Bounded and Compact Integral Operators

Mathematics and Its Applications

Volume 543

Bounded and Compact Integral Operators

by

David E. Edmunds

Centre for Mathematical Analysis and its Application,
University of Sussex, Sussex, United Kingdom

Vakhtang Kokilashvili

A. Razmadze Mathematical Institute,
Georgian Academy of Sciences, Tbilisi, Georgia

and

Alexander Meskhi

A. Razmadze Mathematical Institute,
Georgian Academy of Sciences, Tbilisi, Georgia

KLUWER ACADEMIC PUBLISHERS
DORDRECHT / BOSTON / LONDON

A C.I.P. Catalogue record for this book is available from the Library of Congress.

ISBN 978-90-481-6018-1

Published by Kluwer Academic Publishers,
P.O. Box 17, 3300 AA Dordrecht, The Netherlands.

Sold and distributed in North, Central and South America
by Kluwer Academic Publishers,
101 Philip Drive, Norwell, MA 02061, U.S.A.

In all other countries, sold and distributed
by Kluwer Academic Publishers,
P.O. Box 322, 3300 AH Dordrecht, The Netherlands.

Printed on acid-free paper

Contents

Preface

The monograph presents some of the authors' recent and original results concerning boundedness and compactness problems in Banach function spaces both for classical operators and integral transforms defined, generally speaking, on nonhomogeneous spaces. It focuses on integral operators naturally arising in boundary value problems for PDE, the spectral theory of differential operators, continuum and quantum mechanics, stochastic processes etc. The book may be considered as a systematic and detailed analysis of a large class of specific integral operators from the boundedness and compactness point of view. A characteristic feature of the monograph is that most of the statements proved here have the form of criteria. These criteria enable us, for example, to give various explicit examples of pairs of weighted Banach function spaces governing boundedness/compactness of a wide class of integral operators.

The book has two main parts.

The first part, consisting of Chapters 1–5, covers the investigation of classical operators: Hardy–type transforms, fractional integrals, potentials and maximal functions. Our main goal is to give a complete description of those Banach function spaces in which the above–mentioned operators act boundedly (compactly). When a given operator is not bounded (compact), for example in some Lebesgue space, we look for weighted spaces where boundedness (compactness) holds. We develop the ideas and the techniques for the derivation of appropriate conditions, in terms of weights, which are equivalent to boundedness (compactness). It should be stressed that there is a wide range of problems of Mathematical Physics whose solutions are closely connected to the subject matter of the book. We emphasize the very profound impact of trace inequalities on spectral problems of differential operators, and in particular on eigenvalue estimates for Schrödinger operators (see [87]); and the close connection with the solubility of certain semilinear differential operators with minimal restrictions on the regularity of the coefficients and data. In fact, the existence of positive solutions of certain nonlinear differential equations is equivalent to the validity of a certain two–weighted inequality for a potential–type operator, in which the weights are expressed in terms of coefficients and data (Cf. [120], [13], [3], [197], [303] etc).

One of the most challenging problems of the spectral theory of differential operators is the derivation of eigenvalue and singular value estimates of integral operators in terms of their kernels. The works [34], [235], [164] mark an important stage in the development of this theory; see also [66], [79]. Until recently the list of non–trivial cases in which sharp two–sided estimates are available was rather short. Here we present two–sided estimates of the singular numbers for a large class of operators.

The subjects of our investigation (fractional integrals, potentials, maximal functions, singular integrals) are important tools for solving a variety of problems in several areas of mathematics and its applications. Some examples are worth mentioning here. The Hardy–type transforms are closely related to the solubility problems of nonlinear PDE (see, e.g., [63]). They are of considerable importance in the prediction of stock price futures in an equity market solely on the base of past performance of the stock price or market evaluation respectively. Integral equations involving generalized Hardy operators arise in the theory of automatic complex control systems (see Notes to Chapter 1). Concerning fractional integrals, from the historical point of view we recall the then completely new method of M. Riesz for the solution of the Cauchy problem for PDE of hyperbolic type by means of a semi–group of operators corresponding to the Riemann–Liouville transform. We also mention the close connection between potentials and integral representations for the solution of differential equations; potential and singular integral methods in Mathematical Physics, etc.

The considerable interest in fractional calculus in recent years has been stimulated by applications in different fields of science, including Stochastic Analysis of Long Memory Processes, Numerical Analysis, Physics, Chemistry, Engineering, Biology, Economics and Finance. For the theory of fractional integration and differentiation we refer to the well–known monograph [255]. For multidimensional fractional integrals see [247]. For a wide range of applications of modern Fractional Calculus see the references from [107], [248].

In the second part of the book our attention is concentrated on the investigation of integral transforms on general measure spaces from the boundedness and compactness viewpoint. Topological spaces endowed with a locally finite complete measure and quasi–metric are considered. By a nonhomogeneous space we mean a measure space, with a quasi–metric, in which the doubling condition is not assumed and so may fail. If the measure satisfies the doubling condition, then the measure space is said to be of homogeneous type (SHT). For these spaces we refer to the well–known monographs [56], [286]. SHT in all their generalities naturally arise when studying boundary value problems for PDE with variable coefficients, for instance, when the quasi–metric might be induced by a differential operator, or tailored to fit kernels of integral operators. A weight theory for a wide class of integral transforms on SHT was developed in the monograph [100]. We observe that the general setting of the problem enabled new results to be obtained even in some classical cases. Moreover, it should be emphasized that in that monograph a novel concept was developed, namely a two–weight inequality for integral transforms with arbitrary positive kernels in nonhomogeneous spaces, that is, in spaces where the doubling condition need not hold.

In the late 1990s, in the papers [219–221], [292–294], it was shown that for the validity of almost all statements concerning Calderón–Zygmund singular integrals on an arbitrary metric space with nonatomic measure the doubling condition is not needed. In the present book we are interested in the development of this approach to integral transforms with positive kernels and singular integrals in the framework of the theory of two–weighted estimates.

Now we shall briefly describe the contents of the book.

Chapter 1 deals with Hardy transforms on measure spaces: two–weighted boundedness and compactness criteria are established. Here the exact values of the norms of integral transforms with positive kernels on certain cones of functions are explored. Results relating to the measure of noncompactness and two–sided estimates for the approximation numbers of integral transforms are distributed in various chapters (Chapters 1,2,4,5), corresponding to the study of particular operators.

Chapters 2 and 4 contain an exposition of authors' latest results on the boundedness and compactness of fractional integrals on the line and also in the multidimensional case. Our approach is to do with criteria of boundedness and compactness from L_w^p to L_v^q $(1 < p \leq q < \infty)$. For the Riemann–Liouville and Weyl operators criteria the trace inequalities are established. The "diagonal case" $p = q$ is essentially treated. The latter enables us to give a new criteria for the trace inequality for potentials as well.

The conditions which emerge are transparent and easy to verify. Experience of dealing with fractional integrals with specific kernels enables us to extend these results to a certain class of integral operators. Applications to the solvability problems of Abel's integral equation and certain superlinear inhomogeneous integral equations are presented.

Chapter 3 is devoted to fractional maximal functions, one–sided maximal functions and potentials on the line.

In Chapter 5, our attention is concentrated on boundedness and compactness problems for Riesz potentials and truncated potentials in a half–space. We are concerned with necessary and sufficient conditions for boundedness (compactness) in weighted Lebesgue spaces.

In Chapter 6, we develop the study of integral transforms with positive kernels on nonhomogeneous spaces. We present a complete description of those measures on which potential–type integral transforms generate bounded operators in Lebesgue spaces. Here theorems of Sobolev and Adams type in nonhomogeneous spaces are proved.

In Chapter 7, on the basis of the results of the previous chapters, two–sided estimates are derived for the singular numbers of operators of Volterra and potential type. We establish necessary and sufficient conditions for these operators to belong to Schatten–von Neumann ideals. Asymptotic formulae for singular and entropy numbers of some Volterra–type operators are presented.

Chapter 8 deals with two–weighted estimates for singular integrals in nonhomogeneous spaces. We provide a special treatment of the problem in SHT. In another section focussing on Heisenberg groups, two–weighted estimates for Cauchy–Szegö projections are derived. For the Hilbert transform, two–weighted criteria are given for a certain class of pairs of weights. In the same chapter two–weight estimates for higher–dimensional singular integrals defined on Lipschitz surfaces in Clifford algebras are presented.

Chapter 9 is devoted to two–weight problems for Fourier multipliers. Concerning one–weight estimates with A_p–weights we recall [169], [268], [140], and with exponential weights– [267], [170], [42]. A feature to which we draw attention is that estimates for Fourier multipliers with different weights on the two sides are presented in this book for the first time. We prove two–weighted inequalities for multipliers in Triebel–Lizorkin spaces. The setting of the problem in the framework of two–weight theory enables us to determine new classes of multipliers that it was known in unweighted case. The results of this direction are heavily based on the criteria of boundedness from L_w^p to L_v^q ($1 < p \leq q < \infty$) for fractional and singular integrals derived in previous chapters.

Finally we provide a list of problems which were open at the time of completion of the book. We hope that this list will be useful in stimulating further research.

A few words about the organization of the book are necessary. The enumeration of theorems, lemmas, formulas, etc. follows the natural three–digit system, where after the chapter and the section a consecutive number within the section is used. There are three categories for numbering: theorems, lemmas, propositions and remarks, and then come the formulae.

The bibliography contains about 310 titles; we have collected the most relevent ones related to the topics presented here and it is by no means complete; our concern was to provide a basic orientation rather than a database.

The book is aimed at a rather wide audience, ranging from researchers in Functional and Harmonic Analysis to experts in Applied Mathematics and prospective students.

Acknowledgments

We wish to express our sincere gratitude to Dr . Maia Kvinikadze, Dr. Lida Gogolauri and Dr. Inga Gabisonija from A. Razmadze Mathematical Institute of Georgian Academy of Sciences for the energy and patience they devoted to the production of the TEX master of this book.

We would also like to thank Mr. Tom Armour from the University of Sussex for helping us with solving various technical problems.

The work on this book started in 1997, when the second author visited the Centre for Mathematical Analysis and its Applications, University of Sussex. The collaboration was developed during the visits of the second and third authors to Sussex. They express their deep gratitude to the Centre for support and warm hospitality.

Partial support of Grant No GR/N33034/01 of EPSRC, Grant of the Royal Society, an INTAS Fellowship Grant for Young Scientists (Fellowship Reference No YSF 01/1–8) and Grant No. 1.7 of the Georgian Academy of Sciences is gratefully acknowledged.

Basic notation

R: real line

$R_+ = [0, \infty)$

R^n: n–dimensional Euclidean space

N: set of all positive integers

$Z_+ = N \cup \{0\}$

Z: set of all integers

$n! = 1 \cdot 2 \cdots n$

$\hat{X} = X \times R_+$

$R_+^{n+1} = R^n \times R_+$

$\tilde{R}_+^2 = R_+ \times R_+$

$dist\,(a, A)$: distance from a to A.

$B(x, r)$: open ball in R^n with centre x and radius r

$\overline{B}(x, r)$: closed ball in R^n with centre x and radius r

C^n: n- dimensional complex plane

Ω: domain in R^n

$a_n \uparrow a$: a_n is increasing and $\lim\limits_{n\to\infty} a_n = a$

$a_n \downarrow a$: a_n is decreasing and $\lim\limits_{n\to\infty} a_n = a$

σ_m: volume of the unit ball in R^n, i.e. $\sigma_m = \frac{\pi^{m/2}}{\Gamma(1+m/2)}$

$f \approx g$: there exist positive constants c_1 and c_2 such that $c_1 f \leq g \leq c_2 f$

$a_n \approx b_n$: there exist positive constants c_1 and c_2 such that $c_1 a_n \leq b_n \leq c_2 a_n$
for all $n \in N$

$\mu \times \mu$: product measure

$[r]$: integer part of the constant r

$C^k(\Omega)$: set of k times differentiable functions

$\hat{\varphi}, F(\varphi)$: Fourier transform

$\check{\varphi}, F^{-1}(\varphi)$: inverse Fourier transform

SHT: space of homogemeous type

BFS: Banach function space

RD : reverse doubling condition

PDE: partial differential equatins

$\square$ denotes the end of the proof

Chapter 1

HARDY–TYPE OPERATORS

Being an efficient tool with many applications in various fields of analysis, the Hardy operator

$$Hf(x) = \int_0^x f(y)dy, \quad x > 0,$$

for the past eighty years has been the subject of investigations by many mathematicians.

This chapter deals with boundedness/compactness criteria in weighted Banach function spaces for Hardy–type transforms defined on a measure space endowed with a quasi–metric. Our interest in these extensions of the classical Hardy operator arises from the significant role which they play in the study of fractional and singular integrals defined on such spaces.

Along with compactness criteria, estimates of the measure of non–compactness and approximation numbers for extended Hardy–type transforms are derived. For positive integral transforms on measure spaces the exact value of norms are explored on the cone of functions which is analogous to the set of monotonic functions on the line.

1.1. Boundedness and compactness in *BFS*

In this section we find weight criteria for the boundedness and compactness of generalized Hardy–type operators on Banach function spaces (*BFS*) defined on measure spaces.

Let (X, μ) be a σ– finite measure space and let $\varphi, \psi : X \to [0, \infty)$ be μ– measurable functions such that for every t_1 and t_2 with $0 < t_1 < t_2 < \infty$ the conditions:

$$0 < \mu\{x \in X \ : \ t_1 < \varphi(x) < t_2\} < \infty,$$

1

$$0 < \mu\{x \in X \ : \ t_1 < \psi(x) < t_2\} < \infty$$

are fulfilled.

Let w be a μ–measurable positive function on X. By definition (see [186]), the Banach function space (BFS) $X_1 = (X_1, \mu, w)$ is a normed linear space for which the following conditions are satisfied:

1) the norm $\|f\|_{X_1} = \|f\|_{X_1,\mu,w}$ is defined for every μ–measurable function f and $\|f\|_{X_1} = 0$ if and only if $f = 0$ μ–almost everywhere (μ – a.e.);

2) $\|f\|_{X_1} = \|\,|f|\,\|_{X_1}$ for every $f \in X_1$;

3) if $0 \le f \le g$ μ–a.e., then $\|f\|_{X_1} \le \|g\|_{X_1}$;

4) if $0 \le f_n \uparrow f$ μ–a.e., then $\|f_n\|_{X_1} \uparrow \|f\|_{X_1}$;

5) if E is a μ–measurable subset of X with $w(E) = \int_E w(x)d\mu < \infty$, then $\|\chi_E\|_{X_1} < \infty$, where χ_E is the characteristic function of the set E;

6) for every measurable $E \subset X$ with $w(E) < \infty$ there exists a positive constant c_E such that $\int_E f(x)w(x)d\mu \le c_E\|f\|_{X_1}$ for every $f \in X_1$.

We refer to [20] for a clear presentation of the fundamental properties of Banach function spaces.

For the BFS $X_1 = (X_1, \mu, w)$ its associate space $X_1' = (X_1', \mu, w)$ is given by

$$X_1' = (X_1', \mu, w) = \{f : \int_X f\,g\,wd\mu < \infty \text{ for all } g \in X_1\},$$

and endowed with the associate norm

$$\|f\|_{X_1'} = \sup\left\{ \int_X fgwd\mu \ : \ \|g\|_{X_1} \le 1\right\}$$

it is also a Banach function space.

Both X_1 and X_1' are complete linear spaces, and $X_1'' = X_1$. Moreover, for every $f \in X_1$ and $g \in X_1'$ the Hölder's inequality is fulfilled:

$$\int_X f(x)g(x)w(x)d\mu \le \|f\|_{X_1} \cdot \|g\|_{X_1'}.$$

Now we define a Hardy–type operator on X as follows:

$$Tf(x) = u_1(x) \int_{\{y:\varphi(y)\le\psi(x)\}} f(y)u_2(y)w(y)d\mu,$$

where φ, ψ, u_1 and u_2 are positive μ–measurable functions.

In the same way we shall consider the operator of the type

$$T_1f(x) = u_1(x) \int_{\{y:\varphi(y)<\psi(x)\}} f(y)u_2(y)w(y)d\mu.$$

Both T and T_1 are generalizations of the Hardy operator $Hf(x) = \int\limits_0^x f(t)dt$.

Let $X_1 = (X_1, \mu, w)$ and $X_2 = (X_2, \mu, v)$ be Banach function spaces and let

$$T'g(x) = u_2(x) \int\limits_{\{y:\psi(y)\geq\varphi(x)\}} g(y)u_1(y)v(y)d\mu,$$

$$T_1'g(x) = u_2(x) \int\limits_{\{y:\psi(y)>\varphi(x)\}} g(y)u_1(y)v(y)d\mu.$$

Then the following lemma holds.

Lemma 1.1.1. *The operator T is bounded from X_1 to X_2 if and only if T' is bounded from X_2' to X_1', and the norm of T' is equal to the norm of T.*

Proof. Let the operator T be bounded from X_1 to X_2. Then we obtain

$$\|T'g\|_{X_1'} = \sup_{\|f\|_{X_1}\leq 1} \left| \int\limits_X (T'g)(x)f(x)d\mu(x) \right| \leq$$

$$\leq \sup_{\|f\|_{X_1}\leq 1} \int\limits_X |f(x)|u_2(x) \Big(\int\limits_{\{y:\psi(y)\geq\varphi(x)\}} |g(y)| \times$$

$$\times u_1(y)v(y)d\mu(y) \Big) w(x)d\mu(x) =$$

$$= \sup_{\|f\|_{X_1}\leq 1} \int\limits_X |g(y)|u_1(y) \Big(\int\limits_{\{x:\varphi(x)\leq\psi(y)\}} |f(x)| \times$$

$$\times u_2(x)w(x)d\mu(x) \Big) v(y)d\mu(y) \leq$$

$$\leq \sup_{\|f\|_{X_1}\leq 1} \|Tf\|_{X_2} \cdot \|g\|_{X_2'} \leq \|T\| \, \|g\|_{X_2'}.$$

The inverse assertion is proved analogously. Moreover, $\|T\| = \|T'\|$. $\square$

The lemma below is proved similarly.

Lemma 1.1.2. *The operator T_1 is bounded from X_1 to X_2 if and only if T_1' is bounded from X_2' to X_1', and $\|T_1\| = \|T_1'\|$.*

The corresponding connection as regard compactness of maps is given by the next lemma.

Lemma 1.1.3. *The operator T (resp. T_1) is compact from X_1 to X_2 if and only if T' (resp. T_1') is compact from X_2' to X_1'.*

Proof. We shall prove the lemma for the operator T.

Let T be compact from X_1 to X_2. The boundedness of T' follows from Lemma 1.1.1. Let S_1 and S_2' be the closed unit balls in X_1 and X_2' respectively.

We take an arbitrary sequence $\{f_n\} \subset S_2'$ and consider the functions F_n,

$$F_n(g) = \int_X g(y) f_n(y) w(y) d\mu,$$

on X_2. By Hölder's inequality, we have

$$|F_n(g)| \leq \|g\|_{X_2} \|f_n\|_{X_2'} \leq \|g\|_{X_2}$$

and

$$|F_n(g_1 - g_2)| \leq \|g_1 - g_2\|_{X_2}.$$

Let $\overline{T(S_1)}$ be the closure of $T(S_1)$. Then $\overline{T(S_1)}$ is compact and closed in X_2, while the functions F_n are uniformly bounded and equicontinuous (with respect to n) on $\overline{T(S_1)}$. Therefore by the Arzelá–Ascoli theorem, there exists a subsequence $\{F_{n_k}\}$ converging uniformly on $\overline{T(S_1)}$ and hence $G_{n_k} = F_{n_k} \circ T$ converges uniformly on S_1. But if $h \in S_1$, then

$$\int_X T'(f_{n_k} - f_{n_m})(x) h(x) v(x) d\mu(x) =$$

$$\int_X (f_{n_k} - f_{n_m})(y)(Th)(y) w(y) d\mu(y) = (G_{n_k} - G_{n_m})(h)$$

(Here the change of the integration order is justified, since

$$\int_X |(f_{n_k} - f_{n_m})(y)| |T(|h|)(y) w(y) d\mu(y) \leq \|f_{n_k} - f_{n_m}\|_{X_2'} \|T(|h|)\|_{X_2} < \infty).$$

This implies that

$$\|T'f_{n_k} - T'f_{n_m}\|_{X_1'} = \sup_{\|h\|_{X_1} \leq 1} |(G_{n_k} - G_{n_m})(h)|$$

and hence $\{T'f_{n_k}\}$ converges in X_1'.

If T' is compact from X_2' to X_1', then taking into account the arguments above, we find that T is compact from X_1 to X_2. $\square$

Now we establish weighted criteria for the boundedness of the Hardy–type operators in Banach function spaces.

We shall start from the necessary condition for the boundedness of the operator T.

Theorem 1.1.1. *Suppose there exists a positive constant c such that for any $f \in X_1$ the inequality*

$$\|Tf\|_{X_2} \leq c\|f\|_{X_1} \tag{1.1.1}$$

holds. Then the condition

$$B = \sup_{t \geq 0} B(t) = \sup_{t \geq 0} \|u_1 \chi_{\{y : \psi(y) \geq t\}}\|_{X_2} \|u_2 \chi_{\{y : \varphi(y) \leq t\}}\|_{X_1'} < \infty \tag{1.1.2}$$

is satisfied. Moreover, $B \leq c$.

Proof. Let $\|f\|_{X_1} \leq 1$ and $t > 0$. Then we obtain

$$c \geq c\|f\|_{X_1} \geq \|Tf\|_{X_2} \geq \|(Tf)\chi_{\{y:\psi(y)\geq t\}}\|_{X_2} \geq$$

$$\geq \int_{\{x:\varphi(x)\leq t\}} f(x)u_2(x)w(x)d\mu \|u_1\chi_{\{y:\psi(y)\geq t\}}\|_{X_2}.$$

If we take the supremum with respect to all f and t, then (1.1.2) follows. $\square$

The theorem below is proved analogously.

Theorem 1.1.2. *Let the inequality*

$$\|T_1 f\|_{X_2} \leq c\|f\|_{X_1}, \quad f \in X_1, \tag{1.1.3}$$

hold, where the positive constant c does not depend on f. Then

$$B_1 = \sup_{t>0} B_1(t) = \sup_{t>0} \|u_1\chi_{\{y:\psi(y)>t\}}\|_{X_2}\|u_2\chi_{\{y:\varphi(y)\leq t\}}\|_{X_1'} < \infty. \tag{1.1.4}$$

Moreover, $B_1 \leq c$.

We shall sometimes require that X_1 and X_2 satisfy the following condition: Let c_1, c_2 be fixed positive numbers. Then there exist a number α, $\alpha > 1$, and positive constants k_1 and k_2 such that for any sequences of measurable sets $\{E_k\}$, $\{F_k\}$ with the conditions $\sum_k \chi_{E_k} \leq c_1$, $\sum_k \chi_{F_k} \leq c_2$ and for any measurable functions f and g the following inequalities are valid:

$$\sum_k \|\chi_{E_k} f\|_{X_1}^{\alpha} \leq k_1 \|f\chi_{\cup_k E_k}\|_{X_1}^{\alpha},$$
$$\sum_k \|\chi_{F_k} g\|_{X_2'}^{\alpha'} \leq k_2 \|g\chi_{\cup_k F_k}\|_{X_2'}^{\alpha'}, \tag{1.1.5}$$

where $\alpha' = \frac{\alpha}{\alpha-1}$.

We give examples of spaces X_1 and X_2 for which this condition holds at the end of this section. Its importance is that it enables us to obtain sufficiency condition for boundedness.

Theorem 1.1.3. *Let X_1 and X_2 satisfy condition (1.1.5) and let the condition (1.1.2) be fulfilled. Then inequality (1.1.1) holds. Moreover, $\|T\| \leq kB$, where $k = 1 + 4k_1^{\frac{1}{\alpha}} k_2^{\frac{1}{\alpha'}}$ (see (1.1.5)).*

Proof. By Lemma 1.1.1, it suffices to show that

$$\|T'f\|_{X_1'} \leq c\|f\|_{X_2'}. \tag{1.1.6}$$

Let

$$I(s) = \int_{\{y:\psi(y)\geq s\}} fu_1 v d\mu, \quad f \geq 0, \quad f \in X_2';$$

then for every $s > 0$ we have

$$I(s) \leq \|u_1 \chi_{\{y:\psi(y) \geq s\}}\|_{X_2} \|f\|_{X_2'} < \infty.$$

The function $I(s)$ is left–continuous and $\lim_{s \to \infty} I(s) = 0$.
Let

$$\int_{\{y:\psi(y) \neq 0\}} f u_1 v d\mu \in (2^m, 2^{m+1}]$$

for some $m \in Z$. If $s_j = \inf\{s : I(s) \leq 2^j\}$, $j \leq m$, then $I(s_j) \geq 2^j$ and $I(s) \leq 2^j$ for $s > s_j$. Let $s_{m+1} = 0$; then the sequence $\{s_j\}_{j=-\infty}^{m+1}$ is non-increasing. Moreover, the inequality

$$\int_{\{y:s_j \leq \psi(y) \leq s_{j-1}\}} f u_1 v d\mu \geq 2^{j-1}, \quad \text{for } j \leq m,$$

holds.

Let $J_m = \{j \leq m+1 : s_j < s_{j-1}\}$; then $J_m \neq \emptyset$, since $s_{m+1} < s_m$, and hence $m + 1 \in J_m$.

Let $\beta = \lim_{j \to -\infty} s_j$. Then

$$(0, \infty) = (\cup_{j \in J_m} (s_j, s_{j-1}]) \cup [\beta, +\infty),$$

if J_m is infinite and

$$(0, \infty) = (\cup_{j \in J_m} (s_j, s_{j-1}]) \cup (\beta, +\infty),$$

if J_m is finite.
Clearly,

$$\{x \in X : \varphi(x) \neq 0\} = (\cup_{j \in J_m} E_j) \cup F,$$

where $E_j = \{x : s_j < \varphi(x) \leq s_{j-1}\}$, $F = \{x : \varphi(x) \geq \beta\}$ if J_m is infinite and $F = \{\varphi(x) > \beta\}$, if J_m is finite.
If $\int_{\{y: \psi(y) \neq 0\}} f u_1 v d\mu = \infty$, then we take $m = \infty$ and

$$\{x \in X : \varphi(x) \neq 0\} = (\cup_{j \in J} E_j) \cup F,$$

where $J = \{j \in Z : s_j < s_{j-1}\}$.
If $s \in (s_j, s_{j-1}]$, then $I(s) \leq 2^j$, and if $y \in F$, then $I(\varphi(y)) \leq 2^j$ for every j. Hence $I(\varphi(y)) = 0$.
Let g be a measurable function on X with $\|g\|_{X_1} \leq 1$. Then we obtain

$$\int_X (T'f)(x)g(x)w(x)d\mu = \int_{\{y:\varphi(y)=0\}} (T'f)(x)g(x)w(x)d\mu +$$

$$+ \int_{\{y:\varphi(y) \neq 0\}} (T'f)(x)g(x)w(x)d\mu = I_1 + I_2.$$

Now we estimate I_1 and I_2.

$$I_1 = \left(\int_{\{x:\psi(x)\geq 0\}} fu_1 v d\mu \right) \left(\int_{\{y:\varphi(y)=0\}} gu_2 w d\mu \right) \leq$$

$$\|u_1\chi_{\{y:\psi(y)\geq 0\}}\|_{X_2}\|f\|_{X_2'}\|u_2\chi_{\{y:\varphi(y)=0\}}\|_{X_1'}\|g\chi_{\{y:\varphi(y)=0\}}\|_{X_1} \leq$$

$$\leq B\|f\|_{X_2'}\|g\|_{X_1},$$

$$I_2 = \sum_{j\in J_m} \int_{E_j} (T'f(x))g(x)w(x)d\mu \leq \sum_{j\in J_m} 2^j \int_{E_j} g(x)u_2(x)w(x)d\mu =$$

$$= 4 \sum_{j\in J_m} 2^{j-2} \int_{E_j} g(x)u_2(x)w(x)d\mu \leq$$

$$\leq 4 \sum_{j\in J_m} \int_{\{y:s_{j-1}\leq \psi(y)\leq s_{j-2}\}} f(y)u_1(y)v(y)d\mu \times$$

$$\times \int_{E_j} g(x)u_2(x)w(x)d\mu \leq$$

$$\leq 4 \sum_{j\in J_m} \|u_1\chi_{\{\psi(y)\geq s_{j-1}\}}\|_{X_2}\|f\chi_{\{s_{j-1}\leq \psi(y)\leq s_{j-2}\}}\|_{X_2'} \times$$

$$\times \|u_2\chi_{\{\varphi(y)\leq s_{j-1}\}}\|_{X_1'}\|g\chi_{\{s_j\leq \varphi(y)\leq s_{j-1}\}}\|_{X_1} \leq$$

$$\leq 4B \left(\sum_{j\in J_m} \|f\chi_{\{s_{j-1}\leq \psi(y)\leq s_{j-2}\}}\|_{X_2'}^{\alpha'} \right)^{\frac{1}{\alpha'}} \times$$

$$\times \left(\sum_{j\in J_m} \|g\chi_{\{s_j\leq \varphi(y)\leq s_{j-1}\}}\|_{X_1}^{\alpha} \right)^{\frac{1}{\alpha}} \leq 4Bk_2^{\frac{1}{\alpha'}}k_1^{\frac{1}{\alpha}}\|f\|_{X_2'}\|g\|_{X_1},$$

where α is from (2.4). In the last inequality we used the fact that

$$\sum_{j\in J_m} \chi_{\{x\in X:\, s_{j-1}\leq \psi(x)\leq s_{j-2}\}} \leq 2, \quad \sum_{j\in J_m} \chi_{\{x\in E:\, s_j\leq \varphi(x)\leq s_{j-1}\}} \leq 2.$$

We have

$$I_1 + I_2 \leq (4k_2^{\frac{1}{\alpha'}}k_1^{\frac{1}{\alpha}} + 1)B\|f\|_{X_2'}.$$

Taking the supremum with respect to all such g, we obtain inequality (1.1.6). $\square$

Theorem 1.1.4. *Let X_1 and X_2 satisfy condition* (1.1.5), *suppose that* $\mu\{x : \varphi(x) = 0\} = 0$, *and let condition* (1.1.4) *be fulfilled. Then inequality* (1.1.3) *holds, and if c is the best constant in* (1.1.3), *then $c \leq 4k_1^{\frac{1}{\alpha}}k_2^{\frac{1}{\alpha'}}B_1$.*

Proof. Let $f \geq 0$, $f \in X_1$ and $I(s) = \int_{\{y:\varphi(y)<s\}} f(y)u_2(y)w(y)d\mu$. Then by condition (1.1.4), we have that $I(s) < \infty$ for any $s > 0$. The function $I(s)$ increases and $\lim_{s\to 0} I(s) = 0$. Moreover, the function $I(s)$ is left-continuous.

Suppose that $\int_X f u_2 w d\mu \in (2^m, 2^{m+1}]$, for some $m \in Z$ and put $s_j = \sup\{s : I(s) \le 2^j\}$, $j \le m+1$. The sequence $\{s_j\}_{j=-\infty}^{m+1}$ is non-decreasing. It is clear that $s_{m+1} = \infty$. Moreover, $I(s_j) \le 2^j$, $I(s) \ge 2^j$ if $s > s_j$, and

$$\int_{\{x : s_j \le \varphi(y) \le s_{j+1}\}} f u_2 w d\mu \ge 2^j.$$

Let $\beta = \lim_{j \to -\infty} s_j$. Then $[0, \infty) = \cup_{j \in J_m}(s_j, s_{j+1}] \cup [0, \beta]$, where $J_m = \{j \le m+1, \ s_j < s_{j+1}\}$. This implies that $X = (\cup_{j \in J_m} E_j) \cup F$, where $E_j = \{x : s_j < \psi(y) \le s_{j+1}\}$, $F = \{y : \psi(y) \le \beta\}$.

If $\int_X f u_2 w d\mu = \infty$, then we take $m = \infty$. Thus

$$X = (\cup_{j \in J} E_j) \cup F,$$

where $J = \{j \in Z : s_j < s_{j+1}\}$.

If $s \in (s_j, s_{j+1}]$, then $I(s) \le 2^{j+1}$ for $j \le m$, and if $s \in [0, \beta]$, then $I(s) = 0$.

Arguing as in the proof of Theorem 1.1.3, we easily obtain inequality (1.1.3) with the constant $4k_1^{\frac{1}{\alpha}} k_2^{\frac{1}{\alpha'}} B_1$. $\square$

Our ultimate aim is to obtain information about the operators which is more precise than mere boundedness. With this in mind, it is convenient to introduce various auxiliary operators and to establish their properties. We now set about this.

Further, let $0 < a < b < \infty$ and define

$$T_a f(x) = \chi_{\{y : \psi(y) < a\}}(x) u_1(x) \int_{\{y : \varphi(y) < \psi(x)\}} \chi_{\{z : \varphi(z) < a\}}(y) f(y) u_2(y) d\mu,$$

$$T_b f(x) = \chi_{\{y : \psi(y) > b\}}(x) u_1(x) \int_{\{y : \varphi(y) < \psi(x)\}} \chi_{\{z : \varphi(z) > b\}}(y) f(y) u_2(y) d\mu,$$

$$T_{ab} f(x) = \chi_{\{y : a < \psi(y) \le b\}}(x) u_1(x) \times$$
$$\times \int_{\{y : \varphi(y) < \psi(x)\}} \chi_{\{z : a < \varphi(z) < b\}}(y) f(y) u_2(y) d\mu.$$

The following lemmas hold.

Lemma 1.1.4. *Suppose that* $\mu\{x : \psi(x) = t\} = \mu\{x : \varphi(x) = t\} = 0$ *for every* $t \in [0, a]$. *If* X_1 *and* X_2 *satisfy the condition* (1.1.5) *and*

$$B_a = \sup_{0 < t < a} B_a(t) = \sup_{0 < t < a} \|u_1 \chi_{\{y : t < \psi(y) < a\}}\|_{X_2} \times$$
$$\times \|u_2 \chi_{\{y : \varphi(y) < t\}}\|_{X_1'} < \infty, \tag{1.1.7}$$

then the inequality

$$\|T_a f\|_{X_2} \le c\|f \cdot \chi_{\{y:\varphi(y)<a\}}\|_{X_1}, \quad f \in X_1, \tag{1.1.8}$$

holds, where the constant c does not depend on f, and

$$\|T_a\| \le 4k_1^{\frac{1}{\alpha}} k_2^{\frac{1}{\alpha'}} B_a.$$

Conversely, if inequality (1.1.8) is fulfilled, then the condition (1.1.7) is satisfied, and

$$B_a \le \|T_a\|.$$

Lemma 1.1.5. *Let* $\mu\{x : \psi(x) = t\} = \mu\{x : \varphi(x) = t\} = 0$ *for all* $t \in [a, b]$. *If* X_1 *and* X_2 *satisfy the condition* (1.1.5) *and*

$$B_{ab} = \sup_{a<t<b} B_{ab}(t) = \sup_{a<t<b} \|u_1 \chi_{\{x:t<\psi(x)<b\}}\|_{X_2} \times$$
$$\times \|u_2 \chi_{\{a<\varphi(x)<t\}}\|_{X_1'} < \infty \tag{1.1.9}$$

then there exists a positive constant c *such that for every* $f \in X_1$ *the inequality*

$$\|T_{ab} f\|_{X_2} \le c\|f \chi_{\{a<\varphi(y)<b\}}\|_{X_1} \tag{1.1.10}$$

is fulfilled, and

$$\|T_{ab}\| \le 4k_1^{\frac{1}{\alpha}} k_2^{\frac{1}{\alpha'}} B_{ab}.$$

Conversely, if for X_1 and X_2 inequality (1.1.10) holds, then condition (1.1.9) is satisfied, and

$$B_{ab} \le \|T_{ab}\|.$$

Lemma 1.1.6. *Suppose that* $\mu\{x : \psi(x) = t\} = \mu\{x : \varphi(x) = t\} = 0$ *for every* $t \ge b$. *If for* X_1 *and* X_2 *condition* (1.1.5) *is fulfilled and*

$$B_b = \sup_{t>b} B_b(t) = \sup_{t>b} \|u_1 \chi_{\{x:t<\psi(x)\}}\|_{X_2} \times$$
$$\times \|u_2 \chi_{\{x:b<\varphi(x)<t\}}\|_{X_1'} < \infty, \tag{1.1.11}$$

then the inequality

$$\|T_b f\|_{X_2} \le c\|f \cdot \chi_{\{\varphi(y)>t\}}\|_{X_1}, \quad f \in X_1, \tag{1.1.12}$$

holds, where the positive constant c does not depend on f, and

$$\|T_b\| \le 4k_1^{\frac{1}{\alpha}} k_2^{\frac{1}{\alpha'}} B_b.$$

Conversely, if inequality (1.1.12) *is fulfilled, then the condition* (1.1.11) *is satisfied, and*

$$B_b \leq \|T_b\|.$$

These lemmas are proved in the same way as Theorem 1.1.4. (see also [69], the proof of Theorem 3).

Let $0 < a < b < \infty$, $\lambda > 0$ and write

$$Pf(x) = u_1(x) \int_{\{a\psi(x)<\varphi(y)<b\psi(x)\}} f(y)u_2(y)w(y)d\mu,$$

$$P_1f(x) = \chi_{\{y:a<\psi(y)<b\}}(x)u_1(x) \int_{\{\lambda a<\varphi(y)<\lambda\psi(x)\}} f(y)u_2(y)w(y)d\mu,$$

$$P_2f(x) = \chi_{\{y:a<\psi(y)<b\}}(x)u_1(x) \int_{\{\lambda\psi(x)<\varphi(y)<\lambda b\}} f(y)u_2(y)w(y)d\mu.$$

Then the following lemma holds.

Lemma 1.1.7. *Let φ and ψ satisfy the conditions:*

$$\mu\{x : \varphi(x) = t\} = 0 \quad \text{for every} \quad t \in [\lambda a, \lambda b],$$
$$\mu\{x : \psi(x) = t\} = 0 \quad \text{for every} \quad t \in [a, b].$$

If X_1 and X_2 satisfy condition (1.1.5) *and*

$$A_{1ab} = \sup_{a<t<b} \|u_1\chi_{y:t<\psi(y)<b}\|_{X_2}\|u_2\chi_{\{y:\lambda a<\varphi(y)<\lambda t\}}\|_{X_1'} < \infty, \quad (1.1.13)$$

then the inequality

$$\|P_1f\|_{X_2} \leq c\|f \cdot \chi_{\{a\lambda<\varphi(y)<b\lambda\}}\|_{X_1}, \quad (1.1.14)$$

holds, and

$$\|P_1\| \leq 4k_1^{\frac{1}{\alpha}}k_2^{\frac{1}{\alpha'}}A_{1ab}.$$

Conversely, if inequality (1.1.14) *holds, then condition* (1.1.13) *is satisfied.*
Proof. It can be easily seen that

$$P_1f(x) = \chi_{\{y:\lambda a<\lambda\psi(y)<\lambda b\}}(x)u_1(x) \times$$

$$\times \int_{\{y:\varphi(y)<\lambda\psi(x)\}} \chi_{\{\lambda a<\varphi(z)<\lambda b\}}(y)u_2(y)f(y)w(y)d\mu.$$

Let $\lambda\psi = \psi_1$. Then

$$P_1f(x) = \chi_{\{y:\lambda a<\psi_1(y)<\lambda b\}}(x)u_1(x) \times$$

$$\times \int_{\{y:\varphi(y)<\psi_1(x)\}} \chi_{\{\lambda a<\varphi(z)<\lambda b\}}(y)u_2(y)w(y)f(y)d\mu.$$

Moreover,

$$A_{1ab} = \sup_{\lambda a<s<\lambda b} \|u_1\chi_{\{y:s<\psi_1(y)<\lambda b\}}\|_{X_2}\|u_2\chi_{\{\lambda a<\varphi(y)<s\}}\|_{X_1'},$$

and Lemma 1.1.5 gives what was to be shown. □

The next Lemma follows analogously.

Lemma 1.1.8. *Let φ and ψ satisfy the conditions of Lemma* 1.1.7. *If for X_1 and X_2 condition* (1.1.5) *is fulfilled and*

$$A_{2ab} = \sup_{a<t<b} \|u_1\chi_{a<\psi(y)<t}\|_{X_2}\|u_2\chi_{\{\lambda t<\varphi(y)<\lambda b\}}\|_{X_1'} < \infty, \qquad (1.1.15)$$

then for every $f \in X_1$ the inequality

$$\|P_2f\|_{X_2} \le c\|f \cdot \chi_{\{\lambda a<\varphi(y)<\lambda b\}}\|_{X_1}, \qquad (1.1.16)$$

holds, and

$$\|P_2\| \le 4k_1^{\frac{1}{\alpha}} k_2^{\frac{1}{\alpha'}} A_{2ab}.$$

Conversely, if inequality (1.1.16) *is valid, then the condition* (1.1.15) *is fulfilled.*

In the sequel we shall need the following notation:

$$A_{1k} = \sup_{(\frac{b}{a})^k<t<(\frac{b}{a})^{k+1}} \|u_1\chi_{\{t<\psi(y)<(\frac{b}{a})^{k+1}\}}\|_{X_2}\|u_2\chi_{\{b(\frac{b}{a})^k<\varphi(y)<bt\}}\|_{X_1'},$$

$$A_{2k} = \sup_{(\frac{b}{a})^k<t<(\frac{b}{a})^{k+1}} \|u_1\chi_{\{(\frac{b}{a})^k<\psi(y)<t\}}\|_{X_2}\|u_2\chi_{\{at<\varphi(y)<a(\frac{b}{a})^{k+1}\}}\|_{X_1'}$$

$$A_1 = \sup_{k\in Z}\{A_{1k}\}, A_2 = \sup_{k\in Z}\{A_{2k}\}, A = \max\{A_1, A_2\}.$$

We now have the next theorem.

Theorem 1.1.5. *Let X_1 and X_2 satisfy condition* (1.1.5), *suppose that* $\mu\{x : \varphi(x) = t\} = \mu\{x : \psi(x) = t\} = 0$ *for any $t \in [0, \infty)$, and assume that*

$$A < \infty.$$

Then the operator P is bounded from X_1 to X_2, and

$$\|P\| \le 8k_1^{\frac{2}{\alpha}} k_2^{\frac{2}{\alpha'}} A.$$

Conversely, if P is bounded from X_1 to X_2, then $A < \infty$ and $A \le \|P\|$.

Proof. Let $F_k = \{x : (\frac{b}{a})^k \leq \psi(x) < (\frac{b}{a})^{k+1}\}$. Then $X = \cup_{k \in Z} F_k$. Let us take a function g with $\|g\|_{X_2'} \leq 1$. We obtain

$$\int\limits_X (Pf)(x)g(x)v(x)d\mu = \sum_k \int\limits_{F_k} Pf(x)g(x)v(x)d\mu =$$

$$= \sum_k \int\limits_{F_k} u_1(x) \Bigg[\int\limits_{\{a\psi(x)<\varphi(y)<b(\frac{b}{a})^k\}} f(y)u_2(y)v(y)d\mu +$$

$$+ \int\limits_{\{a(\frac{b}{a})^{k+1}<\varphi(y)<b\psi(x)\}} f(y)u_2(y)v(y)d\mu \Bigg] g(x)v(x)d\mu =$$

$$= \sum_k \int\limits_{F_k} u_1(x) \Bigg(\int\limits_{\{y:a\psi(x)<\varphi(y)<b(\frac{b}{a})^k\}} f(y)u_2(y)v(y)d\mu \Bigg) g(x)v(x)d\mu +$$

$$+ \sum_k \int\limits_{F_k} u_1(x) \Bigg(\int\limits_{\{y:a(\frac{b}{a})^{k+1}<\varphi(y)<b\psi(x)\}} f(y)u_2(y)v(y)d\mu \Bigg) g(x)v(x)d\mu =$$

$$= \sum_k \tau_{1k} + \sum_k \tau_{2k}.$$

By Lemma 1.1.8 we derive

$$\sum_k \tau_{1k} \leq \sum_k \|u_1 \chi_{F_k} \int\limits_{\{a\psi(x)<\varphi(y)<b(\frac{b}{a})^k\}} u_2 f w d\mu\|_{X_2} \|g\chi_{F_k}\|_{X_2'} \leq$$

$$\leq 4k_1^{\frac{1}{\alpha}} k_2^{\frac{1}{\alpha'}} \sum_k A_{2k} \|f\chi_{\{a(\frac{b}{a})^k<\varphi(y)<a(\frac{b}{a})^{k+1}\}}\|_{X_1} \|g\chi_{F_k}\|_{X_2'} \leq$$

$$\leq 4k_1^{\frac{1}{\alpha}} k_2^{\frac{1}{\alpha'}} A \Big(\sum_k \|f\chi_{\{a(\frac{b}{a})^k<\varphi(y)<a(\frac{b}{a})^{k+1}\}}\|_{X_1}^\alpha \Big)^{\frac{1}{\alpha}} \Big(\sum_k \|g\chi_{F_k}\|_{X_2'}^{\alpha'} \Big)^{\frac{1}{\alpha'}} \leq$$

$$\leq 4k_1^{\frac{2}{\alpha}} k_2^{\frac{2}{\alpha'}} A \|g\|_{X_2'} \|f\|_{X_1} \leq 4k_1^{\frac{2}{\alpha}} k_2^{\frac{2}{\alpha'}} A \|f\|_{X_1}.$$

Similarly, using Lemma 1.1.7, we get

$$\sum_k \tau_{2k} \leq 4k_1^{\frac{2}{\alpha}} k_2^{\frac{2}{\alpha'}} A \|f\|_{X_1},$$

and

$$\int\limits_X Pf(x)g(x)v(x)d\mu \leq 8k_1^{\frac{2}{\alpha}} k_2^{\frac{2}{\alpha'}} A \|f\|_{X_1}.$$

Taking the supremum with respect to all g, we arrive at the required result.

Conversely, let the operator P be bounded from X_1 to X_2 and let $\|f\|_{X_1} \leq 1$, $k \in Z$, and $(\frac{b}{a})^k < t < (\frac{b}{a})^{k+1}$.

$$c \geq c\|f\|_{X_1} \geq \|Pf\|_{X_2} \geq \|\chi_{\{t<\psi(y)<(\frac{b}{a})^{k+1}\}}Pf\|_{X_2} \geq$$

$$\geq \left(\int_{\{y:b(\frac{b}{a})^k<\varphi(y)<bt\}} fu_2wd\mu \right) \|\chi_{\{t<\psi(y)<(\frac{b}{a})^{k+1}\}}u_1f\|_{X_2}.$$

If we take the supremum with respect to all f, then we find that $A_{1k} \leq c, k \in Z$. This implies that $A_1 \leq c$. Moreover, we obtain

$$c \geq c\|f\|_{X_1} \geq \|Pf\|_{X_2} \geq \|\chi_{\{(\frac{b}{a})^k<\psi(y)<t\}}Pf\|_{X_2} \geq$$

$$\geq \left(\int_{\{at<\varphi(y)<b(\frac{b}{a})^k\}} fu_2wd\mu \right) \|u_1\chi_{\{(\frac{b}{a})^k<\psi(y)<t\}}\|_{X_2}.$$

which implies that $A_{2k} \leq c$ for every k, and $A_2 \leq c$. Finally we have $A \leq c$.
□

An analogous theorem for the classical Hardy operator in Lebesgue spaces has been proved in [15].

Let σ and ν be σ–finite measures on X and let $X_1(\sigma)$ and $X_2(\nu)$ be Banach function space with measures σ and ν respectively (see the definition of BFS).

Define Hardy–type operators:

$$Kf(x) = u_1(x) \int_{\{y:\varphi(y)\leq\psi(x)\}} f(y)u_2(y)d\sigma,$$

$$K_1f(x) = u_1(x) \int_{\{y:\varphi(y)<\psi(x)\}} f(y)u_2(y)d\sigma,$$

and

$$Gf(x) = u_1(x) \int_{\{y:a\psi(x)<\varphi(y)<b\psi(x)\}} f(y)u_2(y)d\sigma, \quad 0 < a < b < \infty,$$

where u_1 (respectively u_2) is a ν–measurable (respectively σ–measurable) positive function defined on X and $\psi : X \to [0,\infty)$ (respectively $\varphi : X \to [0,\infty)$) is a ν measurable (respectively σ–measurable) function such that for every t_1 and t_2 $(0 < t_1 < t_2 < \infty)$ the following condition $0 < \sigma\{x \in X : \varphi(x) \in (t_1,t_2)\} < \infty$ (respectively $0 < \nu\{x \in X : \psi(x) \in (t_1,t_2)\} < \infty$) holds.

As above, we sometimes require that $X_1 = X_1(\sigma)$ and $X_2 = X_2(\nu)$ satisfy condition (1.1.5).

The following theorems are proved analogously to those given earlier:

Theorem 1.1.6. *Suppose that operator K is bounded from X_1 to X_2. Then condition* (1.1.2) *holds. If X_1 and X_2 satisfy condition* (1.1.5) *and condition* (1.1.2) *holds, then K is bounded from X_1 to X_2.*

Theorem 1.1.7. *Suppose that operator K_1 is bounded from X_1 to X_2. Then condition* (1.1.4) *is satisfied. Conversely, if $\sigma\{x : \varphi(x) = 0\} = 0$ and X_1 and X_2 satisfy* (1.1.5), *then condition* (2.3) *implies the boundedness of the operator K_1 from X_1 to X_2.*

Theorem 1.1.8. *Let $\sigma\{x \in X : \varphi(x) = t\} = \nu\{x \in X : \psi(x) = t\} = 0$ for any $t \geq 0$. If operator G is bounded from X_1 to X_2 then $A < \infty$; and if X_1 and X_2 satisfy condition* (1.1.5) *and $A < \infty$, then G is bounded from X_1 to X_2.*

We can now obtain necessary and sufficient conditions for the compactness of the operator T from X_1 to X_2 .

We assume that for every $t \geq 0$ the following condition is satisfied:

$$\mu\{x \in X : \psi(x) = t\} = \mu\{x \in X : \varphi(x) = t\} = 0.$$

In this case $Tf(x) = T_1 f(x)$ for every $x \in X$ and $f \in X_1$.

Let $0 < a < b < \infty$ and

$$T_{ab}f(x) = \chi_{\{y:a<\psi(y)\leq b\}}(x)u_1(x) \times$$

$$\times \int_{\{y:\varphi(y)<\psi(x)\}} \chi_{\{z:a<\varphi(z)<b\}}(y)u_2(y)f(y)w(y)d\mu.$$

Lemma 1.1.9. *Let*

$$\|u_1\chi_{\{y:a<\psi(y)<b\}}\|_{X_2} = c_\psi < \infty, \quad \|u_2\chi_{\{y:a<\varphi(y)<b\}}\|_{X_1'} = c_\varphi < \infty.$$

Then the operator T_{ab} is compact from X_1 to X_2 if and only if for every $\alpha \in [a, b)$,

$$\lim_{s\to\alpha+} \|u_1\chi_{\{\alpha<\psi(y)<s\}}\|_{X_2} = 0 \ or \ \lim_{s\to\alpha+} \|u_2\chi_{\{\alpha<\varphi(y)<s\}}\|_{X_1'} = 0 \quad (1.1.17)$$

and for every $\alpha \in (a, b]$

$$\lim_{s\to\alpha-} \|u_1\chi_{\{s<\psi(y)<\alpha\}}\|_{X_2} = 0 \ or \ \lim_{s\to\alpha-} \|u_2\chi_{\{s<\varphi(y)<\alpha\}}\|_{X_1'} = 0. \quad (1.1.18)$$

Proof. First let us prove that the operator T_{ab} is bounded from X_1 to X_2. Indeed,

$$\|T_{ab}f\|_{X_2} \leq$$

$$\leq \|u_1 \chi_{\{y:a<\psi(y)<b\}}\|_{X_2} \int\limits_{\{y:a<\varphi(y)<b\}} u_2(y)f(y)w(y)d\mu \leq$$

$$\leq \|u_1 \chi_{\{y:a<\psi(y)<b\}}\|_{X_2} \|u_2 \chi_{\{y:a<\varphi(y)<b\}}\|_{X_1'} \|f\|_{X_1} \leq c_\psi c_\varphi \|f\|_{X_1}.$$

Let $\varepsilon > 0$. Then for every $\alpha \in [a,b]$ there exist c and d with $c < \alpha < d$, such that

$$\|u_1 \chi_{\{c<\psi(y)<\alpha\}}\|_{X_2} < \varepsilon \quad \text{or} \quad \|u_2 \chi_{\{c<\varphi(y)<\alpha\}}\|_{X_1'} < \varepsilon$$

and

$$\|u_1 \chi_{\{\alpha<\psi(y)<d\}}\|_{X_2} < \varepsilon \quad \text{or} \quad \|u_2 \chi_{\{\alpha<\varphi(y)<d\}}\|_{X_1'} < \varepsilon.$$

Thus we obtain an open covering of the segment $[a,b]$ by such intervals (c,d) from which there follows a finite subcovering $\{(c_i,d_i)\}$ with appropriate interior points α_i. The points c_i, α_i and d_i form a partition of $[a,b]$, and we obtain closed intervals $I_j = [\beta_j, \beta_{j+1}]$, $j = 0, 1, \ldots, N$ ($\beta_0 = a, \beta_{N+1} = b$) with the conditions $(\beta_j, \beta_{j+1}) \cap (\beta_i, \beta_{i+1}) = \emptyset$ for $i \neq j$, and $\cup_{j=0}^{N}[\beta_j, \beta_{j+1}] = [a,b]$. Moreover

$$\|u_1 \chi_{\{\beta_j<\psi(y)<\beta_{j+1}\}}\|_{X_2} < \varepsilon \quad \text{or} \quad \|u_2 \chi_{\{\beta_j<\varphi(y)<\beta_{j+1}\}}\|_{X_1'} < \varepsilon.$$

Let

$$Sf(x) = \sum_{j=0}^{N} \chi_{\{\beta_j<\psi(y)\leq\beta_{j+1}\}}(x)u_1(x) \int\limits_{\{a<\varphi(y)<\beta_j\}} u_2(y)f(y)w(y)d\mu.$$

Then

$$T_{ab}f(x) - Sf(x) =$$

$$= \sum_{j=0}^{N} \chi_{\{y:\beta_j<\psi(y)\leq\beta_{j+1}\}}(x)u_1(x) \int\limits_{\{y:\beta_j<\varphi(y)<\psi(x)\}} u_2(y)f(y)w(y)d\mu.$$

This implies that

$$\|T_{ab} - S\| = \sup_{\|f\|_{X_1}\leq 1} \|(T_{ab} - S)f\|_{X_2} =$$

$$= \sup_{\|f\|_{X_1}\leq 1} \sup_{\|g\|_{X_2'}\leq 1} \int_X (T_{ab} - S)f(x)g(x)v(x)d\mu =$$

$$= \sup_{\|f\|_{X_1}\leq 1} \sup_{\{g\}_{X_2'}\leq 1} \sum_{j=0}^{N} \int\limits_{\{x:\beta_j<\psi(x)\leq\beta_{j+1}\}} u_1(x)g(x)v(x) \times$$

$$\times \left(\int\limits_{\{y:\beta_j<\varphi(y)<\psi(x)\}} u_2(y)f(y)w(y)d\mu(y) \right) d\mu(x) \le$$

$$\le \sup_{\|f\|_{X_1}\le 1} \sup_{\|g\|_{X_2'}\le 1} \sum_{j=0}^{N} \left(\int\limits_{\{x:\beta_j<\psi(x)\le\beta_{j+1}\}} u_1(x)g(x)v(x)d\mu \right) \times$$

$$\times \left(\int\limits_{\{y:\beta_j<\varphi(y)<\beta_{j+1}\}} u_2(y)f(y)w(y)d\mu \right) =$$

$$= \sup_{\|f\|_{X_1}\le 1} \sup_{\|g\|_{X_2'}\le 1} \sum_{j=0}^{N} I_{\psi\varphi}.$$

If

$$A_1 = \{j : \|u_1\chi_{\{y:\beta_j<\psi(y)<\beta_{j+1}\}}\|_{X_2} < \varepsilon\}$$

and

$$A_2 = \{j : \|u_2\chi_{\{y:\beta_j<\varphi(y)<\beta_{j+1}\}}\|_{X_1'} < \varepsilon\},$$

then we obtain

$$\sup_{\|f\|_{X_1}\le 1} \sup_{\|g\|_{X_2'}} \sum_{j\in A_1} I_{\psi\varphi} \le \varepsilon c_\varphi$$

and

$$\sup_{\|f\|_{X_1}\le 1} \sup_{\|g\|_{X_2'}} \sum_{j\in A_2} I_{\psi\varphi} \le \varepsilon c_\psi.$$

Finally we have

$$\|T_{ab} - S\| \le \varepsilon(c_\varphi + c_\psi),$$

and since ε was chosen arbitrarily, we conclude that T_{ab} is a limit of finite rank operators. Consequently, T_{ab} is compact.

Thus the sufficiency is proved.

To prove necessity, assume the contrary. Let there exist numbers $\alpha \in [a, b)$ and $\varepsilon > 0$ and a sequence $\{t_n\}$, $t_n \to \alpha+$ such that

$$\|u_1\chi_{\{\alpha<\psi(y)<t_n\}}\|_{X_2} \ge \varepsilon \quad \text{and} \quad \|u_2\chi_{\{\alpha<\varphi(y)<t_n\}}\|_{X_1'} \ge \varepsilon.$$

For $\gamma \in (0,1)$ there exist functions f_n, g_n with support in $[\alpha, t_n]$ such that $\|f_n\|_{X_1} \le 1$, $\|g_n\|_{X_2'} \le 1$ and

$$\int\limits_{\{y:\alpha<\varphi(y)<t_n\}} u_2(y)f_n(y)w(y)d\mu \ge \gamma\|u_2\chi_{\{\alpha<\varphi(y)<t_n\}}\|_{X_1'},$$

$$\int\limits_{\{y:\alpha<\psi(y)<t_n\}} u_1(y)g_n(y)v(y)d\mu \ge \gamma\|u_1\chi_{\{\alpha<\psi(y)<t_n\}}\|_{X_2}.$$

By the condition

$$\mu\{x : \varphi(x) = t\} = \mu\{x : \psi(x) = t\} = 0, \quad t \in [0, \alpha],$$

there exist numbers $\beta_n \in (\alpha, t_n)$ such that

$$\int_{\{y : \beta_n < \varphi(y) < t_n\}} u_2(y) f_n(y) w(y) d\mu \geq \gamma^2 \|u_2 \chi_{\{\alpha < \varphi(y) < t_n\}}\|_{X_1'}$$

and

$$\int_{\{y : \beta_n < \psi(y) < t_n\}} u_1(y) g_n(y) v(y) d\mu \geq \gamma^2 \|u_1 \chi_{\{\alpha < \psi(y) < t_n\}}\|_{X_2}.$$

Let $F_n = f_n \chi_{\{\beta_n < \varphi(y) < t_n\}}$, and let m, k and n be natural numbers such that the condition $t_m < \beta_k < t_k < \beta_n$ is fulfilled for them. Then we obtain

$$\|T_{ab} F_m - T_{ab} F_n\|_{X_2} \geq$$

$$\geq \|\chi_{\{\beta_k < \psi(y) < t_k\}} (T_{ab} F_n - T_{ab} F_m)\|_{X_2} =$$

$$= \|\chi_{\{\beta_k < \psi(y) < t_k\}} u_1 \int_{\{y : \beta_m < \varphi(y) < t_m\}} u_2(y) f_m(y) w(y) d\mu\|_{X_2} =$$

$$= \left(\int_{\{y : \beta_m < \varphi(y) < t_m\}} u_2(y) f_m(y) w(y) d\mu \right) \|u_1 \chi_{\{\beta_k < \psi(y) < t_k\}}\|_{X_2} \geq$$

$$\geq \left(\int_{\{y : \beta_m < \varphi(y) < t_m\}} u_2(y) f_m(y) w(y) d\mu \right) \times$$

$$\times \left(\int_{\{y : \beta_k < \varphi(y) < t_k\}} u_1(y) g_k(y) w(y) d\mu \right) \geq$$

$$\geq \gamma^4 \epsilon^2 > 0.$$

Since $\{T_{ab} F_k\}$ does not converge in X_2, we conclude that T_{ab} is non–compact.

Arguing analogously for

$$T_{ab}' f(x) = u_2(x) \chi_{\{y : a < \varphi(y) < b\}}(x) \times$$

$$\times \int_{\{y : \psi(y) > \varphi(x)\}} \chi_{\{z : a < \psi(z) \leq b\}}(y) f(y) u_1(y) v(y) d\mu$$

which is compact if and only if T_{ab} is compact, we arrive at the condition (1.1.18). $\square$

Theorem 1.1.19. *Let X_1 and X_2 satisfy condition (1.1.5) and let*

$$\sup_{t>0} \|u_1\chi_{\{y:\psi(y)>t\}}\|_{X_2} \|u_2\chi_{\{y:\varphi(y)<t\}}\|_{X_1'} < \infty. \tag{1.1.19}$$

If

$$\lim_{a\to 0+} B_a = 0, \tag{1.1.20}$$

$$\lim_{b\to\infty} B_b = 0, \tag{1.1.21}$$

and conditions (1.1.17) and (1.1.18) are satisfied for every $\alpha \in (0,\infty)$, then the operator T is compact from X_1 to X_2.

Conversely, if the operator T is compact from X_1 to X_2, then the conditions (1.1.19), (1.1.20), (1.1.21) are fulfilled and (1.1.17), (1.1.18) hold for every $\alpha \in (0,\infty)$.

Proof. *Sufficiency.* Let us take numbers a, b with $0 < a < b < \infty$. Then it can be easily verified that

$$Tf(x) = T_a f(x) + T_b f(x) + T_{ab} f(x) + (H_1 + H_2 + H_3),$$

where H_1, H_2, H_3 are finite rank operators. By Lemmas 1.1.3 and 1.1.5 we obtain

$$\|T - T_{ab} - (H_1 + H_2 + H_3)\| \le \|T_a\| + \|T_b\| \le 4k_1^{\frac{1}{\alpha}} k_2^{\frac{1}{\alpha'}} (B_a + B_b).$$

By conditions (1.1.20), (1.1.21) and Lemma 1.1.9 we conclude that the operator T is compact, since it is the limit of compact operators.

Necessity. Let the condition (1.1.20) be not fulfilled. Then there exist $\varepsilon > 0$, a sequence $\{a_n\}$, $a_n \to 0+$ and numbers $t_n \in (0, a_n)$ such that

$$\|u_1\chi_{\{t_n<\psi(y)<a_n\}}\|_{X_2} \|u_2\chi_{\{\varphi(y)<t_n\}}\|_{X_1'} \ge \varepsilon.$$

Moreover, for $\gamma \in (0,1)$ there exist functions f_n with $\|f_n\|_{X_1} \le 1$, such that

$$\int\limits_{\{y:\varphi(y)<t_n\}} u_2(y)f(y)w(y)dw \ge \gamma\|u_2\chi_{\{\varphi(y)<t_n\}}\|_{X_1'}.$$

Since the integral is continuous (the integral continuity is guaranteed by the condition $\mu\{x : \varphi(x) = t\} = 0$, $t \ge 0$) there exist numbers $\beta_n \in (0, t_n)$ such that

$$\int\limits_{\{y:\beta_n<\varphi(y)<t_n\}} u_2(y)f(y)w(y)d\mu \ge \gamma^2\|u_2\chi_{\{\varphi(y)<t_n\}}\|_{X_1'}.$$

let $F_n = f_n \chi_{\{\beta_n < \varphi(y) < t_n\}}$. Then for m and n with $a_m < \beta_n$ we obtain

$$\|TF_m - TF_n\|_{X_2} \geq \|\chi_{\{t_m < \psi(y) < a_m\}}(TF_m - TF_n)\|_{X_2} =$$
$$= \|\chi_{\{t_m < \psi(y) < a_m\}} TF_m\|_{X_2} =$$
$$= \int\limits_{\{y:\beta_m < \varphi(y) < t_m\}} u_2(y) f(y) w(y) d\mu \|u_1 \chi_{\{t_m < \psi(y) < a_m\}}\|_{X_2} \geq$$
$$\geq \gamma^2 \|u_2 \chi_{\{\varphi(y) < t_m\}}\|_{X_1'} \|u_1 \chi_{\{t_m < \psi(y) < a_m\}}\|_{X_2} \geq \gamma^2 \varepsilon$$

which implies that the operator T is non–compact. Using the fact that T' is compact from X_1' to X_2' if and only if T is compact from X_1 to X_2 and repeating the arguments used above, we obtain condition (1.1.21).

Lemma 1.1.9 shows that conditions (1.1.17) and (1.1.18) are fulfilled for every $\alpha \in (0, \infty)$. We now introduce an important quantity, the distance of our operator T from the finite rank operator. Let $\square$

Let

$$\alpha(T) = \inf\{\|T - F\| : \operatorname{rank} F < \infty\}.$$

Two–sided estimates of $\alpha(T)$ are given by the following theorem.

Theorem 1.1.20. *Let X_1 and X_2 satisfy condition* (1.1.5)*, let condition* (1.1.19) *be fulfilled and suppose that conditions* (1.1.17) *and* (1.1.18) *hold for every $\alpha \in (0, \infty)$. Moreover, assume that μ–measurable functions whose supports are sets of the type $\{y \in X : c < \psi(y) < d\}$ (respectively $\{y \in X : c < \varphi(y) < d\}$) are everywhere dense in X_2 (respectively in X_1'). If*

$$h_1 = \lim_{a \to 0+} B_a, \quad h_2 = \lim_{b \to \infty} B_b$$

and

$$h = h_1 + h_2,$$

then

$$\frac{1}{4} h \leq \alpha(T) \leq 4 k_1^{\frac{1}{\alpha}} k_2^{\frac{1}{\alpha}} h. \tag{1.1.22}$$

Proof. As we already know (see the proof of Theorem 1.1.19),

$$T = T_a + T_b + T_{ab} + H$$

where the operator H is of finite rank and T_{ab} is the limit of finite rank operators. Therefore

$$\alpha(T) \leq \alpha(T_a + T_b) \leq \|T_a\| + \|T_b\|$$

and the right–hand side of inequality (1.1.22) follows from Lemmas 1.1.3 and 1.1.4.

To prove the left–hand side of inequality (1.1.22), we need the assumption that functions with supports of the type $\{y \in X : c < \psi(y) < d\}$ are everywhere dense in X_2. Let $\lambda > \alpha(T)$. Then there exists a finite rank operator $G : X_1 \to X_2$, $Gf(x) = \sum_{i=1}^{N} \alpha_i(f)g_i(x)$ $(\alpha_i(f) \in R, g_i \in X_2)$, such that $\alpha(T) \leq \|T - G\| < \lambda$. Moreover, there exists a finite rank operator G_0 such that $\text{supp}G_0 f \subset \{c_0 < \psi(x) < d_0\}$ for some c_0 and d_0 with $0 < c_0 < d_0 < \infty$ and all f, and $\|T - G_0\| < \lambda$.

Let $f \in X_1, \|f\|_{X_1} \leq 1$. Then for $a \in (0, c_0)$ and $t \in (0, a)$ we have

$$\lambda \geq \lambda \|f\chi_{\{y:\varphi(y)<a\}}\|_{X_1} \geq \|(T - G_0)(f\chi_{\{y:0<\varphi(y)<a\}})\chi_{\{y:\psi(y)<a\}}\|_{X_2} =$$

$$= \|T_a f\|_{X_2} \geq \left(\int\limits_{\{y:\phi(y)<t\}} u_2(y)f(y)w(y)d\mu \right) \|u_1\chi_{\{y:t<\psi(y)<a\}}\|_{X_2}.$$

If we take the supremum with respect to all f and t, we get $\lambda \geq h_1$, and since λ does not depend on a, we obtain $\lambda \geq B_a$. Repeating the above arguments, we arrive at $\lambda \geq h_2$, and thus $\lambda \geq \frac{1}{2}(h_1 + h_2)$.

If functions with supports of the form $\{y : c < \varphi(y) < d\}$ are everywhere dense in X_1', we obtain the similar result using the fact that $\alpha(T') \leq 2\alpha(T)$. □

We now assume that σ–finite measures ν and σ defined on X satisfy the following condition

$$\sigma\{x \in X : \varphi(x) = t\} = \nu\{x \in X : \psi(x) = t\} = 0$$

for all $t \geq 0$, where φ and ψ are respectively σ–measurable and ν–measurable non–negative functions on X. In this case we have that $Kf(x) = K_1 f(x)$ for every $x \in X$ and f.

As above we use the following notation:

$$X_1(\sigma) = X_1, \quad X_2(\nu) = X_2.$$

The following theorems are proved analogously to those just given:

Theorem 1.1.21. *Let operator K be compact from X_1 to X_2. Then conditions (1.1.19), (1.1.20), (1.1.21) are satisfied and (1.1.17) and (1.1.18) hold for every $\alpha \in (0, \infty)$. Conversely, if X_1 and X_2 satisfy condition (1.1.5), conditions (1.1.19), (1.1.20), (1.1.21) hold and (1.1.17) and (1.1.18) are fullfilled for every $\alpha \in (0, \infty)$, then K is compact from X_1 to X_2.*

Theorem 1.1.22. *Let X_1 and X_2 satisfy condition (1.1.5). Suppose that condition (1.1.19) is fulfilled and (1.1.17) and (1.1.18) hold for every $\alpha > 0$. Moreover, assume that ν– measurable (respectively σ–measurable) functions whose supports are sets of the type $\{y \in X : c < \psi(y) < d\}$ (respectively $\{y \in X : c < \varphi(y) < d\}$) are everywhere dense in X_2 (respectively in X_1'). Then*

$$\frac{1}{4}h \leq \alpha(K) \leq 4k_1^{\frac{1}{\alpha}} k_2^{\frac{1}{\alpha'}} h,$$

where $h = h_1 + h_2$, $h_1 = \lim\limits_{a \to 0+} B_a$, $h_2 = \lim\limits_{b \to \infty} B_b$.

We now turn our attention to the situation in which the operators act between spaces of Lebesgue or Lorentz type.

Let (X, μ) be a σ–finite measure space. Denote by $L_w^{rs}(X)$ the Lorentz space representing the class of all μ–measurable functions $f : X \to R^1$ for which

$$\|f\|_{L_w^{rs}(X)} = \left(s \int\limits_0^\infty \lambda^{s-1}(w\{x \in X : |f(x)| > \lambda\})^{\frac{s}{r}} d\lambda \right)^{\frac{1}{s}} < \infty \quad (1.1.23)$$

if $1 \le r < \infty$, $1 \le s < \infty$, and

$$\|f\|_{L_w^{r\infty}(X)} = \sup\limits_{\lambda > 0} \lambda(w\{x \in X : |f(x)| > \lambda\})^{\frac{1}{r}} < \infty \quad (1.1.24)$$

for $1 \le r < \infty$.

Note that $L_w^{rr}(X) = L_w^r(X)$ is the usual weighted Lebesgue space, and
$$\|f\|_{L_w^{rr}(X)} = \|f\|_{L_w^r(X)} = \left(\int\limits_X |f(x)|^r w(x) d\mu \right)^{\frac{1}{r}}.$$
The Lorentz space L_w^{rs} is a Banach function space (see [130], [50]) if and only if $r = s = 1$ or $r = s = \infty$ or $r \in (1, \infty)$ and $s \in [1, r]$. If $1 < r < s \le \infty$, then (1.1.23) ((1.1.24)) is a quasi–norm. But with respect to another norm equivalent to (1.1.23) ((1.1.24)) $L_w^{rs}(X)$ is a Banach function space (see [20], [130] for details). Moreover, there exists a positive constant b such that

$$b^{-1}\|f\|_{L_w^{rs}(X)} \le$$
$$\sup\left\{ \int\limits_X fgwd\mu : \|g\|_{L_w^{r's'}(X)} \le 1 \right\} \le b\|f\|_{L_w^{rs}(X)}. \quad (1.1.25)$$

Some important properties of these spaces is in the following lemmas:

Lemma A ([261], [50], [239]) . *Let $1 \le r < \infty$, $1 \le s < \infty$ and $c_0 > 0$. Then there exists a number $k = k(c_0)$ such that*

$$\sum_k \|\chi_{E_k} f\|_{L_w^{rs}(X)}^\sigma \le k \|\chi_{\cup_k E_k} f\|_{L_w^{rs}(X)}^\sigma$$

for all sequences of meaurable sets $\{E_k\}$ with $\sum_k \chi_{E_k} \le c_0$ and for all f, where $\sigma \ge \max\{r, s\}$.

In particular, if $c_0 = 1$, then $k = 1$, and if $1 < s \le r < \infty$ and $\sigma = r$, then $k = c_0$.

Lemma 1.1.10. *Let $1 < \max\{r, s\} \le \min\{p, q\} < \infty$ and $X_1 = L_w^{rs}(X)$, $X_2 = L_v^{pq}(X)$. Then for X_1 and X_2 the condition (1.1.5) is fulfilled.*

Proof. Let $\max\{r, s\} \leq \alpha \leq \min\{p, q\}$. Then $\alpha' \geq \max\{p', q'\}$. By (1.1.25) and Lemma A we conclude that for $L_w^{rs}(X)$ and $L_v^{pq}(X)$ the condition (1.1.5) is fulfilled. $\square$

Corollary 1.1.1. *Let $r = s = 1$ or $r \in (1, \infty)$ and $s \in [1, \infty]$, $p = q = 1$ or $p \in (1, \infty)$ and $q \in [1, \infty]$, $\max\{r, s\} \leq \min\{p, q\}$. Then the operator T is bounded from $L_w^{rs}(X)$ to $L_v^{pq}(X)$ if and only if*

$$B = \sup_{t \geq 0} \|u_1 \chi_{\{\psi(y) \geq t\}}\|_{L_v^{pq}(X)} \|u_2 \chi_{\{\varphi(y) \leq t\}}\|_{L_w^{r's'}(X)} < \infty.$$

Proof. The proof follows from Lemma 1.1.10 and Theorem 1.1.3 for $1 < r, s, p, q < \infty$. The remaining parts can be derived in the same way as in Theorems 1.1.1 and 1.1.3. $\square$

The next corollaries can be establishes in the same way:

Corollary 1.1.2. *Let r, s, p, q satisfy the conditions of Corollary 1.1.1, and let $\mu\{x : \varphi(x) = 0\} = 0$. Then the operator T_1 is bounded from $L_w^{rs}(X)$ to $L_v^{pq}(X)$ if and only if the condition*

$$B_1 = \sup_{t > 0} \|u_1 \chi_{\{\psi(y) > t\}}\|_{L_v^{pq}(X)} \|u_2 \chi_{\{\varphi(y) \leq t\}}\|_{L_w^{r's'}(X)} < \infty$$

is satisfied.

Corollary 1.1.3. *Let r, s, p, q satisfy the condition of Corollary 1.1.1, and let*

$$\mu\{x : \varphi(x) = t\} = \mu\{x : \psi(x) = t\} = 0$$

Then the inequality

$$\|P_{ab}f\|_{L_v^{pq}(X)} \leq c\|f\|_{L_w^{rs}(X)},$$

where the positive constant c does not depend on f, holds if and only if $A < \infty$, where

$$A = \max\{A_1, A_2\}, \quad A_1 = \sup_k\{A_{1k}\}, \quad A_2 = \sup_k\{A_{2k}\},$$

$$A_{1k} = \sup_{(\frac{b}{a})^k < t < (\frac{b}{a})^{k+1}} \|u_1 \chi_{\{t < \psi(y) < (\frac{b}{a})^{k+1}\}}\|_{L_v^{pq}(X)} \times$$

$$\times \|u_2 \chi_{\{b(\frac{b}{a})^k < \varphi(y) < bt\}}\|_{L_w^{r's'}(X)},$$

and

$$A_{2k} = \sup_{(\frac{b}{a})^k < t < (\frac{b}{a})^{k+1}} \|u_1 \chi_{\{(\frac{b}{a})^k < \psi(y) < t\}}\|_{L_v^{pq}(X)} \times$$

$$\times \|u_2\chi_{\{at<\varphi(y)<a(\frac{b}{a})^{k+1}\}}\|_{L_w^{r's'}(X)}.$$

Corollary 1.1.4. *Let r, s, p, q satisfy the conditions of Corollary 1.1.1, and let $\mu\{x : \varphi(x) = t\} = \mu\{x : \psi(x) = t\} = 0$ for all $t \geq 0$. The operator T is bounded from $L_w^{rs}(X)$ to $L_v^{pq}(X)$ if and only if*

$$B = \sup_{t>0} \|u_1\chi_{\{\psi(y)>t\}}\|_{L_v^{pq}(X)}\|u_2\chi_{\{\phi(y)<t\}}\|_{L_w^{r's'}(X)} < \infty. \qquad (1.1.26)$$

Moreover $\|T\| \leq 4B$.

For the map T we also have the criterion for compactness and the estimates for $\alpha(T)$ given in the next two corollaries.

Corollary 1.1.5. *Let r, s p, q satisfy the conditions of Corollary 1.1.4 and let φ, and ψ satisfy the condition of Corollary 1.1.4. The operator T is compact from $L_w^{rs}(X)$ to $L_v^{pq}(X)$ if and only if the condition (1.1.26) is fulfilled and one of the following conditions is also fulfilled:*
(i) *$q < \infty$ or $s > 1$ and*

$$\lim_{a\to 0} \sup_{0<t<a} \|u_1\chi_{\{t<\psi(y)<a\}}\|_{L_v^{pq}(X)}\|u_2\chi_{\{\varphi(y)<t\}}\|_{L_w^{r's'}(X)} = 0, \qquad (1.1.27)$$

$$\lim_{b\to\infty} \sup_{t>b} \|u_1\chi_{\{\psi(y)>t\}}\|_{L_v^{pq}(X)}\|u_2\chi_{\{b<\varphi(y)<t\}}\|_{L_w^{r's'}(X)} = 0 \qquad (1.1.28)$$

or (ii) *$q = \infty$, $s = 1$, and $r \neq 1$, the conditions (1.1.27), (1.1.28) are fulfilled and for every $\alpha \in (0, \infty)$,*

$$\lim_{s\to\alpha+} \|u_1\chi_{\{\alpha<\psi(y)<s\}}\|_{L_v^{p\infty}(X)} = 0 \quad or \quad \lim_{s\to\alpha+} \|u_2\chi_{\{s<\psi(y)<\alpha\}}\|_{L_w^{r'\infty}(X)} = 0$$

and

$$\lim_{s\to\alpha-} \|u_1\chi_{\{s<\psi(y)<\alpha\}}\|_{L_v^{p\infty}(X)} = 0 \quad or \quad \lim_{s\to\alpha-} \|u_2\chi_{\{s<\varphi(y)<\alpha\}}\|_{L_w^{r'\infty}(X)} = 0$$

hold.

Corollary 1.1.6. *Let r, s, p, q satisfy the conditions of Corollary 1.1.1, suppose that φ and ψ satisfy the conditions of Corollary 1.1.4 and let condition (1.1.6) be fulfilled. Then for $X_1 = L_w^{rs}(X)$ and $X_2 = L_v^{pq}(X)$, Theorem 1.1.20 is valid.*

Analogous results for the operator $H_{u_1u_2}f(x) = u_1(x)\int_0^x u_2(y)f(y)w(y)dy$ have been obtained in [Sa6].

Using Lemmas 1.1.1–1.1.3 and Theorems 1.1.1–1.1.4, then we obtain the following results for the dual operators:

$$T^*f(x) = u_1(x)\int_{\{\varphi(y)\geq\psi(x)\}} f(y)u_2(y)w(y)d\mu,$$

and

$$T_1^* f(x) = u_1(x) \int\limits_{\{\varphi(y) > \psi(x)\}} f(y) u_2(y) w(y) d\mu.$$

Corollary 1.1.7. *Let* $1 < \max\{r, s\} \le \min\{p, q\} < \infty$. *Then the operator* T^* *is bounded from* $L_w^{rs}(X)$ *into* $L_v^{pq}(X)$ *if and only if*

$$B = \sup_{t \ge 0} \|u_1 \chi_{\{y:\psi(y) \le t\}}\|_{L_v^{pq}(X)} \|u_2 \chi_{\{y:\varphi(y) \ge t\}}\|_{L_w^{r's'}(X)} < \infty.$$

Moreover, $\|T^*\| \approx B$.

Corollary 1.1.8. *Let* $1 < \max\{r, s\} \le \min\{p, q\} < \infty$. *Assume that* $\mu\{x : \psi(x) = 0\} = 0$. *Then the operator* T_1^* *is bounded from* $L_w^{rs}(X)$ *into* $L_v^{pq}(X)$ *if and only if*

$$B_1 = \sup_{t > 0} \|u_1 \chi_{\{y:\psi(y) \le t\}}\|_{L_v^{pq}(X)} \|u_2 \chi_{\{y:\varphi(y) > t\}}\|_{L_w^{r's'}(X)} < \infty.$$

Moreover, $\|T_1^*\| \approx B_1$.

We conclude this section by introducing the notion of a Hölder's inequality. This provides a general framework within which many of our results can be established.

Definition 1.1.1. A space of homogeneous type (SHT) (X, d, μ) is a topological space X with a complete measure μ such that:

(a) the space of continuous functions with compact supports is everywhere dense in $L_\mu^1(X)$;

(b) there exists a nonnegative real function (quasi–metric) $d : X \times X \to R$ which satisfies the following conditions:

(i) $d(x, x) = 0$ for all $x \in X$;

(ii) $d(x, y) > 0$ for all $x \ne y$, $x, y \in X$;

(iii) there exists a positive constant a_0 such that $d(x, y) \le a_0 d(y, x)$ for every $x, y \in X$;

(iv) there exists a constant a_1 such that $d(x, y) \le a_1(d(x, z) + d(z, y))$ for every $x, y, z \in X$;

(v) for every neighbourhood V of the point $x \in X$ there exists $r > 0$ such that the ball $B(x, r) = \{y \in X : d(x, y) < r\}$ is contained in V;

(vi) the ball $B(x, r)$ is measurable for every $x \in X$ and for arbitrary $r > 0$;

(vii) there exists a constant $b > 0$ such that $\mu B(x, 2r) \le b\mu(B(x, r)) < \infty$ for every $x \in X$ and r, $0 < r < \infty$.

We shall also suppose that there exists a point x_0 such that for all numbers t_1, t_2 with $0 < t_1 < t_2 < \infty$ we have

$$\mu(B(x_0, t_2) \setminus B(x_0, t_1)) > 0$$

(for the definition and some properties of SHT, see [56], [100]).

Let u_1, u_2, v and w be positive μ–measurable functions on X and $x_0 \in X$. Assume that

$$Tf(x) = u_1(x) \int_{\{d(x_0,y) \leq \psi(x)\}} f(y)u_2(y)w(y)d\mu,$$

$$T_1f(x) = u_1(x) \int_{\{d(x_0,y) < \psi(x)\}} f(y)u_2(y)w(y)d\mu$$

and

$$Pf(x) = u_1(x) \int_{\{a\psi(x) < d(x_0,y) < b\psi(x)\}} f(y)u_2(y)w(y)d\mu, \quad 0 < a < b < \infty,$$

where $\psi : X \to [0, \infty)$ is a measurable function with the condition

$$0 < \{x \in X : t_1 < \psi(x) < t_2\} < \infty, \quad 0 < t_1 < t_2 < \infty.$$

Then for the operators T, T_1 and P we can get the boundedness and compactness criteria in Banach function spaces defined on (X, d, μ).

If $X = [0, \infty)$, $d\mu = dx$, $u_1 \equiv 1$, $u_2 \equiv \frac{1}{w}$, $x_0 = 0$ and $\psi(x) = x$, then T is a classical Hardy–type operator.

1.2. An extension of the Hardy transform

In this section we establish two–weighted criteria for some other extensions of Hardy–type transforms on an arbitrary measure space (X, μ) with a σ–finite measure μ. In the sequel we shall assume that for a μ–measurable a.e. positive function $\varphi : X \to [0, \infty)$ the following condition

$$0 < \mu\{x : t_1 < \varphi(x) < t_2\} < \infty$$

holds for every t_1 and t_2 satisfying the condition $0 < t_1 < t_2 < a$, where $a = \sup\{\varphi(x) : x \in X\}$.

Let $w, u_2 : X \to R^1$ (resp. $v, u_1 : (0, a) \to R^1$) be μ–measurable, μ–a.e. positive (resp. Lebesgue–measurable, a.e. positive) functions and let $X_1 = (X_1, \mu, w)$ (resp. $X_2 = (X_2, dx, v)$) be a weighted Banach function space with elements defined on X (on $(0, a)$).

For any μ–measurable f put

$$Hf(t) = u_1(t) \int_{\{y : \varphi(y) < t\}} f(y)u_2(y)w(y)d\mu.$$

We also consider the operator

$$H'g(x) = u_2(x) \int_{\varphi(x)}^{a} g(t)u_1(t)v(t)dt,$$

where $g : (0, a) \to R^1$ is Lebesgue–measurable.

As in the previous section, we shall again assume that X_1 and X_2 satisfy the following condition: for some fixed positive constants d_1 and d_2 there exist a number α, $\alpha > 1$, and positive constants $\overline{k}_1$ and $\overline{k}_2$ such that for the sequences of μ– (resp. Lebesgue–) measurable sets $\{E_k\}$ and $\{F_k\}$, with $\sum_k \chi_{E_k} \le d_1$, $\sum_k \chi_{F_k} \le d_2$, and for μ– (resp. Lebesgue–) measurable function f (resp. g) the inequalities

$$\sum_k \|\chi_{E_k}(\cdot)f(\cdot)\|_{X_1}^{\alpha} \le \overline{k}_1 \left\| f(\cdot)\chi_{\bigcup_k E_k(\cdot)} \right\|_{X_1}^{\alpha}$$

$$\sum_k \|\chi_{F_k}(\cdot)g(\cdot)\|_{X_2'}^{\alpha'} \le \overline{k}_2 \left\| g(\cdot)\chi_{\bigcup_k F_k(\cdot)} \right\|_{X_2'}^{\alpha'} \tag{1.2.1}$$

hold, where $\alpha' = \alpha/(\alpha - 1)$.

The following two lemmas will be of immediate use below in our study of the boundedness of operators.

Lemma 1.2.1. *The operator H is bounded from X_1 to X_2 if and only if the operator H' is bounded from X_2' to X_1'. Moreover, $\|H\| = \|H'\|$.*

Proof.
Let the operator H be bounded from X_1 to X_2. Then we have

$$\|H'g\|_{X_1'} = \sup_{\|f\|_{X_1} \le 1} \left| \int_X (H'g)(x)f(x)d\mu \right| \le$$

$$\le \sup_{\|f\|_{X_1} \le 1} \int_X |f(x)|u_2(x) \left(\int_{\varphi(x)}^{a} |g(y)|u_1(y)v(y)dy \right) w(x)d\mu =$$

$$= \sup_{\|f\|_{X_1} \le 1} \int_0^a |g(y)|u_1(y) \left(\int_{\{x:\, \varphi(x)<y\}} |f(x)|u_2(x)w(x)d\mu \right) v(y)dy$$

$$= \sup_{\|f\|_{X_1} \le 1} \|Hf\|_{X_2} \|g\|_{X_2'} \le \|H\|\|g\|_{X_2'}.$$

Hence $\|H'\| \le \|H\|$. Analogously, we can show the reverse inequality. $\square$

Now let

$$H_1 f(t) = u_1(t) \int\limits_{\{y:\ \varphi(y)>t\}} f(y)u_2(y)w(y)d\mu$$

and

$$H_1' g(x) = u_2(x) \int\limits_0^{\varphi(x)} g(t)u_1(t)v(t)dt.$$

Lemma 1.2.2. *The operator H_1 is bounded from X_1 to X_2 if and only if the operator H_1' is bounded from X_2' to X_1'.*

We omit the proof, since it is similar to that of the previous lemma.

With the help of these lemmas we can now give concrete characterizations of the boundedness of H and H_1.

Theorem 1.2.1. *Let X_1 and X_2 satisfy the condition (1.2.1) and let*

$$\mu\{x:\ \varphi(x) = 0\} = 0.$$

Then the inequality

$$\|Hf\|_{X_2} \le c\|f\|_{X_1}, \quad f \in X_1, \tag{1.2.2}$$

with some positive constant c, holds if and only if

$$D = \sup_{t>0} \left\| u_1 \chi_{(t,a)} \right\|_{X_2} \left\| u_2 \chi_{\{x:\ \varphi(x)<t\}} \right\|_{X_1'} < \infty. \tag{1.2.3}$$

Moreover, if c is the best constant in (1.2.2), then

$$D \le c \le 4(\overline{k_1})^{\frac{1}{\alpha}} (\overline{k_2})^{\frac{1}{\alpha'}} D.$$

Proof. First we shall show that (1.2.3) imples (1.2.2). Thanks to Lemma 1.2.1 it is sufficient to prove that under condition (1.2.3) we have the validity of the inequality

$$\|H' g\|_{X_1'} \le c\|g\|_{X_2'},$$

with some constant c independent of $g \in X_2'$.

Let $g \ge 0$, $g \in X_2'$. For arbitrary $t \in (0, a)$ we have

$$\int\limits_t^a g(y)u_1(y)v(y)dy \le \|g\|_{X_2'} \|\chi_{(t,a)}u_1\|_{X_2} < \infty.$$

Evidently the function

$$I(t) = \int_t^a g(y)u_1(y)v(y)dy$$

is continous and decreasing on $(0, a)$. Moreover, $I(t) \to 0$ as $t \to a$. Now suppose

$$\int_0^a g(y)u_1(y)v(y)dy \in \left(2^m, 2^{m+1}\right]$$

for some integer m. Then for each integer $k \le m$ there exists $t_k \in (0, a)$ such that

$$2^k = \int_{t_k}^a g(y)u_1(y)v(y)dy = \int_{t_k+1}^{t_k} g(y)u_1(y)v(y)dy$$

and

$$2^m = \int_{t_m}^a g(y)u_1(y)v(y)dy.$$

The sequence $\{t_k\}_{k=-\infty}^m$ decreases. Let $\alpha = \lim_{k \to -\infty} t_k$. Then

$$[0, a] = \left(\cup_{k=-\infty}^m [t_{k+1}, t_k) \right) \cup [\alpha, a],$$

where $t_{m+1} = 0$. Thus

$$X = \{x : 0 \le \varphi(x) \le a\} = \bigcup_{k=-\infty}^m \{x \in X : t_{k+1} \le \varphi(x) < t_k\}$$

$$\bigcup \{x : \alpha \le \varphi(x) \le a\}.$$

If

$$\int_0^a g(y)u_1(y)v(y)dy = \infty,$$

then in this case $m = \infty$ and

$$X = \bigcup_{k \in Z} \{x : t_{k+1} \le \varphi(x) < t_k\} \bigcup \{\alpha \le \varphi(x) \le a\} \bigcup$$

$$\{x : \varphi(x) = 0\}.$$

Let $t \in [t_{k+1}, t_k)$; then $I(t) \le \int_{t_{k+1}}^a g(y)u_1(y)v(y)dy \le 2^{k+1}$ as $k \le m$. For $t \in [\alpha, a]$ we have $I(t) \le \int_\alpha^a g(y)u_1(y)v(y)dy \le I(t_k) = 2^k$ for arbitrary $k \le$

m and consequently $I(t) = 0$. Further, let f be a μ–measurable non–negative function on X satisfing $\|f\|_{X_1} \leq 1$. Obviously, we have the representation

$$\int_X (H'g)(x)f(x)w(x)d\mu = \int_{\{x:\,\varphi(x)=0\}} (H'g)(x)f(x)w(x)d\mu +$$

$$+ \int_{\{x:\,0<\varphi(x)\leq a\}} (H'g)(x)f(x)w(x)d\mu =$$

$$= I_1 + I_2$$

Since $\mu\{x:\,\varphi(x)=0\} = 0$, we immediately obtain that $I_1 = 0$.
For I_2 we derive

$$I_2 = \sum_{k\leq m} \int_{\{x:\,t_{k+1}\leq\varphi(x)<t_k\}} (H'g)(x)f(x)w(x)d\mu =$$

$$= \sum_{k\leq m} \int_{\{x:\,t_{k+1}\leq\varphi(x)<t_k\}} u_2(x)f(x)I(\varphi(x))w(x)d\mu \leq$$

$$\leq \sum_{k\leq m} I(t_{k+1}) \int_{\{x:\,t_{k+1}\leq\varphi(x)<t_k\}} u_2(x)w(x)f(x)d\mu =$$

$$= \sum_{k\leq m} 2^{k+1} \int_{\{x:\,t_{k+1}\leq\varphi(x)<t_k\}} u_2(x)w(x)f(x)d\mu =$$

$$= 4 \sum_{k\leq m} \left(\int_{t_k}^{t_{k-1}} g(y)u_1(y)v(y)dy \right) \times$$

$$\times \left(\int_{\{x:t_{k+1}\leq\varphi(x)<t_k\}} u_2(x)w(x)f(x)d\mu \right) \leq$$

$$\leq 4 \sum_{k\leq m} \|g\cdot\chi_{(t_k,t_{k-1})}\|_{X_2'} \|u_1\chi_{(t_k,t_{k-1})}\|_{X_2} \times$$

$$\times \|u_2\chi_{\{x:\,t_{k+1}\leq\varphi(x)<t_k\}}\|_{X_1'} \|f\chi_{\{x:\,t_{k+1}\leq\varphi(x)<t_k\}}\|_{X_1} \leq$$

$$\leq 4D \sum_{k\leq m} \|g\chi_{(t_k,t_{k-1})}\|_{X_2'} \|f\chi_{\{x:\,t_{k+1}\leq\varphi(x)<t_k\}}\|_{X_1} \leq$$

$$\leq 4D \left(\sum_{k\leq m} \|g\chi_{(t_k,t_{k-1})}\|_{X_2'}^{\alpha'} \right)^{\frac{1}{\alpha'}} \left(\sum_{k\leq m} \|f\chi_{\{x:\,t_{k+1}\leq\varphi(x)<t_k\}}\|_{X_1}^{\alpha} \right)^{\frac{1}{\alpha}} \leq$$

$$\leq 4D(\overline{k_1})^{\frac{1}{\alpha}}(\overline{k_2})^{\frac{1}{\alpha'}} \|g\|_{X_2'} \|f\|_{X_1} \leq$$

$$\leq 4D(\overline{k_1})^{\frac{1}{\alpha}}(\overline{k_2})^{\frac{1}{\alpha'}} \|g\|_{X_2'}.$$

Taking the supremum with respect to all such f, we obtain the validity of inequality (1.2.2).

Now it remains to prove that (1.2.2) imples (1.2.3). Let $t \in (0, a)$ and let $f \geq 0$ with $\|f\|_{X_1} \leq 1$. Then

$$c \geq c \|f\|_{X_1} \geq C \|Hf\|_{X_2} \geq \left\| (Hf) \chi_{(t,a)} \right\|_{X_2} \geq$$
$$\geq \left\| u_1 \chi_{(t,a)} \right\|_{X_2} \int_{\{x:\ \varphi(x) < t\}} u_2(x) w(x) f(x) d\mu.$$

If we take the supremum with respect to all f and t, then we obtain the condition (1.2.3). $\square$

Theorem 1.2.2. *Let* $\mu \{x : \varphi(x) = a\} = 0$. *Assume that* X_1 *and* X_2 *satisfy condition* (1.2.1). *Then the inequality*

$$\|H_1 f\|_{X_2} \leq c \|f\|_{X_1}, \tag{1.2.4}$$

holds with some constant c independent of f if, and only if

$$D_1 = \sup_{0 < t < a} \left\| \chi_{(0,t)} u_1 \right\|_{X_2} \left\| \chi_{\{x:\ \varphi(x) > t\}} u_2 \right\|_{X_1'} < \infty \tag{1.2.5}.$$

Moreover, if c is the best constant in (1.2.4) *then*

$$D_1 \leq c \leq 4 (\overline{k_1})^{1/\alpha} (\overline{k_2})^{1/\alpha'} D_1.$$

Proof. Let us show that (1.2.5) implies (1.2.4). By Lemma 1.2.2, it is sufficient to prove that (1.2.5) guarantees the validity of the inequality

$$\|H_1' g\|_{X_1'} \leq c \|g\|_{X_2'},$$

for arbitrary $g \in X_2'$, where c is independent of g.

Let $g \geq 0$, $g \in X_2'$. Then it is easy to see that

$$J(t) = \int_0^t g(y) u_1(y) v(y) dy < \infty$$

for any $t \in (0, a)$. Clearly, the function J is continuous and increasing on $(0, a)$ and $\lim_{t \to 0} J(t) = 0$. Let

$$\int_0^a g(y) u_1(y) v(y) dy \in \left(2^m, 2^{m+1} \right]$$

for some integer m. Then by continuity of J, given any integer $k, k \leq m$, there exists $t_k, t_k \in (0, a)$, such that

$$2^k = J(t_k) = \int\limits_{t_k}^{t_{k+1}} g(y)u_1(y)v(y)dy$$

for $k \leq m - 1$ and

$$2^m = J(t_m).$$

The sequence $\{t_k\}_{k=-\infty}^{m}$ increases. Put $\lim\limits_{k \to -\infty} t_k = \beta$. Then

$$X = \{x \in X : 0 \leq \varphi(x) \leq a\} =$$

$$= \{x : 0 \leq \varphi(x) \leq \beta\} \bigcup \left(\bigcup_{j=-\infty}^{m} \{x : t_j < \varphi(x) \leq t_{j+1}\} \right),$$

where $t_{m+1} = a$. If $\int\limits_0^a g(y)u_1(y)v(y)dy = \infty$, then $m = \infty$ and

$$X = \{x \in X : 0 \leq \varphi(x) \leq \beta\} \bigcup$$

$$\bigcup \left(\bigcup_{j=-\infty}^{+\infty} \{x : t_j < \varphi(x) \leq t_{j+1}\} \right) \bigcup \{x : \varphi(x) = a\}.$$

If $t \in [0, \beta]$, then $J(t) \leq J(t_k) = 2^k$ for arbitrary $k, k \leq m$. Thus $J(t) = 0$. If $t \in (t_j, t_{j+1}]$ for $j \leq m$ then

$$J(t) \leq J(t_{j+1}) = 2^{j+1}.$$

Now for μ– measurable non–negative function f, $\|f\|_{X_1} \leq 1$, we have

$$\int\limits_X (H_1'g(x))f(x)w(x)d\mu = \int\limits_{\{x:\varphi(x)=a\}} (H_1'g(x))f(x)w(x)d\mu +$$

$$+ \int\limits_{\{x:0\leq\varphi(x)<a\}} (H_1'g(x))f(x)w(x)d\mu = I_1 + I_2.$$

Since $\mu\{x \in X : \varphi(x) = a\} = 0$, we have that $I_1 = 0$.
As to I_2, we find that

$$I_2 = \sum_{j\leq m} \int\limits_{\{x:t_j<\varphi(x)\leq t_{j+1}\}} (H_1'g(x))f(x)w(x)d\mu =$$

$$= \sum_{j \leq m} \int_{\{x:\, t_j < \varphi(x) \leq t_{j+1}\}} u_2(x) J(\varphi(x)) f(x) w(x) d\mu \leq$$

$$\leq \sum_{j \leq m} J(t_{j+1}) \int_{\{x:\, t_j < \varphi(x) \leq t_{j+1}\}} u_2(x) f(x) w(x) d\mu =$$

$$= \sum_{j \leq m} 2^{j+1} \int_{\{x:\, t_j < \varphi(x) \leq t_{j+1}\}} u_2(x) f(x) w(x) d\mu =$$

$$= 4 \sum_{j \leq m} 2^{j-1} \int_{\{x:\, t_j < \varphi(x) \leq t_{j+1}\}} u_2(x) f(x) w(x) d\mu =$$

$$= 4 \sum_{j \leq m} \left(\int_{t_{j-1}}^{t_j} g(y) u_1(y) v(y) dy \right) \times$$

$$\times \left(\int_{\{x:\, t_j < \varphi(x) \leq t_{j+1}\}} u_2(x) f(x) w(x) d\mu \right) \leq$$

$$\leq 4 \sum_{j \leq m} \left\| g\chi_{(t_{j-1}, t_j)} \right\|_{X_2'} \left\| u_1 \chi_{(0, t_j)} \right\|_{X_2} \times$$

$$\times \left\| u_2 \chi_{\{x:\, t_j < \varphi(x) \leq t_{j+1}\}} \right\|_{X_1'} \left\| f\chi_{\{x:\, t_j < \varphi(x) \leq t_{j+1}\}} \right\|_{X_1} \leq$$

$$\leq 4 D_1 \sum_{j \leq m} \left\| g\chi_{(t_{j-1}, t_j)} \right\|_{X_2'} \left\| f\chi_{\{x:\, t_j < \varphi(x) \leq t_{j+1}\}} \right\|_{X_1} \leq$$

$$\leq 4 D_1 \left(\sum_{j \leq m} \left\| g\chi_{(t_{j-1}, t_j)} \right\|_{X_2'}^{\alpha'} \right)^{\frac{1}{\alpha'}} \times$$

$$\times \left(\sum_{j \leq m} \left\| f\chi_{\{x:\, t_j < \varphi(x) \leq t_{j+1}\}} \right\|_{X_1}^{\alpha} \right)^{\frac{1}{\alpha}} \leq$$

$$\leq 4 D_1 (\overline{k_1})^{\frac{1}{\alpha}} (\overline{k_2})^{\frac{1}{\alpha'}} \|g\|_{X_2'} \|f\|_{X_1} \leq 4 D_1 (\overline{k_1})^{\frac{1}{\alpha}} (\overline{k_2})^{\frac{1}{\alpha'}} \|g\|_{X_2'} .$$

If we take the supremum with respect to all such f, we obtain (1.2.4).

In order to prove necessity, we take $f \geq 0$ with $\|f\|_{X_1} \leq 1$ and $t \in (0, a)$. Then from (1.2.4) we have

$$c \geq c \|f\|_{X_1} \geq \|H_1 f\|_{X_2} \geq \left\| (H_1 f) \chi_{(0,t)} \right\|_{X_2} \geq$$

$$\geq \left\| u_1 \chi_{(0,t)} \right\|_{X_2} \int_t^a u_2(x) f(x) w(x) d\mu.$$

Taking the supremum with respect to all f and $t \in (0, a)$ we obtain (1.2.5). $\square$

Theorems 1.2.1 and 1.2.2 imply the following results in the setting of Lorentz spaces.

Theorem 1.2.3. *Let* $\mu\{x : \varphi(x) = 0\} = 0, r = s = 1$ *or* $r \in (1, \infty)$ *and* $s \in [1, \infty]$, $p = q = 1$ *or* $p \in (1, \infty)$ *and* $q \in [1, \infty]$. *Suppose that* $\max\{r, s\} \le \min\{p, q\}$. *Then the inequality*

$$\|Hf\|_{L_v^{pq}(0,a)} \le c\|f\|_{L_w^{rs}(X)} \quad f \in L_w^{rs}(X),$$

with the positive constant c independent of f, is fulfilled if and only if

$$\overline{D} = \sup_{0<t<a} \left\|u_1\chi_{(t,a)}\right\|_{L_v^{pq}(0,a)} \left\|u_2\chi_{\{x:\,\varphi(x)<t\}}\right\|_{L_w^{r's'}(X)} < \infty.$$

Moreover, $\overline{D} \le \|H\| \le 4\overline{D}$.

Theorem 1.2.4. *Let* $\mu\{x : \varphi(x) = 0\} = 0$. *Assume that that* r, s, p *and* q *satisfy the conditions of Theorem 1.2.3. Then the operator* H_1 *is bounded from* $L_w^{rs}(X)$ *to* $L_v^{pq}(0, a)$ *if and only if*

$$\overline{D}_1 = \sup_{0<t<a} \left\|u_1\chi_{(0,t)}\right\|_{L_v^{pq}(0,a)} \left\|u_2\chi_{\{x:\,\varphi(x)>t\}}\right\|_{L_w^{r's'}(x)} < \infty.$$

Moreover, $\overline{D}_1 \le \|H\| \le 4\overline{D}_1$.

Further, specialization of Theorems 1.2.3 and 1.2.4 gives the next corollaries, which are set in Lebesgue spases.

Corollary 1.2.1. *Let* $1 < p \le q < \infty$ *and* $\mu\{x : \varphi(x) = 0\} = 0$. *Then the inequality*

$$\left(\int_0^a v(t) \left|\int_{\{x:\,\varphi(x)<t\}} f(x)d\mu\right|^q dt\right)^{\frac{1}{q}} \le c\left(\int_X |f(x)|^p w(x)d\mu\right)^{\frac{1}{p}}, \quad (1.2.6)$$

holds with some $c > 0$ *independent of* f, $f \in L_w^p(X)$, *if and only if*

$$\tilde{D} = \sup_{0<t<a} \left(\int_t^a v(\tau)d\tau\right)^{\frac{1}{q}} \left(\int_{\{x:\,\varphi(x)<t\}} w^{1-p'}(x)d\mu\right)^{\frac{1}{p'}} < \infty.$$

Moreover, if c is the best constant in (1.2.6), *then* $\tilde{D} \le c \le 4\tilde{D}$.

Corollary 1.2.2. *Let* $1 < p \le q < \infty$ *and* $\mu\{x : \varphi(x) = a\} = 0$. *Then the inequality*

$$\left(\int_0^a v(t) \left|\int_{\{x:\,\varphi(x)>t\}} f(x)d\mu\right|^q dt\right)^{\frac{1}{q}} \le c\left(\int_X |f(x)|^p w(x)d\mu\right)^{\frac{1}{p}}, \quad (1.2.7)$$

holds with some constant c independent of f, $f \in L_w^p(X)$, if and only if

$$\tilde{D}_1 = \sup_{0 < t < a} \left(\int_0^t v(\tau)d\tau \right)^{\frac{1}{q}} \left(\int_{\{x: \varphi(x) > t\}} w^{1-p'}(x)d\mu \right)^{\frac{1}{p'}} < \infty.$$

Moreover, if c is the best constant in (1.2.7) then we have that $\tilde{D}_1 \le c \le 4\tilde{D}_1$.

Below we assume that (X, μ) and (Y, ν) are measure spaces and $u_{1,\psi} : Y \to R^1$ and $u_{2,\varphi} : X \to R^1$ are pairs of a.e. positive functions which are respectively $\nu-$ measurable and $\mu-$ measurable. Put

$$H_{ab}f(x) = u_1(x) \int_{\{y: a < \varphi(y) < \psi(x) < b\}} f(y)u_2(y)d\mu$$

and

$$H_{1ab}f(x) = u_1(x) \int_{\{y: b > \varphi(y) > \psi(x) > a\}} f(y)u_2(y)d\mu,$$

where $-\infty \le a < b \le \infty$.

We also assume that for every a and b, $-\infty \le a < b \le \infty$ the inequalities $\nu\{y \in Y : \psi(y) \in (a, b)\} > 0$, $\mu\{x \in X : \varphi(x) \in (a, b)\} > 0$ hold.

Denote by Y_{ab} (X_{ab}) the set $\{y \in Y : \psi(y) \in (a, b)\}$ $(\{x \in X : \varphi(y) \in (a, b)\})$. Let $X_1 = X_1(X_{ab}, \mu)(X_2 = X_2(Y_{ab}, \nu))$ be the Banach function space with elements defined on the set X_{ab} and measure μ $(Y_{ab}$ and measure $\nu)$.

Characterizations of the boundedness of H_{ab} and H_{1ab} are provided by the next two theorems.

Theorem 1.2.5. *Let $X_1 = X_1(X_{ab}, \mu)$ and $X_2 = X_2(Y_{ab}, \nu)$ satisfy the condition (1.2.1). Then for the boundedness of the operator H_{ab} from X_1 to X_2 it is necessary and sufficient that*

$$B_{ab} = \sup_{a < t < b} \left\| u_1 \chi_{\{x: t < \psi(x) < b\}} \right\|_{X_2} \left\| u_2 \chi_{\{y: a < \varphi(x) \le t\}} \right\|_{X_1'} < \infty. \quad (1.2.8)$$

Moreover, $B_{ab} \le \|H_{ab}\| \le 4(\overline{k}_1)^{\frac{1}{\alpha}}(\overline{k}_2)^{\frac{1}{\alpha'}} B_{ab}$.

Proof. Let $f \ge 0$, $f \in X_1$ and let

$$I_{ab}(s) = \int_{\{y: a < \varphi(x) < s\}} f u_2 d\mu.$$

By condition (1.2.8) we have that $I(s) < \infty$ for any $s \in (a, b)$. Moreover, the function $I(s)$ increases and $\lim_{s \to a+} I(s) = 0$. It is easy to verify that $I(s)$ is left–continuous. Suppose that $\int_{X_{ab}} fu_2 d\mu \in (2^m, 2^{m+1}]$, for some $m \in Z$ and let $s_j = \sup\{s : I(s) \leq 2^j\}$, where $j \leq m - 1$. The sequence $\{s_j\}_{j=-\infty}^{m+1}$ is non–decreasing and $s_{m+1} = b$. Moreover, $I(s_j) \leq 2^j, I(s) \leq 2^j$ if $s > s_j$ and

$$\int_{\{y: s_j \leq \varphi(y) \leq s_{j+1}\}} fu_2 d\mu \geq 2^j.$$

Let $\beta = \lim_{j \to -\infty} s_j$. Then $(a, b) = \left(\bigcup_{j \in J_m} (s_j, s_{j+1}]\right) \cup (a, \beta]$, where

$$J_m = \{j \leq m + 1, s_j < s_{j+1}\}.$$

Consequently $Y_{ab} = \left(\bigcup_{j \in J_m} E_j\right) \cup F$, where

$$E_j = \{x \in Y_{ab} : s_j < \psi(x) \leq s_{j+1}\}$$

and $F = \{x : a < \psi(x) \leq \beta\}$. If $\int_{X_{ab}} fu_2 d\mu = \infty$, then we take $m = \infty$ and in this case $Y_{ab} = \left(\bigcup_{j \in J} E_j\right) \cup F$, where $J = \{j \in Z, s_j < s_{j-1}\}$. It is easy to see that if $s \in (s_j, s_{j+1}]$, then $I(s) \leq 2^{j+1}$ for $j \leq m$; and if $s \in [0, \beta]$, then $I(s) = 0$. Let $g \in X_2', g \geq 0$ and $\|g\|_{X_2'} \leq 1$. We have

$$S = \int_{Y_{ab}} (H_{ab}f(x)) g(x) d\nu = \sum_j \int_{E_j} (H_{ab}f(x)) g(x) d\nu \leq$$

$$\leq \sum_j 2^{j+1} \int_{E_j} u_1(x)g(x) d\nu = 4 \sum_j 2^{j-1} \int_{E_j} u_1(x)g(x) d\nu =$$

$$= 4 \sum_j \left(\int_{\{y: s_{j-1} \leq \varphi(y) \leq s_j\}} fu_2 d\mu\right)\left(\int_{E_j} u_1(x)g(x) d\nu\right).$$

Arguing as in the proof of Theorem 1.2.1, we easily obtain the sufficiency.

Now let $t \in (a, b)$ and let $f \in X_1, f \geq 0$, with $\|f\|_{X_1} \leq 1$. Then from (1.2.1) we get

$$c \geq c\|f\|_{X_1} \geq \|H_{ab}f\|_{X_2} \geq \left\|\chi_{\{x: t<\psi(x)<b\}} H_{ab}f\right\|_{X_2} \geq$$

$$\geq \int_{\{y: a<\varphi(y) \leq t\}} fu_2 d\mu \left\|u_1 \chi_{\{x: t<\psi(x)<b\}}\right\|_{X_2}.$$

If we take the supremum with respect to all f and t, then we obtain the condition (1.2.8). □

Theorem 1.2.6. *Let $X_1 = X_1(X_{ab}, \mu)$ and $X_2 = X_2(Y_{ab}, \nu)$ satisfy the condition (1.2.1). Then the operator H_{1ab} is bounded from X_1 to X_2 if and only if*

$$\overline{B_{1ab}} = \sup_{a<t<b} \left\| u_2 \chi_{\{y: \, t<\varphi(y)<b\}} \right\|_{X_1'} \left\| u_1 \chi_{\{x: \, a<\psi(x)\leq t\}} \right\|_{X_2} < \infty.$$

Moreover, $\overline{B_{1ab}} \leq \|H_{1ab}f\| \leq 4(\overline{k_1})^{\frac{1}{\alpha}}(\overline{k_2})^{\frac{1}{\alpha'}}\overline{B_{1ab}}$.

The proof is similar to that of the previous theorem.

Now let $-\infty \leq a < b \leq \infty$ and suppose that μ and ν are non–negative Borel measures on (a,b). Let u_1 and v (resp. u_2 and w) be a μ–measurable and μ–a.e. positive (resp. ν– measurable and ν– a.e.positive) pair of functions on (a,b). From Theorem 1.2.5 we obtain

Theorem 1.2.7. *Let $r = s = 1$ or $r \in (1,\infty)$ and $s \in [1,\infty]$; $p = q = 1$ or $p \in (1,\infty)$ and $q \in [1,\infty]$. Assume that $\max\{r,s\} \leq \min\{p,q\}$. Then the operator*

$$K_{ab}f(x) = u_1(x) \int_a^x u_2(y) f(y) w(y) d\mu$$

is bounded from $L_w^{rs}(\mu, (a,b))$ to $L_v^{pq}(\nu, (a,b))$ if and only if

$$D_{ab} = \sup_{a<t<b} \left\| u_1 \chi_{(t,b)} \right\|_{L_v^{pq}(\nu,(a,b))} \left\| u_2 \chi_{(a,t]} \right\|_{L_w^{rs}(\mu,(a,b))} < \infty.$$

Moreover, $D_{ab} \leq \|K_{ab}\| \leq dD_{ab}$ for some positive constant d.

For μ– measurable f we now put

$$K'_{ab}f(x) = u_1(x) \int_x^b u_2(y) f(y) w(y) d\mu.$$

Then from Theorem 1.2.6 we can obtain the following

Theorem 1.2.8. *Let r, s, p and q satisfy the conditions of Theorem 1.2.7. Then the operator K'_{ab} is bounded from $L_w^{rs}(\mu, (a,b))$ to $L_v^{pq}(\nu, (a,b))$ if and only if*

$$D_{1ab} = \sup_{a<t<b} \left\| u_2 \chi_{(t,b)} \right\|_{L_w^{r's'}(\mu,(a,b))} \left\| u_1 \chi_{(a,t]} \right\|_{L_v^{pq}(\nu,(a,b))} < \infty.$$

Moreover, $D_{1ab} \leq \|K'_{ab}\| \leq d_1 D_{1ab}$, where d_1 is a positive constant.

Now we discuss weak–type inequalities for the Hardy–type operators. Let (X, d, μ) be a measure space with quasi–metric d. We shall assume that:

$$a \equiv \sup\{d(x_0, x) : x \in X\},$$

$$\overline{P}_\eta f(x) = \frac{1}{(\mu B(x_0, d(x_0, x)))^\eta} \int\limits_{B(x_0, d(x_0, x))} f(y) d\mu(y),$$

$$\overline{Q}_\eta f(x) = \frac{1}{(\mu B(x_0, d(x_0, x)))^\eta} \int\limits_{\{y:\, d(x_0, y) > d(x_0, x)\}} f(y) d\mu(y),$$

where $\eta \in R$ and $x_0 \in X$.

We also assume that whenever all $0 < \tau < t < \infty$ the following condition is satisfied

$$\mu(B(x_0, t) \setminus B(x_0, \tau)) > 0.$$

Theorem 1.2.9. *Let (X, D, μ) be an SHT, $1 \le p \le q < \infty$, $\mu\{x_0\} = 0$ and $\mu(X) = \infty$. Suppose that $\eta > 0$ and v and w are μ– measurable, μ–a.e. positive functions on X. Then the operator $\overline{P}_\eta$ is bounded from $L^p_w(X)$ to $L^{q\infty}_v(X)$ if and only if*

$$A_\eta \equiv \sup_{\tau > t} A_\eta(t, \tau) \equiv$$

$$\equiv \sup_{\tau > t} \frac{1}{(\mu B(x_0, \tau))^\eta} \left(\int\limits_{t < d(x_0, x) < \tau} v(y) d\mu \right)^{\frac{1}{q}} \times$$

$$\times \left(\int\limits_{\{d(x_0, x) \le t\}} w^{1-p'}(y) d\mu \right)^{\frac{1}{p'}} < \infty$$

for $p > 1$ and

$$A_{1,\eta} \equiv \sup_{\tau > t} A_{1,\eta}(t, \tau) \equiv \sup_{\tau > t} \frac{1}{(\mu B(x_0, \tau))^\eta} \times$$

$$\times \left(\int\limits_{t < d(x_0, y) < \tau} v(y) d\mu \right)^{\frac{1}{q}} \operatorname*{ess\,sup}_{d(x_0, x) \le t} \frac{1}{w(y)} < \infty$$

for $p = 1$. Moreover, $\left\| \overline{P}_\eta \right\| \approx A_\eta$ when $p > 1$ and $\left\| \overline{P}_\eta \right\| \approx A_{1,\eta}$ when $p = 1$.

Proof. Let us deal with the case $p > 1$. From Corollary 1.1.2 we deduce that the boundness of the operator $\overline{P}_\eta$ from $L^p_w(X)$ to $L^q_v(X)$ is equivalent to the validity of the condition

$$\overline{A}_\eta = \sup_{t > 0} \overline{A}_\eta(t) = \sup_{t > 0} \left\| (\mu B(x_0, d(x_0, \cdot)))^{-\eta} \chi_{\{y:\, d(x_0, y) > t\}}(\cdot) \right\|_{L^{q\infty}_v(X)} \times$$

$$\times \left\| \frac{1}{w(\cdot)} \chi_{\{y:\, d(x_0, y) \le t\}}(\cdot) \right\|_{L_w^p(X)} < \infty$$

(note that from the condition $\mu(X) = \infty$ it follows that $a = \infty$, as otherwise $X \subset B(x_0, a)$ and $\mu(X) < \mu B(x_0, a) < \infty$).

Let

$$h(t) \equiv \left\| \chi_{\{y:\, d(x_0, y) > t\}}(\cdot)(\mu B(x_0, d(x_0, \cdot)))^{-\eta} \right\|_{L_v^{q\infty}(X)}.$$

From the definition of $L_v^{q\infty}(X)$ we have

$$h(t) = \sup_{\lambda < (\mu B(x_0, t))^{-\eta}} \lambda \left(\int_{\{x:\, d(x_0, x) > t, (\mu B(x_0, d(x_0, x))^{-\eta} > \lambda\}} v(x) d\mu \right)^{\frac{1}{q}} \equiv$$

$$\equiv \sup_{\lambda < (\mu B(x_0, t))^{-\eta}} G(\lambda).$$

Assume that $0 < \lambda < (\mu B(x_0, t))^{-\eta}$. Then there exists $\tau > 0$ such that $\mu\left(B(x_0, \frac{\tau}{2})\right) \le \lambda^{-\frac{1}{\eta}} < \mu\left(B(x_0, \tau)\right)$ and

$$G(\lambda) \le \mu\left(B\left(x_0, \frac{\tau}{2}\right)\right)^{-\eta} \times$$

$$\times \left(\int_{\{x:\, d(x_0, t) < \mu B(x_0, d(x_0, x) < \mu(B(x_0, \tau))\}} v(x) d\mu \right)^{\frac{1}{q}} \le$$

$$\le b\left(\mu B(x_0, \tau)\right)^{-\eta} \left(\int_{\{t < d(x_0, x) < \tau\}} v(x) d\mu \right)^{\frac{1}{q}} \le$$

$$\le b \sup_{\tau > t} \left(\mu B(x_0, \tau)\right)^{-\eta} \left(\int_{\{t < d(x_0, x) < \tau\}} v(x) d\mu \right)^{\frac{1}{q}}.$$

Consequently, we have

$$A_\eta \le b \sup_{\tau > t} \frac{1}{(\mu B(x_0, \tau))^\eta} \left(\int_{\{x \in X:\, t < d(x_0, x) < \tau\}} v(x) d\mu \right)^{\frac{1}{q}} \times$$

$$\times \left(\int_{\{x \in X:\, d(x_0, x) \le t\}} w^{1-p'}(x) d\mu \right)^{\frac{1}{p'}} < \infty.$$

The proof when $p = 1$ is analogous.

Now we proceed to the proof of necessity. Let $1 < p \leq q < \infty$ and put

$$\overline{B}(x_0, t) \equiv \{x \in X : d(x_0, x) \leq t\}.$$

If $H(t) = \int_{\overline{B}(x_0,t)} w^{1-p'}(y)dy = 0$ for some $t > 0$, then $A_\eta(t, \tau) = 0$. If $H(t) = \infty$, then $w^{-1/p} \notin L^{p'}(\overline{B}(x_0, t))$. Consequently, there exists a non-negative function $g \in L^p(\overline{B}(x_0, t))$ such that $gw^{-1/p} \in L^1(\overline{B}(x_0, t))$. Let $f(x) = g(x)w^{-\frac{1}{p}}(x)\chi_{\overline{B}(x_0,t)}$. If $d(x_0, x) > t$, then

$$\overline{P}_\eta f(x) \geq (\mu B(x_0, d(x_0, x)))^{-\eta} \int_{B(x_0,t)} f d\mu = \infty$$

and using the boundness of the operator $\overline{P}_\eta$ we find that

$$\left(\int_{\{x:\, d(x_0,x)>t\}} v d\mu \right)^{\frac{1}{q}} \leq \left(\int_{\{x:\, \overline{P}_\eta f(x)>\lambda\}} v d\mu \right)^{\frac{1}{q}} \leq$$

$$\leq \frac{c}{\lambda}\left(\int_{\overline{B}(x_0,t)} f^p w d\mu \right)^{\frac{1}{p}} < \infty$$

for all $\lambda > 0$. Therefore

$$\left(\int_{\{x:\, d(x_0,x)>t\}} v d\mu \right)^{\frac{1}{q}} = 0$$

and thus $v = 0$ μ– a.e. on the set $\{x : d(x_0, x) > t\}$. Hence we can assume that $0 < H(t) < \infty$ for all $t > 0$. Let $f(x) = w^{1-p'}(x)\chi_{\overline{B}(x_0,t)}$. Then for any x satisfying $t < d(x_0, x) < \tau$ we have

$$\overline{P}_\eta f(x) \geq \frac{1}{(\mu B(x_0, \tau))^\eta} \int_{\overline{B}(x_0,t)} w^{1-p'}(x)d\mu \equiv H_1(t, \tau).$$

Using the boundedness of the operator $\overline{P}_\eta$ we derive

$$\left(\int_{\{x:\, t<d(x_0,x)<\tau\}} v d\mu \right)^{\frac{1}{q}} \leq \left(\int_{\{x:\, \overline{P}_\eta f(x)\geq H_1(t,\tau)\}} v d\mu \right)^{\frac{1}{q}} \leq$$

$$\leq \frac{c}{H_1(t,\tau)}\left(\int\limits_{\overline{B}(x_0,t)} w^{1-p'}(x)d\mu\right)^{\frac{1}{p}} =$$

$$= c\left(\mu B(x_0,\tau)\right)^{\eta}\left(\int\limits_{\overline{B}(x_0,t)} w^{1-p'}(x)d\mu\right)^{-\frac{1}{p'}}$$

and hence $A_\eta < \infty$.

Now let $p = 1$. Let us take a positive ε and let E be a set of positive μ measure such that $E \subset \overline{B}(x_0,t)$ and

$$w(x) \leq \varepsilon + \operatorname*{ess\,inf}_{x\in\overline{B}(x_0,t)} w(x)$$

on E. Let $f(x) = \chi_E(x)$ and let $t < d(x_0,x) < \tau$. Then

$$\overline{P}_\eta f(x) \geq \frac{\mu(E)}{\mu B(x_0,\tau)^\eta}$$

and by the boundedness of the operator $\overline{P}_\eta$ we obtain

$$\left(\int\limits_{\{x:\,t<d(x_0,x)<\tau\}} vd\mu\right)^{\frac{1}{q}} \leq c_1\mu B(x_0,\tau)^\eta\left(\varepsilon + \operatorname*{ess\,inf}_{x\in\overline{B}(x_0,t)} w(x)\right).$$

Finally, we have $A_{1,\eta} < \infty$. $\square$

The following theorem we can obtain analogously:

Theorem 1.2.10. *Let (X,D,μ) be an SHT, $1 \leq p \leq q < \infty$, $\eta < 0$ and $\mu(X) = \infty$. Then for the boundedness of the operator $\overline{Q}_\eta$ from $L_w^p(X)$ to $L_v^{q\infty}(X)$ it is necessary and sufficient that*

$$B_\eta \equiv \sup_{\tau>t}\left(\mu B(x_0,t)\right)^{-\eta}\left(\int\limits_{\{t<d(x_0,x)<\tau\}} v(x)d\mu\right)^{\frac{1}{q}} \times$$

$$\times\left(\int\limits_{\{d(x_0,x)>\tau\}} w^{1-p'}(x)d\mu\right)^{\frac{1}{p'}} < \infty$$

for $p > 1$ and

$$B_{1,\eta} \equiv \sup_{\tau>t}\left(\mu B(x_0,t)\right)^{-\eta} \times$$

$$\times\left(\int\limits_{\{t<d(x_0,x)<\tau\}} v(x)d\mu\right)^{\frac{1}{q}} \operatorname*{ess\,sup}_{d(x_0,x)>\tau} \frac{1}{w(x)} < \infty$$

for p = 1. Moreover, $\|\tilde{Q}_\eta\| \approx B_\eta$, *when p > 1 and* $\|\tilde{Q}_\eta\| \approx B_{1,\eta}$ *when p = 1.*

Specialising these results we now introduce operators $\tilde{P}_\eta$, $\tilde{Q}_\eta$ defined for μ–measurable f by

$$\tilde{P}_\eta f(x) = \frac{1}{(d(x_0, x))^\eta} \int_{B(x_0, d(x_0, x))} f(y) d\mu(y)$$

and

$$\tilde{Q}_\eta f(x) = \frac{1}{(d(x_0, x))^\eta} \int_{d(x_0, y) > d(x_0, x)} f(y) d\mu(y),$$

where $x_0 \in X$ and $\eta \in R$.

Arguing as in the proof of Theorem 1.2.9, we easily obtain the following

Theorem 1.2.11. *Let* $1 \leq p \leq q < \infty, \mu\{x_0\} = 0$ *and* $a = \infty$. *Assume that* $\eta > 0$. *Then the operator* $\tilde{P}_\eta$ *is bounded from* $L^p_w(X)$ *to* $L^{q\infty}_v(X)$ *if and only if*

$$D_\eta \equiv \sup_{\tau > t} \frac{1}{\tau^\eta} \left(\int_{\{t < d(x_0, x) < \tau\}} v(x) d\mu \right)^{\frac{1}{q}} \times$$

$$\times \left(\int_{\{d(x_0, x) \leq t\}} w^{1-p'}(x) d\mu \right)^{\frac{1}{p'}} < \infty$$

for p > 1, and

$$D_{1\eta} \equiv \sup_{\tau > t} \frac{1}{\tau^\eta} \left(\int_{\{t < d(x_0, x) < \tau\}} v(x) d\mu \right)^{\frac{1}{q}} \operatorname*{ess\,sup}_{d(x_0, x) < t} \frac{1}{w(x)} < \infty$$

for p = 1. Moreover, $\|\tilde{P}_\eta\| \approx D_\eta$, *when p > 1, and* $\|\tilde{P}_\eta\| \approx D_{1\eta}$ *when p = 1.*

Theorem 1.2.12. *Let* $1 \leq p \leq q < \infty$, *and let* $\eta < 0$. *Then for the boundness of the operator* $\tilde{Q}_\eta$ *from* $L^p_w(X)$ *to* $L^{q\infty}_v(X)$ *it is necessary and sufficient that*

$$\tilde{D}_\eta \equiv \sup_{\tau > t} t^{-\eta} \left(\int_{\{t < d(x_0, x) \leq \tau\}} v(x) d\mu \right)^{\frac{1}{q}} \times$$

$$\times \left(\int_{\{d(x_0, x) > \tau\}} w^{1-p'}(x) d\mu \right)^{\frac{1}{p'}} < \infty$$

for $p > 1$, and

$$\tilde{D}_{1,\eta} \equiv \sup_{\tau > t} t^{-\eta} \left(\int\limits_{\{ t < d(x_0,x) \leq \tau \}} v(x)d\mu \right)^{\frac{1}{q}} \times$$

$$\times \operatorname*{ess\,sup}_{d(x_0,x) > \tau} \frac{1}{w(x)} < \infty$$

for $p = 1$. Moreover, $\|\tilde{Q}_\eta\| \approx \tilde{D}_\eta$ if $p > 1$, and $\left\|\tilde{Q}_\eta\right\| \approx \tilde{D}_{1,\eta}$ for $p = 1$.

Theorem 1.2.13. $(X, d\mu)$ *be an SHT (see Definition 1.1.1.), $1 \leq p \leq q < \infty, \mu\{x_0\} = 0$ and $a = \infty$. Suppose that $\eta \leq 0$. Then the operator $\overline{P}_\eta$ is bounded from $L_w^p(X)$ into $L_v^{q\infty}(X)$ if and only if*

$$E_\eta(p, q) \equiv \sup_{t > 0} \frac{1}{(\mu B(x_0, t))^\eta} \left(\int\limits_{\{ d(x_0,x) > t \}} v(x)d\mu \right)^{\frac{1}{q}} \times$$

$$\times \left(\int\limits_{\{ d(x_0,x) \leq t \}} w^{1-p'}(x)d\mu \right)^{\frac{1}{p'}} < \infty$$

for $p > 1$, and

$$E_\eta(1, q) \equiv \sup_{t > 0} \frac{1}{(\mu B(x_0, t))^\eta} \left(\int\limits_{\{ d(x_0,x) > t \}} v(x)d\mu \right)^{\frac{1}{q}} \operatorname*{ess\,sup}_{d(x_0,x) \leq t} \frac{1}{w(x)} < \infty$$

for $p = 1$. Moreover, $\left\|\overline{P}_\eta\right\| \approx E_\eta(p, q)$ if $p > 1$, and $\left\|\overline{P}_\eta\right\| \approx E_\eta(1, q)$ when $p = 1$.

Proof. Sufficiency can be easily obtained from Corollary 1.1.2 using the estimate of the expression

$$\|\chi_{\{d(x_0,x) > t\}}(\cdot)(\mu B(x_0, d(x_0, \cdot)))^{-\eta}\|_{L_v^{q\infty}(X)} \|\chi_{\{d(x_0,x) \leq t\}}(\cdot)w^{-1}(\cdot)\|_{L_w^{p'}(X)}$$

by $E_\eta(p, q)$. We also need to use the doubling condition. The necessity follows in the standard way. $\square$

In the following theorem we do not need the doubling condition on μ.

Theorem 1.2.14. *Let $1 \leq p \leq q < \infty$, $\mu\{x_0\} = 0$ and $a = \infty$. Suppose that $\eta \leq 0$. Then the operator $\tilde{P}_\eta$ is bounded from $L_w^p(X)$ to $L_v^{q\infty}(X)$ if and only if*

$$\tilde{E}_\eta(p, q) \equiv \sup_{t > 0} \frac{1}{t^\eta} \left(\int\limits_{\{ d(x_0,x) > t \}} v(x)d\mu \right)^{\frac{1}{q}} \times$$

$$\times \left(\int\limits_{\{ d(x_0,x)\leq t\}} w^{1-p'}(x)d\mu \right)^{\frac{1}{p'}} < \infty$$

for $p > 1$, and

$$\tilde{E}_\eta(1,q) = \sup_{t>0} \frac{1}{t^\eta} \left(\int\limits_{\{ t<d(x_0,x)<\tau\}} v(x)d\mu \right)^{\frac{1}{q}} \operatorname*{ess\,sup}_{d(x_0,x)<t} \frac{1}{w(x)} < \infty$$

for $p = 1$. Moreover, $\left\| \tilde{P}_\eta \right\| \approx \tilde{E}_\eta(p,q)$ when $p > 1$, and $\left\| \tilde{P}_\eta \right\| \approx \tilde{E}_\eta(1,q)$ when $p = 1$.

We next consider the operators $\overline{Q}_\eta$ and $\tilde{Q}_\eta$ when $\eta \geq 0$. The calculation of the norm, taking into account $\| \cdot \|_{L_v^{q\infty}(X)}$ with the doubling condition, leads us to the following result:

Theorem 1.2.15. *Let (X,d,μ) be an SHT, $1 \leq p \leq q < \infty$ and $a = \infty$. Suppose that $\eta \geq 0$. Then for the boundedness of $\overline{Q}_\eta$ from $L_w^p(X)$ to $L_v^{q\infty}(X)$ it is necessary and sufficient that*

$$F_\eta(p,q) \equiv \sup_{t>0} \frac{1}{(\mu B(x_0,t))^\eta} \left(\int\limits_{\{ d(x_0,x)\leq t\}} v(x)d\mu \right)^{\frac{1}{q}} \times$$

$$\times \left(\int\limits_{\{ d(x_0,x)>t\}} w^{1-p'}(x)d\mu \right)^{\frac{1}{p'}} < \infty$$

for $p > 1$, and

$$F_\eta(1,q) \equiv \sup_{t>0} \frac{1}{(\mu B(x_0,t))^\eta} \left(\int\limits_{\{ d(x_0,x)\leq t\}} v(x)d\mu \right)^{\frac{1}{q}} \operatorname*{ess\,sup}_{d(x_0,x)>t} \frac{1}{w(x)} < \infty$$

for $p = 1$. Moreover, $\left\| \overline{Q}_\eta \right\| \approx F_\eta(p,q)$ when $p > 1$, and $\left\| \overline{Q}_\eta \right\| \approx F_\eta(1,q)$ when $p = 1$.

The following theorem can be derived in a similar way without assuming the doubling condition.

Theorem 1.2.16. *Let p, q and η satisfy the conditions of the previous theorem and $a = \infty$. Then $\tilde{Q}_\eta$ is bounded from $L_w^p(X)$ to $L_v^{q\infty}(X)$ if and only if*

$$\tilde{F}_\eta(p,q) \equiv \sup_{t>0} t^{-\eta} \left(\int\limits_{\{ d(x_0,x)\leq t\}} v(x)d\mu \right)^{\frac{1}{q}} \times$$

$$\times \left(\int\limits_{\{ d(x_0,x)>t \}} w^{1-p'}(x)d\mu \right)^{\frac{1}{p'}} < \infty$$

for $p > 1$, and

$$\widetilde{F}_\eta(1,q) \equiv \sup_{t>0} t^{-\eta} \left(\int\limits_{\{ d(x_0,x)\le t \}} v(x)d\mu \right)^{\frac{1}{q}} \operatorname*{ess\,sup}_{d(x_0,x)>t} \frac{1}{w(x)} < \infty$$

for $p = 1$.

Moreover, $\left\| \widetilde{Q}_\eta \right\| \approx \widetilde{F}_\eta(p,q)$ when $p > 1$, and $\left\| \widetilde{Q}_\eta \right\| \approx \widetilde{F}_\eta(1,q)$ when $p = 1$.

To conclude this section we mention that two–weighted (p,q), $1 \le p \le q < \infty$, weak type inequalities for the classical Hardy operator were established in [8]. Analogous problems for the Hardy–type operator defined on measure space (R_+, μ), where μ is a Borel measure, were studied in [5].

1.3.　Estimates for approximation numbers

In the present section the Volterra–type integral operator

$$(Tf)(x) = u_1(x) \int\limits_{A(x)} u_2(y)f(y)d\mu(y)$$

is considered, where u_1 and u_2 are weight functions on X, (X,μ) is a σ–finite measure space, $A(x) \equiv \{y \in X : \varphi(y) < \psi(x)\}$ for non–negative measurable functions φ and ψ defined on X satisfying the following conditions:

a)
$$\mu\{x \in X : \varphi(x) = t\} = \mu\{x \in X, \psi(x) = t\} = 0$$

for all $t \ge 0$;

b)
$$0 < \mu\{x \in X : r < \varphi(x) < R\} < \infty,$$

$$0 < \mu\{x \in X : r < \psi(x) < R\} < \infty$$

for all r and R with $0 < r < R < \infty$.

Under suitable conditions on u_1 and u_2 we establish upper and lower estimates for the approximation numbers of T

$$\alpha_n(T) = \inf \{\|T - R\|, \ \operatorname{rank} R < n\}, \ n \in N.$$

It is clear that

$$\|T\| = \alpha_1(T) \ge \alpha_2(T) \ge \cdots$$

and

$$\lim_{n \to \infty} \alpha_n(T) = \alpha(T),$$

where

$$\alpha(T) = \inf\{\|T - R\|, \ \text{rank } R < \infty\}, \ n \in N.$$

We also need the following notation:

$$F(s) = \int_{\{y:\varphi(y)<s\}} f(y)u_2(y)d\mu(y), \ f \in L^p(X), \ s \geq 0$$

and

$$F_{I_\psi} = \frac{1}{\nu(I_\psi)} \int_{I_\psi} F(\psi(x))d\nu(x), \ \nu(I_\psi) \neq 0,$$

where

$$\nu(I_\psi) = \int_{I_\psi} (u_1(y))^q d\mu(y),$$

$$I_\psi = \{x \in X : \ \psi(x) \in I, \ I \subset R^+\}.$$

Let $I \subset R^+$. We define the operator

$$l(I_\psi, f) = \int_{I_\psi \times I_\psi} (u_1(x)u_1(y))^q |F(\psi(x)) - F(\psi(y))|^q d(\mu \times \mu),$$

where $\mu \times \mu$ is a product measure defined on $X \times X$ and $f \in L^p(X)$.
Below we need the following lemmas:

Lemma 1.3.1. *Let $\lambda > 0$. Then for an arbitrary $\varepsilon > 0$ there exists $\delta > 0$ such that for any function f with $\|f\|_{L^p(X)} \leq 1$ and for any pair of intervals $I_1, I_2 \subset [0, \lambda]$ satisfying $(\mu \times \mu)((I_{1\psi}^2 \cup I_{2\psi}^2) \setminus (I_{1\psi}^2 \cap I_{2\psi}^2)) < \delta$ we have $|l(I_{1\psi}, f) - l(I_{2\psi}, f)| < \varepsilon$.*
Proof. Let $J_\psi \equiv (I_{1\psi}^2 \cup I_{2\psi}^2) \setminus (I_{1\psi}^2 \cap I_{2\psi}^2)$. Then

$$|l(I_{1\psi}, f) - l(I_{2\psi}, f)| =$$

$$= \left| \int_{I_{1\psi}^2} (u_1(x)u_1(y))^q |F(\psi(x) - F(\psi(y))|^q d(\mu \times \mu) - \right.$$

$$\left. - \int_{I_{2\psi}^2} (u_1(x)u_1(y))^q |F(\psi(x) - F(\psi(y))|^q d(\mu \times \mu) \right| \leq$$

$$\leq \int_{J_\psi} (u_1(x)u_1(y))^q |F(\psi(x) - F(\psi(y))|^q d(\mu \times \mu) \leq$$

$$\leq \left(\int_{J_\psi} (u_1(x)u_1(y))^q d(\mu \times \mu) \right) \left(\int_{\{0<\varphi(y)<\lambda\}} u_2(y)|f(y)|d\mu \right)^q \leq$$

$$\leq \left(\int_{J_\psi} (u_1(x)u_1(y))^q d(\mu \times \mu) \right) \|f\|^q_{L^p(X)} \|u_2\|^q_{L^{p'} \{y: \, 0<\varphi(y)<\lambda\}} \leq$$

$$\leq \left(\int_{J_\psi} (u_1(x)u_1(y))^q d(\mu \times \mu) \right) \|u_2\|^q_{L^{p'} \{y: \, 0<\varphi(y)<\lambda\}}.$$

By virtue of the absolute continuity of the integral the desired result follows. $\square$

Lemma 1.3.2. *Let* $0 \leq a < b < \infty$, $a_n \to a$, $b_n \to b$ *as* $n \to \infty$ *and* $0 < a_n < b_n$. *Assume that* $I_n = (a_n, b_n)$ *and* $I = (a, b)$. *Then* $(\mu \times \mu)(J_{n\psi}) \to 0$, *as* $n \to \infty$, *where* $J_{n\psi} = (I_\psi^2 \cup I_{n\psi}^2) \setminus (I_\psi^2 \cap I_{n\psi}^2)$.

Proof. Since $\mu\{x : \psi(x) = t\} = 0$ for $t \geq 0$, for arbitrary sequences $\{t_n\}$ and $\{\tau_n\}$ such that $t_n \to t+$, $\tau_n \to t-$ we have

$$\lim_{n\to\infty} \mu\{x : \, t \leq \psi(x) \leq t_n\} = 0, \quad \lim_{n\to\infty} \mu\{x : \, \tau_n \leq \psi(x) \leq t\} = 0.$$

First assume that $a_n \to a-$ and $b_n \to b-$. Then it is easy to verify that

$$J_{n\psi} \subset \{a_n \leq \psi(y) \leq a\} \times \{a_n \leq \psi(y) \leq b_n\} \cup$$
$$\{a \leq \psi(y) \leq b_n\} \times \{a_n \leq \psi(y) \leq a\}$$
$$\cup\{b_n \leq \psi(y) \leq b\} \times \{a \leq \psi(y) \leq b\}$$
$$\cup\{a \leq \psi(y) \leq b_n\} \times \{b_n \leq \psi(y) \leq b\}.$$

Therefore,

$$(\mu \times \mu)(J_{n\psi}) \leq (\mu \times \mu)(\{a_n \leq \psi(y) \leq a\} \times \{a_n \leq \psi(y) \leq b_n\}) +$$
$$+ (\mu \times \mu)(\{a \leq \psi(y) \leq b_n\} \times \{a_n \leq \psi(y) \leq a\}) +$$
$$+ (\mu \times \mu)(\{b_n \leq \psi(y) \leq b\} \times \{a \leq \psi(y) \leq b\}) +$$
$$(\mu \times \mu)(\{a \leq \psi(y) \leq b_n\} \times \{b_n \leq \psi(y) \leq b\}) =$$
$$= \mu\{a_n \leq \psi(y) \leq a\} \cdot \mu\{a_n \leq \psi(y) \leq b_n\} + \mu\{a \leq \psi(y) \leq b_n\} \times$$
$$\times \mu\{a_n \leq \psi(y) \leq a\} + \mu\{b_n \leq \psi(y) \leq b\} \cdot \mu\{a \leq \psi(y) \leq b\} +$$
$$+ \mu\{a \leq \psi(y) \leq b_n\} \cdot \mu\{b_n \leq \psi(y) \leq b\}$$

and consequently,

$$(\mu \times \mu)(J_{n\psi}) \to 0, \quad n \to \infty.$$

The proof in the other cases is analogous.
$\square$

Next we define the function

$$G(a, b) = \sup \{l((a, b)_\psi, f) : \|f\|_{L^p(\{x: a<\psi(x)<b\})} \leq 1\}.$$

It is easy to see that

$$G(a, b) = \sup \{l((a, b)_\psi, f) : \|f\|_{L^p(X)} \leq 1\}.$$

Lemma 1.3.3. *For arbitrary finite interval* $I \subset R^+ (I = (a, b))$ *the function* $G(a, b)$ *is a continuous as a function of two variables.*

The proof follows using Lemma 1.3.1.

For $I = (a, b)$ we denote

$$L(a, b) = \Big\{ \frac{1}{\nu(I_\psi)} \sup \{l(I_\psi, f) : \|f\|_{L^p(I_\psi)} \leq 1\} \Big\}^{1/q}.$$

Lemma 1.3.4. $L(a, b)$ *is continuous as a function of two variables. Moreover,* $L(a, b)$ *is decreasing as* a *increases and increasing as* b *increases.*

Proof. The continuity of $L(a, b)$ follows from Lemma 1.3.2 and from the continuity of $\nu(I_\psi)$. Let us show that $L(a, b)$ is decreasing as a increases. Put

$$g(s, b) = \int\limits_{\{s < \psi(y) < b\}} |F(s) - F(\psi(y))|^q d\nu(y) =$$

$$= \int\limits_{\{s < \psi(y) < b\}} \Big| \int\limits_{\{y : s < \varphi(y) < \psi(y)\}} u_2(y) f(y) d\mu(y) \Big|^q d\nu(y),$$

It is clear that

$$l((a, b)_\psi, f) = 2 \int\limits_{\{y : a < \psi(x) < b\}} g(\psi(x), b) d\nu(x).$$

In addition, if $f \geq 0$, then $g(s, b)$ is decreasing when s increases. Assume that $a \leq a_1 < a_2 < b$ and $\nu((a_2, b)_\psi) > 0$. Then

$$\frac{l((a_1, b)_\psi, f)}{2\nu((a_1, b)_\psi, f)} = \frac{1}{\nu((a_1, b)_\psi)} \Big(\int\limits_{\{x : a_1 < \psi(x) < a_2\}} g(\psi(x), b) d\nu(x) +$$

$$+ \int\limits_{\{x : a_2 < \psi(x) < b\}} g(\psi(x), b) d\nu(x) \Big) \geq$$

$$\geq \frac{1}{\nu((a_1, b)_\psi)} \Big(\frac{\nu((a_1, a_2)_\psi)}{\nu((a_2, b)_\psi)} \int\limits_{\{x : a_2 < \psi(x) < b\}} g(\psi(x), b) d\nu(x) +$$

$$+ \int\limits_{\{a_2 < \psi(x) < b\}} g(\psi(x), b) d\nu(x) \Big) =$$

$$= \frac{1}{\nu((a_2, b)_\psi)} \int\limits_{\{a_2 < \psi(x) < b\}} g(\psi(x), b) d\nu(x) =$$

$$= \frac{l((a_2, b)_\psi, f)}{2\nu((a_2, b)_\psi)}.$$

From this we conclude that $L(a, b)$ decreases as a increases. Analogously, we obtain that $L(a, b)$ increases as b increases. $\square$

Lemma 1.3.5. *Let $1 \leq p \leq q < \infty$, $0 \leq a < c < b < \infty$. Then the inequalities*

$$\left(\int\limits_{\{a < \psi(x) < c\}} (u_1(x))^q \left| \int\limits_{\{y : \psi(x) < \varphi(y) < c\}} u_2(y) f(y) d\mu(y) \right|^q d\mu(x) \right)^{1/q} \leq$$

$$\leq 4A_c \left(\int\limits_{\{a < \varphi(y) < c\}} |f(y)|^p d\mu(y) \right)^{1/p}$$

and

$$\left(\int\limits_{\{c < \psi(x) < b\}} (u_1(x))^q \left| \int\limits_{\{y : c < \varphi(y) < \psi(x)\}} u_2(y) f(y) d\mu(y) \right|^q d\mu(x) \right)^{1/q} \leq$$

$$\leq 4B_c \left(\int\limits_{\{c < \varphi(y) < b\}} |f(y)|^p d\mu \right)^{1/p}$$

hold, where

$$A_c = \sup_{a < t < c} \|u_1 \chi_{\{y : a < \psi(x) < t\}}\|_{L^q(X)} \|u_2 \chi_{\{y : t < \varphi(y) < c\}}\|_{L^{p'}(X)}$$

and

$$B_c = \sup_{c < t < b} \|u_1 \chi_{\{y : t < \psi(x) < b\}}\|_{L^q(X)} \|u_2 \chi_{\{y : a < \varphi(y) < t\}}\|_{L^{p'}(X)}.$$

The lemma is proved in the same manner as Theorems 1.1.3 and 1.1.4. Again put $I = (a, b)$. It is possible to select $c \in (a, b)$ such that

$$\int\limits_{\{a < \psi(x) < c\}} (u_1(x))^q d\mu = \frac{1}{2} \int\limits_{\{a < \psi(x) < b\}} (u_1(x))^q d\mu. \qquad (1.3.1)$$

Put $J(I) = \max\{A_c, B_c\}$, where A_c and B_c are defined as in Lemma 1.3.5 and

$$K(I) = \sup\{\|F(\psi(\cdot)) - F_{I\psi}\|_{L^q_\nu(I_\psi)} / \|f\|_{L^p(I_\varphi)} : f \in L_p(I_\varphi) f \neq 0\}$$

if $\nu(I_\psi) \neq 0$ and $K(I) = 0$ if $\nu(I_\psi) = 0$.

Lemma 1.3.6. *Let* $1 \leq p \leq q < \infty$, $I = (a, b)$. *Then*

$$2^{1/q'} J(I) \leq K(I) \leq 8J(I),$$
$$\frac{1}{2}L(I) \leq K(I) \leq L(I).$$

Proof. Let $c \in (a, b)$. According to Lemma 1.3.5, we obtain

$$\|F(\psi(\cdot)) - F(c)\|_{L^q_\nu(a,c)_\psi} =$$

$$= \left(\int\limits_{\{a<\psi(y)<c\}} |F(\psi(x)) - F(c)|^q (u_1(x))^q d\mu \right)^{1/q} =$$

$$= \left(\int\limits_{\{a<\psi(x)<c\}} (u_1(x))^q \left| \int\limits_{\{y:\psi(x)<\varphi(y)<c\}} u_2(y)f(y)d\mu(y) \right|^q d\mu(x) \right)^{1/q} \leq$$

$$\leq 4A_c \left(\int\limits_{\{a<\varphi(y)<c\}} |f(y)|^p d\mu \right)^{1/p}.$$

Analogously, we find that

$$\|F(\psi(\cdot)) - F(c)\|_{L^q_\nu((c,b)_\psi)} \leq 4B_c \|f\|_{L^p((c,b)_\varphi)}.$$

Consequently,

$$\|F(\psi(\cdot)) - F(c)\|^p_{L^q_\nu(I_\psi)} \leq$$

$$\leq \left(\int\limits_{\{a<\psi(x)<c\}} |F(\psi(x)) - F(c)|^q (u_1(x))^q d\mu(x) \right)^{p/q} +$$

$$+ \left(\int\limits_{\{c<\psi(x)<b\}} |F(\psi(x)) - F(c)|^q (u_1(x))^q d\mu(x) \right)^{p/q} \leq$$

$$\leq 4^p \left(A^p_c \int\limits_{\{a<\varphi(y)<c\}} |f(y)|^p d\mu + B^p_c \int\limits_{\{c<\varphi(y)<b\}} |f(y)|^p d\mu \right) \leq$$

$$\leq 4^p (J(I))^p \int\limits_{\{a<\varphi(y)<b\}} |f(y)|^p d\mu.$$

On the other hand,

$$\|F(\psi(\cdot)) - F_{I_\psi}\|_{L^q_\nu(I_\psi)} \leq \|F(\psi(\cdot)) - F(c)\|_{L^q_\nu(I_\psi)} +$$
$$+ \|F_{I_\psi} - F(c)\|_{L^q_\nu(I_\psi)} \leq 2\|F(\psi(\cdot)) - F(c)\|_{L^q_\nu(I_\psi)}.$$

Therefore

$$K(I) \le 8J(I).$$

Now we estimate $K(I)$ from below. First we observe that

$$\int_{I_\psi \times I_\psi} |F(\psi(x)) - F(\psi(y))|^q (u_1(x))^q (u_1(y))^q d(\mu \times \mu) \le$$

$$\le 2^{q-1} \int_{I_\psi \times I_\psi} \left(|F(\psi(x)) - F_{I_\psi}|^q + |F(\psi(y) - F_{I_\psi}|^q \right) \times$$

$$\times (u_1(x) u_1(y))^q d(\mu \times \mu) =$$

$$= 2^{q-1} \int_{I_\psi \times I_\psi} |F(\psi(x)) - F_{I_\psi}|^q (u_1(x) u_1(y))^q d(\mu \times \mu) +$$

$$+ 2^{q-1} \int_{I_\psi \times I_\psi} |F(\psi(y)) - F_{I_\psi}|^q (u_1(x) u_1(y))^q d(\mu \times \mu) =$$

$$= 2^{q-1} \nu(I_\psi) \int_{I_\psi} |F\psi(x)) - F_{I_\psi}|^q (u_1(x))^q d\mu(x) +$$

$$+ 2^{q-1} \nu(I_\psi) \int_{I_\psi} |F(\psi(y)) - F_{I_\psi}|^q (u_1(y))^q d\mu(y) =$$

$$= 2^q \nu(I_\psi) \int_{I_\psi} |F(\psi(x)) - F_{I_\psi}|^q (u_1(x))^q d\mu(x).$$

Hence

$$\int_{I_\psi \times I_\psi} |F(\psi(x)) - F(\psi(y))|^q (u_1(x))^q (u_1(y))^q d(\mu \times)\mu \le$$

$$\le 2^q \nu(I_\psi) \int_{I_\psi} |F(\psi(x)) - F_{I_\psi}|^q (u_1(x))^q d\mu(x). \tag{1.3.2}$$

Let $a \le \lambda < \eta \le b$ and $f(y) = (u_2(y))^{p'/p} \chi_{\{\lambda < \varphi(y) < \eta\}}(y)$.
By virtue of (1.3.2) we obtain

$$2^q \nu(I_\psi) \|F(\psi(\cdot)) - F_{I_\psi}\|^q_{L^q_\nu(I_\psi)} \ge$$

$$\ge \int_{\{a < \psi(x) < \lambda\} \times \{\eta < \psi(y) < b\}} |F(\psi(x)) - F(\psi(y))|^q \times$$

$$\times (u_1(x) u_1(y))^q d(\mu \times \mu) =$$

$$= \int_{\{a < \psi(x) < \lambda\}} (u_1(x))^q \left(\int_{\{\eta < \psi(y) < b\}} (u_1(y))^q \left| \int_{\{\psi(x) < \varphi(z) < \psi(y)\}} u_2(z) \times \right. \right.$$

$$\times f(z)d\mu(z)\Big|^q d\mu(y)\Big)d\mu(x) \geq$$

$$\geq \int\limits_{\{a<\psi(x)<\lambda\}} (u_1(x))^q \Big(\int\limits_{\{\eta<\psi(y)<b\}} (u_1(y))^q \times$$

$$\times \Big(\int\limits_{\{\lambda<\varphi(z)<\eta\}} (u_2(z))^{p'}d\mu(z)\Big)^q d\mu(y)\Big)d\mu(x) =$$

$$= \Big(\int\limits_{\{a<\psi(x)<\lambda\}} (u_1(x))^q d\mu(x)\Big)\Big(\int\limits_{\{\eta<\psi(y)<b\}} (u_1(y))^q d\mu(y)\Big) \times$$

$$\times \Big(\int\limits_{\{\lambda<\varphi(z)<\eta\}} (u_2(z))^{p'}d\mu(z)\Big)^{q/p'}\Big(\int\limits_{I_\varphi} |f(z)|^p d\mu(z)\Big)^{q/p}.$$

Hence we deduce that

$$\frac{\|F(\psi(\cdot)) - F_{I_\psi}\|_{L^q_\nu(I_\psi)}}{\|f\|_{L^p(I_\varphi)}} \geq 2\Big(\int\limits_{\{a<\psi(x)<\lambda\}} (u_1(x))^q d\mu(x)\Big)^{1/q} \times$$

$$\times \Big(\int\limits_{\{\eta<\psi(y)<b\}} (u_1(y))^q d\mu(y)\Big)^{1/q} \times$$

$$\times \Big(\int\limits_{\{\lambda<\varphi(z)<\eta\}} (u_2(z))^{p'}d\mu(z)\Big)^{1/p'}\Big(\int\limits_{\{a<\psi(y)<b\}} (u_1(y))^q d\mu(y)\Big)^{-1/q}.$$

Choose $c \in (a, b)$ such that the equality (1.3.1) is fulfilled.
For $\eta = c$ and $\lambda \in (a, c)$ we have

$$\frac{\|F(\psi(\cdot)) - F_{I_\psi}\|_{L^q_\nu(I_\psi)}}{\|f\|_{L^p(I_\varphi)}} \geq 2^{1/q'}\Big(\int\limits_{\{a<\psi(x)<\lambda\}} (u_1(x))^q d\mu(x)\Big)^{1/q} \times$$

$$\times \Big(\int\limits_{\{\lambda<\varphi(z)<c\}} (u_2(z))^{p'}d\mu(z)\Big)^{1/p'}.$$

Therefore

$$K(I) \geq 2^{1/q'} A_c.$$

Analogously,

$$K(I) \geq 2^{1/q'} B_c.$$

Thus the first part of lemma is proved.

On the other hand, from (1.3.2) it follows that

$$L(I) \leq 2\,K(I).$$

In addition, we have

$$|F(\psi(x)) - F_{I_\varphi}| = \frac{1}{\nu(I_\psi)}\left| \int_{I_\psi} (F(\psi(x)) - F(\psi(y)))d\nu(y) \right| \leq$$

$$\leq \nu(I_\psi)^{-1/q}\left(\int_{I_\psi} |F(\psi(x)) - F(\psi(y))|d\nu(y) \right)^{1/q}.$$

Finally, we conclude that

$$K(I) \leq L(I).$$

□

Lemma 1.3.7. *Let* $1 \leq p \leq q < \infty$, $\varepsilon > 0$ *and assume that there exist* $n \in N$ *and constants* c_k $(0 < c_0 < c_1 < \cdots c_{n+1} = \infty)$ *such that* $L((I_k)_\varphi) \leq \varepsilon$, *where* $I_k = (c_k, c_{k+1})$ *and* $k = 0, 1, \cdots, n$. *Then the estimate*

$$a_{n+2}(T) \leq \varepsilon$$

holds.

Proof. Put

$$Pf(x) = \sum_{k=0}^{n} P_{I_k} f(x),$$

where

$$f \in L^p(X), \quad \|f\|_{L^p(X)} = 1, \quad P_I f(x) = \chi_{I_\psi}(x) u_1(x) F_{I_\psi}.$$

Evidently, operator P is linear and $rank\, P < n+1$. Further,

$$\|Tf - Pf\|_{L^q(X)}^q = \sum_{k=0}^{n} \|Tf - P_{I_k} f\|_{L^q((I_k)_\psi)}^q =$$

$$= \sum_{k=0}^{n} \|F(\psi(\cdot)) - F_{I_k}\|_{L^q_\nu((I_k)_\psi)}^q \leq \sum_{k=0}^{n} K^q(I_k)\|f\|_{L^p((I_k)_\varphi)}^q.$$

By Lemma 1.3.6,

$$\|Tf - Pf\|_{L^q(X)}^q \leq \varepsilon^q \sum_{k=0}^{n} \|f\|_{L^p((I_k)_\varphi)}^q \leq \varepsilon^q \|f\|_{L^p(X)}^q.$$

Hence

$$a_{n+2}(T) \le \varepsilon.$$

$\square$

Lemma 1.3.8. *Let* $1 \le p \le q < \infty$, $\varepsilon > 0$ *and let* c_i ($i = 0, 1, \cdots, n$) *satisfying* $0 = c_0 < c_1 < \cdots < c_n < \infty$ *be such that* $L((I_k)_\varphi)) \ge \epsilon$, *where* $I_k = (c_k, c_{k+1})$ *for* $k = 0, 1, \cdots, n-1$.
Then

$$a_n(T) \ge \frac{1}{4} n^{\frac{1}{q} - \frac{1}{p}} \varepsilon.$$

Proof. Let $\eta \in (0, 1)$. According to Lemma 1.3.6, there exist $\gamma_k \in L^p((I_k)_\varphi)$ ($k = 0, 1, \cdots, n-1$) such that

$$\frac{\|\Gamma_k(\psi(\cdot)) - (\Gamma_k)_{(I_k)_\psi}\|_{L^q_\nu((I_k)_\psi)}}{\|\gamma_k\|_{L^p((I_k)_\varphi)}} > \frac{1}{2} \eta L((I_k)_\psi) \ge \frac{1}{2} \epsilon \eta,$$

where

$$\Gamma_k(s) = \int\limits_{\{\varphi(y) < s\}} u_2(y) \gamma_k(y) d\mu.$$

Let $\gamma_k(x) = 0$ when $x \notin (I_k)_\varphi$ and let $rank \, P \le n - 1$. Then there exist constants $\lambda_0, \cdots, \lambda_{n-1}$, such that at least one of them is non–zero and

$$P(\sum_{k=0}^{n-1} \lambda_k \gamma_k) = 0.$$

Put

$$\gamma = \sum_{k=0}^{n-1} \lambda_k \gamma_k$$

and let

$$\Gamma(\psi(x)) = \int\limits_{\{\varphi(y) < \psi(x)\}} u_2(y) \gamma(y) d\mu.$$

For any $x \in I_{k_\psi}$ we have

$$\Gamma(\psi(x)) = \lambda_k \Gamma_k(\psi(x)) + \mu_k,$$

where μ_k is some constant. Thus the following estimates hold for arbitrary c.

$$\|F(\psi(\cdot)) - F_{I_\psi}\|_{L^q_\nu(I_\psi)} \le \|F(\psi(\cdot)) - c\|_{L^q_\nu(I_\psi)} +$$
$$+ \|c - F_{I_\psi}\|_{L^q_\nu(I_\psi)} \le 2\|F(\psi(\cdot)) - c\|_{L^q_\nu(I_\psi)}$$

hold. Hence

$$\|F(\psi(\cdot)) - F_{I_\psi}\|_{L^q_\nu(I_\psi)} \leq 2 \inf \|F(\psi(\cdot)) - c\|_{L^q_\nu(I_\psi)},$$

$$\|T\gamma - P\gamma\|^q_{L^q(X)} \|T\gamma\|^q_{L^q(X)} = \sum_{k=0}^{n-1} \|\Gamma(\psi(\cdot))\|^q_{L^q_\nu((I_k)_\psi)} =$$

$$= \sum_{k=0}^{n-1} \|\lambda_k \Gamma_k(\psi(\cdot)) + \mu_k\|^q_{L^q_\nu((I_k)_\psi)} \geq$$

$$\geq 2^{-q} \sum_{k=0}^{n-1} \|\lambda_k \Gamma_k(\psi(\cdot)) - (\lambda_k \Gamma_k)_{(I_k)_\psi}\|^q_{L^q_\nu((I_k)_\psi)} =$$

$$= 2^{-q} \sum_{k=0}^{n-1} |\lambda_k|^q \|\Gamma_k(\psi(\cdot)) - (\Gamma_k)_{((I_k)_\psi)}\|^q_{L^q_\nu((I_k)_\psi)} \geq$$

$$\geq 2^{-q} (\frac{1}{2}\varepsilon\eta)^q \sum_{k=0}^{n-1} |\lambda_k|^q \|\gamma_k\|^q_{L^p((I_k)_\psi)} \geq$$

$$\geq 2^{-2q}\varepsilon^q\eta^q \sum_{k=0}^{n-1} \|\lambda_k\gamma_k\|^q_{L^p((I_k)_\varphi)} \geq$$

$$\geq 4^{-q}\varepsilon^q\eta^q n^{1-q/p} (\sum_{k=0}^{n-1} \|\lambda_k\gamma_k\|^p_{L^p((I_k)_\varphi)})^{q/p} =$$

$$= 4^{-q}\varepsilon^q\eta^q n^{1-q/p} \|\gamma\|^q_{L^p(X)}.$$

Therefore,

$$\frac{\|T\gamma - P\gamma\|_{L^q(X)}}{\|\gamma\|_{L(X)}} \geq \frac{1}{4} n^{1/q-1/p} \varepsilon\eta$$

and

$$a_n(T) \geq \frac{1}{4} n^{1/q-1/p} \eta\varepsilon.$$

As the constant η can be chosen arbitrarily near to 1, we have

$$a_n(T) \geq \frac{1}{4} n^{1/q-1/p} \varepsilon.$$

Hence the lemma is proved. $\square$

Let ε be an arbitrary positive number and let the constants c_k ($k = 0, 1, \cdots$) be defined by the recurrence relations

$$c_0 = 0, \quad c_{k+1} = \inf\{s : L((c_k, s)_\psi) > \varepsilon\}.$$

We call this sequence an (ε, L)– sequence.

There are two possibilities:
(i) The (ε, L)– sequence is finite. In this case there exists $n \in N$ such that

$$c_0 < c_1 < c_n < c_{n+1} = \infty.$$

By the continuity of L we have

$$L((c_k, c_{k+1})_\psi) = \varepsilon, \quad \text{when} \quad k = 0, \ldots, n - 1$$

and

$$L((c_n, c_{n+1})_\psi) \leq \varepsilon.$$

In this case the length of the (ε, L)– sequence will be defined to equal to $n + 1$.
(ii) The (ε, L)– sequence is infinite. In this case we have $L(c_k, c_{k+1}) = \varepsilon$ for arbitrary $k \in N$.

According to Lemmas 1.3.7 and 1.3.8 we obtain the validity of the following statements:

Theorem 1.3.1. *Let* $1 \leq p \leq q < \infty$ *and let the length of the* (ε, L)– *sequence be equal to* $n + 1$. *Then*

$$a_{n+2}T \leq \varepsilon \quad \text{and} \quad a_n(T) \geq \frac{1}{4}\varepsilon n^{1/q-1/p}.$$

If $p = q$ *then*

$$a_n(T) \geq \frac{1}{4}\varepsilon.$$

Theorem 1.3.2. *Let* $1 \leq p \leq q < \infty$. *If the* (ε, L)–*sequence is infinite then*

$$a_n(T) \geq \frac{1}{4}n^{1/q-1/p}\varepsilon$$

for any n.
If $p = q$ *then*

$$a_n(T) \geq \frac{1}{4}\varepsilon.$$

Remark. If (ε, L) is infinite then $c_k \to \infty$ as $k \to \infty$. Indeed, otherwise $\lim_{k\to\infty} |c_k - c_{k+1}| = 0$, and consequently $\lim_{k\to\infty} L(c_k, c_{k+1}) = 0$, by the continuity of L. This contradicts the condition $L((c_k, c_{k+1})_\psi) = \varepsilon$ for arbitrary k.

In the sequel, we shall assume that

$$L((a, \infty)_\psi) = \lim_{b\to\infty} L((a, b)_\psi).$$

It may heppen that $L((a,\infty)_\psi) = \infty$.

Theorem 1.3.3. *Let* $1 \leq p \leq q < \infty$. *Then the following statements are valid:*

1) *operator* T *is bounded if and only if*

$$L((0,\infty)_\psi) < \infty,$$

2) *if* T *is bounded then*

$$a(T) \leq \lim_{s\to\infty} L((s,\infty)_\psi) \equiv M$$

and $a(T) = 0$ *if and only if* $M = 0$. *Moreover,* $a(T) \geq \frac{1}{4}M$ *when* $p = q$;

3) *Let* $\varepsilon \in (M, L((0,\infty)_\psi)$ *and let the* $(\varepsilon, L)-$ *sequence has length* $n + 1$. *Then*

$$a_{n+2}(T) \leq \varepsilon \quad and \quad a_n(T) \geq \frac{1}{4}n^{1/q-1/p}\varepsilon.$$

For $p = q$ *we have*

$$a_n(T) \geq \frac{1}{4}\varepsilon.$$

Proof. Let $L((0,\infty)_\psi) < \infty$ and $\varepsilon = L((0,\infty)_\psi)$. Then the length of (ε, L) is 1. According to Lemma (1.3.7), we obtain

$$a_2(T) = \inf\{\|T - P\| : rank\ P = 1\} \leq \varepsilon.$$

Hence T is bounded.

Conversely, if T is bounded then $K(I) \leq 2\|T\|$ for arbitrary $I \subset R^+$. By Lemma (1.3.6), $L(I) \leq 4\|T\|$ for any I and we conclude that

$$L((0,\infty)_\psi) \leq 4\|T\|.$$

2) Let $\varepsilon \geq M$, then $L(s,\tau) < \varepsilon$ for sufficiently large s and τ, let (ε, L) be of length $N + 1$. From Lemma (1.3.7), we have $a_{N+2} \leq \varepsilon$ and consequently, $a(T) \leq a_{N+2} \leq \varepsilon$. As a number ε can be chosen arbirarily close to M, we conclude that $a(T) \leq M$.

Assume now that $a(T) = 0$. In this case T is compact and, consequently (see Theorem 1.1.19), $\lim_{a\to\infty} J_a = 0$, where

$$J_a = \sup_{s>a} \left(\int\limits_{\{\psi(y)>s\}} (u_1(y))^q d\mu \right)^{1/q} \left(\int\limits_{\{a<\psi(y)<s\}} (u_2(y))^{p'} d\mu \right)^{1/p'}.$$

Hence $\lim_{a\to\infty} J((a,\infty)) = 0$ and therefore, by Lemma 1.3.5, we conclude that $M = 0$.

Now let $p = q$. Assume that $M > 0$ and $\eta \in (0, M)$. Then $L(t, \tau) > \eta$ for sufficiently large t and τ. This shows that (η, L)– sequence is infinite. Otherwise $L(c_n, s) \leq \eta$ for all $s \geq c_n$, which is impossible. Using Lemma 1.3.8 we see that $a_n(T) \geq 1/4M$ for all $n \in N$. Thus $a(T) \geq 1/4M$.

3) Let $\varepsilon \in (M, L(0, \infty))$. Then the (ε, L)– sequence is of length $n + 1$ and by virtue of Lemma 1.3.7 we have $a_{n+2} \leq \varepsilon$. Finally, according to Lemma 1.3.8, we conclude that $a_n(T) \geq \frac{1}{4} n^{1/q - 1/p} \varepsilon$. $\square$

Remark 1.3.2. Theorem 1.3.3 can be formulated in terms of J rather than L.

1.4. Norms of positive operators

In the present section we calculate norms of positive operators on some cones of functions defined on measure spaces. In the classical case (the case of the real axis) analogous problems have been studied in [41].

Necessary and sufficient conditions for some classical operators to be bounded on cones of monotone functions can be found in [12], [222], [264] etc.

The best constants in the Hardy inequalities are found in [21], [23], [40], [234], [217], [171].

For the criteria for the boundedness and compactness of Hardy type operators in the Banach function spaces defined on measure spaces see Sections 1.1 and 1.2.

In this section, as an example, we calculate norms of Hardy–type operators defined on measure spaces.

Let (Y_1, β) and (Y_2, γ) be measure spaces and let $S(Y_1; \beta)$ (resp. $S(Y_2, \gamma)$) be the linear space of all real–valued β– measurable (resp. γ– measurable) functions which are, almost everywhere finite with respect to the nonnegative measure β (resp. γ) on Y_1 (resp. on Y_2). Recall that $L_\beta^p(Y_1)$ is the space of all functions $f \in S(Y_1, \beta)$ for which

$$\|f\|_{p, \beta, Y_1} = \left(\int_{Y_1} |f|^p d\beta \right)^{1/p} < \infty, \quad 0 < p < \infty.$$

We consider a quasi–Banach space $X \subset S(Y_2, \gamma)$ with quasi-norm $\| \cdot \|_X$. Recall that for a quasi-norm, the triangle inequality for arbitrary $f, g \in X$ has the form

$$\|f + g\|_X \leq c(\|f\|_X + \|g\|_X),$$

where $c \geq 1$ is a constant. If $c = 1$, then X is a normed (Banach) space.

A quasi–Banach space $X \subset S(Y_2, \gamma)$ is said to be ideal, if

$$\{f \in S(Y_2, \gamma), g \in X, |f| \leq |g| \ \gamma\text{–a.e.}\} \Rightarrow \{f \in X \quad \text{and} \quad \|f\|_X \leq \|g\|_X\}.$$

An ideal space X will be called l_q–convex $(0 < q \le \infty)$, if

$$\left\|\left(\sum_{k\in Z} |x_k|^q\right)^{1/q}\right\|_X \le \left(\sum_{k\in Z} \|x_k\|_X^q\right)^{1/q}.$$

We note that every normed ideal space is l_1–convex and from the l_q–convexity $(0 < q < 1)$ follows the triangle inequality with the constant $c = 2^{\frac{1}{q}-1}$ (for details, see [41]).

Let us mention that the space $X = L_\gamma^s(Y_2)$ will be l_q–convex for any $q \in (0, s]$. Let D be a cone of nonnegative functions on (a, b), $-\infty \le a < b \le +\infty$, i.e.,

$$\{f, g \in D, \alpha_1, \alpha_2 \ge 0\} \Rightarrow \alpha_1 f + \alpha_2 g \in D.$$

Further, let ρ be the mapping from $Y_1 \times Y_1 \to (a, b)$ and let there exist an element $y_0 \in Y$ such that $\rho(y_0, \cdot) : Y_1 \to (a, b)$ $(x \to \rho(y_0, x))$ is β–measurable. In addition, we suppose that for every α_1, α_2 with $a < \alpha_1 < \alpha_2 < b$ the condition

$$\beta\{x \in Y_1 : \alpha_1 < \rho(y_0, x) < \alpha_2\} > 0$$

is fulfilled.

Denote by D_1 the set of functions:

$$D_1 \equiv \{f : \exists \widetilde{f} \in D, f(x) = \widetilde{f}(\rho(y_0, x)) \quad \text{for all } x \in Y_1\}.$$

An operator $T : D_1 \to X$ is said to be l_r–convex $(0 < r < \infty)$ if for every $f_k \in D_1$ such that $(\sum_{k\in Z} f_k^r)^{1/r} \in D_1$ the inequality

$$T\left[\left(\sum_{k\in Z} f_k^r\right)^{1/r}\right] \le \left(\sum_{k\in Z} |Tf_k|^r\right)^{1/r}$$

is fulfilled almost everywhere on Y_2, and moreover, for every $f \in D_1$ and $\alpha \ge 0$,

$$T(\alpha f) = \alpha T f.$$

Note that the l_1– convexity of the operator T coincides with its sublinearity

$$T\left(\sum_{k\in Z} f_k\right) \le \sum_{k\in Z} Tf_k.$$

The operator T is said to be positive, if from the condition $f, g \in D_1$ and $0 \le f \le g$ β–almost everywhere it follows that $0 \le Tf \le Tg$ γ–almost everywhere.

As an example of an l_r– convex positive operator $(0 < r < \infty)$ we can take an operator of the type $Tf = (Lf^r)^{1/r}$, where L is a countably sublinear positive operator. Let for $H \subset D_1$

$$\|T\|_H \equiv \sup_{h\in H, h\neq 0} \frac{\|Th\|_X}{\|h\|_{L_\beta^p(Y_1)}}.$$

We consider operators on the following cones:

$$G \equiv \{g \in L_\beta^p(Y_1) : \exists \tilde{g} : (a,b) \to R_+^1, \ \tilde{g} \downarrow, \ g(x) = \tilde{g}(\rho(y_0,x)) \forall x \in Y_1\};$$

$$\overset{\circ}{G} \equiv \{g \in G : \lim_{u \to a+} \tilde{g}(u) = 0\};$$

$$G_0 \equiv \{g : g(\cdot) \equiv \chi_{(a,t)}(\rho(y_0, \cdot)), \quad a < t \le b\}$$

$$\overset{\circ}{G_0} \equiv \{g : g(\cdot) \equiv \chi_{(a,t)}(\rho(y_0, \cdot)), \quad a < t < b\}.$$

Now we formulate the basic results and then prove them. We begin with the cone G.

Theorem 1.4.1. *Let* $0 < p \le q \le r < \infty$ *and let* $X \subset S(Y_2, \gamma)$ *be an ideal* l_q- *convex space. Suppose* $T : G \to X$ *is a* l_r- *convex positive operator and the condition*

$$\beta\{y : a < \rho(y_0, y) < b\} = \beta(Y_1) < \infty \tag{1.4.1}$$

is fulfilled. Then

$$\|T\|_G = \|T\|_{G_0},$$

where

$$\|T\|_{G_0} \equiv \sup_{a < t \le b} [\|F(\cdot, t)\|_X (\beta\{y : a < \rho(y_0, y) < t\})^{-1/p}]$$

and

$$F(x, t) = T[\chi_{(a,t)}(\rho(y_0, \cdot))](x).$$

Theorem 1.4.2. *Let* $0 < p \le q \le r < \infty$ *and let* $X \subset S(Y_2, \gamma)$ *be an ideal* l_q- *convex space. Suppose* $T : \overset{\circ}{G} \to X$ *is a* l_r- *convex positive operator. Then*

$$\|T\|_{\overset{\circ}{G}} = \|T\|_{\overset{\circ}{G_0}},$$

where

$$\|T\|_{\overset{\circ}{G_0}} \equiv \sup_{a < t < b} [\|F(\cdot, t)\|_X (\beta\{y : a < \rho(y_0, y) < t\})^{-1/p}].$$

Remark 1.4.1. If $\beta(Y_1) = \beta\{y : a < \rho(y_0, y) < b\} = \infty$, then $G = \overset{\circ}{G}$, because if $\tilde{g} \ge 0$, $\tilde{g} \downarrow$ and $\lim_{u \to b-} \tilde{g}(u) > 0$, then $g(\cdot) \equiv \tilde{g}(\rho(y_0, \cdot)) \notin L_\beta^p(Y_1)$. If (1.4.1) is fulfilled, then $G_0 \subset G$, since in this case $f(\cdot) \equiv \chi_{(a,b)}(\rho(y_0, \cdot)) \in L_\beta^p(Y_1)$.

Remark 1.4.2. It follows from (1.4.1) that $f(\cdot) \equiv \chi_{(a,b)}(\rho(y_0, \cdot)) \in G$. Therefore $F(\cdot, b) = Tf \in X$. Hence $F(x, b) < \infty$ for almost all $x \in Y_2$.

Taking into account the fact that the operator T is positive, the condition $0 < t < \tau < b$ results in $F(x,t) \leq F(x,\tau) \leq F(x,b)$, and we conclude that the function $F(x,t)$ increases with t and has the limit

$$F_0(x,b) \equiv \lim_{t \to b-} F(x,t) \leq F(x,b). \tag{1.4.2}$$

In many cases we have

$$F_0(x,b) = F(x,b) \quad \text{a.e. on} \quad Y_2. \tag{1.4.2'}$$

This implies that $\|T\|_{G_0} = \|T\|_{\dot{G}}$. The condition (1.4.2′) is, for example, fulfilled for any linear bounded operator. Indeed,

$$\|F(\cdot,b) - F(\cdot,t)\|_X =$$
$$= \|T[\chi_{(a,b)}\rho(y_0,\cdot) - \chi_{(a,t)}(\rho(y_0,\cdot))]\|_X \leq$$
$$\leq \|T\|\|\chi_{(a,b)}(\rho(y_0,\cdot)) - \chi_{(a,t)}(\rho(y_0,\cdot))\|_{p,\beta,Y_1} =$$
$$= \|T\|(\beta\{y : t \leq \rho(y_0,y) < b\})^{1/p} \to 0,$$

as $t \to b-$, since $\lim_{t \to b-} \beta\{y : t \leq \rho(y_0,y) < b\} = 0$. But equality (1.4.2′) is not always satisfied and consequently the equality $\|T\|_{G_0} = \|T\|_{\dot{G}_0}$ does not always hold (see [41], Remark 4).

Let φ be a positive continuous function on (a,b). We consider the following cones of functions:

$$G \equiv G(p,\beta,\varphi,Y_1) \equiv \{f \in L^p_\beta(Y_1) : \exists \tilde{f} : (a,b) \to R^1_+, \frac{\tilde{f}}{\varphi} \downarrow,$$

$$f(x) = \tilde{f}(\rho(y_0,x)) \ \forall x \in Y_1\};$$

$$\dot{G} \equiv \dot{G}(p,\beta,\varphi,Y_1) \equiv \{f \in G : \lim_{u \to b-} \frac{\tilde{f}(u)}{\varphi(u)} = 0\}.$$

The following theorem holds.

Theorem 1.4.3. *Let the conditions of Theorem 1.4.2 be fulfilled with* $\dot{G}(p,\beta,\varphi,Y_1)$ *instead of* $\dot{G}$. *Then*

$$\|T\|_{\dot{G}(p,\beta,\varphi,Y_1)} \equiv \sup_{a<t<b} \left[\|T(\varphi\chi_{(a,t)}(\rho(y_0,\cdot)))\|_X \times \right.$$

$$\left. \times \left(\int\limits_{\{y:a<\rho(y_0,y)<t\}} \varphi^p(\rho(y_0,y))d\beta \right)^{-1/p} \right].$$

If the condition

$$\int\limits_{Y_1} \varphi^p(\rho(y_0,y))d\beta < \infty \tag{1.4.3}$$

is violated, then $G(p,\beta,\varphi,Y_1) = \overset{\cdot}{G}(p,\beta,\varphi,Y_1)$; if, however, it is satisfied, then we have the following generalization of Theorem 1.4.1.

Theorem 1.4.4. *Let the conditions of Theorem 1.4.2 be fulfilled with* $G(p,\beta,\varphi,Y_1)$ *instead of* $\overset{\cdot}{G}$ *and condition (1.4.3). Then*

$$\|T\|_{G(p,\beta,\varphi,Y_1)} \sup_{a<t\le b}\left[\|T(\varphi\chi_{(a,t)}(\rho(y_0,\cdot)))\|_X \times \right.$$
$$\left.\left(\int\limits_{\{a<\rho(y_0,y)<t\}} \varphi^p(\rho(y_0,y))^p d\beta\right)^{-1/p}\right].$$

Similar results are also valid on cones of nondecreasing functions:

$$H \equiv \{f \in L^p_\beta(Y_1) : \exists \tilde{f} : (a,b) \to \tilde{R}^1_+, \ \tilde{f}\uparrow, f(x) \equiv \tilde{f}(\rho(y_0,x)) \ \forall x \in Y_1\};$$
$$\overset{\cdot}{H} \equiv \{f \in H : \lim_{u\to a+} \tilde{f}(u) = 0\}.$$

We also consider the following cones of functions:

$$H(p,\beta,\varphi,Y_1) \equiv \{f \in L^p_\beta(Y_1) : \exists \tilde{f} : (a,b) \to \tilde{R}^1_+,$$
$$\frac{\tilde{f}}{\varphi}\uparrow, f(x) = \tilde{f}(\rho(y_0,x)) \ \forall x \in Y_1\};$$
$$\overset{\cdot}{H}(p,\beta,\varphi,Y_1) \equiv \{f \in H : \lim_{u\to a+} \frac{\tilde{f}(u)}{\varphi(u)} = 0\}.$$

Denote

$$H_0 \equiv \{\chi_{(t,b)}(\rho(y_0,y)) : a \le t < b\};$$
$$\overset{\cdot}{H_0} \equiv \{\chi_{(t,b)}(\rho(y_0,y)) : a < t < b\}.$$

Theorem 1.4.5. *Let the conditions of Theorem 1.4.2 with* $\overset{\cdot}{H}$ *instead of* $\overset{\cdot}{G}$ *be fulfilled. Then*

$$\|T\|_{\overset{\cdot}{H}} = \sup_{a<t<b}\left[T[\chi_{(t,b)}(\rho(y_0,\cdot))]\|_X \left(\beta\{y : t < \rho(y_0,y) < b\}\right)^{-1/p}\right].$$

Theorem 1.4.6. *Let the conditions of Theorem 1.4.2 with H instead of G and condition (1.4.1) be fulfilled. Then*

$$\|T\|_H = \sup_{a \le t < b} \left[\|T[(\chi_{(t,b)}(\rho(y_0, \cdot)))]\|_X \left(\beta\{y : t < \rho(y_0, y) < b\} \right)^{-1/p} \right].$$

The following theorems are also valid.

Theorem 1.4.7. *Let the conditions of Theorem 1.4.2 with $\dot{H}$ (p, β, φ, Y_1) instead of G be fulfilled. Then*

$$\|T\|_{\dot{H}(p,\beta,\varphi,Y_1)} \equiv \sup_{a < t < b} \left[\left\| T[\varphi \chi_{(t,b)}(\rho(y_0, \cdot))] \right\|_X \times \right.$$
$$\left. \times \left(\int\limits_{\{t < \rho(y_0, y) < b\}} \varphi^p(\rho(y_0, y)) d\beta \right)^{-1/p} \right].$$

Theorem 1.4.8. *Let the conditions of Theorem 1.4.2 with $H(p, \beta, \varphi, Y_1)$ instead of $\dot{G}$ and condition (1.4.3) be fulfilled. Then*

$$\|T\|_{H(p,\beta,\varphi,Y_1)} \equiv \sup_{a \le t < b} \left[\left\| T[\varphi \chi_{(t,b)}(\rho(y_0, \cdot))] \right\|_X \times \right.$$
$$\left. \times \left(\int\limits_{\{t < \rho(y_0, y) < b\}} \varphi^p(\rho(y_0, y)) d\beta \right)^{-1/p} \right].$$

To prove these results we need the following Lemma (see, e.g., [41]).

Lemma 1.4.1. *Let a_k, $b_k \ge 0$, $k \in Z$, $a_k \uparrow$, $b_k \downarrow$. Suppose that $\lim\limits_{k \to +\infty} a_k = A$, $\lim\limits_{k \to +\infty} b_k = B$. Let $C = AB$ (with the convention that $\infty \cdot 0 = 0$). If $C < \infty$, then for $0 < q < r$ we have*

$$\left\{ C^r + \sum_k a_k^r (b_k^r - b_{k+1}^r) \right\}^{1/r} \le \left\{ C^q + \sum_k a_k^q (b_k^q - b_{k+1}^q) \right\}^{1/q}. \quad (1.4.4)$$

Moreover, inequality (1.4.4) remains valid even if we replace C by any number $D \ge C$.

We also shall need

Lemma 1.4.2. *Let $a_k, b_k \ge 0$, $k \in Z$, $a_k \downarrow$, $b_k \uparrow$, $\lim\limits_{k \to -\infty} a_k = A$, $\lim\limits_{k \to -\infty} b_k = B$, $C = AB$. Assume that $0 < q < r$. If $C < \infty$, then*

$$\left\{ C^r + \sum_k a_k^r (b_{k+1}^r - b_k^r) \right\}^{1/r} \le \left\{ C^q + \sum_k a_k^q (b_{k+1}^q - b_k^q) \right\}^{1/q}. \quad (1.4.5)$$

Inequality (1.4.5) *remains valid even if we replace* C *by any number* $D \geq C$.

Proof. First we shall prove the integral analogue of inequality (1.4.5) for $B = 0$ (i.e., $C = 0$ in (1.4.5)) and $r = 1$. Let the functions $a = a(x)$, $b = b(x)$ and $b'(x)$ be continuous on R_+^1, $a(x) \geq 0$, $b(x) \geq 0$, $a(x) \downarrow$, $b(x) \uparrow$, $b(0+) \equiv \lim_{x \to 0+} b(x) = 0$. Then

$$I \equiv \int_0^\infty a(x)d(b(x)) \leq \left[\int_0^\infty a(x)^q d(b(x)^q) \right]^{1/q}, \quad 0 < q < 1. \qquad (1.4.6)$$

Indeed, we have

$$I = \frac{1}{q} \int_0^\infty a(x)b(x)^{1-q}d(b(x)^q)$$

$$= \frac{1}{q} \int_0^\infty [a(x)b(x)]^{1-q}a(x)^q d(b(x)^q). \qquad (1.4.7)$$

But $a(t) \downarrow$, $b(0+) = 0$ and therefore

$$\int_0^x a(t)^q d(b(t)^q) \geq a(x)^q \int_0^x d(b(t)^q) =$$

$$= a(x)^q [b(x)^q - b(0)^q] = a(x)^q b(x)^q.$$

Consequently from (1.4.7) we get

$$I \leq \frac{1}{q} \int_0^\infty \left[\int_0^x a(t)^q d(b(t)^q) \right]^{1/q-1} a(x)^q d(b(x)^q) =$$

$$= \int_0^\infty d\left[\int_0^x a(t)^q d(b(t)^q) \right]^{1/q} = \left(\int_0^\infty a(t)^q d(b(t)^q) \right)^{1/q}.$$

which gives (1.4.6).

Now let $0 \leq a_k \downarrow$; $0 \leq b_k \uparrow$, $\lim_{k \to -\infty} b_k = 0$. We choose the function $b = b(x) \in C^1(R_+)$ so that $b(2^k) = b_k$, $k \in Z$, and the functions $a_n = a_n(x) \in C(R_+)$, $n \in N$, so that

$$a_n(x) = \begin{cases} a_k & \text{for } 2^k \leq x \leq 2^{k+1} - \frac{2^{k-1}}{n}, \\ \text{linear for } 2^{k+1} - \frac{2^{k-1}}{n} \leq x \leq 2^{k+1}, & k \in Z. \end{cases}$$

Then by (1.4.6) we have

$$I_n \equiv \int_0^\infty a_n(x)d(b(x)) \leq J_n \equiv \left[\int_0^\infty a_n(x)^q d(b(x)^q) \right]^{1/q}. \qquad (1.4.8)$$

But $a_n(x) \uparrow a(x)$, $a(x) = a_k$, for $2^k \leq x < 2^{k+1}$, and hence by the Beppo Levi theorem as $n \to +\infty$ we obtain

$$\lim_{n \to +\infty} I_n = \int_0^\infty a(x) d(b(x)) = \sum_k \int_{2^k}^{2^{k+1}} a(x) d(b(x)) = \sum_k a_k(b_{k+1} - b_k),$$

$$\lim_{n \to +\infty} J_n = \left[\int_0^\infty a(x)^q d(b(x)^q) \right]^{1/q} = \left[\sum_k a_k^q(b_{k+1}^q - b_k^q) \right]^{1/q}.$$

Therefore from (1.4.8) we have inequality (1.4.5) for $r = 1$, $0 < q < 1$, $B = 0$ (i.e., $C = 0$).

Further, let $A < \infty$ and $B > 0$. Suppose that for $N \in Z$

$$\tilde{b}_k = b_k, \quad \text{for} \quad K > N, \quad \tilde{b}_k = 0, \quad \text{for} \quad k \leq N.$$

Then $\tilde{b}_k$ is increasing and $\lim_{k \to -\infty} \tilde{b}_k = 0$. According to the above argument,

$$\sum_k a_k(\tilde{b}_{k+1} - \tilde{b}_k) \leq \left\{ \sum_k a_k^q(\tilde{b}_{k+1}^q - \tilde{b}_k^q) \right\}^{1/q},$$

that is,

$$\sum_{k > N} a_k(b_{k+1} - b_k) = \sum_{k \geq N+1} a_k(b_{k+1} - b_k) +$$

$$+ a_N(b_{N+1} - b_N) = \sum_{k \geq N+1} a_k(b_{k+1} - b_k) + a_N b_{N+1} \leq$$

$$\leq \left\{ \sum_{k \geq N+1} a_k^q(b_{k+1}^q - b_k^q) + a_N^q b_{N+1}^q \right\}^{1/q}.$$

But $\lim_{N \to -\infty} a_N b_{N+1} = AB = C$. Passing to the limit as $N \to -\infty$ we obtain (1.4.5) for $r = 1$, $0 < q < 1$.

Let $C < \infty$. We shall show that in (1.4.5) we can replace C by $D \geq C$. For $u, v \in R_+$, $0 < q < 1$ we consider the function

$$\varphi(x) = \varphi_{u,v}(x) = \frac{(u + x)^q}{\varphi_{u,v} + x^q}.$$

As is shown in [41], the implication

$$\{x_0 > 0, \varphi(x_0) \leq 1\} \Rightarrow \{\varphi(x) \leq 1, \quad \forall x \geq x_0\} \qquad (1.4.9)$$

is valid. Put $x_0 \equiv C$, $u \equiv \sum_k a_k(b_{k+1} - b_k)$, $v \equiv \left[\sum_k a_k^q(b_{k+1}^q - b_k^q) \right]^{1/q}$. Then inequality (1.4.5) has the form $\varphi(x_0) \leq 1$. By (1.4.9), $\varphi(D) \leq 1$,

$\forall D \le C = x_0$. This means that inequality (1.4.5) for $r = 1, 0 < q < 1$ remains valid even if we replace C by $D \ge C$. In the general case, for $0 < q < r$ we have $0 < q/r < 1$, and if $\tilde{a}_k \downarrow, \tilde{b}_k \uparrow$, $\lim\limits_{k \to -\infty} \tilde{a}_k = A$, $\lim\limits_{k \to -\infty} \tilde{b}_k = B$ and $\tilde{C} = \tilde{A}\tilde{B}$, then for any $\tilde{D} \ge \tilde{C}$ (according to the above argument)

$$\tilde{D} + \sum_k \tilde{a}_k(\tilde{b}_{k+1} - \tilde{b}_k) \le \left\{ \tilde{D}^{q/r} + \sum_k \tilde{a}_k^{q/r}(\tilde{b}_{k+1}^{q/r} - \tilde{b}_k^{q/r}) \right\}^{r/q}.$$

Putting $\tilde{a}_k = a_k^r$, $\tilde{b}_k = b_k^r$ here we see that $\tilde{A} = A^r$, $\tilde{B} = B^r$, $\tilde{C} = C^r$ and for any $D \ge C$ we have $\tilde{D} = D^r \ge C^r = \tilde{C}$.

Hence for any $D \ge C$,

$$D^r + \sum_k a_k^r(b_{k+1}^r - b_k^r) \le \left\{ D^q + \sum_k a_k^q(b_{k+1}^q - b_k^q) \right\}^{r/q}.$$

□

We now set about proving the basic results. First we shall show that Theorem 1.4.1 holds.

Let $a < t_k < t_{k+1} < b$, $k \in Z$ where

$$\lim_{k \to -\infty} t_k = a, \qquad \lim_{k \to +\infty} t_k = b.$$

For $g \in G$ we consider the function $\bar{g}(u) \equiv \sum_k \tilde{g}(t_k)\chi_{[t_k,t_{k+1})}(u)$, where $g(x) \equiv \tilde{g}(\rho(y_0, \cdot))$ and $\tilde{g} \downarrow$ on (a, b).

It is clear that $\tilde{g}(u) \le \bar{g}(u)$ for all $u \in (a, b)$. Note that for any $s > 0$

$$\bar{g}(u) = \left\{ \sum_k \tilde{g}(t_k)^s \chi_{[t_k,t_{k+1}]}^s(u) \right\}^{1/s}. \tag{1.4.10}$$

Denote $B \equiv \lim\limits_{u \to +b} \tilde{g}(u)$, $c_k(s) \equiv [\tilde{g}(t_{k-1})^s - \tilde{g}(t_k)^s]^{1/s}$. Then for any $u \in (a, b)$ and $s > 0$ we have

$$\bar{g}(u) = \{\sum_k c_k^s(s)\chi_{(a,t_k)}^s(u) + B^s \chi_{(a,b)}^s(u)\}^{1/s}. \tag{1.4.11}$$

Equality (1.4.11) follows from (1.4.10) using Abel's transform:

$$\sum_{K=M}^{N} e_k(d_{k+1} - d_k) =$$

$$= \sum_{K=M}^{N} (e_k - e_{k+1})d_{k+1} + e_{N+1}d_{N+1} - e_M d_M \tag{1.4.12}$$

putting in it

$$e_k = \tilde{g}(t_k)^s, \quad d_k = \chi^s_{(a,t_k)}(u) = \chi_{(a,t_k)}(u)$$

and taking into account that

$$\lim_{N \to +\infty} e_{N+1} d_{N+1} =$$
$$= B^s \chi_{(a,b)}(u),$$
$$\lim_{M \to -\infty} e_M d_M = \lim_{M \to -\infty} \tilde{g}(t_M)^s \chi_{(a,t_M)}(u) = 0.$$

Thus passing in (1.4.12) to the limit as $M \to -\infty$, $N \to +\infty$ we obtain

$$\sum_k e_k(d_{k+1} - d_k) = \sum_k (e_k - e_{k+1}) d_{k+1} + B^s \chi^s_{(a,b)}(u)$$

which coincides with (1.4.11). Further, as $0 \le \tilde{g}(u) \le \overline{g}(u)$, therefore

$$0 \le \tilde{g}(\rho(y_0, y)) \le \overline{g}(\rho(y_0, y))$$

for any $y \in Y_1$. Moreover, the equality (1.4.11) yields

$$\overline{g}(\rho(y_0, y)) =$$
$$= \left\{ \sum_k c_k^s(s) \chi_{(a,t_k)}(\rho(y_0, y)) + B^s \chi^s_{(a,b)}(\rho(y_0, y)) \right\}^{1/s}, \qquad (1.4.13)$$

and due to the positivity and l_r- convexity of the operator T, we obtain

$$0 \le T[\tilde{g}(\rho(y_0, \cdot))] \le T[\overline{g}(\rho(y_0, \cdot))] \le$$
$$\le \left(\sum_k c_k^s(s) F(x, t_k)^r + B^s F(x, b)^r \right)^{1/r}. \qquad (1.4.14)$$

We now apply Lemma 1.4.1 for $0 < q < r$ and assume that $0 \le a_k \equiv F(x, t_k) \uparrow, 0 \le b_k \equiv \overline{g}(t_{k-1}) \downarrow$. Taking into account that

$$A = \lim_{k \to +\infty} a_k = F_0(x, b), \quad B = \lim_{k \to +\infty} b_k$$
$$C = B F_0(x, b), \quad D = B F(x, b) \ge C$$

we see the fact that (1.4.14) implies

$$0 \le T[\tilde{g}(\rho(y_0, y))] \le \left\{ \sum_k [\tilde{g}(t_{k-1})^q - \tilde{g}(t_k)] F(x, t_k)^q + B^q F(x, b)^q \right\}^{1/q}$$

from which and from the l_q- convexity of the space X we obtain

$$\|T[g]\|_X \le$$
$$\le \left\{ \sum_k [\tilde{g}(t_{k-1})^q - \tilde{g}(t_k)^q] \|F(\cdot, t_k)\|_X^q + B^q \|F(\cdot, b)\|_X^q \right\}^{1/q}.$$

Thus

$$\|T[\tilde{g}(\rho(y_0, \cdot))]\|_X \leq$$

$$\leq \|T\|_{G_0} \Big\{ \sum_k [\tilde{g}(t_{k-1})^q - \tilde{g}(t_k)^q] \big(\beta\{a < \rho(y_0, y) < t_k\} \big)^{q/p} +$$

$$+ B^s (\beta(Y_1))^{q/p} \Big\}^{1/p}.$$

We again use Lemma 1.4.1 with q instead of r and p instead of q $(0 < p < q)$ and put

$$0 \leq a_k \equiv \big(\beta\{a < \rho(y_0, y) < t_k\} \big)^{1/p} \uparrow, \quad 0 \leq b_k \equiv g(t_{k-1}) \downarrow.$$

In this case

$$A = \lim_{k \to +\infty} a_k \leq (\beta(Y_1))^{1/p},$$

$$B = \lim_{k \to +\infty} b_k, \quad D = B(\beta(Y_1))^{1/p} \geq C = AB.$$

In notation inequality (1.4.4) gives

$$\|T[\tilde{g}(\rho(y_0, \cdot))]\|_X \leq \|T\|_{G_0} \Big\{ \sum_k [\tilde{g}(t_{k-1})^p - \tilde{g}(t_k)^p] \times$$

$$\times \beta\{a < \rho(y_0, y) < t_k\} + B^p \beta(Y_1) \Big\}^{1/p}. \tag{1.4.15}$$

We now assume $s = p$ in (1.4.13). After integration we obtain

$$\int_{Y_1} \tilde{g}(\rho(y_0, y))^p = \sum_k c_k^p(p) \beta\{a < \rho(y_0, y) < t_k\} + B^p \beta(Y_1)$$

which together with (1.4.15) gives the inequality

$$\|T[\tilde{g}(\rho(y_0, \cdot))]\|_X \leq \|T\|_{G_0} \|\tilde{g}(\rho(y_0, \cdot))\|_{p, \beta, Y_1}.$$

Next, for every $n \in N$ we construct sequences $\{t_k(n)\}_{k \in Z}$ such that the corresponding functions $\bar{g}_n(u)$ of type (1.4.10) form for $t_k = t_k(n)$ a nonincreasing sequence which converges everywhere to the given function $\tilde{g}$, i.e., $\bar{g}_n(\rho(y_0, y))$ converges to $g(\rho(y_0, y))$ everywhere on Y_1. Then by the Beppo Levi theorem,

$$\lim_{n \to \infty} \|\bar{g}_n(\rho(y_0, \cdot))\|_{p, \beta, Y_1} = \|\tilde{g}(\rho(y_0, \cdot))\|_{p, \beta, Y_1}.$$

Hence, passing to the limit as $n \to \infty$ in the inequality

$$\|T[\bar{g}_n(\rho(y_0, \cdot))]\|_X \leq \|T\|_{G_0} \|\bar{g}(\rho(y_0, \cdot))\|_{p, \beta, Y_1}$$

we obtain

$$\|Tg\|_X \leq \|T\|_{G_0}\|g\|_{p,\beta,Y_1}, \quad \text{for all} \quad g \in G_0.$$

This implies $\|T\|_G \leq \|T\|_{G_0}$. The reverse inequality follows from the inclusion $G_0 \subset G$.

The proof of Theorem 1.4.2 is analogous. To prove Theorem 1.4.3 we first verify that the condition $f \in G\,(p,\beta,\varphi,Y_1)$ is equivalent to the condition $g(\cdot) \equiv (\tilde{f}/\varphi)(\rho(y_0,\cdot)) \in G(p,\beta_\varphi,1,Y_1)$, where $d\beta_\varphi(y) = \varphi^p(\rho(y_0,y))d\beta(y)$. Moreover,

$$\|f\|_{p,\beta,Y_1} = \|g\|_{p,\beta_\varphi,Y_1}.$$

For T we define the l_r-convex positive operator T_φ by the formula

$$T_\varphi(g) = T[\varphi(\rho(y_0,\cdot))g].$$

Then

$$\|T\|_{G(p,\beta,\varphi,Y_1)} = \sup_{f \in G(p,\beta,\varphi,Y_1)} \frac{\|T[f]\|_X}{\|f\|_{p,\beta,Y_1}} =$$

$$= \sup_{g \in G(p,\beta_\varphi,1,Y_1)} \frac{\|T_\varphi(g)\|_X}{\|g\|_{p,\beta_\varphi,Y_1}} = \|T_\varphi\|_{G(p,\beta_\varphi,1,Y_1)}.$$

Now we apply Theorem 1.4.2 to the operator T_φ on the cone $\dot{G} = \dot{G}\,(p,\beta_\varphi,1,Y_1)$ and obtain

$$\|T_\varphi\|_{\dot{G}(p,\beta_\varphi,1,Y_1)} =$$

$$= \sup_{a<t<b}\left[\|T_\varphi[\chi_{(a,t)}(\rho(y_0,\cdot))]\|_X \left(\beta_\varphi\{a < \rho(y_0,y) < t\}\right)^{-1/p}\right].$$

Reasoning as above, from Theorem 1.4.1 we obtain Theorem 1.4.4. Using Lemma 1.4.2 we can prove Theorem 1.4.6 in the same way as Theorem 1.4.1, and so we omit the proof.

Theorems 1.4.5, 1.4.7 and 1.4.8 follows in a similar way to Theorems 1.4.2, 1.4.3 and 1.4.4 respectively.

Let φ and ψ be a.e. positive measurable functions on $(0,\infty)$ such that $\varphi(x) < \psi(x)$ a.e. on $(0,\infty)$. Suppose that for any measurable function $f : (0,\infty) \to R$,

$$T_{\varphi\psi}(f)(x) = \int\limits_{\varphi(x)}^{\psi(x)} f(y)dy.$$

As before, we shall use the following notation:

$$G = \{g \in L^p_\beta(0,\infty) : g \geq 0, g \downarrow\},$$

$$\dot{G} = \{g \in G : \lim_{u \to +\infty} g(u) = 0\},$$

$$H = \{f \in L^p_\beta(0,\infty) : g \geq 0, g \uparrow\},$$

$$\dot{H} = \{f \in H : \lim_{u \to 0+} g(u) = 0\}, F(x,t) = T_{\varphi\psi}(\chi_{(0,t)})(x),$$

$$\Phi(x,t) = T_{\varphi\psi}(\chi_{(t,\infty)})(x), \quad x, t > 0.$$

It can be easily seen that

$$F(x,t) = \begin{cases} 0, & t \leq \varphi(x), \\ t - \varphi(x), & t \in (\varphi(x), \psi(x)) \\ \psi(x) - \varphi(x), & t \geq \psi(x). \end{cases}$$

Let $0 < p \leq \min\{s,1\}$ and $X = L^s_\gamma(0,\infty)$. Then

$$\|T_{\varphi\psi}\|_G \equiv \|T_{\varphi\psi}\|_{\dot{G}} = \sup_{t>0} \|F(\cdot,t)\|_{L^s_\gamma(0,\infty)} (\beta(0,t))^{-1/p} =$$

$$= \sup_{t>0} \left(\int_{\{x:\psi(x)\leq t\}} (\psi(x) - \varphi(x))^s d\gamma(x) + \right.$$

$$\left. + \int_{\{x:\varphi(x)<t<\psi(x)\}} (t - \varphi(x))^s d\gamma(x) \right)^{1/s} (\beta(0,t))^{-1/p}.$$

Analogously we have

$$\|T_{\varphi\psi}\|_H \equiv \|T_{\varphi\psi}\|_{\dot{H}} = \sup_{t>0} \|\Phi(\cdot,t)\|_{L^s_\gamma(0,\infty)} (\beta(t,\infty))^{-1/p} =$$

$$= \sup_{t>0} \left(\int_{\{x:\varphi(x)\geq t\}} (\psi(x) - \varphi(x))^s d\gamma(x) + \right.$$

$$\left. + \int_{\{x:\varphi(x)<t<\psi(x)\}} (t - \varphi(x))^s d\gamma(x) \right)^{1/s} (\beta(t,\infty))^{-1/p}.$$

Let us consider the particular case of the operator $T_{\varphi\psi}$. Let $0 < a < b < \infty$ and let $\varphi(x) = ax, \psi(x) = bx$. In this case we put $T_{\varphi\psi} \equiv T_{ab}$. It is not difficult to see that if $0 < p \leq 1$, $X = L^1_\gamma(0,\infty)$, $d\gamma(x) = x^\sigma dx$, $d\beta(u) = u^\alpha du$, then

$$\|T_{ab}\|_G = \|T_{ab}\|_{\dot{G}} = \sup_{t>0} C_0 t^{\sigma+2-\alpha/p-1/p}, \quad \alpha > -1, \sigma > -2,$$

where

$$C_0 = \left\{ \frac{b-a}{(\sigma+2)b^{\sigma+2}} + \frac{a^{-\sigma-1}-b^{-\sigma-1}}{\sigma+1} - a\frac{a^{-\sigma-2}-b^{-\sigma-2}}{\sigma+2} \right\}(\alpha+1)^{1/p}.$$

Hence the operator T_{ab} is bounded from G to $L^1_\gamma(0,\infty)$ if and only if $\sigma = \alpha/p + 1/p - 2$, and $\|T_{ab}\|_G = C_0$.

In a similar way we obtain

$$\|T_{ab}\|_H = \|T_{ab}\|_{\cdot_H} = \sup_{t>0} C_1 t^{\sigma+2-\alpha/p-1/p}, \quad \alpha < -1, \sigma < -2,$$

where

$$C_1 = \left\{ \frac{a-b}{(\sigma+2)a^{\sigma+2}} + \frac{a^{-\sigma-1}-b^{-\sigma-1}}{\sigma+2} - a\frac{a^{-\sigma-2}-b^{-\sigma-2}}{\sigma+2} \right\}(-\alpha-1)^{1/p}.$$

Consequently the operator T_{ab} acts boundedly from H to $L^1_\gamma(0,\infty)$ if and only if $\sigma = \alpha/p + 1/p - 2$, and in this case we have $\|T_{ab}\|_H = \|T_{ab}\|_{\cdot_H} = C_1$.

Now consider the Riemann–Liouville operator

$$R_\alpha f(x) = \int\limits_0^x f(y)(x-y)^{\alpha-1}dy, \quad x > 0, \quad \alpha > 0.$$

As above, we assume that $0 < p \le \min\{1,s\}$, $X = L^s_\gamma(0,\infty)$ and in this case we have

$$\|R_\alpha\|_G = \|R_\alpha\|_{\cdot_G} = \sup_{t>0} \frac{1}{\alpha} \left\{ \int\limits_0^t x^{\alpha s} d\gamma(x) + \right.$$

$$\left. + \int\limits_t^\infty [x^\alpha - (x-t)^\alpha]^s d\gamma(x) \right\}^{1/s} \left(\int\limits_0^t d\beta(y) \right)^{-1/p};$$

$$\|R_\alpha\|_H = \|R_\alpha\|_{\cdot_H} = \sup_{t>0} \frac{1}{\alpha} \left\{ \int\limits_t^\infty (x-t)^{\alpha s} d\gamma(x) \right\}^{1/s} \left(\int\limits_t^\infty d\beta(y) \right)^{-1/p}.$$

Next let (Y, d, μ) be a Hölder's inequality (SHT) (for the definition of SHT see Section 1.1.).

Assume that measures β and γ are defined on Y with the property that all balls in Y are measurable with respect to β and γ.

Suppose that $x_0 \in Y$ and let $a = \sup\{d(x_0,x) : x \in Y\}$. It will be assumed that $\mu\{x_0\} = \mu\{x \in Y : d(x_0,x) = a\} = 0$ and $\beta(B(x_0,R)\backslash B(x_0,r)) > 0$

for all $r, R, 0 < r < R < a$. We consider the following cones of functions:

$$G = \{g \in L^p_\beta(Y) : \exists \tilde{g} : (0, a) \to R^1_+, \tilde{g} \downarrow, g(x) = \tilde{g}(d(x_0, x)) \forall x \in Y\};$$

$$\overset{.}{G} = \{g \in G : \lim_{u \to a} \tilde{g}(u) = 0)\}$$

$$H = \{f \in L^p_\beta(Y) : \exists \tilde{g} : (0, a) \to R^1_+, \tilde{g} \uparrow, g(x) = \tilde{g}(d(x_0, x)) \forall x \in Y\};$$

$$\overset{.}{H} = \{f \in H : \lim_{u \to 0} \tilde{f}(u) = 0\},$$

where $0 < p < \infty$. Consider on Y the Hardy–type operator

$$\mathcal{H}f(x) = \left(\int_{B(x_0, d(x_0, x))} (f(y))^r d\mu(y) \right)^{1/r}, \quad r > 0.$$

(For the operator $\mathcal{H}$ for $r = 1$, see [73], [155] and also [100], Section 9.4).

Let $0 < p \leq q \equiv \min\{r, s\}$ and $X = L^s(\gamma, Y)$. According to the theorems proven above, we have

$$\|\mathcal{H}\|_G = \|\mathcal{H}\|_{\overset{.}{G}} = \sup_{0 < t < a} \left\{ \int_{B(x_0, t)} \beta(B(x_0, d(x_0, x)))^{s/r} d\gamma(x) + \right.$$

$$\left. + (\beta B(x_0, t))^{s/r} \gamma(X \setminus B(x_0, t)) \right\}^{1/s} (\mu B(x_0, t))^{-1/p},$$

$$\|\mathcal{H}\|_H = \|\mathcal{H}\|_{\overset{.}{H}} =$$

$$= \sup_{0 < t < a} \left\{ \int_{\{x : d(x_0, x) > t\}} (\beta\{t < d(x_0, y) < d(x_0, x)\})^{s/r} d\gamma(x) \right\}^{1/s} \times$$

$$\times (\mu\{x \in Y : d(x_0, x) > t\})^{-1/p}.$$

Consider now a particular case of the space (Y, d, μ).

Definition 1.4.1. A homogeneous group is a nilpotent, connected, simply connected Lie group $\mathcal{G}$ whose Lie algebra g is endowed with a one–parameter group of extensions $\delta_t = \exp(A \ln t)$, $t > 0$, where A is a diagonalized linear operator on g with positive eigenvalues (see, for example, [86], p. 5).

The mappings $\exp \circ \delta_t \circ \exp^{-1}$ are the group automorphisms. We denote them again by δ_t.

The number $Q = tr A$ is said to be the homogeneous dimension of the group $\mathcal{G}$.

As examples of homogeneous groups we may take an n– dimensional Euclidean space with a fixed diagonal matrix $A = (a_{ij})$, $i, j = 1, 2$, the Heisenberg groups H^n, and so on.

A homogeneous norm on $\mathcal{G}$ is a continuous function $r : \mathcal{G} \to [0, \infty)$ of class C^∞ on $\mathcal{G}\backslash\{e\}$ possessing the following properties:

(i) $r(x) = r(x^{-1})$ for arbitrary $x \in \mathcal{G}$;

(ii) $r(x) = 0$ if and only if $x = e$;

(iii) $r(\delta_t x) = tr(x)$ for arbitrary $x \in \mathcal{G}$ and $t > 0$;

(iv) there exists a constant $c_0 > 0$ such that

$$r(xy) \le c_0(r(x) + r(y)) \quad \text{for all} \quad x, y \in \mathcal{G}.$$

For $x \in \mathcal{G}$ we assume that

$$B(x, \rho) \equiv \{y \in \mathcal{G} : r(xy^{-1}) < \rho\}.$$

Fix on $\mathcal{G}$ a normed Haar measure so that the measure of the unit ball $B(e, 1)$ is equal to 1. The Haar measure of any measurable set $E \subset \mathcal{G}$ will be denoted by $|E|$.

It is known that $|B(x, \rho)| = \rho^Q$ for all $x \in \mathcal{G}$ and $\rho > 0$.

Theorem A ([86], p. 14). *Let $S_\mathcal{G} \equiv \{x \in \mathcal{G} : r(x) = 1\}$. There exists a measure σ concentrated on $S_\mathcal{G}$ such that for all $u \in L^1(\mathcal{G})$ the equality*

$$\int_\mathcal{G} u(x)dx = \int_0^\infty \left(\int_{S_\mathcal{G}} u(\delta_\rho(y))\rho^{Q-1}d\sigma(y) \right) d\rho$$

holds.

Consider on $\mathcal{G}$ the following Hardy–type operator:

$$Kf(x) = \left(\int_{B(e,r(x))} (f(y))^r dy \right)^{1/r}, \quad r > 0.$$

Let $d\gamma(x) = v(x)dx$, $d\beta(x) = w(x)dx$ where v and w are a.e. positive, measurable functions on $\mathcal{G}$. Furthermore, let $0 < p \le \min\{s, r\}$ and $X = L_v^s(\mathcal{G})$. The norm of the operator K will be calculated on the following cones:

$$
\begin{aligned}
G &= \{f \in L_w^p(\mathcal{G}) : \exists \tilde{f} : R_+^1 \to R_+^1, \quad \tilde{f} \downarrow, \quad f(x) = \tilde{f}(r(x)) \forall x \in \mathcal{G}\}; \\
\dot{G} &= \{f \in G : \lim_{u \to +\infty} \tilde{f}(u) = 0\}; \\
H &= \{f \in L_w^p(\mathcal{G}) : \exists \tilde{f} : R_+^1 \to R_+^1, \quad \tilde{f} \uparrow, \quad f(x) = \tilde{f}(r(x)) \forall x \in \mathcal{G}\}; \\
\dot{H} &= \{f \in H : \lim_{u \to 0} \tilde{f}(u) = 0\}.
\end{aligned}
$$

The formulae below follow from routine calculations;

$$\|K\|_G = \|K\|_{\dot{G}} =$$

$$= \sup_{t>0} \left\{ \int_{B(e,t)} (r(x))^{Qs/r} v(x)dx + v(\mathcal{G}\backslash B(e,t))t^{Qs/r} \right\}^{1/s} \times$$

$$\times (w(B(e,t)))^{-1/p},$$

$$\|K\|_H = \|K\|_{\overset{\cdot}{H}} =$$

$$= \sup_{t>0} \left\{ \int_{\mathcal{G}\backslash B(e,t)} (r(x)^Q - t^Q)^{s/r} v(x)dx \right\}^{1/s} (w(\mathcal{G}\backslash B(e,t)))^{-1/p},$$

where $v(E) = \int_E v(x)dx$, $w(E) = \int_E w(x)dx$.

Consider now the particular cases of weights. Let v and w be radial functions, i.e., $v(x) = v(r(x))$ and $w(x) = w(r(x))$. Then by Theorem A we obtain

$$\|K\|_G = \|K\|_{\overset{\cdot}{G}} = c_1 c_2 \sup_{t>0} \left\{ \int_0^t \tau^{Qs/r+Q-1} v(\tau)d\tau + \right.$$

$$\left. + t^{Qs/r} \int_t^\infty v(\tau)\tau^{Q-1}d\tau \right\}^{1/s} \left(\int_0^t w(\tau)\tau^{(Q-1)}d\tau \right)^{-1/p},$$

$$\|K\|_H = \|K\|_{\overset{\cdot}{H}} = c_1 c_2 \sup_{t>0} \left\{ \int_t^\infty (\tau^Q - t^Q)^{s/r}\tau^{Q-1} \times \right.$$

$$\left. \times v(\tau)d\tau \right\}^{1/s} \left(\int_t^\infty w(\tau)\tau^{Q-1}d\tau \right)^{-1/p},$$

where $c_1 = \left(\int_{S_{\mathcal{G}}} d\sigma(\zeta) \right)^{1/s}$, $c_2 = \left(\int_{S_{\mathcal{G}}} d\sigma(\zeta) \right)^{-1/p}$.

In particular, if $v(x) = r(x)^{\lambda_2}$ and $w(x) = r(x)^{\lambda_1}$, then

$$\|K\|_G = \|K\|_{\overset{\cdot}{G}} = c_1 c_2 \sup_{t>0} \left\{ \int_0^t \tau^{Qs/r+Q-1+\lambda_2}d\tau + \right.$$

$$\left. + t^{Qs/r} \int_t^\infty \tau^{\lambda_2+Q-1}d\tau \right\}^{1/s} \left(\int_0^t \tau^{\lambda_1+Q-1}d\tau \right)^{-1/p}.$$

We can distinguish the following cases:

(i) if $\lambda_1 \leq Q$, then $G = \overset{\cdot}{G} = \{0\}$ and $\|K\|_G = \|K\|_{\overset{\cdot}{G}} = 0$;

(ii) if $\lambda_1 > -Q$ and, moreover, $\lambda_2 \geq Q$ or $\lambda_2 \leq -Qs/r - Q$, then

$$\|K\|_G = \|K\|_{\dot{G}} = \infty;$$

(iii) for $\lambda_1 > -Q$ and $-Qs/r - Q < \lambda_2 < -Q$ we have

$$\|K\|_G = \|K\|_{\dot{G}} = c_1 c_2 A \sup_{t>0} t^{\frac{Q}{r} + \frac{Q}{s} + \frac{\lambda_2}{s} - \frac{\lambda_1}{s} - \frac{Q}{p}},$$

where

$$A = \left\{ \frac{1}{\frac{Qs}{r} + Q + \lambda_2} - \frac{1}{\lambda_2 + Q} \right\}^{1/s} (-(\lambda_2 + Q))^{1/p}.$$

Consequently, $\|K\|_G = \|K\|_{\dot{G}} < \infty$ if and only if

$$Q/r + \lambda_2/s + Q/s = (\lambda_1 + Q)/P,$$

and in this case

$$\|K\|_G = \|K\|_{\dot{G}} = A.$$

Let us now pass to the case of cones H and $\dot{H}$. We easily calculate that

$$\|K\|_H = \|K\|_{\dot{H}} = c_1 c_2 \sup_{t>0} \left\{ \int_t^\infty (\tau^Q - t^Q)^{s/r} \tau^{Q-1+\lambda_2} d\tau \right\}^{1/s} \times$$

$$\times \left(\int_t^\infty \tau^{\lambda_1 + Q - 1} d\tau \right)^{-1/p}.$$

We consider the following cases:

(i) $\lambda_1 \geq -Q$. Then $H = \dot{H} = \{0\}$ and hence $\|K\|_H = \|K\|_{\dot{H}} = 0$;

(ii) $\lambda_1 < -Q$ and $\lambda_2 \geq -\frac{Qs}{r} - Q$. Then $\|K\|_H = \|K\|_{\dot{H}} = \infty$;

(iii) $\lambda_1 < -Q$ and $\lambda_2 < -\frac{Qs}{r} - Q$. Then we have $\|K\|_H = \|K\|_{\dot{H}} < \infty$ if and only if $(\frac{Qs}{r} + \lambda_2 + Q - 1)/s = \lambda_1/p + Q/p$, and in this case

$$\|K\|_H = \|K\|_{\dot{H}} = c_1 c_2 \left\{ \int_0^1 (1 - u^Q)^{s/r} u^{-Qs/r - \lambda_2 - Q - 1} du \right\}^{1/s} \times$$

$$\times (-(\lambda_1 + Q))^{1/p}.$$

1.5. Notes and comments on Chapter 1

The first fundamental results concerning the mapping properties of Hardy transforms appeared in the 1920s and 1930s (see e.g. [119]). These transforms on the line in Lebesgue spaces studied in [38], [139], [195], [212], [290], [296], [272] and etc.

Weak-type inequalities for classical Hardy operator were obtained in [8] for the case of $p \leq q$ and in [190] when $q < p$.

There is considerable interest in the critical case of Hardy's inequality

$$\int_0^\infty \left| \frac{1}{x} \int_{|t| \leq x} f(t)dt \right|^p x^\alpha dx \leq c_{p,\alpha} \int_{-\infty}^{+\infty} |f(x)|^p |x|^\alpha dx,$$

where $1 < p < \infty$ and $\alpha = p - 1$.

Let ${}^y f(x) = f(x)$ if $|f(x)| > y$ and ${}^y f(x) = 0$ if $|f(x)| \leq y$.

Theorem ([115]). *Let $f \in L_{loc} \cap L^p_{|x|^{p-1}}(-\infty, +\infty)$ and let $1 < p < \infty$. Then ${}^y f \in L(-\infty, +\infty)$ for arbitrary $y > 0$, and*

$$\int_0^\infty \left| \int_{|t| \leq x} f(t)dt \right|^p \frac{dx}{x} \leq c_p \left(\int_{-\infty}^{+\infty} |f(x)|^p |x|^{p-1} dx + \int_0^{+\infty} \left| \int_{-\infty}^{+\infty} {}^y f(t)dt \right|^p \frac{dy}{y} \right),$$

$$\int_0^\infty \left| \int_{-\infty}^{+\infty} {}^y f(t)dt \right|^p \frac{dy}{y} \leq c_p \left(\int_{-\infty}^{+\infty} |f(x)|^p |x|^{p-1} dx + \int_0^{+\infty} \left| \int_{|t| \leq x} f(t)dt \right|^p \frac{dx}{x} \right).$$

Hardy transforms on trees were studied in [83]. For two-dimensional Hardy operators two-weighted boundedness criteria are established in [260]. For several approaches to the Hardy transforms on the line we refer to the monographs [227], [124] and the survey [123], and [166].

For compactness in weighted Lebesgue spaces see [242]. More general results i.e. with estimation of measure of non–compactness can be found in [67], [133], [134], [289]. In BFS similar problems for Hardy transforms on the line, including two–sided estimates of the measure of non–compactess, was investigated in [69] and [184].

For estimates of approximation numbers see [48], [284]. Generalizations of these results are given in [67–68] and [121]. Similar problems in L^1 and L^∞ are considered in [81]. Two–sided estimates of approximation numbers in Lorentz spaces are obtained in [184] (see also [185]).

Necessary and sufficient conditions for some classical operators to be bounded on cones of monotone functions can be found in [12], [222], [264], [6], [39],

[106], [123], [124], [283], etc. The best constants in the Hardy inequalities on some cones of functions were derived in [21], [23], [40], [234], [24], [217], [171].

The exact values of norms of positive operators on some cones of functions defined on the line were studied in [41].

Note that the weight inequalities derived in this chapter, in particular, involve estimates for moving averaging operators. The introduction of weights and the replacement of estimates by mean estimates for the above–mentioned operator with good control of the constants may be useful for "technical analysis" to get additional quantitive information in the study of financial markets, namely, to predict the future of the stock price or the future in equity markets (see Remark in [168], p.4).

Integral equations involving general Hardy transforms arise in nonlinear mathematical models of certain technological processes from various areas of manufacturing. The models are described by integral equations with Hardy–type transforms which serve as the design and control of these processes. In connection with applications in this area we refer to the monograph [14] and the paper [253]. In particular, in this last paper using the mapping properties of Hardy–type transforms and fractional integrals in weighted Lebesgue spaces, the solution of the integral equations which arise in automated closed cycles occurring in granulation processes, is presented (see Chapter 2, Section 2.12 for details).

This chapter is partly based on the papers [73], [74], [154–156], [158].

Chapter 2

FRACTIONAL INTEGRALS ON THE LINE

In this chapter, boundedness and compactness problems are investigated for various fractional integrals defined on the real line. Our main objective is to give complete descriptions of those pairs of weight functions for which these fractional integrals generate operators which are bounded or compact from one weighted Banach function space into another. This problem was studied earlier by many authors, for instance, for fractional Riemann–Liouville operators R_α when $\alpha \geq 1$. Here the problem is studied in a more general setting. Transparent, easy to verify criteria are presented for a wider range of fractional integral orders. At the end of the chapter, some applications to nonlinear Volterra–type integral equations are given.

2.1. Fractional integrals

In this section, we give necessary and sufficient conditions on the weight function v for the boundedness and compactness of the Riemann–Liouville transform

$$R_\alpha f(x) = \int_0^x \frac{f(t)}{(x-t)^{1-\alpha}} dt$$

from $L^p(0,\infty)$ to $L_v^q(0,\infty)$ ($L_v^{q\infty}(0,\infty)$), where $0 < p,\ q < \infty, p > 1$ and $\alpha > \frac{1}{p}$.

The corresponding problem for the Weyl operator is also solved. In addition, we deal with the two-weight case.

Let (X,μ) be a σ–finite measure space. Recall that by $L_\mu^{pq}(X), p \in (0,\infty)$, $q \in (0,\infty]$, we denote the Lorentz space consisting of all functions f satisfying

$\|f\|_{L_\mu^{pq}(X)} < \infty$, where

$$\|f\|_{L_\mu^{pq}(X)} = \left(q \int_0^\infty \lambda^{q-1}(\mu\{x \in X : |f(x)| > \lambda\})^{q/p} d\lambda \right)^{1/q}$$

for $q \in (0, \infty)$ and

$$\|f\|_{L_\mu^{p\infty}(X)} = \sup_{\lambda>0} \lambda(\mu\{x \in X : |f(x)| > \lambda\})^{1/p}.$$

If $X = R^n$, dx is Lebesgue–measure and $d\mu(x) = w(x)dx$, where w is a Lebesgue–measurable a.e. positive function on R^n, then we write $L_\mu^{pq}(X) \equiv L_w^{pq}(X)$. It is obvious that $L_\mu^{pp}(X) \equiv L_\mu^p(X)$ is a Lebesgue space, i.e. it is the set of all μ–measurable functions with finite norm (quasinorm if $p \in (0,1)$)

$$\|f\|_{L_\mu^p(X)} = \left(\int_X |f(x)|^p d\mu(x) \right)^{1/p}.$$

We shall assume that v and w are Lebesgue–measurable, a.e. positive functions on $(0, \infty)$.

First, let us recall some familiar results (see [212], [290], [296] for $p = q$ and [38], [139], [195] for $p \le q$):

Theorem A. *Let* $1 \le p \le q < \infty$. *The inequality*

$$\left(\int_0^\infty \left| \int_0^x f(t)dt \right|^q v(x)dx \right)^{\frac{1}{q}} \le c \left(\int_0^\infty |f(x)|^p w(x)dx \right)^{\frac{1}{p}}, \qquad (2.1.1)$$

where the positive constant c does not depend on f, $f \in L_w^p(0, \infty)$, holds if and only if

$$A = \sup_{t>0} \left(\int_t^\infty v(x)dx \right)^{\frac{1}{q}} \left(\int_0^t w^{1-p'}(x)dx \right)^{\frac{1}{p'}} < \infty \quad \left(p' = \frac{p}{p-1} \right).$$

Moreover, if c is the best constant in (2.1.1.), then $c \approx A$.

The following theorem is also well–known (see [195] for $1 \le q < p < \infty$ and [272–273] for $q \le 1 \le p < \infty$). In the case $q < 1 < p < \infty$ we shall assume that $w^{1-p'}$ is locally integrable on $(0, \infty)$.

Theorem B. *Let* $0 < q < p < \infty$ *and* $p > 1$. *Then inequality (2.1.1) for all $f \in L_w^p(0, \infty)$ holds if and only if*

$$A_1 = \left(\int_0^\infty \left[\left(\int_t^\infty v(x)dx \right) \left(\int_0^t w^{1-p'}(x)dx \right)^{q-1} \right]^{\frac{p}{p-q}} w^{1-p'}(t)dt \right)^{\frac{p-q}{pq}} < \infty.$$

Moreover, if c is the best constant in (2.1.1), then $c \approx A_1$.

Definition 2.1.1. Let $1 \leq p, q, r, s < \infty$ and let (X, μ) and (Y, ν) be a σ–finite measure space. Denote by $L_\mu^{pq}(X)[L_\nu^{rs}(Y)]$ the spase of all $\mu \times \nu$–measurable functions $k(s, t)$ satisfying the conditions:
1) the function $s \to k(s, t)$ belongs to $L_\nu^{rs}(Y)$;
2) the function $W_Y(k)(t) = \|k(\cdot, t)\|_{L_\nu^{rs}(Y)}$ belongs to $L_\mu^{pq}(X)$.

As for the the Lorentz norm $\| \cdot \|_{L_\mu^{pq}(X)}$ defined on σ– finite measure space (X, μ) we have that $\|f_n\|_{L_\mu^{pq}(X)} \downarrow 0$ when $f_n \downarrow 0$ a.e., therefore the following statement follows from [135], Chapter XI, Section 4, Lemma 2.

Proposition A. *Let $1 \leq r, p < \infty$, $1 \leq s, q \leq \infty$ and let (X, μ) and (Y, ν) be σ–finite measure spaces. Then the set of all functions of the form*

$$k_n(s, t) \equiv \sum_{i=1}^n \eta_i(s)\lambda_i(t), \quad s \in X, \ t \in Y,$$

is dense in the space $L_\mu^{pq}(X)[L_\nu^{rs}(Y)]$ where $\lambda_i \equiv \chi_{B_i}$, $\chi_{B_i} \in L_\nu^{rs}(Y)$ (B_i are ν–measurable pointwise disjoint sets) and $\eta_i \in L_\mu^{pq}(X) \cap L_\mu^\infty(X)$.

Now we prove the following statement:

Theorem C. *Let (X, μ) and (Y, ν) be σ–finite measure spaces and let $1 < r, p < \infty$, $1 \leq s, q < \infty$. Then the condition*

$$M \equiv \|\|k(x, y)\|_{L_\nu^{r's'}(Y)}\|_{L_\mu^{pq}(X)} < \infty$$

implies the compactness of the operator

$$Kf(x) = \int_Y k(x, y)f(y)d\nu(y), \quad x \in X,$$

from $L_\nu^{rs}(Y)$ into $L_\mu^{pq}(X)$.

Proof. By Proposition A we find that the set of all functions of the type

$$k_m(s, t) \equiv \sum_{i=1}^m \eta_i(s)\lambda_i(t), \quad s \in X, \ t \in Y,$$

is dense in $L_\mu^{pq}(X)[L_\nu^{r's'}(Y)]$.
First we show the boundedness of K. By Hölder's inequality we have

$$|Kf(x)| = \left| \int_Y k(x, y)f(y)d\nu(y) \right| \leq$$
$$\leq \|f\|_{L_\nu^{rs}(Y)}\|k(x, y)\|_{L_\nu^{r's'}(Y)}.$$

Hence

$$\|Kf\|_{L_\mu^{pq}(X)} \leq \|f\|_{L_\nu^{rs}(Y)}\|\|k(x, y)\|_{L_\nu^{r's'}(Y)}\|_{L_\mu^{pq}(X)} \leq$$
$$\leq M\|f\|_{L_\nu^{rs}(Y)}.$$

Moreover, $\|K\| \leq M$.

Now we prove the compactness of K. For each $n \in N$ let

$$(K_n\varphi)(x) = \int_Y k_n(x, y)\varphi(y)d\nu(y).$$

Note that

$$(K_n\varphi)(x) = \int_Y k_n(x, y)\varphi(y)d\nu(y) =$$

$$= \sum_{i=1}^{n} \eta_i(x) \int_Y \lambda_i(y)\varphi(y)d\nu(y) \equiv \sum_{i=1}^{n} \eta_i(x)b_i,$$

where

$$b_i = \int_Y \lambda_i(y)\varphi(y)d\nu(y).$$

This means that K_n is a finite rank operator, i.e., it is compact. Further, let $\epsilon > 0$. Then using above–mentioned arguments, there exists $n \in N$ such that

$$\|K - K_n\| \leq \| \|k(x, y) - k_n(x, y)\|_{L_\nu^{r's'}(Y)}\|_{L_\mu^{pq}(X)} < \epsilon.$$

Thus K can be represented as a limit of finite rank operators. Hence K is compact. □

We shall also need the following statement (see, e.g., [10], [165], Sections 5.3 and 5.4):

Theorem D.*Let $0 < q < \infty$, $1 < p < \infty$ and $q < p$. Suppose that v and w are Lebesgue–measurable a.e. positive functions on $\Omega \subseteq R^n$. Let the operator*

$$Tf(x) = \int_\Omega k(x, y)f(y)dy, \quad k \geq 0, \quad x \in \Omega,$$

act from $L_w^p(\Omega)$ into $L_v^q(\Omega)$. Then T is compact.

Armed with these results we can characterise the boundedness and compactness of various important operators.

Theorem 2.1.1. *Let $1 < p \leq q < \infty$, $\alpha > 1/p$. Then the following conditions are equivalent:*
(i) *R_α is bounded from $L^p(0, \infty)$ into $L_v^q(0, \infty)$;*
(ii) *R_α is bounded from $L^p(0, \infty)$ into $L_v^{q\infty}(0, \infty)$;*
(iii)

$$B \equiv \sup_{t>0} \left(\int_t^\infty \frac{v(x)}{x^{(1-\alpha)q}}dx \right)^{1/q} t^{1/p'} < \infty;$$

(iv)

$$B_1 \equiv \sup_{k \in Z} \left(\int_{2^k}^{2^{k+1}} v(x)x^{(\alpha-1/p)q}dx \right)^{1/q} < \infty.$$

Moreover, $\|R_\alpha\|_{L^p \to L^q_v} \approx \|R_\alpha\|_{L^p \to L^{q,\infty}_v} \approx B \approx B_1$.

Proof. Denoting

$$I_{1\alpha}f(x) \equiv \int_0^{\frac{x}{2}} \frac{f(t)}{(x-t)^{1-\alpha}} dt$$

and

$$I_{2\alpha}f(x) \equiv \int_{\frac{x}{2}}^x \frac{f(t)}{(x-t)^{1-\alpha}} dt$$

for $f \in L^p(0,\infty)$ we write $R_\alpha f$ as

$$R_\alpha f(x) = I_{1\alpha}f(x) + I_{2\alpha}f(x).$$

We obtain

$$\|R_\alpha f\|_{L^q_v(0,\infty)}^q \le c_1 \int_0^\infty |I_{1\alpha}f(x)|^q v(x)dx + c_1 \int_0^\infty |I_{2\alpha}f(x)|^q v(x)dx = S_1 + S_2.$$

If $0 < t < \frac{x}{2}$, then $(x-t)^{\alpha-1} \le bx^{\alpha-1}$, where the positive constant b depends only on α. Consequently, using Theorem A with $w \equiv 1$, we have

$$S_1 \le c_2 \int_0^\infty \frac{v(x)}{x^{(1-\alpha)q}} \left(\int_0^x |f(t)|dt \right)^q dx \le c_3 B^q \|f\|_{L^p(0,\infty)}^q.$$

Next we shall estimate S_2. Using the Hölder inequality and the condition $1/p < \alpha$, we obtain

$$S_2 = c_1 \int_0^\infty v(x) \left| \int_{\frac{x}{2}}^x \frac{f(t)}{(x-t)^{1-\alpha}} dt \right|^q dx \le$$

$$\le c_1 \int_0^\infty v(x) \left(\int_{\frac{x}{2}}^x |f(t)|^p dt \right)^{q/p} \left(\int_{\frac{x}{2}}^x \frac{dt}{(x-t)^{(1-\alpha)p'}} \right)^{q/p'} dx =$$

$$= c_4 \sum_{k \in Z} \int_{2^k}^{2^{k+1}} v(x) \cdot x^{(\alpha-1)q+q/p'} \left(\int_{\frac{x}{2}}^x |f(t)|^p dt \right)^{q/p} dx \le$$

$$\le c_4 \sum_{k \in Z} \left(\int_{2^{k-1}}^{2^{k+1}} |f(t)|^p dt \right)^{q/p} \left(\int_{2^k}^{2^{k+1}} v(x) \cdot x^{(\alpha-1)q+q/p'} dx \right) \le$$

$$\le c_5 \sum_{k \in Z} \left(\int_{2^{k-1}}^{2^{k+1}} |f(t)|^p dt \right)^{q/p} \left(\int_{2^k}^{2^{k+1}} v(x) \cdot x^{(\alpha-1)q} dx \right) \cdot 2^{kq/p'} \le$$

$$\le c_5 B^q \sum_{k \in Z} \left(\int_{2^{k-1}}^{2^{k+1}} |f(t)|^p dt \right)^{q/p} \le c_6 B^q \|f\|_{L^p(0,\infty)}^q.$$

which proves the sufficiency. Hence (iii) $\Rightarrow$ (i).

Now we show that (ii) $\Rightarrow$ (iv). Let $f_k(x) = \chi_{(0,2^{k-1})}(x)$. Note that if $0 < y < 2^{k-1}$ and $x \in (2^k, 2^{k+1})$, then $(x - y)^{\alpha-1} \geq b_1 x^{\alpha-1}$, where the positive constant b_1 depends only on α. We have

$$\|R_\alpha f\|_{L_v^{q\infty}(0,\infty)} \geq c_7 \|\chi_{[2^k,2^{k+1})} x^{\alpha-1}\|_{L_v^{q\infty}(0,\infty)} 2^k \geq$$

$$\geq c_8 \|\chi_{[2^k,2^{k+1})}\|_{L_v^{q\infty}(0,\infty)} 2^{k\alpha} = c_8 \left(\int_{2^k}^{2^{k+1}} v(x) dx \right)^{1/q} 2^{k\alpha}.$$

On the other hand, $\|f_k\|_{L^p(0,\infty)} = c_9 2^{k/p}$ and by virtue of the boundedness of R_α from $L^p(0,\infty)$ into $L_v^{q\infty}(0,\infty)$ we find that $B_1 \leq \infty$.

Next we prove that (iv) implies (iii). Let

$$B(t) \equiv \left(\int_t^\infty \frac{v(x)}{x^{(1-\alpha)q}} dx \right)^{1/q} t^{1/p'}$$

and $t \in (0,\infty)$. Then $t \in (2^m, 2^{m+1}]$ for some $m \in Z$. We have

$$B(t)^q = \left(\int_t^\infty \frac{v(x)}{x^{(1-\alpha)q}} dx \right) t^{q/p'} \leq \left(\int_{2^m}^\infty \frac{v(x)}{x^{(1-\alpha)q}} dx \right) 2^{\frac{(m+1)q}{p'}} =$$

$$= b_1 2^{mq/p'} \sum_{k=m}^\infty \left(\int_{2^k}^{2^{k+1}} \frac{v(x)}{x^{(1-\alpha)q}} dx \right) \leq$$

$$b_1 2^{mq/p'} \sum_{k=m}^\infty 2^{-kq/p'} \int_{2^k}^{2^{k+1}} \frac{v(x) x^{q/p'}}{x^{(1-\alpha)q}} dx \leq$$

$$\leq b_2 B_1^q 2^{mq/p'} \sum_{k=m}^\infty 2^{-kq/p'} \leq b_3 B_1^q$$

and therefore $B \leq b_4 B_1 < \infty$.

As (i) implies (ii) we finally have that (i) $\Rightarrow$ (ii) $\Rightarrow$ (iv) $\Rightarrow$ (iii) $\Rightarrow$ (i). $\square$

A somewhat more complicated proof of a similar theorem in the strong type case is given in [223] for $p = q = 2$.

Let

$$W_{\alpha,w} f(x) = \int_x^\infty f(y)(y - x)^{\alpha-1} w(y) dy,$$

where w is a Lebesgue-measurable, a.e. positive function on $(0,\infty)$.

By a duality argument and Theorem 2.1.1 we obtain

Theorem 2.1.2 *Let $1 < p \leq q < \infty$, $\alpha > 1/q'$. Then the following conditions are equivalent:*
(i) *$W_{\alpha,w}$ is bounded from $L_w^p(0,\infty)$ into $L^q(0,\infty)$;*
(ii)

$$\tilde{B} \equiv \sup_{r>0} \left(\int_t^\infty \frac{w(x)}{x^{(1-\alpha)p'}} dx \right)^{1/p'} t^{1/q} < \infty;$$

(iii)

$$\tilde{B}_1 \equiv \sup_{k \in Z} \left(\int_{2^k}^{2^{k+1}} w(x) x^{(\alpha - 1/q')p'} dx \right)^{1/p'} < \infty.$$

Moreover, $\|W_{\alpha,w}\| \approx \tilde{B} \approx \tilde{B}_1$.

Now we consider the case $q < p$.

Theorem 2.1.3. *Let* $0 < q < p < \infty$, $p > 1$, $\alpha > 1/p$. *The operator* R_α *is bounded from* $L^p(0,\infty)$ *into* $L_v^q(0,\infty)$ *if and only if*

$$D = \left(\int_0^\infty \left(\int_x^\infty \frac{v(t)}{t^{(1-\alpha)q}} dt \right)^{p/(p-q)} x^{(q-1)p/(p-q)} dx \right)^{\frac{p-q}{pq}} < \infty.$$

Moreover, $\|R_\alpha\| \approx D$.

Proof. *Sufficiency.* With the notation introduced in the proof of Theorem 2.1.1 we have

$$\|R_\alpha f\|_{L_v^q(0,\infty)}^q \leq S_1 + S_2.$$

Using Theorem B with $w \equiv 1$ and the argument from the proof of Theorem 2.1.1, we obtain

$$S_1 \leq c_2 D^q \|f\|_{L^p(0,\infty)}^q.$$

Applying Hölder's inequality twice and the fact that $1/p < \alpha$, we have

$$S_2 \leq c_1 \int_0^\infty \left(\int_{\frac{x}{2}}^x |f(t)|^p dt \right)^{q/p} \left(\int_{\frac{x}{2}}^x \frac{dt}{(x-t)^{(1-\alpha)p'}} \right)^{q/p'} v(x) dx =$$

$$= c_3 \sum_{k \in Z} \int_{2^k}^{2^{k+1}} \left(\int_{\frac{x}{2}}^x |f(t)|^p dt \right)^{q/p} v(x) x^{(\alpha-1)q + q/p'} dx \leq$$

$$\leq c_3 \sum_{k \in Z} \left(\int_{2^{k-1}}^{2^{k+1}} |f(t)|^p dt \right)^{q/p} \left(\int_{2^k}^{2^{k+1}} v(x) x^{(\alpha-1)q + q/p'} dx \right) \leq$$

$$\leq c_3 \left(\sum_{k \in Z} \int_{2^{k-1}}^{2^{k+1}} |f(t)|^p dt \right)^{q/p} \times$$

$$\left(\sum_{k \in Z} \left(\int_{2^k}^{2^{k+1}} v(x) x^{(\alpha-1)q + q/p'} dx \right)^{p/(p-q)} \right)^{(p-q)/p} \leq$$

$$\leq c_4 \|f\|_{L^p(0,\infty)}^q \left(\sum_{k \in Z} \left(\int_{2^k}^{2^{k+1}} v(x) x^{(\alpha-1)q + q/p'} dx \right)^{p/(p-q)} \right)^{(p-q)/p} =$$

$$= c_4 \|f\|_{L^p(0,\infty)}^q \left(\sum_{k \in Z} S_{2,k} \right)^{(p-q)/p},$$

where

$$S_{2,k} \equiv \left(\int_{2^k}^{2^{k+1}} v(x) x^{(\alpha-1)q + q/p'} dx \right)^{p/(p-q)}.$$

For $S_{2,k}$ we have

$$S_{2,k} \le 2^{\frac{(k+1)qp}{p'(p-q)}} \left(\int_{2^k}^{2^{k+1}} v(x) x^{(\alpha-1)q} dx \right)^{p/(p-q)} \le$$

$$\le c_5 \int_{2^{k-1}}^{2^k} \left(\int_{2^k}^{2^{k+1}} v(y) y^{(\alpha-1)q} dy \right)^{p/(p-q)} x^{p(q-1)/(p-q)} dx \le$$

$$\le c_5 \int_{2^{k-1}}^{2^k} \left(\int_x^{\infty} v(y) y^{(\alpha-1)q} dy \right)^{p/(p-q)} x^{p(q-1)/(p-q)} dx.$$

Thus

$$S_2 \le c_6 \|f\|_{L^p(0,\infty)}^q D^q$$

and finally we obtain (i).

Necessity. Put $v_n(t) = v(t) \cdot \chi_{(1/n,n)}(t)$, where $n \in N$, $n \ge 2$ and let

$$f(x) = \left(\int_x^{\infty} \frac{v_n(t)}{t^{(1-\alpha)q}} dt \right)^{1/(p-q)} x^{(q-1)/(p-q)}.$$

Then using integration by parts we have

$$\|f\|_{L^p(0,\infty)} = \left(\int_0^{\infty} \left(\int_x^{\infty} \frac{v_n(t)}{t^{(1-\alpha)q}} dt \right)^{p/(p-q)} x^{(q-1)p/(p-q)} dx \right)^{1/p} =$$

$$= c_7 \left(\int_0^{\infty} \left(\int_x^{\infty} \frac{v_n(t)}{t^{(1-\alpha)q}} dt \right)^{q/(p-q)} \times \right.$$

$$\left. \times x^{(p-1)q/(p-q)} v_n(x) x^{(\alpha-1)q} dx \right)^{1/p} < \infty.$$

On the other hand, integration by parts also shows that

$$\|R_\alpha f\|_{L^q_v(0,\infty)} = \left(\int_0^{\infty} v(x) \left(\int_0^x \frac{f(t)}{(x-t)^{1-\alpha}} dt \right)^q dx \right)^{1/q} \ge$$

$$\ge c_8 \left(\int_0^{\infty} \frac{v_n(x)}{x^{(1-\alpha)q}} \left(\int_0^{x/2} \left(\int_t^{\infty} \frac{v_n(y)}{y^{(1-\alpha)q}} dy \right)^{1/(p-q)} \times \right. \right.$$

$$\left. \left. \times t^{(q-1)/(p-1)} dt \right)^q dx \right)^{1/q} \ge$$

$$\ge c_9 \left(\int_0^{\infty} \frac{v_n(x)}{x^{(1-\alpha)q}} \left(\int_x^{\infty} \frac{v_n(y)}{y^{(1-\alpha)q}} dy \right)^{q/(p-q)} x^{q(p-1)/(p-q)} dx \right)^{1/q} =$$

$$= c_{10} \left(\int_0^{\infty} \left(\int_x^{\infty} \frac{v_n(y)}{y^{(1-\alpha)q}} dy \right)^{p/(p-q)} x^{p(q-1)/(p-q)} dx \right)^{1/q}.$$

From the boundedness of R_α we obtain

$$\left(\int_0^\infty \left(\int_x^\infty \frac{v_n(y)}{y^{(1-\alpha)q}} dy \right)^{p/(p-q)} x^{p(q-1)/(p-q)} dx \right)^{\frac{p-q}{pq}} \le c_{11}$$

where c_{11} does not depend on n. By Fatou's lemma we finally see that $D < \infty$.
□

By a duality argument and Theorem 2.1.3 we have

Theorem 2.1.4. *Let* $1 < q < p < \infty$, $\alpha > 1/q'$. *The operator* $W_{\alpha,w}$ *is bounded from* $L_w^p(0,\infty)$ *into* $L^q(0,\infty)$ *if and only if*

$$\tilde{D} = \left(\int_0^\infty \left(\int_x^\infty \frac{w(t)}{t^{(1-\alpha)p'}} dt \right)^{q(p-1)/(p-q)} x^{q/(p-q)} dx \right)^{\frac{p-q}{pq}} < \infty.$$

Moreover, $\|W_{\alpha,w}\| \approx \tilde{D}$.

Let us now investigate the compactness of the operators R_α and W_α.

Theorem 2.1.5. *Let* $1 < p \le q < \infty$, $\alpha > \frac{1}{p}$. *Then the following conditions are equivalent:*
(i) R_α *is compact from* $L^p(0,\infty)$ *into* $L_v^q(0,\infty)$;
(ii) R_α *is compact from* $L^p(0,\infty)$ *into* $L_v^{q\infty}(0,\infty)$;
(iii)

$$B \equiv \sup_{t>0} \left(\int_t^\infty \frac{v(x)}{x^{(1-\alpha)q}} dx \right)^{\frac{1}{q}} t^{\frac{1}{p'}} < \infty$$

and $\lim_{a\to 0} B^{(a)} = \lim_{b\to\infty} B^{(b)} = 0$, where

$$B^{(a)} \equiv \sup_{0<t<a} \left(\int_t^a \frac{v(x)}{x^{(1-\alpha)q}} dx \right)^{\frac{1}{q}} t^{\frac{1}{p'}} ;$$

$$B^{(b)} \equiv \sup_{t>b} \left(\int_t^\infty \frac{v(x)}{x^{(1-\alpha)q}} dx \right)^{\frac{1}{q}} (t-b)^{\frac{1}{p'}} ;$$

(iv) $B < \infty$ and $\lim_{t\to 0} B(t) = \lim_{t\to\infty} B(t) = 0$, where

$$B(t) \equiv \left(\int_t^\infty \frac{v(x)}{x^{(1-\alpha)q}} dx \right)^{\frac{1}{q}} t^{\frac{1}{p'}} ;$$

(v)

$$B_1 \equiv \sup_{k\in Z} \left(\int_{2^k}^{2^{k+1}} v(x) x^{(\alpha-1/p)q} dx \right)^{\frac{1}{q}} < \infty$$

and $\lim\limits_{k\to-\infty} B_1(k) = \lim\limits_{k\to+\infty} B_1(k) = 0$, where

$$B_1(k) \equiv \Big(\int_{2^k}^{2^{k+1}} \frac{v(x)}{x^{(1/p-\alpha)q}} dx \Big)^{\frac{1}{q}}.$$

Proof. Let $0 < a < b < \infty$. We write $R_\alpha f$ as

$$R_\alpha f = \chi_{[0,a]} R_\alpha(f \cdot \chi_{(0,a)}) + \chi_{(a,b)} R_\alpha(f \cdot \chi_{(0,b)}) + \chi_{[b,\infty)} R_\alpha(f \cdot \chi_{(0,\frac{b}{2})}) +$$

$$+\chi_{[b,\infty)} R_\alpha(f \cdot \chi_{(\frac{b}{2},\infty)}) = P_{1\alpha} f + P_{2\alpha} f + P_{2\alpha} f + P_{4\alpha} f.$$

For $P_{2\alpha} f$ we have $P_{2\alpha} f(x) = \int_0^\infty k_1(x,y) f(y) dy$, with $k_1(x,y) = (x - y)^{\alpha-1} \chi_{(a,b)}(x)$ for $y < x$ and $k_1(x,y) = 0$ for $y \geq x$. Consequently

$$\int_0^\infty v(x) \Big(\int_0^\infty (k_1(x,y))^{p'} dy \Big)^{\frac{q}{p'}} dx \leq \Big(\int_a^b \frac{v(x)}{x^{(1-\alpha)q}} dx \Big) b^{\frac{q}{p'}} < \infty$$

and by Theorem C we conclude that $P_{2\alpha}$ is compact from $L^p(0,\infty)$ to $L^q_v(0,\infty)$. In a similar manner we show that $P_{3\alpha}$ is compact, too.
Using Theorem 2.1.1 for the operators $P_{1\alpha}$ and $P_{4\alpha}$, we obtain

$$\|P_{1\alpha}\| \leq c_1 B^{(a)} \quad \text{and} \quad \|P_{4\alpha}\| \leq c_2 B^{(b/2)}.$$

Consequently

$$\|R_\alpha - P_{2\alpha} - P_{3\alpha}\| \leq \|P_{1\alpha}\| + \|P_{4\alpha}\| \leq c_1 B^{(a)} + c_2 B^{(b/2)} \to 0$$

as $a \to 0$ and $b \to \infty$.
Thus the operator R_α is compact, since it is a limit of compact operators. Hence (iii) $\Rightarrow$ (i).
Now we prove that (i) $\Rightarrow$ (iv).
Note that the fact $B < \infty$ follows from Theorem 2.1.1. Thus we need to prove the remaining part. Let $f_t(x) = \chi_{(0,t)}(x) t^{-1/p}$. Then the sequence f_t is weakly convergent to 0. Indeed, assuming that $\varphi \in L^{p'}$, we obtain

$$\Big| \int_0^\infty f_t(x) \varphi(x) dx \Big| \leq \Big(\int_0^t |\varphi(x)|^{p'} dx \Big)^{\frac{1}{p'}} \to 0 \quad \text{as } t \to 0.$$

On the other hand, we have

$$\|R_\alpha f_t\|_{L^q_v(0,\infty)} \geq \Big(\int_t^\infty v(x) \Big(\int_0^t \frac{f_t(y)}{(x-y)^{1-\alpha}} dy \Big)^q dx \Big)^{\frac{1}{q}} \geq$$

$$\geq c_3 \Big(\int_t^\infty \frac{v(x)}{x^{(1-\alpha)q}} dx \Big)^{\frac{1}{q}} t^{\frac{1}{p'}} = c_3 B(t).$$

Using the fact that a compact operator maps a weakly convergent sequence into a strongly convergent one, we find that $B(t) \to 0$ as $t \to 0$. Let

$$g_t(y) = \chi_{(t,\infty)}(y) y^{(\alpha-1)(q-1)} \left(\int_t^\infty x^{(\alpha-1)q} v(x) dx \right)^{-1/q'}.$$

Then it is easy to verify that g_t converges weakly to 0. On the other hand, using the fact that the compactness of $W_{\alpha,v}$ is compact from $L_v^q(0,\infty)$ into $L^{p'}(0,\infty)$ is equivalent to the compactness of R_α from $L^p(0,\infty)$ into $L_v^q(0,\infty)$, we obtain $\lim_{b\to\infty} B^{(b)} = 0$.

Let now $a > 0$. Then $a \in [2^m, 2^{m+1})$ for some $m \in Z$. Therefore $B^{(a)} \le \sup_{0<t<2^m} B_{2^m,t} = B^{(2^m)}$, where

$$B_{2^m,t} \equiv \left(\int_t^{2^m} \frac{v(x)}{x^{(1-\alpha)q}} dx \right)^{\frac{1}{q}} t^{\frac{1}{p'}}.$$

If $t \in [0, 2^m)$, then $t \in [2^{j-1}, 2^j)$ for some $j \in Z, j \le m$. Furthermore,

$$B_{2^m,t}^q \le 2^{jq/p'} \sum_{k=j}^m \int_{2^{k-1}}^{2^k} x^{(\alpha-1)q} v(x) dx \le$$

$$\le c_2 \left(\sup_{k\le m} B_1(k-1) \right)^q.$$

Hence we have $B^{(2^m)} \le c_3 B_1^{(m)}$, where $B_1^{(m)} \equiv \left(\sup_{k\le m} B_1(k-1) \right)^q$. If $a \to 0$, then $m \to -\infty$ and consequently $B_1^{(m)} \to 0$. Therefore $\lim_{a\to 0} B^{(a)} = 0$.

Now let $r \in (0,\infty)$. Then $r \in [2^m, 2^{m+1})$ for some integer m. We have

$$B^q(r) \le c_4 B^q(2^m) = c_4 \left(\sum_{k=m}^{+\infty} \int_{2^k}^{2^{k+1}} v(x) x^{(\alpha-1)q} dx \right) 2^{mq/p'} \le$$

$$\le c_5 \left(\sup_{k\ge m} B_1(k) \right)^q.$$

If $r \to +\infty$, then $m \to +\infty$. Consequently $\sup_{k\ge m} B_1(k) \to 0$. Hence $\lim_{r\to +\infty} B(r) = 0$. and so (v) $\Rightarrow$ (iii).

Let $k \in Z$ and $f_k(x) = \chi_{(2^{k-2},2^{k-1})} 2^{-k/p}$. Then it is easy to verify that f_k converges weakly to 0 when $k \to +\infty$ or $k \to -\infty$. On the other hand,

$$\|R_\alpha f_k\|_{L_v^{q\infty}(0,\infty)} \ge c_6 B_1(k).$$

as $k \to +\infty$ or $k \to -\infty$. Therefore (ii) implies (v). It is obvious that (iv) $\Rightarrow$ (iii) and (i) $\Rightarrow$ (ii) Finally we have (iv) $\Leftrightarrow$ (i), (i) $\Rightarrow$ (ii) $\Rightarrow$ (v) $\Rightarrow$ (iii) $\Rightarrow$ (i). $\square$

In [223] an analogous theorem (in the strong–type case) was proved for $p = q = 2$.

Example 2.1.1. Let $1 < p < \infty$, $\alpha > 1/p$, $v(x) = e^{Ax}x^{p-\alpha p}$, where $A < 0$. Then the operator R_α is compact $L^p(0, \infty)$ into $L^p_v(0, \infty)$. Indeed, let $t > 0$. Then

$$B(t) \equiv \left(\int_t^\infty \frac{v(x)}{x^{(1-\alpha)p}} dx \right)^{\frac{1}{p}} t^{\frac{1}{p'}} =$$

$$= \left(\int_t^\infty e^{Ax} dx \right)^{1/p} t^{1/p'} = c e^{At/p} t^{1/p'}.$$

Moreover, $B = \sup_{t>0} B(t) < \infty$ and $\lim_{t \to 0} B(t) = \lim_{t \to \infty} B(t) = 0$.

From a duality argument we have the following result:

Theorem 2.1.6. *Let* $1 < p \leq q < \infty$, $\alpha > 1/q'$. *Then the following conditions are equivalent:*
(i) $W_{\alpha,w}$ *is compact from* $L^p_w(0, \infty)$ *into* $L^q(0, \infty)$;
(ii) $\tilde{B} < \infty$ *and* $\lim_{a \to 0} \tilde{B}^{(a)} = \lim_{b \to \infty} \tilde{B}^{(b)} = 0$, *where*

$$\tilde{B}^{(a)} \equiv \sup_{0<t<a} \left(\int_t^a \frac{w(x)}{x^{(1-\alpha)p'}} dx \right)^{\frac{1}{q}} t^{\frac{1}{q}};$$

$$B^{(b)} \equiv \sup_{t>b} \left(\int_t^\infty \frac{w(x)}{x^{(1-\alpha)p'}} dx \right)^{\frac{1}{p'}} (t - b)^{\frac{1}{q}};$$

(iii) $\tilde{B} < \infty$ *and* $\lim_{t \to 0} \tilde{B}(t) = \lim_{t \to \infty} \tilde{B}(t) = 0$, *where*

$$\tilde{B}(t) \equiv \left(\int_t^\infty \frac{w(x)}{x^{(1-\alpha)p'}} dx \right)^{\frac{1}{p'}} t^{\frac{1}{q}};$$

(iv) $\tilde{B}_1 < \infty$ *and* $\lim_{k \to -\infty} \tilde{B}_1(k) = \lim_{k \to +\infty} \tilde{B}_1(k) = 0$, *where*

$$\tilde{B}_1(k) \equiv \left(\int_{2^k}^{2^{k+1}} \frac{w(x)}{x^{(1/q'-\alpha)p'}} dx \right)^{\frac{1}{p'}}.$$

The next theorem follows directly from Theorem D.

Theorem 2.1.7 *Let* $0 < q < p < \infty$, $p > 1$ *and* $\alpha > 1/p$. *The operator* R_α *is compact from* $L^p(0, \infty)$ *to* $L^q_v(0, \infty)$ *if and only if* $D < \infty$.

By a duality argument we have

Theorem 2.1.8 *Let* $1 < q < p < \infty$, $\alpha > 1/q'$. *The operator* $W_{\alpha,w}$ *is compact from* $L^p_w(0, \infty)$ *to* $L^q(0, \infty)$ *iff* $\tilde{D} < \infty$.

In the non-compact case it is useful to estimate the distance of the operator K from the space of compact operators.

Let X and Y be Banach function spaces. Denote by $\mathcal{B}(X,Y)$ the space of all bounded linear operators from X to Y. Let $\mathcal{K}(X,Y)$ be the class of all compact linear operators from X to Y. Suppose that $\mathcal{F}_r(X,Y)$ is the space of linear operators with finite rank.

We shall assume that v is a Lebesgue–measurable a.e. positive function on (a,b), where $-\infty < a < b \leq +\infty$.

The following Lemma holds (see [226] and [66], Corollary V.5.4).

Lemma 2.1.1. *Let* $1 \leq p \leq \infty$, $-\infty < a < b \leq +\infty$ *and let* $P \in \mathcal{B}(X,Y)$, *where* $Y = L^p_v(a,b)$. *Then*

$$dist(P, \mathcal{K}(X,Y)) = dist(P, \mathcal{F}_r(X,Y)).$$

We also need (see [226], [66], Lemma V.5.4)

Lemma 2.1.2. *Let* $1 \leq p < \infty$, $-\infty < a < b \leq +\infty$ *and let* $Y = L^p_v(a,b)$. *Suppose that* $P \in \mathcal{F}_r(X,Y)$ *and* $\epsilon > 0$. *Then there exist* $T \in \mathcal{F}_r(X,Y)$ *and* $[\alpha, \beta] \subset (a,b)$ *such that*

$$\|P - T\| < \epsilon \tag{2.1.2}$$

and

$$supp\, Tf \subset [\alpha, \beta] \tag{2.1.3}$$

for every $f \in X$.

Now we give an estimate of the measure of non–compactness $dist(R_\alpha, \mathcal{K}(\mathcal{X},\mathcal{Y}))$ for the Riemann–Liouville operator R_α. This is contained in the following theorem.

Theorem 2.1.9. *Let* $1 < p \leq q < \infty$ *and let* $\alpha > 1/p$; *suppose that* $B < \infty$. *Assume that* $X = L^p(0,\infty)$, $Y = L^q_v(0,\infty)$. *Then*

$$b_1 I \leq dist(R_\alpha, \mathcal{K}(X,Y)) \leq b_2 I,$$

where $I = \lim_{c \to 0} I_c + \lim_{d \to \infty} I_d$,

$$I_c = \sup_{0 < t < c} \left(\int_t^c \frac{v(x)}{x^{(1-\alpha)q}} dx \right)^{1/q} t^{1/p'},$$

$$I_d = \sup_{t > d} \left(\int_t^\infty \frac{v(x)}{x^{(1-\alpha)q}} dx \right)^{1/q} (t-d)^{1/p'}$$

and the positive constants b_1 *and* b_2 *depend only on* p, q *and* α.

Proof. If we repeat the arguments used in the proof of Theorem 2.1.5, then we find

$$dist(R_\alpha, \mathcal{K}(X,Y)) \leq b_2 I.$$

Now let $\lambda > dist(R_\alpha, \mathcal{K}(X,Y))$. Then by Lemma 2.1.1. there exists $P \in \mathcal{F}_r(X,Y)$ such that $\|R_\alpha - P\| < \lambda$. By virtue of Lemma 4 for $\epsilon = (\lambda - \|R_\alpha - P\|)/2$ there are $T \in \mathcal{F}_r(X,Y)$ and $[\alpha, \beta] \subset (0, \infty)$ such that (2.1.2) and (2.1.3) hold. From (2.1.2) we obtain

$$\|R_\alpha f - Tf\|_Y \le \lambda \|f\|_X \qquad (2.1.4)$$

for every $f \in X$. Further, from (2.1.2), (2.1.3) and (2.1.4) we have

$$\int_0^\alpha |R_\alpha f(x)|^q v(x)dx + \int_\beta^\infty |R_\alpha f(x)|^q v(x)dx \le \lambda^q \|f\|_{L^p(0,\infty)}^q.$$

Let $d \ge \beta$ and let $t \in (d, \infty)$. Then for $f_t(y) = \chi_{(d/2, t/2)}(y)$ we have

$$\int_t^\infty |R_\alpha f_t(x)|^q v(x)dx \ge \int_t^\infty \Big(\int_{d/2}^{t/2} \frac{f_t(y)}{(x-y)^{1-\alpha}} dy \Big)^q v(x)dx \ge$$

$$\ge c_1 \Big(\int_t^\infty x^{(\alpha-1)q} v(x)dx \Big)(t-d)^q.$$

On the other hand,

$$\|f\|_{L^p(0,\infty)}^q = c_2 (t-d)^{q/p},$$

whence

$$\lambda \ge c_3 \Big(\int_t^\infty x^{(\alpha-1)q} v(x)dx \Big)^{1/q} (t-d)^{1/p'}$$

for $t > d$. Consequently $\lambda \ge c_3 I_d$ for every $d, d > \beta$. From the last inequality we have

$$c_3 \lim_{d \to \infty} I_d \le \lambda.$$

As λ is an arbitrary number greater than $dist(R_\alpha, \mathcal{K}(X,Y))$, we conclude that

$$c_4 \lim_{d \to \infty} I_d \le dist(R_\alpha, \mathcal{K}(X,Y)).$$

Analogously we can show that

$$c_2 \lim_{c \to 0} I_c \le dist(R_\alpha, \mathcal{K}(X,Y)).$$

Consequently

$$b_1 I \le dist(R_\alpha, \mathcal{K}(X,Y)).$$

Now we establish sufficient and in some cases necessary conditions for the validity of two–weight inequalities for the Riemann–Liouville and Weyl operators.

Theorem 2.1.10. *Let* $1 < p \le q < \infty$ *and let* $\alpha > 0$. *Assume that* ν *is a non–negative Borel measure on* $(0, \infty)$ *and* w *is a Lebesgue–measurable a.e. positive function on* $(0, \infty)$. *If the following two conditions are satisfied:*

(i)

$$B_1 \equiv \sup_{t>0} \left(\int_{[2t,\infty)} \frac{d\nu(x)}{x^{(1-\alpha)q}} \right)^{1/q} \left(\int_0^t w^{1-p'}(x) dx \right)^{1/p'} < \infty \qquad (2.1.5)$$

(ii)

$$B_2 \equiv \sup_{t>0} \left(\nu[t/2, 2t) \right)^{1/q} \left(\int_{t/2}^t \frac{w^{1-p'}(x)}{(t-x)^{(1-\alpha)p'}} dx \right)^{1/p'} < \infty,$$

then R_α is bounded from $L_w^p(0, \infty)$ into $L_\nu^q(0, \infty)$. Conversely, the boundedness of R_α from $L_w^p(0, \infty)$ into $L_\nu^q(0, \infty)$ implies the validity of (2.1.5). Moreover, there exist positive constants b_1 and b_2 such that $b_1 B_1 \le \|R_\alpha\| \le b_2 \max(B_1, B_2)$.

Proof. Let $f \in L_w^p(0, \infty)$ and let $f \ge 0$. Then

$$\|R_\alpha f\|_{L_\nu^q([0,\infty))} \le \left(\int_{[0,\infty)} \left(\int_0^{x/2} f(t)(x-t)^{\alpha-1} dt \right)^q d\nu(x) \right)^{1/q} +$$

$$+ \left(\int_{[0,\infty)} \left(\int_{x/2}^x f(t)(x-t)^{\alpha-1} dt \right)^q d\nu(x) \right)^{1/q} \equiv S_1 + S_2.$$

If $t < x/2$, then $x/2 \le x - t$ and using Theorem 1.1.6 we obtain

$$S_1 \le c_1 \left(\int_{(0,\infty)} \left(\int_0^x f(t) dt \right)^q d\nu(x) \right)^{1/q} \le c_2 \|f\|_{L_w^p(0,\infty)}.$$

Now we estimate S_2. Using Hölder's inequality and condition (ii), we find that

$$S_2^q \le \int_{(0,\infty)} \left(\int_{x/2}^x w(t)(f(t))^p dt \right)^{q/p} \times$$

$$\times \left(\int_{x/2}^x w^{1-p'}(t)(x-t)^{(\alpha-1)p'} dt \right)^{q/p'} d\nu(x) =$$

$$= \sum_{k \in Z} \int_{(2^k, 2^{k+1}]} \left(\int_{x/2}^x w(t)(f(t))^p dt \right)^{q/p} \times$$

$$\left(\int_{x/2}^x w^{1-p'}(t)(x-t)^{(\alpha-1)p'} dt \right)^{q/p'} d\nu(x) =$$

$$\sum_{k \in Z} \left(\int_{2^{k-1}}^{2^{k+1}} w(t)(f(t))^p dt \right)^{q/p} \times$$

$$\times \left(\int_{(2^k,2^{k+1}]} \left(\int_{x/2}^x w^{1-p'}(t)(x-t)^{(\alpha-1)p'} dt \right)^{q/p'} d\nu(x) \right) \le$$

$$\le c_3 B_2^q \sum_{k \in Z} \left(\int_{2^{k-1}}^{2^{k+1}} w(t)(f(t))^p dt \right)^{q/p} \times$$

$$\times \left(\int_{(2^k,2^{k+1}]} \left(\nu[x/2,2x] \right)^{-1} d\nu(x) \right) \le$$

$$\le c_4 B_2^q \sum_{k \in Z} \left(\int_{2^{k-1}}^{2^{k+1}} w(t)(f(t))^p dt \right)^{q/p} \times$$

$$\times \left(\nu(2^k,2^{k+1}] \right) \left(\nu(2^k,2^{k+1}] \right)^{-1} =$$

$$= c_4 \sum_{k \in Z} \left(\int_{2^{k-1}}^{2^{k+1}} w(t)(f(t))^p dt \right)^{q/p} \le c_5 B_2^q \|f\|_{L_w^p(0,\infty)}^q.$$

Next suppose that R_α is bounded and that $f_t(x) = \chi_{(0,t)}(x)w^{1-p'}(x)$ for $t > 0$. Then it is easy to show that

$$\int_0^t w^{1-p'} < \infty.$$

Further,

$$\|f_t\|_{L_w^p(0,\infty)} = \left(\int_0^t w^{1-p'}(x) dx \right)^{1/p}.$$

On the other hand,

$$\left(\int_0^\infty \left(R_\alpha f(x) \right)^q d\nu(x) \right)^{1/q} \ge$$

$$\ge c_7 \left(\int_{2t}^\infty \frac{d\nu(x)}{x^{(1-\alpha)q}} \right)^{1/q} \left(\int_0^t w^{1-p'}(x) dx \right).$$

Consequently, $c_8 B_1 \le \|R_\alpha\| < \infty.$ □

The following theorem is proved analogously

Theorem 2.1.11. Let $1 < p \le q < \infty$ and $\alpha > 0$. If
(i)

$$D_1 = \sup_{t>0} \left((\nu(0,t/2))^{1/q} \left(\int_t^\infty w(x)x^{(\alpha-1)p'} dx \right)^{1/p'} \right) < \infty$$

(ii)

$$D_2 = \sup_{t>0} \left(\nu[t/2,2t] \right)^{1/q} \left(\int_t^{2t} \frac{w(x)}{(x-t)^{(1-\alpha)p'}} dx \right)^{1/p'} < \infty,$$

then $W_{\alpha,w}$ is bounded from $L_w^p(0,\infty)$ into $L_\nu^q(0,\infty)$. Conversely, from the boundedness of R_α from $L_w^p(0,\infty)$ into $L_\nu^q(0,\infty)$ the validity of (i) follows. Moreover, there exist positive constants b_1 and b_2 such that

$$b_1 D_1 \leq \|W_{\alpha,w}\| \leq b_2 \max(D_1, D_2).$$

From Theorems 2.1.10 and 2.1.11 the next results follow directly:

Theorem 2.1.12. *Let* $1 < p \leq q < \infty$, $\alpha > 0$. *Assume that the weight function* w *satisfies the following condition: there exists a positive constant* c *such that for all* $t > 0$ *the following inequality holds*

$$\int_{t/2}^t \frac{w^{1-p'}(x)}{(t-x)^{(1-\alpha)p'}} dx \leq ct^{(\alpha-1)p'} \int_0^{t/4} w^{1-p'}(x)dx. \tag{2.1.6}$$

Then for the boundedness of R_α *from* $L_w^p(0,\infty)$ *into* $L_\nu^q(0,\infty)$ *it is necessary and sufficient that* (2.1.5) *holds. Moreover,* $\|R_\alpha\| \approx B_1$.

Note that if $1 < p < \infty$, $\alpha > 1/p$ and either (i) w is an increasing function on $(0,\infty)$, or (ii) w is a decreasing function on $(0,\infty)$ satisfying the condition

$$\int_t^{2t} w^{1-p'}(x)dx \leq b \int_0^{t/4} w^{1-p'}(x)dx \tag{2.1.7}$$

with a positive constant b independent of t, then condition (2.1.6) is satisfied.

The two-weight problem for R_α has been solved when $\alpha > 1$ (see [192], [284]), but as a corollary of Theorem 2.1.10 we have a two–weight boundedness criterion, when the right–hand weight w satisfies the additional condition:

$$\int_0^{2t} w^{1-p'}(x)dx \leq b_1 \int_0^t w^{1-p'}(x)dx, \tag{2.1.8}$$

with b_1 independent of $t > 0$. This is given in the next theorem.

Theorem 2.1.13. *Let* $1 < p \leq q <$ *and* $\alpha \geq 1$. *Assume that* w *satisfies the condition* (2.1.8). *Then* R_α *is bounded from* $L_w^p(0,\infty)$ *into* $L_\nu^q(0,\infty)$ *if and only if* (2.1.5) *is fulfilled. Moreover,* $\|R_\alpha\| \approx B_1$.

Now we shall deal with the case $q < p$. The following theorem is proved in the same way as Theorem 2.1.3.

Theorem 2.1.14. *Let* $0 < q < p < \infty$, $p > 1$ *and* $\alpha > \frac{1}{p}$. *Assume also that the weight* w *is increasing on* $(0,\infty)$. *Then the operator* R_α *is bounded from*

$L^p(0, \infty)$ into $L_v^q(0, \infty)$ if and only if

$$D = \left(\int_0^\infty \left(\int_{2x}^\infty \frac{v(t)}{t^{(1-\alpha)q}} dt \right)^{\frac{p}{p-q}} \left(\int_0^x w^{1-p'}(t)dt \right)^{\frac{(q-1)p}{p-q}} \times \right.$$
$$\left. \times w^{1-p'}(x)dx \right)^{\frac{p-q}{pq}} < \infty. \tag{2.1.9}$$

Moreover, $\|R_\alpha\| \approx D.$

From a duality argument we have

Theorem 2.1.15. *Let* $1 < q < p < \infty$, $\alpha > 1/q'$ *and let* v *be a decreasing function on* $(0, \infty)$. *Then* $W_{\alpha,w}$ *is bounded from* $L_w^p(0, \infty)$ *into* $L_v^q(0, \infty)$ *if and only if*

$$D_1 = \left(\int_0^\infty \left(\int_{2x}^\infty \frac{w(t)}{t^{(1-\alpha)p'}} dt \right)^{\frac{q(p-1)}{p-q}} \times \right.$$
$$\left(\int_0^x v(t)dt \right)^{\frac{q}{p-q}} v(x)dx \right)^{\frac{p-q}{pq}} < \infty. \tag{2.1.10}$$

Moreover, $\|W_{\alpha,w}\| \approx D_1.$

Next we are going to discuss the compactness of R_α and W_α. The following theorem can be obtained in the same way as Theorem 2.1.5; we therefore omit the proof.

Theorem 2.1.16. *Let* $1 < p \le q < \infty$ *and* $\alpha > 0$. *Suppose the weight* w *satisfies* $(2.1.6)$. *Then* R_α *is compact from* $L_w^p(0, \infty)$ *into* $L_v^q(0, \infty)$ *if and only if* $(2.1.5)$ *is satisfied and*

$$\lim_{t \to 0} B_1(t) = \lim_{t \to \infty} B_1(t) = 0,$$

where

$$B_1(t) = \left(\int_{[2t,\infty)} \frac{d\nu(x)}{x^{(1-\alpha)q}} \right)^{1/q} \left(\int_0^t w^{1-p'}(x)dx \right)^{1/p'}.$$

From duality arguments we have the following result for W_α:

Theorem 2.1.17. *Let* $1 < p \le q < \infty$ *and* $\alpha > 1/q'$. *Assume that* v *and* w *are Lebesgue–measurable a.e. positive functions on* $(0, \infty)$. *Suppose also that either* (i) v *is a decreasing function on* $(0, \infty)$, *or* (ii) v *is an increasing function satisfying the condition*

$$\int_t^{2t} v(x)dx \le b \int_0^{t/4} v(x)dx.$$

Then $W_{\alpha,w}$ is compact from $L_w^p(0,\infty)$ into $L_v^q(0,\infty)$ if and only if

$$D_1 = \sup_{t>0} (v(0,t/2))^{1/q} \left(\int_t^\infty w(x) x^{(\alpha-1)p'} dx \right)^{1/p'} < \infty$$

and

$$\lim_{t\to 0} D_1(t) = \lim_{t\to\infty} D_1(t) = 0,$$

where

$$D_1(t) \equiv \left(v(0,t/2) \right)^{1/q} \left(\int_t^\infty w(x) x^{(\alpha-1)p'} dx \right)^{1/p'}$$

and

$$v(0,\tau) \equiv \int_0^\tau v(x) dx.$$

Moreover, we have

Theorem 2.1.18. *Let* $1 < q < p < \infty$ *and* $\alpha > \frac{1}{p}$; *assume also that the weight* w *is increasing on* $(0,\infty)$. *Then the operator* R_α *is compact from* $L_w^p(0,\infty)$ *into* $L_v^q(0,\infty)$ *if and only if* (2.1.9) *is fulfilled.*

This follows immediately from Theorem D.

From duality arguments we obtain the next result.

Theorem 2.1.19. *Let* $1 < q < p < \infty$, $\alpha > 1/q'$ *and let* v *be a decreasing function on* $(0,\infty)$. *Then* $W_{\alpha,w}$ *is compact from* $L_w^p(0,\infty)$ *into* $L_v^q(0,\infty)$ *if and only if* (2.1.10) *holds.*

The two–weight compactness problem for R_α has been settled when $\alpha > 1$, see [Ste4]. We can obtain a two–weight compactness criterion, when the right–hand weight w satisfies the additional condition (2.1.8).

Theorem 2.1.20. *Let* $1 < p \le q < \infty$ *and* $\alpha > 1$. *Assume that* w *satisfies the condition* (2.1.8). *Then* R_α *is compact from* $L_w^p(0,\infty)$ *into* $L_v^q(0,\infty)$ *if and only if* (2.1.5) *is fulfilled and*

$$\lim_{t\to 0} B_1(t) = \lim_{t\to\infty} B_1(t) = 0,$$

where

$$B_1(t) = \left(\int_{[2t,\infty)} \frac{d\nu(x)}{x^{(1-\alpha)q}} \right)^{1/q} \left(\int_0^t w^{1-p'}(x) dx \right)^{1/p'} < \infty.$$

This follows easily from 2.1.16.

The next theorem is well–known (see e.g. [195], Section 1.3.2) and will be useful for us.

Theorem E. *Let $1 \leq q < p < \infty$ and let μ be a nonnegative Borel measure on $(0, \infty)$. The inequality*

$$\left(\int_0^\infty \Big| \int_0^x f(t)dt \Big|^q d\mu(x) \right)^{1/q} \leq c \left(\int_0^\infty |f(x)|^p w(x)dx \right)^{1/p}, \quad (2.1.11)$$

where the positive constant c does not depend on f, $f \in L_w^p(0, \infty)$, is fulfilled if and only if

$$A_2 = \left(\int_0^\infty \left[(\mu[t, \infty)) \left(\int_0^t w^{1-p'}(x)dx \right)^{q-1} \right]^{\frac{p}{p-q}} w^{1-p'}(t)dt \right)^{\frac{p-q}{pq}} < \infty.$$

Moreover, if c is the best constant in (2.1.11), then $c \approx A_2$.

Now we shall deal with the operators R_α and W_α.

Theorem 2.1.21. *Let $1 \leq q < p < \infty$ and $\alpha > 1/p$. Assume that ν is a non–negative Borel measure on $(0, \infty)$. Then the following statements are equivalent:*
   (i) R_α *is bounded from* $L^p(0, \infty)$ *into* $L_\nu^q(0, \infty)$;
   (ii) R_α *is compact from* $L^p(0, \infty)$ *into* $L_\nu^q(0, \infty)$;
   (iii)

$$D_1 = \left(\int_0^\infty \left(\int_{[x,\infty)} y^{(1-\alpha)q} d\nu(y) \right)^{\frac{p}{p-q}} x^{\frac{(q-1)p}{p-q}} dx \right)^{\frac{p-q}{pq}} < \infty.$$

Moreover, $\|R_\alpha\| \approx D_1$.

Proof. The implication (iii) $\Rightarrow$ (i) follows in the same way as in the proof of Theorem 2.1.3 (we need to use Theorem E instead of Theorem B). The fact (i) $\Leftrightarrow$ (ii) can be obtained as above (see also Theorem D). It remains to show that (i) $\Rightarrow$ (iii). Let R_α be bounded from $L^p(0, \infty)$ into $L_\nu^q(0, \infty)$ and for each $n \in N$ choose the function f_n as follows: $f_n(x) = S_k$, $-n \leq k \leq n$, where

$$S_k \equiv 2^{\frac{k(q-1)}{p-q}} \left(\int_{[2^k, 2^{k+1})} y^{(\alpha-1)q} d\nu(y) \right)^{\frac{1}{p-q}},$$

when $x \in [2^{k-2}, 2^{k-1})$ and $f_n(x) = 0$, when x does not belong to the interval $[2^{-n-2}, 2^{n-1})$. Then

$$\|f_n\|_{L^p(0,\infty)} = c_1 \left(\sum_{k=-n}^n 2^{\frac{k(p-1)q}{p-q}} \times \right.$$

$$\times \left(\int_{[2^k,2^{k+1})} y^{(\alpha-1)q} d\nu(y) \right)^{\frac{p}{p-q}} dx \right)^{1/p} < \infty.$$

On the other hand,

$$\|R_\alpha f_n\|_{L^q_\nu(0,\infty)} \geq \left(\int_{[-n,n+1)} \left(\int_0^x f_n(y)(x-y)^{\alpha-1} dy \right)^q d\nu(x) \right)^{1/q} =$$

$$= \left(\sum_{k=-n}^{n} \int_{[2^k,2^{k+1})} \left(\int_{2^{k-2}}^{2^{k-1}} f_n(y)(x-y)^{\alpha-1} dy \right)^q d\nu(x) \right)^{1/q} \geq$$

$$\geq c_2 \left(\sum_{k=-n}^{n} \left(\int_{[2^k,2^{k+1})} x^{(\alpha-1)q} d\nu(x) \right)^{\frac{p}{p-q}} 2^{\frac{k(p-1)q}{p-q}} \right)^{1/q}.$$

From the boundedness of R_α we see that

$$\left(\sum_{k=-n}^{n} \left(\int_{[2^k,2^{k+1})} x^{(\alpha-1)q} d\nu(x) \right)^{\frac{p}{p-q}} 2^{\frac{k(p-1)q}{p-q}} \right)^{\frac{p-q}{pq}} \leq c_3 \|R_\alpha\|,$$

where c_3 does not depend on n. Consequently

$$\tilde{D}_1 \equiv \left(\sum_{k=-\infty}^{+\infty} \left(\int_{[2^k,2^{k+1})} x^{(\alpha-1)q} d\nu(x) \right)^{\frac{p}{p-q}} 2^{\frac{k(p-1)q}{p-q}} \right)^{\frac{p-q}{pq}} \leq c_3 \|R_\alpha\| < \infty.$$

Now we show that $D_1 \leq c_4 \tilde{D}_1$ for some constant c_4. Indeed, using the Hardy inequality in discrete form, we obtain that

$$D_1^{\frac{pq}{p-q}} \leq c_5 \sum_{k=-\infty}^{+\infty} \left(\int_{[2^k,\infty)} x^{(\alpha-1)q} d\nu(x) \right)^{\frac{p}{p-q}} 2^{\frac{k(p-1)q}{p-q}} \leq$$

$$\leq c_6 \sum_{k=-\infty}^{+\infty} \left(\sum_{j=k}^{+\infty} \int_{[2^j,2^{j+1})} x^{(\alpha-1)q} d\nu(x) \right)^{\frac{p}{p-q}} 2^{\frac{k(p-1)q}{p-q}} \leq$$

$$\leq c_7 \sum_{k=-\infty}^{+\infty} \left(\int_{[2^k,2^{k+1})} x^{(\alpha-1)q} d\nu(x) \right)^{\frac{p}{p-q}} 2^{\frac{k(p-1)q}{p-q}} = c_7 \tilde{D}_1^{\frac{pq}{p-q}}.$$

The theorem is proved. $\square$

Duality arguments easily give the next theorem.

Theorem 2.1.22. *Let* $1 < q < p < \infty$, $\alpha > \frac{1}{q}$. *Then the following statements are equivalent:*

(i) $W_{\alpha,\nu}$ is bounded from $L_\nu^p(0,\infty)$ into $L^q(0,\infty)$, where

$$W_{\alpha,\nu}g(x) = \int_{[x,\infty)} g(z)(z-x)^{\alpha-1}d\nu(z);$$

(ii) $W_{\alpha,\nu}$ is bounded from $L_\nu^p(0,\infty)$ into $L^q(0,\infty)$;
(iii)

$$D_2 \equiv \left(\int_0^\infty \left(\int_{[x,\infty)} y^{(1-\alpha)p'}d\nu(y) \right)^{\frac{q(p-1)}{p-q}} x^{\frac{q}{p-q}} dx \right)^{\frac{p-q}{pq}} < \infty.$$

Moreover, $\|W_{\alpha,\nu}\| \approx D_2$.

Example 2.1.2. Let $0 < q < \infty$, $1 < p < \infty$ and let $q < p$. We assume that $\alpha > 1/p$ and that $v(x) = x^\gamma$ if $x > 1$ and $v(x) = x^\beta$ if $x \leq 1$, where $\gamma < -\alpha q + q/p - 1$, $-\alpha q + q/p - 1 < \beta < (1-\alpha)q - 1$. Then

$$D \equiv \left(\int_0^\infty \left(\int_x^\infty v(y)y^{(\alpha-1)q}dy \right)^{p(p-q)} x^{p(q-1)/(p-q)}dx \right)^{(p-q)/(pq)} < \infty$$

and by Theorem 2.1.3 (Theorem 2.1.7) we conclude that R_α is bounded (compact) from $L^p(0,\infty)$ into $L_v^q(0,\infty)$.

In what follows we shall investigate the mapping properties of the Riemann–Liouville and Weyl operators defined on bounded intervals, and give some appropriate examples of weight functions.

We assume that v and w are measurable a.e. positive functions on $(0,\infty)$.

Proposition 2.1.1. *Let $0 < q < \infty$, $1 < p < \infty$, $0 < a < \infty$ and $\alpha > 0$. Assume that the weight function w satisfies the following condition: there exists a positive constant c such that for all $t \in (0,a)$,*

$$\int_{t/2}^t \frac{w^{1-p'}(x)}{(t-x)^{(1-\alpha)p'}} dx \leq ct^{(\alpha-1)p'} \int_0^{t/4} w^{1-p'}(x)dx. \qquad (*)$$

(a) If $1 < p \leq q < \infty$, then for the boundedness of R_α from $L_w^p(0,a)$ into $L_v^q(0,a)$ it is necessary and sufficient that the condition

$$B_a \equiv \sup_{0<t<a} B_a(t) \equiv \sup_{0<t<a} \left(\int_t^a \frac{v(x)dx}{x^{(1-\alpha)q}} \right)^{1/q} \left(\int_0^{t/2} w^{1-p'}(x)dx \right)^{1/p'} < \infty$$

be satisfied. Moreover, $\|R_\alpha\| \approx B_a$.

(b) Let $1 < p \leq q < \infty$. Then R_α is compact from $L_w^p(0,a)$ into $L_v^q(0,a)$ if and only if $B_a < \infty$ and $\lim_{t\to 0} B_a(t) = 0$.

(c) Let $q < p$, $1 < p < \infty$ and $1/p < \alpha < 1$. Then R_α is bounded (compact) from $L^p(0, a)$ into $L_v^q(0, a)$ if and only if

$$D_a = \left(\int_0^a \left(\int_x^a \frac{v(t)}{t^{(1-\alpha)q}} dt \right)^{\frac{p}{p-q}} \times \right.$$
$$\left. \times \left(\int_0^x w^{1-p'}(t) dt \right)^{\frac{(q-1)p}{p-q}} w^{1-p'}(x) dx \right)^{\frac{p-q}{pq}} < \infty.$$

Moreover, $\|R_\alpha\| \approx D$.

Proof. The proof follows in the same way as in the case of $a = \infty$. $\square$

Now we give some examples of appropriate weights satisfying the above–mentioned conditions.

Example 2.1.3. Let $1 < p \le q < \infty$, $\alpha > 1/p$, $v(x) = x^\beta$ and let $w(x) = x^\eta \ln^\gamma \frac{2a}{x}$, where $\beta = q - \alpha q - 1$, $\eta = p - 1$, $\gamma = p - 1 + p/q$ and $a = e^{\gamma/\eta}$. Then the condition

$$G \equiv \sup_{0<t<1} G(t) \equiv \sup_{0<t<1} \left(\int_t^1 \frac{v(v) dx}{x^{(1-\alpha)q}} \right)^{1/q} \left(\int_0^{t/2} w^{1-p'}(x) dx \right)^{1/p'} < \infty$$

is satisfied. Further by Proposition 2.1.1 (part (a)) we conclude that R_α is bounded from $L_w^p(0, 1)$ into $L_v^q(0, 1)$.

Example 2.1.4. Let $1 < p \le q < \infty$, $\alpha > 1/p$, $v(x) = x^\beta$ and let $w(x) = x^\eta \ln^\gamma \frac{2a}{x}$, where $\beta = q - \alpha q - 1$, $\eta = p - 1$, $\gamma > p - 1 + p/q$ and $a = e^{\gamma/\eta}$. Then for the pair (v, w) the conditions $G < \infty$ and $\lim_{t\to 0} G(t) = 0$ are satisfied and by Proposition 2.1.1 (part (b)) we conclude that R_α is compact from $L_w^p(0, 1)$ into $L_v^q(0, 1)$.

Example 2.1.5. Let $1 < p \le q < \infty$ and $\alpha > 1/p$. Then R_α is bounded (compact) from $L^p(0, 1)$ into $L_v^q(0, 1)$, where $v(x) = x^\mu(1 - x)^\nu$ with $\mu = -\alpha q + q$ and $\nu > -1$.

Taking into account a duality argument we give some examples of weights guaranteeing the boundedness (compactness) of the Weyl–type operator

$$\overline{W}_\alpha f(x) = \int_x^1 (y - x)^{\alpha-1} f(y) dy, \quad x \in (0, 1), \quad \alpha > 0.$$

Example 2.1.6. Let $1 < p \le q < \infty$ and $\alpha > 1/q'$. Then
(a)
If $v(x) = x^{-1} \ln^\gamma \frac{2a}{x}$, $w(x) = x^\beta$, where $\beta = \alpha p - 1$, $\gamma = -1 - q/p$ and $a = e^{-\gamma}$, then $\overline{W}_\alpha$ is bounded from $L_w^p(0, 1)$ into $L_v^q(0, 1)$.

(b) Let $v(x) = x^{-1} \ln^\gamma \frac{2a}{x}$ and $w(x) = x^\beta$, where $\beta = \alpha p - 1$, $\gamma < -1 - q/p$ and $a = e^{-\gamma}$. Then $\overline{W}_\alpha$ is compact from $L_w^p(0, 1)$ into $L_v^q(0, 1)$.

In the sequel we shall use the following notation:

$$w_1(x) \equiv w\Big(\frac{1}{x+1}\Big)(x+1)^{(\alpha+1)p-2},$$

$$v_1(x) \equiv v\Big(\frac{1}{x+1}\Big)(x+1)^{(1-\alpha)q-2},$$

$$\overline{w}(x) \equiv w\Big(\frac{1}{x+1}\Big)(x+1)^{-2},$$

$$\overline{v}(x) \equiv v\Big(\frac{1}{x+1}\Big)(x+1)^{-2}.$$

Proposition 2.1.2. *Let* $0 < p,q < \infty$ *and* $\alpha > 0$. *Then the opertor* $\overline{W}_\alpha$ *is bounded from* $L^p_w(0,1)$ *into* $L^q_v(0,1)$ *if and only if* R_α *is bounded from* $L^p_{w_1}(0,\infty)$ *into* $L^q_{v_1}(0,\infty)$.

Proof. Let $f \geq 0$ and let $\overline{W}_\alpha$ act boundedly from $L^p_w(0,1)$ into $L^q_v(0,1)$. If we change the variable : $y = 1/\tau - 1$, $x = 1/t - 1$, then we obtain

$$\Big(\int_0^\infty v\Big(\frac{1}{x+1}\Big)(x+1)^{(1-\alpha)q-2} \times$$

$$\times \Big(\int_0^x f(y)(x-y)^{\alpha-1}(y+1)^{-\alpha-1}dy\Big)^q dx\Big)^{1/q} =$$

$$= \Big(\int_0^1 v(t)\Big(\int_t^1 f(1/\tau - 1)(\tau - t)^{\alpha-1}d\tau\Big)^q dy\Big)^{1/q} \leq$$

$$\leq c\Big(\int_0^1 w(t)\Big(f(1/t - 1)\Big)^p dt\Big)^p =$$

$$= c\Big(\int_0^\infty w\Big(\frac{1}{x+1}\Big)(f(x))^p(x+1)^{-2}dx\Big)^{1/p}.$$

Analogously, we have

$$\Big(\int_0^1 v(x)\Big(\int_x^1 f(y)(y-x)^{\alpha-1}dy\Big)^q dx\Big)^{1/q} =$$

$$= \Big(\int_0^\infty v\Big(\frac{1}{t+1}\Big)(t+1)^{(1-\alpha)q-2}\times$$

$$\times\Big(\int_0^t f\Big(\frac{1}{\tau+1}\Big)(\tau+1)^{-\alpha-1}(t-\tau)^{\alpha-1}d\tau\Big)^q dt\Big)^{1/q} \leq$$

$$\leq c\Big(\int_0^\infty w\Big(\frac{1}{t+1}\Big)(t+1)^{-2}\Big(f\Big(\frac{1}{t+1}\Big)\Big)^p dt\Big)^{1/p} =$$

$$= c\Big(\int_0^\infty w(x)(f(x))^p dx\Big)^{1/p}.$$

The theorem is proved. □

Proposition 2.1.3. *Let* $0 < p, q < \infty$ *and* $\alpha > 0$. *Then the compactness of* $\overline{W}_\alpha$ *from* $L^p_w(0,1)$ *into* $L^q_v(0,1)$ *is equivalent to the compactness of* R_α *from* $L^p_{w_1}(\mathcal{R}_+)$ *into* $L^q_{v_1}(\mathcal{R}_+)$.

Proof. It is easy to verify that the compactness of R_α from $L^p_{w_1}(0,1)$ into $L^q_{v_1}(0,1)$ is equivalent to the compactness of the operator

$$\overline{R}_\alpha(x) = (x+1)^{1-\alpha} \int_0^x f(\tau)(x-\tau)^{\alpha-1}(\tau+1)^{-1-\alpha}d\tau$$

from $L^p_{\overline{w}}(\mathcal{R}_+)$ into $L^q_{\overline{v}}(\mathcal{R}_+)$. Thus it suffices to show that the latter is equivalent to the compactness of $\overline{W}_\alpha$ from $L^p_w(0,1)$ into $L^q_v(0,1)$.

Next let $\|f_n\|_{L^p_{\overline{w}}(\mathcal{R}_+)} \leq M$. Then using a change of variable $t = \frac{1}{x+1}$ we derive that $\|\tilde{f}_n\|_{L^p_w(0,1)} \leq 1$, where $\tilde{f}_n(x) = f_n(1/x - 1)$. Further, if $\overline{W}_\alpha$ is compact from $L^p_w(0,1)$ into $L^q_v(0,1)$, then there exist a subsequence $\{\tilde{f}_{n_k}\}$ and $f \in L^q_v(0,1)$ such that $\|\overline{W}_\alpha \tilde{f}_{n_k} - f\|_{L^q_v(0,1)} \to \infty$ as $k \to \infty$. But

$$\|\overline{W}_\alpha \tilde{f}_{n_k} - f\|_{L^q_v(0,1)} = \left(\int_0^1 \left|(\overline{W}_\alpha \tilde{f}_{n_k})(x) - f(x)\right|^q v(x) dx \right)^{1/q} =$$

$$= \left(\int_0^\infty \left|(\overline{W}_\alpha \tilde{f}_{n_k})\left(\frac{1}{t+1}\right) - f\left(\frac{1}{t+1}\right)\right|^q \times \right.$$

$$\left. \times v\left(\frac{1}{t+1}\right)(t+1)^{-2}dt \right)^{1/q} =$$

$$= \left(\int_0^\infty \left|(t+1)^{1-\alpha} \int_0^t f_{n_k}(\tau)(\tau-t)^{\alpha-1}(\tau+1)^{-1-\alpha}d\tau - f\left(\frac{1}{t+1}\right)\right|^q \times \right.$$

$$\left. \times v\left(\frac{1}{t+1}\right)(t+1)^{-2}dt \right)^{1/q} =$$

$$= \left(\int_0^\infty \left|\overline{R}_\alpha f_{n_k}(t) - F(t)\right|^q \overline{v}(t)dt \right)^{1/q}$$

which tends to 0 as $k \to +\infty$, where $F(t) = f\left(\frac{1}{t+1}\right)$. Moreover, $F \in L^q_{\overline{v}}(\mathcal{R}_+)$. Finally we conclude that $\overline{R}_\alpha$ is compact from $L^p_{\overline{w}}(\mathcal{R}_+)$ into $L^q_{\overline{v}}(\mathcal{R}_+)$. Analogously we find that the compactness of $\overline{R}_\alpha$ from $L^p_{\overline{w}}(\mathcal{R}_+)$ into $L^q_{\overline{v}}(\mathcal{R}_+)$ implies that of $\overline{W}_\alpha$ from $L^p_w(0,1)$ into $L^q_v(0,1)$. □

The next proposition can be established in a similar way.

Proposition 2.1.4. *Let* $0 < p, q < \infty$ *and* $\alpha > 0$.

(a) The following conditions are equivalent:
(i) R_α is bounded from $L^p_w(0,1)$ into $L^q_v(0,1)$;

(ii) W_α is bounded from $L^p_{w_1}(R_+)$ into $L^q_{v_1}(R_+)$.

(b) The following two conditions are equivalent:

(i) R_α is compact from $L^p_w(0,1)$ into $L^q_v(0,1)$;

(ii) W_α is compact from $L^p_{w_1}(R_+)$ into $L^q_{v_1}(R_+)$.

Now we give some examples of weights:

Example 2.1.7. Let $1 < p \leq q < \infty$ and $\alpha > 1/q'$. Then:

(a) R_α is bounded from $L^p_{w_1}(R_+)$ into $L^q_{v_1}(R_+)$, where v and w are as in Example 2.1.6 (a).

(b) R_α is compact from $L^p_{w_1}(R_+)$ into $L^q_{v_1}(R_+)$, where v and w are as in Example 2.1.6 (b).

Example 2.1.8. Let $1 < p \leq q < \infty$ and $\alpha > 1/p$. Then

(a) W_α is bounded from $L^p_{w_1}(R_+)$ into $L^q_{v_1}(R_+)$, where the weights v and w are from Example 2.1.3.

W_α is compact from $L^p_{w_1}(R_+)$ into $L^q_{v_1}(R_+)$, with weight functions v and w as in Example 2.1.4.

Example 2.1.9. Let $0 < q \leq \infty$, $1 < p < \infty$, $q < p$ and $\alpha > 1/p$. We assume that $v(x) = x^\gamma$, where $q/p - 1 - \alpha q < \gamma < q - 1 - \alpha q$. Then taking into account Proposition 2.1.1 (part (b)), we conclude that R_α is bounded (compact) from $L^p(0,1)$ into $L^q_v(0,1)$.

Example 2.1.10. Let $0 < q \leq \infty$, $1 < p < \infty$, $q < p$ and $\alpha > 1/q'$. If $w(x) = x^\eta$, where $\alpha p - 1 < \eta < \alpha p + p/q - 1$, then $\overline{W}_\alpha$ is bounded (compact) from from $L^p_w(0,1)$ into $L^q(0,1)$.

Further examples of weights are given in the next two examples.

Example 2.1.11. Let $0 < q < \infty$, $1 < p < \infty$, $q < p$ and let $\alpha > 1/q'$. Then R_α is bounded (compact) from $L^p_{w_1}(0,\infty)$ into $L^q_u(0,\infty)$, where $u(x) = (x+1)^{(1-\alpha)q-2}$ and w is from Example 2.1.10.

Example 2.1.12. Let $0 < q < \infty$, $1 < p < \infty$, $q < p$ and let $\alpha > 1/p$. Then W_α is bounded (compact) from $L^p_{u_1}(0,\infty)$ into $L^q_{v_1}(0,\infty)$, where $u_1(x) = (x+1)^{(1-\alpha)p-2}$ and v is from Example 2.1.9.

2.2.   Two–weight problems

In this section we give a complete description of pairs of weights governing the two–weight, strong–type inequalities for the Riemann–Liouville and Weyl transforms on the line, when the order of integration is less than one.

Let

$$Kf(x) = \int\limits_{-\infty}^{+\infty} k(x,y)f(y)dy$$

for measurable $f : R \to R$, where $k : R \times R \to R$ is a positive measurable kernel.

We shall need the following

Theorem A. *Let $1 < p < q < \infty$ and let v, w be weights. Then the operator K is bounded from $L_w^p(R)$ into $L_v^{q\infty}(R)$ if*

$$\equiv \sup_{x \in R, h > 0} \left(\int\limits_{x-h}^{x+h} v(x)dx \right)^{1/q} \left(\int\limits_{R \backslash (x-h,x+h)} k^{p'}(x,y)w^{1-p'}(y)dy \right)^{1/p'} < \infty.$$

The proof of this theorem in a more general setting is given in [100] (see Theorem 3.1.1).

For measurable $f : R \to R$ put

$$\mathcal{R}_\alpha f(x) = \int\limits_{-\infty}^{x} (x-y)^{\alpha-1} f(y)dy,$$

$$\mathcal{W}_\alpha f(x) = \int\limits_{x}^{+\infty} (y-x)^{\alpha-1} f(y)dy,$$

where $x \in R$ and $\alpha > 0$.

Theorem 2.2.1. *Let $1 < p < q < \infty$, $0 < \alpha < 1$ and let v, w be weights. Then the inequality*

$$\|\mathcal{R}_\alpha f\|_{L_v^{q\infty}(R)} \le c\|f\|_{L_w^p(R)}, \quad f \in L_w^p(R), \tag{2.2.1}$$

with a positive constant c independent of f, holds if and only if

$$C_1 \equiv \sup_{x \in R, h > 0} \left(\int\limits_{x-h}^{x+h} v(y)dy \right)^{1/q} \left(\int\limits_{-\infty}^{x-h} \frac{w^{1-p'}(y)}{(x-y)^{(1-\alpha)p'}} dy \right)^{1/p'} < \infty. \tag{2.2.2}$$

Proof. From Theorem A we see that if $C_1 < \infty$, then the two-weight weak type inequality (2.2.1) holds. Now we show that the condition $C_1 < \infty$ is also necessary.

First we show that

$$I(x,h) \equiv \int\limits_{-\infty}^{x-h} \frac{w^{1-p'}(y)}{(x-y)^{(1-\alpha)p'}}dy < \infty$$

for every x and h. Indeed, if we assume on the contrary that for some $x \in R$ and $h > 0$, $I(x,h) = \infty$, then there exists a non-negative $g : R \to R$ such that

$$\int\limits_{-\infty}^{x-h} g^p(y)w(y)dy \leq 1$$

and

$$\int\limits_{-\infty}^{x-h} g(y)(x-y)^{\alpha-1}dy = \infty.$$

On the other hand, if $z \in (x-h, x+h)$, then

$$\mathcal{R}_\alpha g(z) \geq \int\limits_{-\infty}^{x-h} g(y)(z-y)^{\alpha-1}dy \geq$$

$$c_1 \int\limits_{-\infty}^{x-h} g(y)(x-y)^{\alpha-1}dy = \infty.$$

Consequently

$$(x-h, x+h) \subset \{z : \mathcal{R}_\alpha g(z) > \lambda\}$$

for every $\lambda > 0$. From (2.2.1) we have

$$\int\limits_{x-h}^{x+h} v(y)dy \leq c\lambda^{-q}.$$

As λ is an arbitrary positive number, we conclude that $\int\limits_{x-h}^{x+h} v(x)dx = 0$, which is absurd. Hence $I(x,h) < \infty$.

Now let $f \geq 0$, $x \in R$, $h > 0$ and $z \in (x-h, x+h)$. We have

$$\mathcal{R}_\alpha f(z) = \int\limits_{-\infty}^{z} (z-y)^{\alpha-1}f(y)dy \geq$$

$$\geq c_2 \int\limits_{-\infty}^{x-h} (x-y)^{\alpha-1} f(y) dy.$$

From the two-weight weak type inequality (2.2.1) we obtain

$$\int\limits_{x-h}^{x+h} v(y) dy \leq \int\limits_{\{z:\mathcal{R}_\alpha f(z) \geq c_2 \int\limits_{-\infty}^{x-h} (x-y)^{\alpha-1} f(y) dy\}} v(z) dz \leq$$

$$\leq cc_2^{-q} \left(\int\limits_{-\infty}^{x-h} (x-y)^{\alpha-1} f(y) dy \right)^{-q} \left(\int\limits_{-\infty}^{+\infty} f^p(y) w(y) dy \right)^{q/p},$$

where the constants c and c_2 are independent of $x \in R$, $h > 0$ and $f \geq 0$. If we put here

$$f(y) = \chi_{(-\infty, x-h)}(y) w^{1-p'}(y) (x-y)^{(p'-1)(\alpha-1)},$$

then we obtain

$$\int\limits_{x-h}^{x+h} v(y) dy \leq c_3 \left(\int\limits_{-\infty}^{x-h} (x-y)^{(\alpha-1)p'} w^{1-p'}(y) dy \right)^{-q/p'}$$

and finally we see that $C_1 < \infty$. $\square$

Theorem 2.2.2. *Let* $1 < p < q < \infty$, $\alpha \in (0,1)$ *and let* v, w *be weights. Then* $\mathcal{W}_\alpha$ *is bounded from* $L^p_w(R)$ *into* $L^{q\infty}_v(R)$ *if and only if*

$$C_2 \equiv \sup_{a \in R, h > 0} \left(\int\limits_{a-h}^{a+h} v(y) dy \right)^{1/q} \left(\int\limits_{a+h}^{+\infty} \frac{w^{1-p'}(y)}{(y-a)^{(1-\alpha)p'}} dy \right)^{1/p'} < \infty.$$

The proof is similar to that of the previous theorem.

From the last statement obviously we have that $\mathcal{W}_\alpha$ acts boundedly from $L^{q'}_{v^{1-q'}}(R)$ into $L^{p'\infty}_{w^{1-p'}}(R)$ if and only if

$$\sup_{a \in R, h > 0} \left(\int\limits_{a-h}^{a+h} w^{1-p'}(y) dy \right)^{1/p'} \left(\int\limits_{a+h}^{+\infty} \frac{v(y)}{(y-a)^{(1-\alpha)q}} dy \right)^{1/q} < \infty. \quad (2.2.3)$$

In the papers [181–182], criteria of Sawyer type were derived for the operators $\mathcal{R}_\alpha$ and $\mathcal{W}_\alpha$. In fact, the above–mentioned results lead to

Theorem B. *Let* $1 < p \leq q < \infty$ *and let* $0 < \alpha < 1$; *let* v, w *be weights. Then* $\mathcal{R}_\alpha$ *is bounded from* $L^p_w(R)$ *to* $L^q_v(R)$ *if and only if* $\mathcal{R}_\alpha$ *is bounded from* $L^p_w(R)$ *to* $L^{q\infty}_v(R)$ *and simultaneously* $\mathcal{W}_\alpha$ *acts boundedly from* $L^{q'}_{v^{1-q'}}(R)$ *into* $L^{p'\infty}_{w^{1-p'}}(R)$

Now combining the results stated above we have

Theorem 2.2.3. *Let* $1 < p < q < \infty$ *and* $0 < \alpha < 1$. *For the boundedness of* $\mathcal{R}_\alpha$ *from* $L^p_w(R)$ *into* $L^q_v(R)$ *it is necessary and sufficient that the conditions* (2.2.2) *and* (2.2.3) *are satisfied simultaneously.*

From duality arguments we have the following result for $\mathcal{W}_\alpha$:

Theorem 2.2.4. *Let* $1 < p < q < \infty$ *and let* $\alpha \in (0,1)$. *Then* $\mathcal{W}_\alpha$ *is bounded from* $L^p_w(R)$ *into* $L^q_v(R)$ *if and only if*

$$\sup_{a \in R, h > 0} \left(\int_{a-h}^{a+h} w^{1-p'}(y)dy \right)^{1/p'} \left(\int_{-\infty}^{a-h} \frac{v(y)}{(a-y)^{(1-\alpha)q}}dy \right)^{1/q} < \infty \quad (2.2.4)$$

and

$$\sup_{a \in R, h > 0} \left(\int_{a-h}^{a+h} v(y)dy \right)^{1/q} \left(\int_{a+h}^{+\infty} \frac{w^{1-p'}(y)}{(y-a)^{(1-\alpha)p'}}dy \right)^{1/p'} < \infty. \quad (2.2.5)$$

Now let k be a positive measurable function, lower semicontinuous with support in $(0, \infty)$, nonincreasing in $(0, \infty)$, with $\lim_{x \to +\infty} k(x) = 0$ and satisfying the condition

$$k(x) \leq ck(2x), \quad x \in (0, \infty)$$

where c does not depend on x.

Assume that

$$\mathcal{K}f(x) = \int_{-\infty}^{x} k(x-y)f(y)dy$$

for measurable f defined on R.

We denote by $\mathcal{K}^*$ the dual operator to $\mathcal{K}$, i.e.

$$\mathcal{K}^*f(x) = \int_{x}^{+\infty} k(y-x)f(y)dy.$$

As conditions of Sawyer type, ensuring the boundedness of operators K and K^*, are also known (see [181]) and results similar to Theorems A and B for above–mentioned operators hold, we are led to the following criteria:

Theorem 2.2.5. *Let $1 < p < q < \infty$ and let $0 < \alpha < 1$. Then $\mathcal{K}$ is bounded from $L^p_w(R)$ into $L^q_v(R)$ if and only if*

$$\sup_{a \in R, h > 0} \left(\int_{a-h}^{a+h} v(y)dy \right)^{1/q} \left(\int_{-\infty}^{a-h} w^{1-p'}(y)k^{p'}(a-y)dy \right)^{1/p'} < \infty$$

and

$$\sup_{a \in R, h > 0} \left(\int_{a-h}^{a+h} w^{1-p'}(y)dy \right)^{1/p'} \left(\int_{a+h}^{+\infty} v(y)k^q(y-a)dy \right)^{1/q} < \infty.$$

Theorem 2.2.6. *Let α, p, q satisfy the conditions of Theorems 2.2.5. Then $\mathcal{K}^*$ is bounded from $L^p_w(R)$ into $L^q_v(R)$ if and only if*

$$\sup_{a \in R, h > 0} \left(\int_{a-h}^{a+h} w^{1-p'}(y)dy \right)^{1/p'} \left(\int_{-\infty}^{a-h} v(y)k^q(a-y)dy \right)^{1/q} < \infty$$

and

$$\sup_{a \in R, h > 0} \left(\int_{a-h}^{a+h} v(y)dy \right)^{1/q} \left(\int_{a+h}^{\infty} w^{1-p'}(y)k^{p'}(a-y)dy \right)^{1/p'} < \infty.$$

2.3. Trace inequalities ("Diagonal case"). Examples

In the present section we derive necessary and sufficient conditions for the boundedness of the Riemann–Liouville and Weyl operators R_α ($\mathcal{R}_\alpha$) and W_α ($\mathcal{W}_\alpha$) from L^p_w into L^p_v in the case when $0 < \alpha < 1/p$. Some examples of weight pairs are presented. For the definition and some results concerning the operator R_α see Section 2.1.

First we are going to establish trace inequalities for the Riemann–Liouville and Weyl operators R_α and W_α (see Section 2.1 for the definitions of R_α and W_α).

Theorem 2.3.1. *Let $1 < p < \infty$ and let $0 < \alpha < \frac{1}{p}$. Then the inequality*

$$\int_0^\infty |R_\alpha f(x)|^p v(x)dx \leq c_0 \int_0^\infty |f(x)|^p dx, \quad f \in L^p(R_+), \qquad (2.3.1)$$

holds if and only if $W_\alpha v \in L^{p'}_{loc}(R_+)$ and

$$W_\alpha [W_\alpha v]^{p'}(x) \leq c W_\alpha v(x) \quad a.e. \qquad (2.3.2)$$

To prove this theorem we need

Proposition 2.3.1. *Let $1 < p < \infty$ and let $0 < \alpha < \frac{1}{p}$. If (2.3.1) holds, then*

$$\int_{x}^{x+h} v(y)dy \le c\, h^{1-\alpha p} \tag{2.3.3}$$

for all positive x and h.

Proof. By the duality argument (2.3.1) is equivalent to the inequality

$$\|W_\alpha f\|_{L^{p'}(R_+)} \le c_0^{1/p} \|f\|_{L^{p'}_{v^{1-p'}}(R_+)}. \tag{2.3.4}$$

Replacing here $f(y) = \chi_{(x,x+h)}(y)v(y)$ for $0 < h \le x$, we get

$$\int_{x-h}^{x} \left(\int_{x}^{x+h} v(z)(z-y)^{\alpha-1}dz \right)^{p'} dy \le c_0^{p'-1} \int_{x}^{x+h} v(y)dy.$$

Hence (2.3.3) holds for all x and h with the condition $0 < h \le x$. Now let $0 < x < h < \infty$. Then taking into account the condition $0 < \alpha < \frac{1}{p}$ we obtain

$$\int_{x}^{x+h} v(y)dy = \sum_{k=0}^{\infty} \int_{x+\frac{h}{2^{k+1}}}^{x+\frac{h}{2^k}} v(y)dy =$$

$$= \sum_{k=0}^{\infty} \left(\int_{x+\frac{h}{2^{k+1}}}^{x+\frac{h}{2^k}} v(y)dy \right) \left(\frac{h}{2^{k+1}} \right)^{\alpha p-1} \left(\frac{h}{2^{k+1}} \right)^{1-\alpha p} \le$$

$$\le \sup_{t \le a} \left(\left(\int_{a}^{a+t} v(y)dy \right) t^{\alpha p-1} \right) h^{1-\alpha p} \sum_{k=0}^{\infty} 2^{(k+1)(\alpha p-1)} \le c\, h^{1-\alpha p}.$$

Hence (2.3.3) holds. Note that $c = c_0 2^{(1-\alpha)p} \max\{1, \dfrac{2^{\alpha p-1}}{1-2^{\alpha p-1}}\}$ in (2.3.3).

$\square$

Proof of Theorem 1.1.5. Necessity. Let us first show that from (2.3.1) it follows that $W_\alpha v \in L^{p'}_{loc}(R_+)$. For $f(y) = v(y)\chi_{(x,x+h)}(y)$ ($x \in R_+$ and $h > 0$), from (2.3.1) we have

$$\int_{x}^{x+h} (W_\alpha(v\chi_{(x,x+h)}))^{p'}(y)dy \le c \int_{x}^{x+h} v(y)dy. \tag{2.3.5}$$

Let $v_1(y) = \chi_{(x,x+2h)}v(y)$ and $v_2(y) = \chi_{R_+\setminus(x,x+2h)}(y)v(y)$, where $x \in R_+$ and $h > 0$. We have

$$\int_x^{x+h} (W_\alpha v)^{p'}(y)dy \le c(I_1(x) + I_2(x)),$$

where

$$I_1(x) = \int_x^{x+h} (W_\alpha v_1)^{p'}(y)dy \text{ end } I_2(x) = \int_x^{x+h} (W_\alpha v_2)^{p'}(y)dy.$$

From (2.3.5) we

$$I_1(x) \le c \int_x^{x+2h} v(y)dy < \infty.$$

Thus $W_\alpha v_1 \in L_{loc}^{p'}(R_+)$.

Note that for $z > x + 2h$ and $x < y < x + h$ we have $z - x \le 2(z - y)$. Hence using (2.3.3) we arrive at the estimate

$$(W_\alpha v_2)(y) = \int_{x+2h}^\infty (z-y)^{\alpha-1}v(z)dz \le 2^{1-\alpha}\int_{x+h}^\infty (z-x)^{\alpha-1}v(z)dz =$$

$$= 2^{1-\alpha}\int_h^\infty t^{\alpha-2}\left(\int_x^{x+t} v(z)dz\right)dt \le c2^{1-\alpha}\int_h^\infty t^{\alpha-1-\alpha p}dt < \infty.$$

Therefore $W_\alpha v_2 \in L_{loc}^{p'}(R_+)$, and $W_\alpha v \in L_{loc}^{p'}(R_+)$.

Now we prove that (2.3.1) yields (2.3.2).

We shall use the equality

$$W_\alpha v(x) = (1-\alpha)\int_0^\infty \tau^{\alpha-1}\left(\int_x^{x+\tau} v(y)dy\right)\frac{d\tau}{\tau} \qquad (2.3.6).$$

Thus

$$W_\alpha[(W_\alpha v)^{p'}](x) = (1-\alpha)\int_0^\infty \tau^{\alpha-1}\left(\int_x^{x+\tau} (W_\alpha v)^{p'}(y)dy\right)\frac{d\tau}{\tau}. \qquad (2.3.7)$$

Let v_1 and v_2 be defined as before. By (2.3.5) we have

$$\int_x^{x+h} (W_\alpha v_1)^{p'}(y)dy \le c \int_x^{x+2h} v(y)dy. \qquad (2.3.8)$$

Then from (2.3.7) and (2.3.8) we derive the estimate

$$W_\alpha[(W_\alpha v_1)^{p'}](x) \le c \int_0^\infty \tau^{\alpha-1} \left(\int_x^{x+2\tau} v(t)dt \right) \frac{d\tau}{\tau} = cW_\alpha v(x). \qquad (2.3.9)$$

It is easy to see that for $t \in (x, x+h)$

$$(W_\alpha v_2)(t) \le c \int_h^\infty r^{\alpha-1} \left(\int_x^{x+r} v(y)dy \right) \frac{dr}{r}.$$

Therefore (2.3.7) yields

$$W_\alpha[(W_\alpha v_2)^{p'}](x) \le$$
$$\le c \int_0^\infty t^\alpha \left(\int_t^\infty r^{\alpha-1} \left(\int_x^{x+r} v(y)dy \right) \frac{dr}{r} \right)^{p'} \frac{dt}{t}.$$

Integration by parts on the right–hand side of the last inequality leads to the estimate

$$W_\alpha[(W_\alpha v_2)^{p'}](x) \le c \int_0^\infty r^\alpha \left(\int_r^\infty \tau^{\alpha-1} \left(\int_x^{x+\tau} v(y)dy \right) \frac{d\tau}{\tau} \right)^{p'-1} \times$$
$$\times \left(r^\alpha \int_x^{x+r} v(y)dy \right) \frac{dr}{r}. \qquad (2.3.10)$$

Now recall that estimate (2.3.3) holds by Proposition 2.3.1 . From the inequality (2.3.3) by a simple computation we obtain

$$W_\alpha[(W_\alpha v_2)^{p'}](x) \le c \int_0^\infty h^{\alpha-1} \left(\int_x^{x+h} v(y)dy \right) \frac{dh}{h}.$$

Thus (11)
$$W_\alpha[(W_\alpha v_2)^{p'}](x) \le c(W_\alpha v)(x) \quad \text{a.e.} \qquad (2.3.11)$$
Finally (2.3.9) and (2.3.11) imply (2.3.2).

Remark 2.3.1. It follows from the proof of necessity in Theorem 2.3.1. that if c_0 is the best constant in (2.3.2), then for c from (2.3.3) we have

$$c = c_0^{p'-1} 2^{p'-\alpha} + 2^{p'-1}(1-\alpha)^{p'} \frac{p'c_1}{\alpha(\alpha p - \alpha)^{p'-1}},$$

where $c_1 = c_0 2^{(1-\alpha)p} \max\{1, \frac{2^{\alpha p-1}}{1-2^{\alpha p-1}}\}$.

Sufficiency of Theorem 1.1.5. To show sufficiency, we shall need the following lemmas:

Lemma 2.3.1. *Let* $1 < p < \infty$ *and* $0 < \alpha < 1$. *Then there exists a positive constant c such that for all* $f \in L^1_{loc}(R_+)$, $f \geq 0$, *and for arbitrary* $x \in R_+$ *the following inequality holds:*

$$(R_\alpha f(x))^p \leq c\, R_\alpha \left((R_\alpha f)^{p-1} f \right)(x) \qquad (2.3.12)$$

(*for c we have:* $c = 2^{\frac{1}{p-1}}$ *if* $p \leq 2$ *and* $c = 2^{p(p-1)}$ *if* $p > 2$).

Proof. First we assume that $R_\alpha f(x) < \infty$ and prove (2.3.12) for such x. We also assume that

$$V_\alpha f(x) \leq (R_\alpha f(x))^p,$$

where $V_\alpha f(x) \equiv R_\alpha \left((R_\alpha f)^{p-1} f \right)(x)$, otherwise (2.3.12) is obvious for $c = 1$. Now let us assume that $1 < p \leq 2$. Then we have

$$(R_\alpha f(x))^p = \int_0^x (x-y)^{\alpha-1} f(y) \left(\int_0^x (x-z)^{\alpha-1} f(z)dz \right)^{p-1} dy \leq$$

$$\leq \int_0^x (x-y)^{\alpha-1} f(y) \left(\int_0^y (x-z)^{\alpha-1} f(z)dz \right)^{p-1} dy +$$

$$+ \int_0^x (x-y)^{\alpha-1} f(y) \left(\int_y^x (x-z)^{\alpha-1} f(z)dz \right)^{p-1} dy \equiv I_1(x) + I_2(x).$$

It is obvious that if $z < y < x$, then $y - z \leq x - z$. Consequently

$$I_1(x) \leq \int_0^x (x-y)^{\alpha-1} f(y) \left(\int_0^y (y-z)^{\alpha-1} f(z)dz \right)^{p-1} dy = V_\alpha f(x).$$

Now we use Hölder's inequality with respect to the exponents $\frac{1}{p-1}$, $\frac{1}{2-p}$ and measure $d\sigma(y) = (x-y)^{\alpha-1} f(y)dy$. We have

$$I_2(x) \leq \left(\int_0^x (x-y)^{\alpha-1} f(y)dy \right)^{2-p} \times$$

$$\times \left(\int_0^x \left(\int_y^x (x-z)^{\alpha-1} f(z)dz \right)(x-y)^{\alpha-1} f(y)dy \right)^{p-1} =$$

$$= (R_\alpha f(x))^{2-p} (J(x))^{p-1},$$

where

$$J(x) \equiv \int_0^x \left(\int_y^x (x-z)^{\alpha-1} f(z) dz \right) (x-y)^{\alpha-1} f(y) dy.$$

Using Tonelli's theorem we have

$$J(x) = \int_0^x (x-z)^{\alpha-1} f(z) \left(\int_0^z (x-y)^{\alpha-1} f(y) dy \right) dz.$$

Further, the simple inequality

$$\int_0^z (x-y)^{\alpha-1} f(y) dy \le \left(\int_0^z (x-y)^{\alpha-1} f(y) dy \right)^{p-1} (R_\alpha f(x))^{2-p} \le$$

$$\le (R_\alpha f(z))^{p-1} (R_\alpha f(x))^{2-p}$$

is clear, where $z < x$.

Taking into account the last estimate we obtain

$$J(x) \le \left(\int_0^x (x-z)^{\alpha-1} f(z) (R_\alpha f(z))^{p-1} dz \right) (R_\alpha f(x))^{2-p} =$$

$$= (V_\alpha f(x)) (R_\alpha f(x))^{2-p}.$$

Thus

$$I_2(x) \le (R_\alpha f(x))^{2-p} (R_\alpha f(x))^{(2-p)(p-1)} (V_\alpha f(x))^{p-1} =$$

$$= (R_\alpha f(x))^{p(2-p)} (V_\alpha f(x))^{p-1}.$$

Combining the estimates for I_1 and I_2 we derive

$$(R_\alpha f(x))^p \le V_\alpha f(x) + (R_\alpha f(x))^{p(2-p)} (V_\alpha f(x))^{p-1}.$$

As we have assumed that $V_\alpha f(x) \le (R_\alpha f(x))^p$, we obtain

$$V_\alpha f(x) = (V_\alpha f(x))^{2-p} (V_\alpha f(x))^{p-1} \le (V_\alpha f(x))^{p-1} (R_\alpha f(x))^{p(2-p)}.$$

Hence

$$(R_\alpha f(x))^p \le (V_\alpha f(x))^{p-1} (R_\alpha f(x))^{p(2-p)} +$$
$$+ (V_\alpha f(x))^{p-1} (R_\alpha f(x))^{p(2-p)}$$

$$= 2 \left(V_\alpha f(x) \right)^{p-1} \left(R_\alpha f(x) \right)^{p(2-p)} .$$

Since $R_\alpha f(x) < \infty$ we deduce that

$$\left(R_\alpha f(x) \right)^{p-1} \leq 2^{\frac{1}{p-1}} \left(V_\alpha f(x) \right) .$$

Now we shall deal with the case $p > 2$. Let us assume again that

$$V_\alpha f(x) \leq \left(R_\alpha f(x) \right)^p ,$$

where

$$V_\alpha f(x) \equiv R_\alpha \left[(R_\alpha f)^{p-1} f \right] (x).$$

As $p > 2$ we have

$$\left(R_\alpha f(x) \right)^p = \int\limits_0^x f(y)(x-y)^{\alpha-1} \left(\int\limits_0^x (x-z)^{\alpha-1} f(z) dz \right)^{p-1} dy \leq$$

$$\leq 2^{p-1} \int\limits_0^x f(y)(x-y)^{\alpha-1} \left(\int\limits_0^y (x-z)^{\alpha-1} f(z) dz \right)^{p-1} dy +$$

$$+ 2^{p-1} \int\limits_0^x f(y)(x-y)^{\alpha-1} \left(\int\limits_y^x (x-z)^{\alpha-1} f(z) dz \right)^{p-1} dy \equiv$$

$$2^{p-1} I_1(x) + 2^{p-1} I_2(x).$$

It is clear that if $z < y < x$, then $(x-z)^{\alpha-1} \leq (y-z)^{\alpha-1}$. Therefore $I_1(x) \leq V_\alpha f(x)$. Now we estimate $I_2(x)$. We obtain

$$\left(\int\limits_y^x (x-z)^{\alpha-1} f(z) dz \right)^{p-1} =$$

$$= \left(\int\limits_y^x (x-z)^{\alpha-1} f(z) dz \right)^{p-2} \left(\int\limits_y^x (x-z)^{\alpha-1} f(z) dz \right) \leq$$

$$\leq \left(R_\alpha f(x) \right)^{p-2} \int\limits_y^x (x-z)^{\alpha-1} f(z) dz.$$

Using Tonelli's theorem and the last estimate we have

$$I_2(x) \leq \left(R_\alpha f(x) \right)^{p-2} \int\limits_0^x f(y)(x-y)^{\alpha-1} \left(\int\limits_y^x (x-z)^{\alpha-1} f(z) dz \right) dy =$$

$$= (R_\alpha f(x))^{p-2} \int_0^x f(z)(x-z)^{\alpha-1} \left(\int_0^z (x-y)^{\alpha-1} f(y) dy \right) dz \le$$

$$\le (R_\alpha f(x))^{p-2} \int_0^x f(z)(x-z)^{\alpha-1} \left(\int_0^z (z-y)^{\alpha-1} f(y) dy \right) dz.$$

Hölder's inequality with respect to the exponents $p-1$ and $\frac{p-1}{p-2}$ gives

$$\int_0^x (x-z)^{\alpha-1} f(z) \left(\int_0^z (z-y)^{\alpha-1} f(y) dy \right) dz \le$$

$$\le \left(\int_0^x (x-z)^{\alpha-1} f(z) dz \right)^{\frac{p-2}{p-1}} \times$$

$$\times \left(\int_0^x \left(\int_0^z (z-y)^{\alpha-1} f(y) dy \right)^{p-1} (x-z)^{\alpha-1} f(z) dz \right)^{\frac{1}{p-1}} =$$

$$= (R_\alpha f(x))^{\frac{p-2}{p-1}} (V_\alpha f(x))^{\frac{1}{p-1}}.$$

Combining these estimates we obtain

$$(R_\alpha f(x))^p \le 2^{p-1} V_\alpha f(x) + 2^{p-1} (R_\alpha f(x))^{\frac{p(p-2)}{p-1}} (V_\alpha f(x))^{\frac{1}{p-1}}.$$

From the inequality $V_\alpha f(x) \le (R_\alpha f(x))^p$ it follows that

$$V_\alpha f(x) = (V_\alpha f(x))^{\frac{1}{p-1}} (V_\alpha f(x))^{\frac{p-2}{p-1}} \le (V_\alpha f(x))^{\frac{1}{p-1}} (R_\alpha f(x))^{\frac{p(p-2)}{p-1}}.$$

Hence

$$(R_\alpha f(x))^p \le 2^{p-1} (V_\alpha f(x))^{\frac{1}{p-1}} (R_\alpha f(x))^{\frac{p(p-2)}{p-1}} +$$

$$+ (V_\alpha f(x))^{\frac{1}{p-1}} (R_\alpha f(x))^{\frac{p(p-2)}{p-1}} =$$

$$= 2^p (V_\alpha f(x))^{\frac{1}{p-1}} (R_\alpha f(x))^{\frac{p(p-2)}{p-1}}.$$

The last estimate yields

$$(R_\alpha f(x))^p \le 2^{p(p-1)} (V_\alpha f(x)),$$

where $2 < p < \infty$.

Now we show (2.3.12) for any x which satisfies the condition $R_\alpha f(x) = \infty$.

Let $k_n(x,y) = \chi_{(0,x)}(y)\min\{(x-y)^{\alpha-1},n\}$, where $n \in N$. It is easy to verify that (2.3.12) holds if we replace $k(x,y) = \chi_{(0,x)}(y)(x-y)^{\alpha-1}$ by $k_n(x,y)$. Let $I = (a,b)$, where $0 < a < b < \infty$. Then

$$\int_I k_n(x,y)f(y)dy < \infty \quad \text{and} \quad \sup_{I,n}\int_I k_n(x,y)f(y)dy = \infty.$$

Taking into account the arguments used above we obtain

$$\left(\int_0^x \chi_I(y)k_n(x,y)f(y)dy\right)^p \leq$$

$$\leq c\left(\int_0^x \chi_I(y)\left(\int_0^y \chi_I(z)k_n(y,z)f(z)dz\right)^{p-1} f(z)k_n(x,z)dy\right).$$

(In the last inequality we can assume that f has support in I. In this case $\int_0^x k_n(x,y)f(y)dy \leq n\int_0^x \chi_I(y)f(y)dy \leq n\int_a^b f(y)dy < \infty$). The constant c is defined as follows: $c = 2^{\frac{1}{p-1}}$ if $1 < p \leq 2$ and $c = 2^{p(p-1)}$ if $p > 2$. Taking the supremum with respect to all I and making n tent to $+\infty$, we obtain (2.3.12) for all x. $\square$

Remark 2.3.2. Let $1 < p < \infty, 0 < \alpha < 1, k_n(x,y) \equiv \min\{n,(x-y)^{\alpha-1}\}$. Then for all $f \in L^1_{loc}(R_+)$ $(f \geq 0)$ and for all $x \in R_+$ we have the inequality

$$\left(\int_0^x k_n(x,y)f(y)dy\right)^p \leq c\int_0^x k_n(x,y)\left(\int_0^y k_n(x,y)f(z)dz\right)^{p-1} f(y)dy,$$

where c is the same as in inequlity (2.3.12).

Lemma 2.3.2. *Let* $0 < \alpha < 1$, v *be a locally integrable a.e. positive function on* R_+. *Let there exist a constant* $c > 0$ *such that the inequality*

$$\|R_\alpha f\|_{L^p_{v_1}(R_+)} \leq c_1 \|f\|_{L^p(R_+)}, \quad v_1(x) = [(W_\alpha v)(x)]^{p'} \tag{2.3.13}$$

holds for all $f \in L^p(R_+)$. *Then*

$$\|R_\alpha f\|_{L^p_v(R_+)} \leq c_2 \|f\|_{L^p(R_+)}, \quad f \in L^p(R_+),$$

where $c_2 = c_1^{1/p'}c^{1/p}$ *and* c *is the same as in* (2.3.12).

Proof. Let $f \geq 0$. Using Lemma 2.3.1, Tonelli's theorem and Hölder's inequality we have

$$\int_0^\infty (R_\alpha f(x))^p v(x)dx \leq c\int_0^\infty R_\alpha\left[f(R_\alpha f)^{p-1}\right](x)v(x)dx =$$

$$= c \int\limits_0^\infty (R_\alpha f)^{p-1}(y) f(y) (W_\alpha v)(y) dy \le$$

$$\le c \left(\int\limits_0^\infty (f(y))^p \, dy \right)^{\frac{1}{p}} \left(\int\limits_0^\infty (R_\alpha f(y))^p v_1(y) dy \right)^{\frac{1}{p'}} =$$

$$= c \|f\|_{L^p(R_+)} \|R_\alpha f\|^{p-1}_{L^p_{v_1}(R_+)} \le c_1^{p-1} c \|f\|_{L^p(R_+)} \|f\|^{p-1}_{L^p(R_+)} =$$

$$= c_1^{p-1} c \|f\|^p_{L^p(R_+)} \, .$$

Hence

$$\|R_\alpha f\|_{L^p_v(R_+)} \le c_1^{\frac{1}{p'}} c^{\frac{1}{p}} \|f\|_{L^p(R_+)} \, .$$

□

Lemma 2.3.3. Let $1 < p < \infty$, $0 < \alpha < 1$, $W_\alpha v \in L^{p'}_{loc}$, and let

$$W_\alpha (W_\alpha v)^{p'}(x) \le c_3 (W_\alpha v)(x) \quad \text{a.e.}$$

Then

$$\|R_\alpha f\|_{L^p_{v_1}(R_+)} \le c_4 \|f\|_{L^p(R_+)}, \quad f \in L^p(R_+), \tag{2.3.14}$$

where $v_1(x) = [(W_\alpha v)(x)]^{p'}$ and $c_4 = cc_3$ (c is from (2.3.12)).

Proof. Let $f \ge 0$ be supported in $I \subset R_+$, where I has a form $I = (a, b)$, $0 < a < b < \infty$. Let $k_n(x, y) = \min\left\{(x - y)^{\alpha-1}, n\right\}$. Then using Lemma 2.3.1, Remark 2.3.1 and Tonelli's theorem we have

$$\int\limits_0^\infty \left(\int\limits_0^x k_n(x, y) f(y) dy \right)^p v_1(x) dx \le$$

$$\le c \int\limits_0^\infty \left(\int\limits_0^x k_n(x, y) \left(\int\limits_0^y k_n(y, z) f(z) dz \right)^{p-1} f(y) dy \right) v_1(x) dx =$$

$$= c \int\limits_0^\infty f(y) \left(\int\limits_0^y k_n(y, z) f(z) dz \right)^{p-1} \left(\int\limits_y^\infty k_n(x, y) v_1(x) dx \right) dy \le$$

$$\le c \|f\|_{L^p(R_+)} \left(\int\limits_I \left(\int\limits_0^y k_n(y, z) f(z) dz \right)^p \left(W_\alpha \left[(W_\alpha v)^{p'} \right] (y) \right)^{p'} dy \right)^{1/p'} \le$$

$$\leq cc_3 \, \|f\|_{L^p(R_+)} \left(\int\limits_I \left(\int\limits_0^y k_n(y,z) f(z) dz \right)^p [(W_\alpha v)(y)]^{p'} \, dy \right)^{1/p'}.$$

The last expression is finite

$$\int\limits_0^y k_n(y,z) f(z) dz \leq n \int\limits_I f(z) dz \leq n \, |I|^{\frac{1}{p'}} \|f\|_{L^p(R_+)} < \infty$$

and

$$\int\limits_I ((W_\alpha v)(y))^{p'} \, dy < \infty.$$

Consequently

$$\left(\int\limits_I \left(\int\limits_0^x k_n(x,y) f(y) dy \right)^p v_1(x) dx \right)^{1/p} \leq c_3 c \, \|f\|_{L^p(R_+)} ,$$

where c is from (2.3.12). Finally, we have (2.3.14). $\square$

Combining these Lemmas we obtain the sufficiency part of Theorem 2.3.1.

The next theorem concerns the boundedness of W_α and is proved just in the same way as the previous result.

Theorem 2.3.2. *Let $1 < p < \infty$. Then the inequality*

$$\|W_\alpha f\|_{L^p_v(R_+)} \leq c\|f\|_{L^p(R_+)}, \quad f \in L^p(R_+),$$

holds if and only if $R_\alpha v \in L^{p'}_{loc}(R_+)$ and

$$R_\alpha[R_\alpha v]^{p'}(x) \leq c R_\alpha v(x)$$

for a.a. $x \in R_+$.

As the following statements follow analogously, proofs are omitted.

Theorem 2.3.3. *Let $1 < p < \infty$ and $0 < \alpha < 1/p$. Assume that v is a weight on R. Then $\mathcal{R}_\alpha$ is bounded from $L^p(R)$ to $L^p_v(R)$ if and only if $\mathcal{W}_\alpha v \in L^{p'}_{loc}(R)$ and*

$$\mathcal{W}_\alpha[\mathcal{W}_\alpha v]^{p'}(x) \leq c \mathcal{W}_\alpha v(x)$$

for a.a. $x \in R$.

Theorem 2.3.4. *Let $1 < p < \infty$ and $0 < \alpha < 1/p$. Then the inequality*

$$\|\mathcal{W}_\alpha f\|_{L^p_v(R)} \leq c\|f\|_{L^p(R)}, \quad f \in L^p(R),$$

holds if and only if $\mathcal{R}_\alpha \in L^{p'}_{loc}(R)$ *and*

$$\mathcal{R}_\alpha[\mathcal{R}_\alpha v]^{p'}(x) \leq c\mathcal{R}_\alpha v(x)$$

for a.a. $x \in R$.

Let us now consider the case of two weights.
Let μ and ν be locally finite Borel measures on R_+ and let

$$R_{\alpha,\mu}f(x) = \int\limits_{[0,x]} f(y)(x-y)^{\alpha-1} d\mu(y),$$

$$W_{\alpha,\nu}f(x) = \int\limits_{[x,\infty)} f(y)(y-x)^{\alpha-1} d\nu(y),$$

where $x \in R_+$ and $\alpha \in (0,1)$.

Theorem 2.3.5. *Let* $1 < p < \infty$ *and* $0 < \alpha < 1$. *Assume that the measures* μ *and* ν *satisfy the conditions* $W_{\alpha,\nu}(1) \in L^{p'}_{\mu,loc}(R_+)$ *and*

$$B \equiv \sup_{x>0,r>0} \left(\int\limits_r^\infty \frac{\nu(I_t(x))}{t^{1-\alpha}} \frac{dt}{t} \right)^{1/p} \left(\int\limits_0^r \frac{\mu(I_t(x))}{t^{\alpha-1}} \frac{dt}{t} \right)^{1/p'} < \infty, \quad (2.3.15)$$

where $I_t(x)$ *is the interval* $[x, x+t)$. *Then the inequality*

$$\int\limits_0^\infty |R_{\alpha,\mu}f(x)|^p d\nu(x) \leq c_0 \int\limits_0^\infty |f(x)|^p d\mu(x), \quad f \in L^p_\mu(R_+), \quad (2.3.16)$$

holds if and only if

$$W_{\alpha,\mu}[W_{\alpha,\nu}(1)]^{p'}(x) \leq cW_{\alpha,\nu}(1)(x) \quad (2.3.17)$$

for μ–*a.a.* x.

The sufficiency part of this theorem is a consequence of the following Lemmas:

Lemma 2.3.4. *Let* $1 < p < \infty$ *and* $0 < \alpha < 1$. *Then there exists a positive constant c such that for all* $f \in L^1_{\mu,loc}(R_+)$, $f \geq 0$, *and for arbitrary* $x \in R_+$ *the inequality*

$$(R_{\alpha,\mu}f(x))^p \leq c\,R_{\alpha,\mu}\left((R_{\alpha,\mu}f)^{p-1} f \right)(x) \quad (2.3.18)$$

holds (for c we have: $c = 2^{\frac{1}{p-1}}$ *if* $p \leq 2$ *and* $c = 2^{p(p-1)}$ *if* $p > 2$).

Lemma 2.3.5. *Let* $0 < \alpha < 1$. *Assume that there exists a positive constant* $c_1 > 0$ *such that the inequality*

$$\|R_{\alpha,\mu}f\|_{L^p_{\nu_1}(R_+)} \leq c_1 \|f\|_{L^p_\mu(R_+)}, \quad d\nu_1(x) = [(W_{\alpha,\nu}(1))(x)]^{p'} d\mu(x)$$

holds for all $f \in L^p_\mu(R_+)$. *Then*

$$\|R_{\alpha,\mu}f\|_{L^p_\nu(R_+)} \leq c_2 \|f\|_{L^p_\mu(R_+)}, \quad f \in L^p_\mu(R_+).$$

where $c_2 = c_1^{1/p'} c^{1/p}$ *and* c *is the same as in* (2.3.18).

Lemma 2.3.6. *Let* $1 < p < \infty$, $0 < \alpha < 1$, $W_{\alpha,\nu}(1) \in L^{p'}_{\mu,loc}(R_+)$ *and let*

$$W_{\alpha,\mu}(W_{\alpha,\nu}(1))^{p'}(x) \leq c_3(W_{\alpha,\nu}(1))(x) \quad \mu - a.e.$$

Then

$$\|R_{\alpha,\mu}f\|_{L^p_{\nu_1}(R_+)} \leq c_4\|f\|_{L^p_\mu(R_+)}, \quad f \in L^p_\mu(R_+),$$

where $d\nu_1(x) = [(W_{\alpha,\nu}(1))(x)]^{p'}$ *and* $c_4 = cc_3$ (*c is from* (2.3.18)).
These lemmas can be proved in the same way as the corresponding propositions derived above.

Taking into account the proof of Theorem 2.3.1, we easily obtain necessity. Moreover, for the constant c in condition (2.3.17) we have

$$c = 2^{p'-1}(c_0^{p'-1}2^{1-\alpha} + (1-\alpha)^{p'}2^{(1-\alpha)p'} B^{p'} p'),$$

where c_0 and B are from (2.3.16) and (2.3.15) respectively.

For the Riesz potential

$$I_\alpha f(x) = \int_R f(y)|x - y|^{\alpha-1} dy$$

we have

Proposition 2.3.2. *Let* $1 < p < \infty$ *and* $0 < \alpha < 1/p$. *Then the inequality*

$$\|I_\alpha f\|_{L^p_v(R)} \leq c\|f\|_{L^p(R)},$$

where the positive constant c does not depend on f, $f \in L^p(R)$, *holds if and only if*

(i) $\mathcal{W}_\alpha v \in L^{p'}_{loc}(R)$ *and*

$$\mathcal{W}_\alpha[\mathcal{W}_\alpha v]^{p'}(x) \leq c\mathcal{W}_\alpha v(x), \quad a.a. \ x \in R,$$

(ii) $\mathcal{R}_\alpha v \in L^{p'}_{loc}(R)$ *and*

$$\mathcal{R}_\alpha[\mathcal{R}_\alpha v]^{p'}(x) \leq c\mathcal{R}_\alpha v(x), \quad a.a. \ x \in R,$$

where the constant c_1 and c_2 are independent of x.

Proof. Sufficiency follows from Theorems 2.3.3- 2.3.4 and from the equality

$$I_\alpha f(x) = \mathcal{R}_\alpha f(x) + \mathcal{W}_\alpha f(x).$$

To prove necessity we observe that

$$\mathcal{R}_\alpha f(x) \le I_\alpha f(x), \quad \mathcal{W}_\alpha f(x) \le I_\alpha f(x), \quad f \ge 0.$$

Theorems 2.3.3–2.3.4 completes the proof. $\square$

Further, we note that the following proposition holds for the Volterra–type integral operator

$$K_\mu f(x) = \int_{[0,x]} f(y) k(x,y) d\mu(y),$$

where μ is a locally finite Borel measure on R_+.

Proposition 2.3.3. *Let $1 < p < \infty$ and the kernel k satisfy the condition : there exists a positive constant b such that for all x, y, z, with $0 < y < z < x < \infty$, the inequality*

$$k(x,y) \le bk(z,y)$$

is fulfilled. Let ν and μ be locally finite Borel measures on R_+. Suppose that $K'_\nu(1) \in L^{p'}_\mu(R_+)$, where

$$K'_\nu g(x) = \int_{[x,\infty)} g(y) k(y,x) d\nu(y).$$

Then the condition

$$K'_\mu[K'_\nu(1)]^{p'}(x) \le c K'_\nu(1)(x), \quad \mu - a.e.$$

implies it follows the boundedness of the operator K_μ from $L^p_\mu(R_+)$ to $L^p_\nu(R_+)$.

The proof of this statement follows in the same way as the sufficiency part of Theorem 2.3.5. We omit the proof.

We shall now need the following

Lemma 2.3.7. *Let $1 < p < \infty$, $0 < \alpha < 1/p$. If v and w are positive increasing functions on $(0, \infty)$ and*

$$B \equiv \sup_{t>0} \Big(\int_t^\infty v(x) x^{(\alpha-1)p} dx \Big)^{\frac{1}{p}} \Big(\int_0^{t/2} w^{1-p'}(x) dx \Big)^{\frac{1}{p'}} < \infty \qquad (2.3.19)$$

then there exists a positive constant b_1 such that for all $t > 0$ the following inequality holds:

$$t^{\alpha p} v(t) \le b w(t/4),$$

where $b = 2^{3p-\alpha p-2}B^p$.

Proof. Let

$$B(t) \equiv \Big(\int\limits_t^\infty v(x)x^{(\alpha-1)p}dx \Big)^{\frac{1}{p}} \Big(\int\limits_0^{t/2} w^{1-p'}(x)dx \Big)^{\frac{1}{p'}}.$$

Then for all $t > 0$ we have

$$B \geq B(t) \geq \Big(\int\limits_t^{2t} v(x)x^{(\alpha-1)p}dx \Big)^{\frac{1}{p}} \Big(\int\limits_0^{t/4} w^{1-p'}(x)dx \Big)^{\frac{1}{p'}} \geq$$

$$\geq cv^{1/p}(t)t^{\alpha-1/p'}w^{-1/p}(t/4)t^{1/p'} = c\Big(\frac{v(t)}{w(t/4)} \Big)^{1/p} t^\alpha,$$

where $c = 2^{\alpha-1-2/p'}$. $\square$

Theorem 2.3.6.*Let* $1 < p < \infty$ *and* $0 < \alpha < 1/p$. *Suppose that* v *and* w *are positive increasing functions on* $(0,\infty)$. *Then for the boundedness of* R_α *from* $L_w^p(0,\infty)$ *to* $L_v^p(0,\infty)$ *it is necessary and sufficient that* (2.3.19) *holds. Moreover, there exist positive constants* b_1 *and* b_2 *depending only on* α *and* p *such that* $b_1 B \leq \|R_\alpha\| \leq b_2 B$.

Proof. Let $f \geq 0$. Represent $R_\alpha f$ as follows:

$$R_\alpha f(x) = R_\alpha^{(1)}(x) + R_\alpha^{(2)}(x),$$

where

$$R_\alpha^{(1)} f(x) = \int\limits_0^{x/2} f(y)(x-y)^{\alpha-1}dy, \quad R_\alpha^{(2)} f(x) = \int\limits_{x/2}^x f(y)(x-y)^{\alpha-1}dy.$$

If $y < x/2$, then $x/2 \leq x - y$ and using Theorem A of Section 1.2 we obtain

$$\|R_\alpha^{(1)}f\|_{L_v^p(0,\infty)} \leq c_1 \Big(\int\limits_0^\infty v(x)x^{(\alpha-1)p} \Big(\int\limits_0^{x/2} f(y)dy \Big)^p dx \Big)^{1/p} \leq$$

$$\leq c_2 B \Big(\int\limits_0^\infty w(x)(f(x))^p dx \Big)^{1/p}.$$

Using Hölder's inequality we get

$$I \equiv \int\limits_0^\infty v(x) \Big(\int\limits_{x/2}^x f(y)(x-y)^{\alpha-1}dy \Big)^p dx =$$

$$= \sum_{k \in Z} \int_{2^k}^{2^{k+1}} v(x) \left(\int_{x/2}^{x} f(y)(x-y)^{\alpha-1} dy \right)^p dx \leq$$

$$\leq \sum_{k \in Z} \left(\int_{2^k}^{2^{k+1}} v^{1/(\alpha p)}(x) dx \right)^{\alpha p} \left(\int_{2^k}^{2^{k+1}} \left(\int_{x/2}^{x} f(y)(x-y)^{\alpha-1} dy \right)^{p^*} dx \right)^{p/p^*},$$

where $p^* = p/(1 - \alpha p)$. Moreover, using Lemma 1.3.1 we obtain

$$\left(\int_{2^k}^{2^{k+1}} v^{1/(\alpha p)}(x) dx \right)^{\alpha p} \leq v(2^{k+1}) 2^{k\alpha p} \leq c_3 B^p w(2^{k-1}).$$

On the other hand, using the boundedness of R_α from $L^p(0, \infty)$ to $L^{p^*}(0, \infty)$ we have

$$\left(\int_{2^k}^{2^{k+1}} \left(\int_{x/2}^{x} f(y)(x-y)^{\alpha-1} dy \right)^{p^*} dx \right)^{p/p^*} \leq$$

$$\leq \left(\int_{2^k}^{2^{k+1}} (R_\alpha(f \chi_{(2^{k-1}, 2^{k+1})})(x))^{p^*} dx \right)^{p/p^*} \leq c_4 \int_{2^{k-1}}^{2^{k+1}} (f(x))^p dx.$$

Consequently

$$I \leq c_5 B^p \sum_{k \in Z} w(2^{k-1}) \int_{2^{k-1}}^{2^{k+1}} (f(x))^p dx \leq c_5 B^p \sum_{k \in Z} \int_{2^{k-1}}^{2^{k+1}} (f(x))^p w(x) dx \leq$$

$$\leq c_6 B^p \int_0^\infty (f(x))^p w(x) dx.$$

Finally we have that

$$\|R_\alpha f\|_{L_v^p(0,\infty)} \leq c_7 B \|f\|_{L_w^p(0,\infty)}.$$

Now we prove necessity. First we are going to show that the boundedness of R_α from $L_w^p(0, \infty)$ to $L_v^p(0, \infty)$ implies that

$$J(t) = \int_0^t w^{1-p'}(x) dx < \infty.$$

for all $t > 0$.

Suppose that $J(t) = \infty$ for some $t > 0$. This means that $\|w^{-1}\|_{L^{p'}_w(0,t)} = \infty$.

Consequently, there exists $g \in L^p_w(0,t)$ such that $\int_0^t g(x)dx = \infty$. Now let $f(x) = \chi_{(0,t)}(x)g(x)$. Then we have that $\|f\|_{L^p_w(0,\infty)} = \|g\|_{L^p_w(0,t)} < \infty$. On the other hand,

$$\|R_\alpha\|_{L^p_v(0,\infty)} \geq \left(\int_t^\infty v(x)\left(\int_0^t g(y)(x-y)^{\alpha-1}dy \right)^p dx \right)^{1/p} \geq$$

$$\geq c_8 \left(\int_t^\infty v(x)x^{(\alpha-1)p}dx \right)^{1/p} \int_0^t g(y)dy = \infty.$$

From the boundedness of R_α we conclude that $J(t) < \infty$ for all $t > 0$.

Now let $f_t(x) = \chi_{(0,t/2)}(x)w^{1-p'}(x)$. Then we have

$$\|f_t\|_{L^p_w(0,\infty)} = \left(\int_0^{t/2} w^{1-p'}(x)dx \right)^{1/p}.$$

On the other hand,

$$\|R_\alpha f_t\|_{L^p_v(0,\infty)} \geq \left(\int_t^\infty v(x)\left(\int_0^{t/2} f(y)(x-y)^{\alpha-1}dy \right)^p dx \right)^{1/p} \geq$$

$$\geq c_9 \left(\int_t^\infty v(x)x^{(\alpha-1)p}dx \right)^{1/p} \left(\int_0^{t/2} w^{1-p'}(y)dy \right).$$

Since R_α is bounded we see that $B < \infty$. □

Lemma 2.3.8. *Let $1 < p < \infty$, $0 < \alpha < 1/p$. If v and w are positive increasing functions on $(0, \infty)$ and*

$$B_1 = \sup_{\tau > t > 0} \tau^{\alpha-1}\left(\int_t^\tau v(x)dx \right)^{\frac{1}{p}}\left(\int_0^t w^{1-p'}(x)dx \right)^{\frac{1}{p'}} < \infty, \qquad (2.3.20)$$

then there exists a constant b_1 such that for all $t > 0$ the following inequality holds

$$t^{\alpha p}v(t) \leq bw(t/4),$$

where $b = 2^{3p-\alpha p-2}$.

Proof. Assume that

$$B_1(t,\tau) = \tau^{\alpha-1} \Big(\int\limits_t^\tau v(x)dx \Big)^{1/p} \Big(\int\limits_0^t w^{1-p'}(x)dx \Big)^{\frac{1}{p'}}.$$

Then for all $t > 0$ we have

$$B_1 \geq B_1(t,2t) = (2t)^{\alpha-1} \Big(\int\limits_t^{2t} v(x)dx \Big)^{1/p} \Big(\int\limits_0^{t/4} w^{1-p'}(x)dx \Big)^{\frac{1}{p'}} \geq$$

$$\geq ct^{\alpha-1}v^{1/p}(t)t^{1/p}w^{-1/p}(t/4)t^{1/p'} = ct^\alpha \Big(\frac{v(t)}{w(t/4)} \Big)^{1/p},$$

where $c = 2^{\alpha-1-2/p'}$. $\square$

Theorem 2.3.7. *Let* $1 < p < \infty$, $0 < \alpha < 1/p$. *Suppose that* v *and* w *are positive increasing functions on* $(0,\infty)$. *Then the operator* R_α *is bounded from* $L_w^p(0,\infty)$ *to* $L_v^{p\infty}(0,\infty)$ *if and only if* (2.3.20) *holds. Moreover,* $b_1B \leq \|R_\alpha\| \leq b_2B$, *where positive constants* b_1 *and* b_2 *depend only on* α *and* p.

Proof. As in the proof of Theorem 2.3.6 we represent $R_\alpha f$ as follows

$$R_\alpha f(x) = R_\alpha^{(1)}(x) + R_\alpha^{(2)}(x),$$

where $f \geq 0$. Using the weak- type inequality for the Hardy operator (see [8] and also Theorem 1.2.9) we obtain

$$\|R_\alpha^{(1)}f\|_{L_v^{p\infty}(0,\infty)} \leq c_1\|x^{\alpha-1}\int\limits_0^x f(y)dy\|_{L_v^{p\infty}(0,\infty)} \leq$$

$$\leq c_2 B_1 \|f\|_{L_w^p(0,\infty)}.$$

For R_α^2 we have

$$\|R_\alpha^{(2)}f\|_{L_v^{p\infty}(0,\infty)}^p \leq \|R_\alpha^{(2)}f\|_{L_v^p(0,\infty)}^p.$$

Using the Hölder inequality, Lemma 2.3.8 and the boundedness of R_α from $L^p(0,\infty)$ to $L^{p^*}(0,\infty)$ (see the proof of Theorem 2.3.6) we get

$$\|R_\alpha^{(2)}f\|_{L_v^{p\infty}(0,\infty)}^p \leq c_3 B_1^p \|f\|_{L_w^p(0,\infty)}^p.$$

To prove necessity we show that $I(t) = \int\limits_0^t w^{1-p'}(x)dx < \infty$ for all $t > 0$. Let

$I(t) = \infty$ for some $t > 0$. Then there exists $g \in L_w^p(0,t)$ such that $\int\limits_0^t g(x)dx =$

∞. Let $f_t(x) = \chi_{(0,t)}(x)g(x)$, then $\|f_t\|_{L^p_w(0,\infty)} = \|g\|_{L^p_w(0,t)} < \infty$. On the other hand,

$$\|R_\alpha f\|_{L^{p\infty}_v(0,\infty)} \geq \|\chi_{(t,2t)}(x)\int_0^t g(y)(x-y)^{\alpha-1}dy\|_{L^{p\infty}_v(0,\infty)} \geq$$

$$\geq c_4\|\chi_{(t,2t)}(x)x^{\alpha-1}\|_{L^{p\infty}_v(0,\infty)}\int_0^t g(y)dy \geq$$

$$\geq c_5 t^{\alpha-1}\left(\int_t^{2t} v(x)dx\right)^{1/p}\int_0^t g(y)dy = \infty.$$

Consequently $I(t) < \infty$ for all $t > 0$.

Now let R_α be bounded, $0 < t < \tau < \infty$ and let $f_t(x) = \chi_{(0,t)}(x)w^{1-p'}(x)$. If $x \in (t,\tau)$, then

$$R_\alpha f_t(x) \geq \int_0^t f(y)(x-y)^{\alpha-1}dy \geq c_6 x^{\alpha-1}\int_0^t f(y)dy \geq$$

$$\geq c_6 \tau^{\alpha-1}\int_0^t w^{1-p'}(y)dy \equiv \eta_{t,\tau}.$$

Consequently we have

$$\int_t^\tau v(x)dx \leq \int_{\{x:R_\alpha f_t(x)>\eta_{t,\tau}\}} v(x)dx \leq$$

$$\leq \frac{c_7}{(\eta_{t,\tau})^p}\left(\int_0^t w^{1-p'}(x)dx\right) = c_8\tau^{(1-\alpha)p}\left(\int_0^t w^{1-p'}(x)dx\right)^{1-p}.$$

Hence $B_1 \leq c_8$. $\square$

Example 2.3.1. Let $1 < p < \infty$, $0 < \alpha < \min\{\frac{1}{p}, \frac{1}{p'}\}$ and suppose that

$$v(x) = \begin{cases} x^{-\alpha p+p-1} & \text{if } x \in (0, e^{-p'}), \\ e^{\alpha pp'-pp'+p'+\beta p'}x^\beta & \text{if } x \in [e^{-p'}, \infty) \end{cases}$$

$$w(x) = \begin{cases} x^{p-1}\ln^p\frac{1}{x} & \text{if } x \in (0, e^{-p'}), \\ e^{-p+\gamma p'}(p')^p x^\gamma & \text{if } x \geq e^{-p'}, \end{cases}$$

where $0 < \beta \leq \gamma - \alpha p$, $\alpha p < \gamma \leq \min\{1, p - 1\}$. Then it is easy to verify that the weight pair (v, w) satisfies the condition

$$B = \sup_{t>0} \left(\int_t^\infty \frac{v(x)}{x^{(1-\alpha)p}} dx \right)^{\frac{1}{p}} \left(\int_0^{t/2} w^{1-p'}(x) dx \right)^{\frac{1}{p'}} < \infty$$

and consequently the operator R_α is bounded from $L_w^p(0, \infty)$ to $L_v^p(0, \infty)$.

Example 2.3.2. Let $1 < p < \infty$, $0 < \alpha < \min\{\frac{1}{p}, \frac{1}{p'}\}$. Suppose that

$$v(x) = \begin{cases} x^{-\alpha p + p - 1} \ln \frac{1}{x} & \text{if } x \in (0, e^\lambda), \\ (-\lambda) e^{-\alpha p \lambda + \frac{\lambda}{p} - \lambda - \lambda \beta} x^\beta & \text{if } x \geq e^\lambda, \end{cases}$$

$$w(x) = \begin{cases} x^{p-1} \ln^p \frac{1}{x} & \text{if } x \in (0, e^\lambda), \\ e^{\lambda(p-1) - \gamma\lambda} (-\lambda)^p x^\gamma & \text{if } x \geq e^\lambda, \end{cases}$$

where $0 < \beta \leq \gamma - \alpha p$, $\alpha p < \gamma < \min\{1, p - 1\}$, $\lambda = \min\{-p', \frac{1}{\alpha p - p + 1}\}$. Then for v and w the condition (2.3.20) is satisfied, but $B = \infty$. Thus the operator R_α is bounded from $L_w^p(0, \infty)$ to $L_v^{p\infty}(0, \infty)$, but for these weights $B = \infty$ and therefore R_α is not bounded from $L_w^p(0, \infty)$ to $L_v^p(0, \infty)$.

Corollary 2.3.1.. *Let $1 < p < \infty$ and $\alpha \in (0, 1)$. Then the class of weight pairs governing the boundedness of R_α from $L_w^p(0, \infty)$ to $L_v^{p\infty}(0, \infty)$ is wider than the class of weight pairs ensuring the boundedness of R_α from $L_w^p(0, \infty)$ to $L_v^p(0, \infty)$.*

For Lebesgue-measurable $f : R_+ \to R$ let

$$W_\alpha f(x) = \int_x^\infty f(y)(y - x)^{\alpha-1} dy, \quad \alpha > 0, \ x > 0.$$

The following theorems are proved analogously:

Theorem 2.3.8. *Let $1 < p < \infty$, $0 < \alpha < 1/p$. Assume that v and w are positive decreasing functions on $(0, \infty)$. Then the following statements are equivalent:*

(i) W_α *is bounded from* $L_w^p(0, \infty)$ *to* $L_v^p(0, \infty)$;
(ii) W_α *is bounded from* $L_w^p(0, \infty)$ *to* $L_v^{p\infty}(0, \infty)$;
(iii)

$$\overline{B} \equiv \sup_{t>0} \left(\int_0^{t/2} v(x) dx \right)^{1/p} \left(\int_t^\infty w^{1-p'}(x) x^{(\alpha-1)p'} dx \right)^{1/p'} < \infty.$$

Moreover, $\|W_\alpha\|_{L^p_w \to L^p_v} \approx \|W_\alpha\|_{L^p_w \to L^{p\infty}_v} \approx \overline{B}$.

Example 2.3.3. Let $1 < p < \infty$, $0 < \alpha < \min\{\frac{1}{p}, \frac{1}{p'}\}$ and suppose that

$$v(x) = \begin{cases} x^{-1} \ln^{-p} \frac{1}{x} & \text{if } x \in (0, e^{-p'}), \\ e^{p+\gamma p} p^{-p} x^\gamma & \text{if } x \in [e^{-p'}, \infty) \end{cases}$$

$$w(x) = \begin{cases} x^{\alpha p - 1} & \text{if } x \in (0, e^{-p'}), \\ e^{-p^2\alpha + p + \beta p} x^\beta & \text{if } x \geq e^{-p'}, \end{cases}$$

where $-1 < \gamma \leq \beta - \alpha p$, $\alpha p - 1 < \beta \leq 0$. Then for the weight pair (v, w) the condition $\overline{B} < \infty$ is satisfied and consequently W_α is bounded from $L^p_w(0, \infty)$ to $L^q_v(0, \infty)$; thus R_α is bounded from $L^p_w(0, \infty)$ to $L^p_v(0, \infty)$ (to $L^{p\infty}_v(0, \infty)$).

Now we establish sufficient conditions for the weight v which guarantee the boundedness of R_α from $L^p(0, \infty)$ to $L^p_v(0, \infty)$.

Lemma 2.3.9 *Let $1 < p < \infty$, $0 < \alpha < 1/p$. If the following condition is satisfied*

$$B \equiv \sup_{k \in Z} \left(\int_{2^k}^{2^{k+1}} v^{1/(\alpha p)}(x) dx \right)^\alpha < \infty, \qquad (2.3.21)$$

then

$$B_1 \equiv \sup_{t > 0} \left(\int_t^\infty v(x) x^{(\alpha-1)p} dx \right)^{1/p} t^{1/p'} < \infty.$$

Moreover, $B_1 \leq bB$.

Proof. Let $t > 0$. Then $t \in [2^m, 2^{m+1})$ for some integer m. Using Hölder's inequality we obtain

$$\cdot \left(\int_t^\infty v(x) x^{(\alpha-1)p} dx \right) t^{p-1} \leq c_1 2^{m(p-1)} \int_{2^m}^\infty v(x) x^{(\alpha-1)p} dx =$$

$$= c_1 2^{m(p-1)} \sum_{k=m}^{+\infty} \int_{2^k}^{2^{k+1}} v(x) x^{(\alpha-1)p} dx \leq$$

$$\leq c_1 2^{m(p-1)} \sum_{k=m}^{+\infty} \left(\int_{2^k}^{2^{k+1}} (v(x))^{1/(\alpha p)} dx \right)^{\alpha p} \left(\int_{2^k}^{2^{k+1}} x^{\frac{(\alpha-1)p}{1-\alpha p}} dx \right)^{1-\alpha p} \leq$$

$$\leq c_2 B^p 2^{m(p-1)} \sum_{k=m}^{+\infty} 2^{k(1-p)} \leq c_3 B^p.$$

☐

Theorem 2.3.9. *Let* $1 < p < \infty$, $0 < \alpha < 1/p$. *If* (2.3.21) *holds, then* R_α *is bounded from* $L^p(0, \infty)$ *to* $L^p_v(0, \infty)$. *Moreover,* $\|R_\alpha\| \leq bB$, *where* b *depends only on* α *and* p.

Proof. Suppose that $f \geq 0$. We have

$$\|R_\alpha\|_{L^p_v(0,\infty)} \leq \Big(\int\limits_0^\infty \Big(\int\limits_0^{x/2} f(y)(x-y)^{\alpha-1} dy \Big)^p v(x) dx \Big)^{1/p} +$$

$$+ \Big(\int\limits_0^\infty \Big(\int\limits_{x/2}^x f(y)(x-y)^{\alpha-1} dy \Big)^p v(x) dx \Big)^{1/p} \equiv I_1 + I_2.$$

Using Theorem 2.3.7 and Lemma 2.3.9 for I_1 we obtain

$$I_1 \leq c_1 \Big(\int\limits_0^\infty v(x) \Big(\int\limits_0^x (x-y)^{\alpha-1} dy \Big)^p dx \Big)^{1/p} \leq c_2 B \|f\|_{L^p(0,\infty)}.$$

Using Hölder's inequality we get

$$I_2^p = \sum_{k \in Z} \int\limits_{2^k}^{2^{k+1}} \Big(\int\limits_{x/2}^x f(y)(x-y)^{\alpha-1} dy \Big)^p v(x) dx \leq$$

$$\leq \sum_{k \in Z} \Big(\int\limits_{2^k}^{2^{k+1}} v^{1/(\alpha p)}(x) dx \Big)^{\alpha p} \Big(\int\limits_{2^k}^{2^{k+1}} \Big(\int\limits_{x/2}^x f(y)(x-y)^{\alpha-1} dy \Big)^{p^*} dx \Big)^{p/p^*} \leq$$

$$\leq B^p \sum_{k \in Z} \Big(\int\limits_0^\infty ((R_\alpha f_k)(x))^{p^*} dx \Big)^{p/p^*},$$

where $f_k(x) = \chi_{[2^{k-1}, 2^{k+1}]}(x)$.

By the boundedness of R_α from $L^p(0, \infty)$ to $L^{p^*}(0, \infty)$, where $p^* = \frac{p}{1-\alpha p}$, we obtain

$$I_2^p \leq c_3 B^p \sum_{k \in Z} \int\limits_{2^{k-1}}^{2^{k+1}} (f(y))^p dy \leq c_4 B^p \int\limits_0^\infty (f(y))^p dy.$$

☐

We need the following

Lemma 2.3.10. *Let* $1 < p < \infty$, $0 < \alpha < 1/p$. *If* (1.3.1) *holds, then*

$$B_2 \equiv \sup_{t>0} \Big(\int\limits_0^t v(x)dx \Big)^{1/p} t^{\alpha - 1/p} < \infty.$$

Moreover, $B_2 \leq bB$, *where B is from (2.3.21) and b depends only on p and α.*

Proof. Let $t > 0$. Then $t \in [2^m, 2^{m+1})$ for some $m \in Z$. We have

$$\Big(\int\limits_0^t v(x)dx \Big) t^{\alpha p - 1} \leq 2^{m(\alpha p - 1)} \int\limits_0^{2^{m+1}} v(x)dx =$$

$$= 2^{m(\alpha p - 1)} \sum_{k=-\infty}^{m} \int\limits_{2^k}^{2^{k+1}} v(x)dx \leq$$

$$\leq 2^{m(\alpha p - 1)} \sum_{k=-\infty}^{m} \Big(\int\limits_{2^k}^{2^{k+1}} v^{1/(\alpha p)} v(x)dx \Big)^{\alpha p} 2^{k(1-\alpha p)} \leq$$

$$\leq B^p 2^{m(\alpha p - 1)} \sum_{k=-\infty}^{m} 2^{k(1-\alpha p)} \leq cB^p.$$

Consequently, $B_2 \leq bB$. $\square$

Theorem 2.3.10. *Let α and p satisfy the conditions of Lemma 2.3.10. If* (2.3.21) *holds, then the operator W_α is bounded from $L^p(0,\infty)$ to $L_v^p(0,\infty)$. Moreover, $\|W_\alpha\| \leq bB$ for some positive constant b depending only on p and α.*

Proof. Assume that $f \geq 0$. Then

$$\|W_\alpha\|_{L_v^p(0,\infty)} \leq \Big(\int\limits_0^\infty \Big(\int\limits_{2x}^\infty \frac{f(y)}{(y-x)^{1-\alpha}}dy \Big) v(x)dx \Big)^{1/p} +$$

$$+ \Big(\int\limits_0^\infty \Big(\int\limits_x^{2x} \frac{f(y)}{(y-x)^{1-\alpha}}dy \Big) v(x)dx \Big)^{1/p} \equiv I_1 + I_2.$$

If $y > 2x$, then $y/2 \leq y - x$. Consequently, by Theorem 2.3.8. and Lemma 2.3.10 we have

$$I_1 \leq c_1 \Big(\int\limits_0^\infty \Big(\int\limits_x^\infty \frac{f(y)}{(y-x)^{1-\alpha}}dy \Big) v(x)dx \Big)^{1/p} \leq$$

$$\leq c_2 \Big(\int\limits_0^\infty (f(y))^p dy \Big)^{1/p}.$$

Using Hölder's inequality we obtain

$$I^p = \sum_{k \in Z} \int\limits_{2^k}^{2^{k+1}} \Big(\int\limits_x^{2x} \frac{f(y)}{(y-x)^{1-\alpha}} dy \Big) v(x) dx \leq$$

$$\leq \sum_{k \in Z} \Big(\int\limits_{2^k}^{2^{k+1}} v^{1/(\alpha p)}(x) dx \Big)^{\alpha p} \Big(\int\limits_{2^k}^{2^{k+1}} \Big(\int\limits_x^{2x} \frac{f(y)}{(y-x)^{1-\alpha}} dy \Big)^{p^*} dx \Big)^{p/p^*} \leq$$

$$\leq B^p \sum_{k \in Z} \Big(\int\limits_0^\infty (W_\alpha f_k)^{p^*} dx \Big)^{p/p^*},$$

where $f_k(x) = f(x)\chi_{[2^k, 2^{k+2})}(x)$.

Together with the boundedness of W_α from $L^p(0, \infty)$ to $L^{p^*}(0, \infty)$, this shows that

$$I^p \leq c_3 B^p \sum_{k \in Z} \int\limits_{2^k}^{2^{k+2}} (f(y))^p dy \leq c_4 B^p \int\limits_0^\infty (f(y))^p dy.$$

□

From duality arguments we can easily obtain the following theorems:

Theorem 2.3.11. *Let* $1 < p < \infty$, $0 < \alpha < (p-1)/p$. *If*

$$\overline{B} = \sup \Big(\int\limits_{2^k}^{2^{k+1}} w^{-1/(\alpha p)} dx \Big)^\alpha < \infty, \qquad (2.3.22)$$

then the operator W_α *is bounded from* $L_w^p(0, \infty)$ *to* $L^p(0, \infty)$. *Moreover,* $\|W_\alpha\| \leq b\overline{B}$ *for some positive constant* b *which depends only on* p *and* α.

Theorem 2.3.12. *Let* $1 < p < \infty$, $0 < \alpha < \frac{p-1}{p}$. *If* (2.3.22) *is satisfied, then* R_α *is bounded from* $L_w^p(0, \infty)$ *to* $L^p(0, \infty)$. *Moreover, there exists a positive constant* b *such that* $\|R_\alpha\| \leq b\overline{B}$, *where* b *depends only on* p *and* α.

Theorem 2.3.13. *Let* $0 < a < \infty$, $1 < p < \infty$. *Suppose that* v *and* w *are positive increasing functions on* $(0, a)$. *Suppose the following two conditions hold:*

(i)

$$B_a \equiv \sup_{0 < t < a} \left(\int_t^a v(y) y^{-\alpha p} dy \right)^{1/p} \left(\int_0^{t/2} w^{1-p'}(y) dy \right)^{1/p'} < \infty;$$

(ii) there exist positive constants b such that for all $t \in (0, a]$ the following inequality holds

$$t^{\alpha p} v(t) \leq b w(t/4).$$

Then the operator R_α is bounded from $L_w^p(0, a)$ into $L_v^p(0, a)$. Conversely, if R_α is bounded from $L_w^p(0, a)$ into $L_w^p(0, a)$, then condition (i) is satisfied.

Example 2.3.4. Let $0 < a < \infty$, $1 < p < \infty$ and $0 < \alpha < \min\{\frac{1}{p}, \frac{1}{p'}\}$. Assume that $v(x) = x^{-\alpha p + p - 1}$, $w(x) = x^{p-1} \ln^p \frac{e^{p'} a}{x}$. Then it is easy to verify that v and w are increasing functions on $(0, a)$, $B_a < \infty$ and $v(t) t^{\alpha p} \leq c w(t/4)$ for all $t \in (0, a]$.

The boundedness criterion for R^α from $L_w^p(0, \infty)$ into $L_v^q(0, \infty)$ is already known (see Section 2.2.). Here we consider the case where $w \equiv 1$; for this we have

Theorem 2.3.14. *Let* $1 < p < q < \infty$, $0 < \alpha < 1/p$. *Then the following statements are equivalent:*

(i) There exists a positive constant b_1 such that for all $f \in L^p(0, \infty)$,

$$\left(\int_0^\infty \left| \int_0^x \frac{f(y)}{(x-y)^{1-\alpha}} dy \right|^q d\nu(x) \right)^{1/q} \leq b_1 \left(\int_0^\infty |f(y)|^p dy \right)^{1/p};$$

(ii) There exists a positive constant b_2 such that for all $f \in L^p(0, \infty)$,

$$\left(\int_0^\infty \left| \int_x^\infty \frac{f(y)}{(y-x)^{1-\alpha}} dy \right|^q d\nu(x) \right)^{1/q} \leq b_1 \left(\int_0^\infty |f(y)|^p dy \right)^{1/p};$$

(iii) The inequality

$$\left(\int_0^\infty \left| \int_0^\infty \frac{f(y)}{|y-x|^{1-\alpha}} dy \right|^q d\nu(x) \right)^{1/q} \leq b_1 \left(\int_0^\infty |f(y)|^p dy \right)^{1/p};$$

is fulfilled, where the positive constant b_3 is independent of f, $f \in L^p(0, \infty)$.

(iv)

$$B \equiv \sup_{0 \leq h \leq a} (\nu[a; a+h])^{1/q} h^{\alpha - 1/p} < \infty;$$

(v)

$$B_1 \equiv \sup_{I \subset (0,\infty)} (\nu(I))^{1/q} |I|^{\alpha - 1/p} < \infty.$$

Proof. The implications (iii) $\Leftrightarrow$ (v) follow from [1], (see also [98]). Now we show that (v) $\Rightarrow$ (i). Indeed, let $f \geq 0$. Then using the boundedness of the potential operator, we heve

$$\left(\int\limits_{(0,\infty)} \Big| \int\limits_0^x \frac{f(y)}{(x-y)^{1-\alpha}} dy \Big|^q d\nu(x) \right)^{1/q} \leq$$

$$\leq \left(\int\limits_{(0,\infty)} \Big| \int\limits_0^\infty \frac{f(y)}{|x-y|^{1-\alpha}} dy \Big|^q d\nu(x) \right)^{1/q} \leq$$

$$\leq c \|f\|_{L^p(0,\infty)}.$$

Analogously, (v) $\Rightarrow$ (ii). Now we prove that (iv) $\Rightarrow$ (v). Let $0 \leq a < b < \infty$. Using the fact that $0 < \alpha < 1/p$, we obtain

$$\left(\int\limits_{(a,b)} d\nu \right)(b-a)^{q(\alpha-1/p)} = \sum_{k=0}^{+\infty} \left(\int\limits_{[s_{k+1}, s_k)} d\nu \right)(b-a)^{q(\alpha-1/p)} =$$

$$= \sum_{k=0}^{+\infty} \left(\int\limits_{[s_{k+1}, s_k)} d\nu \right) \left(\frac{(b-a)}{2^{k+1}} \right)^{q(\alpha-1/p)} \left(2^{k+1} \right)^{q(\alpha-1/p)} \leq$$

$$\leq B^q \sum_{k=0}^\infty 2^{(k+1)q(\alpha-1/p)} \leq c_1 B^q,$$

where $s_k = a + \frac{(b-a)}{2^k}$. Thus $B_1 \leq c_2 B$.

The inequality $B \leq c_3 B_1$ is clear. Next we are going to show that (i) $\Rightarrow$ (iv) and (ii) $\Rightarrow$ (v). Let $0 \leq h \leq a < \infty$ and let $f(y) = \chi_{(a-h,a)}(y)$. Then

$$\left(\int\limits_{(0,\infty)} \Big(\int\limits_0^x \frac{f(y)}{|x-y|^{1-\alpha}} dy \Big)^q d\nu(x) \right)^{1/q} \geq$$

$$\geq \left(\int\limits_{[a,a+h)} \Big(\int\limits_{a-h}^a \frac{1}{(x-y)^{1-\alpha}} dy \Big)^q d\nu(x) \right)^{1/q} \geq c_4 (\nu[a, a+h))^{1/q} h^\alpha.$$

Further, $\|f\|_{L^p(0,\infty)} = h^{1/p}$. Consequently, $B < \infty$. Now let $f(y) = \chi_{(a,a+h)}(y)$. Then

$$\left(\int\limits_{(0,\infty)} \left(\int\limits_{x}^{\infty} \frac{f(y)}{(y-x)^{1-\alpha}} dy \right)^q d\nu(x) \right)^{1/q} \geq$$

$$\geq \left(\int\limits_{[a-h,a)} \left(\int\limits_{a}^{a+h} \frac{1}{(y-x)^{1-\alpha}} dy \right)^q d\nu(x) \right)^{1/q} \geq$$

$$\geq c_4 h^{\alpha} (\nu[a-h,a))^{1/q}.$$

On the other hand, $\|f\|_{L^p(0,\infty)} = h^{1/p}$, whence $B_1 < \infty$. $\square$

The following theorems are proved analogously.

Theorem 2.3.15. *Let* $1 < p < q < \infty$, $0 < \alpha < 1/p$. *Then the following statements are equivalent:*

(i) There exists a positive constant c_1 such that for all $f \in L^p(R)$,

$$\left(\int\limits_{-\infty}^{+\infty} \left| \int\limits_{-\infty}^{x} \frac{f(y)}{(x-y)^{1-\alpha}} dy \right|^q d\nu(x) \right)^{1/q} \leq c_1 \left(\int\limits_{-\infty}^{\infty} |f(y)|^p dy \right)^{1/p};$$

(ii) There exists a positive constant c_2 such that for all $f \in L^p(R)$,

$$\left(\int\limits_{-\infty}^{\infty} \left| \int\limits_{x}^{+\infty} \frac{f(y)}{(y-x)^{1-\alpha}} dy \right|^q d\nu(x) \right)^{1/q} \leq c_2 \left(\int\limits_{-\infty}^{\infty} |f(y)|^p dy \right)^{1/p};$$

(iii) The inequality

$$\left(\int\limits_{-\infty}^{+\infty} \left| \int\limits_{-\infty}^{\infty} \frac{f(y)}{|x-y|^{1-\alpha}} dy \right|^q d\nu(x) \right)^{1/q} \leq c_2 \left(\int\limits_{-\infty}^{+\infty} |f(y)|^p dy \right)^{1/p};$$

holds, where the positive constant c_3 is independent of f, $f \in L^p(R)$.

(iv)
$$D \equiv \sup_{I \subset R} (\nu(I))^{1/q} |I|^{\gamma - 1/p} < \infty.$$

Theorem 2.3.16. *Let* $1 < p < q < \infty$, $0 < \alpha < 1/p$. *Assume that* $0 < a < \infty$. *Then the following statements are equivalent:*

(i) There exists a positive constant c_1 such that for all $f \in L^p(0,a)$,

$$\left(\int\limits_{(0,a)} \left| \int\limits_{0}^{x} \frac{f(y)}{(x-y)^{1-\alpha}} dy \right|^q d\nu(x) \right)^{1/q} \leq c_1 \left(\int\limits_{(0,a)} |f(y)|^p dy \right)^{1/p};$$

(ii) There exists a positive constant c_2 such that for all $f \in L^p(0, a)$,

$$\left(\int\limits_{(0,a)} \left| \int\limits_x^a \frac{f(y)}{(y-x)^{1-\alpha}} dy \right|^q d\nu(x) \right)^{1/q} \leq c_2 \left(\int\limits_0^a |f(y)|^p dy \right)^{1/p};$$

(iii) The inequality

$$\left(\int\limits_{(0,a)} \left| \int\limits_0^a \frac{f(y)}{|x-y|^{1-\alpha}} dy \right|^q d\nu(x) \right)^{1/q} \leq c_3 \left(\int\limits_0^a |f(y)|^p dy \right)^{1/p};$$

holds, where the positive constant c_3 is independent of f, $f \in L^p(0, a)$.
(iv)
$$B_1 \equiv \sup_{h<z<a-h} (\nu[z, z+h))^{1/q} h^{\alpha-1/p} < \infty,$$

(v)
$$B_2 \equiv \sup_{h<z<a-h} (\nu[z-h; z))^{1/q} h^{\alpha-1/p} < \infty,$$

(vi)
$$\sup_{I \subset (0,a)} (\nu(I)))^{1/q} |I|^{\alpha-1/p} < \infty.$$

Despite the fact that the two-weight boundedness problem for R_α and W_α from L_w^p into L_v^q, where $1 < p < q < \infty$, is already solved (see Section 2.2, [100]), it is sometimes important to establish more easily verifiable criteria for the two-weight inequality for the operators mentioned above.

Taking into account the proofs of Theorems 2.3.6 - 2.3.8 and using the Hardy - Littlewood theorem for R_α and W_α (see, e.g., [119], Section 10.17, Theorem 383), we can easily obtain the following results:

Theorem 2.3.17. Let $1 < p < \infty$, $0 < \alpha < 1/p$, $1/p - 1/q = \alpha$. Assume that v and w are positive increasing functions on $(0, \infty)$. Then R_α is bounded from $L_w^p(0, \infty)$ into $L_v^q(0, \infty)$ if and only if

$$B \equiv \sup_{t>0} \left(\int\limits_t^\infty v(x) x^{(\alpha-1)q} dx \right)^{\frac{1}{q}} \left(\int\limits_0^{t/2} w^{1-p'}(x) dx \right)^{\frac{1}{p'}} < \infty.$$

Moreover, $\|R_\alpha\| \approx B$.

Theorem 2.3.18. Let $1 < p < \infty$, $0 < \alpha < 1/p$, $1/p - 1/q = \alpha$. Assume also that v and w are positive increasing functions on $(0, \infty)$. Then R_α is *bounded from $L_w^p(0,\infty)$ into $L_v^{q\infty}(0,\infty)$ if and only if*

$$B_1 = \sup_{\tau>t} \tau^{\alpha-1}\left(\int\limits_t^\tau v(x)dx\right)^{\frac{1}{q}}\left(\int\limits_0^t w^{1-p'}(x)dx\right)^{\frac{1}{p'}} < \infty,$$

Moreover, $\|R_\alpha\| \approx B_1$.

Theorem 2.3.19. Let *p, q and α satisfy the conditions of Theorem 2.3.17. Assume that v and w are positive decreasing functions on $(0,\infty)$. Then the following statements are equivalent:*
   (i) W_α *is bounded from* $L_w^p(0,\infty)$ *to* $L_v^q(0,\infty)$;
   (ii) W_α *is bounded from* $L_w^p(0,\infty)$ *to* $L_v^{q\infty}(0,\infty)$;
   (iii)

$$\overline{B} \equiv \sup_{t>0}\left(\int\limits_0^{t/2} v(x)dx\right)^{1/q}\left(\int\limits_t^\infty w^{1-p'}(x)x^{(\alpha-1)p'}dx\right)^{1/p'} < \infty.$$

Moreover, $\|W_\alpha\|_{L_w^p \to L_v^q} \approx \|W_\alpha\|_{L_w^p \to L_v^{q\infty}} \approx \overline{B}$.

2.4. Weak–type inequalities. Examples

Let g be a Lebesgue-measurable, a.e. positive function on $(0,\infty)$. Assume that

$$T_g f(x) = g(x)\int\limits_0^x f(y)dy.$$

We shall need the following known result.

Theorem A ([190]). *Let $1 \le q < p < \infty$, $1/r = 1/q - 1/p$ and assume that g is a decreasing function on $(0,\infty)$. Then for the boundedness of T from $L^p(0,\infty)$ into $L_v^{q\infty}(0,\infty)$ it is necessary and sufficient that $\Phi \in L_v^{r\infty}(0,\infty)$, where*

$$\Phi(x) = \sup_{b>x} g(b)\left(\int\limits_x^b v(y)dy\right)^{1/p} x^{1/p'}.$$

Suppose that

$$R_\alpha f(x) = \int\limits_0^x \frac{f(y)}{(x-y)^{1-\alpha}}dy, \quad \alpha > 0.$$

Theorem 2.4.1. *Let* $1 \leq q < p < \infty$, $1/r = 1/q - 1/p$. *Suppose that,* $1/p < \alpha \leq 1$. *Then* R_α *is bounded from* $L^p(0, \infty)$ *into* $L_v^{q\infty}(0, \infty)$ *if and only if* $\Psi \in L_v^{r\infty}(0, \infty)$, *where*

$$\Psi(x) = \sup_{t > x} t^{\alpha - 1} \left(\int_x^t v(y) dy \right)^{1/p} x^{1/p'}.$$

Proof. Represent $R_\alpha f$ as follows:

$$R_\alpha f(x) = \int_0^{x/2} \frac{f(y)}{(x-y)^{1-\alpha}} dy + \int_{x/2}^x \frac{f(y)}{(x-y)^{1-\alpha}} dy \equiv$$

$$\equiv R_\alpha^{(1)} f(x) + R_\alpha^{(2)} f(x).$$

We have

$$\|R_\alpha f(\cdot)\|_{L_v^{q\infty}(0,\infty)} \leq \|R_\alpha^{(1)} f(\cdot)\|_{L_v^{q\infty}(0,\infty)} +$$

$$+ \|R_\alpha^{(2)} f(\cdot)\|_{L_v^{q\infty}(0,\infty)} \equiv I_1 + I_2.$$

If $y < x/2$, then $x/2 \leq x - y$ and using Theorem A we obtain

$$I_1 \leq c_1 \|x^{\alpha - 1} \int_0^x f(y) dy\|_{L_v^{q\infty}(0,\infty)} \leq c_2 \|f\|_{L^p(0,\infty)}.$$

Let $\|f\|_{L^p(0,\infty)} = 1$. Put

$$\{x : R_\alpha^{(2)} f(x) > \lambda\} \equiv F_\lambda.$$

We have

$$\int_{\{x:|R_\alpha^{(2)} f(x)| > \lambda\}} v(x) dx = \int_{F_\lambda \cap \{x \in (0,\infty): \Psi(x) > \lambda^{q/r}\}} v(x) dx +$$

$$\int_{F_\lambda \cap \{x \in (0,\infty): \Psi(x) \leq \lambda^{q/r}\}} v(x) dx \equiv \tilde{I}_{2,1} + \tilde{I}_{2,2}.$$

For $\tilde{I}_{2,1}$ we obtain

$$\tilde{I}_{2,1} \leq \int_{\{\Psi(x) > \lambda^{q/r}\}} v(x) dx = \lambda^{-q} \sup_{t > 0} t^q \left(\int_{\{\Psi(x) > t^{q/r}\}} v(x) dx \right) =$$

$$= \lambda^{-q} \|\Psi(\cdot)\|_{L_v^{r\infty}(0,\infty)}^r.$$

Now let $E_k \equiv (2^k, 2^{k+1}) \cap F_\lambda \cap \{x : \Psi(x) \leq \lambda^{q/r}\}$, $\alpha_k = \inf E_k$ and $\beta_k = \sup E_k$. It is clear that $(\alpha_k, \beta_k) \subset (2^k, 2^{k+1})$. Moreover, there exists a sequence $\{x_k^{(n)}\}$, $x_k^{(n)} \in E_k$ such that $\lim\limits_{n \to \infty} x_k^{(n)} = \alpha_k$ and $x_k^{(n)}$ is decreasing for fixed k.

If $x \in E_k$, then using Hölder's inequality we have

$$\lambda < R_\alpha^{(2)} f(x) \leq c_3 \left(\int_{x/2}^{x} (f(y))^p dy \right)^{1/p} x^{((\alpha-1)p'+1)1/p'} \leq$$

$$\leq c_3 \left(\int_{2^{k-1}}^{2^{k+1}} (f(y))^p dy \right)^{1/p} x^{\alpha-1+1/p'} \leq$$

$$\leq c_4 \left(\int_{2^{k-1}}^{2^{k+1}} (f(y))^p dy \right)^{1/p} 2^{(k+1)(\alpha-1)} 2^{k/p'} \leq$$

$$\leq c_4 \left(\int_{2^{k-1}}^{2^{k+1}} (f(y))^p dy \right)^{1/p} (x_k^{(n)})^{1/p'} (\beta_k)^{\alpha-1}.$$

Hence

$$\lambda \left(\int_{x_k^{(n)}}^{\beta_k} v(x) dx \right)^{1/p} \leq c_4 \left(\int_{2^{k-1}}^{2^{k+1}} (f(y))^p dy \right)^{1/p} \left(\int_{x_k^{(n)}}^{\beta_k} v(x) dx \right)^{1/p} \times$$

$$\times (x_k^{(n)})^{1/p'} (\beta_k)^{\alpha-1} \leq c_4 \Psi(x_k^{(n)}) \left(\int_{2^{k-1}}^{2^{k+1}} (f(y))^p dy \right)^{1/p}.$$

Further, as $x_k^{(n)} \in E_k$, we get $\Psi(x_k^{(n)}) \leq \lambda^{q/r}$ and so

$$\lambda \left(\int_{x_k^{(n)}}^{\beta_k} v(x) dx \right)^{1/p} \leq c_4 \lambda^{q/r} \left(\int_{2^{k-1}}^{2^{k+1}} (f(y))^p dy \right)^{1/p}.$$

Consequently

$$\lambda \left(\int_{\alpha_k}^{\beta_k} v(x) dx \right)^{1/p} \leq c_4 \lambda^{q/r} \left(\int_{2^{k-1}}^{2^{k+1}} (f(y))^p dy \right)^{1/p}.$$

Hence

$$\lambda^p \int_{\alpha_k}^{\beta_k} v(x)dx \le c_5 \lambda^{\frac{qp}{r}} \int_{2^{k-1}}^{2^{k+1}} (f(y))^p dy.$$

From the last inequality we have

$$\int_{E_k} v(x)dx \le c_5 \lambda^{-q} \int_{2^{k-1}}^{2^{k+1}} (f(y))^p dy,$$

whence

$$\int_{E_\lambda \cap \{x: \Psi(x) \le \lambda^{q/r}\}} v(x)dx \le c_6 \lambda^{-q}.$$

Finally we obtain

$$\int_{E_\lambda} v(x)dx \le \frac{c_7}{\lambda^q}\Big(1 + \|\Psi(\cdot)\|_{L_v^{r\infty}(0,\infty)}^r\Big) \le \frac{c_8}{\lambda^q}.$$

Sufficiency is proved.

To prove necessity we take $\lambda > 0$. Let $S_\lambda \equiv \{x : \Psi(x) > \lambda\}$ and $z \in S_\lambda$. Then there exist a_z and b_z with $0 < a_z < z < b_z$ such that

$$\Big(\int_{a_z}^{b_z} v(x)dx\Big)^{1/p} a_z^{1/p'} b_z^{\alpha-1} > \lambda. \qquad (2.4.1)$$

Indeed, if $z \in S_\lambda$, then $\Psi(z) > \lambda$. Consequently

$$\sup_{t>z} t^{\alpha-1}\Big(\int_z^t v(y)dy\Big)^{1/p} z^{1/p'} > \lambda.$$

Hence, there exists b_z such that

$$b_z^{\alpha-1}\Big(\int_z^{b_z} v(y)dy\Big)^{1/p} z^{1/p'} > \lambda.$$

Further, for fixed z we have

$$\lim_{a_z \to z-} b_z^{\alpha-1}\Big(\int_{a_z}^{b_z} v(y)dy\Big)^{1/p} a_z^{1/p'} > \lambda.$$

From the last inequality we see that there exists a_z with

$$b_z^{\alpha-1}\left(\int_{a_z}^{b_z} v(y)dy\right)^{1/p} a_z^{1/p'} > \lambda.$$

Let $K \subset S_\lambda$ be a compact set. Then we have that for every $z \in K$ there exists an interval (a_z, b_z) such that (2.4.1) holds. Hence there are intervals $(a_{z_1}, b_{z_1}), \cdots, (a_{z_k}, b_{z_k})$ such that $\sum_{j=1}^k \chi_{(a_{z_j}, b_{z_j})} \le 2\chi_{\cup_{j=1}^k (a_{z_j}, b_{z_j})}$ and $K \subset \cup_{j=1}^k (a_{z_j}, b_{z_j})$.

Let

$$f(x) = \left(\sum_{j=1}^k ((b_{z_j})^{1-\alpha}(a_{z_j})^{-1})^p \chi_{(0, a_{z_j})}(x)\right)^{1/p}.$$

If $z \in (a_{z_j}, b_{z_j})$, then we obtain

$$R_\alpha f(z) \ge c_1 z^{\alpha-1}\left(\int_0^z f(y)dy\right) \ge c_1 z^{\alpha-1}\left(\int_0^{a_{z_j}} dy\right) \times$$

$$\times a_{z_j}^{-1}(b_{z_j})^{1-\alpha} \ge c_1,$$

where the positive constant c_1 depends only on α. Hence the set $\{x \in (0, \infty) : R_\alpha f(x) \ge c_1\}$ contains $\cup_{j=1}^k (a_{z_j}, b_{z_j})$. From the inequality (2.4.1) and from the weak-type inequality we obtain

$$\int_{\cup_{j=1}^k (a_{z_j}, b_{z_j})} v(x)dx \le \int_{\{R_\alpha f(x) \ge c_1\}} v(x)dx \le c_2\left(\int_0^\infty (f(x))^p dx\right)^{q/p} =$$

$$= c_2\left(\sum_{j=1}^k \int_0^{a_{z_j}} (a_{z_j}^{-1} b_{z_j}^{1-\alpha})^p dx\right)^{q/p} = c_2\left(\sum_{j=1}^k (b_{z_j})^{(1-\alpha)p}(a_{z_j})^{1-p}\right)^{q/p} \le$$

$$\le \frac{c_2}{\lambda^q}\left(\sum_{j=1}^k \int_{a_{z_j}}^{b_{z_j}} v(x)dx\right)^{q/p} \le \frac{c_3}{\lambda^q}\left(\int_{\cup_{j=1}^k (a_{z_j}, b_{z_j})} v(x)dx\right)^{q/p}.$$

Hence

$$\lambda\left(\int_{\cup_{j=1}^k (a_{z_j}, b_{z_j})} v(x)dx\right)^{1/r} \le c.$$

Since $K \subset (\cup_{j=1}^k (a_{z_j}, b_{z_j}))$ we obtain

$$\lambda\left(\int\limits_K v(x)dx\right)^{1/r} \leq c$$

for all compact K contained in S_λ. Consequently

$$\lambda\left(\int\limits_{S_\lambda} v(x)dx\right)^{1/r} \leq c.$$

The theorem is proved. □

Example 2.4.1. Let $1 \leq q < p < \infty$, $1/r = 1/q - 1/p$, $1/p < \alpha \leq 1$. Assume that $v(y) = y^\beta$, where $\beta = -\frac{p+\alpha rp}{p+r}$. Then the function

$$\Psi(x) = \sup_{t>x} t^{\alpha-1}\left(\int\limits_x^t v(y)dy\right)^{1/p} x^{1/p'}$$

belongs to $L_v^{r\infty}(0,\infty)$, but

$$B = \left(\int\limits_0^\infty \left(\int\limits_x^\infty v(y)y^{(\alpha-1)q}dy\right)^{\frac{p}{p-q}} x^{\frac{p(q-1)}{p-q}} dx\right)^{\frac{p-q}{pq}} = \infty \qquad (2.4.2).$$

Corollary 2.4.1. *Let p, q and α satisfy the conditions of Example 2.4.1, $v(y) = y^\beta$, where $\beta = -\frac{p+\alpha rp}{p+r}$. Then R_α is bounded from $L^p(0,\infty)$ into $L_v^{q\infty}(0,\infty)$, but R_α is not bounded from $L^p(0,\infty)$ into $L_v^q(0,\infty)$.*

Another known result that we shall need is

Theorem B[190]. *Let $1 \leq q < p < \infty$, $1/r = 1/q - 1/p$. Assume that g is an increasing function on $(0,\infty)$. Then T_g is bounded from $L^p(0,\infty)$ into $L_v^{q\infty}(0,\infty)$ if and only if $\Phi_1 \in L_v^{r\infty}(0,\infty)$, where*

$$\Phi_1(x) = g(x)\left(\int\limits_x^\infty v(y)dy\right)^{1/p} x^{1/p'}.$$

The following theorem can be proved in the same way as the previous theorem, using Theorem B this time.

Theorem 2.4.2. *Let $1 \leq q < p < \infty$, $1/r = 1/q - 1/p$. Assume that $\alpha > 1$. Then R_α is bounded from $L^p(0,\infty)$ into $L_v^{q\infty}(0,\infty)$ if and only if $\Psi_1 \in L_v^{r\infty}(0,\infty)$, where*

$$\Psi_1(x) = x^{\alpha-1/p}\left(\int\limits_x^\infty v(y)dy\right)^{1/p}.$$

Example 2.4.2. Let $1 \leq q < p < \infty$, $1/r = 1/q - 1/p$, $\alpha > 1$. Assume that $v(y) = y^\beta$, where $\beta = -\frac{p+\alpha rp}{p+r}$. Then the function $\Psi_1(x)$ belongs to $L_v^{r\infty}(0, \infty)$, but (2.4.2) holds.

Corollary 2.4.2. Let p, q and α satisfy the conditions of Example 2.4.2, $v(y) = y^\beta$, where $\beta = -\frac{p+\alpha rp}{p+r}$. Then R_α is bounded from $L^p(0, \infty)$ into $L_v^{q\infty}(0, \infty)$, but R_α is not bounded from $L^p(0, \infty)$ into $L_v^q(0, \infty)$.

2.5. Integral transforms with power–logarithmic kernels

In the present section we discuss the boundedness and compactness of integral operators with power–logarithmic kernels

$$I_{\alpha,\beta}(f)(x) = \int_0^x (x - t)^{\alpha-1} \ln^\beta \frac{\gamma}{x - t} f(t) dt$$

from $L^p(0, a)$ to $L_v^q(0, a)$ (or to $L_v^{q\infty}(0, a)$), where $0 < a \leq \gamma < \infty$, $1 < p, q < \infty$, $\alpha > \frac{1}{p}$ and $\beta \geq 0$ are given.
The corresponding problem is studied for the operator

$$I'_{\alpha,\beta}(g)(y) = \int_y^a (x - y)^{\alpha-1} \ln^\beta \frac{\gamma}{x - y} g(x) d\nu.$$

The measure of non–compactness is also estimated.
For weighted criteria for the boundedness and compactness of the Riemann–Liouville operator see Sections 2.1–2.3.
In the sequel we shall assume that ν is a non–negative σ- finite Borel measure on $(0, a)$.
Let $H(f)(x) = \int_0^x f(t) dt$ be the Hardy operator. The following theorems are known (see e.g. [195], Section 1.3 for $1 < p < \infty$, $1 \leq q < \infty$ and [272] when $0 < q < 1 < p < \infty$.):

Theorem A. Let $1 < p \leq q < \infty$ and μ be a non–negative Borel measure on $(0, \infty)$. The operator H is bounded from $L^p(0, a)$ to $L_v^q(0, a)$ if and only if

$$A = \sup_{0 < t < a} (\mu[t, a))^{1/q} t^{1/p'} < \infty,$$

where $p' = \frac{p}{p-1}$. Moreover, $A \leq \|H\| \leq (q')^{1/p'} q^{1/q} A$.

Theorem B. Let $0 < q < p < \infty$, $p > 1$ and let v be a positive, Lebesgue-measurable function on $(0, a)$. The operator H is bounded from $L^p(0, a)$ to

$L_v^q(0,a)$ *if and only if*

$$\tilde{A} = \left(\int\limits_0^a \left(\int\limits_x^a v(t)dt \right)^{\frac{p}{p-q}} x^{\frac{p(q-1)}{p-q}} dx \right)^{\frac{p-q}{pq}} < \infty.$$

Moreover there exists positive constants b_1 and b_2 which depend only on p and q such that $b_1 \tilde{A} \leq \|H\| \leq b_2 \tilde{A}$.

With the help of Theorem A we establish

Theorem 2.5.1. *Let $1 < p \leq q < \infty$, $\alpha > \frac{1}{p}$, $\beta \geq 0$. The following statements are equivalent:*

(i) $I_{\alpha,\beta}$ *is bounded from $L^p(0,a)$ to $L_v^q(0,a)$;*

(ii) $I_{\alpha,\beta}$ *is bounded from $L^p(0,a)$ to $L_v^{q\infty}(0,a)$;*

(iii)

$$B = \sup_{0<t<a} B(t) = \sup_{0<t<a} \left(\int\limits_{[t,a)} \frac{1}{x^{(1-\alpha)q}} \ln^{\beta q} \frac{2\gamma}{x} d\nu \right)^{\frac{1}{q}} t^{\frac{1}{p'}} < \infty;$$

(iv)

$$B_1 = \sup_{k \in Z, k \leq 0} \left(\int\limits_{[2^{k-1}a, 2^k a)} \frac{1}{x^{q(1/p-\alpha)}} \ln^{\beta q} \frac{2\gamma}{x} d\nu \right)^{\frac{1}{q}} < \infty.$$

Moreover, there exist constants $b_i, i = 1, \cdots, 6$, depending only on p, q and α (not depending on a), such that the inequalities

$$b_1 B \leq \|I_{\alpha,\beta}\|_{L^p(0,a) \to L_v^q(0,a)} \leq b_2 B, \quad b_3 B_1 \leq B \leq b_4 B_1,$$
$$b_5 B_1 \leq \|I_{\alpha,\beta}\|_{L^p(0,a) \to L_v^{q\infty}(0,a)} \leq b_6 B_1$$

hold.

Proof. First let us show that (iii) implies (i). Let $f \geq 0$. Then we have

$$\|I_{\alpha,\beta}f\|_{L_v^q(0,a)}^q \leq c_1 \int\limits_0^a \left(\int\limits_0^{x/2} f(y)(x-y)^{\alpha-1} \ln^{\beta} \frac{\gamma}{x-y} dy \right)^q d\nu +$$

$$+ c_1 \int\limits_0^a \left(\int\limits_{x/2}^x f(y)(x-y)^{\alpha-1} \ln^{\beta} \frac{\gamma}{x-y} dy \right)^q d\nu = S_1 + S_2.$$

If $0 < y < x/2$, then $(x-y)^{\alpha-1} \ln^{\beta} \frac{\gamma}{x-y} \leq \tilde{b} x^{\alpha-1} \ln^{\beta} \frac{2\gamma}{x}$, where the positive constant $\tilde{b}$ does not depend on x and y. By Theorem A (here $d\mu =$

$x^{(\alpha-1)q}\ln^{\beta q}\frac{2\gamma}{x}d\nu)$ we obtain

$$S_1 \le c_2 \int\limits_0^a \Big(\int\limits_0^{x/2} f(y)dy \Big)^q x^{(\alpha-1)q}\ln^{\beta q}\frac{2\gamma}{x}d\nu \le c_3 B^q \Big(\int\limits_0^a (f(y))^p dy \Big)^{q/p},$$

where the positive constant c_3 does not depend on a.

Let us now estimate S_2. Using Hölder's inequality we obtain

$$S_2 \le c_1 \int\limits_0^a \Big(\int\limits_{x/2}^x (f(y))^p dy \Big)^{q/p} \Big(\int\limits_{x/2}^x \frac{1}{(x-y)^{(1-\alpha)p'}}\ln^{\beta p'}\frac{\gamma}{x-y}dy \Big)^{q/p'}d\nu,$$

where $p' = \frac{p}{p-1}$. Moreover using the facts that $\alpha > 1/p$ and $\beta \ge 0$ we have

$$\int\limits_{x/2}^x \frac{1}{(x-y)^{(1-\alpha)p'}}\ln^{\beta p'}\frac{\gamma}{x-y}dy = d_1 \int\limits_0^{x/2}\frac{1}{t^{(1-\alpha)p'}}\ln^{\beta p'}\frac{\gamma}{t}dt =$$

$$= d_2 x \int\limits_0^1 \frac{1}{(xu)^{(1-\alpha)p'}}\ln^{\beta p'}\frac{2\gamma}{ux}du =$$

$$= d_2 x^{(\alpha-1)p'+1} \int\limits_0^1 \frac{1}{u^{(1-\alpha)p'}}(\ln\frac{2\gamma}{x}+\ln\frac{1}{u})^{\beta p'}du \le$$

$$\le d_3 \Big(x^{(\alpha-1)p'+1}\ln^{\beta p'}\frac{2\gamma}{x}\int\limits_0^1\frac{1}{u^{(1-\alpha)p'}}du +$$

$$+x^{(\alpha-1)p'+1}\int\limits_0^1\frac{1}{u^{(1-\alpha)p'}}\ln^{\beta p'}\frac{1}{u}du \Big) \le d_4 x^{(\alpha-1)p'+1}\ln^{\beta p'}\frac{2\gamma}{x}$$

for every $x \in (0,a)$. Using the last estimates, we find that

$$S_2 \le c_4 \int\limits_0^a \Big(\int\limits_{x/2}^x (f(y))^p dy \Big)^{q/p} x^{(\alpha-1)q+q/p'}\ln^{\beta q}\frac{2\gamma}{x}d\nu =$$

$$= c_4 \sum\limits_{k\in Z, k\le 0} \int\limits_{[2^{k-1}a,2^k a)} \Big(\int\limits_{x/2}^x (f(y))^p dy \Big)^{q/p} \times$$

$$\times x^{(\alpha-1)q+q/p'}\ln^{\beta q}\frac{2\gamma}{x}d\nu \le$$

$$\leq c_5 \sum_{k \in Z, k \leq 0} \Big(\int_{2^{k-2}a}^{2^k a} (f(y))^p dy \Big)^{q/p} \times$$

$$\times \Big(\int_{[2^{k-1}a, 2^k a)} x^{(\alpha-1)q} \ln^{\beta q} \frac{2\gamma}{x} d\nu \Big) (2^k a)^{q/p'} =$$

$$= c_6 \sum_{k \in Z, k \leq 0} \Big(\int_{2^{k-2}a}^{2^k a} (f(y))^p dy \Big)^{q/p} \times$$

$$\times \Big(\int_{[2^{k-1}a, 2^k a)} x^{(\alpha-1)q} \ln^{\beta q} \frac{2\gamma}{x} d\nu \Big) (2^{k-1} a)^{q/p'} \leq$$

$$\leq c_7 B^q \Big(\int_0^a (f(y))^p dy \Big)^{q/p}.$$

We show that (ii) implies (iv). Indeed, let $k \in Z, k \leq 0, f_k(y) = \chi_{(0, 2^k a)}(y)$ and $x \in [2^{k-1}a, 2^k a)$. Then

$$I_{\alpha,\beta}(f)(x) \geq \int_{3x/4}^{x} (x-y)^{\alpha-1} \ln^{\beta} \frac{\gamma}{x-y} dy \geq c_8 x^{\alpha} \ln^{\beta} \frac{4\gamma}{x} \geq$$

$$\geq c_8 (2^{k-1}a)^{\alpha} \ln^{\beta} \frac{\gamma}{2^{k-2}a}.$$

Since the operator $I_{\alpha,\beta}$ is bounded from $L^p(0, a)$ to $L_\nu^{q\infty}(0, a)$, we have

$$\int_{[2^{k-1}a, 2^k a)} d\nu \leq \int_{\{x \in (0,a): I_{\alpha,\beta}(f)(x) > c_8(2^{k-1}a)^{\alpha} \ln^{\beta} \frac{\gamma}{2^{k-2}a}\}} d\nu \leq$$

$$\leq \frac{c_9 (2^k a)^{q/p}}{(2^{k-1}a)^{\alpha q} \ln^{\beta q} \frac{\gamma}{2^{k-2}a}} =$$

$$= c_{10} \frac{1}{(2^k a)^{q(\alpha-1/p)} \ln^{\beta q} \frac{\gamma}{2^{k-2}a}}.$$

Hence

$$c_{11} \geq (2^k a)^{q(\alpha-1/p)} \ln^{\beta q} \frac{\gamma}{2^{k-2}a} \int_{[2^{k-1}a, 2^k a)} d\nu \geq$$

$$\geq c_{12} \int_{[2^{k-1}a, 2^k a)} x^{q(\alpha-1/p)} \ln^{\beta q} \frac{2\gamma}{x} d\nu,$$

where the positive constants c_{11} and c_{12} do not depend on a and k.

Next we prove that (iv) implies (iii). Let $t \in (0, a)$. Then there exists an integer $m \leq 0$ such that $t \in [2^{m-1}a, 2^m a)$.

We see that

$$B^q(t) \leq \left(\int_{[2^{m-1}a, a)} \frac{1}{x^{(1-\alpha)q}} \ln^{\beta q} \frac{2\gamma}{x} d\nu \right)(2^m a)^{q/p'} =$$

$$= c_{13}(2^m a)^{q/p'} \sum_{k=m}^{0} \int_{[2^{k-1}a, 2^k a)} \frac{1}{x^{(1-\alpha)q}} \ln^{\beta q} \frac{2\gamma}{x} d\nu \leq$$

$$\leq c_{13}(2^m a)^{q/p'} \sum_{k=m}^{0} (2^{k-1}a)^{-q/p'} \int_{[2^{k-1}a, 2^k a)} \frac{1}{x^{q(1/p-\alpha)}} \ln^{\beta q} \frac{2\gamma}{x} d\nu \leq$$

$$\leq c_{14} B_1^q (2^m a)^{q/p'} \sum_{k=m}^{0} (2^k a)^{-q/p'} \leq c_{15} B_1^q.$$

Thus $B \leq c_{15} B_1$. Since (i) implies (ii), we obtain (iii) $\Rightarrow$ (i) $\Rightarrow$ (ii) $\Rightarrow$ (iv) $\Rightarrow$ (iii). $\square$

From dual considerations, by virtue of Theorem 1 we easily have

Theorem 2.5.2. *Let* $1 < p \leq q < \infty$, $\alpha > \frac{q-1}{q}$, $\beta \geq 0$. *The following conditions are equivalent:*

(i) $I'_{\alpha,\beta}$ *is bounded from* $L^p_\nu(0, a)$ *to* $L^q(0, a)$;

(ii)

$$\tilde{B} = \sup_{0 < t < a} \left(\int_{[t, a)} x^{(\alpha-1)p'} \ln^{\beta p'} \frac{2\gamma}{x} d\nu \right)^{\frac{1}{p'}} t^{\frac{1}{q}} < \infty;$$

(iii)

$$\tilde{B}_1 = \sup_{k \in Z, k \leq 0} \tilde{B}_1(k) = \sup_{k \in Z, k \leq 0} \left(\int_{[2^{k-1}a, 2^k a)} x^{p'(\alpha-1/q')} \ln^{\beta p'} \frac{2\gamma}{x} d\nu \right)^{\frac{1}{p'}} < \infty.$$

Moreover, there exist constants $\tilde{b}_1, \tilde{b}_2, \tilde{b}_3, \tilde{b}_4$, depending only on p, q and α, such that the inequalities

$$\tilde{b}_1 \tilde{B} \leq \|I_{\alpha,\beta}\| \leq \tilde{b}_2 \tilde{B}, \quad \tilde{b}_3 \tilde{B} \leq \tilde{B}_1 \leq \tilde{b}_4 \tilde{B}$$

hold.

We now consider the case $1 < q < p < \infty$. We shall assume that v and w are positive Lebesgue–integrable functions on (a_1, a) for every $a_1, 0 < a_1 < a$.

Theorem 2.5.3. *Let* $1 < q < p < \infty$, $\alpha > \frac{1}{p}$, $\beta \geq 0$. *The operator* $I_{\alpha,\beta}$ *is bounded from* $L^p(0,a)$ *to* $L^q_v(0,a)$ *if and only if*

$$D = \left(\int\limits_0^a \left(\int\limits_x^a v(t) t^{(\alpha-1)q} \ln^{\beta q} \frac{2\gamma}{t} dt \right)^{\frac{p}{p-q}} x^{\frac{(q-1)p}{p-q}} dx \right)^{\frac{p-q}{pq}} < \infty.$$

Moreover, there exist positive constants b_1 *and* b_2, *which depend only on* p *and* q (*do not depend on* a), *such that the inequality*

$$b_1 D \leq \|I_{\alpha,\beta}\| \leq b_2 D$$

holds.

Proof. *Sufficiency.* Let $f \geq 0$. We have

$$\|I_{\alpha,\beta}f\|^q_{L^q_v(0,a)} \leq c_1 \int\limits_0^a \left(\int\limits_0^{x/2} f(y)(x-y)^{\alpha-1} \ln^\beta \frac{\gamma}{x-y} dy \right)^q v(x) dx +$$

$$+ c_1 \int\limits_0^a \left(\int\limits_{x/2}^x f(y)(x-y)^{\alpha-1} \ln^\beta \frac{\gamma}{x-y} dy \right)^q v(x) dx = \widetilde{S_1} + \widetilde{S_2}.$$

Using Theorem B and the argument from the proof of Theorem 2.5.1 we obtain

$$\widetilde{S_1} \leq c_2 D^q \|f\|^q_{L^p(0,a)}.$$

Applying Hölder's inequality twice and the facts that $\frac{1}{p} < \alpha$ and $\beta \geq 0$, we have

$$\widetilde{S_2} \leq c_1 \int\limits_0^a \left(\int\limits_{\frac{x}{2}}^x (f(y))^p dy \right)^{\frac{q}{p}} \left(\int\limits_{\frac{x}{2}}^x (x-y)^{(\alpha-1)p'} \ln^{\beta p'} \frac{\gamma}{x-y} dy \right)^{\frac{q}{p'}} v(x) dx =$$

$$= c_3 \sum_{k \in Z, k \leq 0} \int\limits_{2^{k-1}a}^{2^k a} \left(\int\limits_{\frac{x}{2}}^x (f(y))^p dy \right)^{\frac{q}{p}} \frac{v(x)}{x^{(1-\alpha)q - \frac{q}{p'}}} \ln^{\beta q} \frac{2\gamma}{x} dx \leq$$

$$\leq c_3 \sum_{k \in Z, k \leq 0} \left(\int\limits_{2^{k-2}a}^{2^k a} (f(y))^p dy \right)^{\frac{q}{p}} \left(\int\limits_{2^{k-1}a}^{2^k a} \frac{v(x)}{x^{(1-\alpha)q - \frac{q}{p'}}} \ln^{\beta q} \frac{2\gamma}{x} dx \right) \leq$$

$$\leq c_3 \left(\sum_{k \in Z, k \leq 0} \int\limits_{2^{k-2}a}^{2^k a} (f(y))^p dy \right)^{\frac{q}{p}} \times$$

$$\times \left(\sum_{k \in Z, k \le 0} \left(\int_{2^{k-1}a}^{2^k a} \frac{v(x)}{x^{(1-\alpha)q - \frac{q}{p'}}} \ln^{\beta q} \frac{2\gamma}{x} dx \right)^{\frac{p}{p-q}} \right)^{\frac{p-q}{p}} \le$$

$$\le c_4 \|f\|_{L^p(0,a)}^q \left(\sum_{k \in Z, k \le 0} \left(\int_{2^{k-1}a}^{2^k a} v(x) x^{(\alpha-1)q + \frac{q}{p'}} \ln^{\beta q} \frac{2\gamma}{x} dx \right)^{\frac{p}{p-q}} \right)^{\frac{p-q}{p}} \equiv$$

$$\equiv c_4 \|f\|_{L^p(0,a)}^q \left(\sum_{k \in Z, k \le 0} \widetilde{S_{2k}} \right)^{\frac{p-q}{p}}.$$

Moreover

$$\widetilde{S_{2k}} \le (2^k a)^{\frac{qp}{p'(p-q)}} \left(\int_{2^{k-1}a}^{2^k a} v(x) x^{(\alpha-1)q} \ln^{\beta q} \frac{2\gamma}{x} dx \right)^{\frac{p}{p-q}} =$$

$$= c_5 \int_{2^{k-2}a}^{2^{k-1}a} x^{\frac{p(q-1)}{p-q}} \left(\int_{2^{k-1}a}^{2^k a} \frac{v(y) \ln^{\beta q} \frac{2\gamma}{y}}{y^{(1-\alpha)q}} dy \right)^{\frac{q}{p-q}} dx \le$$

$$\le c_5 \int_{2^{k-2}a}^{2^{k-1}a} x^{\frac{p(q-1)}{p-q}} \left(\int_x^a \frac{v(y)}{y^{(1-\alpha)q}} \ln^{\beta q} \frac{2\gamma}{y} dy \right)^{\frac{q}{p-q}} dx.$$

Hence

$$\widetilde{S_2} \le c_6 D^q \|f\|_{L^p(0,a)}^q.$$

Necessity. Let $I_{\alpha,\beta}$ be bounded from $L^p(0,a)$ to $L_v^q(0,a)$. Then

$$\int_x^a \frac{v(t)}{t^{(1-\alpha)q}} \ln^{\beta q} \frac{2\gamma}{t} dt < \infty$$

for all $x \in (0,a)$. Put $v_k(t) = v(t) \cdot \chi_{(2^{k-1}a, a)}(t)$, where $k \le 0$, and define

$$f(x) = \left(\int_x^a \frac{v_k(t)}{t^{(1-\alpha)q}} \ln^{\beta q} \frac{2\gamma}{t} dt \right)^{\frac{1}{p-q}} x^{\frac{q-1}{p-q}}.$$

Then we have

$$\|f\|_{L^p(0,a)} = \left(\int_0^a \left(\int_x^a \frac{v_k(t)}{t^{(1-\alpha)q}} \ln^{\beta q} \frac{2\gamma}{t} dt \right)^{\frac{p}{p-q}} x^{\frac{(q-1)p}{p-q}} dx \right)^{\frac{1}{p}} =$$

$$= c_7 \left(\int_{2^k a}^a \left(\int_x^a \frac{v_k(t)}{t^{(1-\alpha)q}} \ln^{\beta q} \frac{2\gamma}{t} dt \right)^{\frac{q}{p-q}} \frac{v(x)}{x^{(1-\alpha)q}} \left(\ln^{\beta q} \frac{2\gamma}{x} \right) x^{\frac{(p-1)q}{p-q}} dx \right)^{\frac{1}{p}} <$$

$$< \infty$$

On the other hand,

$$\|I_{\alpha,\beta}f\|_{L^q_v(0,a)} \geq \left(\int\limits_0^a v(x) \left(\int\limits_{x/2}^{\frac{3x}{4}} \frac{f(t)}{(x-t)^{1-\alpha}} \ln^{\beta q} \frac{2\gamma}{t} dt \right)^q dx \right)^{\frac{1}{q}} \geq$$

$$\geq c_8 \left(\int\limits_0^a \frac{v(x)}{x^{(1-\alpha)q}} \ln^{\beta q} \frac{2\gamma}{x} \left(\int\limits_{x/2}^{\frac{3x}{4}} f(t)dt \right)^q dx \right)^{\frac{1}{q}} \geq$$

$$\geq c_9 \left(\int\limits_0^a \frac{v_k(x)}{x^{(1-\alpha)q}} \ln^{\beta q} \frac{2\gamma}{x} \left(\int\limits_x^a \frac{v_k(y)}{y^{(\alpha-1)q}} \ln^{\beta q} \frac{2\gamma}{y} dy \right)^{\frac{q}{p-q}} \times \right.$$

$$\left. \times \left(\int\limits_{x/2}^{\frac{3x}{4}} t^{\frac{q-1}{p-q}} dt \right)^q dx \right)^{\frac{1}{q}} \geq$$

$$\geq c_9 \left(\int\limits_0^a \frac{v_k(x)}{x^{(1-\alpha)q}} \ln^{\beta q} \frac{2\gamma}{x} \left(\int\limits_x^a \frac{v_k(y)}{y^{(1-\alpha)q}} \ln^{\beta q} \frac{2\gamma}{y} dy \right)^{\frac{q}{p-q}} x^{\frac{(p-1)q}{p-q}} dx \right)^{\frac{1}{q}} =$$

$$= c_{10} \left(\int\limits_0^a \left(\int\limits_x^a \frac{v_k(t)}{t^{(1-\alpha)q}} \ln^{\beta q} \frac{2\gamma}{t} dt \right)^{\frac{p}{p-q}} x^{\frac{(q-1)p}{p-q}} dx \right)^{\frac{1}{q}}$$

As $I_{\alpha,\beta}$ is bounded, we obtain

$$\left(\int\limits_0^a \left(\int\limits_x^a \frac{v_k(t)}{t^{(1-\alpha)q}} \ln^{\beta q} \frac{2\gamma}{t} dt \right)^{\frac{p}{p-q}} x^{\frac{(q-1)p}{p-q}} dx \right)^{\frac{p-q}{pq}} \leq c,$$

By Fatou's lemma we finally see that $D < \infty$. $\square$

By a duality argument and Theorem 2.5.3 we have

Theorem 2.5.4. *Let* $1 < q < p < \infty$, $\alpha > \frac{q}{q-1}$, $\beta \geq 0$. *The operator*

$$J_{\alpha,\beta}f(x) = \int\limits_x^a (y-x)^{\alpha-1} \ln^\beta \frac{\gamma}{y-x} f(y)w(y)dy$$

is bounded from $L^p_w(0,a)$ *to* $L^q(0,a)$ *if and only if*

$$\overline{D} = \left(\int\limits_0^a \left(\int\limits_x^a w(t)t^{(\alpha-1)p'} \ln^{\beta p'} \frac{2\gamma}{t} dt \right)^{\frac{q(p-1)}{p-q}} x^{\frac{q}{p-q}} dx \right)^{\frac{p-q}{pq}} < \infty.$$

Moreover, there exist positive constants b_1 *and* b_2, *not depending on* a, *such that the inequality*

$$b_1 \overline{D} \leq \|J_{\alpha,\beta}\| \leq b_2 \overline{D}$$

holds.

Consider now the compactness of the above-mentioned operators.

Theorem 2.5.5. *Let p, q, α and β satisfy the conditions of Theorem 2.5.1. Then the following conditions are equivalent:*

(i) $I_{\alpha,\beta}$ *is compact from* $L^p(0, a)$ *to* $L^q_\nu(0, a)$;

(ii) $I_{\alpha,\beta}$ *is compact from* $L^p(0, a)$ *to* $L^{q\infty}_\nu(0, a)$;

(iii) $B < \infty$ *and* $\lim_{b \to 0} B_b = 0$, *where*

$$B_b = \sup_{0<t<b} B_b(t) = \sup_{0<t<b} \left(\int_{[t,b)} \frac{1}{x^{(1-\alpha)q}} \ln^{\beta q} \frac{2\gamma}{x} d\nu \right)^{\frac{1}{q}} t^{\frac{1}{p'}};$$

(iv) $B_1 < \infty$ *and* $\lim_{k \to -\infty} B_1(k) = 0$, *where*

$$B_1(k) = \left(\int_{[2^{k-1}a, 2^k a)} \frac{1}{x^{q(1/p-\alpha)}} \ln^{\beta q} \frac{2\gamma}{x} d\nu \right)^{\frac{1}{q}}.$$

Proof. First we shall show that (iii) implies (i). Let $0 < b < a$ and represent $I_{\alpha,\beta}$ as follows:

$$I_{\alpha,\beta} = \chi_{(0,b)} I_{\alpha,\beta} + \chi_{[b,a)} I_{\alpha,\beta} = P_{1b} + P_{2b}.$$

For the operator P_{2b} we have

$$P_{2b} f(x) = \chi_{[b,a)}(x) \int_0^a k_1(x, y) dy,$$

where $k_1(x, y) = (x - y)^{\alpha-1} \ln^\beta \frac{\gamma}{x-y}$ for $0 < y < x < a$, and $k_1(x, y) = 0$ for $0 < x \leq y < a$. Thus we have

$$\int_{[b,a)} \left(\int_0^a (k_1(x,y))^{p'} dy \right)^{q/p'} d\nu =$$

$$= \int_{[b,a)} \left(\int_0^x (x - y)^{(\alpha-1)p'} \ln^{\beta p'} \frac{\gamma}{x - y} dy \right)^{q/p'} d\nu =$$

$$= c_1 \int_{[b,a)} \left(x \int_0^1 (xu)^{(\alpha-1)p'} \ln^{\beta p'} \frac{\gamma}{xu} du \right)^{q/p'} d\nu \leq$$

$$\leq c_2 \int_{[b,a)} x^{(\alpha-1)q + q/p'} \ln^{\beta q} \frac{2\gamma}{x} d\nu \leq$$

$$\le c_2 \Big(\int\limits_{[b,a)} x^{(\alpha-1)q} \ln^{\beta q} \frac{2\gamma}{x} d\nu \Big) a^{q/p'} < \infty.$$

for every b, $0 < b < a$.

By Theorem C of Section 2.1 we find that the operator P_{2b} is compact from $L^p(0,a)$ to $L^q_\nu(0,a)$ for every b with $0 < b < a$.

Repeating the arguments appearing in the proof of Theorem 2.5.1, we readily obtain $\|P_{1b}\| \le c_3 B_b$, where the positive constant c_3 does not depend on b. Hence

$$\|I_{\alpha,\beta} - P_{2b}\| \le c_3 B_b \to 0$$

as $b \to 0$. Thus the operator $I_{\alpha,\beta}$, being the limit of compact operators, is compact.

One can easily see that (i) implies (ii).

We now show that (ii) implies (iv). Let $k \in Z$, $k \le 0$ and $f_k(y) = \chi_{(0,2^k a)}(y)(2^k a)^{-1/p}$. Then the sequence $\{f_k\}$ converges weakly to 0. Indeed, let $\varphi \in L^{p'}(0,a)$. Then we have

$$\Big| \int\limits_0^a f_k(y)\varphi(y)dy \Big| \le \Big(\int\limits_0^{2^k a} |f_k(y)|^p dy \Big)^{1/p} \Big(\int\limits_0^{2^k a} |\varphi(y)|^{p'} dy \Big)^{1/p'} =$$

$$= \Big(\int\limits_0^{2^k a} |\varphi(y)|^{p'} dy \Big)^{1/p'} \to 0$$

as $k \to -\infty$.

On the other hand, we have

$$\|I_{\alpha,\beta} f_k\|_{L^{q\infty}_\nu(0,a)} \ge \|\chi_{[2^{k-1}a,2^k a)} I_{\alpha,\beta} f_k\|_{L^{q\infty}_\nu(0,a)} \ge$$

$$\ge \|\chi_{[2^{k-1}a,2^k a)}(x) \Big(\int\limits_{3x/4}^x \frac{f_k(y)}{(x-y)^{1-\alpha}} \ln^\beta \frac{\gamma}{x-y} dy \Big)\|_{L^{q\infty}_\nu(0,a)} \ge$$

$$\ge c_4 \|\chi_{[2^{k-1}a,2^k a)}(x) x^{\alpha-1} (\ln^\beta 4\gamma x^{-1}) x\|_{L^{q\infty}_\nu(0,a)} (2^k a)^{-1/p} \ge$$

$$\ge c_5 \|\chi_{[2^{k-1}a,2^k a)}(x)\|_{L^{q\infty}_\nu(0,a)} (2^{k-1}a)^{\alpha-1/p} \ln^\beta \frac{\gamma}{2^{k-2}a} =$$

$$= c_6 \Big(\int\limits_{[2^{k-1}a,2^k a)} d\nu \Big)^{1/q} (2^k a)^{\alpha-1/p} \ln^\beta \frac{2\gamma}{2^{k-1}a} \ge$$

$$\ge c_6 \Big(\int\limits_{[2^{k-1}a,2^k a)} x^{q(\alpha-1/p)} \ln^{\beta q} \frac{2\gamma}{x} d\nu \Big)^{1/q}.$$

Since c_6 does not depend on k and the compact operator maps a weakly convergent sequence into a strongly convergent one, we obtain $B_1(k) \to 0$ as $k \to -\infty$.

Now we show that (iv) implies (iii). Let $0 < b < a$. Then there exists an integer $m \le 0$, depending on b, such that $2^{m-1}a \le b < 2^m a$. Therefore we have

$$B_b \le \sup_{0<t<2^m a} \left(\int_{[t,2^m a)} \frac{1}{x^{(1-\alpha)q}} \ln^{\beta q} \frac{2\gamma}{x} d\nu \right)^{\frac{1}{q}} t^{\frac{1}{p'}} =$$

$$= \sup_{0<t<2^m a} B_{2^m a}(t) = B_{2^m a}$$

Now let $0 < t < 2^m a$. Then $t \in [2^{n-1}a, 2^n a)$ for some integer $n \le m$. We have

$$B_{2^m a}^q(t) \le \left(\int_{[2^{n-1}a,2^m a)} x^{(\alpha-1)q} \ln^{\beta q} \frac{2\gamma}{x} d\nu \right) (2^n a)^{q/p'} =$$

$$= (2^n a)^{q/p'} \sum_{k=n}^{m} \int_{[2^{k-1}a,2^k a)} x^{(\alpha-1)q} \ln^{\beta q} \frac{2\gamma}{x} d\nu \le$$

$$\le (2^n a)^{q/p'} \sum_{k=n}^{m} (2^{k-1}a)^{-q/p'} \int_{[2^{k-1}a,2^k a)} x^{q(\alpha-1/p)} \ln^{\beta q} \frac{2\gamma}{x} d\nu \le$$

$$\le (\sup_{k\le m} B_1(k))^q (2^n a)^{q/p'} \sum_{k=n}^{m} (2^k a)^{-q/p'} \le c_7 (\sup_{k\le m} B_1(k))^q = c_7 B_{1m}^q$$

Thus

$$B_{2^m a} \le c_7 B_{1m}.$$

If $b \to 0$, then $m \to -\infty$, and hence $B_{1m} \to 0$, because $\lim_{k\to-\infty} B_1(k) = 0$. This implies that $\lim_{b\to 0} B_b = 0$. From Theorem 2.5.1 we find that if $B_1 < \infty$, then $B < \infty$. Finally we obtain (iii) $\Rightarrow$ (i) $\Rightarrow$ (ii) $\Rightarrow$ (iv) $\Rightarrow$ (iii). $\square$

From dual considerations and Theorem 2.5.5 we obtain

Theorem 2.5.6. *Let p, q, α and β satisfy the conditions of Theorem 2.5.2. Then the following conditions are equivalent:*
(i) *operator $I'_{\alpha,\beta}$ is compact from $L_\nu^p(0,a)$ to $L^q(0,a)$;*
(ii) *$\widetilde{B} < \infty$ and $\lim_{b\to 0} \widetilde{B}_b = 0$, where*

$$\widetilde{B}_b = \sup_{0<t<b} \widetilde{B}_b(t) = \sup_{0<t<b} \left(\int_{[t,b)} \frac{1}{x^{(1-\alpha)p'}} \ln^{\beta p'} \frac{2\gamma}{x} d\nu \right)^{\frac{1}{p'}} t^{\frac{1}{q}};$$

(iii) $\widetilde{B}_1 < \infty$ and $\lim\limits_{k \to -\infty} \widetilde{B}_1(k) = 0$,

The following Theorem is proved in the same way as Theorem 2.5.5.

Theorem 2.5.7. *Let p, q α, β and v satisfy the conditions of Theorem 2.5.3. $I_{\alpha,\beta}$ is compact from $L^p(0,a)$ to $L_v^q(0,a)$ if and only if $D < \infty$.*

By a duality argument we have

Theorem 2.5.8. *Let p, q α, β and w satisfy the conditions of Theorem 2.5.4. $J_{\alpha,\beta}$ is compact from $L_w^p(0,a)$ to $L^q(0,a)$ if and only if $\overline{D} < \infty$.*

In the non-compact case we estimate the distance between the operator $I_{\alpha,\beta}$ and the space of compact operators.

In the sequel the notation of Section 2.1 will be used.

We shall assume that v is a Lebesgue-measurable a.e. positive function on $(0,a)$.

Theorem 2.5.9. *Let $1 < p \le q < \infty$. Suppose that $B < \infty$ for $d\nu(x) = v(x)dx$, $X = L^p(0,a)$ and $Y = L_v^q(0,a)$. Then the inequality*

$$b_1 J \le dist(I_{\alpha,\beta}, \mathcal{K}(X,Y)) \le b_2 J \qquad (2.5.1)$$

is fulfilled, where the positive constants b_1 and b_2 depend only on p, q, α and β, $J = \lim\limits_{c \to 0} R_c$ and

$$R_c = \sup_{0 < t < c} \left(\int\limits_t^c x^{(\alpha-1)q} \ln^{\beta q} \frac{2\gamma}{x} v(x)dx \right)^{\frac{1}{q}} t^{\frac{1}{p'}}.$$

Proof. It follows from the proof of Theorem 2.5.5 that

$$\|I_{\alpha,\beta} - K_c\| \le c_1 R_c,$$

where K_c is a compact operator for every c. Therefore we have

$$dist(I_{\alpha,\beta}, \mathcal{K}(X,Y)) \le c_1 J,$$

where c_1 depends only on p, q, α and β. Now we establish the lower estimate of (2.5.1).

Let $\lambda > dist(I_{\alpha,\beta}, \mathcal{K}(X,Y))$. Then using Lemma 2.1.1 there exists $P \in \mathcal{F}_r(X,Y)$ such that $\|I_{\alpha,\beta} - P\| < \lambda$. On the other hand, taking into account Lemma 2.1.2 with $\epsilon = (\lambda - \|I_{\alpha,\beta} - P\|)/2$ there exist $T \in \mathcal{F}_r(X,Y)$ and $[\alpha,\eta] \subset (0,a)$ such that

$$\|P - T\| < \epsilon \qquad (2.5.2)$$

and

$$supp \, Tf \subset [\alpha,\eta].$$

Using inequality (2.5.2) we obtain

$$\|I_{\alpha,\beta}f - Tf\|_Y \leq \lambda\|f\|_X$$

for every $f \in X$. Hence we have

$$\int\limits_0^\alpha |I_{\alpha,\beta}f(x)|^q v(x)dx + \int\limits_\eta^b |I_{\alpha,\beta}f(x)|^q v(x)dx \leq \lambda^q\|f\|_X^q \qquad (2.5.3)$$

for every $f \in X$.

Let us choose $n \in Z$ such that $2^n a < \alpha$. Assume that $j \in Z$, $j \leq n$ and $f_j(y) = \chi_{(0,2^j a)}(y)$. Then we obtain

$$\int\limits_{2^{j-1}a}^{2^j a} |I_{\alpha,\beta}f_j(x)|^q v(x)dx \geq$$

$$\geq \int\limits_{2^{j-1}a}^{2^j a} \left(\int\limits_{x/2}^x (x-y)^{\alpha-1}\ln^\beta \frac{\gamma}{x-y}f(y)dy \right)^q v(x)dx \geq$$

$$\geq c_2 \int\limits_{2^{j-1}a}^{2^j a} x^{(\alpha-1)q}\left(\ln^{\beta q}\frac{2\gamma}{x} \right)x^q v(x)dx.$$

On the other hand,

$$\|f_j\|_X^q = b(2^j a)^{q/p}$$

and according to (2.5.3) we find

$$c_3 R(j) \equiv c_3 \left(\int\limits_{2^{j-1}a}^{2^j a} x^{(\alpha-1)q}\left(\ln^{\beta q}\frac{2\gamma}{x} \right)x^{q/p'} v(x)dx \right)^{1/q} \leq \lambda$$

for every integer j, $j \leq n$. Consequently $\sup_{j \leq n} R(j) \leq c_4\lambda$ for every integer n satisfying the condition $2^n a < \alpha$. Therefore we have

$$\lim_{n \to -\infty} \sup_{j \leq n} R(j) \leq c_4\lambda.$$

Let $c \in (0,\alpha)$; then $c \in [2^{m-1}a, 2^m a)$ for some $m = m(c)$, $m \leq 0$. We have (see the proof of Theorem 2.5.5)

$$R_c \leq c_5 \sup_{n \leq m} R(n) \equiv c_5\overline{R}_m.$$

From the last inequality it follows that

$$\lim_{c \to 0+} R_c \le c_5 \lim_{m \to -\infty} \overline{R}_m \le c_6 \lambda,$$

where c_6 is independent of a. Finally, we obtain the lower estimate of (2.5.1) and the proof is complete. $\square$

2.6.   Erdelyi–Köber   operators

For measurable $f : [0, \infty) \to R^1$ let

$$J^\alpha_{\sigma,\eta} f(x) = x^{-\sigma(\alpha+\eta)} \int\limits_0^x (x^\sigma - y^\sigma)^{\alpha-1} y^{\sigma\eta+\sigma-1} f(y) dy,$$

where $\sigma > 0, 0 < \alpha < 1$.

Theorem 2.6.1. *Let* $1 < p \le q < \infty$, $1/p < \alpha < 1$. *Then the following statements are equivalent:*
(i) $J^\alpha_{\sigma,\eta}$ *is bounded from* $L^p(0, \infty)$ *into* $L^q_\nu(0, \infty)$;
(ii) $J^\alpha_{\sigma,\eta}$ *is bounded from* $L^p(0, \infty)$ *into* $L^{q\infty}_\nu(0, \infty)$;
(iii) $\eta > \frac{1}{\sigma p} - 1$ *and*

$$D \equiv \sup_{t>0} \left(\int\limits_{[t,\infty)} x^{-\sigma q(1+\eta)} d\nu(x) \right)^{1/q} t^{\sigma\eta+\sigma-1/p} < \infty;$$

(iv) $\eta > \frac{1}{\sigma p} - 1$ *and*

$$D_1 \equiv \sup_{k \in Z} \left(\int\limits_{[2^k, 2^{k+1})} x^{-q/p} d\nu(x) \right)^{1/q} < \infty.$$

Moreover, $\|J^\alpha_{\sigma,\eta}\|_{L^p \to L^q_\nu} \approx \|J^\alpha_{\sigma,\eta}\|_{L^p \to L^{q\infty}_\nu} \approx D \approx D_1$.
Proof. The fact that (iii) $\Rightarrow$ (i) can be proved in the same way as in Theorem 2.1.1. Now we prove that (ii) implies the condition $\eta > \frac{1}{\sigma p} - 1$. Indeed, let $\eta \le \frac{1}{\sigma p} - 1$. Then there exists a function $g \in L^p(0, t)$ such that

$$I(t) \equiv \int\limits_0^t g(y) y^{\sigma\eta+\sigma-1} dy = \infty.$$

Let $f_t(y) = g(y)\chi_{(0,t)}(y)$. Then $\|f\|_{L^p(R_+)} = \|g\|_{L^p(0,t)} < \infty$. On the other hand,

$$\|J^\alpha_{\sigma,\eta} f_t(x)\|_{L^{q\infty}_\nu(0,\infty)} \ge$$

$$\geq \|\chi_{[2t,\infty)}(\cdot)J^\alpha_{\sigma,\eta}f_t(\cdot)\|_{L^{q\infty}_\nu(0,\infty)} \geq$$

$$\geq c_1\|\chi_{[2t,\infty)}(x)x^{-\sigma(\eta+1)}I(t)\|_{L^{q\infty}_\nu(0,\infty)} = \infty.$$

Consequently the condition $\eta > 1/(\sigma p) - 1$ follows from (ii).
Now let $k \in Z$ and let $f_k(y) = \chi_{(2^{k-2},2^{k-1})}(y)$. Then we have

$$\|f_k\|_{L^p(0,\infty)} = c_2 2^{k/p}.$$

On the other hand,

$$\|J^\alpha_{\sigma,\eta}f_k(\cdot)\|_{L^{q\infty}_\nu(0,\infty)} \geq c_3\|\chi_{[2^k,2^{k+1})}(x)x^{-\sigma(\eta+1)}\|_{L^{q\infty}_\nu(0,\infty)} 2^{k(\sigma\eta+\sigma)} \geq$$

$$\geq c_4\left(\nu([2^k, 2^{k+1}))\right)^{1/q}.$$

From (ii) it follows that

$$\sup_{k\in Z}\left(\nu[2^k, 2^{k+1})\right)^{1/q}2^{-k/p}.$$

Hence $D_1 < \infty$.
The fact that (iv) $\Rightarrow$ (iii) follows in the same manner as in the previous sections. As (i) implies (ii), finally we have (iii) $\Rightarrow$ (i) $\Rightarrow$ (ii) $\Rightarrow$ (iv) $\Rightarrow$ (iii) $\square$

Theorem 2.6.2. *Let* $0 < q < p < \infty, p > 1$ *and* $\alpha > 1/p$. *Then the operator* $J^\alpha_{\sigma,\eta}$ *is bounded from* $L^p(0,\infty)$ *into* $L^q_\nu(0,\infty)$ *if and only if* $\eta > \frac{1}{\sigma p} - 1$ *and*

$$\overline{D} = \left(\int\limits_0^\infty \left(\int\limits_x^\infty v(y)y^{-\sigma q(1+\eta)}dy\right)^{\frac{p}{p-q}} x^{\frac{\sigma q p(\eta+1)-p}{p-q}}dx\right)^{\frac{p-q}{pq}}.$$

Moreover, $\|J^\alpha_{\sigma\eta}\| \approx \overline{D}$.
Proof. That the boundedness of $J^\alpha_{\sigma\eta}$ implies the condition $\eta > \frac{1}{\sigma p} - 1$ can be shown as in the proof of Theorem 3.2.1, and the remaing part of the theorem is proved as in Section 2.1. $\square$

Now we consider the compactness of $J^\alpha_{\sigma,\eta}$.

Theorem 2.6.3. *Let* $1 < p \leq q < \infty$ *and* $\alpha > 1/p$. *Then the following conditions are equivalent:*
(i) $J^\alpha_{\sigma,\eta}$ *is compact from* $L^p(0,\infty)$ *into* $L^q_\nu(0,\infty)$;
(ii) $J^\alpha_{\sigma,\eta}$ *is compact from* $L^p(0,\infty)$ *into* $L^{q\infty}_\nu(0,\infty)$;
(iii) $\eta > \frac{1}{\sigma p} - 1, D < \infty$ *and* $\lim\limits_{a\to 0} D^{(a)} = \lim\limits_{b\to\infty} D^{(b)} = 0$, *where*

$$D^{(a)} \equiv \sup_{0<t<a}\left(\int\limits_{[t,a)} x^{-\sigma q(1+\eta)}d\nu(x)\right)^{1/q} t^{\sigma\eta+\sigma-1/p};$$

$$D^{(b)} \equiv \sup_{t>b} \left(\int\limits_{[t,\infty)} x^{-\sigma q(1+\eta)} d\nu(x) \right)^{1/q} \times$$

$$\times \left(t^{(\sigma\eta+\sigma-1)p'+1} - b^{(\sigma\eta+\sigma-1)p'+1} \right)^{1/p'};$$

(iv) $\eta > \frac{1}{\sigma p} - 1$, $D < \infty$ and $\lim\limits_{t\to 0} D_t = \lim\limits_{t\to\infty} D_t = 0$, where

$$D_t \equiv \left(\int\limits_{[t,\infty)} x^{-\sigma q(1+\eta)} d\nu(x) \right)^{1/q} t^{\sigma\eta+\sigma-1/p};$$

(v) $\eta > \frac{1}{\sigma p} - 1$, $D_1 < \infty$ and $\lim\limits_{k\to -\infty} D_1(k) = \lim\limits_{k\to +\infty} D_1(k) = 0$, where

$$D_1(k) \equiv \sup_{k\in Z} \left(\int\limits_{[2^k,2^{k+1})} x^{-q/p} d\nu(x) \right)^{1/q}.$$

Proof. First we prove that (iii) $\Rightarrow$ (i). Represent $J^\alpha_{\sigma,\eta} f$ as follows:

$$J^\alpha_{\sigma,\eta} f = \chi_{[0,a)} J^\alpha_{\sigma,\eta}(f\chi_{[0,a)}) +$$
$$+\chi_{[a,b]} J^\alpha_{\sigma,\eta}(f\chi_{[0,b)}) + \chi_{(b,\infty)} J^\alpha_{\sigma,\eta}(f\chi_{[0,b/2)}) +$$
$$+\chi_{[b,\infty)} J^\alpha_{\sigma,\eta}(f\chi_{[b/2,\infty)}) \equiv P_1 f + P_2 f + P_3 f + P_4 f.$$

As in the proof of Theorem 2.1.5 it follows that P_2 and P_3 are compact. Moreover,

$$S \equiv \|J^\alpha_{\sigma,\eta} - P_2 - P_3\| \le b_1 D^{(a)} + b_2 D^{(b/2)}.$$

If (iii) holds, then $S \to 0$ and consequently $J^\alpha_{\sigma,\eta}$ is compact as it is a limit of compact operators. Hence (iii) $\Rightarrow$ (i).

From the inequalities:

$$D^{(a)} \le \sup_{0<t<a} D_t$$

and

$$D^{(b/2)} \le \sup_{t\ge b/2} D_t$$

we see that (iv) implies (iii). To prove that (i) $\Rightarrow$ (iv), we take $f_t(x) = \chi_{(0,t)}(x) t^{-1/p}$. Then $f_t \to 0$ weakly. In addition

$$\|J^\alpha_{\sigma,\eta} f_t\|_{L^q_\nu(0,\infty)} \ge c_1 \left(\int\limits_{t}^{\infty} x^{-\sigma(\alpha+\eta)q+\sigma(\alpha-1)q} \times \right.$$

$$\left. \times \left(\int\limits_{0}^{t} f(y) y^{\sigma\eta+\sigma-1} dy \right)^q d\nu(x) \right)^{1/q} \ge c_2 D_t \to 0$$

when $t \to 0$. To prove the equality $\lim\limits_{t\to\infty} D_t = 0$, we use the fact that $J^\alpha_{\sigma,\eta}$ is compact from $L^p(0,\infty)$ into $L^q_\nu(0,\infty)$ if and only if $\overline{J}^\alpha_{\sigma,\eta}$ is compact from $L^{q'}_\nu(0,\infty)$ into $L^{p'}(0,\infty)$, where

$$\overline{J}^\alpha_{\sigma,\eta}g(y) = y^{\sigma\eta+\sigma-1}\int\limits_{[y,\infty)} g(x)(x^\sigma - y^\sigma)^{\alpha-1}x^{-\sigma(\alpha+\eta)}d\nu(x).$$

Now let

$$g_t(y) = \chi_{[t,\infty)}(y)y^{-\sigma(\eta+1)(q-1)}\left(\int\limits_{[t,\infty)} x^{-\sigma q(1+\eta)}d\nu(x)\right)^{-1/q'}.$$

Then $g_t \to 0$ weakly as $t \to +\infty$. Indeed, let $\varphi \in L^{q'}(0,\infty)$. Then

$$\left|\int\limits_{[0,\infty)} g_t(y)\varphi(y)d\nu(y)\right| \le$$

$$\le \left(\int\limits_{[t,\infty)} |\varphi(x)|^q d\nu(x)\right)^{1/q} \times$$

$$\times \left(\int\limits_{[t,\infty)} |g_t(y)|^{q'} d\nu(y)\right)^{1/q'} =$$

$$= \left(\int\limits_{[t,\infty)} |\varphi(x)|^q d\nu(x)\right)^{1/q} \to 0$$

as $t \to \infty$. On the other hand,

$$\left(\int\limits_0^\infty (\overline{J}^\alpha_{\sigma,\eta}g(y))^{p'} dy\right)^{1/p'} \ge$$

$$\ge \left(\int\limits_0^t y^{(\sigma\eta+\sigma-1)p'}\left(\int\limits_{[y,\infty)} x^{-\sigma(1+\eta)(q-1)}\left(\int\limits_{[t,\infty)} z^{-\sigma q(1+\eta)}d\nu(z)\right)^{-1/q'} \times\right.\right.$$

$$\left.\left.\times (x^\sigma - y^\sigma)^{\alpha-1}x^{-\sigma(\alpha+\eta)}d\nu(x)\right)^{p'} dy\right)^{1/p'} \ge$$

$$\ge c_3\left(\int\limits_{[t,\infty)} x^{-\sigma q(1+\eta)}d\nu(x)\right)^{1/q} t^{\sigma\eta+\sigma-1/p} \to 0$$

as $t \to \infty$. Thus (i) $\Rightarrow$ (iv). Now we show that (ii) $\Rightarrow$ (v). Let $f_k(y) = \chi_{[2^{k-2},2^{k-1})}(y)2^{-k/p}$, where $k \in Z$. Then it is easy to verify that $f_k(y)$ converges weakly to 0, when $k \to +\infty$ or $k \to -\infty$. On the other hand,

$$\|J^\alpha_{\sigma,\eta} f_k(\cdot)\|_{L^{q\infty}_\nu(0,\infty)} \geq \|\chi_{[2^k,2^{k+1})}(\cdot) J^\alpha_{\sigma,\eta} f_k(\cdot)\|_{L^{q\infty}_\nu(0,\infty)} \geq$$

$$\geq c_4 \|\chi_{[2^k,2^{k+1})}(x) x^{-\sigma(\eta+1)}\|_{L^{q\infty}_\nu(0,\infty)} 2^{k(\sigma\eta+\sigma-1/p)} \geq$$

$$\geq c_5 (\nu[2^k, 2^{k+1}))^{1/q} 2^{-k/p} \geq c_6 D_1(k).$$

As f_k converges strongly to 0 as $k \to -\infty$ or $k \to +\infty$, we see that

$$\lim_{k\to-\infty} D_1(k) = \lim_{k\to+\infty} D_1(k) = 0.$$

That $\eta > 1/(\sigma p) - 1$ and $D_1 < \infty$ follows from Theorem 2.6.1.

To prove that (v) $\Rightarrow$ (iii), let $a > 0$, so that $a \in [2^m, 2^{m+1})$ for some integer m. Consequently

$$D^{(a)} \leq \sup_{0<t<2^m} D_{2^m,t} = D^{(2^m)},$$

where

$$D_{2^m,t} = \left(\int\limits_{[t,2^m)} x^{-\sigma q(1+\eta)} d\nu(x) \right)^{1/q} t^{\sigma\eta+\sigma-1/p}.$$

Now if $t \in [0, 2^m)$, then $t \in [2^{i-1}, 2^i)$ for some $i \leq m$. Further,

$$D^q_{2^m,t} \leq 2^{i(\sigma\eta q+\sigma q-q/p)} \left(\int\limits_{[2^{i-1},2^m)} x^{-\sigma q(1+\eta)} d\nu(x) \right) =$$

$$= 2^{i(\sigma\eta q+\sigma q-q/p)} \sum_{k=i}^{m} \int\limits_{[2^{k-1},2^k)} x^{-\sigma q(1+\eta)} d\nu(x) \leq$$

$$\leq c_7 \left(\sup_{k\leq m} D_1(k-1) \right)^q.$$

Thus $D^{(2^m)} \leq c_8 D_1^{(m)}$, where $D_1(m) = \sup_{k\leq m} D_1(k-1)$. If $a \to 0$, then $m \to -\infty$. Hence $D_1^{(m)} \to 0$. Therefore $\lim_{a\to 0} D^{(a)} = 0$. In the same way we can show that $\lim_{b\to+\infty} D^{(b)} = 0$. Thus (v) $\Rightarrow$ (iii).

As (i) $\Rightarrow$ (ii), finally we have: (i) $\Rightarrow$ (ii) $\Rightarrow$ (v) $\Rightarrow$ (iii) $\Rightarrow$ (i), (i) $\Leftrightarrow$ (iv). $\square$

We can establish the following theorem in a similar way. It can also be proved also from Ando's theorem (see Theorem D from Section 2.1).

Theorem 2.6.4. *Let* $0 < q < p < \infty$, $p > 1$ *and* $\alpha > 1/p$. *Then* $J^\alpha_{\sigma,\eta}$ *is compact from* $L^p(0,\infty)$ *into* $L^q_\nu(0,\infty)$ *if and only if* $\overline{D} < \infty$.

Our next aim is to give an estimate of the distance between $J^\alpha_{\sigma,\eta}$ and the space of compact operators.

We shall assume that v is a Lebesgue-measurable a.e. positive function on $(0, \infty)$.

Theorem 2.6.5. *Let $1 < p \le q < \infty$ and let $\alpha > 1/p$; suppose that $D < \infty$ for $d\nu = v(x)dx$ and assume that $X = L^p(0, \infty)$, $Y = L_v^q(0, \infty)$. Then*

$$b_1 J \le dist(R_\alpha, \mathcal{K}(X, Y)) \le b_2 J,$$

where $J = \lim_{a \to 0} J^a + \lim_{d \to \infty} J^{(d)}$,

$$J^{(a)} = \sup_{0 < t < a} \left(\int_t^a \frac{v(x)}{x^{(1+\eta)\sigma q}} dx \right)^{1/q} t^{\sigma\eta + \sigma - 1/p},$$

$$J^{(d)} = \sup_{t > d} \left(\int_t^\infty \frac{v(x)}{x^{(1+\eta)q\sigma}} dx \right)^{1/q} \left(t^{(\sigma\eta+\sigma-1)p'+1} - d^{(\sigma\eta+\sigma-1)p'+1} \right)^{1/p'},$$

and the positive constants b_1 and b_2 depend only on p, q, α, σ and η.

Proof. Let $\lambda > I$, where $I = dist(J_{\sigma,\eta}^\alpha, \mathcal{K}(X, Y))$. By Lemma 2.1.1 there exists $P \in \mathcal{F}_r(X, Y)$ such that

$$\| J_{\sigma,\eta}^\alpha - P \| < \lambda.$$

Using Lemma 2.1.2 we see that for $\epsilon = (\lambda - \| J_{\sigma,\eta}^\alpha - P \|)/2$ there are $T \in \mathcal{F}_r(X, Y)$ and $[\alpha, \beta] \subset (0, \infty)$ such that $\| T - P \| < \epsilon$ and $suppTf \subset [\alpha, \beta]$. Hence for all $f \in X$ we have

$$\| J_{\sigma,\eta}^\alpha f - Tf \|_{L_v^q(0,\infty)} \le \lambda \| f \|_{L^p(0,\infty)}.$$

Further

$$\int_0^\alpha v(x) |J_{\sigma,\eta}^\alpha f(x)|^q dx + \int_\beta^\infty v(x) |J_{\sigma,\eta}^\alpha f(x)|^q dx \le$$

$$\le \lambda^q \| f \|_{L^p(0,\infty)}^q.$$

Now let $d > \beta$ and $t \in (d, \infty)$. Suppose that $f_t(y) = \chi_{(0,t)}(y)$. Then $\| f_t \|_{L^p(0,\infty)} = t^{1/p}$. On the other hand,

$$\| J_{\sigma,\eta}^\alpha \|_{L_v^q(0,\infty)} \ge$$

$$\ge \left(\int_t^\infty v(x) \left(\int_0^t (x^\sigma - y^\sigma)^{\alpha-1} y^{\sigma\eta+\sigma-1} dy \right)^q x^{-\sigma q(\alpha+\eta)} dx \right)^{1/q} \ge$$

$$\ge c_1 \left(\int_t^\infty v(x) x^{-\sigma q(1+\eta)} dx \right)^{1/q} t^{\sigma\eta+\sigma}.$$

Consequently

$$c_1 \sup_{t>d} D(t) \le \lambda,$$

where

$$D(t) = \left(\int\limits_t^\infty x^{-\sigma q(1+\eta)} v(x) dx \right)^{1/q} t^{\sigma\eta + \sigma - 1/p}.$$

Hence

$$c_1 J^{(d)} \le \sup_{t>d} D(t) \le \lambda.$$

From the last inequality we see that

$$c_1 \lim_{d \to +\infty} J^{(d)} \le \lambda.$$

Since λ is arbitrarily close to I, we conclude that $c_1 \lim\limits_{d \to +\infty} J^{(d)} \le I$. Analogously we find that $c_2 \lim\limits_{a \to 0} J^{(a)} \le I$. Finally $b_1 J \le I$. The inequality $I \le b_2 J$ follows from the proof of Theorem 2.6.3. $\square$

2.7. Integral transforms with positive kernels

In the present section we find necessary and sufficient conditions for the boundedness and compactness of the operator

$$K(f)(x) = \int\limits_a^x k(x,y) f(y) dy$$

from $L^p(a,b)$ to $L^q_\nu(a,b)$ $(p,q \in (1,\infty)$ or $0 < q \le 1 < p < \infty$, $-\infty < a < b \le \infty$ and ν is a non–negative σ–finite Borel measure on $(a,b))$.

Analogous problems for the Riemann–Liouville type operator

$$R_\alpha f(x) = \int\limits_0^x \frac{f(y)}{(x-y)^{1-\alpha}} dy,$$

with $a = 0$, $b = +\infty$, $p,q \in (1,\infty)$ and $\alpha > 1/p$ were solved in Section 2.1. For boundedness and compactness criteria for operators with power–logarithmic kernels

$$I_{\alpha,\beta}(f)(y) = \int\limits_0^x (x-y)^{\alpha-1} \ln^\beta \frac{\gamma}{x-y} f(y) dy$$

with $0 < b \le \gamma < \infty$, $\alpha > 1/p$ and $\beta \ge 0$, see Section 2.3.

In the non-compact case we give the upper and the lower bounds for the distance of K from the subspace of compact operators from $L^p(a, b)$ to $L_v^q(a, b)$ when $1 < p \leq q < \infty$.

The following lemma is known for the case $a = 0$ and $b = \infty$ (see [195], Section 1.3), but we give the proof in the case $-\infty < a < b \leq +\infty$ for completeness.

Lemma 2.7.1. *Let $-\infty < a < b \leq +\infty$, $1 < p \leq q < \infty$ and let μ be a non-negative Borel measure on (a, b). The inequality*

$$\left(\int_{(a,b)} \Big| \int_a^x f(y) dy \Big|^q d\mu \right)^{\frac{1}{q}} \leq c \left(\int_a^b |f(y)|^p dy \right)^{\frac{1}{p}}, \qquad (2.7.1)$$

where the positive constant c does not depend on $f \in L^p(a, b)$, holds if and only if

$$A = \sup_{a < t < b} \left(\mu([t, b)) \right)^{\frac{1}{q}} (t - a)^{1/p'} < \infty,$$

where, $p' = p/(p - 1)$. Moreover, if c is the best constant in (2.7.1), then $A \leq c \leq 4A$.

Proof. Let $f \geq 0$, $f \in L^p(a, b)$ and let

$$\int_a^b f(y) dy \in (2^m, 2^{m+1}]$$

for some integer m. Put

$$\int_a^x f(y) dy \equiv I(x);$$

then for every $x \in (a, b)$ we have $I(x) \leq \|f\|_{L^p(a,b)} (x - a)^{1/p'} < \infty$. The function I is continuous on (a, b). Therefore for every $k \in Z$, with $k \leq m$, there exists t_k such that $2^k = I(t_k) = \int_{t_k}^{t_{k+1}} f(y) dy$ for $k \leq m - 1$ and $2^m = I(t_m)$.

It is easy to verify that the sequence $\{t_k\}$ is increasing. Let $\alpha = \lim_{k \to -\infty} t_k$. Then we have $(a, b) = (a, \alpha] \cup (\cup_{k \leq m} E_k)$, where $E_k = [t_k, t_{k+1})$ and $t_{m+1} = b$.

When $\int_a^b f(y) dy = \infty$ we have $(a, b) = (a, \alpha] \cup (\cup_{k=-\infty}^{+\infty} E_k)$ (i.e. $m = +\infty$).

If $t \in (a, \alpha)$, then $I(t) = 0$; and if $t \in E_k$, then $I(t) \leq I(t_{k+1}) \leq 2^{k+1}$.

We have

$$\left(\int\limits_{(a,b)} \Big(\int\limits_a^x f(y)dy \Big)^q d\mu \right)^{p/q} = \left(\sum_{k \le m} \int\limits_{E_k} (I(x))^q d\mu \right)^{p/q} \le$$

$$\le \sum_{k \le m} \left(\int\limits_{E_k} (I(x))^q d\mu \right)^{p/q} \le \sum_{k \le m} 2^{(k+1)p} \left(\int\limits_{E_k} d\mu \right)^{p/q} =$$

$$= 4^p \sum_{k \le m} 2^{(k-1)p} (\mu(E_k))^{p/q} = 4^p \sum_{k \le m} \left(\int\limits_{t_{k-1}}^{t_k} f(y)dy \right)^p (\mu(E_k))^{p/q} \le$$

$$\le 4^p \sum_{k \le m} \left(\int\limits_{t_{k-1}}^{t_k} (f(y))^p dy \right) (t_k - t_{k-1})^{p-1} (\mu(E_k))^{p/q} \le$$

$$\le 4^p A^p \|f\|^p_{L^p(a,b)}.$$

To prove necessity, we put $f(y) = \chi_{(a,t)}(y)$ in (2.7.1), where $t \in (a,b)$. Then we have $\|f\|_{L^p(a,b)} = (t - a)^{1/p}$. On the other hand,

$$\left(\int\limits_{(a,b)} \Big(\int\limits_a^x f(y)dy \Big)^q d\mu \right)^{1/q} \ge \Big(\mu([t,b)) \Big)^{1/q} (t - a)$$

and consequently we obtain $A \le c$. $\square$

We also need

Lemma 2.7.2. *Let* $-\infty < a < b \le +\infty$, $0 < q < p < \infty$ *and let* $p > 1$. *Then the inequality*

$$\left(\int\limits_a^b \Big| \int\limits_a^x f(y)dy \Big|^q v(x)dx \right)^{\frac{1}{q}} \le c \left(\int\limits_a^b |f(y)|^p dy \right)^{\frac{1}{p}}, \tag{2.7.2}$$

where the positive constant c *does not depend on* $f \in L^p(a,b)$, *is fulfilled if and only if*

$$\overline{A} = \left(\int\limits_a^b \Big(\int\limits_x^b v(t)dt \Big)^{\frac{p}{p-q}} (x - a)^{\frac{p(q-1)}{p-q}} dx \right)^{\frac{p-q}{pq}} < \infty.$$

Moreover, there exist positive constants c_1 *and* c_2 *depending only on* p *and* q *such that if* c *is the best constant in* (2.7.2), *then*

$$c_1 \overline{A} \le c \le c_2 \overline{A}.$$

This lemma can be proved in the same way as Lemma 1.3.2 of [195] (for the case $0 < q < 1 < p < \infty$, see for example [273]).

Definition 2.7.1. Let $-\infty < a < b \leq +\infty$. A kernel $k : \{(x,y) : a < y < x < b\} \to (0,\infty)$ belongs to V ($k \in V$) if there exists a positive constant d_1 such that for all x, y, z with $a < y < z < x < b$ the inequality

$$k(x,y) \leq d_1 k(x,z)$$

holds.

Definition 2.7.2 Let $-\infty < a < b \leq +\infty$. We say that k belongs to V_λ ($k \in V_\lambda$) ($1 < \lambda < \infty$) if there exists a positive constant d_2 such that for all x, $x \in (a,b)$, the inequality

$$\int\limits_{a+\frac{x-a}{2}}^{x} k^{\lambda'}(x,y)dy \leq d_2(x-a)k^{\lambda'}(x, a + \frac{x-a}{2}),$$

is fulfilled, where $\lambda' = \lambda/(\lambda - 1)$.

Let k_1 be a positive measurable function on $(0, b-a)$

Definition 2.7.3 Let $-\infty < a < b \leq +\infty$. We say that k_1 belongs to $V_{1\lambda}$ ($k_1 \in V_{1\lambda}$) ($1 < \lambda < \infty$) if there exists a positive constant d_3 such that the inequality

$$\int\limits_{0}^{\frac{x-a}{2}} k_1^{\lambda'}(y)dy \leq d_3(x-a)k_1^{\lambda'}(\frac{x-a}{2}), \quad \lambda' = \lambda/(\lambda - 1).$$

is fulfilled for all x, $x \in (a,b)$.

It is easy to verify that if k_1 is a nonincreasing function on $(0, b-a)$ and $k_1 \in V_{1\lambda}$, then the kernel $k(x,y) \equiv k_1(x-y)$ belongs to $V \cap V_\lambda$.

Now we give some examples of kernels satisfying the above-mentioned conditions.

Let $-\infty < a < b \leq +\infty$ and let $k_1(y) = y^{\alpha-1}$, where $\alpha > 0$. If $1 < \lambda < \infty$ and $\frac{1}{\lambda} < \alpha \leq 1$, then $k_1 \in V_{1\lambda}$ and consequently the kernel $k(x,y) \equiv k_1(x-y)$ belongs to $V \cap V_\lambda$.

Assume that $-\infty < a < b < +\infty, b-a \leq \gamma < \infty, 1/\lambda < \alpha \leq 1$ and $\beta \geq 0$. Let $k_1(y) = y^{\alpha-1}\ln^\beta \frac{\gamma}{y}$. Then $k_1 \in V_{1\lambda}$ and therefore $k(x,y) \equiv k_1(x-y)$ belongs to $V \cap V_\lambda$.

Now suppose that $-\infty < a < b \leq +\infty$, $k(x,y) = (x-y)^{\alpha-1}\ln^{\beta-1}\frac{x-a}{y-a}$, where $1/\lambda < \alpha \leq 1$ and $1 - \alpha + 1/\lambda < \beta \leq 1$. Then $k \in V \cap V_\lambda$

Let $a = 0, 0 < b \leq +\infty$ and let $k(x,y) = x^{-\sigma(\alpha+\eta)}(x^\sigma - y^\sigma)^{\alpha-1}y^{\sigma\eta+\sigma-1}$ be the Erdelyi–Köber kernel, where $\sigma > 0$ and $0 < \alpha \leq 1$. It easy to see that if $1/\lambda < \alpha \leq 1$ and $\eta > 1/\sigma - 1$, then $k \in V \cap V_\lambda$.

Some results about integral transforms with these kernels can be found in [255].

We turn to the compactness of the operators K and K'.

Theorem 2.7.1. *Let $-\infty < a < b \leq +\infty$. Suppose that $1 < p \leq q < \infty$ and $k \in V \cap V_p$. Then the operator K is bounded from $L^p(a,b)$ to $L^q_\nu(a,b)$ if and only if*

$$B \equiv \sup_{a<t<b} \left(\int\limits_{[t,b)} k^q(x, a + (x-a)/2)d\nu \right)^{\frac{1}{q}} (t-a)^{\frac{1}{p'}} < \infty.$$

Moreover, there exist positive constants b_1 and b_2 depending only on d_1, d_2, p and q such that the inequality

$$b_1 B \leq \|K\| \leq b_2 B$$

holds fulfilled (If the constants d_1 and d_2 from the Definitions 2.7.1 and 2.7.2 do not depend on a and b, then the constants b_1, b_2 are independent of a and b).

Proof. First we prove the theorem when $b = \infty$. Let $f \geq 0$. Then we have

$$\|Kf\|_{L^q_\nu(a,\infty)} \leq \left(\int\limits_{(a,\infty)} \left(\int\limits_a^{a+\frac{x-a}{2}} k(x,y)f(y)dy \right)^q d\nu \right)^{1/q} +$$

$$+ \left(\int\limits_{(a,\infty)} \left(\int\limits_{a+\frac{x-a}{2}}^x k(x,y)f(y)dy \right)^q d\nu \right)^{1/q} \equiv I_1 + I_2.$$

If $a < y < a + \frac{x-a}{2}$, then $k(x,y) \leq k(x, a + \frac{x-a}{2})$ and consequently, using Lemma 2.7.1, we obtain

$$I_1 \leq c_1 \left(\int\limits_{(a,\infty)} k^q(x, a + \frac{x-a}{2}) \left(\int\limits_a^x f(y)dy \right)^q d\nu \right)^{1/q} \leq$$

$$\leq c_2 B \|f\|_{L^p(a,\infty)}.$$

Using Hölder's inequality, the condition $k \in V_p$ and the notation $s_j \equiv a + 2^j$, we find that

$$I_2^q \leq \int\limits_{(a,\infty)} \left(\int\limits_{a+\frac{x-a}{2}}^x (f(y))^p dy \right)^{q/p} \left(\int\limits_{a+\frac{x-a}{2}}^x k^{p'}(x,y)dy \right)^{q/p'} d\nu \leq$$

$$\leq c_3 \int\limits_{(a,\infty)} \left(\int\limits_{a+\frac{x-a}{2}}^{x} (f(y))^p dy \right)^{q/p} \times$$

$$\times (x-a)^{q/p'} k^q(x, a+(x-a)/2) d\nu \leq$$

$$\leq c_3 \sum_{j\in Z} \int\limits_{[s_j, s_{j+1})} \left(\int\limits_{a+\frac{x-a}{2}}^{x} (f(y))^p dy \right)^{q/p} \times$$

$$\times (x-a)^{q/p'} k^q(x, a+(x-a)/2) d\nu \leq$$

$$\leq c_3 \sum_{j\in Z} \left(\int\limits_{s_{j-1}}^{s_{j+1}} (f(y))^p dy \right)^{q/p} \int\limits_{[s_j, s_{j+1})} \times$$

$$\times (x-a)^{q/p'} k^q(x, a+(x-a)/2) d\nu \leq$$

$$\leq c_4 B^q \sum_{j\in Z} \left(\int\limits_{s_{j-1}}^{s_{j+1}} (f(y))^p dy \right)^{q/p} \leq c_5 B^q \|f\|^q_{L^p(a,\infty)}$$

Now we prove necessity. First we show that from the boundedness of the operator K the following condition can be obtained:

$$\tilde{B} \equiv \sup_{j\in Z} \left(\int\limits_{[s_j, s_{j+1})} k^q(x, a+(x-a)/2)(x-a)^{q/p'} d\nu \right)^{\frac{1}{q}} < \infty. \quad (2.7.3)$$

Let $f_j(y) = \chi_{(a,s_{j+1})}(y)$, where $j \in Z$. Then we have that

$$\|Kf_j\|_{L^q_\nu(a,\infty)} \geq \left(\int\limits_{[s_j, s_{j+1})} (Kf(x))^q d\nu \right)^{1/q} \geq$$

$$\geq \left(\int\limits_{[s_j, s_{j+1})} \left(\int\limits_{a+\frac{x-a}{2}}^{x} f_j(y) k(x,y) dy \right)^q d\nu \right)^{1/q} \geq$$

$$\geq c_6 \left(\int\limits_{[s_j, s_{j+1})} k^q(x, a+(x-a)/2)(x-a)^q d\nu \right)^{1/q}.$$

Consequently, using the boundedness of K, we obtain $\tilde{B} < \infty$. Now we show that $B \leq c_7 \tilde{B}$. Denote

$$\left(\int\limits_{[t,\infty)} k^q(x, a+(x-a)/2) d\nu \right)^{\frac{1}{q}} (t-a)^{\frac{1}{p'}} \equiv B(t).$$

Let $t \in (a, \infty)$; then $t \in [s_m, s_{m+1})$ for some $m \in Z$.
We have

$$B^q(t) \leq \Big(\int\limits_{[s_m, \infty)} k^q(x, a + (x-a)/2) d\nu \Big) 2^{(m+1)q/p'} =$$

$$= c_8 2^{\frac{mq}{p'}} \sum_{j=m}^{+\infty} \int\limits_{[s_j, s_{j+1})} k^q(x, a + (x-a)/2) d\nu \leq$$

$$\leq c_9 \widetilde{B}^q 2^{\frac{mq}{p'}} \sum_{j=m}^{+\infty} 2^{\frac{-jq}{p'}} = c_{10} \widetilde{B}^q,$$

where c_{10} depends only on q and p.

The case $b \leq \infty$ can be proved analogously. In this case we take $s_j = a + (b-a)2^j$ (It is clear that $(a, b) = \cup_{j \leq 0}[s_{j-1}, s_j)$). $\square$

Remark 2.7.1. There exist positive constants a_1, a_2, a_3 and a_4 depending only on p and q such that

$$a_1 B \leq \widetilde{B} \leq a_2 B,$$

if $b = \infty$, where $\widetilde{B}$ is from (2.7.3) and

$$a_3 B \leq \overline{B} \leq a_4 B,$$

if $b < \infty$, where

$$\overline{B} = \sup_{j \leq 0} \Big(\int\limits_{[a+(b-a)2^{j-1}, \, a+(b-a)2^j)} k^q(x, a + (x-b)/2)(x-a)^{q/p'} d\nu \Big)^{\frac{1}{q}}.$$

Indeed, let $b = \infty$. Then the inequality $a_1 B \leq \widetilde{B}$ follows from the proof of Theorem 2.7.1. Moreover,

$$\Big(\int\limits_{[a+2^j, a+2^{j+1})} k^q(x, a + (x-a)/2)(x-a)^{q/p'} d\nu \Big)^{\frac{1}{q}} \leq$$

$$\leq c_1 \Big(\int\limits_{[a+2^j, a+2^{j+1})} k^q(x, a + (x-a)/2) d\nu \Big)^{\frac{1}{q}} 2^{j/p'} \leq c_1 B$$

for every $j \in Z$. Consequently $\widetilde{B} \leq a_2 B$, where a_2 depends only on p and q. We have an analogous result for $\overline{B}$.

Let g be a ν- measurable positive function on (a, b) and let

$$K'g(y) = \int\limits_y^b k(x, y)g(x)d\nu(x),$$

where $y \in (a, b)$.

From a duality arguments the next result follows.

Theorem 2.7.2. *Let* $-\infty < a < b \le +\infty$ *and let* $1 < p \le q < \infty$. *Suppose that* $k \in V \cap V_{q'}$. *Then the operator* K' *is bounded from* $L_v^p(a, b)$ *to* $L^q(a, b)$ *if and only if*

$$B' = \sup_{a < t < b} \Big(\int\limits_{[t,b)} k^{p'}(x, a + (x - a)/2) d\nu \Big)^{\frac{1}{p'}} (t - a)^{\frac{1}{q}} < \infty.$$

Moreover, there exist positive constants b_1 *and* b_2 *depending only on* d_1, d_2, p *and* q *such that*

$$b_1 B' \le \|K'\| \le b_2 B'.$$

Now we shall deal the case $q < p$. We shall assume that v and w are Lebesgue- measurable, a.e. positive functions on (a, b).

Theorem 2.7.3. *Let* $-\infty < a < b \le +\infty$, $0 < q < p < \infty$ *and let* $p > 1$. *Suppose that* $k \in V \cap V_p$. *Then the operator* K *is bounded from* $L^p(a, b)$ *to* $L_v^q(a, b)$ *if and only if*

$$B_1 = \Big(\int\limits_a^b \Big(\int\limits_x^b k^q(t, a + \frac{t - a}{2}) v(t) dt \Big)^{\frac{p}{p-q}} (x - a)^{\frac{p(q-1)}{p-q}} dx \Big)^{\frac{p-q}{pq}} < \infty.$$

Moreover, there exist positive constants b_1 *and* b_2 *such that*

$$b_1 B_1 \le \|K\| \le b_2 B_1.$$

Proof. We prove the theorem when $b = \infty$. The case $b < \infty$ can be proved similarly. Let $f \ge 0$. Then

$$\|Kf\|_{L_v^q(a,\infty)}^q \le c_1 \int\limits_a^\infty \Big(\int\limits_a^{a+\frac{x-a}{2}} f(y) k(x, y) dy \Big)^q v(x) dx +$$

$$+ c_1 \int\limits_a^\infty \Big(\int\limits_{a+\frac{x-a}{2}}^x f(y) k(x, y) dy \Big)^q v(x) dx = \overline{I}_1 + \overline{I}_2.$$

Using Lemma 2.7.2, we obtain $\overline{I}_1 \le c_2 B_1^q \|f\|_{L^p(a,\infty)}^q$, where c_2 depends only on p, q and d_1. By Hölder's inequality and the condition $k \in V_p$ we find that

$$\overline{I}_2 \le c_3 \int\limits_a^\infty \left(\int\limits_{a+\frac{x-a}{2}}^x (f(y))^p dy \right)^{\frac{q}{p}} \times$$

$$\times (x-a)^{\frac{q}{p'}} k^q(x, a + (x-a)/2) v(x) dx =$$

$$= c_3 \sum_{j\in Z} \int\limits_{s_j}^{s_{j+1}} \left(\int\limits_{a+\frac{x-a}{2}}^x (f(y))^p dy \right)^{\frac{q}{p}} \times$$

$$\times (x-a)^{\frac{q}{p'}} k^q(x, a + (x-a)/2) v(x) dx \le$$

$$\le c_3 \sum_{j\in Z} \left(\int\limits_{s_{j-1}}^{s_{j+1}} (f(y))^p dy \right)^{\frac{q}{p}} \int\limits_{s_j}^{s_{j+1}} (x-a)^{\frac{q}{p'}} k^q(x, a + (x-a)/2) v(x) dx,$$

where $s_j = a + 2^j$. Using Hölder's inequality again, we have

$$\overline{I}_2 \le c_3 \left(\sum_{j\in Z} \int\limits_{s_{j-1}}^{s_{j+1}} (f(y))^p dy \right)^{q/p} \times$$

$$\times \left(\sum_{j\in Z} \left(\int\limits_{s_j}^{s_{j+1}} (x-a)^{q/p'} k^q(x, a + (x-a)/2) v(x) dx \right)^{\frac{p}{p-q}} \right)^{\frac{p-q}{p}} \le$$

$$\le c_4 \overline{B}_1^q \|f\|_{L^p(a,\infty)}^q,$$

where

$$\overline{B}_1 \equiv \left(\sum_{j\in Z} \left(\int\limits_{s_j}^{s_{j+1}} (x-a)^{q/p'} k^q(x, a + (x-a)/2) v(x) dx \right)^{\frac{p}{p-q}} \right)^{\frac{p-q}{pq}}.$$

Moreover,

$$\overline{B}_1^{\frac{pq}{p-q}} \le c_5 \sum_{j\in Z} 2^{\frac{jq(p-1)}{p-q}} \left(\int\limits_{s_j}^{s_{j+1}} k^q(x, a + (x-a)/2) v(x) dx \right)^{\frac{p}{p-q}} \le$$

$$\le c_5 \sum_{j\in Z} \int\limits_{s_{j-1}}^{s_j} (y-a)^{\frac{p(q-1)}{p-q}} \left(\int\limits_y^{s_{j+1}} k^q(x, a + (x-a)/2) v(x) dx \right)^{\frac{p}{p-q}} dy \le$$

$$\le c_5 \int\limits_a^\infty (y-a)^{\frac{p(q-1)}{p-q}} \left(\int\limits_y^\infty k^q(x, a + (x-a)/2) v(x) dx \right)^{\frac{p}{p-q}} dy = c_5 B_1^{\frac{pq}{p-q}}.$$

Consequently $\bar{I}_2 \leq c_6 B_1^q \|f\|_{L^p(a,\infty)}^q$, where the positive constant c_6 depends only on d_2, p and q.

Now we prove necessity. Let the operator K be bounded from $L^p(a,\infty)$ to $L_v^q(a,\infty)$. If we repeat the arguments used in the proof of Theorem 2.7.1, then we see that for every $x \in (a,\infty)$

$$\int_x^\infty v(t)k^q(t, a+(t-a)/2)dt < \infty.$$

Let $v_n(t) = v(t)\chi_{(a+1/n,a+n)}(t)$, where $n \in Z$ and $n \geq 2$. Suppose that

$$f_n(x) = \left(\int_x^\infty v_n(t)k^q(t, a+\frac{t-a}{2})dt\right)^{\frac{1}{p-q}}(x-a)^{\frac{q-1}{p-q}}.$$

Then by integration by parts we obtain

$$\|f_n\|_{L^p(a,\infty)} =$$

$$= \left(\int_a^\infty \left(\int_x^\infty v_n(t)k^q(t, a+\frac{t-a}{2})dt\right)^{\frac{p}{p-q}}(x-a)^{\frac{(q-1)p}{p-q}}dx\right)^{1/p} =$$

$$= c_7 \left(\int_a^\infty \left(\int_x^\infty v_n(t)k^q(t, a+\frac{t-a}{2})dt\right)^{\frac{q}{p-q}} \times \right.$$

$$\left. \times (x-a)^{\frac{(p-1)q}{p-q}} v_n(x)k^q(x, a+(x-a)/2)dx\right)^{1/p} < \infty.$$

On the other hand,

$$\|Kf_n\|_{L_v^q(a,\infty)} \geq c_8 \left(\int_a^\infty v(x)\left(\int_{a+(x-a)/2}^x f_n(t)k(x,t)dt\right)^q dx\right)^{1/q} \geq$$

$$\geq c_9 \left(\int_a^\infty v_n(x)k^q(x, a+(x-a)/2) \times \right.$$

$$\times \left(\int_x^\infty v_n(t)k^q(x, a+(t-a)/2)dt\right)^{\frac{q}{p-q}} \times$$

$$\left. \times \left(\int_{a+(x-a)/2}^x (t-a)^{\frac{q-1}{p-q}}dt\right)^q dx\right)^{1/q} \geq$$

$$\geq c_{10}\left(\int_a^\infty v_n(x)k^q(x, a+(x-a)/2) \times \right.$$

$$\times \left(\int\limits_{x}^{\infty} v_n(t) k^q(t, a + (t-a)/2) dt \right)^{\frac{-q}{p-q}} (x-a)^{\frac{(p-1)q}{p-q}} dx \right)^{1/q} =$$

$$= c_{11} \left(\int\limits_{a}^{\infty} \left(\int\limits_{x}^{\infty} v_n(t) k^q(t, a + (t-a)/2) dt \right)^{\frac{p}{p-q}} (x-a)^{\frac{p(q-1)}{p-q}} dx \right)^{1/q}.$$

Since K is bounded,

$$\left(\int\limits_{a}^{\infty} \left(\int\limits_{x}^{\infty} k^q(t, a + (t-a)/2) v_n(t) dt \right)^{\frac{p}{p-q}} (x-a)^{\frac{p(q-1)}{p-q}} dx \right)^{\frac{p-q}{pq}} \leq c,$$

where the positive constant c does not depend on n. By Fatou's lemma we finally obtain $B_1 < \infty$. $\square$

Now let us consider the operator

$$\widetilde{K} f(x) = \int\limits_{x}^{b} f(y) k(y, x) w(y) dy,$$

where w is a Lebesgue-measurable a.e. positive function on (a, b). From duality arguments and Theorem 3.7.3 we can obtain

Theorem 2.7.4. *Let* $-\infty < a < b \leq +\infty$ *and let* $1 < q < p < \infty$. *Suppose that* $k \in V \cap V_{q'}$. *Then the operator* $\widetilde{K}$ *is bounded from* $L_w^p(a, b)$ *to* $L^q(a, b)$ *if and only if*

$$\widetilde{B}_1 = \left(\int\limits_{a}^{b} \left(\int\limits_{x}^{b} k^{p'}(t, (t-a)/2) w(t) dt \right)^{\frac{q(p-1)}{p-q}} (x-a)^{\frac{q}{p-q}} dx \right)^{\frac{p-q}{pq}} < \infty.$$

Moreover, there exist positive constants $\widetilde{b}_1$ *and* $\widetilde{b}_2$ *such that*

$$\widetilde{b}_1 \widetilde{B}_1 \leq \|\widetilde{K}\| \leq \widetilde{b}_2 \widetilde{B}_1.$$

Now we prove theorems concerning the compactness of K and K'. First we have

Theorem 2.7.5. *Let* $-\infty < a < b \leq +\infty$, $1 < p \leq q < \infty$ *and let* $k \in V \cap V_p$. *Then the following statements are equivalent:*
(i) *the operator* K *is compact from* $L^p(a, b)$ *to* $L_v^q(a, b)$;

(ii) $B < \infty$ and $\lim\limits_{c \to a+} B^{(c)} = \lim\limits_{d \to +\infty} B^{(d)} = 0$ for $b = \infty$; $B < \infty$ and $\lim\limits_{c \to a+} B^{(c)} = 0$ for $b < \infty$, where

$$B^{(c)} \equiv \sup_{a < t < c} \left(\int_{[t,c)} k^q(x, a + (x-a)/2) d\nu \right)^{\frac{1}{q}} (t-a)^{\frac{1}{p'}},$$

$$B^{(d)} \equiv \sup_{t > d} \left(\int_{[t,\infty)} k^q(x, a + (x-a)/2) d\nu \right)^{\frac{1}{q}} (t-a)^{\frac{1}{p'}};$$

(iii) $\overline{B} < \infty$ and $\lim\limits_{j \to -\infty} \overline{B}(j) = \lim\limits_{j \to +\infty} \overline{B}(j) = 0$ if $b = \infty$; $\overline{B} < \infty$ and $\lim\limits_{j \to -\infty} \overline{B}(j) = 0$ if $b < \infty$, where

$$\overline{B}(j) = \left(\int_{[s_j, s_{j+1})} k^q(x, a + (x-a)/2)(x-a)^{q/p'} d\nu \right)^{\frac{1}{q}},$$

$s_j = a + 2^j$ if $b = \infty$ and $s_j = a + (b-a)2^{j-1}$ if $b < \infty$.

Proof. Let us assume that $b = \infty$. First we prove that (ii) implies (i). Let $a < c < d < \infty$ and represent Kf as follows:

$$Kf = \chi_{(a,c)} K(f\chi_{(a,c)}) + \chi_{[c,d]} K(f\chi_{(a,d)}) +$$
$$+ \chi_{(d,\infty)} K(f\chi_{(a,(a+d)/2]}) + \chi_{(d,\infty)} K(f\chi_{((a+d)/2,\infty)}) \equiv$$
$$\equiv P_1 + P_2 + P_3 + P_4.$$

For P_2 we have

$$P_2 f(x) = \chi_{[c,d]}(x) \int_a^\infty T_1(x,y) dy,$$

where $T_1(x,y) = k(x,y)$ when $a < y < x < \infty$ and $T_1(x,y) = 0$ if $a < x \le y < \infty$. Consequently

$$S \equiv \int_{[c,d]} \left(\int_a^\infty (T_1(x,y))^{p'} dy \right)^{q/p'} d\nu =$$

$$= \int_{[c,d]} \left(\int_a^x (k(x,y))^{p'} dy \right)^{q/p'} d\nu \le$$

$$\le c_1 \int_{[c,d]} \left(\int_a^{a+\frac{x-a}{2}} (k(x,y))^{p'} dy \right)^{q/p'} d\nu +$$

$$+c_1 \int\limits_{[c,d]} \left(\int\limits_{a+\frac{x-a}{2}}^{x} (k(x,y))^{p'} dy \right)^{q/p'} d\nu \equiv S_1 + S_2.$$

If $a < y < a + \frac{x-a}{2}$, then $k(x,y) \le d_1 k(x, a + \frac{x-a}{2})$ and therefore

$$S_1 \le c_2 \int\limits_{[c,d]} k^q(x, a + \frac{x-a}{2}) \left(\frac{x-a}{2} \right)^{q/p'} d\nu \le$$

$$\le c_2 \left(\int\limits_{[c,d]} k^q(x, a + \frac{x-a}{2}) d\nu \right) \left(\frac{d-a}{2} \right)^{q/p'} < \infty.$$

Using the condition $k \in V_p$, for S_2 we obtain

$$S_2 \le c_3 \int\limits_{[c,d]} k^q(x, a + \frac{x-a}{2})(x-a)^{q/p'} d\nu < \infty.$$

Finally, we have $S < \infty$ and by Theorem C of Section 2.1 we conclude that P_2 is compact for every c and d. Analogously, we obtain that P_3 is compact. In addition, by virtue of Theorem 2.7.1 we have

$$\|P_1\| \le c_4 B^{(c)}; \quad \|P_4\| \le c_5 B^{(d)},$$

where the positive constants c_4 and c_5 do not depend on c and d. Consequently

$$\|K - P_2 - P_3\| \le c_4 B^{(c)} + c_5 B^{(d)} \to 0$$

as $c \to a$ and $d \to +\infty$. Hence the operator K is compact as it is a limit of compact operators.

To prove that (i) implies (iii), let $j \in Z$ and put $f_j(y) = \chi_{(s_{j-1}, s_{j+1})}(y) 2^{-j/p}$. Then for $\varphi \in L^{p'}(a, \infty)$ we have

$$\left| \int\limits_{a}^{\infty} f_j(y) \varphi(y) dy \right| \le$$

$$\le \left(\int\limits_{s_{j-1}}^{s_{j+1}} |f_j(y)|^p dy \right)^{1/p} \left(\int\limits_{s_{j-1}}^{s_{j+1}} |\varphi(y)|^{p'} dy \right)^{1/p'} =$$

$$= c_6 \left(\int\limits_{s_{j-1}}^{s_{j+1}} |\varphi(y)|^{p'} dy \right)^{1/p'} \to 0$$

as $j \to -\infty$ or $j \to +\infty$. On the other hand,

$$\|Kf_j\|_{L^q_\nu(a,\infty)} \geq \left(\int_{[s_j,s_{j+1})} (Kf_j(x))^q d\nu \right)^{1/q} \geq$$

$$\geq \left(\int_{[s_j,s_{j+1})} k^q(x,a + \frac{x-a}{2}) \left(\int_{a+\frac{x-a}{2}}^x f_j(y)dy \right)^q d\nu \right)^{1/q} \geq$$

$$\geq c_7 \left(\int_{[s_j,s_{j+1})} k^q(x,a + \frac{x-a}{2})(x-a)^q d\nu \right)^{1/q} 2^{-j/p} \geq c_8 \overline{B}(j).$$

As a compact operator maps a weakly convergent sequence into a strongly convergent one, we have that (iii) is satisfied. That $\overline{B} < \infty$ follows from Remark 2.7.1 and Theorem 2.7.1.

Now we prove that (ii) follows from (iii). Let $c > a$. Then there exists an integer m such that $c \in [s_m, s_{m+1})$ and

$$B^{(c)} \leq \sup_{a<t<s_{m+1}} \left(\int_{[t,s_{m+1})} k^q(x,a + (x-a)/2)d\nu \right)^{\frac{1}{q}} (t-a)^{\frac{1}{p'}} \equiv B_{s_m}.$$

Put

$$B_{s_m}(t) \equiv \left(\int_{[t,s_m)} k^q(x,a + (x-a)/2)d\nu \right)^{\frac{1}{q}} (t-a)^{\frac{1}{p'}}.$$

Let $t \in (a, s_m)$; then $t \in [s_{n-1}, s_n)$ for some integer $n \leq m$. We obtain

$$B^q_{s_m}(t) \leq \left(\int_{[s_{n-1},s_m)} k^q(x,a + (x-a)/2)d\nu \right) 2^{nq/p'} =$$

$$= 2^{nq/p'} \sum_{j=n}^{m} \int_{[s_{j-1},s_j)} k^q(x,a + (x-a)/2)d\nu \leq$$

$$\leq c_9 2^{nq/p'} \left(\sum_{j=n}^{m} 2^{-jq/p'} \right) \left(\sup_{j\leq m} \overline{B}(j-1) \right)^q \leq$$

$$\leq c_{10} \left(\sup_{j\leq m} \overline{B}(j-1) \right)^q \equiv c_{10} \overline{B}^q_m.$$

Thus

$$B_{s_m} \leq c_{11} \overline{B}_m.$$

If $c \to a$, then $s_m \to a$. Therefore $B^{(c)} \to 0$ as $\lim_{j \to -\infty} \overline{B}(j) = 0$. Analogously we obtain that $\lim_{d \to +\infty} B^{(d)} = 0$. The condition $B < \infty$ follows from Theorem 2.7.1. Hence (ii) $\Longrightarrow$ (i) $\Longrightarrow$ (iii) $\Longrightarrow$ (ii).

The case $b < \infty$ is proved in a similar manner. $\square$

From a duality argument we obtain

Theorem 2.7.6. *Let* $-\infty < a < b \leq +\infty$ *and let* $1 < p \leq q < \infty$. *Then the following statements are equivalent:*

(i) *the operator* K' *is compact from* $L_v^p(a,b)$ *to* $L^q(a,b)$;

(ii) $B' < \infty$ *and* $\lim_{c \to a+} (B')^{(c)} = \lim_{d \to +\infty} (B')^{(d)} = 0$ *if* $b + \infty$; $B' < \infty$ *and* $\lim_{c \to a+} (B')^{(c)} = 0$ *if* $b = \infty$, *where*

$$(B')^{(c)} \equiv \sup_{a<t<c} \left(\int\limits_{[t,c)} k^{p'}(x, a + (x-a)/2) d\nu \right)^{\frac{1}{p'}} (t-a)^{\frac{1}{q}};$$

$$(B')^{(d)} \equiv \sup_{t>d} \left(\int\limits_{[t,\infty)} k^{p'}(x, a + (x-a)/2) d\nu \right)^{\frac{1}{p'}} (t-a)^{\frac{1}{q}};$$

(iii) $\overline{B}' \equiv \sup_{j \in Z} \overline{B}'(j) < \infty$ *and* $\lim_{j \to -\infty} \overline{B}'(j) = \lim_{j \to +\infty} \overline{B}'(j) = 0$ *if* $b = \infty$; $\overline{B}' \equiv \sup_{j \leq 0} \overline{B}'(j) < \infty$ *and* $\lim_{j \to -\infty} \overline{B}'(j) = 0$ *if* $b < \infty$, *where*

$$\overline{B}'(j) \equiv \left(\int\limits_{[s_j, s_{j+1})} k^{p'}(x, a + (x-a)/2)(x-a)^{p'/q} d\nu \right)^{\frac{1}{p'}},$$

$s_j = a + 2^j$ *if* $b = \infty$ *and* $s_j = a + (b-a)2^{j-1}$ *if* $b < \infty$.

Theorem 2.7.7. *Let* $-\infty < a < b \leq +\infty$, $0 < q < p < \infty$ *and let* $p > 1$. *Suppose that* $k \in V \cap V_p$. *Then the operator* K *is compact from* $L^p(a,b)$ *to* $L_v^q(a,b)$ *if and only if* $B_1 < \infty$.

Proof. The sufficiency of the theorem can be derived in the same way as in the proof of Theorem 2.7.5 (It also follows from Theorem D of Section 2.1). Theorem 2.7.3 gives the necessity. $\square$

As an immediate consequence we have

Theorem 2.7.8. *Let* $-\infty < a < b \lesssim +\infty$ *and let* $1 < q < p < \infty$. *Suppose that* $k \in V \cap V_{q'}$. *Then the operator* $\widetilde{K}$ *is compact from* $L_w^p(a,b)$ *to* $L^q(a,b)$ *if and only if* $\widetilde{B}_1 < \infty$.

We shall assume that v is a Lebesgue-measurable a.e. positive function on (a,b), where $-\infty < a < b \leq +\infty$.

Theorem 2.7.9. *Let* $1 < p \leq q < \infty$, $-\infty < a < b \leq +\infty$ *and let* $k \in V \cap V_p$. *Assume that K is bounded from X to Y, where $X \equiv L^p(a,b)$ and $Y \equiv L^q_v(a,b)$. Then there exist positive constants b_1 and b_2 depending only on p, q, d_1 and d_2 such that the inequality*

$$b_1 J \leq dist(K, \mathcal{K}(X,Y)) \leq b_2 J \qquad (2.7.4)$$

holds, where $J = \lim\limits_{c \to a+} R^{(c)} + \lim\limits_{d \to +\infty} R^{(d)}$ *if $b = \infty$ and $J = \lim\limits_{c \to a+} R^{(c)}$ if* $b < \infty$,

$$R^{(c)} \equiv \sup_{a<t<c} \left(\int_t^c k^q(x, a + (x-a)/2)v(x)dx \right)^{\frac{1}{q}} (t-a)^{\frac{1}{p'}},$$

$$R^{(d)} \equiv \sup_{t>d} \left(\int_t^\infty k^q(x, a + (x-a)/2)v(x)dx \right)^{\frac{1}{q}} (t-a)^{\frac{1}{p'}}.$$

Proof. Let us prove the theorem when $b = \infty$. The case $b < \infty$ follows analogously. From the proof of Theorem 2.7.5 we see that

$$\|K - \overline{P}_1 - \overline{P}_2\| \leq c_1(R^{(c)} + R^{(d)}),$$

where $\overline{P}_1$ and $\overline{P}_2$ are compact operators for every c and d with the condition $a < c < d < \infty$. Consequently

$$dist(K, \mathcal{K}(X,Y)) \leq b_2 J,$$

where b_2 depends only on p, q, d_1 and d_2. To obtain the lower bound

$$dist(K, \mathcal{K}(X,Y)) \geq b_1 J,$$

let $\lambda > dist(K, \mathcal{K}(X,Y))$. Then by Lemma 2.1.1 there exists $P \in \mathcal{F}_r(X,Y)$ such that $\|K - P\| < \lambda$. On the other hand, using Lemma 2.1.2, for $\epsilon = (\lambda - \|K - P\|)/2$ there exist $T \in \mathcal{F}_r(X,Y)$ and $[\alpha, \beta] \subset (a,b)$ such that

$$\|P - T\| < \epsilon \qquad (2.7.5)$$

and

$$supp\, Tf \subset [\alpha, \beta].$$

From (2.7.5) we obtain

$$\|Kf - Tf\|_Y \leq \lambda \|f\|_X$$

for every $f \in X$. Thus

$$\int\limits_a^\alpha |Kf(x)|^q v(x)dx + \int\limits_\beta^b |Kf(x)|^q v(x)dx \leq \lambda^q \|f\|_X^q \qquad (2.7.6)$$

for every $f \in X$.

Let us choose $n \in Z$ such that $a + 2^n < \alpha$. Assume that $j \in Z, j \leq n$ and $f_j(y) = \chi_{(s_{j-1}, s_{j+1})}(y)$. Then

$$\int\limits_{s_j}^{s_{j+1}} |Kf_j(x)|^q v(x)dx \geq \int\limits_{s_j}^{s_{j+1}} \left(\int\limits_{a+(x-a)/2}^x k(x,y)f_j(y)dy \right)^q v(x)dx \geq$$

$$\geq c_2 \int\limits_{s_j}^{s_{j+1}} k^q(x, a + (x-a)/2)(x-a)^q v(x)dx.$$

On the other hand, $\|f_j\|_X^q = c_3 2^{jq/p}$ and by (2.7.6) we find that

$$c_3 R(j) \equiv c_3 \left(\int\limits_{s_j}^{s_{j+1}} k^q(x, a + \frac{x-a}{2})(x-a)^{q/p'} v(x)dx \right)^{1/q} \leq \lambda$$

for every integer $j, j \leq n$. Hence $\sup_{j \leq n} R(j) \leq c_4 \lambda$ for every integer n with the condition $s_n < \alpha$. Therefore $\lim\limits_{n \to -\infty} \sup_{j \leq n} R(j) \leq c_4 \lambda$.

Let $c \in (a, \alpha)$; then $c \in [s_{m-1}, s_m)$ for some $m = m(c), m \in Z$. We obtain (see the proof of Theorem 2.7.5)

$$R^{(c)} \leq c_5 \sup_{n \leq m} R(n).$$

From the last inequality we have

$$\lim\limits_{c \to a+} R_c \leq c_5 \overline{\lim}_{m \to -\infty} R_m \leq c_6 \lambda,$$

where c_6 does not depend on a and b.

Now we take $m, \in Z$ such that $s_m > \beta$. Then for $f_j(y) = \chi_{(0, s_{j+1})}(y)$ $(j \geq m)$ we obtain

$$\int\limits_{s_j}^{s_{j+1}} |Kf_j(x)|^q v(x)dx \geq c_7 \int\limits_{s_j}^{s_{j+1}} k^q(x, a + (x-a)/2)(x-a)^q v(x)dx.$$

On the other hand, $\|f_j\|_X^q = c_8 2^{jq/p}$. Hence $\sup_{j \geq m} R(j) \leq c_9 \lambda$, where c_9 depends only on p, q and d_1. Consequently $\lim\limits_{m \to +\infty} \sup_{j \geq m} R(j) \leq c_9 \lambda$.

Further, it is easy to verify that

$$\lim_{d \to +\infty} R^{(d)} \le c_{10} \lim_{m \to +\infty} \sup_{j \ge m} R(j) \le c_{11}\lambda,$$

where c_{11} depends only on p, q and d_1. As λ is an arbitrary number greater than $dist(K, \mathcal{K}(X, Y))$, finally we obtain (2.7.4). $\square$

Remark 2.7.2. It follows from the proof of Theorem 2.7.9 that there exist positive constants a_1 and a_2, depending only on p, q d_1 and d_2, such that the inequality

$$a_1 I \le dist(K, \mathcal{K}(X, Y)) \le a_2 I$$

holds, where $I = \overline{\lim}_{j \to +\infty} R(j) + \overline{\lim}_{j \to -\infty} R(j)$ if $b = \infty$, $I = \overline{\lim}_{j \to -\infty} R(j)$ if $b < \infty$ and

$$R(j) \equiv \left(\int_{s_j}^{s_{j+1}} k^q(x, a + \frac{x-a}{2})(x-a)^{q/p'} v(x)dx \right)^{1/q} \le \lambda$$

($s_j = a + 2^j$ for $b = \infty$ and $s_j = a + (b-a)2^{j-1}$ for $b < \infty$).

Let $\omega_\lambda(x) \equiv e^{\lambda x}$ ($x \in R$). We define operators of Riemann-Liouville and Weyl type on R in the following way :

$$\mathcal{R}_{\alpha,\omega_\lambda} f(x) = \int_{-\infty}^{x} (x-y)^{\alpha-1} f(y)\omega_\lambda(y)dy,$$

$$\mathcal{W}_{\alpha,\omega_\alpha} f(x) = \int_{x}^{+\infty} (y-x)^{\alpha-1} f(y)\omega_\lambda(y)dy,$$

where $\alpha > 0$ and $\lambda \in R$.

We are going to establish boundedness and compactness criteria for $\mathcal{R}_{\alpha,\omega_\lambda}$ and $\mathcal{W}_{\alpha,\omega_\lambda}$ in weighted Lebesgue spaces. First we have

Theorem 2.7.10. *Let* $\lambda > 0$, $1 < p \le q < \infty$ *and let* $1/p < \alpha < 1$. *We assume that* $\lambda > 0$. *Then for the boundedness of* $\mathcal{R}_{\alpha,\omega_\lambda}$ *from* $L^p_{\omega_\lambda}(R)$ *into* $L^q_{v \cdot \omega_\lambda}(R)$ *it is necessary and sufficient that*

$$B_\lambda \equiv \sup_{r \in R} B_\lambda(t) \equiv \sup_{r \in R} \left(\int_{r}^{+\infty} v(x)\omega_\lambda(x)dx \right)^{1/q} (\omega_\lambda(r))^{1/p'} < \infty.$$

Moreover, $\|\mathcal{R}_{\alpha,\omega_\lambda}\| \approx B_\lambda$.

Proof. Using the change of variable : $x = \frac{1}{\lambda}\ln t$, we easily obtain that $\mathcal{R}_{\alpha,\omega_\lambda}$ is bounded from $L^p_{\omega_\lambda}(R)$ into $L^q_{v\cdot\omega_\lambda}(R)$ if and only if the operator

$$A_\alpha f(x) = \lambda^{-\alpha} \int\limits_0^x (\ln \frac{x}{y})^{\alpha-1} f(y) dy$$

is bounded from $L^p(0,\infty)$ into $L^q_{\bar{v}}(0,\infty)$, where $\bar{v}(x) \equiv v(\frac{1}{\lambda}\ln x)$. On the other hand, the boundedness of A_α is equivalent to the condition (see Theorem 2.7.1)

$$\bar{B}_\lambda \equiv \sup_{t>0} \bar{B}_\lambda \equiv \sup_{t>0} \left(\int\limits_t^\infty \bar{v}(x) dx \right)^{\frac{1}{q}} t^{\frac{1}{p'}} < \infty.$$

Using again a change of variable, we finally see that the condition $B_\lambda < \infty$ is equivalent to the condition $\bar{B}_\lambda < \infty$. $\square$

In a similar way the next result can be established.

Theorem 2.7.11. *Let* λ, p, q *and* α *satisfy the conditions of Theorem* 2.7.10. *Then* $\mathcal{R}_{\alpha,\omega_\lambda}$ *is compact from* $L^p_{\omega_\lambda}(R)$ *into* $L^q_{v\omega_\lambda}(R)$ *if and only if* $B_\lambda < \infty$ *and* $\lim\limits_{t\to-\infty} B_\lambda(t) = \lim\limits_{t\to+\infty} B_\lambda(t) = 0$.

Repeating the arguments used in the proof of Theorem 2.7.10 we easily obtain the following result for $\mathcal{W}_{\alpha,\omega_\lambda}$:

Theorem 2.7.12. *Let* $\lambda < 0$, $1 < p \le q < \infty$ *and let* $\alpha > 1/p$. *Then the following statements hold:*

(a) $\mathcal{W}_{\alpha,\omega_\lambda}$ *is bounded from* $L^p_{\omega_\lambda}(R)$ *into* $L^q_{v\omega_\lambda}(R)$ *if and only if*

$$D_\lambda \equiv \sup_{r\in R} D_\lambda(r) \equiv \sup_{r\in R} \left(\int\limits_{-\infty}^r v(x)\omega_\lambda(x) dx \right)^{1/q} \omega_\lambda(r)^{1/p'} < \infty;$$

(b) $\mathcal{W}_{\alpha,\omega_\lambda}$ *is compact from* $L^p_{\omega_\lambda}(R)$ *into* $L^q_{v\cdot\omega_\lambda}(R)$ *if and only if* $D_\lambda < \infty$ and $\lim\limits_{t\to-\infty} D_\lambda(t) = \lim\limits_{t\to+\infty} D_\lambda(t) = 0$.

Examples of appropriate weights satisfying the conditions above are given below.

Example 2.7.1. Let $1 < p \le q < \infty$, $\lambda > 0$ and $\eta = -\lambda(1 + \frac{q}{p'})$. Then the condition $B_\lambda < \infty$ is satisfied for the weight $v(x) = \omega_\eta(x)$. Thus $\mathcal{W}_{\alpha,\omega_\lambda}$ is bounded from $L^p_{\omega_\lambda}(R)$ into $L^q_{\omega_{\eta+\lambda}}(R)$.

Example 2.7.2. let $1 < p \le q < \infty$ and let $\lambda < 0$. We assume that $\eta = -\lambda(1 + \frac{q}{p'})$. If $v(x) = \omega_\eta(x)$, then the condition $D_\lambda < \infty$ is satisfied and using Theorem 2.7.12 we have that $\mathcal{W}_{\alpha,\omega_\lambda}$ is bounded from $L^p_{\omega_\lambda}(R)$ into $L^q_{v\cdot\omega_{\eta+\lambda}}(R)$.

Two- sided estimates of the distance between the operator $\mathcal{R}_{\alpha,\omega_\lambda}$ and the class of all compact linear operators from $L^p_{\omega_\lambda}(R)$ into $L^q_{v\omega_\lambda}(R)$ ($\mathcal{K}(L^p_{\omega_\lambda}(R)$, $L^q_{v\omega_\lambda}(R))$) are provided in

Theorem 2.7.13. *Let $1 < p \le q < \infty$ and $\lambda > 0$. Suppose that $\alpha \in (\frac{1}{p}, 1)$ and that $\mathcal{R}_{\alpha,\omega_\lambda}$ is bounded from $L^p_{\omega_\lambda}(R)$ into $L^q_{v\omega_\lambda}(R)$. Then there exist positive constants b_1 and b_2, depending only on p, q, α and λ, such that*

$$b_1 J \le dist(\mathcal{R}_{\alpha,\omega_\lambda}, \mathcal{K}(L^p_{\omega_\lambda}(R), L^q_{v\omega_\lambda}(R))) \le b_2 J,$$

where $J = \lim\limits_{c \to -\infty} J_c + \lim\limits_{d \to +\infty} J^{(d)}$ and

$$J_c \equiv \sup_{-\infty < r < c} \left(\int_r^c v(x)\omega_\lambda(x)dx \right)^{1/q} (\omega_\lambda(r))^{1/p'},$$

$$J^{(d)} \equiv \sup_{r>d} \left(\int_r^\infty v(x)\omega_\lambda(x)dx \right)^{1/q} (\omega_\lambda(r) - \omega_\lambda(d))^{1/p'}.$$

Proof. Let us introduce the following notation: $\bar{g}(x) \equiv g(\frac{1}{\lambda}\ln x)$, $\widetilde{\varphi}(x) \equiv \varphi(\omega_\lambda(x))$, where g and φ are measurable functions respectively on R_+ and R. Let B be a linear operator from $L^p_{\omega_\lambda}(R)$ into $L^q_{v\cdot\omega_\lambda}(R)$. For B we define an operator $(Af)(x) \equiv (\overline{B\widetilde{f}})(x)$. It is easy to verify that B acts boundedly (compactly) from $L^p_{\omega_\lambda}(R)$ into $L^q_{v\cdot\omega_\lambda}(R)$ if and only if A is bounded (compact) from $L^p(R_+)$ into $L^q_{\bar{v}}(R_+)$. In addition, it is easy to see that

$$dist\,(\mathcal{R}_{\alpha,\omega_\lambda}, \mathcal{K}(L^p_{\omega_\lambda}(R), L^q_{v\cdot\omega_\lambda}(R))) \approx$$
$$\approx dist(A_\alpha, \mathcal{K}(L^p(R_+), L^q_{\bar{v}}(R_+))) \equiv S,$$

where A_α is from the proof of Theorem 2.7.10.

By Theorem 2.7.9 we see that $S \approx \bar{J}$, where $\bar{J} = \lim\limits_{a \to 0} \bar{J}_a + \lim\limits_{b \to \infty} \bar{J}^{(b)}$,

$$\bar{J}_a \equiv \sup_{0 < t < a} \left(\int_t^a \bar{v}(x)dx \right)^{1/q} t^{1/p'},$$

$$\bar{J}^{(b)} \equiv \sup_{t>b} \left(\int_t^\infty \bar{v}(x)dx \right)^{1/q} (t - b)^{1/p'}.$$

Using a change of variable, we easily derive the following equalities:

$$\bar{J}_a = c_3 J_c \quad \text{and} \quad \bar{J}^{(b)} = c_4 J^{(d)},$$

where $c = \frac{1}{\lambda}\ln a$, $d = \frac{1}{\lambda}\ln b$. The theorem is proved. $\square$

For $q < p$ we have the next theorem which follows in the same manner as Theorem 2.7.10.

Theorem 2.7.14. *Let* $0 < q < \infty$, $1 < p < \infty$ *and* $q < p$. *We assume that* $\lambda > 0$ *and* $\frac{1}{p} < \alpha < 1$. *The following statements hold:*
(a) *If* $\lambda > 0$, *then the following three conditions are equivalent:*

(i) $\mathcal{R}_{\alpha,\omega_\lambda}$ is bounded from $L^p_{\omega_\lambda}(R)$ into $L^q_{v\cdot\omega_\lambda}(R)$;
(ii) $\mathcal{R}_{\alpha,\omega_\lambda}$ is compact from $L^p_{\omega_\lambda}(R)$ into $L^q_{v\cdot\omega_\lambda}(R)$;
(iii)

$$A \equiv \left(\int\limits_{-\infty}^{+\infty} \left(\int\limits_{x}^{+\infty} v(y)\omega_\lambda(y)dy \right)^{\frac{p}{p-q}} (\omega_\lambda(x))^{\frac{q(p-1)}{p-q}} dx \right)^{\frac{p-q}{pq}} < \infty.$$

Moreover, $\|\mathcal{R}_{\alpha,\omega_\lambda}\| \approx A$.
(b) *If* $\lambda < 0$, *then then the following conditions are equivalent:*

(i) $\mathcal{W}_{\alpha,\omega_\lambda}$ is bounded from $L^p_{\omega_\lambda}(R)$ into $L^q_{v\cdot\omega_\lambda}(R)$;
(ii) $\mathcal{W}_{\alpha,\omega_\lambda}$ is compact from $L^p_{\omega_\lambda}(R)$ into $L^q_{v\cdot\omega_\lambda}(R)$;
(iii)

$$C \equiv \left(\int\limits_{-\infty}^{+\infty} \left(\int\limits_{-\infty}^{x} v(y)\omega_\lambda(y)dy \right)^{\frac{p}{p-q}} (\omega_\lambda(x))^{\frac{q(p-1)}{p-q}} dx \right)^{\frac{p-q}{pq}} < \infty.$$

Moreover, $\|\mathcal{W}_{\alpha,\omega_\lambda}\| \approx C$.

From Theorems 2.7.10, 2.7.11 and 2.7.15 we obtain the next corollaries for the Riemann-Liouville operator

$$\mathcal{R}_\alpha f(x) = \int\limits_{-\infty}^{x} f(y)(x-y)^{\alpha-1}dy, \quad \alpha > 0$$

(as before we use the following notation: $\omega_\mu(x) = e^{\mu x}$, where $\mu \in R$):

Corollary 2.7.1. *Let* $\mu < 0$, $1 < p < \infty$, $0 < q < \infty$ *and* $\alpha \in (\frac{1}{p}, 1)$. *Then the following criteria hold:*
(a) *If* $p \le q$, *then* $\mathcal{R}_\alpha$ *is bounded from* $L^p_{\omega_\mu}(R)$ *into* $L^q_v(R)$ *if and only if*

$$F \equiv \sup_{t\in R} R(t) \equiv \sup_{t\in R} \left(\int\limits_{t}^{+\infty} v(x)dx \right)^{\frac{1}{q}} \omega_\mu^{-\frac{1}{p}}(t) < \infty;$$

(b) *If* $q < p$, *then* $\mathcal{R}_\alpha$ *is bounded from* $L^p_{w_\mu}(R)$ *into* $L^q_v(R)$ *if and only if*

$$G \equiv \left(\int\limits_{-\infty}^{+\infty} \left(\int\limits_{x}^{+\infty} v(t)dt \right)^{\frac{p}{p-q}} \omega_\mu^{-\frac{p'(q-1)}{p-q}} (x)dx \right)^{\frac{p-q}{pq}} < \infty.$$

Moreover, $\|\mathcal{R}_\alpha\| \approx F$ for $p \le q$ and $\|\mathcal{R}_\alpha\| \approx G$ for $q < p$.

Corollary 2.7.2. *Let $\mu < 0$, $1 < p < \infty$, $0 < q < \infty$ and $\alpha \in (\frac{1}{p}, 1)$. Then the following criteria hold:*

(a) If $p \le q$, then $\mathcal{R}_\alpha$ acts compactly from $L^p_{\omega_\lambda}(R)$ into $L^q_v(R)$ if and only if $F < \infty$ and $\lim_{t \to -\infty} F(t) = \lim_{t \to +\infty} F(t) = 0$;

(b) If $q < p$, then $\mathcal{R}_\alpha$ is compact from $L^p_{\omega_\lambda}(R)$ into $L^q_v(R)$ if and only if $G < \infty$.

Let

$$\mathcal{W}_\alpha f(x) = \int_x^{+\infty} f(y)(y - x)^{\alpha-1} dy, \quad \alpha > 0$$

be the Weyl-type transform. Theorems 2.7.12 and 2.7.16 yield the next corollaries:

Corollary 2.7.3. *Let $\mu > 0$, $1 < p < \infty$, $0 < q < \infty$ and $\alpha \in (\frac{1}{p}, 1)$. Then $\mathcal{W}_\alpha$ acts boundedly from $L^p_{\omega_\mu}(R)$ into $L^q_v(R)$ if and only if*

$$F_1 \equiv \sup_{t \in R} F_1(t) \equiv \sup_{t \in R} \left(\int_{-\infty}^t v(x)dx \right)^{\frac{1}{q}} \omega_\mu^{-\frac{1}{p}}(t) < \infty$$

if $p \le q$ and

$$G_1 \equiv \left(\int_{-\infty}^{+\infty} \left(\int_{-\infty}^x v(t)dt \right)^{\frac{p}{p-q}} (\omega_\mu(x))^{-\frac{p'(q-1)}{p-q}} dx \right)^{\frac{p-q}{pq}} < \infty$$

if $q < p$.

Moreover, $\|\mathcal{W}_\alpha\| \approx F_1$ for $p \le q$ and $\|\mathcal{W}_\alpha\| \approx G_1$ for $q < p$.

Corollary 2.7.4. *Let $\mu > 0$, $1 < p < \infty$, $0 < q < \infty$ and $\alpha \in (\frac{1}{p}, 1)$. Then $\mathcal{W}_\alpha$ is compact from $L^p_{\omega_\mu}(R)$ into $L^q_v(R)$ if and only if (i) $F_1 < \infty$ and $\lim_{t \to -\infty} F_1(t) = \lim_{t \to +\infty} F_1(t) = 0$ for $p \le q$; (ii) $G_1 < \infty$ for $q < p$.*

From Theorem 2.7.13 we have the next statement concerning the measure of non–compactness for $\mathcal{R}_\alpha$.

Corollary 2.7.5. *Let $1 < p \le q < \infty$, $\alpha \in (\frac{1}{p}, 1)$ and $\mu < 0$. We assume that $\mathcal{R}_\alpha$ is bounded from $L^p_{\omega_\mu}(R)$ into $L^q_v(R)$. Then there exist positive constants c_1 and c_2 depending only on α, p, q and μ such that*

$$c_1 S \le dist(\mathcal{R}_\alpha, \mathcal{K}(L^p_{\omega_\mu}(R), L^q_v(R))) \le c_2 S,$$

where $S = \lim_{a \to -\infty} S_a + \lim_{b \to +\infty} S^{(b)}$,

$$S_a \equiv \sup_{-\infty < t < a} \left(\int_t^a v(x)dx \right)^{1/q} (\omega_\mu(t))^{-1/p},$$

$$S^{(b)} \equiv \sup_{t > b} \left(\int_t^\infty v(x)dx \right)^{1/q} (\omega_\mu^{1-p'}(t) - \omega_\mu^{1-p'}(b))^{1/p'}.$$

Now we consider the case $0 < \alpha < 1/p$.

Lemma 2.7.3. *Let* $1 < p < \infty$ *and* $u(x) \equiv \frac{1}{x}$. *Then the operator*

$$L_\alpha f(x) = \int_0^x \left(\ln \frac{x}{y} \right)^{\alpha-1} f(y) \frac{dy}{y}$$

is bounded from $L_u^p(R_+)$ *into* $L_u^{p^*}(R_+)$ *if and only if* $0 < \alpha < \frac{1}{p}$ *and* $p^* = \frac{p}{1-\alpha p}$.

Proof. It is easy to see that L_α is bounded from $L_u^p(R_+)$ into $L_u^{p^*}(R_+)$ if and only if $\mathcal{R}_\alpha$ is bounded from $L^p(R)$ into $L^{p^*}(R)$. This follows by using a change of variable. The latter is equivalent to the conditions: $\alpha \in (0, 1/p)$ and $p^* = \frac{p}{1-\alpha p}$ (see, e.g., [119]). $\square$

Corollary 2.7.6. *Let* $1 < p < \infty$, $w(x) = x^{p-1}$ *and* $u(x) = \frac{1}{x}$. *Then the operator*

$$A_\alpha f(x) = \int_0^x \left(\ln \frac{x}{y} \right)^{\alpha-1} f(y) dy$$

is bounded from $L_w^p(R_+)$ *into* $L_u^p(R_+)$ *if and only if* $\alpha \in (0, \frac{1}{p})$ *and* $p^* = \frac{p}{1-\alpha p}$.

As before we shall use the notation: $\omega_\lambda(x) \equiv e^{\lambda x}$, where $\lambda \in R$. The following statement holds:

Theorem 2.7.15. *Let* $1 < p < \infty$, $\lambda > 0$ *and let* v *be a monotone function on* R. *We assume that* $\alpha \in (0, \frac{1}{p})$. *Then the operator* $\mathcal{R}_{\alpha, \omega_\lambda}$ *is bounded from* $L_{\omega_\lambda}^p(R)$ *into* $L_{v \cdot \omega_\lambda}^p(R)$ *if and only if*

$$B \equiv \sup_{r \in R} \left(\int_r^{+\infty} v(x) e^{\lambda x} dx \right)^{1/p} (\omega_\lambda(r))^{1/p'} < \infty.$$

Proof. Using a change of variable $x = \frac{1}{\lambda} \ln y$ we easily obtain that $\mathcal{R}_{\alpha, \omega_\lambda}$ is bounded from $L_{\omega_\lambda}^p(R)$ into $L_{v \cdot \omega_\lambda}^p(R)$ if and only if A_α is bounded from

$L^p(R_+)$ into $L^p_{\bar{v}}(R_+)$, where $\bar{v}(x) = v(\frac{1}{\lambda}\ln x)$. Thus it suffices to show that the boundedness of A_α is equivalent to the condition

$$\sup_{t>0}\left(\int_t^\infty \bar{v}(x)dx\right)^{1/p} t^{1/p'} < \infty. \tag{2.7.7}$$

Let $f \geq 0$. We have

$$\|A_\alpha f\|^p_{L^p_{\bar{v}}(R_+)} \leq c_1 \int_0^{+\infty} \bar{v}(x)\left(\int_0^{x/2}\left(\ln\frac{x}{y}\right)^{\alpha-1}f(y)dy\right)^p dx +$$

$$+c_1 \int_0^{+\infty} \bar{v}(x)\left(\int_{x/2}^{x}\left(\ln\frac{x}{y}\right)^{\alpha-1}f(y)dy\right)^p dx \equiv I_1 + I_2.$$

It is obvious that if $y < x/2$, then $(\ln\frac{x}{y})^{\alpha-1} \leq (\ln 2)^{\alpha-1}$. Consequently by Theorem A of Section 2.1 we obtain

$$I_1 \leq c_2 \int_0^{+\infty} \bar{v}(x)\left(\int_0^x f(y)dy\right)^p dx \leq c_3 \int_0^{+\infty}(f(y))^p dy.$$

Let $p^* \equiv \frac{p}{1-\alpha p}$. Then taking into account Hölder's inequality ($\frac{p^*}{p} > 1$), Corollary 2.7.6 and the monotonicity of $\bar{v}$ we easily obtain

$$I_2 = \sum_{k \in Z} \int_{2^k}^{2^{k+1}} \bar{v}(x)\left(\int_{x/2}^x\left(\ln\frac{x}{y}\right)^{\alpha-1}f(y)dy\right)^p dx \leq$$

$$\leq \sum_{k \in Z}\left(\int_{2^k}^{2^{k+1}}(x\bar{v}(x))^{\frac{1}{\alpha p}}\frac{dx}{x}\right)^{\alpha p} \times$$

$$\times \left(\int_{2^k}^{2^{k+1}}\left(\int_{x/2}^x\left(\ln\frac{x}{y}\right)^{\alpha-1}f(y)dy\right)^{p^*}\frac{dx}{x}\right)^{p/p^*} \leq$$

$$\leq \sum_{k \in Z}\left(\int_{2^k}^{2^{k+1}}(x\bar{v}(x))^{\frac{1}{\alpha p}}\frac{dx}{x}\right)^{\alpha p}\int_{2^{k-1}}^{2^{k+1}}y^{p-1}(f(y))^p dy \leq$$

$$\leq c_4 \sum_{k \in Z}\int_{2^{k-1}}^{2^{k+1}}(f(y))^p dy = c_5 \int_0^{+\infty}(f(y))^p dy.$$

Here we used the inequality $\bar{v}(2^k)2^{kp} \leq c, k \in Z$, which follows from condition (7.2.7).

Let A_α be bounded from $L^p(R_+)$ into $L^p_{\bar{v}}(R_+)$ and let

$$f_k(x) = \chi_{(2^{k-1},2^{k+1})}(x),$$

where $k \in Z$. Then we have

$$\|A_\alpha f_k\|^p_{L^p_{\bar{v}}(R_+)} \geq \int\limits_{2^k}^{2^{k+1}} \bar{v}(x)\left(\int\limits_{x/2}^{x} \left(\ln\frac{x}{y}\right)^{\alpha-1} f(y)dy\right)^p dx \geq$$

$$\geq c_6 \int\limits_{2^k}^{2^{k+1}} \bar{v}(x)x^p dx \geq c_6 \left(\int\limits_{2^k}^{2^{k+1}} \bar{v}(x)dx\right) 2^{kp}.$$

Moreover, $\|f_k\|^p_{L^p(R_+)} = c_7 2^k$. Thus

$$D \equiv \sup_{k\in Z}\left(\int\limits_{2^k}^{2^{k+1}} \bar{v}(x)dx\right)^{1/p} 2^{k/p'} < \infty.$$

Further, it is easy to verify that $B \leq c_8 D$ for some positive constant c_8. $\square$

Theorem 2.7.17. *Let* $1 < p < \infty$, $\lambda < 0$ *and let* v *be a monotone function on* R. *We assume that* $\alpha \in (0, 1/p)$. *Then for the boundedness of* $\mathcal{W}_{\alpha,\omega_\lambda}$ *from* $L^p_{\omega_\lambda}(R)$ *into* $L^p_{v\cdot\omega_\lambda}(R)$ *it is necessary and sufficient that*

$$D \equiv \sup_{r\in R}\left(\int\limits_{-\infty}^{r} v(y)\omega_\lambda(y)dy\right)^{1/p}(\omega_\lambda(r))^{1/p'}.$$

This follows in the same way as the previous theorem.

Corollary 2.7.7. *Let* $1 < p < \infty$ *and* $\alpha \in (0, \frac{1}{p})$. *We assume that* v *is a monotone function on* R. *Then the following criteria hold:*

(a) $\mathcal{R}_\alpha$ *is bounded from* $L^p_{\omega_\mu}(R)$ *into* $L^p_v(R)$ *if and only if*

$$D \equiv \sup_{r\in R}\left(\int\limits_{r}^{+\infty} v(x)dx\right)^{1/p}(\omega_\mu(r))^{-1/p} < \infty,$$

where $\mu < 0$;

(b) $\mathcal{W}_\alpha$ *is bounded from* $L^p_{\omega_\mu}(R)$ *into* $L^p_v(R)$ *if and only if*

$$D_1 \equiv \sup_{r\in R}\left(\int\limits_{-\infty}^{r} v(x)dx\right)^{1/p}(\omega_\mu(r))^{-1/p} < \infty,$$

where $\mu > 0$.

2.8. Extended Erdelyi–Köber operators

Let φ be a positive function on $(a, +\infty)$, $-\infty < a < +\infty$, such that $\varphi'(x)$ is continuous and $\varphi(a) = \lim_{t \to a+} \varphi(t) \neq -\infty$. Assume that

$$I^\alpha_{\varphi,a} f(x) = \int_a^x (\varphi(x) - \varphi(a))^{\alpha-1} f(s)\varphi'(s)ds, \quad x > a, \ \alpha > 0$$

for measurable $f : (a, \infty) \to R^1$. We shall suppose that v is a Lebesgue-measurable a.e. positive function on $(a, +\infty)$ and that $\varphi'(x) \neq 0$ for $x \in (a, +\infty)$. For some properties of the operator $I^\alpha_{\varphi,a}$ see for example [255], Section 18.2.

A criterion for boundedness is given by

Theorem 2.8.1. *Let* $1 < p \leq q < \infty$ *and* $\alpha > 1/p$. *The* $I^\alpha_{\varphi,a}$ *is bounded from* $L^p_{\varphi'}(a, \infty)$ *into* $L^q_v(a, +\infty)$ *if and only if*

$$B_{\varphi,a} \equiv \sup_{t>a} \left(\int_t^\infty v(x)(\varphi(x) - \varphi(a))^{(\alpha-1)q}dx \right)^{1/q} (\varphi(t) - \varphi(a))^{1/p'} < \infty.$$

Moreover, $\|I^\alpha_{\varphi,a}\| \approx B_{\varphi,a}$.

Proof. The theorem will be proved if we can show that the boundedness of $I^\alpha_{\varphi,a}$ from $L^p_{\varphi'}(0, \infty)$ into $L^q_{\overline{v}\varphi'}(a, +\infty)$ is equivalent to the condition

$$B'_{\varphi,a} \equiv \sup_{t>a} B'_{\varphi,a}(t) \equiv \sup_{t>a} \left(\int_t^\infty \overline{v}(x)(\varphi(x) - \varphi(a))^{(\alpha-1)q} \times \right.$$
$$\left. \times \varphi'(x)dx \right)^{1/q} (\varphi(t) - \varphi(a))^{1/p'} < \infty, \tag{2.8.1}$$

where $\overline{v}$ is a Lebesgue-measurable a.e. positive function on (a, ∞). If we change a variable, then we can easily see that $I^\alpha_{\varphi,a}$ is bounded from $L^p_{\varphi'}(a, \infty)$ into $L^q_{\overline{v}\varphi'}(a, \infty)$ if and only if the operator

$$R^\alpha_{\varphi,a} g(x) = \int_{\varphi(a)}^x (x - s)^{\alpha-1}g(s)ds$$

is bounded from $L^p(\varphi(a), \varphi(\infty))$ into $L^q_{\tilde{v}}(\varphi(a), \varphi(\infty))$, where

$$\tilde{v}(x) = \overline{v}(\varphi^{-1}(x)).$$

Using Theorem 2.1.2 (or 2.7.1) the latter is equivalent to the condition

$$\overline{B}_{\varphi,a} \equiv \sup_{\varphi(a)<t<\varphi(+\infty)} \overline{B}_{\varphi,a}(t) \equiv$$

$$\sup_{\varphi(a)<t<\varphi(+\infty)} \left(\int_t^\infty \tilde{v}(x)(x - \varphi(a))^{(\alpha-1)q} dx \right)^{1/q} (t - \varphi(a))^{1/p'} < \infty.$$

$$(2.8.2)$$

But if we put $t = \varphi(r)$, then

$$\overline{B}_{\varphi,a}(t) = \left(\int_{\varphi(r)}^{\varphi(\infty)} \tilde{v}(y)(y - \varphi(a))^{(\alpha-1)q} dy \right)^{1/q} (\varphi(r) - \varphi(a))^{1/p'} =$$

$$= \left(\int_r^\infty \overline{v}(x)(x - \varphi(a))^{(\alpha-1)q} \varphi'(x) dx \right)^{1/q} (\varphi(r) - \varphi(a))^{1/p'}.$$

Consequently $B'_{\varphi,a} = \overline{B}_{\varphi,a}$. The theorem is proved. $\square$

The next theorem is established in a similar way.

Theorem 2.8.2. *Let* $0 < q < p < \infty, p > 1$ *and* $\alpha > 1/p$. *Then the operator* $I^\alpha_{\varphi,a}$ *is bounded from* $L^p_{\varphi'}(a, \infty)$ *into* $L^q_v(a, \infty)$ *if and only if*

$$D_{\varphi,a} \equiv \left(\int_a^\infty \left(\int_x^\infty v(y)(\varphi(y) - \varphi(a))^{(\alpha-1)q} dy \right)^{\frac{p}{p-q}} \times \right.$$

$$\left. \times (\varphi(x) - \varphi(a))^{\frac{p(q-1)}{p-q}} \varphi'(x) dx \right)^{\frac{p-q}{p}} < \infty.$$

Moreover, $\|I^\alpha_{\varphi,a}\| \approx D_{\varphi,a}$.

Now we consider the compactness of $I^\alpha_{\varphi,a}$.

Theorem 2.8.3. *Let* $1 < p \le q < \infty$ *and* $\alpha > 1/p$. *Assume that* $\varphi(+\infty) = +\infty$. *Then the following conditions are equivalent:*

(i) $I^\alpha_{\varphi,a}$ *is compact from* $L^p_{\varphi'}(a, \infty)$ *into* $L^q_v(a, \infty)$;

(ii) $B_{\varphi,a} < \infty$ *and* $\lim_{t\to a} B_{\varphi,a}(t) = \lim_{t\to\infty} B_{\varphi,a}(t) = 0$, *where*

$$B_{\varphi,a}(t) = \left(\int_t^\infty v(x)(\varphi(x) - \varphi(a))^{(\alpha-1)q} dx \right)^{1/q} (\varphi(t) - \varphi(a))^{1/p'};$$

(iii) $B_{\varphi,a} < \infty$ *and* $\lim_{c\to a} B^{(c)}_{\varphi,a} = \lim_{b\to\infty} B^{(b)}_{\varphi,a} = 0$, *where*

$$B^{(c)}_{\varphi,a} \equiv \sup_{a<t<c} \left(\int_t^c v(x)(\varphi(x) - \varphi(a))^{(\alpha-1)q} dx \right)^{1/q} (\varphi(t) - \varphi(a))^{1/p'};$$

$$B_{\varphi,a}^{(b)} \equiv \sup_{t>b} \left(\int_t^\infty v(x)(\varphi(x) - \varphi(a))^{(\alpha-1)q} dx \right)^{1/q} (\varphi(t) - \varphi(b))^{1/p'}.$$

Proof. It is easy to verify that the operator $I_{\varphi,a}^\alpha$ is compact from $L_\varphi^p(a,\infty)$ into $L_{\overline{v}\varphi'}^q(a,\infty)$, where $\overline{v}$ is a Lebesgue-measurable a.e. positive function on $(0,\infty)$, if and only if the operator $R_{\varphi,a}^\alpha$ is compact $L^p(\varphi(a),\varphi(\infty))$ into $L_{\widetilde{v}}^q(\varphi(a),\varphi(\infty))$, where $\widetilde{v}(x) = \overline{v}(\varphi^{-1}(x))$. The latter is equivalent to the following conditions: (2.8.2) holds and

$$\lim_{r \to \varphi(a)} \overline{B}_{\varphi,a}(r) = \lim_{r \to \infty} \overline{B}_{\varphi,a}(r) = 0.$$

Using a change of variable, we find that these conditions are equivalent to the conditions: (2.8.1) is fulfilled and $\lim_{t \to a} B_{\varphi,a}'(t) = \lim_{t \to \infty} B_{\varphi,a}'(t) = 0$.

In a similar way we can show that the compactness of $R_{\varphi,a}^\alpha$ from $L^p(\varphi(a),\varphi(\infty))$ into $L_{\widetilde{v}}^q(\varphi(a),\varphi(\infty))$ is equivalent to the following conditions: (2.8.2) holds and

$$\lim_{c \to \varphi(a)} \overline{B}_{\varphi,a}^{(c)} = \lim_{b \to \infty} \widetilde{B}_{\varphi,a}^{(b)} = 0,$$

where

$$\overline{B}_{\varphi,a}^{(c)} \equiv \sup_{\varphi(a)<t<c} \left(\int_t^c \widetilde{v}(x)(x - \varphi(a))^{(\alpha-1)q} dx \right)^{1/q} \\ (t - \varphi(a))^{1/p'}, \tag{2.8.3}$$

$$\widetilde{B}_{\varphi,a}^{(b)} \equiv \sup_{t>b} \left(\int_t^\infty \widetilde{v}(x)(x - \varphi(a))^{(\alpha-1)q} dx \right)^{1/q} (t - b)^{1/p'}. \tag{2.8.4}$$

A change of variable completes the proof. □

The next theorem gives another criteria for compactness when $\varphi(+\infty) < \infty$.

Theorem 2.8.4. *Let* $1 < p \le q < \infty$, $\alpha > 1/p$. *Assume that* $\varphi(+\infty) < \infty$. *Then the following statements are equivalent:*

(i) $I_{\varphi,a}^\alpha$ *is compact from* $L^p(a,\infty)$ *into* $L_v^q(a,\infty)$;

(ii) $B_{\varphi,a} < \infty$ *and* $\lim_{t \to a} B_{\varphi,a}(t) = 0$;

(iii) $B_{\varphi,a} < \infty$ *and* $\lim_{c \to a} B_{\varphi,a}^{(c)} = 0$.

Proof. The proof is similar to that of Theorem 2.8.3. In the present situation the conditions $\lim_{t \to \infty} B_{\varphi,a}(t) = 0$ and $\lim_{b \to \infty} B_{\varphi,a}^{(b)} = 0$ are automatically satisfied.
□

The next theorem is an immediate consequence of Theorem 2.8.2:

Theorem 2.8.5. *Let $0 < q < p < \infty$, $p > 1$ and $\alpha > 1/p$. Then $I_{\varphi,a}^{\alpha}$ is compact from $L_{\varphi'}^{p}(a, \infty)$ into $L_{v}^{q}(a, \infty)$ if and only if $D_{\varphi,a} < \infty$.*

As before, we now estimate the measure of non–compactness of the operator $I_{\varphi,a}^{\alpha}$.

Theorem 2.8.6. *Let $1 < p \leq q < \infty$ and $\alpha > 1/p$. Assume that $B_{\varphi,a} < \infty$, where $\varphi(+\infty) = +\infty$. Then there exist positive constants b_1 and b_2 depending only on p, q, α such that*

$$b_2 I \leq dist(I_{\varphi,a}^{\alpha}, \mathcal{K}(L_{\varphi'}^{p}(a, \infty), L_{v}^{q}(a, \infty))) \leq b_1 I,$$

where

$$I \equiv \lim_{c \to a} B_{\varphi,a}^{(c)} + \lim_{b \to \infty} B_{\varphi,a}^{(b)}.$$

Proof. We put:

$$d \equiv dist(I_{\varphi,a}^{\alpha}, \mathcal{K}(L_{\varphi'}^{p}(a, \infty), L_{\bar{v}\varphi'}^{q}(a, \infty))),$$

$$d_1 \equiv dist(R_{\varphi,a}^{\alpha}, \mathcal{K}(L_{\varphi'}^{p}(\varphi(a), \varphi(\infty)), L_{\bar{v}}^{q}(\varphi(a), \varphi(\infty)))),$$

where $\widetilde{v} = \bar{v}(\varphi^{-1}(x))$.

It is easy to verify that the operator A is compact from $L_{\varphi'}^{p}(a, \infty)$ into $L_{\bar{v}\varphi'}^{q}(a, \infty)$ if and only if $\widetilde{A}$ is compact from $L^{p}(\varphi(a), \varphi(\infty))$ into $L_{\bar{v}}^{q}(\varphi(a), \varphi(\infty))$, where $\widetilde{A}f(x) \equiv A(f \circ \varphi)(\varphi^{-1}(x))$. Using this it follows that $d_1 = d$. By Theorem 2.7.9 we obtain that

$$c_2 S \leq d_1 \leq c_1 S,$$

where $S = \lim_{c_1 \to \varphi(a)} \overline{B}_{\varphi,a}^{(c_1)} + \lim_{b_1 \to \infty} \widetilde{B}_{\varphi,a}^{(b_1)}$, where $\overline{B}_{\varphi}^{(c_1)}$ and $\widetilde{B}_{\varphi}^{(b_1)}$ are defined by (2.8.3) and (2.8.4) and c_1 and c_2 depend only on p, q and α. If we change the variable, then we find that

$$\lim_{c_1 \to \varphi(a)} \overline{B}_{\varphi,a}^{(c_1)} = \lim_{c \to a} A_a^{(c)}$$

and

$$\lim_{b_1 \to \infty} \widetilde{B}_{\varphi,a}^{(b_1)} = \lim_{b \to \infty} A_a^{(b)},$$

where

$$A_a^{(c)} \equiv \sup_{r \in (a,c)} \left(\int_r^c \bar{v}(x)(\varphi(x) - \varphi(a))^{(\alpha-1)q} \varphi'(x)dx \right)^{1/q} (\varphi(r) - \varphi(a))^{1/p'};$$

$$\overline{A}_a^{(b)} \equiv \sup_{r>b} \left(\int\limits_r^\infty \overline{v}(x)(\varphi(x) - \varphi(a))^{(\alpha-1)q} \varphi'(x)dx \right)^{1/q} (\varphi(r) - \varphi(b))^{1/p'}.$$

Finally we derive the inequality

$$b_2 I' \leq d \leq b_1 I'.$$

with positive constants b_1 and b_2 depending only on p, q and α, where

$$I' = \lim_{c \to a} A_a^{(c)} + \lim_{b \to \infty} \overline{A}_a^{(b)}.$$

□

In a similar way we can prove the following Theorem:

Theorem 2.8.7. *Let* $1 < p \leq q < \infty$ *and* $\alpha > 1/p$. *Assume that* $\varphi(+\infty) < +\infty$. *Then there exist positive constants* b_1 *and* b_2 *depending only on* p, q *and* α *such that*

$$b_1 I \leq dist(I_{\varphi,a}^\alpha, \mathcal{K}(L_{\varphi'}^p(a,\infty), L_v^q(a,\infty))) \leq b_1 I,$$

where $I = \lim_{c \to a} \overline{B}_{\varphi,a}^{(c)}$.

For measurable $f : (a, \infty) \to R^1, (a > 0)$ let

$$\mathcal{J}_a f(x) = \int\limits_a^x f(t)t^{-1}(\ln \frac{x}{t})^{\alpha-1}dt, \quad x > a.$$

The operator $\mathcal{J}$ is known as Hadamard's operator (see, e.g., [255], Section 18.2). Let us put

$$\mathcal{L}^p(a, \infty) = \{f : \|f\|_{\mathcal{L}}(a, \infty) = \left(\int\limits_a^\infty |f(t)|^p \frac{dt}{t} \right)^{1/p} < \infty\}.$$

Corollary 2.8.1. *Let* $1 < p \leq q < \infty$ *and* $\alpha > 1/p$. *Then* $\mathcal{J}_1$ *bounded from* $\mathcal{L}^p(1, \infty)$ *into* $L_v^q(1, \infty)$ *if and only if*

$$B \equiv \sup_{t>1} \left(\int\limits_t^\infty v(x)(\ln x)^{(\alpha-1)q}dx \right)^{1/q} (\ln t)^{1/p'} < \infty.$$

Moreover, $\|\mathcal{J}_1\| \approx B$.

Corollary 2.8.2. *Let* α, p *and* q *satisfy the conditions of Corollary 2.8.1. Then the following statements are equivalent:*

(i) $\mathcal{J}_1$ is bounded from $\mathcal{L}^p(1, \infty)$ into $L_v^p(1; \infty)$;

(ii) $B < \infty$ and $\lim\limits_{t \to 1} B(t) = \lim\limits_{t \to \infty} B(t) = 0$, where

$$B(t) = \left(\int\limits_t^\infty v(x)(\ln x)^{(\alpha-1)q} dx \right)^{1/q} (\ln t)^{1/p'};$$

(iii) $B < \infty$ and $\lim\limits_{c \to 1} B^{(c)} \lim\limits_{b \to \infty} B^{(b)} = 0$, where

$$B^{(c)} \equiv \sup\limits_{1 < t < c} \left(\int\limits_t^c v(x)(\ln x)^{(\alpha-1)q} dx \right)^{1/q} (\ln t)^{1/p'};$$

$$B^{(b)} \equiv \sup\limits_{t > b} \left(\int\limits_t^\infty v(x)(\ln x)^{(\alpha-1)q} dx \right)^{1/q} (\ln \frac{t}{b})^{1/p'}.$$

Corollary 2.8.3. *Let* $0 < q < p < \infty$, $p > 1$ *and* $\alpha > 1/p$. *Then the following statements are equivalent:*

(i) $\mathcal{J}_1$ *is bounded from* $\mathcal{L}^p(1, \infty)$ *into* $L_v^q(1, \infty)$.

(ii) $\mathcal{J}_1$ *is compact from* $\mathcal{L}^p(1, \infty)$ *into* $L_v^q(1, \infty)$.

(iii)

$$D \equiv \left(\int\limits_1^\infty \left(\int\limits_x^\infty v(y)(\ln y)^{(\alpha-1)q} dy \right)^{\frac{p}{p-q}} (\ln x)^{\frac{p(q-1)}{p-q}} \frac{dx}{x} \right)^{\frac{p-q}{pq}} < \infty.$$

Moreover, $\| \mathcal{J}_1 \| \approx D$.

Corollary 2.8.4. *Let* $1 < p \leq q < \infty$ *and* $\alpha > 1/p$. *Assume that* $B < \infty$. *Then there exist positive constants* b_1 *and* b_2 *depending only on* p, q, α *such that*

$$b_2 I \leq dist(\mathcal{J}_1, \mathcal{K}(\mathcal{L}^p(a, \infty), L_v^q(a, \infty))) \leq b_1 I,$$

where

$$I \equiv \lim\limits_{c \to 1} B^{(c)} + \lim\limits_{b \to \infty} B^{(b)}.$$

The next operator to be analysed is given for measurable $f : (a, \infty) \to R^1$ $(-\infty < a < \infty)$ by

$$J_{\varphi,a}^\alpha f(x) = \int\limits_x^\infty (\varphi(y) - \varphi(x))^{\alpha-1} f(y) w(y) \varphi'(y) dy, \quad x > a, \quad \alpha > 0,$$

where w is a Lebesgue–measurable a.e. positive function on (a, ∞).

From duality arguments we obtain the following results:

Theorem 2.8.8. *Let* $1 < p \le q < \infty$ *and* $\alpha > \frac{q-1}{q}$. *Then the operator* $J_{\varphi,a}^{\alpha}$ *is bounded from* $L_{w\varphi'}^{p}(a, \infty)$ *into* $L_{\varphi'}^{q}(a, \infty)$ *if and only if*

$$A_{\varphi,a} \equiv \sup_{t>a} \left(\int_{t}^{\infty} w(y)(\varphi(y) - \varphi(a))^{(\alpha-1)p'} \varphi'(y)dy \right)^{1/p'} \times$$

$$\times (\varphi(t) - \varphi(a))^{1/q} < \infty, \quad p' = \frac{p}{p-1}.$$

Moreover, $\|J_{\varphi,a}^{\alpha}\| \approx A_{\varphi,a}$.

Theorem 2.8.9. *Let* $1 < q < p < \infty$ *and* $\alpha > \frac{q}{q-1}$. *Then the operator* $J_{\varphi,a}^{\alpha}$ *is bounded from* $L_{w\varphi'}^{p}(a, \infty)$ *into* $L_{\varphi'}^{q}(a, \infty)$ *if and only if*

$$E \equiv \left(\int_{a}^{\infty} \left(\int_{x}^{\infty} w(y)(\varphi(y) - \varphi(a))^{(\alpha-1)p'} \varphi'(y)dy \right)^{\frac{(p-1)q}{p-q}} \times \right.$$

$$\left. \times (\varphi(x) - \varphi(a))^{\frac{q}{p-q}} \varphi'(x)dx \right)^{\frac{p-q}{pq}} < \infty.$$

Moreover, $\|J_{\varphi,a}^{\alpha}\| \approx E$.

Theorem 2.8.10. *Let* $1 < p \le q < \infty$ *and* $\alpha > 1/q'$. *Assume that* $\varphi(+\infty) = +\infty$. *Then the following statements are equivalent:*

(i) $J_{\varphi,a}^{\alpha}$ *is compact from* $L_{w\varphi'}^{p}(a, \infty)$ *into* $L_{\varphi'}^{q}(a, \infty)$;

(ii) $A_{\varphi,a} < \infty$ *and* $\lim\limits_{t \to a} A_{\varphi,a}(t) = \lim\limits_{t \to \infty} A_{\varphi,a}(t) = 0$, *where*

$$A_{\varphi,a}(t) = \left(\int_{t}^{\infty} w(x)(\varphi(x) - \varphi(a))^{(\alpha-1)p'} \varphi'(x)dx \right)^{1/p'} (\varphi(t) - \varphi(a))^{1/q};$$

(iii) $A_{\varphi,a} < \infty$ *and* $\lim\limits_{c \to a} A_{\varphi,a}^{(c)} = \lim\limits_{b \to \infty} B_{\varphi,a}^{(b)} = 0$, *where*

$$A_{\varphi,a}^{(c)} \equiv \sup_{a<t<c} \left(\int_{t}^{c} w(x)(\varphi(x) - \varphi(a))^{(\alpha-1)p'} \varphi'(y)dy \right)^{1/p'} (\varphi(t) - \varphi(a))^{1/q};$$

$$A_{\varphi,a}^{(b)} \equiv \sup_{t>b} \left(\int_{t}^{\infty} w(x)(\varphi(x) - \varphi(a))^{(\alpha-1)p'} \varphi'(x)dx \right)^{1/p'} (\varphi(t) - \varphi(b))^{1/p'}.$$

Theorem 2.8.11. *Let p, q and α satisfy the conditions of Theorem* 2.8.10. *Assume that* $\varphi(+\infty) < +\infty$. *Then the following statements are equivalent:*

(i) $J_{\varphi,a}^{\alpha}$ *is compact from* $L_{w\varphi'}^p(a,\infty)$ *into* $L_{\varphi}^q(a,\infty)$;

(ii) $A_{\varphi,a} < \infty$ *and* $\lim_{t\to\infty} A_{\varphi,a}(t) = 0$;

(iii) $A_{\varphi,a} < \infty$ *and* $\lim_{c\to a} A_{\varphi,a}^{(c)} = 0$.

Theorem 2.8.12. *Let* $1 < q < p < \infty$ *and* $\alpha > 1/q'$. *Then* $J_{\varphi,a}^{\alpha}$ *is compact from* $L_{\varphi'w}^p(a,\infty)$ *into* $L_{\varphi'}^q(a,\infty)$ *if and only if* $E < \infty$.

Now let

$$H_{\varphi}^a f(x) = \int\limits_{\varphi(a)}^{\varphi(x)} f(y)dy$$

for measurable $f : (\varphi(a), \varphi(\infty)) \to R^1$, $-\infty < a < \infty$.

In the sequel we shall assume that v and w are Lebesgue-measurable a.e. positive functions on (a, ∞).

We need the following Lemma:

Lemma 2.8.1 *Let* $1 < p \leq q < \infty$, μ *be a non-negative Borel measure on* (a, ∞). *Then the operator* H_{φ}^a *is bounded from* $L_w^p(\varphi(a), \varphi(\infty))$ *into* $L_{\mu}^q(a, \infty)$ *if and only if*

$$B \equiv \sup_{t>a}(\mu([t,\infty)))^{1/q} \left(\int\limits_{\varphi(a)}^{\varphi(t)} w^{1-p'}(y)dy \right)^{1/p'} < \infty.$$

Moreover, $\|H_{\varphi}^a\| \approx B$

Proof. Let $f \geq 0$ and $f \in L_w^p(\varphi(a), \varphi(\infty))$. We put $I_{\varphi}(t) = \int\limits_{\varphi(a)}^{\varphi(t)} f(y)dy$.

Let $I_{\varphi}(\infty) = \int\limits_{\varphi(a)}^{\varphi(\infty)} f(y)dy \in [2^m, 2^{m+1})$ for some integer m. As I_{φ} is continuous on (a, ∞), there exists x_k such that

$$2^k = \int\limits_{\varphi(a)}^{\varphi(x_k)} f(y)dy, \quad k \leq m.$$

It is obvious that for $k \leq m - 1$ the equality

$$2^k = \int\limits_{\varphi(x_k)}^{\varphi(x_{k+1})} f(y)dy$$

holds. The sequence x_k increases and $[a, \infty) = [a, \alpha] \cap (\cap_{k \leq m}[x_k, x_{k+1}))$, where $\alpha = \lim\limits_{k \to -\infty} x_k$ and $x_{k+1} = \infty$. If $x \in [a, \alpha]$, then

$$I_\varphi(x) \leq \int\limits_{\varphi(a)}^{\varphi(x)} f(y)dy \leq \int\limits_{\varphi(a)}^{\varphi(x_k)} f(y)dy$$

for every k, $k \leq m$ and consequently $I_\varphi(x) = 0$. If $x \in [x_k, x_{k+1})$, then

$$I_\varphi(x) \leq \int\limits_{\varphi(a)}^{\varphi(x_{k+1})} f(y)dy = 2^{k+1}.$$

For $I_\varphi(\infty)$, we have that

$$[a, \infty) = [a, \alpha] \cap (\cap_{k \in Z}[x_k, x_{k+1})).$$

(In this case $m = \infty$). Further

$$\|H_\varphi^a f\|_{L_\mu^q(a,\infty)}^q = \sum_k \int\limits_{[x_k, x_{k+1})} (H_\varphi^a f(x))^q d\mu(x) \leq$$

$$\leq \sum_k 2^{(k+1)q} \int\limits_{[x_k, x_{k+1})} d\mu(x) =$$

$$= 4^q \sum_k \left(\int\limits_{\varphi(x_{k-1})}^{\varphi(x_k)} f(y)dy \right)^q \mu([x_k, x_{k+1})) \leq$$

$$\leq 4^q \sum_k \left(\int\limits_{\varphi(x_{k-1})}^{\varphi(x_k)} w(y)(f(y))^p dy \right)^{q/p}$$

$$\times \quad \times \left(\int\limits_{\varphi(x_{k-1})}^{\varphi(x_k)} w^{1-p'}(y)dy \right)^{q/p'} \mu([x_k, x_{k+1})) \leq$$

$$\leq 4^q B^q \|f\|_{L_w^p(\varphi(a),\varphi(\infty))}.$$

Sufficiency is proved. To prove necessity we take $t > a$ and let $f_t(x) = w^{1-p'}(y)\chi_{(\varphi(a),\varphi(t))}(y)$. Then we have

$$\|f\|_{L_w^p(\varphi(a),\varphi(\infty))} = \left(\int\limits_{\varphi(a)}^{\varphi(x_k)} w^{1-p'}(y)dy \right)^{1/p'} < \infty.$$

On the other hand,

$$\|H_\varphi^a f\|_{L_\mu^q(a,\infty)} \geq \left(\int_{[t,\infty)} \left(\int_{\varphi(a)}^{\varphi(x)} w^{1-p'}(y)dy \right)^q d\mu(x) \right)^{1/q} \geq$$

$$\geq (\mu[t,\infty))^{1/q} \left(\int_{\varphi(a)}^{\varphi(t)} w^{1-p'}(y)dy \right).$$

From the boundedness of H_φ^a we finally obtain that $B < \infty$. □

The next operator to be studied is

$$T_\varphi^\alpha f(x) = \int_{\varphi(a)}^{\varphi(x)} (\varphi(x) - u)^{\alpha-1} f(u)du, \quad x > a, \ \alpha > 0,$$

for measurable $f : (\varphi(a), \varphi(\infty)) \to R$

Theorem 2.8.13. *Let $1 < p \leq q < \infty$ and $\alpha > 1/p$. Assume that ν ia a non-negative σ-finite Borel measure on $[a,\infty)$. Then the operator T_φ^α is bounded from $L^p(\varphi(a), \varphi(\infty))$ into $L_\nu^q(a,\infty)$ if and only if*

$$F \equiv \sup_{t>a} \left(\int_{[t,\infty)} (\varphi(x) - \varphi(a))^{(\alpha-1)q} d\nu(x) \right)^{1/q} (\varphi(t) - \varphi(a))^{1/p'} < \infty.$$

Moreover, $\|T_\varphi^\alpha\| \approx F$.
Proof. Let $f \geq 0$. Then

$$\|T_\varphi^\alpha f\|_{L_\nu^q(a,\infty)} \leq \left(\int_{[a,\infty)} \left(\int_{\varphi(a)}^{\psi(x)} f(y)(\varphi(x) - y)^{\alpha-1} dy \right)^q d\nu(x) \right)^{1/q} +$$

$$+ \left(\int_{[a,\infty)} \left(\int_{\psi(x)}^{\varphi(x)} f(y)(\varphi(x) - y)^{\alpha-1} dy \right)^q d\nu(x) \right)^{1/q} \equiv S_1 + S_2,$$

where $\psi(x) = (\varphi(a) + \varphi(x))/2$.
If $\varphi(a) < y < \psi(x)$, then

$$\varphi(x) - \varphi(a) \leq \varphi(x) - y + y - \varphi(a) \leq \varphi(x) - y + \frac{\varphi(x) - \varphi(a)}{2}.$$

Hence $(\varphi(x) - \varphi(a))/2 \leq \varphi(x) - y$. Consequently, using Lemma 2.8.1 we obtain

$$
S_1 \leq c_1 \left(\int\limits_{[a,\infty)} (\varphi(x) - \varphi(a))^{(\alpha-1)q} \left(\int\limits_{\varphi(a)}^{\varphi(x)} f(y)dy \right)^q d\nu(x) \right)^{1/q} \leq
$$

$$
\leq c_2 F \|f\|_{L^p(\varphi(a),\varphi(\infty))}.
$$

First we assume that $\varphi(+\infty) = +\infty$. Then using Hölder's inequality we obatain

$$
S_2^q \leq \int\limits_{[a,\infty)} \left(\int\limits_{\psi(x)}^{\varphi(x)} (f(y))^p dy \right)^{q/p} \left(\int\limits_{\psi(x)}^{\varphi(x)} (\varphi(x) - \varphi(a))^{(\alpha-1)p'} dy \right)^{q/p'} =
$$

$$
= \sum_k \int\limits_{[s_k,s_{k+1})} \left(\int\limits_{\psi(x)}^{\varphi(x)} (f(y))^p dy \right)^{q/p} (\varphi(x) - \varphi(a))^{(\alpha-1)q + q/p'} d\nu(x),
$$

where $s_k = \varphi^{-1}(\varphi(a) + 2^k)$. Moreover,

$$
S_2^q \leq c_3 \sum_k \left(\int\limits_{\varphi(a)+2^{k-1}}^{\varphi(a)+2^{k+1}} (f(y))^p dy \right)^{q/p} \times
$$

$$
\times \left(\int\limits_{[s_k,s_{k+1})} (\varphi(x) - \varphi(a))^{(\alpha-1)q} d\nu(x) \right) 2^{kq/p'} \leq
$$

$$
c_3 \leq F^q \|f\|_{L^q(\varphi(a),\varphi(\infty))}.
$$

When $\varphi(+\infty) < \infty$, we take $s_k = \varphi^{-1}\left(\varphi(a) + \frac{\varphi(+\infty)-\varphi(a)}{2^k}\right)$ and consequently $(a,\infty) = \cup_{k \geq 0}[s_{k+1}, s_k)$.

To prove necessity we take $f_t(x) = \chi_{(\varphi(a),\varphi(t))}(x)$ for $t > a$. Then $\|f_t\|_{L^p(\varphi(a),\varphi(\infty))} = (\varphi(t) - \varphi(a))^{1/p}$. On the other hand,

$$
\|T_\varphi^\alpha f\|_{L^q_\nu(a,\infty)} \geq \left(\int\limits_{[t,\infty)} (\varphi(x) - \varphi(a))^{(\alpha-1)q} d\nu(x) \right)^{1/q} (\varphi(t) - \varphi(a)).
$$

Finally we obtain $F < \infty$. $\square$

The next theorem is a direct consequence of Theorem 2.8.2.

Theorem 2.8.14. *Let $0 < q < p < \infty$, $p > 1$ and $\alpha > 1/p$. Then the operator $T_{\varphi,a}^\alpha$ is bounded from $L^p(\varphi(a), \varphi(\infty))$ into $L^q_v(a,\infty)$, where v is a*

Lebesgue-measurable a.e. positive function on (a, ∞), *if and only if*

$$D_{\varphi,a} < \infty$$

(*for the definition of* $D_{\varphi,a}$ *see Theorem 2.8.2*).

The following theorem can be derived in the same way as Theorem 2.1.5.

Theorem 2.8.15. *Let* $1 < p \le q < \infty$ *and* $\alpha > 1/p$. *Assume that* $\varphi(+\infty) = +\infty$. *Then the following conditions are equivalent:*
(i) T_φ^α *is compact from* $L^p(\varphi(a), \varphi(\infty))$ *into* $L_\nu^q(a, \infty)$;
(ii) $F < \infty$ *and* $\lim\limits_{t \to a} F(t) = \lim\limits_{t \to \infty} F(t) = 0$, *where*

$$F(t) = \left(\int\limits_{[t,\infty)} (\varphi(x) - \varphi(a))^{(\alpha-1)q} d\nu(x) \right)^{1/q} (\varphi(t) - \varphi(a))^{1/p'}.$$

Similarly we also have

Theorem 2.8.16. *Let* $1 < p \le q < \infty$ *and* $\alpha > 1/p$. *Assume that* $\varphi(+\infty) < +\infty$. *Then the following conditions are equivalent:*
(i) T_φ^α *is compact from* $L^p(\varphi(a), \varphi(\infty))$ *into* $L_\nu^q(a, \infty)$;
(ii) $F < \infty$ *and* $\lim\limits_{t \to a} F(t) = 0$;
(iii) $F < \infty$ *and* $\lim\limits_{t \to a} F^{(c)} = 0$, *where*

$$F^{(c)} = \sup_{a < t < c} \left(\int\limits_{[t,c)} (\varphi(x) - \varphi(a))^{(\alpha-1)q} d\nu(x) \right)^{1/q} (\varphi(t) - \varphi(a))^{1/p'}.$$

From the previous theorems, using a change of variable we can obtain the following results for I_φ^α.

Theorem 2.8.17. *Let* $1 < p \le q < \infty$ *and* $\alpha > 1/p$. *Then* $I_{\varphi,a}^\alpha$ *is bounded from* $L_{\varphi'}^p(a, \infty)$ *into* $L_\nu^q(a, \infty)$ *if and only if* $F < \infty$. *Moreover,* $\|I_{\varphi,a}^\alpha\| \approx F$.

Theorem 2.8.18. *Let* $1 < p \le q < \infty$ *and* $\alpha > 1/p$. *Suppose that* $\varphi(+\infty) = \infty$. *Then the following conditions are equivalent:*
(i) $I_{\varphi,a}^\alpha$ *is compact from* $L_{\varphi'}^p(a, \infty)$ *into* $L_\nu^q(a, \infty)$;
(ii) $F < \infty$ *and* $\lim\limits_{t \to a} F(t) = \lim\limits_{t \to \infty} F(t) = 0$.

Theorem 2.8.19. *Let* p, q *and* α *satisfy the conditions of Theorem 2.8.18. Suppose that* $\varphi(+\infty) < +\infty$. *Then the following conditions are equivalent:*
(i) $I_{\varphi,a}^\alpha$ *is compact from* $L_{\varphi'}^p(a, \infty)$ *into* $L_\nu^q(a, \infty)$;

(ii) $F < \infty$ and $\lim_{t \to a+} F(t) = 0$;

(iii) $F < \infty$ and $\lim_{c \to a+} F^{(c)} = 0$.

Let $\varphi : (0, +\infty) \to R^1$ be a continuous increasing function such that $\varphi(0) = -\infty$, $\varphi(+\infty) = +\infty$, φ' is continuous and $\varphi'(x) \neq 0$. We shall assume that

$$I^\alpha_{\varphi, -\infty} \equiv I^\alpha_\varphi$$

The next theorem can be derived from Theorem 2.2.2.

Theorem 2.8.20. *Let* $1 < p < q < \infty$. *Then* I^α_φ *is bounded from* $L^p_w(R)$ *into* $L^q_v(R)$ *if and only if*

$$\sup_{h>0, x \in R} \left(\int_{x-h}^{x+h} \widetilde{v_1}(y) dy \right)^{1/q} \left(\int_{-\infty}^{x-h} \frac{\widetilde{w_1}^{1-p'}(y)}{(x-y)^{(1-\alpha)p'}} dy \right)^{1/p'} < \infty$$

$$\sup_{h>0, x \in R} \left(\int_{x-h}^{x+h} \widetilde{w_1}^{1-p'}(y) dy \right)^{1/p'} \left(\int_{x+h}^{+\infty} \frac{\widetilde{v_1}(y)}{(y-x)^{(1-\alpha)q}} dy \right)^{1/q} < \infty,$$

where

$$v_1(x) = v(x)/\varphi'(x), \quad w_1(x) = w(x)/\varphi'(x),$$

$$\widetilde{v_1}(x) = v_1(\varphi^{-1}(x)), \quad \widetilde{w_1}(x) = w_1(\varphi^{-1}(x)).$$

Proof. It is easy to check that the boundedness of I^α_φ from $L^p_{\widetilde{w}\varphi'}(0, \infty)$ into $L^q_{\widetilde{v}\varphi'}(0, \infty)$ is equivalent to the two-weight inequality

$$\left(\int_{-\infty}^{+\infty} \left(\int_{-\infty}^{x} f(y)(x-y)^{-1} dy \right)^p \widetilde{v}(x) dx \right)^{1/q} \leq c \left(\int_{-\infty}^{+\infty} (f(y))^q \widetilde{w}(y) dy \right)^{1/p},$$

where the positive constant c is independent of f and

$$\widetilde{v} = \overline{v}(\varphi^{-1}(x)), \quad \widetilde{w} = \overline{w}(\varphi^{-1}(x)).$$

The last inequality holds if and only if

$$\sup_{h>0, x \in R} \left(\int_{x-h}^{x+h} \widetilde{v}(y) dy \right)^{1/q} \left(\int_{-\infty}^{x-h} \frac{\widetilde{w}^{1-p'}(y)}{(x-y)^{(1-\alpha)p'}} dy \right)^{1/p'} < \infty$$

$$\sup_{h>0, x \in R} \left(\int_{x-h}^{x+h} \tilde{w}^{1-p'}(y) dy \right)^{1/p'} \left(\int_{x+h}^{+\infty} \frac{\tilde{v}(y)}{(y-x)^{(1-\alpha)q}} dy \right)^{1/q} < \infty.$$

The theorem is proved. □

2.9. Generalized one–sided potentials

In this section we establish necessary and sufficient conditions which ensure the boundedness (compactness) of the generalized Riemann-Liouville operator

$$T_\alpha f(x, t) = \int_0^x (x - y + t)^{\alpha - 1} f(y) dy, \quad x, t \in R_+,$$

from $L^p(R_+)$ into $L^q_\nu(\tilde{R}^2_+)$, where $0 < p, q < \infty$, $p > 1$, $\alpha > 1/p$, $R_+ \equiv [0, \infty)$ and ν is a positive σ-finite Borel measure on $\tilde{R}^2_+ \equiv R_+ \times R_+$ (for $q < p$ it will be assumed that $d\nu(x, t) = v(x, t) dx dt$, where v is a Lebesgue-measurable almost everywhere positive function on $\tilde{R}^2_+$).

An analogous problem for the classical Riemann–Liouville operator

$$R_\alpha f(x) = \int_0^x (x - y)^{\alpha - 1} f(y) dy$$

was solved in Section 2.1.

The boundedness problem for the generalized Riesz potential

$$\mathcal{I}_\alpha f(x, t) = \int_{R^n} (|x - y| + t)^{\alpha - n} f(y) dy, \quad 0 < \alpha < n,$$

from $L^p(R^n)$ into $L^q_\nu(R^n \times R_+)$ $(1 < p < q < \infty)$ was solved in [1] (Theorem C) (see [98] for more general case).

A complete description of weight pairs (v, w) ensuring the validity of weak (p, q) $(1 < p < q < \infty)$ type inequality for $\mathcal{I}_\alpha$ was given in [99] (see also [100], Chapter 3). For related Hörmander–type maximal functions see [100], Chapter 4.

Different (Sawyer–type) necessary and sufficient conditions for the validity of two-weight strong (p, q) type inequalities for $\mathcal{I}_\alpha$ and corresponding Hörmander- type fractional maximal functions were established in [266].

Similar operators arise in boundary value problems in PDE, particularly in Polyharmonic Differential Equations. Some applications of operator $\mathcal{I}_\alpha$ in weighted estimates for gradients were presented in [307], p.923.

In this section, criteria for boundedness (compactness) from $L_\nu^p(\tilde{R}_+^2)$ into $L^q(R_+)$ are also established for the operator

$$\tilde{T}_\alpha g(y) = \int_{[y,\infty)\times R_+} g(x,t)(x-y+t)^{\alpha-1} d\nu(x,t).$$

Finally, upper and lower estimates of the distance between the operator T_α and the space of compact operators are derived in the non-compact case.

Let

$$Hf(x) = \int_0^x f(y)dy$$

for a measurable function $f : R_+ \to R^1$.

In the sequel we shall use the notation $U_r \equiv [r,\infty) \times R_+$, where $r > 0$. It is obvious that $[r,R) \times R_+ = U_r \setminus U_R$ for $0 < r < R < \infty$.

To prove our main results, we need

Lemma 2.9.1. *Let $1 < p \le q < \infty$ and μ be a positive Borel measure on $\tilde{R}_+^2$. Then the operator H is bounded from $L^p(R_+)$ into $L_\mu^q(\tilde{R}_+^2)$ if and only if*

$$A \equiv \sup_{r>0}(\mu(U_r))^{1/q} r^{1/p'} < \infty, \quad p' = p/(p-1).$$

Moreover, $A \le \|H\| \le 4A$.

Proof. *Sufficiency.* Let $f \ge 0$, $f \in L^p(R_+)$ and $I(t) \equiv \int_0^t f$. Assume that $\int_0^\infty f \in (2^m, 2^{m+1}]$ for some $m \in Z$. Then there exist x_k ($k \le m$) such that $I(x_k) = 2^k$. It is obvious that $2^k = \int_{x_k}^{x_{k+1}} f$ for $k \le m-1$. The sequence $\{x_k\}$ increases. Moreover, if $\alpha = \lim_{k\to-\infty} x_k$, then $R_+ = [0,\alpha) \cup (\cup_{k\le m}[x_k, x_{k+1}))$, where $x_{k+1} = \infty$. When $\int_0^\infty f = \infty$, we have $R_+ = [0,\alpha] \cup (\cup_{k\in Z}[x_k, x_{k+1}))$ (i.e. $m = \infty$). If $y \in [0,\alpha]$, then $I(y) = 0$, and if $y \in [x_k, x_{k+1})$, then $I(y) \le 2^{k+1}$. We have

$$\|Hf\|_{L_\mu^q(\tilde{R}_+^2)}^p \le \sum_k \|\chi_{U_{x_k}\setminus U_{x_{k+1}}} Hf\|_{L_\mu^q(\tilde{R}_+^2)}^p \le$$

$$\le \sum_k 2^{(k+1)p}\|\chi_{U_{x_k}\setminus U_{x_{k+1}}}\|_{L_\mu^q(\tilde{R}_+^2)}^p =$$

$$= 4^p \sum_k \left(\int_{x_{k-1}}^{x_k} f(y)dy \right)^p (\mu(U_{x_k} \setminus U_{x_{k+1}}))^{p/q} \le$$

$$\le 4^p \left(\int_{x_{k-1}}^{x_k} (f(y))^p dy \right) (x_k - x_{k-1})^{p-1} (\mu(U_{x_k} \setminus U_{x_{k+1}}))^{p/q} \le$$

$$\le 4^p A^p \|f\|_{L^p(R_+)}^p.$$

Necessity. Let $r > 0$ and $f_r(x) = \chi_{[0,r)}(x)$. Then $\|f_r\|_{L^p(R_+)} = r^{1/p}$. On the other hand,

$$\|Hf_r\|_{L^q_\mu(\widetilde{R}^2_+)} \ge \|\chi_{U_r} Hf_r\|_{L^q_\mu(\widetilde{R}^2_+)} \ge (\mu(U_r))^{1/q} r.$$

Hence the boundedness of H implies that $A < \infty$. □

Lemma 2.9.2. *Let* $0 < q < p \le \infty, p > 1$ *and let* v *be an almost everywhere positive measurable function on* $\widetilde{R}^2_+$. *Then the operator* H *is bounded from* $L^p(R_+)$ *into* $L^q_v(\widetilde{R}^2_+)$ *if and only if*

$$A_1 \equiv \left(\int_0^\infty \left(\int_{U_x} v(y,t)dydt \right)^{\frac{p}{p-q}} x^{\frac{(q-1)p}{p-q}} dx \right)^{\frac{p-q}{pq}} < \infty.$$

Moreover, $\lambda_1 A_1 \le \|H\| \le \lambda_2 A_1$, *where* $\lambda_1 = \left(\frac{p-q}{p-1} \right)^{1/q'} q^{1/q}$ *and* $\lambda_2 = (p')^{1/q'} q^{1/q}$ *for* $q > 1$, $\lambda_1 = \lambda_2 = 1$ *for* $q = 1$, $\lambda_1 = (q/p')^{\frac{p-q}{pq}} (p')^{1/p'} q^{1/p} \frac{p-q}{p}$ *and* $\lambda_2 = \left(\frac{p}{p-q} \right)^{\frac{p-q}{pq}} p^{1/p} (p')^{1/p'}$ *for* $0 < q < 1$.

Proof. Applying Lemma 1.3.2 from [195] for $1 \le q < p < \infty$ and using the arguments from [273] for $0 < q < 1 < p < \infty$ we find that the condition $A_1 < \infty$ is equivalent to the boundedness of H from $L^p(R_+)$ into $L^q_{\widetilde{v}}(R_+)$, where

$$\widetilde{v}(y) = \int_0^\infty v(y,t)dt.$$

But

$$\|Hf\|_{L^q_{\widetilde{v}}(R_+)} = \|Hf\|_{L^q_v(\widetilde{R}^2_+)}.$$

Therefore the condition $A_1 < \infty$ is equivalent to the boundedness of H from $L^p(R_+)$ into $L^q_v(\widetilde{R}^2_+)$. The constants λ_1 and λ_2 are from [195] (Section 1.3.2) for $q \ge 1$, and from [273](see Theorem 2.4 and Remark) for $0 < q < 1$. □

Now we establish boundedness criteria for the operators T_α and $\widetilde{T}_\alpha$.

Theorem 2.9.1. *Let $1 < p \leq q < \infty$, $\alpha > 1/p$, ν be a positive σ-finite measure on $\widetilde{R}_+^2$. Then the following conditions are equivalent:*

(i) T_α *is bounded from $L^p(R_+)$ into $L_\nu^q(\widetilde{R}_+^2)$;*

(ii)

$$B \equiv \sup_{r>0} \left(\int\limits_{U_r} (x+t)^{(\alpha-1)q} d\nu(x,t) \right)^{\frac{1}{q}} r^{\frac{1}{p'}} < \infty;$$

(iii)

$$B_1 \equiv \sup_{k \in Z} \left(\int\limits_{U_{2^k} \setminus U_{2^{k+1}}} (x+t)^{(\alpha-1)q} x^{q/p'} d\nu(x,t) \right)^{\frac{1}{q}} < \infty.$$

Moreover, there exist positive constants b_1, b_2, b_3 and b_4 depending only on p, q and α such that

$$b_1 B \leq \|T_\alpha\| \leq b_2 B, \quad b_3 B_1 \leq \|T_\alpha\| \leq b_4 B_1.$$

Proof. First we shall show that (ii) implies (i). Let $f \geq 0$. If $\alpha \geq 1$, then using Lemma 2.9.1 we obtain

$$\|T_\alpha f\|_{L_\nu^q} \leq \left(\int\limits_{\widetilde{R}_+^2} (x+t)^{(\alpha-1)q} \left(\int\limits_0^x f(y)dy \right)^q d\nu(x,t) \right)^{1/q} \leq$$

$$\leq 4B\|f\|_{L^p(R_+)}.$$

Now let $1/p < \alpha < 1$. We have

$$\|T_\alpha f\|_{L_\nu^q(\widetilde{R}_+^2)} \leq \left(\int\limits_{\widetilde{R}_+^2} \left(\int\limits_0^{x/2} f(y)(x-y+t)^{\alpha-1}dy \right)^q d\nu(x,t) \right)^{1/q} +$$

$$+ \left(\int\limits_{\widetilde{R}_+^2} \left(\int\limits_{x/2}^x f(y)(x-y+t)^{\alpha-1}dy \right)^q d\nu(x,t) \right)^{1/q} \equiv S_1 + S_2.$$

If $y < x/2$, then $(x-y+t)^{\alpha-1} \leq 2^{1-\alpha}(x+t)^{\alpha-1}$. By Lemma 2.9.1 we obtain

$$S_1 \leq 2^{1-\alpha} \left(\int\limits_{\widetilde{R}_+^2} (Hf(x))^q (x+t)^{(\alpha-1)q} d\nu(x,t) \right)^{1/q} \leq 2^{3-\alpha} B\|f\|_{L^p(R_+)}.$$

Using Hölder's inequality, we find that

$$S_2^q \leq \int\limits_{\widetilde{R}_+^2} \Big(\int\limits_{x/2}^{x} (f(y))^p dy \Big)^{q/p} (\varphi(x,t))^{q/p'} d\nu(x,t),$$

where

$$\varphi(x,t) \equiv \int\limits_{x/2}^{x} (x - y + t)^{(\alpha-1)p'} dy.$$

Moreover, $\varphi(x,t) \leq c_1 (x+t)^{(\alpha-1)p'} x$, where $c_1 = 2^{(1-\alpha)p'-1} 3((\alpha-1)p' + 1)^{-1}$. Indeed, if $t \leq x$ then $\varphi(x,t) \leq ((\alpha-1)p' + 1)^{-1}(x/2 + t)^{(\alpha-1)p'+1} \leq c_2(x+t)^{(\alpha-1)p'} x$, where $c_2 = 2^{(1-\alpha)p'-1} 3((\alpha - 1)p' + 1)^{-1}$. Let $t > x$. Then $\varphi(x,t) \leq t^{(\alpha-1)p'} x/2 \leq 2^{(1-\alpha)p'-1}(x+t)^{(\alpha-1)p'} x$. Using Minkowski's inequality we obtain

$$S_2^q \leq c_1^{q/p'} \int\limits_{\widetilde{R}_+^2} \Big(\int\limits_{x/2}^{x} (f(y))^p dy \Big)^{q/p} (x+t)^{(\alpha-1)q} x^{q/p'} d\nu(x,t) \leq$$

$$\leq c_1^{q/p'} \Big(\int\limits_{0}^{\infty} (f(y))^p \Big(\int\limits_{U_y \setminus U_{2y}} (x+t)^{(\alpha-1)q} x^{q/p'} d\nu(x,t) \Big)^{p/q} dy \Big)^{q/p} \leq$$

$$\leq 2^{q/p'} c_1^{q/p'} \Big(\int\limits_{0}^{\infty} (f(y))^p \Big(\int\limits_{U_y} (x+t)^{(\alpha-1)q} d\nu(x,t) \Big)^{p/q} y^{p/p'} dy \Big)^{q/p} \leq$$

$$\leq (2c_1)^{q/p'} B^q \|f\|_{L^p(R_+)}^q.$$

Now we shall show that (i) $\Rightarrow$ (iii). Let $k \in Z$ and $f_k(x) = \chi_{[0,2^{k-1})}(x)$. Then $\|f_k\|_{L^p(R_+)} = 2^{(k-1)/p}$. On the other hand,

$$\|T_\alpha f_k\|_{L_\nu^q(\widetilde{R}_+^2)} \geq c_3 \Big(\int\limits_{U_{2^k} \setminus U_{2^{k+1}}} (x+t)^{(\alpha-1)q} 2^{(k-1)q} d\nu(x,t) \Big)^{1/q}.$$

Therefore $c_4 B_1 \leq \|T_\alpha\| < \infty$, where $c_4 = 3^{\alpha-1} 2^{-2/p'+1-\alpha}$ if $1/p < \alpha < 1$ and $c_4 = 2^{1-\alpha-2/p'}$ if $\alpha \geq 1$.

Analogously we can show that $c_5 B \leq \|T_\alpha\|$, where $c_5 = 3^{\alpha-1} 2^{1/p-\alpha}$ if $1/p < \alpha < 1$ and $c_5 = 2^{1/p-\alpha}$ for $\alpha \geq 1$.

Now let $r > 0$. Then $r \in [2^m, 2^{m+1})$ for some $m \in Z$. Therefore

$$\Big(\int\limits_{U_r} (x+t)^{(\alpha-1)q} d\nu(x,t) \Big) r^{q/p'} \leq 2^{(m+1)q/p'} \int\limits_{U_{2^m}} (x+t)^{(\alpha-1)q} d\nu(x,t) =$$

$$= 2^{q/p'} 2^{mq/p'} \sum_{k=m}^{+\infty} \int_{U_{2k} \setminus U_{2k+1}} (x+t)^{(\alpha-1)q} d\nu(x,t) \le$$

$$\le 2^{q/p'} B_1^q 2^{mq/p'} \sum_{k=m}^{+\infty} 2^{-kq/p'} = 2^{q/p'} (1 - 2^{-q/p'})^{-1} B_1^q.$$

Thus (iii) implies (ii). Hence (ii) $\Rightarrow$ (i) $\Rightarrow$ (iii) $\Rightarrow$ (ii). $\square$

Remark 2.9.1. For the constants b_1, b_2, b_3 and b_4 from Theorem 2.9.1 we have: $b_1 = 3^{\alpha-1} 2^{1/p-\alpha}$, $b_2 = 2^{3-\alpha} + 3^{1/p'} 2^{1-\alpha} ((\alpha-1)p'+1)^{-1/p'}$, $b_3 = 3^{\alpha-1} 2^{-2/p'+1-\alpha}$ in the case, where $1/p < \alpha < 1$ and $b_1 = 2^{1/p-\alpha}$, $b_2 = 4$, $b_3 = 2^{-2/p'+1-\alpha}$ if $\alpha \ge 1$. $b_4 = 2^{1/p'} (1 - 2^{-q/p'})^{-1/q} b_2$.

Let us now consider the case $q < p$.

Theorem 2.9.2. *Let $0 < q < p < \infty$, $p > 1$ and $\alpha > 1/p$. Assume that v is an almost everywhere positive Lebesgue-measurable function on $\tilde{R}_+^2$. Then the operator T_α is bounded from $L^p(R_+)$ into $L_v^q(\tilde{R}_+^2)$ if and only if*

$$D \equiv \left(\int_0^\infty \left(\int_{U_x} (y+t)^{(\alpha-1)q} v(y,t) dy dt \right)^{\frac{p}{p-q}} x^{\frac{(q-1)p}{p-q}} dx \right)^{\frac{p-q}{pq}} < \infty.$$

Moreover, there exist positive constants d_1 and d_2 depending only on p, q and α such that

$$d_1 D \le \|T_\alpha\| \le d_2 D.$$

Proof. Let $f \ge 0$ and let $\alpha \ge 1$. Then using Lemma 2.9.2 we obtain

$$\|T_\alpha f\|_{L_v^q} \le \left(\int_{\tilde{R}_+^2} (x+t)^{(\alpha-1)q} \left(\int_0^x f(y) dy \right)^q v(x,t) dx dt \right)^{1/q} \le$$

$$\le \lambda_2 D \|f\|_{L^p(R_+)},$$

where λ_2 is from Lemma 2.9.2. Now let $1/p < \alpha < 1$. Then as in the proof of Theorem 2.9.1, we have

$$\|T_\alpha f\|_{L_v^q(\tilde{R}_+^2)} \le c_1 \left(\int_{\tilde{R}_+^2} \left(\int_0^{x/2} f(y)(x-y+t)^{\alpha-1} dy \right)^q v(x,t) dx dt \right)^{1/q} +$$

$$+ c_1 \left(\int_{\tilde{R}_+^2} \left(\int_{x/2}^x f(y)(x-y+t)^{\alpha-1} dy \right)^q v(x,t) dx dt \right)^{1/q} \equiv I_1 + I_2,$$

where $c_1 = 1$ if $q \geq 1$ and $c_1 = 2^{1/q-1}$ if $0 < q < 1$. By virtue of Lemma 2.9.2, for I_1 we obtain

$$I_1 \leq 2^{1-\alpha}c_1 \Big(\int_{\widetilde{R}_+^2} (Hf(x))^q (x+t)^{(\alpha-1)q} v(x,t)dxdt \Big)^{1/q} \leq$$

$$\leq c_1 \lambda_2 2^{1-\alpha} D \|f\|_{L^p(R_+)}.$$

Applying Hölder's inequality twice, we find

$$I_2^q \leq c_2 \int_{\widetilde{R}_+^2} \Big(\int_{x/2}^{x} (f(y))^p dy \Big)^{q/p} (x+t)^{(\alpha-1)q} x^{q/p'} v(x,t)dxdt \leq$$

$$\leq c_2 \sum_{k \in Z} \Big(\int_{2^{k-1}}^{2^{k+1}} (f(y))^p dy \Big)^{q/p} \times$$

$$\times \Big(\int_{U_{2^k} \backslash U_{2^{k+1}}} (x+t)^{(\alpha-1)q} x^{q/p'} v(x,t)dxdt \Big) \leq$$

$$\leq c_2 \Big(\sum_{k \in Z} \int_{2^{k-1}}^{2^{k+1}} (f(y))^p dy \Big)^{q/p} \Big(\sum_{k \in Z} \Big(\int_{U_{2^k} \backslash U_{2^{k+1}}} (x+t)^{(\alpha-1)q} x^{q/p'} \times$$

$$\times v(x,t)dxdt \Big)^{\frac{p}{p-q}} \Big)^{\frac{p-q}{p}} \leq 2^{q/p} c_2 \|f\|_{L^p(R_+)}^q \widetilde{B}_1,$$

where $c_2 = c_1^q (3 \cdot 2^{(1-\alpha)p'-1}((\alpha-1)p'+1)^{-1})^{q/p'}$ and

$$\widetilde{B}_1 \equiv \Big(\sum_{k \in Z} \Big(\int_{U_{2^k} \backslash U_{2^{k+1}}} (x+t)^{(\alpha-1)q} x^{q/p'} v(x,t)dxdt \Big)^{\frac{p}{p-q}} \Big)^{\frac{p-q}{p}} \equiv$$

$$\equiv \Big(\sum_{k \in Z} \widetilde{B}_{1,k} \Big)^{\frac{p-q}{p}}.$$

For $\widetilde{B}_{1,k}$ we have

$$\widetilde{B}_{1,k} \leq 2^{\frac{(k+1)q(p-1)}{p-q}} \Big(\int_{U_{2^k} \backslash U_{2^{k+1}}} (x+t)^{(\alpha-1)q} v(x,t)dxdt \Big)^{\frac{p}{p-q}} \leq$$

$$\leq c_3 \int_{2^{k-1}}^{2^k} y^{\frac{p(q-1)}{p-q}} \Big(\int_{U_y} (x+t)^{(\alpha-1)q} v(x,t)dxdt \Big)^{\frac{p}{p-q}} dy,$$

where $c_3 = 4^{\frac{(p-1)q}{p-q}} \frac{q(p-1)}{p-q} \left(2^{\frac{(p-1)q}{p-q}} - 1\right)^{-1}$. Therefore $\tilde{B}_1 \leq (c_3)^{\frac{p-q}{p}} D^q$. Finally, we obtain $I_2 \leq c_4 D \|f\|_{L^p(R_+)}$, where $c_4 = 2^{1/p} (c_2)^{1/q} (c_3)^{\frac{p-q}{pq}}$.

Now let us prove necessity. Let T_α be bounded from $L^p(R_+)$ into $L^q_v(\tilde{R}^2_+)$. Then for each $x \in (0, \infty)$ we have

$$\int_{U_x} v(y,t)(y+t)^{(\alpha-1)q} dy dt < \infty.$$

Let $n \in Z$ and

$$f_n(x) = \left(\int_x^\infty \bar{v}_n(y) dy\right)^{\frac{1}{p-q}} x^{\frac{q-1}{p-q}},$$

where

$$\bar{v}_n(x) = \left(\int_0^\infty v(x,t)(x+t)^{(\alpha-1)q} dt\right) \chi_{(1/n,n)}(x).$$

The boundedness of T_α implies that $f_n(x) < \infty$ for each $x \in R_+$. Applying integration by parts, we obtain

$$\|f_n\|_{L^p(R_+)} = \left(\int_0^\infty \left(\int_x^\infty \bar{v}_n(y) dy\right)^{\frac{p}{p-q}} x^{\frac{p(q-1)}{p-q}} dx\right)^{1/p} =$$

$$= \left(\frac{p'}{q} \int_0^\infty \left(\int_x^\infty \bar{v}_n(y) dy\right)^{\frac{q}{p-q}} \bar{v}_n(x) x^{\frac{q(p-1)}{p-q}} dx\right)^{1/p} < \infty.$$

On the other hand,

$$\|T_\alpha\|_{L^q_v(\tilde{R}^2_+)} \geq \left(\int_{\tilde{R}^2_+} \left(\int_0^{x/2} f_n(y)(x-y+t)^{\alpha-1} dy\right)^q v(x,t) dx dt\right)^{1/q} \geq$$

$$\geq \left(\int_{\tilde{R}^2_+} \left(\int_x^\infty \bar{v}_n(y) dy\right)^{\frac{q}{p-q}} \times\right.$$

$$\times \left(\int_0^{x/2} (x-y+t)^{\alpha-1} y^{\frac{q-1}{p-q}} dy\right)^q v(x,t) dx dt\right)^{1/q} \geq$$

$$\geq c_5 \left(\int_{\tilde{R}^2_+} v(x,t) \left(\int_x^\infty \bar{v}_n(y) dy\right)^{\frac{q}{p-q}} (x+t)^{(\alpha-1)q} x^{\frac{q(p-1)}{p-q}} dx dt\right)^{1/q} =$$

$$= c_5 \left(\int\limits_0^\infty \left(\int\limits_0^\infty v(x,t)(x+t)^{(\alpha-1)q} dt \right) \times \right.$$

$$\left. \times \left(\int\limits_x^\infty \bar{v}_n(y) dy \right)^{-\frac{q}{p-q}} x^{\frac{(p-1)q}{p-q}} dx \right)^{1/q} \geq$$

$$\geq c_5 \left(\int\limits_0^\infty \bar{v}_n(x) \left(\int\limits_x^\infty \bar{v}_n(y) dy \right)^{-\frac{q}{p-q}} x^{\frac{(p-1)q}{p-q}} dx \right)^{1/q} =$$

$$= c_6 \left(\int\limits_0^\infty \left(\int\limits_x^\infty \bar{v}_n(y) dy \right)^{\frac{p}{p-q}} x^{\frac{(q-1)p}{p-q}} dx \right)^{1/q},$$

with $c_6 = (q/p')^{1/q} 2^{-\frac{p-1}{p-q}} \frac{p-q}{p-1} c_7$, where $c_7 = (\frac{3}{2})^{\alpha-1}$ if $1/p < \alpha < 1$ and $c_7 = (\frac{1}{2})^{\alpha-1}$ if $\alpha \geq 1$. Therefore

$$c_6 \left(\int\limits_0^\infty \left(\int\limits_x^\infty \bar{v}_n(y) dy \right)^{\frac{p}{p-q}} x^{\frac{(q-1)p}{p-q}} dx \right)^{\frac{p-q}{pq}} \leq \|T_\alpha\|.$$

By virtue of Fatou's lemma we finally conclude that $c_6 D \leq \|T_\alpha\| < \infty$. $\square$

Remark 2.9.2. It follows from the proof of Theorem 2.9.2 that for the constants d_1 and d_2 we have: $d_1 = \left(\frac{q}{p'} \right)^{1/q} 2^{\frac{1-p}{p-q}} \frac{p-q}{p-1} \gamma_1(\alpha)$, where $\gamma_1(\alpha) = (3/2)^{\alpha-1}$ if $1/p < \alpha < 1$ and $\gamma_1(\alpha) = (1/2)^{\alpha-1}$ if $\alpha \geq 1$, $d_2 = \lambda_2$ for $\alpha \geq 1$, and if $1/p < \alpha < 1$, then $d_2 = \lambda_2 \gamma_2(q) 2^{1-\alpha} + 2^{2/p-\alpha} 3^{1/p'} ((\alpha-1)p' + 1)^{-1/p'} 4^{1/p'} \left(\frac{q(p-1)}{p-q} \right)^{\frac{p-q}{pq}} \left(2^{\frac{p-q}{p-q}} - 1 \right)^{-\frac{p-q}{pq}} \gamma_2(q)$, where $\gamma_2(q) = 1$ for $q \geq 1$, $\gamma_2(q) = 2^{1/q-1}$ for $0 < q < 1$.

Using duality arguments, we readily obtain the following theorems:

Theorem 2.9.3. *Let* $1 < p \leq q < \infty$, $\alpha > (q-1)/q$. *Then the following conditions are equivalent:*
(i) $\widetilde{T}_\alpha$ *is bounded from* $L^p_\nu(\widetilde{R}^2_+)$ *into* $L^q(R_+)$;
(ii)

$$\widetilde{B} \equiv \sup_{r>0} \left(\int\limits_{U_r} (x+t)^{(\alpha-1)p'} d\nu(x,t) \right)^{\frac{1}{p'}} r^{\frac{1}{q}} < \infty;$$

(iii)

$$\widetilde{B}_1 \equiv \sup_{k \in Z} \left(\int\limits_{U_{2^k} \setminus U_{2^{k+1}}} (x+t)^{(\alpha-1)p'} x^{p'/q} d\nu(x,t) \right)^{\frac{1}{p'}} < \infty.$$

Moreover, there exist positive contants $\tilde{b}_1, \tilde{b}_2, \tilde{b}_3$ and $\tilde{b}_4$ depending only on p, q and α such that

$$\tilde{b}_1\tilde{B} \leq \|\tilde{T}_\alpha\| \leq \tilde{b}_2\tilde{B}, \quad \tilde{b}_3\tilde{B}_1 \leq \|\tilde{T}_\alpha\| \leq \tilde{b}_4\tilde{B}_1.$$

Theorem 2.9.4. *Let* $1 < q < p < \infty$ *and* $\alpha > (q-1)/q$. *Let* $d\nu(x,y) = w(x,t)dxdt$, *where* w *is a Lebesgue- measurable a.e. positive function on* $\tilde{R}_+^2$. *Then* $\tilde{R}_\alpha$ *is bounded from* $L^p_w(\tilde{R}_+^2)$ *into* $L^q(R_+)$ *if and only if*

$$\tilde{D} \equiv \left(\int\limits_0^\infty \left(\int\limits_{U_x} (y+t)^{(\alpha-1)p'} w(y,t)dydt \right)^{\frac{q(p-1)}{p-q}} x^{\frac{q}{p-q}} dx \right)^{\frac{p-q}{pq}} < \infty.$$

Moreover, $\tilde{d}_1\tilde{D} \leq \|\tilde{T}_\alpha\| \leq \tilde{d}_2\tilde{D}$, *where the positive constants* $\tilde{d}_1$ *and* $\tilde{d}_2$ *depend only on* p, q *and* α.

In the following theorems, criteria for the compactness of the operators T_α and $\tilde{T}_\alpha$ are established. First we shall prove

Lemma 2.9.3. *Let* $1 < p \leq q < \infty$ *and* $\alpha > 1/p$. *If*
(i) $B < \infty$;
(ii) $\lim\limits_{a\to 0} B^{(a)} = \lim\limits_{b\to+\infty} B^{(b)} = 0$, *where*

$$B^{(a)} \equiv \sup_{0<r<a} \left(\int\limits_{U_r \setminus U_a} (x+t)^{(\alpha-1)q}d\nu(x,t) \right)^{1/q} r^{1/p'},$$

$$B^{(b)} \equiv \sup_{r>b} \left(\int\limits_{U_r} (x+t)^{(\alpha-1)q}d\nu(x,t) \right)^{1/q} r^{1/p'},$$

then T_α *is compact from* $L^p(R_+)$ *into* $L^q_\nu(\tilde{R}_+^2)$.

Proof. Let us represent T_α as

$$T_\alpha f = \chi_{V_a} T_\alpha(\chi_{[0,a)}f) + \chi_{V_b \setminus V_a} T_\alpha(\chi_{(0,b)}f) + \chi_{U_b} T_\alpha(\chi_{(0,b/2]}f) +$$
$$+ \chi_{U_b} T_\alpha(\chi_{(b/2,\infty)}f) \equiv P_1f + P_2f + P_3f + P_4f,$$

where $V_r \equiv [0,r) \times R_+$ (It is obvious that $[a,b) \times R_+ = V_b \setminus V_a$).
For P_2 we have

$$P_2f(x,t) = \int\limits_0^\infty \overline{k}(x,t,y)f(y)dy,$$

where $\bar{k}(x,t,y) = \chi_{V_b \setminus V_a}(x,t)\chi_{(0,x)}(y)(x-y+t)^{\alpha-1}$. Moreover, using the inequality

$$\int_0^x (x-y+t)^{(\alpha-1)p'}\,dy \le b(x+t)^{(\alpha-1)p'}x,$$

where the constant $b > 0$ is independent of x and t, we get

$$\||\bar{k}(x,t,y)\|_{L^{p'}(R_+)}\|_{L^q_\nu(\widetilde{R}^2_+)} =$$

$$= \Big(\int_{V_b \setminus V_a} \Big(\int_0^x (x-y+t)^{(\alpha-1)p'}\,dy \Big)^{q/p'} d\nu(x,t) \Big)^{1/q} \le$$

$$\le c_1 \Big(\int_{V_b \setminus V_a} (x+t)^{(\alpha-1)q}x^{q/p'}\,d\nu(x,t) \Big)^{1/q} < \infty.$$

For P_3 we obtain $P_3 f(x,t) = \int_0^\infty \widetilde{k}(x,t,y)f(y)\,dy$, where

$$\widetilde{k}(x,t,y) = \chi_{U_b}(x,t)\chi_{(0,b/2]}(y)(x-y+t)^{\alpha-1}.$$

It can be easily verified that

$$\||\widetilde{k}(x,t,y)\|_{L^{p'}(R_+)}\|_{L^q_\nu(\widetilde{R}^2_+)} < \infty.$$

Using Theorem C of Section 2.1 we conclude that P_2 and P_3 are compact operators.

By Theorem 2.9.1 we have

$$\|P_1\| \le b_2 B^{(a)} < \infty \quad and \quad \|P_4\| \le b_2 B^{(b/2)} < \infty, \qquad (2.9.1)$$

where b_2 is from Theorem 2.9.1. Hence we obtain

$$\|T_\alpha - P_2 - P_3\| \le \|P_1\| + \|P_4\| \to 0 \qquad (2.9.2)$$

as $a \to 0$ and $b \to +\infty$. Therefore T_α is compact as it is a limit of a sequence of compact operators. $\square$

Theorem 2.9.5. *Let p, q and α satisfy the conditions of Lemma* 2.9.3. *Then the following conditions are equivalent:*
(i) *T_α is compact from $L^p(R_+)$ to $L^q_\nu(\widetilde{R}^2_+)$;*
(ii) *$B < \infty$ and $\lim_{a \to 0} B^{(a)} = \lim_{b \to +\infty} B^{(b)} = 0$;*
(iii) *$B < \infty$ and $\lim_{r \to 0} B(r) = \lim_{r \to +\infty} B(r) = 0$, where*

$$B(r) \equiv \Big(\int_{U_r} (x+t)^{(\alpha-1)q}\,d\nu(x,t) \Big)^{\frac{1}{q}} r^{\frac{1}{p'}};$$

(iv) $B_1 < \infty$ and $\lim\limits_{k\to-\infty} B_1(k) = \lim\limits_{k\to+\infty} B_1(k) = 0$, where

$$B_1(k) \equiv \Big(\int\limits_{U_{2k}\setminus U_{2k+1}} (x+t)^{(\alpha-1)q} x^{q/p'}\, d\nu(x,t) \Big)^{\frac{1}{q}}.$$

Proof. By Lemma 2.9.3 we have (ii) $\Rightarrow$ (i). Now let us show that (iii) $\Rightarrow$ (ii). Since

$$B^{(a)} \le \sup\limits_{0<r<a} B(r) \quad and \quad B^{(b)} = \sup\limits_{r>b} B(r),$$

we obtain $B^{(a)} \to 0$ as $a \to 0$ and $B^{(b)} \to +\infty$ as $b \to \infty$. Therefore (iii) $\Rightarrow$ (ii). Now let T_α be compact from $L^p(R_+)$ into $L^q_\nu(\widetilde{R}^2_+)$. Let $r > 0$ and $f_r(x) = \chi_{(0,r/2)}(x)r^{-1/p}$. It can be easily checked that f_r weakly converges to 0 if $r \to 0$. On the other hand,

$$\|T_\alpha f_r\|_{L^q_\nu(\widetilde{R}^2_+)} \ge c_1 B(r) \to 0$$

as $r \to 0$, since $T_\alpha f_r$ converges strongly to 0. If we take

$$g_r(x,t) = \chi_{U_r}(x,t)(x+t)^{(\alpha-1)(q-1)} \Big(\int\limits_{U_r} (y+t)^{(\alpha-1)q} d\nu(y,t) \Big)^{-1/q'},$$

then g_r weakly converges to 0 as $r \to +\infty$. Since $\widetilde{T}_\alpha$ is compact from $L^{q'}_\nu(\widetilde{R}^2_+)$ into $L^{p'}(R_+)$ and

$$\|\widetilde{T}_\alpha g_r\|_{L^{p'}(R_+)} \ge c_2 B(r),$$

we obtain $\lim\limits_{r\to+\infty} B(r) = 0$. Therefore (i) $\Rightarrow$ (iii).

Now we shall prove that (ii) follows from (iv). Using Theorem 2.9.1, we see that $B \le b_1 B_1$. Let $a > 0$. Then $a \in [2^m, 2^{m+1})$ for some $m \in Z$, and so $B^{(a)} \le \sup\limits_{0<r<2^m} B_{2^m,r} \equiv B^{(2^m)}$, where

$$B_{2^m,r} \equiv \Big(\int\limits_{U_r\setminus U_{2m}} (x+t)^{(\alpha-1)q} d\nu(x,t) \Big)^{\frac{1}{q}} r^{\frac{1}{p'}}.$$

If $r \in [0, 2^m)$, then $r \in [2^{j-1}, 2^j)$ for some $j \in Z$, $j \le m$. Furthermore,

$$B^q_{2^m,r} \le 2^{\frac{iq}{p'}} \sum\limits_{k=j}^{m} \int\limits_{U_{2k-1}\setminus U_{2k}} (x+t)^{(\alpha-1)q} d\nu(x,t) \le c_3 \Big(\sup\limits_{k\le m} B_1(k-1) \Big)^q.$$

Hence $B^{(2^m)} \le c_4 B_1^{(m)}$, where $B_1^{(m)} \equiv \sup\limits_{k\le m} B_1(k-1)$. If $a \to 0$, then $m \to -\infty$ and $B_1^{(m)} \to 0$. Therefore $\lim\limits_{a\to 0} B^{(a)} = 0$.

Now let $\tau > 0$. Then $\tau \in [2^m, 2^{m+1})$ and we have

$$B^q(\tau) \leq c_5 B^q(2^m) = c_5 2^{\frac{mq}{p'}} \sum_{k=m}^{+\infty} \int_{U_{2^k} \setminus U_{2^{k+1}}} (x+t)^{(\alpha-1)q} d\nu(x,t) \leq$$

$$\leq c_6 (\sup_{k \geq m} B_1(k))^q.$$

It readily follows that

$$\lim_{\tau \to +\infty} B(\tau) \leq c_7 \lim_{m \to +\infty} \sup_{k \geq m} B_1(k) = 0$$

and $\lim_{b \to +\infty} B^{(b)} = 0$. Thus (iv) $\Rightarrow$ (ii). Now let T_α is compact from $L^p(R_+)$ into $L_\nu^q(\widetilde{R}_+^2)$, $k \in Z$ and

$$f_k(x) = \chi_{[2^{k-2}, 2^{k-1})}(x) 2^{-k/p}.$$

Then the sequence f_k converges weakly to 0 as $k \to -\infty$ or $k \to +\infty$. Moreover, it is easy to show that

$$\|T_\alpha f_k\|_{L_\nu^q(\widetilde{R}_+^2)} \geq c_8 B_1(k).$$

Therefore (iv) is valid. Finally, we obtain (i) $\Leftrightarrow$ (iii), (iv) $\Rightarrow$ (ii) $\Rightarrow$ (i) $\Rightarrow$ (iv). $\square$

Our next theorem is proved in a similar manner. It is also a corollary of the well-known Ando theorem (see, e.g., [165] and [10], §5).

Theorem 2.9.6. *Let p, q, α and v satisfy the condition of Theorem 2.9.6. Then T_α is compact from $L^p(R_+)$ into $L_v^q(\widetilde{R}_+^2)$ if and only if $D < \infty$.*

By duality arguments we obtain the two theorems:

Theorem 2.9.7. *Let $1 < p \leq q < \infty$, $\alpha > \frac{q-1}{q}$. It is assumed that ν is a positive σ-finite measure. Then the following conditions are equivalent:*
(i) $\widetilde{T}_\alpha$ *is compact from $L_\nu^p(\widetilde{R}_+^2)$ into $L^q(R_+)$;*
(ii) $\widetilde{B} < \infty$ *and* $\lim_{a \to 0} \widetilde{B}^{(a)} = \lim_{b \to +\infty} \widetilde{B}^{(b)} = 0$, *where*

$$\widetilde{B}^{(a)} \equiv \sup_{0 < r < a} \left(\int_{U_r \setminus U_a} (x+t)^{(\alpha-1)p'} d\nu(x,t) \right)^{1/p'} r^{1/q},$$

$$\widetilde{B}^{(b)} \equiv \sup_{r > b} \widetilde{B}(r) \equiv \sup_{r > b} \left(\int_{U_r} (x+t)^{(\alpha-1)p'} d\nu(x,t) \right)^{1/p'} r^{1/q};$$

(iii) $\widetilde{B} < \infty$ *and* $\lim_{r \to 0} \widetilde{B}(r) = \lim_{r \to +\infty} \widetilde{B}(r) = 0$;

(iv) $\widetilde{B}_1 < \infty$ and $\lim\limits_{k \to -\infty} \widetilde{B}_1(k) = \lim\limits_{k \to +\infty} \widetilde{B}_1(k) = 0$, where

$$\widetilde{B}_1(k) \equiv \left(\int\limits_{U_{2^k} \setminus U_{2^{k+1}}} (x+t)^{(\alpha-1)p'} x^{p'/q} d\nu(x,t) \right)^{\frac{1}{q}}.$$

Theorem 2.9.8. *Let* $1 < q < p < \infty$ *and* $\alpha > \frac{q-1}{q}$. *Suppose that* $d\nu(x,t) = w(x,t)dxdt$, *where* w *is a measurable a.e. positive function on* $\widetilde{R}_+^2$. *Then* $\widetilde{T}_\alpha$ *is compact from* $L_w^p(\widetilde{R}_+^2)$ *into* $L^q(R_+)$ *if and only if* $\widetilde{D} < \infty$.

As above our next concern is the distance of the operator T_α from a space of compact operators.

It is assumed that v is a Lebesgue-measurable almost everywhere positive function on $\widetilde{R}_+^2$.

We need the following lemmas (see [66], Chapter V, Corollary 5.4).:

Lemma 2.9.4. *Let* $1 \leq q < \infty$ *and* $P \in B(X,Y)$, *where* $Y = L^q(\widetilde{R}_+^2)$. *Then*

$$dist(P, \mathcal{K}(X,Y)) = dist(P, \mathcal{F}_r(X,Y)).$$

Our next lemma is proved in the same way as Lemma V.5.6 in [66] (see also [226], Lemma 2.2).

Lemma 2.9.5. *Let* $1 \leq q < \infty$ *and* $Y = L^q(\widetilde{R}_+^2)$. *It is assumed that* $P \in \mathcal{F}_r(X,Y)$ *and* $\epsilon > 0$. *Then there exist* $T \in \mathcal{F}_r(X,Y)$ *and* $[\alpha,\beta] \subset (0,\infty)$ *such that* $\|P - T\| < \epsilon$ *and* $suppTf \subset [\alpha,\beta] \times R_+$ *for any* $f \in X$.

Let $T'_\alpha (0 < \alpha < 1)$ be an operator of the form $T'_\alpha f(x,t) = v^{1/q}(x,t)T_\alpha f(x,t)$. We denote

$$\widetilde{I} \equiv dist(T_\alpha, \mathcal{K}(X, L_v^q(\widetilde{R}_+^2))),$$

$$\overline{I} \equiv dist(T'_\alpha, \mathcal{K}(X, L^q(\widetilde{R}_+^2))).$$

Lemma 2.9.6. *Let* $1 \leq q < \infty$. *Then* $\widetilde{I} = \overline{I}$.

Proof. Let $E \equiv \{f : \|f\|_X \leq 1\}$ and $P \in \mathcal{K}(X, L_v^q(\widetilde{R}_+^2))$. Then

$$\|T_\alpha - P\| = \sup_E \|(T_\alpha - P)f\|_{L_v^q(\widetilde{R}_+^2)} =$$

$$= \sup_E \|T'_\alpha f - v^{1/q} P f\|_{L^q(\widetilde{R}_+^2)} = \|T'_\alpha - \overline{P}\|,$$

where $\overline{P} = v^{1/q}P$. But $\overline{P} \in \mathcal{K}(X, L^q(R_+^2))$. Therefore $\overline{I} \leq \widetilde{I}$. Similarly we obtain $\widetilde{I} \leq \overline{I}$. $\square$

Theorem 2.9.9. *Let* $1 < p \le q < \infty$, $\alpha > 1/p$ *and let* $X = L^p(\mathcal{R}_+)$, $Y = L^q_v(\widetilde{R}^2_+)$. *Assume that* $B < \infty$ *for* $d\nu(x,t) = v(x,t)dxdt$. *Then there exist positive constants* ϵ_1 *and* ϵ_2 *depending only on* p, q *and* α *such that*

$$\epsilon_1 J \le dist(T_\alpha, \mathcal{K}(X,Y)) \le \epsilon_2 J,$$

where $J = \lim_{a \to 0} J^{(a)} + \lim_{d \to +\infty} J^{(d)}$,

$$J^{(a)} \equiv \sup_{0 < r < a} \left(\int\limits_{U_r \setminus U_a} v(x,t)(x+t)^{(\alpha-1)q}dxdt \right)^{1/q} r^{1/p'},$$

$$J^{(d)} \equiv \sup_{r > d} \left(\int\limits_{U_r} v(x,t)(x+t)^{(\alpha-1)q}dxdt \right)^{1/q} r^{1/p'}.$$

Proof. By the inequalities (2.9.1) and (2.9.2) from the proof of Lemma 2.9.3, we obtain $\widetilde{I} \equiv dist(T_\alpha, \mathcal{K}(X,Y)) \le b_2 J$, where b_2 is from Theorem 2.9.1. Let $\lambda > \widetilde{I}$. By Lemma 2.9.6 we have $\widetilde{I} = \overline{I}$. Using Lemma 2.9.4, we find that there exists an operator of finite rank $P : X \to L^q(\widetilde{R}^2_+)$ such that $\|T'_\alpha - P\| < \lambda$. From Lemma 2.9.5 it follows that for $\epsilon = (\lambda - \|T'_\alpha - P\|)/2$ there are $T \in \mathcal{F}_r(X, L^q(\widetilde{R}^2_+))$ and $[\alpha, \beta] \subset (0, \infty)$ such that $\|P - T\| < \epsilon$ and $suppTf \subset [\alpha, \beta] \times R_+$. Therefore for all $f \in X$ we have

$$\|T'_\alpha f - Tf\|_{L^q(\widetilde{R}^2_+)} \le \lambda \|f\|_X.$$

Moreover,

$$\int\limits_{[0,\alpha] \times R_+} |T'_\alpha f(x,t)|^q dxdt +$$

$$+ \int\limits_{[\beta,\infty) \times R_+} |T'_\alpha f(x,t)|^q dxdt \le \lambda^q \|f\|^q_{L^p(R_+)}. \tag{2.9.3}$$

Now let $d > \beta$ and $r \in (d, \infty)$. Assume that $f_r(y) = \chi_{0,r/2}(y)$. Then $\|f_r\|^q_{L^p(R_+)} = 2^{-q/p} r^{q/p}$. On the other hand,

$$\int\limits_{U_r} |T'_\alpha f_r(x,t)|^q dxdt \ge \int\limits_{U_r} \left(\int\limits_0^{r/2} (x-y+t)^{\alpha-1}dy \right)^q v(x,t)dxdt \ge$$

$$\ge c_1 \left(\int\limits_{U_r} v(x,t)(x+t)^{(\alpha-1)q}dxdt \right) r^q,$$

where $c_1 = 3^{(\alpha-1)q}2^{-\alpha q}$ if $1/p < \alpha < 1$ and $c_1 = 2^{-\alpha q}$ for $\alpha \geq 1$. Therefore

$$\lambda \geq c_1^{1/q}2^{1/p}\Big(\int\limits_{U_r} v(x,t)(x+t)^{(\alpha-1)q}dxdt\Big)^{1/q}r^{1/p'}.$$

for all $r > d$. Hence we have $c_2 J^{(d)} \leq \lambda$ for any $d > \beta$ and, finally, we obtain $c_2 \lim\limits_{d\to+\infty} J^{(d)} \leq \lambda$. Since λ is arbitrarily close to $\tilde{I}$, we conclude that

$c_2 \lim\limits_{d\to+\infty} J^{(d)} \leq \tilde{I}$, where $c_2 = c_1^{1/q}2^{1/p}$.

Let us choose $n \in Z$ such that $2^n < \alpha$. Assume that $j \in Z, j \leq n - 1$ and $f_j(y) = \chi_{(0,2^{j-1})}(y)$. Then we obtain

$$\int\limits_{U_{2j} \setminus U_{2j+1}} |T'_\alpha f(x,t)|^q dxdt \geq$$

$$\geq \int\limits_{U_{2j} \setminus U_{2j+1}} v(x,y)\Big(\int\limits_0^{2^{j-1}} (x - y + t)^{\alpha-1}dy\Big)^q dxdt \geq$$

$$\geq c_3 \int\limits_{U_{2j} \setminus U_{2j+1}} v(x,y)(x+t)^{(\alpha-1)q}2^{(j-1)q}dxdt,$$

where $c_3 = (3/2)^{(\alpha-1)q}$ in the case, where $1/p < \alpha < 1$ and $c_3 = (1/2)^{(\alpha-1)q}$ for $\alpha \geq 1$. On the other hand, $\|f_j\|_X^q = 2^{(j-1)q/p}$. By (2.9.3) we find that

$$c_3^{1/q}4^{-1/p'}\overline{B}_1(j) \leq \lambda$$

for every integer $j, j \leq n - 1$, where

$$\overline{B}(j) \equiv \Big(\int\limits_{U_{2j} \setminus U_{2j+1}} v(x,t)(x+t)^{(\alpha-1)q}x^{q/p'} dxdt\Big)^{1/q}.$$

Consequently $c_3^{1/q}4^{-1/p'} \sup_{j\leq n} \overline{B}_1(j) \leq \lambda$ for every integer n satisfying the condition $2^n < \alpha$. Let $a < 2^n < \alpha$. Then $a \in [2^m, 2^{m+1})$ for some m, $m \leq n - 1$. As in the proof of Theorem 2.9.5 we have that

$$B^{(a)} \leq B^{(2^m)} \leq 2^{1/p'}(1 - 2^{-q/p'})^{-1/q} \sup_{j\leq m} \overline{B}_1(j),$$

where

$$B^{(2^m)} \equiv \sup_{0<r<2^m}{}' \Big(\int\limits_{U_r \setminus U_{2m}} v(x,t)(x+t)^{(\alpha-1)q}dxdt\Big)^{1/q}r^{1/p'}.$$

Therefore $c_4 \lim_{a \to 0} B^{(a)} \leq \lambda$ with $c_4 = 2^{-3/p'} c_3^{1/q} (1 - 2^{-q/p'})^{1/q}$. Finally we obtain $c_5 J \leq \tilde{I}$, where $c_5 = 1/2 \min\{c_2, c_4\}$. $\square$

An analogous theorem for the classical Riemann-Liouville operator R_α is proved for $\alpha > 1/p$ in Section 2.7. Estimates of the distance of R_α from the class of compact operators in the case of two weights for $\alpha > 1$ are obtained in [77], [266] (for the case $\alpha = 1$ see [67]).

Remark 2.9.3. For the constants ϵ_1 and ϵ_2 from Theorem 2.9.9 we have: $\epsilon_2 = b_2$, $\epsilon_1 = 1/2 \min\{\beta_1, \beta_2\}$, where $\beta_1 = 2^{1/p} \gamma_3$, $\beta_2 = 2^{-3/p'}(1 - 2^{-q/p'})^{1/q} \gamma_4$ with $\gamma_3 = 3^{\alpha-1} 2^{-\alpha}$ for $1/p < \alpha < 1$, $\gamma_3 = 2^{-\alpha}$ for $\alpha \geq 1$ and $\gamma_4 = (3/2)^{\alpha-1}$ for $1/p < \alpha < 1$, $\gamma_4 = (1/2)^{\alpha-1}$ if $\alpha \geq 1$.

At the end of this section we discuss the boundedness snd compactness problem for T_α in the case of two weights.

The following lemma can be derived in the same way as Lemma 2.9.1.

Lemma 2.9.7. *Let* $1 < p \leq q < \infty$ *and let* $a \in (0, \infty)$. *Then the operator*

$$H_a f(x) = \int\limits_0^{ax} f(y) dy$$

is bounded from $L_w^p(R_+)$ *into* $L_\nu^q(\tilde{R}_+)$ *if and only if*

$$\tilde{A} \equiv \sup_{r>0} \left(\nu(U_r) \right)^{1/q} \left(\int\limits_0^{at} w^{1-p'}(x) dx \right)^{1/p'} < \infty.$$

Moreover, $\|H_a\| \approx A$.

The following lemma can be obtained in the same mannaer as Lemma 2.9.2, using a change of variable and Theorem B from Section 2.1.

Lemma 2.9.8. *Let* $0 < q < p < \infty$, $p > 1$. *Then* H_a *is bounded from* $L_w^p(R_+)$ *into* $L_\nu^q(\tilde{R}_+^2)$ *if and only if*

$$\overline{A} \equiv \left(\int\limits_0^\infty \left[\left(\int\limits_{U_{t/a}} v(y,x) dy dx \right) \left(\int\limits_0^t w^{1-p'}(x) dx \right)^{q-1} \right]^{\frac{p}{p-q}} w^{1-p'}(t) dt \right)^{\frac{p-q}{pq}} < \infty.$$

Moreover, $\|H_a\| \approx \overline{A}$.

Now we prove

Theorem 2.9.10. *Let $1 < p \le q < \infty$, $\alpha > 0$. Assume that the weight w satisfies the condition*

$$\int\limits_{r/2}^{r} \frac{w^{1-p'}(x)}{(r-x+t)^{(1-\alpha)p'}}dx \le c(r+t)^{(\alpha-1)p'}\int\limits_{0}^{r/4} w^{1-p'}(x)dx, \qquad (2.9.4)$$

where c is independent of $r > 0$ and $t > 0$. Then for the boundedness of T_α from $L_w^p(R_+)$ into $L_v^q(\widetilde{R}_+^2)$ it is necessary and sufficient that

$$B \equiv \sup_{r>0}\left(\int\limits_{U_r} \frac{d\nu(x,t)}{(x+t)^{(1-\alpha)q}}\right)^{1/q}\left(\int\limits_{0}^{r/2} w^{1-p'}(x)dx\right)^{1/p'} < \infty. \qquad (2.9.5)$$

Moreover, $\|T_\alpha\| \approx B$.

Proof. Representing T_α as

$$T_\alpha f(x,t)\int\limits_{0}^{x/2} f(y)(x-y+t)^{\alpha-1}dy \int\limits_{x/2}^{x} f(y)(x-y+t)^{\alpha-1}dy \equiv$$

$$\equiv T_\alpha^{(1)}f(x,t) + T_\alpha^{(2)}f(x,t),$$

we obtain

$$\|T_\alpha f\|_\nu \le \|T_\alpha^{(1)}f\|_{L_v^q([0,\infty))} + \|T_\alpha^{(2)}f\|_{L_v^q([0,\infty))} \equiv I_1 + I_2.$$

Using Lemma 2.9.7 we have

$$I_1 \le c_1\left(\int\limits_{\widetilde{R}_+^2}\left(\int\limits_{0}^{x/2} f(y)dy\right)^q \frac{d\nu(x,t)}{(x+t)^{(1-\alpha)q}}\right)^{1/q} \le c_2\|f\|_{L_w^p(0,\infty)}.$$

For I_2 we obtain

$$I_2^q \le \int\limits_{\widetilde{R}_+^2}\left(\int\limits_{x/2}^{x} w(y)(f(y))^p dy\right)^{q/p} \times$$

$$\times\left(\int\limits_{x/2}^{x} w^{1-p'}(y)(x-y+t)^{(\alpha-1)p'}dy\right)^{q/p'} d\nu(x,t) =$$

$$= \sum_{k\in Z}\int\limits_{U_{2k}\setminus U_{2k+1}}\left(\int\limits_{x/2}^{x} w(y)(f(y))^p dt\right)^{q/p} \times$$

$$\times \left(\int_{x/2}^{x} w^{1-p'}(t)(x - y + t)^{(\alpha-1)p'} dy \right)^{q/p'} d\nu(x,t) =$$

$$\sum_{k \in Z} \left(\int_{2^{k-1}}^{2^{k+1}} w(y)(f(y))^p dt \right)^{q/p} \times$$

$$\times \int_{U_{2^k} \setminus U_{2^{k+1}}} \left(\int_{x/2}^{x} w^{1-p'}(y)(x - y + t)^{(\alpha-1)p'} dy \right)^{q/p'} d\nu(x,t) \leq$$

$$\leq c_3 \sum_{k \in Z} \left(\int_{2^{k-1}}^{2^{k+1}} w(y)(f(y))^p dt \right)^{q/p} \times$$

$$\times \int_{U_{2^k} \setminus U_{2^{k+1}}} (x + t)^{(\alpha-1)q} \left(\int_{0}^{x/2} w^{1-p'}(y) dy \right)^{q/p'} d\nu(x,t) \leq$$

$$\leq c_4 B^q \sum_{k \in Z} \left(\int_{2^{k-1}}^{2^{k+1}} w(y)(f(y))^p dy \right)^{q/p} \left(\nu(2^k, 2^{k+1}]\right)(x + t)^{(\alpha-1)q} \times$$

$$\times \left(\int_{U_{x/2}} (y + t)^{(\alpha-1)q} d\nu(y,t) \right)^{-1} d\nu(x,t) \leq c_5 B^q \|f\|_{L^p_w(0,\infty)}.$$

To prove necessity we take $f_r(x) = w^{1-p'}(x)\chi_{(0,r/2)}(x)$, where $r > 0$. Then

$$\|f_r\|_{L^p_w(0,\infty)} = \left(\int_{0}^{r/2} w^{1-p'}(x) dx \right)^{1/p} < \infty.$$

On the other hand,

$$\|T_\alpha\|_{L^q_\nu(\widetilde{R}^2_+)} \geq c_6 \left(\int_{U_r} (x + t)^{(1-\alpha)q} d\nu(x,t) \right)^{1/q} \left(\int_{0}^{r/2} w^{1-p'}(x) dx \right).$$

From the boundedness of T_α we obtain (2.9.5). $\square$

Example 2.9.1. If $1 < p < \infty$ and $\alpha > 1/p$ and either (i) the weight function w is increasing on $(0,\infty)$, or (ii) w is decreasing on $(0,\infty)$ satisfying the condition

$$\int_r^{2r} w^{1-p'}(x)dx \le b \int_0^{r/4} w^{1-p'}(x)dx \qquad (2.9.6)$$

with a positive constant b independent of t, then the condition (2.9.4) is satisfied.

From Theorem 2.9.10 we easily obtain

Theorem 2.9.11. *Let* $1 < p \le q <$ *and* $\alpha \ge 1$. *Assume that* w *satisfies the condition*

$$\int_0^{2r} w^{1-p'}(x)dx \le b_1 \int_0^r w^{1-p'}(x)dx, \qquad (2.9.7)$$

where b_1 *does not depend on* r. *Then* T_α *is bounded from* $L_w^p(0,\infty)$ *into* $L_\nu^q(\widetilde{R}_+^2)$ *if and only if* (2.9.5) *is fulfilled. Moreover,* $\|T_\alpha\| \approx B$.

Now we discuss the compactness of T_α. The following Theorem is proved in the same way as Theorem 2.9.5.

Theorem 2.9.12. *Let* $1 < p \le q < \infty$ *and* $\alpha > 1/p$. *Suppose that the weight* w *satisfies* (2.9.4). *Then* T_α *is compact from* $L_w^p(0,\infty)$ *into* $L_\nu^q(\widetilde{R}_+^2)$ *if and only if* (2.9.5) *is satisfied and*

$$\lim_{t\to 0} B(r) = \lim_{t\to\infty} B(r) = 0,$$

where

$$B(r) = \left(\int_{U_r} (x+t)^{(1-\alpha)q} d\nu(x,t)\right)^{1/q} \left(\int_0^{r/2} w^{1-p'}(x)dx\right)^{1/p'}.$$

From Theorem 2.9.12 we can derive the following statement:

Theorem 2.9.13. *Let* $1 < p \le q < \infty$ *and let* $\alpha \ge 1$. *Assume that* w *satisfies* (2.9.7). *Then* T_α *is compact from* $L_w^p(0,\infty)$ *into* $L_\nu^q(\widetilde{R}_+^2)$ *if and only if* (2.9.5) *is satisfied and*

$$\lim_{t\to 0} B(r) = \lim_{t\to\infty} B(r) = 0,$$

where $B(r)$ *is from the previous Theorem.*

Taking into account the proof of Theorem 2.9.2 and using Ando's theorem, we obtain the following result for T_α:

Theorem 2.9.14. *Let* $0 < q < p < \infty$, $p > 1$ *and* $\alpha > 1/p$. *Assume that w is an increasing function on* $(0, \infty)$. *Then the following statements are equivalent:*

(i) T_α *is bounded from* $L^p_w(0, \infty)$ *into* $L^q_v(\widetilde{R}^2_+)$;

(ii) T_α *is compact from* $L^p_w(R_+)$ *into* $L^q_v(\widetilde{R}^2_+)$;

(iii)

$$D = \left(\int\limits_0^\infty \left(\int\limits_{U_{2x}} v(y,t)(y+t)^{(1-\alpha)q} dy dt \right)^{\frac{p}{p-q}} \left(\int\limits_0^x w^{1-p'}(y) dy \right)^{\frac{(q-1)p}{p-q}} \times$$

$$\times w^{1-p'}(x) dx \right)^{\frac{p-q}{pq}} < \infty.$$

Moreover, $\|T_\alpha\| \approx D$.

2.10.   One-sided potentials on the half–space

In this section we establish boundedness and compactness criteria for integral transform with generalized positive kernels

$$Kf(x,t) = \int\limits_0^x k(x,y,t) f(y) dy.$$

We begin with some definitions.

Definition 2.10.1. Let $0 < a \leq +\infty$. A kernel $k : \{(x,y) : 0 < y < x < a\} \times [0, \infty) \to (0, \infty)$ belongs to $\mathcal{V}$ ($k \in \mathcal{V}$) if there exists a positive constant b_1 such that for all x, y, z with $0 < y < z < x < a$ and for all $t > 0$ the inequality

$$k(x, y, t) \leq b_1 k(x, z, t)$$

holds.

Definition 2.10.2. Let $0 < a \leq +\infty$. We say that k belongs to $\mathcal{V}_\lambda$ ($k \in \mathcal{V}_\lambda$) ($1 < \lambda < \infty$) if there exists a positive constant b_2 such that for all x, $x \in (0, a)$, and for all $t > 0$ the inequality

$$\int\limits_{x/2}^x k^{\lambda'}(x, y, t) dy \leq b_2 x k^{\lambda'}(x, x/2, t)$$

is fulfilled, where $\lambda' = \lambda/(\lambda - 1)$.

Let k_1 be a positive measurable function on $(0, a) \times (0, \infty)$.

Definition 2.10.3. Let $0 < a \leq +\infty$. We say that k_1 belongs to $\mathcal{V}_{1\lambda}$ ($k \in \mathcal{V}_{1\lambda}$) ($1 < \lambda < \infty$) if there exists a positive constant b_3 such that the inequality

$$\int_0^{x/2} k_1^{\lambda'}(y,t)dy \leq b_3 x k_1^{\lambda'}(x/2,t)$$

is fulfilled for all x, $x \in (0,a)$ and $t > 0$.

It is easy to verify that if k_1 is a non-increasing function on $(0,a)$ with respect to the first variable and $k_1 \in \mathcal{V}_{1\lambda}$, then the kernel $k(x,y,t) \equiv k_1(x-y,t)$ belongs to $\mathcal{V} \cap \mathcal{V}_\lambda$.

Some examples of kernels satisfying the above-mentioned conditions are desirable.

Let $a \leq +\infty$ and let $k(x,y,t) = (x-y+t)^{\alpha-1}$, where $\alpha > 0$. If $1 < \lambda < \infty$ and $\frac{1}{\lambda} < \alpha \leq 1$, then $k \in \mathcal{V}_{1\lambda}$ and consequently the kernel k_1 belongs to $\mathcal{V} \cap \mathcal{V}_\lambda$.

Indeed, the fact $k \in \mathcal{V}$ is obvious. We show that $k \in \mathcal{V}_\lambda$. Let $x \in [0,a)$ and $t \in [0,\infty)$. If $t < x$, then we have

$$\int_{x/2}^x k^{\lambda'}(x,y,t)dy = \int_{x/2}^x (x-y+t)^{(\alpha-1)\lambda'}dy =$$

$$= \int_t^{x/2+t} y^{(\alpha-1)\lambda'}dy \leq c_1(x/2+t)^{(\alpha-1)\lambda'+1} \leq c_2 x(x/2+t)^{(\alpha-1)\lambda'} =$$

$$= c_2 x k^{\lambda'}(x,x/2,t).$$

If $t \geq x$, then

$$\int_{x/2}^x k^{\lambda'}(x,y,t)dy = \int_{x/2}^x (x-y+t)^{(\alpha-1)\lambda'}dy \leq$$

$$\leq c_3 x t^{(\alpha-1)\lambda'} \leq c_4 x(x+t)^{(\alpha-1)\lambda'} \leq c_5 x k^{\lambda'}(x,x/2,t).$$

Now suppose that $a \leq +\infty$, $k(x,y,t) = (\ln \frac{x}{y}+t)^\beta$, where $-1/\lambda' < \beta \leq 0$. Then $k \in \mathcal{V} \cap \mathcal{V}_\lambda$. For let $t \leq \ln 2$. Then

$$I(x,t) \equiv \int_{x/2}^x k^{\lambda'}(x,y,t)dy = \int_{x/2}^x (\ln \frac{x}{y}+t)^{\beta\lambda'}dy =$$

$$= c_6 x \int_1^2 (\ln u+t)^{\beta\lambda'} \frac{1}{u^2}du \leq c_7 x \int_1^2 (\ln u+t)^{\beta\lambda'} \frac{du}{u} \leq$$

$$\leq c_8 x \int\limits_{t}^{\ln 2 + t} u^{\beta \lambda'} du \leq c_{10} x (\ln 2 + t)^{\beta \lambda' + 1} \leq c_{11} x (\ln 2 + t)^{\beta \lambda'} =$$

$$= c_{11} x k^{\lambda'} (x, x/2, t)$$

If $t > \ln 2$, then

$$I(x, t) \leq c_{12} x t^{\beta \lambda'} \leq c_{13} x (t + \ln 2)^{\beta \lambda'} = c_{13} x k^{\lambda'} (x, x/2, t).$$

Hence $k \in \mathcal{V}_\lambda$.

Let $0 < a \leq +\infty$ and let $k(x, y, t) = x^{-\sigma(\alpha + \eta)} (x^\sigma - y^\sigma + t)^{\alpha - 1} y^{\sigma \eta + \sigma - 1}$ be the Erdelyi–Köber kernel, where $\sigma > 0$ and $0 < \alpha \leq 1$. It easy to see that if $1/\lambda < \alpha \leq 1$ and $\eta > 1/\sigma - 1$, then $k \in \mathcal{V} \cap \mathcal{V}_\lambda$.

We prove that $k \in \mathcal{V}_\lambda$. Let $t \leq x^\sigma$. Then

$$J(x, t) \equiv \int\limits_{x/2}^{x} k^\lambda(x, y, t) dy =$$

$$= x^{-\sigma(\alpha + \eta)\lambda'} \int\limits_{x/2}^{x} (x^\sigma - y^\sigma + t)^{(\alpha - 1)\lambda'} y^{(\sigma \eta + \sigma - 1)\lambda'} dy \leq$$

$$\leq c_3 x^{-\sigma \alpha \lambda' + \sigma \lambda' - \lambda'} \int\limits_{t}^{x^\sigma - (x/2)^\sigma + t} u^{(\alpha - 1)\lambda'} (x^\sigma + t - u)^{1/\sigma - 1} du \leq$$

$$\leq c_4 x^{-\sigma \alpha \lambda' + \sigma \lambda' - \lambda'} x^{\sigma(1/\sigma - 1)} \int\limits_{0}^{x^\sigma - (x/2)^\sigma + t} u^{(\alpha - 1)\lambda'} du =$$

$$= c_4 x^{-\sigma \alpha \lambda' + \sigma \lambda' - \lambda'} x^{1 - \sigma} (x^\sigma - (x/2)^\sigma + t)^{(\alpha - 1)\lambda' + 1} =$$

$$= c_4 x^{-\sigma \alpha \lambda' + \sigma \lambda' - \lambda'} x^{1 - \sigma} (x^\sigma - (x/2)^\sigma + t)^{(\alpha - 1)\lambda'} x^\sigma =$$

$$= c_5 x k^{\lambda'} (x, x/2, t).$$

Now let $t > x^\sigma$. Then we have

$$I(x, t) \leq x^{-\sigma(\alpha + \eta)\lambda' + (\sigma \eta + \sigma - 1)\lambda'} t^{(\alpha - 1)\lambda'} x \leq$$

$$\leq c_6 x^{-\sigma(\alpha + \eta)\lambda' + (\sigma \eta + \sigma - 1)\lambda'} (x^\sigma - (x/2)^\sigma + t)^{(\alpha - 1)\lambda'} x =$$

$$= c_7 x k^{\lambda'} (x, x/2, t).$$

Suppose that ν is a positive σ-finite measure on $[0, a) \times [0, \infty) \equiv \tilde{R}_a$ such that all rectangles in $\tilde{R}_a$ are ν-measurable. Let $U_{a,r} \equiv [r, a) \times [0, \infty), 0 < r < a$.

Theorem 2.10.1. *Let* $1 < p \le q < \infty$, $0 < a \le \infty$, $k \in \mathcal{V} \cup \mathcal{V}_{p'}$. *Then the following conditions are equivalent:*
(i) K *is bounded from* $L^p(0,a)$ *into* $L_\nu^q(\tilde{R}_a)$;
(ii)

$$B_a \equiv \sup_{0 < r < a} \left(\int_{U_{a,r}} k^q(x, x/2, t) d\nu(x,t) \right)^{1/q} r^{1/p'} < \infty.$$

Moreover, there exist positive constants d_1 and d_2 such that

$$d_1 B_a \le \|K\| \le d_2 B_a$$

and d_1 and d_2 do not depend on a if b_1 and b_2 are independent of a.

Proof. Let $a = \infty$, $f \ge 0$. Then

$$I \equiv \|Kf\|_{L_\nu^q(\tilde{R}_\infty)} \le \left(\int_{\tilde{R}_\infty} \left(\int_0^{x/2} k(x,y,t)f(y)dy \right)^q d\nu(x,t) \right)^{1/q} +$$

$$+ \left(\int_{\tilde{R}_\infty} \left(\int_{x/2}^x k(x,y,t)f(y)dy \right)^q d\nu(x,t) \right)^{1/q} \equiv$$

$$\equiv S_1 + S_2.$$

Using the fact that $k \in \mathcal{V}$ and Lemma 2.9.1 we obtain

$$S_1 \le c_1 \left(\int_{\tilde{R}_\infty} k^q(x, x/2, t) \left(\int_0^x f(y)dy \right)^q d\nu(x,t) \right)^{1/q} \le$$

$$\le c_2 B_\infty \|f\|_{L^p(R_+)}.$$

By Hölder's inequality and since $k \in \mathcal{V}_p$ we have

$$S_2^q \le \int_{\tilde{R}_\infty} \left(\int_{x/2}^x (f(y))^p dy \right)^{q/p} \left(\int_{x/2}^x k^{p'}(x,y,t)dy \right)^{q/p'} d\nu(x,t) \le$$

$$\le c_3 \int_{\tilde{R}_\infty} \left(\int_{x/2}^x (f(y))^p dy \right)^{q/p} k^q(x,x/2,t) x^{q/p'} d\nu(x,t) =$$

$$= c_3 \sum_{n \in Z} \int_{U_{\infty,2^n} \setminus U_{\infty,2^{n+1}}} \left(\int_{x/2}^x (f(y))^p dy \right)^{q/p} k^q(x,x/2,t) x^{q/p'} d\nu(x,t) \le$$

$$\leq c_3 B_\infty^q \sum_{n\in Z} \left(\int_{2^{n-1}}^{2^{n+1}} (f(y))^q dy \right)^{q/p} \leq c_4 B_\infty^q \|f\|_{L^p(R_+)}^q.$$

Now let $n \in Z$ and let $f_n(y) = \chi_{[0,2^{n+1})}(y)$. We obtain

$$\|Kf_n\|_{L_\nu^q(\widetilde{R}_\infty)} \geq$$

$$\geq \left(\int_{U_{\infty,2^n} \setminus U_{\infty,2^{n+1}}} \left(\int_{x/2}^{x} f_n(y)k(x,y,t)dy \right)^q d\nu(x,t) \right)^{1/q} \geq$$

$$\geq c_5 \left(\int_{U_{\infty,2^n} \setminus U_{\infty,2^{n+1}}} k^q(x, x/2, t)x^q d\nu(x,t) \right)^{1/q}.$$

On the other hand,

$$\|f\|_{L^p(R_+)} = c_6 2^{n/p}$$

and

$$B_1(n) \equiv \left(\int_{U_{\infty,2^n} \setminus U_{\infty,2^{n+1}}} k^q(x, x/2, t)x^{q/p'} d\nu(x,t) \right)^{1/q} < \infty$$

for every $n \in Z$. Consequently

$$B_1 \equiv \sup_{n\in Z} B_1(n) < \infty.$$

Now we show that $B \leq c_7 B_1$. Let $r \in [0, \infty)$. Then $r \in [2^m, 2^{m+1})$ for some $m \in Z$. Then

$$r^{q/p'} \int_{U_{\infty,r}} k^q(x, x/2, t)d\nu(x,t) \leq c_8 2^{mq/p'} \int_{U_{\infty,2^m}} k^q(x, x/2, t)d\nu(x,t) =$$

$$= c_8 2^{mq/p'} B_1^q \sum_{n=m}^{+\infty} 2^{-nq/p'} \leq c_9 B_1^q.$$

Hence $B \leq c_{10} B_1 < \infty$. $\square$

Remark 2.10.1. It is easy to verify that if $a = \infty$, then the condition $B_\infty < \infty$ is equivalent to the condition

$$B_1 \equiv \sup_{n\in Z} B_1(n) \equiv$$

$$\equiv \left(\int\limits_{U_{\infty,2^n} \backslash U_{\infty,2^{n+1}}} k^q(x, x/2, t) x^{q/p'} \, d\nu(x, t) \right)^{1/q} < \infty,$$

and if $a < \infty$, then the condition (ii) of Theorem 2.10.1 is equivalent to the condition

$$B_{a,1} \equiv \sup_{n \leq 0} B_{a,1}(n) \equiv$$

$$\equiv \sup_{n \leq 0} \left(\int\limits_{U_{a,2^{n-1}a} \backslash U_{a,2^n a}} k^q(x, x/2, t) x^{q/p'} \, d\nu(x, t) \right)^{1/q} < \infty.$$

Theorem 2.10.2. *Let* $1 < p \leq q < \infty, 0 < a \leq \infty$. *Suppose that* $k \in \mathcal{V} \cap \mathcal{V}_p$. *Then the following conditions are equivalent:*

(i) K *is compact from* $L^p(0, a)$ *into* $L^q_\nu(\tilde{R}_a)$;

(ii) $B_\infty < \infty$ *and* $\lim_{c \to 0} B^{(c)}_\infty = \lim_{d \to +\infty} B^{(d)}_\infty = 0$ *if* $a = \infty$ *and* $B_a < \infty$,

$\lim_{c \to 0} B^{(c)}_a = 0$ *if* $a < \infty$, *where*

$$B^{(c)}_a = \sup_{0 < r < c} \left(\int\limits_{U_{a,r} \backslash U_{a,c}} k^q(x, x/2, t) d\nu(x, t) \right)^{1/q} r^{1/p'},$$

$$B^{(d)}_\infty = \sup_{r > d} \left(\int\limits_{\tilde{R}_\infty \backslash U_{\infty,c}} k^q(x, x/2, t) d\nu(x, t) \right)^{1/q} r^{1/p'},$$

(iii) $B_1 < \infty$ *and* $\lim_{n \to -\infty} B_1(n) = \lim_{n \to +\infty} B_1(n) = 0$ *for* $a = \infty, B_{a,1} < \infty$

and $\lim_{n \to -\infty} B_{a,1}(n) = 0$ *for* $a < \infty$.

Proof. We shall assume that $a = \infty$. First we prove that (ii) $\Rightarrow$ (i). Let $0 < c < d < \infty$ and represent Kf as follows:

$$Kf(x, t) = \chi_{\tilde{R}_\infty \backslash U_{\infty,c}}(x, t) \int\limits_0^x k(x, y, t) f(y) dy +$$

$$+ \chi_{U_{\infty,c} \backslash U_{\infty,d}}(x, t) \int\limits_0^x k(x, y, t) f(y) dy +$$

$$+ \chi_{U_{\infty,d}}(x, t) \int\limits_0^{d/2} k(x, y, t) f(y) dy +$$

$$+\chi_{U_{\infty,d}}(x,t)\int\limits_{d/2}^{x} k(x,y,t)f(y)dy \equiv P_1 f(x,t) +$$

$$+P_2 f(x,t) + P_3 f(x,t) + P_4 f(x,t).$$

For P_2 we have

$$S \equiv \int\limits_{U_{\infty,c}\backslash U_{\infty,d}} \left(\int\limits_{0}^{x} k^{p'}(x,y,t)dy\right)^{q/p'} d\nu(x,t) \le$$

$$\le \int\limits_{U_{\infty,c}\backslash U_{\infty,d}} \left(\int\limits_{0}^{x/2} k^{p'}(x,y,t)dy\right)^{q/p'} d\nu(x,t) +$$

$$+ \int\limits_{U_{\infty,c}\backslash U_{\infty,d}} \left(\int\limits_{x/2}^{x} k^{p'}(x,y,t)dy\right)^{q/p'} d\nu(x,t) \equiv S_1 + S_2.$$

Using the fact that $k \in \mathcal{V}$, we obtain

$$S_1 \le c_2 \int\limits_{U_{\infty,c}\backslash U_{\infty,d}} k^q(x,x/2,t)x^{q/p}d\nu(x,t) < \infty.$$

Further, from the condition $k \in \mathcal{V}_p$ we have the following inequality

$$S_2 \le c_3 \int\limits_{U_{\infty,c}\backslash U_{\infty,d}} k^q(x,x/2,t)x^{q/p'}d\nu(x,t) < \infty.$$

Hence $S < \infty$ and by Theorem C of Section 2.1 we conclude that P_2 is compact. Analogously it follows that P_3 is compact.

By Theorem 2.10.1 we have

$$\|K - P_2 - P_3\| \le b(\|P_1\| + \|P_4\|) \to 0 \qquad (2.10.1)$$

as $c \to 0$ and $d \to \infty$. Hence K is compact as it is a limit of compact operators.

Now we show that (i) implies (iii). Suppose that $n \in Z$ and let

$$f_n(y) = \chi_{[2^{n-1},2^{n+1})}(y)2^{-n/p}.$$

Then the sequence f_n is weakly convergent to 0. Indeed, if $\varphi \in L^{p'}(0,a)$, then

$$\left|\int\limits_{0}^{\infty} \varphi(x)f_n(x)dx\right| \le b_1 \left(\int\limits_{2^{n-1}}^{2^{n+1}} |\varphi(x)|^p dx\right)^{1/p} \to 0$$

as $n \to -\infty$ or $n \to +\infty$. On the other hand,

$$\|Kf_n\|_{L^q_\nu(\widetilde{R}_a)} \geq \left(\int\limits_{U_{\infty,2^n} \backslash U_{\infty,2^{n+1}}} \left(\int\limits_{x/2}^x k(x,y,t)f_n(y)dy \right)^q d\nu(x,t) \right)^{1/q} \geq$$

$$\geq c_4 \left(\int\limits_{U_{\infty,2^n} \backslash U_{\infty,2^{n+1}}} x^{q/p'} k^q(x,x/2,t) d\nu(x,t) \right)^{1/q} \to 0$$

when $n \to -\infty$ or $n \to +\infty$.

It remains to show that (ii) follows from (iii). Let $c \in (0,\infty)$. Then $c \in [2^m, 2^{m+1})$ for some $m = m_c \in Z$. We obtain

$$B_\infty^{(d)} \leq \sup_{0<r<2^{m+1}} B_\infty^{(2^{m+1})}(r) = B_\infty^{(2^{m+1})},$$

where

$$B_\infty^{(2^{m+1})}(r) \equiv \left(\int\limits_{U_{\infty,r} \backslash U_{\infty,2^{m+1}}} k^q(x,x/2,t) d\nu(x,t) \right)^{1/q} r^{1/p'}.$$

Now let $r \in (0, 2^{m+1})$. Then $r \in [2^n, 2^{n+1})$ for some integer $n \leq m$. We have

$$\left(B_\infty^{(2^{m+1})}(r) \right)^q \leq \left(\int\limits_{U_{\infty,2^n} \backslash U_{\infty,2^{m+1}}} k^q(x,x/2,t) d\nu(x,t) \right) (2^{n+1})^{q/p'} =$$

$$= (2^{n+1})^{q/p'} \sum_{i=n}^m \int\limits_{U_{\infty,2^i} \backslash U_{a,2^{i+1}}} k^q(x,x/2,t) d\nu(x,t) \leq$$

$$\leq c_3 \left(\sup_{i \leq m} B_1(i) \right)^q \equiv c_3 \widetilde{B}_{m,1}^q.$$

Thus

$$B_\infty^{(2^{m+1})} \leq c_4 \widetilde{B}_{m,1}.$$

If $c \to 0$, then $m \to -\infty$ and $\widetilde{B}_{m,1} \to 0$ because $\lim\limits_{n \to -\infty} B_1(n) = 0$. This implies that $\lim\limits_{c \to 0} B_\infty^{(c)} = 0$.

Analogously we can show that $\lim\limits_{d \to +\infty} B_\infty^{(d)} = 0$.

Finally we have (ii) $\Rightarrow$ (i) $\Rightarrow$ (iii) $\Rightarrow$ (ii). $\square$

Now we consider the case $q < p$. Assume that the measure ν is absolutely continuous, i.e. $d\nu(x,t) = v(x,t)dxdt$.

Theorem 2.10.3. *Let* $0 < q < p < \infty$, $p > 1$ *and* $0 < a \leq \infty$. *Assume that* $k \in \mathcal{V} \cap \mathcal{V}_p$. *Then* K *is bounded from* $L^p(0, a)$ *into* $L^q_v(\tilde{R}_a)$ *if and only if*

$$D_a \equiv \left(\int\limits_0^a \left(\int\limits_{U_{a,x}} k^q(y, y/2, t) v(y, t) dy dt \right)^{\frac{p}{p-q}} x^{\frac{p(q-1)}{p-q}} dx \right)^{\frac{p-q}{pq}} < \infty.$$

Moreover, there exist positive constants d_1 *and* d_2 *such that*

$$d_1 D_a \leq \|K\| \leq d_2 D_a.$$

(If b_1 *and* b_2 *do not depend on* a, *then* d_1 *and* d_2 *are also independent of* a *).*

Proof. Let $a = \infty$. The case $a < \infty$ is proved analogously. Let $f \geq 0$. Then we have

$$\|Kf\|_{L^q_v(\tilde{R}_\infty)} \leq c_0 \left(\int\limits_{\tilde{R}_\infty} \left(\int\limits_0^{x/2} f(y) k(x, y, t) dy \right)^q v(x, t) dx dt \right)^{1/q} +$$

$$+ c_0 \left(\int\limits_{\tilde{R}_\infty} \left(\int\limits_{x/2}^x f(y) k(x, y) dy \right)^q v(x, t) dx dt \right)^{1/q} \equiv S_1 + S_2.$$

Using the fact $k \in \mathcal{V}$ and Theorem B from Section 2.1 we have

$$S_1 \leq c_2 D_a \|f\|_{L^p(0,\infty)},$$

where c_2 depends only on p, q and d_1. By Hölder's inequality and the condition $k \in \mathcal{V}_p$ we obtain

$$S_2^q \leq c_3 \int\limits_{\tilde{R}_\infty} \left(\int\limits_{x/2}^x (f(y))^p dy \right)^{\frac{q}{p}} x^{\frac{q}{p'}} k^q(x, x/2, t) v(x, t) dx dt =$$

$$= c_3 \sum_{j \in Z} \int\limits_{U_{\infty,2^j} \setminus U_{\infty,2^{j+1}}} \left(\int\limits_{x/2}^x (f(y))^p dy \right)^{\frac{q}{p}} x^{\frac{q}{p'}} k^q(x, x/2, t) v(x, t) dx dt \leq$$

$$\leq c_3 \sum_{j \in Z} \left(\int\limits_{2^{j-1}}^{2^{j+1}} (f(y))^p dy \right)^{\frac{q}{p}} \int\limits_{U_{\infty,2^j} \setminus U_{\infty,2^{j+1}}} x^{\frac{q}{p'}} k^q(x, x/2) v(x, t) dx dt.$$

Using Hölder's inequality again, we have

$$S_2^q \leq c_3 \left(\sum_{j \in Z} \int\limits_{2^{j-1}}^{2^{j+1}} (f(y))^p dy \right)^{q/p} \times$$

$$\times \left(\sum_{j \in Z} \left(\int_{U_{\infty,2^j} \setminus U_{\infty,2^j+1}} x^{q/p'} k^q(x, x/2, t) v(x) dx \right)^{\frac{p}{p-q}} \right)^{\frac{p-q}{p}} \le$$

$$\le c_4 \widetilde{I}_1^q \|f\|^q_{L^p(0,\infty)},$$

where

$$\widetilde{I} \equiv \left(\sum_{j \in Z} \left(\int_{U_{\infty,2^j} \setminus U_{\infty,2^j+1}} x^{q/p'} k^q(x, x/2, t) v(x, t) dx dt \right)^{\frac{p}{p-q}} \right)^{\frac{p-q}{pq}}.$$

In addition,

$$\widetilde{I}^{\frac{pq}{p-q}} \le c_5 \sum_{j \in Z} 2^{\frac{jq(p-1)}{p-q}} \left(\int_{U_{\infty,2^j} \setminus U_{\infty,2^j+1}} k^q(x, x/2, t) v(x, t) dx dt \right)^{\frac{p}{p-q}} \le$$

$$\le c_5 \sum_{j \in Z} \int_{2^{j-1}}^{2^j} y^{\frac{p(q-1)}{p-q}} \left(\int_{U_{\infty,y} \setminus U_{\infty,2^j+1}} k^q(x, x/2, t) v(x, t) dx dt \right)^{\frac{p}{p-q}} dy =$$

$$= c_5 \int_0^\infty y^{\frac{p(q-1)}{p-q}} \left(\int_{U_{\infty,y}} k^q(x, x/2, t) v(x, t) dx dt \right)^{\frac{p}{p-q}} dy = c_5 \widetilde{B}^{\frac{pq}{p-q}}.$$

Consequently

$$S_2 \le c_6 D_\infty \|f\|_{L^p(0,\infty)},$$

where the positive constant c_6 depends only on p, q and b_2.

Now let us prove necessity. Let the operator K be bounded from $L^p(0, \infty)$ to $L^q_v(\widetilde{R}_\infty)$. It is easy to verify that for every $x \in (a, \infty)$,

$$\int_{U_{\infty,x}} v(y, t) k^q(y, y/2, t) dy dt < \infty.$$

Let $n \in Z, n \ge 2$ and let

$$\widetilde{v}_n(x) = \left(\int_0^\infty v(x, t) k^q(x, x/2, t) dt \right) \chi_{(1/n,n)}(x).$$

Suppose that

$$f_n(x) = \left(\int_x^\infty \widetilde{v}_n(y) dy \right)^{\frac{1}{p-q}} x^{\frac{q-1}{p-q}}.$$

Using integration by parts we obtain

$$\|f_n\|_{L^p(0,\infty)} = \left(\int\limits_0^\infty \left(\int\limits_x^\infty \tilde{v}_n(y)dy \right)^{\frac{p}{p-q}} x^{\frac{(q-1)p}{p-q}} dx \right)^{1/p} =$$

$$= c_7 \left(\int\limits_0^\infty \left(\int\limits_x^\infty \tilde{v}_n(y)dy \right)^{\frac{q}{p-q}} x^{\frac{(p-1)q}{p-q}} v_n(x)dx \right)^{1/p} < \infty.$$

On the other hand,

$$\|Kf_n\|_{L^q_v(\tilde{R}_\infty)} \geq \left(\int\limits_{\tilde{R}_\infty} v(x,t) \left(\int\limits_{x/2}^x f_n(t)k(x,y,t)dy \right)^q dxdt \right)^{1/q} \geq$$

$$\geq c_8 \left(\int\limits_{\tilde{R}_\infty} v(x,t)k^q(x,x/2,t) \left(\int\limits_x^\infty \tilde{v}_n(y)dy \right)^{\frac{q}{p-q}} \times \right.$$

$$\left. \times x^{\frac{q(p-1)}{p-q}} dxdt \right)^{1/q} = c_8 \left(\int\limits_0^\infty \left(\int\limits_0^\infty v(x,t)k^q(x,x/2,t)dt \right) \times \right.$$

$$\left. \times \left(\int\limits_x^\infty \tilde{v}_n(y)dy \right)^{\frac{q}{p-q}} x^{\frac{(p-1)q}{p-q}} dx \right)^{1/q} \geq$$

$$\geq c_8 \left(\int\limits_0^\infty \tilde{v}_n(x) \left(\int\limits_x^\infty \tilde{v}_n(y)dy \right)^{\frac{q}{p-q}} x^{\frac{q(p-1)}{p-q}} dx \right)^{1/q} =$$

$$= c_9 \left(\int\limits_0^\infty \left(\int\limits_x^\infty \tilde{v}_n(y)dy \right)^{\frac{p}{p-q}} x^{\frac{p(q-1)}{p-q}} dx \right)^{1/q}.$$

Consequently

$$\left(\int\limits_0^\infty \left(\int\limits_x^\infty \tilde{v}_n(y)dy \right)^{\frac{p}{p-q}} x^{\frac{p(q-1)}{p-q}} dx \right)^{\frac{p-q}{pq}} \leq c,$$

where the positive constant c does not depend on n. By Fatou's lemma we finally obtain $D_\infty < \infty$.

The case $0 < a < \infty$ can be proved analogously. $\square$

The following theorem is proved as Theorem 2.10.2. For $a = \infty$ it is proved using Theorem D of Section 2.1.

Theorem 2.10.4. *Let p, q, α and k satisfy the conditions of Theorem 2.10.3. Then K is compact from $L^p(0,a)$ into $L^q_v(\tilde{R}_a)$ if and only if $D_a < \infty$.*

Now we investigate the measure of non–compactness for the operator K. We shall assume that v is a Lebesgue–measurable, a.e. positive function on $\tilde{R}_a$. Let us put $\overline{K}f(x,t) = v^{1/q}(x,t)K(x,t)$.

$$\tilde{I}_a \equiv dist(K, \mathcal{K}(X, L_v^q(\tilde{R}_a)),$$

$$\overline{I}_a \equiv dist(\overline{K}, \mathcal{K}(X, L^q(\tilde{R}_a)),$$

where X is a Banach function space and $\mathcal{K}(X,Y)$ is the space of compact linear operators from X to Y. The next Lemma is immediate:

Lemma 2.10.1. *Let* $0 < a \le \infty$ *and* $1 \le q < \infty$. *Then* $\overline{I}_a = \tilde{I}_a$.

Theorem 2.10.5. *Let* $0 < a \le \infty$, $1 < p \le q < \infty$ *and* $k \in \mathcal{V} \cap \mathcal{V}_p$. *Assume that K is bounded from X into $L_v^q(\tilde{R}_a)$, where $X \equiv L^p(0,a)$. Then there exist positive constants a_1 and a_2 depending only on p, q, b_1 and b_2 such that*

$$a_1 J \le \tilde{I}_a \le a_2 J,$$

where $J = \lim_{c \to 0} B_\infty^{(c)} + \lim_{d \to \infty} B_\infty^{(d)}$ *if* $a = \infty$ *and* $J = \lim_{c \to 0} B_a^{(c)} = 0$ *if* $a < \infty$,

$$B_a^{(c)} = \sup_{0 < r < c < a} \left(\int_{U_{a,r} \setminus U_{a,c}} k^q(x, x/2, t)v(x,t)dxdt \right)^{1/q} r^{1/p'},$$

$$B_\infty^{(d)} = \sup_{r > d} \left(\int_{U_{\infty,d}} k^q(x, x/2, t)v(x,t)dxdt \right)^{1/q} r^{1/p'}.$$

Proof. We shall assume that $a = \infty$. The case $a < \infty$ is similar. The upper estimate follows from the proof of Theorem 2.10.2. To prove the lower estimate, we take $\lambda > \tilde{I}_\infty$. By Lemma 2.10.1 we have that $\lambda > \overline{I}_\infty$. Thus, using Lemma 2.9.4 there exists $P \in \mathcal{F}_r(X,Y)$, such that $\|\overline{K} - T\| < \lambda$. Further, by Lemma 2.9.5 for $\epsilon = (\lambda - \|\overline{K} - P\|)/2$ there exist $T \in \mathcal{F}_r(X,Y)$ and $[\alpha, \beta] \subset (0, \infty)$ such that

$$\|P - T\| < \epsilon \quad \text{and} \quad suppTf \subset [\alpha, \beta].$$

Hence

$$\|\overline{K}f - Tf\|_{L^q(R_\infty)} \le \lambda \|f\|_X$$

for every $f \in X$. Consequently

$$\int_a^\alpha \int_0^\infty |\overline{K}f(x,t)|^q dxdt + \int_\beta^\infty \int_0^\infty |\overline{K}f(x,t)|^q dxdt$$

$$\le \lambda^q \|f\|_X^q$$

for all $f \in X$.

Now let $n \in Z$ be chosen such that $2^n < \alpha$. Suppose that $j \in Z$, where $j \leq n - 1$ and let

$$f_j(y) = \chi_{(0, 2^{j+1})}(y).$$

Then

$$\int\limits_{2^j}^{2^{j+1}} \int\limits_0^\infty |\overline{K}f(x,t)|^q dx dt \geq \int\limits_{2^j}^{2^{j+1}a} \int\limits_0^\infty \left(\int\limits_{x/2}^x k(x,y,t)f(y)dy \right)^q v(x,t) dx dt \geq$$

$$\geq c_1 \int\limits_{2^j}^{2^{j+1}} \int\limits_0^\infty k^q(x, x/2, t)v(x,t)x^q dx dt.$$

On the other hand, $\|f_j\|_X^q = c_2 2^{jq/p}$. Hence

$$B_{\infty,1}(j) \equiv \left(\int\limits_{2^j}^{2^{j+1}} \int\limits_0^\infty k^q(x, x/2, t)v(x,t)x^{q/p'} dx dt \right)^{1/q} \leq c_3 \lambda$$

for all j, $j \leq n$. It follows that

$$\sup_{j \leq n} B_{\infty,1}(j) \leq c_3 \lambda.$$

and from the last inequality we obtain

$$\lim_{n \to -\infty} \sup_{j \leq n} B_{\infty,1}(j) \leq c_3 \lambda.$$

Similarly,

$$\lim_{n \to +\infty} \sup_{j \geq n} B_{\infty,1}(j) \leq c_4 \lambda$$

with a positive constant c_4 depending only on p, q and d_1. As λ is an arbitrary number greater than $\overline{I}_\infty$, we conclude that $a_1 J \leq \overline{I}_\infty = \tilde{I}_\infty$. $\square$

2.11.   Weighted criteria in Lorentz spaces

In this section we establish boundedness and compactness criteria for some integral operators defined on weighted Lorentz spaces.

Let ν be a σ–finite positive Borel measure on R_+. We assume that u_1 and u_2 are ν– measurable a.e. positive functions on R_+.

Now we define the weighted Riemann–Liouville and Weyl operators:

$$R_{\alpha,u_1}f(x) = u_1(x) \int\limits_0^x \frac{f(t)}{(x-t)^{1-\alpha}} dt,$$

$$W_{\alpha,u_2}g(y) = \int\limits_{[y,\infty)} g(x)u_2(x)(x-y)^{\alpha-1}d\nu(x),$$

where $\alpha > 0$.

To prove the main results we need several Lemmas.

Lemma 2.11.1. ([50], [261]). *Let $\{E_k\}$ be a countable family of measurable sets $E_k \subset R^n$ and let μ be a positive $\sigma-$ finite Borel measure defined on R^n. Assume that $\sum_k \chi_{E_k}(\cdot) \leq c\chi_{\cup_k E_k}(\cdot)$ for a fixed constant $c > 0$. Then*

(a)

$$\sum_k \|f(\cdot)\chi_{E_k}(\cdot)\|^\lambda_{L^{rs}_\mu(R^n)} \leq c_1 \|f(\cdot)\chi_{\cup E_k}(\cdot)\|^\lambda_{L^{rs}_\mu(R^n)}$$

whenever $\max\{r, s\} \leq \lambda$ and c_1 does not depend on f;

(b)

$$\Big\|\sum_k f(\cdot)\chi_{E_k}(\cdot)\Big\|^\lambda_{L^{pq}_\mu(R^n)} \leq c_2 \sum_k \|f(\cdot)\chi_{E_k}(\cdot)\|^\lambda_{L^{pq}_\mu(R^n)}$$

whenever $0 < \gamma \leq \min\{p, q\}$ and c_2 is independent of f;

Lemma 2.11.2.. *Let μ be a positive σ- finite Borel measure on R^n. Then the following statements are valid:*

(a) *let $1 < r < \infty, 1 \leq s \leq \infty$. If $E \subset R^n$ is a μ-measurable set, then*

$$\|\chi_E(\cdot)\|_{L^{rs}_\mu(R^n)} = (\mu(E))^{1/r};$$

(b)

$$\|f\|_{L^{ps_2}_\mu(R^n)} \leq \|f\|_{L^{ps_1}_\mu(R^n)}$$

for fixed p, $1 \leq p \leq \infty$, and $s_2 \leq s_1$.

(c)

$$\|f_1 f_2\|_{L^{ps}_\mu(R^n)} \leq c\|f_1\|_{L^{p_1 s_1}_\mu R^n}\|f_2\|_{L^{p_2 s_2}_\mu(R^n)}$$

with $1/p = 1/p_1 + 1/p_2$ and $1/s = 1/s_1 + 1/s_2$.

Lemma 2.11.3.. *Let $1 < r < \infty, 1 \leq s < \infty$ and let $\alpha > 1/r$. Then there exists a positive constant c such that for all $x > 0$ the following inequality holds:*

$$I(x) \equiv \Big\|\chi_{(x/2,x)}(\cdot)\frac{1}{(x-\cdot)^{1-\alpha}}\Big\|_{L^{r's'}(R_+)} \leq cx^{\alpha-1/r}.$$

Proof. As the case $\alpha \geq 1$ is trivial, we consider the case $\alpha \in (1/r, 1)$. Let $s > 1$. Then we have

$$I(x) \leq \Big\|\chi_{(0,x)}(\cdot)\frac{1}{(x-\cdot)^{1-\alpha}}\Big\|_{L^{r's'}(R_+)} =$$

$$= \left(s' \int\limits_0^\infty \lambda^{s'-1} \left(\left| \{ y : y \in (0,x), (x-y)^{\alpha-1} > \lambda \} \right| \right)^{s'/r'} d\lambda \right)^{1/s'} =$$

$$= \left(s' \int\limits_0^{x^{\alpha-1}} \lambda^{s'-1} \left(\left| \{ y : y \in (0,x), (x-y)^{\alpha-1} > \lambda \} \right| \right)^{s'/r'} d\lambda \right)^{1/s'} +$$

$$+ \left(s' \int\limits_{x^{\alpha-1}}^\infty \lambda^{s'-1} \left(\left| \{ y : y \in (0,x), (x-y)^{\alpha-1} > \lambda \} \right| \right)^{s'/r'} d\lambda \right)^{1/s'} \equiv$$

$$\equiv S_1(x) + S_2(x).$$

A simple computation gives

$$S_1(x) \le \left(s' \int\limits_0^{x^{\alpha-1}} \lambda^{s'-1} x^{s'/r'} d\lambda \right)^{1/s'} = c_1 x^{\alpha-1/r},$$

while for $S_2(x)$ we have

$$S_2(x) \le \left(s' \int\limits_{x^{\alpha-1}}^\infty \lambda^{s'-1} \lambda^{s'/(\alpha-1)r'} d\lambda \right)^{1/s'} = c_2 x^{\alpha-1/r}.$$

Finally we have

$$I(x) \le c_3 x^{\alpha-1/r}.$$

The case $s = 1$ is proved analogously. $\square$

Theorem 2.11.1. *Let $1 < r, p < \infty$, $1 \le s < \infty$ and $1 < q \le \infty$. Assume also that $\max\{r, s\} \le \min\{p, q\}$ and $\alpha > 1/r$. Then the following conditions are equivalent:*

(i) R_{α, u_1} *is bounded from $L^{rs}(R_+)$ into $L_\nu^{pq}(R_+)$;*

(ii)

$$B_1 \equiv \sup_{t>0} B_1(t) \equiv \sup_{t>0} \left\| u_1(x) x^{\alpha-1} \chi_{[t,\infty)}(x) \right\|_{L_\nu^{pq}} t^{1/r'} < \infty;$$

(iii)

$$B_2 \equiv \sup_{k \in Z} B_2(k) \equiv \sup_{k \in Z} \left\| u_1(x) x^{\alpha-1/r} \chi_{[2^k, 2^{k+1})}(x) \right\|_{L_\nu^{pq}(R_+)} < \infty.$$

Moreover, $\| R_{\alpha, u_1} \| \approx B_1 \approx B_2$.

Proof. First we show that (ii) $\Rightarrow$ (i). We have

$$\|R_{\alpha,u_1}f(\cdot)\|_{L^{pq}_\nu(R_+)} \le c_1\Big(\|R^{(1)}_{\alpha,u_1}f(\cdot)\|_{L^{pq}_\nu(R_+)} +$$
$$+\|R^{(2)}_{\alpha,u_1}f(\cdot)\|_{L^{pq}_\nu(R_+)}\Big)c_1(I_1 + I_2),$$

where

$$R^{(1)}_{\alpha,u_1}f(x) = u_1(x)\int\limits_0^{x/2} f(t)(x-t)^{\alpha-1}dt,$$

$$R^{(2)}_{\alpha,u_1}f(x) = u_1(x)\int\limits_{x/2}^x f(t)(x-t)^{\alpha-1}dt.$$

If $t \in (0, x/2)$, then $(x-t)^{\alpha-1} \le cx^{\alpha-1}$ and using Theorem 1.1.4 we obtain

$$I_1 \le c_2\|u_1(x)x^{\alpha-1}\int\limits_0^x f(t)dt\|_{L^{pq}_\nu(R_+)} \le$$
$$\le c_3\|f(\cdot)\|_{L^{rs}(R_+)}.$$

Using Hölder's inequality and taking into account Lemma 2.11.3, we find that

$$\int\limits_{x/2}^x f(t)(x-t)^{\alpha-1}dt \le \|\chi_{(x/2,x)}(\cdot)(\cdot-x)^{\alpha-1}\|_{L^{r's'}(R_+)} \times$$
$$\times\|\chi_{(x/2,x)}(\cdot)f(\cdot)\|_{L^{rs}(R_+)} \le c_3 x^{\alpha-1/r}\|\chi_{(x/2,x)}(\cdot)f(\cdot)\|_{L^{rs}(R_+)} \le$$
$$\le c_4 x^{\alpha-1}2^{k/r'}\|\chi_{(2^{k-1},2^{k+1})}(\cdot)f(\cdot)\|_{L^{rs}(R_+)}$$

for all $x \in [2^k, 2^{k+1})$.
Let $\max\{r,s\} \le \lambda \le \min\{p,q\}$. Then using Lemma 2.11.1 and the last estimates we have

$$I_2^\lambda \le \sum_{k\in Z} \|\chi_{[2^k,2^{k+1})}(x)u_1(x)\int\limits_{x/2}^x f(t)(x-t)^{\alpha-1}dt\|^\lambda_{L^{pq}_\nu(R_+)} \le$$
$$\le c_5 \sum_{k\in Z} 2^{k\lambda/r'}\|\chi_{[2^k,2^{k+1})}(x)u_1(x)x^{\alpha-1}\|^\lambda_{L^{pq}_\nu(R_+)} \times$$
$$\times\|\chi_{[2^{k-1},2^{k+1})}(\cdot)f(\cdot)\|^\lambda_{L^{rs}(R_+)} \le$$
$$\le c_5 B_1^\lambda\|f(\cdot)\|^\lambda_{L^{rs}(R_+)}.$$

We show that (i) $\Rightarrow$ (iii). Let $k \in Z$ and let $f_k(x) = \chi_{[2^{k-2}, 2^{k-1})}$. Then by Lemma 2.11.2 (part (a)) we obtain

$$\|f_k\|_{L^{rs}(R_+)} = c_6 2^{k/r}.$$

On the other hand,

$$\|R_{\alpha,u_1} f(\cdot)\|_{L_\nu^{pq}(R_+)} \geq \|\chi_{[2^k, 2^{k+1})}(x) R_{\alpha,u_1} f(x)\|_{L_\nu^{pq}(R_+)} \geq$$

$$\geq c_7 2^k \|\chi_{[2^k, 2^{k+1})}(x) u_1(x) x^{\alpha-1}\|_{L_\nu^{pq}(R_+)}$$

and from the boundedness of R_{α,u_1} we finally derive $B_2 < \infty$. To prove the implication (ii) $\Rightarrow$ (iii), we take $t \in [0, \infty)$. Then $t \in [2^m, 2^{m+1})$ for some integer m. Let us take σ such that $\sigma \leq \min\{p, q\}$. Using Lemma 2.11.1 we have

$$B_1^\sigma(t) \leq c_8 \|u_1(x) \chi_{[2^m, \infty)}(x) x^{\alpha-1}\|_{L_\nu^{pq}(R_+)}^\sigma 2^{m\sigma/r'} \leq$$

$$\leq c_8 \sum_{k=m}^\infty \|u_1(x) \chi_{[2^k, 2^{k+1})}(x) x^{\alpha-1}\|_{L_\nu^{pq}(R_+)}^\sigma 2^{m\sigma/r'} \leq$$

$$\leq c_9 B_2^\sigma 2^{m\sigma/r'} \sum_{k=m}^\infty 2^{-\sigma k/r'} \leq c_{10} B_2^\sigma.$$

Consequently $B_1 \leq c_{10} B_2$. $\square$

From duality arguments we can obtain the following statement:

Theorem 2.11.2. *Let* $1 < r, p < \infty$, $1 \leq s < \infty$ *and* $1 < q \leq \infty$. *Assume also that* $\max\{r, s\} \leq \min\{p, q\}$ *and* $\alpha > 1/p'$. *Then the following conditions are equivalent:*
(i) W_{α,u_2} *is bounded from* $L_\nu^{rs}(R_+)$ *into* $L^{pq}(R_+)$;
(ii)

$$B_1' \equiv \sup_{t>0} B_1'(t) \equiv \sup_{t>0} \|u_2(x) x^{\alpha-1} \chi_{[t,\infty)}(x)\|_{L_\nu^{r's'}(R_+)} t^{1/p} < \infty;$$

(iii)

$$B_2' \equiv \sup_{k \in Z} B_2'(k) \equiv \sup_{k \in Z} \|u_2(x) x^{\alpha-1/p'} \chi_{[2^k, 2^{k+1})}(\cdot)\|_{L_\nu^{r's'}(R_+)} < \infty.$$

Moreover, $\|W_{\alpha,u_2}\| \approx B_1' \approx B_2'$.

Now we deal with the compactness of R_{α,u_1} and W_{α,u_2}.

Lemma 2.11.4. *Let* $1 < r, p, q < \infty$ *and let* $1 \leq s < \infty$. *Assume that* $\max\{r, s\} \leq \min\{p, q\}$ *and* $\alpha > 1/r$. *If the following conditions are satisfied:*
(i) $B_1 < \infty$;

(ii) $\lim\limits_{b\to 0} B_1^{(b)} = \lim\limits_{c\to\infty} \overline{B}_1^{(c)} = 0$, where

$$B_1^{(b)} \equiv \sup_{0<t<b} \|u_1(x)x^{\alpha-1}\chi_{[t,b)}(x)\|_{L_v^{pq}(R_+)} t^{1/r'},$$

$$\overline{B}_1^{(c)} \equiv \sup_{t>c} \|u_1(x)x^{\alpha-1}\chi_{[t,\infty)}(x)\|_{L_v^{pq}(R_+)}(t-c)^{1/r'} < \infty,$$

then R_{α,U_1} is compact from $L^{rs}(R_+)$ to $L_v^{pq}(R_+)$.

Proof. Let $0 < b < c < \infty$ and let us represent R_{α,u_1} in the following way

$$R_{\alpha,u_1}f = \chi_{[0,b)}R_{\alpha,u_1}(f\chi_{[0,b)}) + \chi_{[b,c]}R_{\alpha,u_1}(f\chi_{(0,c/2]}) +$$
$$+\chi_{(c,\infty)}R_{\alpha,u_1}(f\chi_{(c/2,\infty)}) \equiv P_1f + P_2f + P_3f + P_4f.$$

Using Theorem C from Section 2.1 we conclude that P_2 and P_3 are compact operators. On the other hand, taking into account Theorem 2.11.1 we have

$$\|R_{\alpha,u_1} - P_2 - P_3\| \leq \|P_1\| + \|P_4\| \leq$$
$$\leq c_1(B_1^{(b)} + \overline{B}_1^{(c/2)}),$$

which converges to 0 when $b \to 0$ and $c \to \infty$. Finally we conclude that R_{α,u_1} is compact as it is a limit of compact operators. $\square$

Now we are going to prove the following statement:

Theorem 2.11.3. *Let* $1 < r, s, p, q < \infty$ *with* $\max\{r,s\} \leq \min\{p,q\}$ *and let* $\alpha > 1/r$. *Then the following conditions are equivalent:*

(i) R_{α,u_1} *is compact from* $L^{rs}(R_+)$ *to* $L_v^{pq}(R_+)$;

(ii) $B_1 < \infty$ *and* $\lim\limits_{b\to 0} B_1^{(b)} + \lim\limits_{c\to\infty} \overline{B}_1^{(c)} = 0$;

(iii) $B_1 < \infty$ *and* $\lim\limits_{t\to 0} B_1(t) = \lim\limits_{t\to\infty} B_1(t) = 0$;

(iv) $B_2 < \infty$ *and* $\lim\limits_{k\to-\infty} B_2(k) = \lim\limits_{k\to+\infty} B_2(k) = 0$.

Proof. The implication (ii) $\Rightarrow$ (i) follows from the previous Lemma. In addition, it is easy to see that (iii) $\Rightarrow$ (ii). Now we prove the implication (i) $\Rightarrow$ (iv). Let $k \in Z$ and let $f_k(y) = \chi_{[2^{k-2},2^{k-1})}(y)2^{-k/r}$. Then f_k converges weakly to 0 as $k \to -\infty$ or $k \to +\infty$. Indeed, if $\varphi \in L^{r's'}(R_+)$, then

$$\left|\int\limits_0^\infty f_k(y)\varphi(y)dy\right| \leq c_1\|\varphi(\cdot)\chi_{[2^{k-2},2^{k-1})}(\cdot)\|_{L^{r's'}(R_+)},$$

which converges to 0 as $k \to -\infty$ or $k \to +\infty$. On the other hand, we have

$$\|R_{\alpha,u_1}f(\cdot)\|_{L_v^{pq}(R_+)} \geq c_2\|\chi_{[2^k,2^{k+1})}(x)u_1(x)x^{\alpha-1}\|_{L_v^{pq}(R_+)}2^{k/r'} \geq$$
$$\geq c_3B_2(k),$$

which tends to 0 when $k \to -\infty$ or $k \to +\infty$, as R_{α,u_1} converges strongly to 0.

Now we show that (i) $\Rightarrow$ (iii). Let $f_t(y) = \chi_{(0,t/2)}(y)t^{-1/r}$. Then f_t tends weakly to 0 as $t \to 0$. On the other hand,

$$\|R_{\alpha,u_1}f_t(\cdot)\|_{L_\nu^{pq}(R_+)} \geq c_4 B_1(t)$$

and consequently $B_1(t) \to \infty$ as $t \to 0$. We claim $\lim_{t \to \infty} B_1(t) = 0$. Let $t > 0$. Then $t \in [2^m.2^{m+1})$ for some integer m. Consequently if $1 < \sigma \leq \min\{p,q\}$, then

$$B_1^\sigma(t) \leq \|\chi_{[2^m,\infty)}(x)x^{\alpha-1}u_1(x)\|_{L_\nu^{pq}(R_+)}^\sigma 2^{(m+1)\sigma/r'} \leq$$

$$\leq c_5 2^{m\sigma/r'} \sum_{k=m}^{+\infty} \|\chi_{[2^k,2^{k+1})}(x)x^{\alpha-1}u_1(x)\|_{L_\nu^{pq}(R_+)}^\sigma 2^{(m+1)\sigma/r'} \leq$$

$$\leq c_5(\sup_{k \geq m} B_2(k))^\sigma \equiv c_5 B_{2,m}^\sigma.$$

Hence if $t \to \infty$, then $m \to +\infty$ and $B_{2,m} \to 0$. Thus $\lim_{t \to \infty} B_1(t) = 0$. Now we prove that (iv) $\Rightarrow$ (ii). Let $b > 0$. Then there exists an integer $m \in Z$ depending on b such that $2^{m-1} \leq b < 2^m$. Also

$$B_1^{(b)} \leq \sup_{0 < t < 2^m} \|\chi_{[t,2^m)}(x)u_1(x)x^{\alpha-1}\|_{L_\nu^{pq}(R_+)} t^{1/r'} =$$

$$= \sup_{0 < t < 2^m} B_1^{(2^m)}(t) = B_1^{(2^m)},$$

where

$$B_1^{(2^m)}(t) \equiv \|\chi_{[t,2^m)}(x)x^{\alpha-1}u_1(x)\|_{L_\nu^{pq}(R_+)} t^{1/r'}.$$

Let $t \in [0, 2^m)$. Then $t \in [2^{n-1}, 2^n)$ for some integer $n \leq m$. Therefore , if $\sigma \leq \min\{p,q\}$, we have

$$(B_1^{(2^m)}(t))^\sigma \leq \|\chi_{[2^{n-1},2^m)}(x)x^{\alpha-1}u_1(x)\|_{L_\nu^{pq}(R_+)}^\sigma 2^{n\sigma/r'} =$$

$$= 2^{n\sigma/r'} \sum_{k=n}^m \|\chi_{[2^{k-1},2^k)}(x)x^{\alpha-1}u_1(x)\|_{L_\nu^{pq}(R_+)}^\sigma \leq$$

$$\leq c_6 \left(\sup_{k \leq m} B_2(k)\right)^\sigma \equiv \overline{B}_{2,m}.$$

Hence $B_1^{(b)} \leq c_6 \overline{B}_{2,m}$. If $b \to 0$, then $m \to -\infty$. Consequently, $\overline{B}_{2,m} \to 0$ as $m \to -\infty$ and $B_1^{(b)} \to 0$ as $b \to 0$.

We have already shown that $\lim\limits_{k\to+\infty} B_2(k) = 0$ follows from $\lim\limits_{t\to+\infty} B_1(t) = 0$. Hence $\lim\limits_{c\to\infty} \overline{B}^{(c)} = 0$. By Theorem 2.11.1 the condition $B_2 < \infty$ implies $B_1 < \infty$. Finally we have (ii) $\Rightarrow$ (i) $\Rightarrow$ (iv) $\Rightarrow$ (ii) and (i) $\Leftrightarrow$ (iii). $\square$

From duality arguments we get the following theorem:

Theorem 2.11.4. *Let* $1 < r, s, p, q < \infty$. *Assume also that* $\max\{r, s\} \le \min\{p, q\}$ *and* $\alpha > 1/p'$. *Then the following conditions are equivalent:*
(i) W_{α, u_2} *is compact from* $L_\nu^{rs}(R_+)$ *into* $L^{pq}(R_+)$;
(ii) $B_1' < \infty$ *and* $\lim\limits_{t\to 0} B_1'(t) = \lim\limits_{t\to\infty} B_1'(t) = 0$;
(iii) $B_1' < \infty$ *and* $\lim\limits_{b\to 0}(B_1^{(b)})' = \lim\limits_{c\to\infty}(\overline{B}_1^{(c)})' = 0$, *where*

$$(B_1^{(b)})' \equiv \sup_{0<t<b} \|u_2(x)x^{\alpha-1}\chi_{[t,b)}(x)\|_{L_\nu^{r's'}(R_+)} t^{1/p};$$

$$(\overline{B}_1^{(c)})' \equiv \sup_{t>c} \|u_2(x)x^{\alpha-1}\chi_{[t,\infty)}(x)\|_{L_\nu^{r's'}(R_+)} (t-c)^{1/p};$$

(iv) $B_2' < \infty$ *and* $\lim\limits_{k\to-\infty} B_2'(k) = \lim\limits_{k\to+\infty} B_2'(k) = 0$.

Now let us consider the operators: $R_\alpha \equiv R_{\alpha,1}$, $W_\alpha \equiv W_{\alpha,1}$.

Theorem 2.11.5. *Let* p, r, s, q *and* α *satisfy the conditions of Theorem* 2.11.1. *Then the following statements are equivalent:*
(i) R_α *is bounded from* $L^r(R_+)$ *into* $L_\nu^p(R_+)$;
(ii) R_α *is bounded from* $L^{rs}(R_+)$ *into* $L_\nu^{pq}(R_+)$;
(iii)

$$A_1 \equiv \sup_{t>0} A_1(t) \equiv \sup_{t>0} \left(\int\limits_{[t,\infty)} x^{(\alpha-1)p} d\nu(x) \right)^{1/p} t^{1/r'} < \infty;$$

(iv)

$$A_2 \equiv \sup_{k\in Z} A_2(k) \equiv \sup_{k\in Z} \left(\int\limits_{[2^k, 2^{k+1})} x^{(\alpha-1/r)p} d\nu(x) \right)^{1/p} < \infty.$$

Moreover, $\|R_\alpha\|_{L^r\to L_\nu^q} \approx B_1 \|R_\alpha\|_{L^{rs}\to L_\nu^{pq}} \approx A_1 \approx A_2$.

Proof. The facts (i) $\Leftrightarrow$ (iii) $\Leftrightarrow$ (iv) follows from the results of Section 2.1. Now we prove that (ii) $\Leftrightarrow$ (iv). By Theorem 2.11.1, (ii) is equivalent to the condition

$$A_2' \equiv \sup_{k\in Z} \|x^{\alpha-1/r}\chi_{[2^k, 2^{k+1})}(x)\|_{L_\nu^{pq}(R_+)} < \infty.$$

Now using Lemma 2.11.2 (part (b)) we have

$$A_2'(k) \approx \|\chi_{[2^k, 2^{k+1})}(x)\|_{L_\nu^{pq}(R_+)} 2^{k(\alpha-1/r)} =$$
$$= (\nu[2^k, 2^{k+1}))2^{k(\alpha-1/r)} \approx A_2(k).$$

Consequently, $A_2' < \infty$ if and only if $A_2 < \infty$. $\square$

Analogously we can derive the following result:

Theorem 2.11.6. *If p, q, r, s and α satisfy the conditions of Theorem 2.11.3, then the following conditions are equivalent:*
(i) R_α *is compact from* $L^r(R_+)$ *into* $L^p_\nu(R_+)$;
(ii) R_α *is compact from* $L^{rs}(R_+)$ *into* $L^{pq}_\nu(R_+)$;
(iii) $A_1 < \infty$ *and* $\lim\limits_{t \to 0} A_1(t) = \lim\limits_{t \to \infty} A_1(t) = 0$;
(iv) $A_1 < \infty$ *and* $\lim\limits_{b \to 0} A_1^{(b)} = \lim\limits_{c \to \infty} \overline{A}_1^{(c)} = 0$, *where*

$$A_1^{(b)} \equiv \sup_{0 < t < b} \left(\int\limits_{[t,b)} x^{(\alpha-1)p} d\nu(x) \right)^{1/p} t^{1/r'},$$

$$\overline{A}_1^{(c)} \equiv \sup_{t > c} \left(\int\limits_{[t,\infty)} x^{(\alpha-1)p} d\nu(x) \right)^{1/p} (t-c)^{1/r'};$$

v) $A_2 < \infty$ *and* $\lim\limits_{k \to -\infty} A_2(k) = \lim\limits_{k \to +\infty} A_2(k) = 0$.

Theorem 2.11.7. *Let r, s, p, q and α satisfy the conditions of Theorem 2.11.2. Then the following conditions are equivalent:*
(i) W_α *is bounded from* $L^r_\nu(R_+)$ *into* $L^p(R_+)$;
(ii) W_α *is bounded from* $L^{rs}_\nu(R_+)$ *into* $L^{pq}(R_+)$;
(iii)

$$D_1 \equiv \sup_{t > 0} D_1(t) \equiv \sup_{t > 0} \left(\int\limits_{[t,\infty)} x^{(\alpha-1)r'} d\nu(x) \right)^{1/r'} t^{1/p} < \infty;$$

(iv)

$$D_2 \equiv \sup_{k \in Z} D_2(k) \equiv \sup_{k \in Z} \left(\int\limits_{[2^k, 2^{k+1})} x^{(\alpha-1/p')r'} d\nu(x) \right)^{1/r'} < \infty.$$

Moreover, $\|W_\alpha\|_{L^r \to L^p_\nu} \approx \|W_\alpha\|_{L^{rs} \to L^{pq}_\nu} \approx D_1 \approx D_2$.

Theorem 2.11.8. *Let r, s, p, q and α satisfy the conditions of Theorem 2.11.4. Then the following conditions are equivalent:*
(i) W_α *is compact from* $L^r_\nu(R_+)$ *into* $L^p(R_+)$;
(ii) W_α *is compact from* $L^{rs}_\nu(R_+)$ *into* $L^{pq}(R_+)$;
(iii) $D_1 < \infty$ *and* $\lim\limits_{t \to 0} D_1(t) = \lim\limits_{t \to \infty} D_1(t) = 0$;
(iv) $D_1 < \infty$ *and* $\lim\limits_{b \to 0} D_1^{(b)}(t) = \lim\limits_{c \to \infty} \overline{D}_1^{(c)}(t) = 0$, *where*

$$D_1^{(b)} \equiv \sup_{0 < t < b} \left(\int_{[t,b)} x^{(\alpha-1)r'} d\nu(x) \right)^{1/r'} t^{1/p} < \infty;$$

$$\overline{D}_1^{(c)} \equiv \sup_{t > b} \left(\int_{[t,\infty)} x^{(\alpha-1)r'} d\nu(x) \right)^{1/r'} (t-c)^{1/p} < \infty;$$

(v) $D_2 < \infty$ and $\lim\limits_{k \to -\infty} D_2(k) = \lim\limits_{k \to +\infty} D_2(k) = 0$.

Now let us consider the Volterra- type integral transform

$$K_v f(x) = v(x) \int_0^x k(x,y) f(y) dy, \quad k \geq 0,$$

where v is a measurable, ν-a.e. positive function on R_+. We recall the following definition for the kernel $k : \{(x,y) : 0 < y < x < a\} \to R_+$ which was introduced in Section 2.7:

Let $a \leq \infty$. A kernel k belongs to V ($k \in V$) if there exists a positive constant d_1 such that for all x, y, z with $0 < y < z < x < a$ the following inequality is fulfilled:

$$k(x,y) \leq d_1 k(x,z)$$

We also need the following

Definition 2.11.1. We say that k belongs to $V_{\lambda,\eta}$ ($k \in V_{\lambda,\eta}$) ($1 < \lambda < \infty$, $1 \leq \eta < \infty$) if there exists a positive constant d_2 such that for all x, $x \in (0,a)$, the inequality

$$\|k(x)\chi_{(x/2,x)}(\cdot)\|_{L^{\lambda'\eta'}} \leq d_2 x^{1/\lambda'} k(x, x/2)$$

holds, where $\lambda' = \lambda/(\lambda-1)$, $\eta' = \eta/(\eta-1)$, and the positive constant d_2 does not depend on x.

As the next results can be derived in the same way as the previous theorems, we omit the proofs (see Section 2.7 for the case of Lebesgue spaces).

Theorem 2.11.9. *Let* $0 < a \leq \infty$, $1 \leq s < \infty$, $1 < q \leq \infty$ *and let* $k \in V \cap V_{r,s}$. *Then the following conditions are equivalent:*
(i) K_v *is bounded from* $L^{rs}(R_+)$ *into* $L_\nu^{pq}(R_+)$;
(ii)

$$E_1 \equiv \sup_{0 < t < a} E_1(t) \equiv \sup_{0 < t < a} \|v(x)\chi_{[t,a)}(x)k(x,x/2)\|_{L_\nu^{pq}} t^{1/r'} < \infty;$$

(iii)

$$E_2^{(\infty)} \equiv \sup_{n \in Z} E_2^{(\infty)}(k) \equiv \sup_{n \in Z} \|v(x)x^{1/r'} k(x,x/2)\chi_{[2^n, 2^{n+1})}(x)\|_{L_\nu^{pq}(R_+)} < \infty$$

if $a = \infty$ and

$$E_2^{(a)} \equiv \sup_{n \in Z_+} E_2^{(a)}(n) \equiv$$

$$\equiv \sup_{n \in Z_+} \|v(x)x^{1/r'}k(x,x/2)\chi_{[a/2^n,a/2^{n+1})}(x)\|_{L_\nu^{pq}(R_+)} < \infty$$

if $a < \infty$. Moreover, $\|K_v\| \approx E_1 \approx E_2^{(a)}$.

Theorem 2.11.10. *Let* $0 < a \le \infty$, $1 < r, p, q < \infty$, *and let* $1 \le s < \infty$.
Assume that $k \in V \cap V_{r,s}$. *Then the following conditions are equivalent:*
(i) K_v *is bounded from* $L^{rs}(R_+)$ *into* $L_\nu^{pq}(R_+)$;
(ii) $E_1 < \infty$ *and* $\lim_{b \to 0} E_1^{(b)} = \lim_{c \to \infty} \overline{E}_1^{(c)} = 0$, *where*

$$E_1^{(b)} \equiv \sup_{0 < t < b} \|v(x)\chi_{[t,b)}(x)k(x,x/2)\|_{L_\nu^{pq}} t^{1/r'};$$

$$\overline{E}_1^{(c)} \equiv \sup_{t > c} \|v(x)\chi_{[t,\infty)}(x)k(x,x/2)\|_{L_\nu^{pq}} (t - c)^{1/r'}$$

if $a = \infty$; $E_1 < \infty$ *and* $\lim_{b \to 0} E_1^{(b)} = 0$ *if* $a < \infty$;
(iii) $E_2^{(\infty)} < \infty$ *and* $\lim_{n \to -\infty} E_2^{(\infty)}(n) = \lim_{n \to +\infty} E_2^{(\infty)}(n) = 0$ *for* $a = \infty$ *and*
$E_2^{(a)} < \infty$ *and* $\lim_{n \to +\infty} E_2^{(a)}(n) = 0$ *for* $a < \infty$.

Next we consider the case of two weights. We shall deal with the operator

$$R_{\alpha,v,w}f(x) = v(x) \int_0^x f(y)(x - y)^{\alpha-1}w(y)dy, \quad \alpha > 0,$$

where $v(w)$ is a ν-measurable (Lebesgue-measurable) ν-a.e. positive (Lebesgue-a.e. positive) function on $(0, a)$, where $0 < a \le \infty$.

We say that a weight function w belongs to $U_{r,s}$ ($w \in U_{r,s}$), $1 < r < \infty$, $1 \le s < \infty$, if there exists a positive constant c such that for all $x \in (0, a)$ the inequality

$$\left\| \chi_{(x/2,x)}(\cdot) \frac{w(\cdot)}{(x - \cdot)^{1-\alpha}} \right\|_{L^{r's'}(0,a)} \le cx^{\alpha-1} \left\| \chi_{(0,x/4)}(\cdot)w(\cdot) \right\|_{L^{r's'}(0,a)}$$

holds, where $r' = r/(r - 1)$, $s' = s/(s - 1)$.

For example, if w is a decreasing function and $\alpha > 1/r$, then $w \in U_{r,s}$. This follows using Lemmas 2.11.2 and 2.11.3.

Now we formulate the main statement concerning the boundedness and compactness of $R_{\alpha,v,w}$.

Theorem 2.11.11. *Let* $0 < a \le \infty$, $1 < r, p < \infty$, $1 \le s < \infty$, $1 < q \le \infty$ *with* $\max\{r, s\} \le \min\{p, q\}$. *We assume that* $w \in U_{r,s}$ *and* $\alpha > 0$. *Then*
(a) $R_{\alpha,v,w}$ *is bounded from* $L^{rs}(0, a)$ *into* $L_\nu^{pq}(0, a)$ *if and only if*

$$B_a \equiv \sup_{0 < t < a} B_a(t) \equiv \|\chi_{[t,a)}(x)v(x)x^{\alpha-1}\|_{L_\nu^{pq}(0,a)} \times$$

$$\times \|\chi_{(0,t/2)}(x)w(x)\|_{L^{r's'}(0,a)} < \infty;$$

(b) *let* $q < \infty$, *then* $R_{\alpha,v,w}$ *is compact from* $L^{rs}(0, \infty)$ *into* $L_\nu^{pq}(0, \infty)$ *if and only if* $B_\infty < \infty$ *and* $\lim_{t\to 0} B_\infty(t) = \lim_{t\to\infty} B_\infty(t) = 0$ *if* $a = \infty$; $B_a < \infty$ *and* $\lim_{t\to 0} B_a(t) = 0$ *if* $a < \infty$.

The first part of this statement follows as in the case of Lebesgue space (see Theorem 2.1.12) using Theorem 1.1.4. The second part is proved in the same way as Theorem 2.11.3. We omit the proofs.

2.12. Applications to Abel's integral equations

In this section we are going to apply the two-weighted boundedness criteria for fractional integrals to Abel's integral equation

$$\frac{1}{\Gamma(\alpha)} \int\limits_0^x \frac{\varphi(t)dt}{(x-t)^{1-\alpha}} = f(x), \quad x > 0, \qquad (2.12.1)$$

where $0 < \alpha < 1$. This equation will be considered on a finite interval $[0, a]$. The results of Sections 2.1, 2.2 allow us to specify the spaces for the solutions of (2.12.1).

It is well-known (see, e.g.,[255], p.32) that for absolutely continuous f the equation may be solvable in $L(0, a)$ and its solution is represented in the form

$$\varphi(x) = \frac{1}{\Gamma(1-\alpha)} \left(\frac{f(0)}{x^\alpha} + \int\limits_0^x \frac{f'(y)}{(x-y)^\alpha} dy \right).$$

By means of Theorems from Section 2.1 and 2.2 we derive the following statements:

Proposition 2.12.1. *Let* $1 < p \le q < \infty$ *and let* $0 < \alpha < 1/p'$. *Assume that* $f \in AC([0, a])$, $f(0) = 0$ *and* $f' \in L^p(0, a)$. *If a Borel measure* ν *satisfies the condition*

$$\sup_{0 < t < a} \left(\int\limits_t^a x^{-\alpha q} d\nu \right)^{1/q} t^{1/p'} < \infty, \qquad (2.12.2)$$

then the equation (2.12.1) *is solvable in* $L_v^q(0,a) \cap L^1(0,a)$ *and for the solution* φ *the estimate*

$$\|\varphi\|_{L_v^q(0,a)} \leq c \|f'\|_{L^p(0,a)}$$

holds, where the constant c *does not depend on* f.

Proposition 2.12.2. *Let* $0 < q < p < \infty$, $p > 1$ *and* $0 < \alpha < 1/p'$. *Suppose that* $f \in AC([0,a])$, $f(0) = 0$ *and* $f' \in L^p(0,a)$. *If a weight* v *satisfies the condition*

$$\left(\int\limits_0^a \left(\int\limits_x^a v(y) y^{-\alpha q} dy \right)^{\frac{p}{p-q}} x^{\frac{(q-1)p}{p-q}} dx \right)^{\frac{p-q}{pq}} < \infty \qquad (2.12.3)$$

then the equation (2.12.1) *is solvable in* $L_v^q(0,a) \cap L^1(0,a)$ *and for the solution* φ *the estimate*

$$\|\varphi\|_{L_v^q(0,a)} \leq c \|f'\|_{L^p(0,a)}$$

is valid, where c *does not depend on* f.

Proposition 2.12.3. *Let* $1 < p < \infty$ *and* $1/p' < \alpha < 1$. *Assume that* v *and* w *are positive increasing functions on* $(0,a)$ *satisfying the conditions:*
(i)

$$\sup_{0<t<a} \left(\int\limits_t^a v(y) y^{-\alpha p} dy \right)^{1/p} \left(\int\limits_0^{t/2} w^{1-p'}(y) dy \right)^{1/p'} < \infty, \qquad (2.12.4),$$

(ii) there exist positive constants b such that for all $t \in (0,a]$ the inequality

$$t^{\alpha p} v(t) \leq b w(t/4) \qquad (2.12.5)$$

is fulfilled.

Then for $f \in AC([0,a])$ with the conditions $f(0) = 0$ and

$$\int\limits_0^a |f'(x)|^p w(x) dx < \infty,$$

the equation (2.12.1) is solvable in $L_v^q(0,a) \cap L^1(0,a)$. Moreover, the following estimate holds for solution φ:

$$\|\varphi\|_{L_v^q(0,a)} \leq c \|f'\|_{L_w^p(0,a)},$$

where c is independent of f.

Analogously for the dual Abel equation

$$\frac{1}{\Gamma(\alpha)} \int_x^b \frac{\varphi(t)}{(t-x)^{1-\alpha}} dt = f(x), \quad 0 < x < b < \infty \qquad (2.12.6)$$

we can derive the following

Proposition 2.12.5. *Let* $1 < p \leq q < \infty$ *and let* $0 < \alpha < 1/q$. *Assume that the weight function* w *satisfies the condition*

$$\sup_{0<t<b} \left(\int_t^b \frac{w^{1-p'}(x)}{x^{\alpha p'}} dx \right)^{1/p'} t^{1/q} < \infty.$$

Then for arbitrary f *from* $AC([0,b])$ *satisfying the conditions :* $f(0) = 0$ *and* $\int_0^b |f'(x)|^p w(x) dx < \infty$, *the equation* (2.12.6) *is solvable in* $L^p(0,b) \cap L^1(0,b)$. *In addition, the solution* φ *satisfies*

$$\|\varphi\|_{L^q(0,b)} \leq c \|f'\|_{L^p_w(0,b)}$$

with a constant c *independent of* f.

Proposition 2.12.6. *Let* $1 < q < p < \infty$ *and* $0 < \alpha < 1/q$. *Assume that the weight* w *satisfies the condition*

$$\left(\int_0^b \left(\int_x^b w^{1-p'}(y) y^{-\alpha p'} dy \right)^{\frac{q(p-1)}{p-q}} x^{\frac{q}{p-q}} dx \right)^{\frac{p-q}{pq}} < \infty.$$

Then for arbitrary f *from* $AC([0,b])$, *which satisfies the conditions:* $f(b) = 0$ *and* $\int_0^b |f'(x)|^p w(x) dx < \infty$, *the equation* (2.12.6) *is solvable in* $L^p(0,b) \cap L^1(0,b)$. *Moreover, the solution* φ *satisfies*

$$\|\varphi\|_{L^q(0,b)} \leq c \|f'\|_{L^p_w(0,b)},$$

where c *is independent of* f.

In the sequel we consider the case when the right-hand side of the equation (2.12.1) is represented as

$$f(x) = x^{-\nu} f^*(x), \quad 0 < x < a, \qquad (2.12.7)$$

where $0 < \nu \leq 1$ and $f^* \in AC([0,a])$. It is known (see, e.g., [255], §2) that in this case the equation (2.12.1) is solvable in $L^1(0,a)$ and the solution is

$$\varphi(x) = \frac{1}{\Gamma(1-\alpha)} \frac{1}{x} \int_0^x \frac{F(t)}{(x-t)^\alpha} dt,$$

where $F'(t) = (1 - \alpha)f(t) + tf'(t)$.

As above, using Theorems from Sections 2.1, 2.2 we can specify the spaces for the solution of (2.12.1). In fact, we have

Proposition 2.12.7. *Let* $1 < p \le q < \infty$ *and let* $0 < \alpha < 1/p'$. *If a Borel measure* ν *satisfies the condition* (2.12.2), *then for arbitrary* f *of the form* (2.12.7) *such that* $F \in L^p(0, a)$ *the equation* (2.12.1) *is solvable in* $L^1 \cap L^q_\mu$ *and the estimate*

$$\|\varphi\|_{L^q_\mu(0,a)} \le c\|F\|_{L^p(0,a)}$$

holds for the solution φ, *where* $d\mu = x d\nu$.

Now we consider the case $q < p$.

Proposition 2.12.8. *Let* p, q *and* α *satisfy the conditions of Proposition 2.12.2. If a weight function* v *satisfies the condition* (2.12.3), *then for arbitrary* f *of the form* (2.12.7) *such that* $F \in L^p(0, a)$ *the equation* (2.12.1) *is solvable in* $L^1 \cap L^q_{\overline{v}}$ *and the estimate of the solution* φ,

$$\|\varphi\|_{L^q_{\overline{v}}(0,a)} \le c\|F\|_{L^p(0,a)}$$

holds, where $\overline{v}(x) = x v(x)$.

Proposition 2.12.9. *Let* p, q *and* α *satisfy the conditions of Proposition 2.12.3. Assume that* v *satisfies the conditions* (2.12.4) *and* (2.12.5). *Then for* f *of the form* (2.12.7) *such that* $F \in L^p_w(0, a)$ *the equation is solvable in* $L^1 \cap L^q_{\overline{v}}$ *and for the solution* φ *the inequality*

$$\|\varphi\|_{L^q_{\overline{v}}(0,a)} \le c\|F\|_{L^p_w(0,a)}$$

is valid, where c *is independent of* F *and* $\overline{v}(x) = x v(x)$.

2.13.   On some Volterra–type integral equations

In this review section we present certain results concerning integral equations involving generalized Hardy transform and fractional integrals. These observations are due to [253].

At first we consider the integral equation

$$\varphi(x) - \alpha \int_x^1 \frac{x^{\alpha-1}}{y^\alpha}\varphi(y)dy = f(x), \quad x \in (0,1), \tag{2.13.1}$$

where $\alpha \in R$.

Let $\alpha + \beta > 1/2$. Then:

(i) If $\beta > 1/2$ the equation (2.11.8) is uniquely solvable in $L^2_{x^\beta}(0,1)$ and the solution has the form

$$\varphi(x) = f(x) + \frac{\alpha}{x}\int\limits_x^1 f(y)dy;$$

(ii) if $\beta < 1/2$, then (2.13.1) is solvable if and only if

$$\int\limits_0^1 f(x)dx = 0; \qquad\qquad (2.13.2)$$

moreover, the solution is given by the formula

$$\varphi(x) = f(x) - \frac{\alpha}{x}\int\limits_0^x f(y)dy;$$

(iii) if $\beta = 1/2$, then (2.13.1) is not normally solvable in $L^2_{x^\beta}(0,1)$.

One of the central part of the proof is the assertion that for the boundedness of the operator

$$V_\alpha\varphi(x) = \int\limits_x^1 \frac{x^{\alpha-1}}{y^\alpha}\varphi(y)dy$$

in $L^2_{x^\beta}(0,1)$ it is necessary and sufficient that $\alpha + \beta > 1/2$. Moreover,

$$\|V_\alpha\|_{L^2_{x^\beta}} = \frac{2}{2\alpha + 2\beta - 1}.$$

Next we consider the integral equation

$$\varphi(x) - 2\int\limits_x^1 \frac{y-x}{y^2}\varphi(y)dy = f(x), \quad x \in (0,1). \qquad (2.13.3)$$

Let $\beta > -1/2$. Then:

(i) in the case where $\beta > 1/2$, (2.13.3) is uniquely solvable for arbitrary $f \in L^2_{x^\beta}(0,1)$ and the solution has the form

$$\varphi(x) = f(x) + \frac{2}{3x}\int\limits_x^1 f(y)dy - \frac{2x^2}{3}\int\limits_x^1 \frac{f(y)}{y^3}dy;$$

(ii) if $-1/2 < \beta < 1/2$, then (2.13.3) is uniquely solvable if and only if (2.13.2) holds and the unique solution admits the representation

$$\varphi(x) = f(x) + \frac{2}{3x} \int\limits_0^x f(y)dy - \frac{2x^2}{3} \int\limits_x^1 \frac{f(y)}{y^3}dy;$$

(iii) if $\beta = 1/2$, then (2.13.3) is not normally solvable.
In this observation the following statement is important:
The operator

$$N_\alpha f(x) = \int\limits_x^1 \frac{(y-x)^{\alpha-1}}{y^\alpha} \varphi(y)dy, \quad \alpha > 0,$$

is bounded in $L^2_{x^\beta}(R_+)$ if and only if $\beta > -1/2$ and

$$\|N_\alpha\|_{L^2_{x^\beta}} = B(1/2 + \beta, \alpha),$$

where B denotes Euler's Beta function.
Note that in [253] the more general equation

$$\varphi(x) - 2 \int\limits_x^1 \frac{(y-x)^{\alpha-1}}{y^2} \varphi(y)dy = f(x), \quad x \in (0,1), \quad \alpha > 0,$$

is solved in explicit form.
For the details of the proofs we refer to the paper mentioned above.

2.14. Application to the existence of positive solutions of nonlinear integral equations

The goal of this section is to give a characterization of the existence of positive solutions of some Volterra–type nonlinear integral equations.
First let us consider the integral equation

$$\varphi(x) = \int\limits_x^\infty \frac{\varphi^p(t)dt}{(t-x)^{1-\alpha}} + \int\limits_x^\infty \frac{v(t)}{(t-x)^{1-\alpha}}dt, \quad 0 < \alpha < 1, \qquad (1.14.1)$$

with given non–negative $v \in L_{loc}(R_+)$.
Recall that by R_α and W_α we denote the Riemann–Liouville and Weyl operators respectively, i.e.

$$R_\alpha f(x) = \int_0^x f(t)(x-t)^{\alpha-1}dt, \quad x > 0, \quad \alpha > 0,$$

$$W_\alpha f(x) = \int_x^\infty f(t)(t-x)^{\alpha-1}dt, \quad x > 0, \quad \alpha > 0.$$

Theorem 1.14.1. *Let* $1 < p < \infty$, $0 < \alpha < 1$, $p' = \frac{p}{p-1}$, *and* $A_p = (p'-1)(p')^{-p}$.

i) If $W_\alpha v \in L^p_{loc}(R_+)$ and the inequality

$$W_\alpha[W_\alpha v]^p(x) \le A_p W_\alpha v(x) \quad \text{a.e.} \tag{1.14.2}$$

holds, then (1.14.1) has a non-negative solution $\varphi \in L^p_{loc}(R_+)$. Moreover, $(W_\alpha v)(x) \le \varphi(x) \le p'(W_\alpha v)(x)$.

ii) If $0 < \alpha < \frac{1}{p}$ and (1.14.1) has a non-negative solution in $L^p_{loc}(R_+)$, then $W_\alpha v \in L^p_{loc}(R_+)$ and

$$W_\alpha[(W_\alpha v)^p](x) \le c W_\alpha v(x) \quad \text{a.e.} \tag{1.14.3}$$

for some constant $c > 0$.

Proof. We shall use the following iteration procedure. Let $\varphi_0 = 0$, and let for $k = 0, 1, 2, \cdots$

$$\varphi_{k+1}(x) = W_\alpha(\varphi_k^p)(x) + W_\alpha v(x). \tag{1.14.4}$$

By induction it is easy to verify that

$$W_\alpha v(x) \le \varphi_k(x) \le \varphi_{k+1}(x), \quad k = 0, 1, 2, \cdots \tag{1.15.5}$$

From (1.14.4) we shall inductively derive an estimate of $\varphi_k(x)$.

Let

$$\varphi_k(x) \le c_k W_\alpha v(x) \tag{1.14.6}$$

for some $k = 0, 1, \cdots$. It is obvious that $c_1 = 1$. Then (1.14.2), (1.14.3) and (1.14.5) yield

$$\varphi_{k+1}(x) \le (A_p c_k^p + 1)(W_\alpha v)(x),$$

where A_p is the constant in (1.14.2). Thus $c_{k+1} = A_p c_k^p + 1$ for $k = 1, 2, \cdots$. Now by induction and the definition of A_p we deduce that the sequence $(c_k)_k$ is increasing. Indeed, it is obvious that $c_1 < c_2$. Let $c_k < c_{k+1}$. Then

$$c_{k+1} = A_p c_k^p + 1 < A_p c_{k+1} + 1 = c_{k+2}.$$

It is also clear that $(c_k)_k$ is bounded from the above by p' and consequently it converges. As the equation $z = A_p z^p + 1$ has only one solution, $x = p'$, it follows that $\lim_{k \to \infty} c_k = p'$. On the other hand, the sequence $(\varphi_k)_k$ is nondecreasing and by (1.14.6) we get

$$\varphi(x) = \lim_{k \to \infty} \varphi_k(x) \le p'(W_\alpha v)(x).$$

By our assumption $W_\alpha v \in L^p_{loc}(R_+)$ and from the preceding estimate we conclude that $\varphi \in L^p_{loc}(R_+)$. Moreover, $(W_\alpha v)(x) \le \varphi(x) \le p'(W_\alpha v)(x)$.

Now we prove the statement ii). Suppose (1.14.1) has a solution $\varphi \in L_{loc}^p(R^+)$. We have

$$W_\alpha(\varphi^p)(x) \leq \varphi(x) < \infty \quad \text{a.e.} \tag{1.14.7}$$

Hence $W_\alpha(\varphi^p) \in L_{loc}^p(R_+)$. Then from (1.14.7) we get

$$W_\alpha[W_\alpha(\varphi^p)(x)]^p(x) \leq W_\alpha(\varphi^p)(x) \quad \text{a.e.}$$

Applying Theorem 2.3.1, we deduce that

$$\|R_\alpha f\|_{L_\rho^{p'}} \leq \|f\|_{L^{p'}}, \tag{1.14.8}$$

where $\rho(x) = \varphi^p(x)$. But $(W_\alpha v)(x) \leq \varphi(x)$. Due to (1.14.7) we get

$$\|R_\alpha f\|_{L_{\rho_1}^{p'}} \leq c\|f\|_{L^{p'}} \tag{1.14.9}$$

with $\rho_1(x) = (W_\alpha v)^p(x)$. Using Lemma 2.3.2 we arrive at

$$\|R_\alpha f\|_{L_v^{p'}} \leq \|f\|_{L^{p'}}.$$

Applying Theorem 2.3.1 we come to condition (1.14.3). $\square$

Analogously, we can prove

Theorem 1.14.2. Let $1 < p < \infty$, $0 < \alpha < 1$ and let $A_p = (p' - 1)(p')^{-p}$.
(i) If $R_\alpha v \in L_{loc}^p(R_+)$ and the inequality

$$R_\alpha[R_\alpha v]^p(x) \leq A_p R_\alpha v(x) \quad \text{a.e.}$$

holds, then the integral equation

$$\varphi(x) = R_\alpha(\varphi^p)(x) + R_\alpha(v)(x) \tag{1.14.10}$$

has a non-negative solution $\varphi \in L_{loc}^p(R_+)$. Moreover, $R_\alpha v(x) \leq \varphi(x) \leq p'(R_\alpha v)(x)$.
(ii) If $0 < \alpha < 1/p'$ and (1.14.10) has a non-negative solution in $L_{loc}^p(R_+)$, then $R_\alpha v \in L_{loc}^p(R_+)$ and

$$R_\alpha[(R_\alpha v)^p](x) \leq c R_\alpha v(x) \quad \text{a.e.}$$

for some constant $c > 0$.

2.15.   Notes and comments on Chapter 2

The first integral inequalities for fractional integrals were obtained in [118]. For a simple proof of the Hardy–Littlewood inequality in Euclidean spaces see

[276], and for periodic case we refer to [310] (Section 9 of Chapter 12). For those transforms the estimates with power weights come from [118], [85] and in more general form from [245].

Criteria for one-weighted inequalities were derived in [9]. Criteria for two-weighted inequalities for the Riemann-Liouville operators R_α when $\alpha \geq 1$ were the core of the investigation of the series of papers [192], [281], [282], [284], [37]. The compactness criteria can be found in [284] and in a more general setting in [77] and [226]. For a valuable survey of boundedness and compactness problems, concerning the case when the order of intergation is not less than one, see [284].

Two-weighted criteria of Sawyer's type for fractional inegrals of arbitrary order are presented in [181–182]. Integral inequalities with power weights for power–logarithmic fractional integrals can be found in [255] (see §21, Section 3.4). For various applications we refer to the same monograph.

Some sufficient conditions for the boundedness of the Riemann-Liouville operators were derived in [238].

In [136] the Banach space $X_\gamma(a, b)$ consisting of the functions f with finite norm

$$\sup_{r \geq 1}(r^{-\gamma}\|f\|_r)$$

was introduced. Then the boundedness of the Riemann-Liouville operator

$$R_{a,1/p}f(x) = \int_a^x f(y)(x - y)^{1/p-1}dy$$

from $L^p(a, b)$ to $X_\gamma(a, b)$, $\gamma \geq 1/p'$, was established. Moreover, the boundedness between these spaces for $\gamma < 1/p'$ fails.

Note that $BMO(a, b) \not\subset X_\gamma(a, b)$ and $X_\gamma(a, b) \not\subset BMO(a, b)$ for $0 < \gamma < 1$ (see [136]).

An intensive study in weighted Lebesgue spaces of Volterra integral operators on the line may be found in [284], [77], [225]. For Lorentz and, in general, Banach function spaces see [184–185]. All the above–mentioned results involved classical fractional integrals, just when the order of integration is not less than one.

For general kernels in Euclidean space we refer [147], [100].

Interpolation properties of compact operators and measures of non–compactness of bounded linear operators were discussed in the survey [54].

The boundedness problems for discrete analogues of Riemann-Liouville transforms were studied in [210] and in some different form in [237]. It should be noted that in [210] the compactness criteria for the mentioned operator is also obtained.

Finally, concerning related problems we recall the survey [157].

Some parts of this chapter are based on our papers [200–202], [157], [163], [203], [205], [206–207], [76] (the second part of Section 2.12 relies on the paper [253]). The results of Sections 2.4, 2.6, 2.8, 2.10, 2.12 are published here for the first time.

Chapter 3

ONE–SIDED MAXIMAL FUNCTIONS

This chapter partly deals with one–sided fractional maximal functions and their various generalizations, among them one–sided maximal functions of Hörmander type. We establish $L^p \to L_v^q$ $(1 < p \le q < \infty)$ boundedness criteria which are very easy to verify. The proofs depend heavily on the results on the Riemann-Liouville operator which were derived in the previous chapter. Then follows a study, from the point of view of boundedness and compactness, of potentials on the line, or on a bounded interval, involving power–logarithmic kernels and their multidimensional analogues as well.

Exploring the well–known connection between maximal functions and potentials we obtain necessary and sufficient conditions for fractional maximal functions to be bounded from L_w^p to L_v^q when one of the weight functions is of power type.

Finally we establish boundedness criteria in weighted Lebesgue spaces for one–sided fractional maximal functions of Hörmander type.

3.1. One–sided maximal functions

This section deals with one–sided maximal functions. Here we present necessary and sufficient conditions on a weight function v to ensure the boundedness of one–sided fractional maximal operators from L^p into L_v^p, $1 < p < \infty$.

Theorem 3.1.1. *Let $1 < p < \infty$, $\frac{1}{p} < \alpha < 1$. Then there exists a constant A_1 such that the inequality*

$$\left(\int\limits_0^\infty (M_\alpha^- f(x)^p v(x) dx \right)^{\frac{1}{p}} \le A_1 \left(\int\limits_0^\infty |f(x)|^p dx \right)^{\frac{1}{p}} \tag{3.1.1}$$

251

holds for all $f \in L^p(0, \infty)$ if and only if

$$D_1 \equiv \sup_{t>0} \left(\int_t^\infty \frac{v(x)}{x^{(1-\alpha)p}} dx \right)^{1/p} t^{1/p'} < \infty. \tag{3.1.2}$$

In addition, if A_1 is the best constant in (3.1.1), *then we have*

$$A_1 \approx D_1.$$

Proof. Necessity follows from the inequality

$$M_\alpha^- f(x) \le cR_\alpha(|f|)(x)$$

and Theorem 2.1.1. In this inequality the constant c does not depend on x and f.

Let us assume that the condition (3.1.1) is fulfilled. In this inequality put $f(x) = \chi_{(0,2^{k+1})}(x)$. Then

$$\int_0^\infty v(x)(M_\alpha^- f(x))^p dx \ge \int_{2^k}^{2^{k+1}} v(x) \left(\frac{1}{2^{(k-1)(1-\alpha)}} \int_{x-2^{k-1}}^x dx \right)^p dx =$$

$$= c_1 \left(\int_{2^k}^{2^{k+1}} v(x) dx \right) \cdot 2^{k\alpha p}.$$

On the other hand,

$$\int_0^\infty |f(x)|^p dx = 2^{k+1}$$

and from (3.1.1) we see that

$$\sup_{k \in Z} \int_{2^k}^{2^{k+1}} \frac{v(x)}{x^{1-\alpha p}} dx < \infty.$$

By Theorem 2.1.1 we conclude that (3.1.2) holds. $\square$

Theorem 3.1.2. *Let $1 < p < \infty$, $\frac{1}{p'} < \alpha < 1$. The following two statements are equivalent:*
i) there exists $A_2 > 0$ such that

$$\left(\int_0^\infty (M_\alpha^+ f(x))^p dx \right)^{\frac{1}{p}} \le A_2 \left(\int_0^\infty |f(x)|^p w(x) dx \right)^{\frac{1}{p}}, \tag{3.1.3}$$

for all $f \in L_w^p(0, \infty)$;

ii)

$$\overline{D_1} \equiv \sup_{t>0} \Big(\int_t^\infty \frac{w^{1-p'}(x)}{x^{(1-\alpha)p'}} dx \Big)^{\frac{1}{p'}} t^{\frac{1}{p}} < \infty. \tag{3.1.4}$$

Moreover, if A_2 *is the best constant in* (3.1.3) *then*

$$A_2 \approx \overline{D_1}.$$

The proof is similar to that of the last Theorem, and is based on Theorem 2.1.2.

Theorem 3.1.3. *Let* $1 < p \le q < \infty$, $0 < \alpha < 1$. *Then there exists* $A > 0$ *such that for all* $f \in L_w^p(0, \infty)$,

$$\Big(\int_0^\infty (M_\alpha^- f(x)^q v(x) dx \Big)^{1/q} \le A \Big(\int_0^\infty |f(x)|^p w(x) dx \Big)^{1/p} \tag{3.1.5}$$

if the following two conditions are fulfilled:

$$D_1 = \sup_{t>0} \Big(\int_{2t}^\infty \frac{v(x)}{x^{(1-\alpha)q}} dx \Big)^{1/q} \Big(\int_0^t w^{1-p'}(x) dx \Big)^{1/p'} < \infty \tag{3.1.6}$$

and

$$D_2 = \sup_{t>0} \Big(\int_{t/2}^{2t} v(x) dx \Big)^{1/q} \Big(\int_{t/2}^t \frac{w^{1-p'}(x)}{(t-x)^{(1-\alpha)p'}} dx \Big)^{1/p'} < \infty. \tag{3.1.7}$$

Conversely if (3.1.5) *holds, then condition* (3.1.6) *is satisfied. Moreover if* A *is the best constant in* (3.1.5) *then we have*

$$c_1 D_1 \le A \le c_2 \max(D_1, D_2).$$

Proof. The first part follows from Theorem 2.1.10. Now let (3.1.5) be true. In this inequality put $f(x) = w^{1-p'} \chi_{(a,t)}(x)$, where $0 < a < t$. We show that

$$x^{\alpha-1} \int_a^t f(y) dy \le M_\alpha^- f(x)$$

for arbitrary $x \in (2t, \infty)$. In fact, for such an x we have

$$x^{\alpha-1} \int_a^t f(y) dy \le x^{\alpha-1} \int_a^x f(y) dy \le (x-a)^{\alpha-1} \int_a^x f(y) dy \le M_\alpha^- f(x).$$

Therefore

$$\Big(\int_0^\infty (M^- f(x)^q v(x)dx\Big)^{1/q} \geq \Big(\int_{2t}^\infty \frac{v(x)}{x^{(1-\alpha)q}}dx\Big)^{1/q}\Big(\int_a^t w^{1-p'}(y)dy\Big).$$
(3.1.8)

On the other hand

$$\int_0^\infty f^p(x)w(x)dx = \int_a^t w^{1-p'}(y)dy < \infty.$$
(3.1.9)

Consequently from (3.1.5), (3.1.8) and (3.1.9) we conclude that

$$\Big(\int_{2t}^\infty v(x)x^{(\alpha-1)q}dx\Big)^{1/q}\Big(\int_a^t w^{1-p'}(y)dy\Big)^{1/p'} \leq cA,$$

where the positive constant c does not depend on a and t.
Passing now to the limit as $a \to 0$ we obtain (3.1.6). □

The following theorem can be proved in a similar way.

Theorem 3.1.4. *Let* $1 < p \leq q < \infty, 0 < \alpha < 1$. *If the following two
conditions*

$$\overline{D}_1 \equiv \sup_{t>0} \Big(\int_0^{t/2} v(x)dx\Big)^{1/q}\Big(\int_t^\infty \frac{w^{1-p'}(x)}{x^{(1-\alpha)p'}}dx\Big)^{1/p'} < \infty$$
(3.1.10)

and

$$\overline{D}_2 \equiv \sup_{t>0} \Big(\int_{t/2}^{2t} v(x)dx\Big)^{1/q}\Big(\int_t^{2t} \frac{w^{1-p'}(x)}{(x-t)^{(1-\alpha)p'}}dx\Big)^{1/p'} < \infty,$$
(3.1.11)

are satisfied, then there exists a constant $A_1 > 0$ *such that*

$$\Big(\int_0^\infty v(x)(M^+ f(x))^q dx\Big)^{1/q} \leq A_1\Big(\int_0^\infty w(x)|f(x)|^p dx\Big)^{1/p},$$
(3.1.12)

for all $f \in L_w^p(0,\infty)$.
Conversely from the validity of (3.1.12) *we have* (3.1.10).
In addition ,

$$c_2\overline{D}_1 \leq A_1 \leq c_1 \max\{\overline{D}_1, \overline{D}_2\}.$$

From the previous Theorems we obtain the following corollaries:

Corollary 3.1.1. *Let* $1 < p \le q < \infty$, $\frac{1}{p} < \alpha < 1$. *Suppose that either* (i) w *is an increasing function on* $(0, \infty)$, *or* (ii) w *is a decreasing function on* $(0, \infty)$ *satisfying the condition*

$$\int\limits_{t}^{2t} w^{1-p'}(x)dx \le b \int\limits_{0}^{t/4} w^{1-p'}(x)dx, \ \ t > 0, \tag{3.1.13}$$

with a positive constant b independent of t. The inequality (3.1.5) is fulfilled if and only if

$$D_1 \equiv \sup_{t>0} \Big(\int\limits_{2t}^{\infty} \frac{v(x)}{x^{(1-\alpha)p}}dx \Big)^{1/q} \Big(\int\limits_{0}^{t} w^{1-p'}(x)dx \Big)^{1/p'} < \infty.$$

Moreover, $\|M_\alpha^-\| \approx D_1$.

Proof. By Theorem 3.1.3 it is sufficient to show that from the condition

$$D_2 \equiv \sup_{t>0} D_2(t) \equiv \sup_{t>0} \Big(\int\limits_{t/2}^{2t} v(x)dx \Big)^{1/q} \Big(\int\limits_{t/2}^{t} \frac{w^{1-p'}(x)dx}{(t-x)^{(1-\alpha)p'}} \Big)^{1/p'} < \infty$$

it follows that $D_1 < \infty$.

Indeed, let w be increasing and let $t > 0$. Then using the fact $\frac{1}{p} < \alpha < 1$ we derive

$$\int\limits_{t/2}^{t} \frac{w^{1-p'}(x)dx}{(t-x)^{(1-\alpha)p'}} \le w^{1-p'}(t/2) \int\limits_{t/2}^{t} \frac{dx}{(t-x)^{(1-\alpha)p'}} = b_1 w^{1-p'}(t/2)t^{(\alpha-1)p'+1}.$$

Consequently we have

$$D_2(t) \le b_2 \Big(\int\limits_{t/2}^{2t} v(x)dx \Big)^{1/q} w^{-1/p}(t/2)t^{\alpha-1+1/p'} \le$$

$$\le b_2 \Big(\int\limits_{t}^{2t} \frac{v(x)}{x^{(1-\alpha)q}}dx \Big)^{1/q} \cdot w^{-1/p}(t/2)t^{1/p'} \le$$

$$\le b_3 \Big(\int\limits_{t}^{2t} \frac{v(x)}{x^{(1-\alpha)q}}dx \Big)^{1/q} \Big(\int\limits_{0}^{t/4} w^{1-p'}(x)dx \Big)^{1/p'} \le b_3 D_1$$

for every $t > 0$.

Analoguosly we can show that $D_2 < \infty$ in the case when w is a decreasing function satisfying condition (3.1.13). □

The following corollary can be proved similarly.

Corollary 3.1.2. *Let* $1 < p \le q < \infty$, $\frac{q}{q-1} < \alpha < 1$. *Suppose that* (i) w *is a decreasing function on* $(0, \infty)$, *or* (ii) w *is an increasing function satisfying the condition*

$$\int_t^{2t} w^{1-p'}(x)dx \le b \int_{4t}^{6t} w^{1-p'}(x)dx, \quad t > 0,$$

with a positive constant b *independent of* t. *For the validity of* (3.1.12) *it is necessary and sufficient that* (3.1.10) *holds. Moreover,* $\|M_\alpha^+\| \approx \overline{D}_1$.

Corollary 3.1.3. *Let* $1 < p \le q < \infty$, $\frac{1}{p} < \alpha < 1$. *Suppose that* w *satisfies the condition*

$$\int_{t/2}^t \frac{w^{1-p'}(x)}{(t-x)^{(1-\alpha)p'}}dx \le ct^{(\alpha-1)p'} \int_0^{t/4} w^{1-p'}(x)dx,$$

with a positive constant c *independent of* t. *The inequality* (3.1.5) *holds if and only if* (3.1.6) *holds. Moreover,* $\|M_\alpha^-\| \approx D_1$.

Corollary 3.1.4. *Let* $1 < p \le q < \infty$, $\frac{q}{q-1} < \alpha < 1$. *Suppose that* w *satisfies the condition*

$$\int_t^{2t} \frac{w^{1-p'}(x)}{(t-x)^{(1-\alpha)p'}}dx \le ct^{(\alpha-1)p'} \int_{4t}^{6t} w^{1-p'}(x)dx,$$

with a positive constant c *independent of* t. *For the validity of* (3.1.12) *it is necessary and sufficient that* (3.1.10) *be satisfied. Moreover,* $\|M_\alpha^+\| \approx \overline{D}_1$.

3.2. Mapping properties of potentials on the line

In this section we present some two-weight criteria for the fractional integral operator

$$T_\gamma f(x) = \int_{-\infty}^{+\infty} \frac{f(y)}{|x-y|^{1-\gamma}}dy, \quad 0 < \gamma < 1.$$

Theorem 3.2.1. *Let* $1 < p < \infty$, $\frac{1}{p} < \gamma < 1$. *For the validity of the inequality*

$$\Big(\int_{-\infty}^{+\infty} |T_\gamma f(x)|^p v(x) dx \Big)^{1/p} \leq c \Big(\int_{-\infty}^{+\infty} |f(x)|^p |x|^\beta dx \Big)^{1/p}, \qquad (3.2.1)$$

with some positive constant c independent of f, $f \in L^p_{|x|^\beta}(R)$, it is necessary and sufficient that

$$\gamma p - 1 < \beta < p - 1 \qquad (3.2.2)$$

and

$$D_1 = \sup_{t>0} \Big(\int_{|x|>t} \frac{v(x)}{|x|^{(1-\gamma)p}} dx \Big)^{1/p} t^{1-1/p-\beta/p} < \infty. \qquad (3.2.3)$$

Moreover, if c is the best constant in (3.2.1), then $c \approx D_1$.
Proof. Sufficiency. Let $f \geq 0$. Then we have

$$\int_{-\infty}^{+\infty} (T_\gamma f(x))^p v(x) dx \leq c_1 \int_{-\infty}^{+\infty} \Big(\int_{|y|<|x|/2} \frac{f(y)}{|x-y|^{1-\gamma}} dy \Big)^p v(x) dx +$$

$$+ c_1 \int_{-\infty}^{+\infty} \Big(\int_{|x|/2<|y|<2|x|} \frac{f(y)}{|x-y|^{1-\gamma}} dy \Big)^p v(x) dx +$$

$$+ c_1 \int_{-\infty}^{+\infty} \Big(\int_{|y|>2|x|} \frac{f(y)}{|x-y|^{1-\gamma}} dy \Big)^p v(x) dx =$$

$$= S_1 + S_2 + S_3.$$

If $|y| < \frac{|x|}{2}$, then it is clear that $|x| \leq |x - y| + |y| \leq |x - y| + \frac{|x|}{2}$. Consequently we have $\frac{|x|}{2} \leq |x - y|$. Using Theorem 1.1.3 and the fact that $\beta < p - 1$ we find that

$$S_1 \leq c_2 \int_{-\infty}^{+\infty} \frac{v(x)}{|x|^{(1-\gamma)p}} \Big(\int_{|y|<|x|} f(y) dy \Big)^p dx \leq$$

$$\leq c_3 D_1^p \int_{-\infty}^{+\infty} (f(x))^p |x|^\beta dx.$$

Now we estimate S_3. If $|y| > 2|x|$, then $|y| \leq |x - y| + |x| \leq |x - y| + \frac{|y|}{2}$. Therefore $\frac{|y|}{2} \leq |x - y|$ and from Corollary 1.1.7 and the fact $\beta > \gamma p - 1$ we have

$$S_3 \leq c_4 \int_{-\infty}^{+\infty} v(x) \left(\int_{|y|>|x|} \frac{f(y)}{|y|^{1-\gamma}} dy \right)^p dx \leq c_5 D_2^p \int_{-\infty}^{+\infty} (f(y))^p |y|^\beta dy,$$

where

$$D_2 = \sup_{t>0} D_2(t) = \sup_{t>0} \left(\int_{|x|<t} v(x) dx \right)^{1/p} t^{\gamma - 1/p - \beta/p}.$$

Using Hölder's inequality we derive the following estimates:

$$S_2 \leq c_1 \int_{-\infty}^{+\infty} v(x) \left(\int_{|x|/2<|y|<2|x|} (f(y))^p \times \right.$$

$$\times |y|^\beta dy \Big) \left(\int_{|x|/2<|y|<2|x|} \frac{|y|^{\beta(1-p')}}{|x-y|^{(1-\gamma)p'}} dy \right)^{p-1} dx \leq$$

$$\leq c_6 \int_{-\infty}^{+\infty} v(x) |x|^\beta \left(\int_{|x|/2<|y|<2|x|} (f(y))^p |y|^\beta dy \right) \times$$

$$\times \left(\int_{|x|/2<|y|<2|x|} \frac{1}{|x-y|^{(1-\gamma)p'}} dy \right)^{p-1} dx.$$

We claim now that

$$I(x) \equiv \int_{|x|/2<|y|<2|x|} \frac{1}{|x-y|^{(1-\gamma)p'}} dy \leq \int_{|y|<2|x|} \frac{1}{|x-y|^{(1-\gamma)p'}} dy =$$

$$= \int_0^\infty |\{y : |x-y|^{(\gamma-1)p'} > \lambda\} \cap \{y : |y| < 2|x|\}| d\lambda =$$

$$= \int_0^{|x|^{(\gamma-1)p'}} |\{y : |x-y|^{(\gamma-1)p'} > \lambda\} \cap (-2|x|, 2|x|)| d\lambda +$$

$$+ \int_{|x|^{(\gamma-1)p'}}^\infty |\{y : |x-y|^{(\gamma-1)p'} > \lambda\} \cap (-2|x|, 2|x|)| d\lambda =$$

$$= I_1(x) + I_2(x).$$

For $I_1(x)$ we have

$$I_1(x) \le \int_0^{|x|^{(\gamma-1)p'}} |(-2|x|, 2|x|)| d\lambda = b_1|x||x|^{(\gamma-1)p'} = b_1|x|^{(\gamma-1)p'+1}$$

and using the condition $\frac{1}{p} < \gamma < 1$ for $I_2(x)$ we conclude that

$$I_2(x) \le \int_{|x|^{(\gamma-1)p'}}^{\infty} |\{y : |x-y|^{(\gamma-1)p'} > \lambda\}| d\lambda =$$

$$= \int_{|x|^{(\gamma-1)p'}}^{\infty} |\{y : |x-y| < \lambda^{\frac{1}{(\gamma-1)p'}}\}| d\lambda =$$

$$= b_2 \int_{|x|^{(\gamma-1)p'}}^{\infty} \lambda^{\frac{1}{(\gamma-1)p'}} d\lambda = b_3|x|^{(\gamma-1)p'+1}.$$

Hence

$$I(x) \le b_4|x|^{(\gamma-1)p'+1}.$$

From the estimates derived above we obtain

$$S_2 \le c_7 \int_{-\infty}^{+\infty} v(x)|x|^{-\beta+\gamma p-1}\left(\int_{|x|/2<|y|<2|x|} (f(y))^p|y|^\beta dy\right)dx =$$

$$= c_7 \sum_{k\in Z} \int_{2^k<|x|<2^{k+1}} v(x)|x|^{-\beta+\gamma p-1}\left(\int_{|x|/2<|y|<2|x|} (f(y))^p|y|^\beta dy\right)dx \le$$

$$\le c_7 \sum_{k\in Z} \left(\int_{2^{k-1}<|y|<2^{k+2}} (f(y))^p|y|^\beta dy\right)\int_{2^k<|x|<2^{k+1}} v(x)|x|^{-\beta+\gamma p-1}dx \le$$

$$\le c_8 \sum_{k\in Z} \left(\int_{2^{k-1}<|y|<2^{k+2}} (f(y))^p|y|^\beta dy\right) \times$$

$$\times\left(\int_{2^k<|x|<2^{k+1}} v(x)|x|^{(\gamma-1)p}dx\right)2^{k(-\beta+p-1)} \le$$

$$\le c_8 D_1^p \sum_{k\in Z} \left(\int_{2^{k-1}<|y|<2^{k+2}} (f(y))^p|y|^\beta dy\right) \le c_9 D_1^p \int_{-\infty}^{+\infty} (f(y))^p|y|^\beta dy.$$

Therefore

$$\|T_\gamma f\|_{L_v^p(R)} \le c_{10}D_1 + c_{11}D_2.$$

Now we show that $D_2 \leq b_5 D_1$. Let $t > 0$. Then there exists an integer m such that $t \in (2^m, 2^{m+1}]$ and the following estimates hold:

$$D_2^p(t) \leq \Big(\int\limits_{|x|>2^m} \frac{v(x)}{|x|^{(1-\gamma)p}} dx \Big) 2^{(m+1)(p-1-\beta)} =$$

$$= \Big(\sum_{k=m}^{+\infty} \int\limits_{2^k<|x|<2_{k+1}} \frac{v(x)}{|x|^{(1-\gamma)p}} dx \Big) 2^{(m+1)(p-1-\beta)} =$$

$$= b_6 2^{m(p-1-\beta)} \sum_{k=m}^{\infty} 2^{k(1-p+\beta)} \int\limits_{2^k<|x|<2^{k+1}} \frac{v(x)}{|x|^{-\gamma p+1+\beta}} dx \leq$$

$$\leq b_6 \sup_{k\in Z} D^p(k) 2^{m(p-1-\beta)} \sum_{k=m}^{\infty} 2^{k(1-p+\beta)} \leq b_7 \sup_{k\in Z} D^p(k),$$

where

$$D(k) = \Big(\int\limits_{2^k<|x|<2^{k+1}} \frac{v(x)}{|x|^{-\gamma p+1+\beta}} dx \Big)^{1/p}.$$

So we have

$$D_2 \leq b_8 \sup_{k\in Z} D(k).$$

Next let $k \in Z$. Then it is easy to see that

$$D(k) \leq b_{10} \Big(\int\limits_{2^k<|x|<2^{k+1}} \frac{v(x)}{|x|^{(1-\gamma)p}} dx \Big)^{1/p} 2^{k(1-\frac{1}{p}-\frac{\beta}{p})} \leq b_9 D_1.$$

Consequently $\sup_{k\in Z} D(k) \leq b_9 D_1$ and $D_2 \leq b_{10} D_1$.
This proves the inequality (3.2.1).
Necessity. First show that from inequality (3.2.1) follows condition (3.2.3). Let $t > 0$ and $f(x) = |x|^{\beta(1-p')} \chi_{\{|y|<t\}}(x)$. Then we have

$$\int\limits_{-\infty}^{+\infty} v(x) |T_\gamma f(x)|^p dx \geq \int\limits_{|x|>t} v(x) \Big(\int\limits_{|y|<t} \frac{|y|^{\beta(1-p')}}{|x-y|^{1-\gamma}} dy \Big)^p dx \geq$$

$$\geq \Big(\int\limits_{|x|>t} v(x) |x|^{(\gamma-1)p} dx \Big) \Big(\int\limits_{|y|<t} |y|^{\beta(1-p')} dy \Big)^p.$$

On the other hand,

$$\int\limits_{-\infty}^{+\infty} |f(x)|^p |x|^\beta dx = \int\limits_{|x|<t} |x|^{\beta(1-p')} dx$$

and consequently

$$\Big(\int\limits_{|x|>t} v(x)|x|^{(\gamma-1)p}dx\Big)\Big(\int\limits_{|x|<t} |x|^{\beta(1-p')}dx\Big)^{p-1} =$$

$$= d_1\Big(\int\limits_{|x|>t} v(x)|x|^{(\gamma-1)p}dx\Big)t^{-\beta+p-1} \le c$$

for every $t > 0$ and therefore $D_1 \le c$.

Now we derive condition (3.2.2). Let $\beta \ge p - 1$. Then $\int\limits_{|x|<t} |x|^{\beta(1-p')}dx = \infty$ and consequently there exists a function $g \ge 0$, $g \in L^p(-t,t)$, such that $g(x)|x|^{-\beta/p}$ does not belong to $L^1(-t,t)$.

Let $f(x) = g(x)|x|^{-\beta/p} \cdot \chi_{(-t,t)}(x)$. Then the right- hand side of the inequality (3.2.1) is finite, while

$$\int\limits_{-\infty}^{+\infty} v(x)(T_\gamma f(x))^p dx \ge$$

$$\ge d_2\Big(\int\limits_{|x|>t} \frac{v(x)}{|x|^{(1-\gamma)p}}dx\Big)\Big(\int\limits_{|x|<t} g(x)|x|^{-\beta/p}dx\Big)^p = \infty.$$

Hence $\beta < p - 1$.

Now let $\beta \le \gamma p - 1$; then

$$\int\limits_{|x|>t} |x|^{\beta(1-p')-(1-\gamma)p'} dx = \infty.$$

This means that $|x|^{-\beta/p+\gamma-1}$ does not belong to $L^{p'}(|x| > t)$.

Consequently, there exists a function $g \in L^p(|x| > t)$ such that

$$g(x)|x|^{-\beta/p+\gamma-1} \notin L^1(|x| > t).$$

Let $f(x) = g(x)|x|^{-\beta/p}\chi_{\{|y|>t\}}(x)$. Then the right-hand side of the inequality (3.2.1) is finite, while

$$\int\limits_{-\infty}^{+\infty} v(x)\Big(\int\limits_{-\infty}^{+\infty} \frac{f(y)}{|x-y|^{1-\gamma}}dy\Big)^p dx \ge \int\limits_{|x|<t} v(x)\Big(\int\limits_{|y|>t} \frac{g(y)|y|^{-\beta/p}}{|x-y|^{1-\gamma}}dy\Big)^p dx \ge$$

$$\ge d_3\Big(\int\limits_{|x|<t} v(x)dx\Big)\Big(\int\limits_{|y|>t} g(y)|y|^{-\beta/p+\gamma-1}dy\Big)^p = \infty.$$

This completes the proof.

From a duality argument and Theorem 3.2.1 we obtain

Theorem 3.2.2. *Let* $1 < p < \infty$, $\frac{1}{p'} < \gamma < 1$. *The inequality*

$$\int_{-\infty}^{+\infty} |T_\gamma f(x)|^p |x|^\beta dx \le c \int_{-\infty}^{+\infty} |f(x)|^p w(x) dx, \quad f \in L_w^p(R), \qquad (3.2.4)$$

where the positive constant c does not depend on f, holds if and only if

$$-1 < \beta < -\gamma p + p - 1, \qquad (3.2.5)$$

and

$$\overline{D}_1 = \sup_{t>0} \Big(\int_{|x|>t} \frac{w^{1-p'}(x)}{|x|^{(1-\gamma)p'}} dx \Big)^{1/p'} t^{\beta/p+1/p} < \infty. \qquad (3.2.6)$$

Moreover, $\|T_\gamma\| \approx \overline{D}_1$.

Theorem 3.2.3. *Let* $0 < q < p < \infty$, $p > 1$ *and* $\frac{1}{p} < \gamma < 1$. *Suppose that* v *is a weight function on* $(-\infty, +\infty)$. *The inequality*

$$\Big(\int_{-\infty}^{+\infty} |T_\gamma f(x)|^q v(x) dx \Big)^{1/q} \le c \Big(\int_{-\infty}^{+\infty} |f(x)|^p |x|^\beta dx \Big)^{1/p}, \qquad (3.2.7)$$

with a positive constant c independent of f, $f \in L_{|x|^\beta}^p(R)$, *holds if and only if the following conditions are satisfied:*
(i)

$$\gamma p - 1 < \beta < p - 1; \qquad (3.2.8)$$

(ii)

$$\tilde{A}_1 = \Big[\int_{-\infty}^{+\infty} \Big(\int_{|y|>|x|} \frac{v(y)}{|y|^{(1-\gamma)q}} dy \Big)^{\frac{p}{p-q}} |x|^{\frac{(q-1)p-\beta q}{p-q}} dx \Big]^{\frac{p-q}{pq}} < \infty. \qquad (3.2.9)$$

Moreover, $\|T_\gamma\| \approx \tilde{A}_1$.

Proof. First we prove sufficiency. For non-negative f we have

$$\Big(\int_{-\infty}^{+\infty} |T_\gamma f(x)|^q v(x) dx \Big)^{1/q} \le$$

$$\le c_1 \Big(\int_{-\infty}^{+\infty} v(x) \Big(\int_{|y|<\frac{|x|}{2}} \frac{f(y)}{|x-y|^{1-\gamma}} dy \Big)^q dx \Big)^{1/q} +$$

$$+c_1\Big(\int\limits_{-\infty}^{+\infty}v(x)\Big(\int\limits_{\frac{|x|}{2}<|y|<2|x|}\frac{f(y)}{|x-y|^{1-\gamma}}dy\Big)^q dx\Big)^{1/q}+$$

$$+c_1\Big(\int\limits_{-\infty}^{+\infty}v(x)\Big(\int\limits_{|y|>2|x|}\frac{f(y)}{|x-y|^{1-\gamma}}dy\Big)^q dx\Big)^{1/q}\equiv$$

$$\equiv S_1+S_2+S_3.$$

If $|y|<\frac{|x|}{2}$, then $\frac{|x|}{2}\le|x-y|$ and consequently by Theorem 1.1.3 and the assumption that $\beta<p-1$ we derive the estimate

$$S_1\le c_2\Big(\int\limits_{-\infty}^{+\infty}\frac{v(x)}{|x|^{(1-\gamma)q}}\Big(\int\limits_{|y|<|x|}f(y)dy\Big)^q dx\Big)^{1/q}\le$$

$$\le c_3\tilde{A}_1\Big(\int\limits_{-\infty}^{+\infty}(f(y))^p|y|^\beta dy\Big)^{\frac{1}{p}}.$$

If $|y|>2|x|$, then $\frac{|y|}{2}\le|x-y|$ and using Corollary 1.1.7 and the fact that $\beta>\gamma p-1$ we obtain

$$S_3\le c_4\Big(\int\limits_{-\infty}^{+\infty}v(x)\Big(\int\limits_{|y|>|x|}\frac{f(y)}{|y|^{1-\gamma}}dy\Big)^q dx\Big)^{1/q}\le$$

$$\le c_5\tilde{A}_2\Big(\int\limits_{-\infty}^{+\infty}(f(x))^p|x|^\beta dx\Big)^{1/p},$$

where

$$\tilde{A}_2=\Big[\int\limits_{-\infty}^{+\infty}\Big(\int\limits_{|y|<|x|}v(y)dy\Big)^{\frac{p}{p-q}}|x|^{\frac{-\beta q+\gamma pq-p}{p-q}}dx\Big]^{\frac{p-q}{pq}}<\infty. \qquad (3.2.10)$$

Using Hölder's inequality twice and the condition $\frac{1}{p}<\gamma<1$ we have

$$S_2^q\le c_6\int\limits_{-\infty}^{+\infty}v(x)\Big(\int\limits_{|x|/2<|y|<2|x|}(f(y))^p|y|^\beta dy\Big)^{q/p}\times$$

$$\Big(\int\limits_{|x|/2<|y|<|x|}\frac{dy}{|x-y|^{(1-\gamma)p'}|y|^{\beta(p'-1)}}\Big)^{q/p'}dx\le$$

$$\leq c_7 \int\limits_{-\infty}^{+\infty} v(x) \Big(\int\limits_{|x|/2<|y|<2|x|} (f(y))^p |y|^\beta dy \Big)^{q/p} \times$$

$$\times |x|^{\beta(1-p')q/p'} |x|^{[(\gamma-1)p'+1]q/p'} dx \leq$$

$$\leq c_7 \int\limits_{-\infty}^{+\infty} v(x) \Big(\int\limits_{|x|/2<|y|<2|x|} (f(y))^p |y|^\beta dy \Big)^{q/p} \times$$

$$\times |x|^{(\gamma-1)q - \beta q/p + q/p'} dx \leq$$

$$\leq c_7 \Big(\sum_{k\in Z} \int\limits_{2^{k-1}<|x|<2^{k+1}} (f(x))^p |x|^\beta dx \Big)^{\frac{q}{p}} \times$$

$$\times \Big(\sum_{k\in Z} \Big(\int\limits_{2^k<|x|<2^{k+1}} v(x) |x|^{(\gamma-1)q - \beta q/p + q/p'} dx \Big)^{\frac{p}{p-q}} \Big)^{\frac{p-q}{p}} =$$

$$= c_7 \|f\|^q_{L^p_{|y|^\beta}(R)} \Big(\sum_{k\in Z} S_{2k} \Big)^{\frac{p-q}{p}}.$$

For S_{2k} we have

$$S_{2k} \leq c_8 \Big(\int\limits_{2^k<|x|<2^{k+1}} v(x) |x|^{(\gamma-1)q} dx \Big)^{\frac{p}{p-q}} 2^{k(-\frac{\beta q}{p}+\frac{q}{p'})\frac{p}{p-q}} =$$

$$= c_8 2^{\frac{kp}{p-q}(\frac{q}{p'} - \frac{\beta q}{p})} \overline{S}_{2k},$$

where

$$\overline{S}_{2k} = \Big(\int\limits_{2^k<|x|<2^{k+1}} v(x) |x|^{(\gamma-1)q} dx \Big)^{\frac{p}{p-q}}$$

Integration by parts in the expression $\overline{S}_{2k}$ gives

$$\overline{S}_{2k} = \Big(\int\limits_{2^k}^{2^{k+1}} (v(x) + v(-x)) x^{(\gamma-1)q} dx \Big)^{\frac{p}{p-q}} =$$

$$= c_9 \int\limits_{2^k}^{2^{k+1}} \Big(\int\limits_{x}^{2^{k+1}} (v(x) + v(-x)) x^{(\gamma-1)q} dx \Big)^{\frac{p}{p-q}} \frac{v(x) + v(-x)}{x^{(\gamma-1)q}} dx =$$

$$= c_{10} \int\limits_{2^k<|x|<2^{k+1}} \frac{v(x)}{|x|^{(1-\gamma)q}} \Big(\int\limits_{|x|}^{\infty} \frac{v(y)}{|y|^{(1-\gamma)q}} dy \Big)^{\frac{q}{p-q}} dx.$$

Consequently we have

$$S_{2k} \le c_{11} \int\limits_{2^k < |x| < 2^{k+1}} \left(\int\limits_{|y| > |x|} \frac{v(y)}{|y|^{(1-\gamma)q}} dy \right)^{\frac{q}{p-q}} \times$$

$$\times \frac{v(x)}{|x|^{(1-\gamma)q}} |x|^{(-\frac{\beta q}{p} + \frac{q}{p'})\frac{p}{p-q}} dx$$

and therefore

$$S_2 \le c_{12} \|f\|^q_{L^p_{|y|^\beta}(R)} \left[\int\limits_{-\infty}^{+\infty} \left(\int\limits_{|y| > |x|} \frac{v(y)}{|y|^{(1-\gamma)q}} dy \right)^{q(p-q)} \times \right.$$

$$\left. \times \frac{v(x)}{|x|^{(1-\gamma)q}} |x|^{(-\frac{\beta q}{p} + \frac{q}{p'})\frac{p}{p-q}} dx \right]^{\frac{p-q}{p}} =$$

$$= c_{13} \|f\|^q_{L^p_{|y|^\beta}(R)} \left[\int\limits_{-\infty}^{+\infty} \left(\int\limits_{|y| > |x|} \frac{v(y)}{|y|^{(1-\gamma)q}} dy \right)^{\frac{p}{p-q}} |x|^{\frac{-\beta q + p(q-1)}{p-q}} dx \right]^{\frac{p-q}{p}} =$$

$$= c_{14} \tilde{A}_1 \|f\|^q_{L^p_{|y|^\beta}(R)}.$$

It remains to prove that $\tilde{A} \le b\tilde{A}_1$. Indeed, let $\lambda = \frac{(q-1)p - \beta q}{p-q}$. Then using the Hardy inequality in the discrete case (see, e.g., [19], [95]) and the condition (i) we obtain

$$\tilde{A}^p = \sum_{k \in Z} \int\limits_{2^k \le |x| < 2^{k+1}} |x|^\lambda \left(\int\limits_{|y| < |x|} v(y) dy \right)^{p/(p-q)} dx \le$$

$$\le \sum_{k \in Z} \left(\int\limits_{2^k \le |x| < 2^{k+1}} |x|^\lambda dx \right) \left(\int\limits_{|y| < 2^{k+1}} v(y) dy \right)^{p/(p-q)} =$$

$$= b_1 \sum_{k \in Z} 2^{k(\lambda+1)} \left(\sum_{n=-\infty}^{k} \int\limits_{2^n \le |y| < 2^{n+1}} v(y) dy \right)^{p/(p-q)} \le$$

$$\le b_2 \sum_{k \in Z} 2^{k(\lambda+1)} \left(\int\limits_{2^k \le |y| < 2^{k+1}} v(y) dy \right)^{p/(p-q)} =$$

$$= b_3 \sum_{k \in Z} 2^{k(\lambda+1) + \frac{(1-\gamma)pqk}{p-q}} \left(\int\limits_{2^k \le |y| < 2^{k+1}} v(y)|y|^{(\gamma-1)q} dy \right)^{p/(p-q)} =$$

$$= b_4 \sum_{k \in Z} \int\limits_{2^{k-1} \le |x| < 2^k} |x|^{\lambda \frac{pq - \gamma pq}{p-q}} \left(\int\limits_{|y| > |x|} v(y)|y|^{(\gamma-1)q} dy \right)^{p/(p-q)} =$$

$$= b_4 \widetilde{A}_1^{\frac{pq}{p-q}}.$$

Thus (3.2.9) implies (3.2.10).

Summarizing the estimates derived above we have (3.2.7).

Now we prove necessity. The validity of the condition (3.2.8) can be proved as in the proof of Theorem 3.2.1.

Let $0 < a < b < \infty$ and let

$$v_0(y) = v(y) \cdot \chi_{\{a < |z| < b\}}(y),$$

$$w_0(y) = |y|^{\beta(1-p')} \chi_{\{a < |z| < b\}}(y),$$

$$f(x) = \Big(\int\limits_{|y| > |x|} \frac{v_0(y)}{|y|^{(1-\gamma)q}} dy \Big)^{\frac{1}{p-q}} \Big(\int\limits_{|y| < |x|} w_0(y) dy \Big)^{\frac{q-1}{p-q}} \cdot w_0(x).$$

Then we have

$$\|f\|_{L^p_{|y|^\beta}(R^1)} = \Big(\int\limits_{-\infty}^{+\infty} \Big(\int\limits_{|y| > |x|} \frac{v_0(y)}{|y|^{(1-\gamma)q}} dy \Big)^{\frac{p}{p-q}} \times$$

$$\times \Big(\int\limits_{|y| < |x|} w_0(y) dy \Big)^{\frac{(q-1)p}{p-q}} w_0(x) dx \Big)^{\frac{1}{p}} < \infty,$$

as

$$\int\limits_{|y| > t} \frac{v(y)}{|y|^{(1-\gamma)q}} dy < \infty$$

for all $t > 0$ (see the proof of Theorem 3.2.1.).

On the other hand,

$$\Big(\int\limits_{-\infty}^{+\infty} (T_\gamma f(x))^q v(x) dx \Big)^{\frac{1}{q}} \geq$$

$$\geq \Big(\int\limits_{-\infty}^{+\infty} v(x) \Big(\int\limits_{|y| < |x|} \frac{f(y)}{|x-y|^{1-\gamma}} dy \Big)^q dx \Big)^{\frac{1}{q}} \geq$$

$$\geq c_1 \Big(\int\limits_{-\infty}^{+\infty} \frac{v_0(x)}{|x|^{(1-\gamma)q}} \Big(\int\limits_{|y| < |x|} f(y) dy \Big)^q dx \Big)^{1/q} \geq$$

$$\geq c_1 \Big(\int\limits_{-\infty}^{+\infty} \frac{v_0(x)}{|x|^{(1-\gamma)q}} \Big(\int\limits_{|y| < |x|} \Big(\int\limits_{|z| > |y|} \frac{v_0(z)}{|z|^{(1-\gamma)q}} dz \Big)^{\frac{1}{p-q}} \times$$

$$\times\Big(\int\limits_{|z|<|y|} w_0(z)dz\Big)^{\frac{q-1}{p-q}} w_0(y)dy\Big)^q dx\Big)^{1/q} \geq$$

$$\geq c_1\Big(\int\limits_{-\infty}^{+\infty}\frac{v_0(x)}{|x|^{(1-\gamma)q}}\Big(\int\limits_{|z|>|x|}\frac{v_0(z)}{|z|^{(1-\gamma)q}}dz\Big)^{\frac{q}{p-q}} \times$$

$$\times\Big(\int\limits_{|y|<|x|}\Big(\int\limits_{|z|<|y|} w_0(z)dz\Big)^{\frac{q-1}{p-q}} w_0(y)dy\Big)^q dx\Big)^{1/q} =$$

$$= c_2\Big(\int\limits_{-\infty}^{+\infty}\frac{v_0(x)}{|x|^{(1-\gamma)q}}\Big(\int\limits_{|z|<|x|}\frac{v_0(z)}{|z|^{(1-\gamma)q}}dz\Big)^{\frac{q}{p-q}} \times$$

$$\times\Big(\int\limits_{|z|>|x|} w_0(z)dz\Big)^{\frac{q(p-1)}{p-q}} dx\Big)^{1/q} = c_3\Big(\int\limits_{-\infty}^{+\infty}\Big(\int\limits_{|z|>|x|}\frac{v_0(z)}{|z|^{(1-\gamma)q}}dz\Big)^{\frac{p}{p-q}} \times$$

$$\times\Big(\int\limits_{|z|<|x|} w_0(z)dz\Big)^{\frac{p(q-1)}{p-q}} w_0(x)dx\Big)^{1/q}.$$

Consequently we have

$$\Big(\int\limits_{-\infty}^{+\infty}\Big(\int\limits_{|z|>|x|}\frac{v_0(z)}{|z|^{(1-\gamma)q}}dz\Big)^{\frac{p}{p-q}}\Big(\int\limits_{|z|<|x|} w_0(z)dz\Big)^{\frac{p(q-1)}{p-q}} w_0(x)dx\Big)^{\frac{p-q}{pq}} \leq c,$$

where the positive constant c does not depend on a and b. By Fatou's lemma we derive condition (3.2.9). $\square$

3.3.   Potentials T_γ on   the   line   (the   case $0 < \gamma < \frac{1}{p}$)

In this section we derive two–weight inequalities of type (p,p), $p > 1$, for potentials on the real axis.

Definition 3.3.1. Let $1 < p < \infty$, $0 < \gamma < 1$, $p^* = \frac{p}{1-\gamma p}$. We say that the weight function w belongs to $A_{pp^*}(R)$ if

$$\sup\Big(\frac{1}{|I|}\int\limits_I w^{p^*}(x)dx\Big)^{\frac{1}{p^*}}\Big(\frac{1}{|I|}\int\limits_I w^{-p'}(x)dx\Big)^{\frac{1}{p'}} < \infty,$$

where the supremum is taken with respect to all intervals $I \subset R^1$ and $p' = \frac{p}{p-1}$.
   The following lemma is well-known for even weights (see, for example, [71], [100], p. 360). We give the proof in the case of general weights for the completeness

Lemma A. *Let $1 < p < \infty$ and let $\rho \in A_p(R)$ i.e.*

$$\sup \left(\frac{1}{|I|} \int_I \rho(x)dx \right)^{1/p} \left(\frac{1}{|I|} \int_I \rho^{1-p'}(x)dx \right)^{1/p'} < \infty, \quad p' = \frac{p}{p-1},$$

where the supremum is taken over all intervals $I \subset R$. Assume that $0 \le c_1 < c_2 \le c_3 < c_4 < \infty$. Then there exists a positive constant c such that for all $t > 0$ the following inequality holds:

$$\int_{c_3 t < |x| < c_4 t} \rho(x)dx \le c \int_{c_1 t < |x| < c_2 t} \rho(x)dx.$$

Proof. Let

$$\Gamma = \{x \in R : c_1 t < |x| < c_2 t\},$$

$$\Gamma_1 = \{x \in R : c_3 t < |x| < c_4 t\},$$

$$B = \{x \in R : |x| < c_4 t\}.$$

Let Mf be the Hardy–Littlwood maximal function:

$$Mf(x) = \sup \frac{1}{|I|} \int_I |f(y)|dy,$$

where the supremum is taken over all intervals I containing the point x. Then we have

$$M\varphi(x) > \left(\frac{1}{|I|} \int_I |\varphi(y)|dy \right) \cdot \chi_{\Gamma_1}(x) \ge \left(\frac{c}{|\Gamma|} \int_\Gamma |\varphi(y)|dy \right) \chi_{\Gamma_1}(x),$$

where $\varphi \in L^p_\rho(R)$ and the constant $c > 0$ is independent of φ. Using the boundedness of M in $L^p_\rho(R)$, $1 < p < \infty$ (see [213]) we get

$$\int_{\Gamma_1} \left(\frac{1}{|\Gamma|} \int_\Gamma \varphi(y)dy \right)^p \rho(x)dx \le c \int_G |\varphi(y)|^p \rho(y)dy.$$

If we replace the function φ by χ_Γ, then we obtain

$$\int_{\{x : c_3 t < |x| < c_4 t\}} \rho(x)dx \le c \int_{\{x : c_1 t < |x| < c_2 t\}} \rho(x)dx.$$

$\square$

Lemma 3.3.1. *Let* $0 < \gamma < 1$, $1 < p < \frac{1}{\gamma}$, $p^* = \frac{p}{1-\gamma p}$. *Suppose that* $0 \le c_1 < c_2 \le c_3 < c_4 < \infty$ *and* $w \in A_{pp^*}(R^1)$. *Then there exists a positive constant* c *such that for all* $t > 0$ *the following inequality*

$$\int_{c_3 t < |x| < c_4 t} w^{-p'}(x)dx \le c \int_{c_1 t < |x| < c_2 t} w^{-p'}(x)dx \qquad (3.3.1)$$

is fulfilled.

Proof. From the condition $w \in A_{pp^*}(R^1)$ we have that $w^{p^*} \in A_{1+p^*/p'}(R^1)$. Therefore $(w^{p^*})^{(-p'/p^*)} \in A_{1+p'/p^*}(R)$ and by Lemma A we obtain the inequality (3.3.1). $\square$

The following Lemma can be proved analogously.

Lemma 3.3.2. *Let* γ, p, p^*, w *and* c_i, $i = 1, 2, 3, 4$, *satisfy the conditions of the previous Lemma. Then the inequality*

$$\int_{c_3 t < |x| < c_4 t} w^p(x)dx \le c \int_{c_1 t < |x| < c_2 t} w^p(x)dx, \quad t > 0$$

holds, where the positive constant c *does not depend on* t.

Lemma 3.3.3. *Let* p, γ *and* p^* *satisfy the conditions of Lemma 3.3.1. Let* $\rho \in A_{pp^*}(R)$ *and* $\beta \ge 1$. *Suppose that* σ *and* u *are positive increasing functions on* $(0, \infty)$. *We put* $v(x) = \rho^p(x)\sigma(|x|)$, $w(x) = \rho^p(x)u(|x|)$. *If*

$$B \equiv \sup_{t>0} \Big(\int_{|x|>t} \frac{v(x)}{|x|^{(1-\gamma)p}}dx \Big)^{\frac{1}{p}} \Big(\int_{|x|<t/2} w^{1-p'}(x)dx \Big)^{\frac{1}{p'}} < \infty. \qquad (3.3.2)$$

Then there exists a positive constant b *such that for all* $t > 0$ *the inequality*

$$\sigma(\beta t)t^{\gamma p} \le bB^p u(t/2)$$

holds.

Proof. By Hölder's inequality we have

$$\Big[\frac{\sigma(\beta t)t^{\gamma p}}{u(t/2)} \Big]^{\frac{1}{p}} = c_1 \Big[\frac{\sigma(\beta t)t^{\gamma p}}{u(t)} \Big]^{\frac{1}{p}} \Big(\frac{1}{t} \int_{\beta t < |x| < 2\beta t} \rho(x)\rho^{-1}(x)dx \Big) \le$$

$$\le c_1 \Big[\frac{\sigma(\beta t)t^{\gamma p}}{u(t/2)} \Big]^{\frac{1}{p}} \frac{1}{t} \Big(\int_{\beta t < |x| < 2\beta t} \rho^p(x)dx \Big)^{\frac{1}{p}} \Big(\int_{\beta t < |x| < 2\beta t} \rho^{-p'}(x)dx \Big)^{\frac{1}{p'}} \le$$

$$\le c_2 \Big[\frac{\sigma(\beta t)t^{\gamma p}}{u(t/2)} \Big]^{\frac{1}{p}} \frac{1}{t} \Big(\int_{\beta t < |x| < 2\beta t} \rho^p(x)dx \Big)^{\frac{1}{p}} \Big(\int_{|x|<t/2} \rho^{-p'}(x)dx \Big)^{\frac{1}{p'}} \le$$

$$\leq c_2 t^{\gamma-1} \Big(\int\limits_{\beta t<|x|<2\beta t} v(x)dx \Big)^{\frac{1}{p}} \Big(\int\limits_{|x|<t/2} w^{1-p'}(x)dx \Big)^{\frac{1}{p'}} \leq c_3 B.$$

$\square$

Theorem A ([214]). *Let γ, p and p^* satisfy the conditions of Lemma 3.3.1. Then the inequality*

$$\Big(\int\limits_{-\infty}^{+\infty} |T_\gamma f(x)|^{p^*} w^{p^*}(x)dx \Big)^{\frac{1}{p^*}} \leq c \Big(\int\limits_{-\infty}^{+\infty} |f(x)|^p w^p(x)dx \Big)^{\frac{1}{p}},$$

holds if and only if $w \in A_{pp^*}(R)$.

Lemma B. *Let $1 < p < \infty$ and let $\rho \in A_{pp^*}$, where $p^* = \frac{p}{1-\gamma p}$. Then the following condition is fulfilled:*

$$B \equiv \sup_{t>0} \Big(\int\limits_{|x|<t} \rho^{p^*}(x)dx \Big)^{\frac{1}{p^*}} \Big(\int\limits_{|x|>t} \frac{\rho^{-p'}(x)}{|x|^{(1-\gamma)p'}}dx \Big)^{\frac{1}{p'}} < \infty. \qquad (3.3.2)$$

Proof. Due to the condition $\rho \in A_{pp^*}(R)$, there exists $c > 0$ such that

$$\Big(\int\limits_{-\infty}^{+\infty} |T_\gamma f(x)|^{p^*} \rho^{p^*}(x)dx \Big)^{1/p^*} \leq \Big(\int\limits_{-\infty}^{+\infty} |f(x)|^p \rho^p(x)dx \Big)^{1/p},$$

where c is independent of f, $f \in L^p_{\rho^p}(R)$. Then it is easy to verify that for all $t > 0$ we have

$$I(t) \equiv \int\limits_{|x|>t} \frac{\rho^{-p'}}{|x|^{(1-\gamma)p'}}dx < \infty.$$

Indeed, if for some t_0, $I(t_0) = \infty$, then there exists

$$g \in L^p_{\rho^p}(|x| > t_0)$$

such that

$$\int\limits_{|x|>t_0} g(x)|x|^{\gamma-1}dx = \infty.$$

If we choose $f(x) = g(x)$ in the two-weight inequality, then we obtain

$$\Big(\int\limits_{-\infty}^{+\infty} |T_\gamma f(x)|^{p^*} \rho^{p^*}(x)dx \Big)^{1/p^*} \geq$$

$$\geq \Big(\int\limits_{|x|<t_0} \rho^{p^*}(x)dx \Big)^{1/p^*} \Big(\int\limits_{|x|>t_0} g(x)|x|^{\gamma-1}dx \Big) = \infty.$$

On the other hand,

$$\|f\|_{L^p_{\rho p}(R)} = \int\limits_{|x|>t_0} \rho^p(x)g(x)dx < \infty.$$

Consequently, $I(t_0) < \infty$.

Now let $f(y) = \chi_{\{|y|>t\}}(x)|x|^{(\gamma-1)(p'-1)}\rho^{-p'}(x)$. Then

$$\|T_\gamma f\|_{L^{p^*}_{\rho p^*}(R)} \geq \Big(\int\limits_{|x|<t} \rho^{p^*}(x)dx \Big)^{1/p^*} \Big(\int\limits_{|x|>t} |x|^{(\gamma-1)p'}\rho^{-p'}(x)dx \Big).$$

On the other hand,

$$\|f\|_{L^p_{\rho p}(R)} = \Big(\int\limits_{|x|>t} |x|^{(\gamma-1)p'}\rho^{-p'}(x)dx \Big)^{1/p},$$

which gives the condition $B < \infty$. $\square$

Lemma 3.3.4. *Let γ, p, p^* satisfy the conditions of Lemma 3.3.1 and let $\rho \in A_{pp^*}(R)$. Suppose that σ and u are positive increasing functions on $(0,\infty)$. We put $v(x) = \sigma(|x|)\rho^p(x)$, $w(x) = u(|x|)\rho^p(x)$. If v and w satisfy the condition (3.3.2) then*

$$B_1 = \sup_{t>0} B_1(t) = \sup_{t>0} \Big(\int\limits_{|x|<t} v(x)dx \Big)^{\frac{1}{p}} \Big(\int\limits_{|x|>t} \frac{w^{1-p'}(x)}{|x|^{(1-\gamma)p'}}dx \Big)^{\frac{1}{p'}} < \infty.$$

Moreover, $B_1 \leq bB$.

Proof. Using the monotonicity of σ and u and Hölder's inequality we obtain

$$B_1(t) \leq \Big(\frac{\sigma(t)}{u(t)}\Big)^{\frac{1}{p}} \Big(\int\limits_{|x|<t} \rho^p(x)dx \Big)^{\frac{1}{p}} \Big(\int\limits_{|x|>t} \frac{\rho^{-p'}(x)}{|x|^{(1-\gamma)p'}}dx \Big)^{\frac{1}{p'}} \leq$$

$$\leq \Big(\frac{\sigma(t)}{u(t)}\Big)^{\frac{1}{p}} \Big(\int\limits_{|x|<t} \rho^{p^*}(x)dx \Big)^{\frac{1}{p^*}} \Big(\int\limits_{|x|<t} dx \Big)^{\frac{1}{p(p^*/p)'}} \Big(\int\limits_{|x|>t} \frac{\rho^{-p'}(x)}{|x|^{(1-\gamma)p'}}dx \Big)^{\frac{1}{p'}}.$$

Moreover, as $\rho \in A_{pp^*}(R)$, by Lemma B we have that

$$\sup_{t>0} \Big(\int\limits_{|x|<t} \rho^{p^*}(x)dx \Big)^{\frac{1}{p^*}} \Big(\int\limits_{|x|>t} \frac{\rho^{-p'}(x)}{|x|^{(1-\gamma)p'}}dx \Big)^{\frac{1}{p'}} < \infty.$$

is satisfied. Therefore by Lemma 3.3.3 we obtain

$$B_1(t) \leq c_1 \Big(\frac{\sigma(t)}{u(t)}\Big)^{\frac{1}{p}} t^\gamma \leq c_2 B$$

for all $t > 0$.

□

After this proposition we can give criteria for boundedness of T_γ.

Theorem 3.3.1. *Let* $0 < \gamma < 1$, $1 < p < \frac{1}{\gamma}$ *and* $p^* = \frac{p}{1-\gamma p}$. *Suppose that* $\rho \in A_{pp^*}(R)$ *and* σ *and* u *are positive increasing functions on* $(0, \infty)$. *We put* $v(x) = \sigma(|x|)\rho^p(x)$, $w(x) = u(|x|)\rho^p(x)$. *Then the operator* T_γ *is bounded from* $L^p_w(R)$ *to* $L^p_v(R)$ *if and only if* (3.3.2) *holds. Moreover,* $\|T_\gamma\| \approx B$.

Proof. We have

$$\int\limits_{-\infty}^{+\infty} |T_\gamma f(x)|^p v(x) dx \leq$$

$$\leq c_1 \int\limits_{-\infty}^{+\infty} \Big| \int\limits_{|y| < \frac{|x|}{2}} \frac{f(y)}{|x-y|^{1-\gamma}} dy \Big|^p v(x) dx +$$

$$+ c_1 \int\limits_{-\infty}^{+\infty} \Big| \int\limits_{\frac{|x|}{2} < |y| < 2|x|} \frac{f(y)}{|x-y|^{1-\gamma}} dy \Big|^p v(x) dx +$$

$$+ c_1 \int\limits_{-\infty}^{+\infty} \Big| \int\limits_{|y| > 2|x|} \frac{f(y)}{|x-y|^{1-\gamma}} dy \Big|^p v(x) dx \equiv$$

$$\equiv S_1 + S_2 + S_3.$$

If $|y| < \frac{|x|}{2}$, then $|x - y| \geq \frac{|x|}{2}$ and consequently using Theorem 1.1.3 we conclude that

$$S_1 \leq c_2 \int\limits_{-\infty}^{+\infty} \Big(\int\limits_{|y| < \frac{|x|}{2}} |f(y)| dy \Big)^p v(x) |x|^{(\gamma-1)p} dx \leq$$

$$\leq c_3 B^p \int\limits_{-\infty}^{+\infty} |f(y)|^p w(x) dx.$$

If $|y| > 2|x|$, then $\frac{|y|}{2} \leq |x - y|$ and we have

$$S_3 \leq c_4 \int\limits_{-\infty}^{+\infty} \Big(\int\limits_{|y|>|x|} \frac{|f(y)|}{|y|^{1-\gamma}} dy \Big)^p v(x) dx.$$

By Lemma 3.3.4 and Corollary 1.1.7 we obtain

$$S_3 \leq c_5 B^p \int\limits_{-\infty}^{+\infty} |f(x)|^p w(x) dx.$$

For S_2, using the Hölder inequality we derive the estimate

$$S_2 = c_1 \sum_{k \in Z} \int\limits_{2^k < |x| < 2^{k+1}} v(x) \Big| \Big(\int\limits_{\frac{|x|}{2} < |y| < 2|x|} \frac{f(y)}{|x-y|^{1-\gamma}} dy \Big| ^p dx \leq$$

$$\leq \sum_{k \in Z} \Big(\int\limits_{2^k < |x| < 2^{k+1}} \sigma^{\frac{1}{\gamma p}}(|x|) dx \Big)^{\gamma p} \Big(\int\limits_{2^k < |x| < 2^{k+1}} \rho^{p^*}(x) \times$$

$$\times \Big(\int\limits_{\frac{|x|}{2} < |y| < 2|x|} \frac{f(y)}{|x-y|^{1-\gamma}} dy \Big)^{p^*} dx \Big)^{\frac{p}{p^*}} =$$

$$= \sum_{y \in Z} S_{2,1,k} \cdot S_{2,2,k}.$$

By Lemma 3.3.3

$$S_{2,1,k} \leq c_6 \sigma(2^{k+1}) 2^{k\alpha p} \leq c_7 B^p u(2^{k-1}).$$

According to Theorem A for $S_{2,2,k}$ we obtain

$$S_{2,2,k} \leq \Big(\int\limits_{-\infty}^{+\infty} \rho^{p^*}(x) \Big(\int\limits_{-\infty}^{+\infty} \frac{|f(y)| \chi_{\{2^{k-1} < |y| < 2^{k+2}\}}(y)}{|x-y|^{1-\gamma}} dy \Big)^{p^*} dx \Big)^{\frac{p}{p^*}} \leq$$

$$\leq c_8 \int\limits_{2^{k-1} < |x| < 2^{k+2}} \rho^p(x) |f(x)|^p dx$$

and consequently for S_2 we have

$$S_2 \leq c_9 B^p \sum_{k \in Z} u(2^{k-1}) \int\limits_{2^{k-1} < |x| < 2^{k+2}} \rho^p(x) |f(x)|^p dx \leq$$

$$\leq c_9 B^p \sum_{k \in Z} \int\limits_{2^{k-1} < |x| < 2^{k+2}} w(x) |f(x)|^p dx \leq c_{10} B^p \int\limits_{-\infty}^{+\infty} |f(x)|^p w(x) dx.$$

Finally we obtain

$$\int\limits_{-\infty}^{+\infty} |T_\gamma f(x)|^p v(x)dx \le c_{10}B^p \int\limits_{-\infty}^{+\infty} |f(x)|^p w(x)dx.$$

Now we prove necessity. First we show that for all $t > 0$

$$\int\limits_{|x|<t} w^{1-p'}(x)dx < \infty. \tag{3.3.3}$$

Indeed, suppose that for some $t > 0$ we have that

$$\int\limits_{|x|<t} w^{1-p'}(x)dx = \infty.$$

Let g be a function that is in $L^p(-t, t)$, but such that $g \cdot w^{-1/p}$ is not integrable on $(-t, t)$. Let $f(x) = |g(x)|w^{-1/p}(x)$ on the interval $(-t, t)$ and $f(x) = 0$ otherwise. Then we have

$$\|f\|_{L^p_w(R)} = \Big(\int\limits_{|x|<t} |g(x)|^p dx \Big)^{\frac{1}{p}} < \infty.$$

On the other hand, we have

$$\|T_\gamma f\|^p_{L^p_w(R)} \ge \Big(\int\limits_{|x|>t} \Big(\int\limits_{|y|<t} \frac{|g(x)|w^{-\frac{1}{p}}(x)}{|x-y|^{1-\gamma}}dx \Big)^p v(x)dx \Big) \ge$$

$$\ge c_{11} \Big(\int\limits_{|x|>t} v(x)|x|^{(\gamma-1)p}dx \Big) \Big(\int\limits_{|y|<t} |g(x)|w^{-\frac{1}{p}}(x)dx \Big)^p = \infty.$$

Hence $v(x) = 0$ almost everywhere in $R \setminus (-t, t)$, and consequently (3.3.3) holds.

Now we show that $B < \infty$. Let $f(x) = w^{1-p'}(x)\chi_{\{|y|<t/2\}}(x)$ for $t > 0$. Then we have

$$\|f\|_{L^p_w(R)} = \Big(\int\limits_{|x|<t/2} w^{1-p'}(x)dx \Big)^{\frac{1}{p}} < \infty.$$

On the other hand,

$$\|T_\gamma f\|_{L^p_w(R)} \ge \Big(\int\limits_{|x|>t} \Big(\int\limits_{|y|<t/2} \frac{w^{1-p'}(y)}{|x-y|^{1-\gamma}}dy \Big)^p v(x)dx \Big)^{\frac{1}{p}} \ge$$

$$\ge \Big(\int\limits_{|x|>t} \frac{v(x)}{|x|^{(1-\gamma)p}}dx \Big)^{\frac{1}{p}} \Big(\int\limits_{|y|<t/2} w^{1-p'}(y)dy \Big).$$

Due to the boundedness of the operator T_γ from $L^p_w(R)$ to $L^p_v(R)$ it follows that $B < \infty$. $\square$

3.4. Weak–type inequalities

In this section we shall deal with weak- type inequalities for potential operators.

First we prove the following

Lemma 3.4.1. *Let* $0 < \gamma < 1$, $1 < p < \frac{1}{\gamma}$, $\frac{1}{p} - \frac{1}{p^*} = \gamma$. *Assume that* $\rho \in A_{pp^*}(R)$ *and that* σ *and* u *are positive increasing functions on* $(0, \infty)$. *We put* $v(x) = \rho^p(x)\sigma(|x|)$, $w(x) = \rho^p(x)u(|x|)$. *If the condition*

$$\overline{B} \equiv \sup_{\tau > t} \frac{1}{\tau^{1-\gamma}} \Big(\int\limits_{t < |x| < \tau} v(x)dx \Big)^{\frac{1}{p}}$$
$$\Big(\int\limits_{|x| < t} w^{1-p'}(x)dx \Big)^{\frac{1}{p'}} < \infty \tag{3.4.1}$$

is satisfied, then there is a positive constant b *such that for all* $t > 0$ *the following inequality holds:*

$$\sigma(\beta t)t^{\gamma p} \leq b\overline{B}^p u(t).$$

Proof. Using the monotonicity of σ and u, Hölder's inequality and Lemma A of Section 3.3 we have

$$\Big(\frac{\sigma(\beta t)t^{\gamma p}}{u(t)} \Big)^{\frac{1}{p}} = c_1 \Big(\frac{\sigma(\beta t)t^{\gamma p}}{u(t)} \Big)^{\frac{1}{p}} \Big(\frac{1}{t} \int\limits_{\beta t < |x| < 2\beta t} \rho(x)\rho^{-1}(x)dx \Big) \leq$$

$$\leq c_1 \Big(\frac{\sigma(\beta t)t^{\gamma p}}{u(t)} \Big)^{\frac{1}{p}} \frac{1}{t} \Big(\int\limits_{\beta t < |x| < 2\beta t} \rho^p(x)dx \Big)^{\frac{1}{p}} \Big(\int\limits_{\beta t < |x| < 2\beta t} \rho^{-p'}(x)dx \Big)^{\frac{1}{p'}} \leq$$

$$\leq c_2 \Big(\frac{\sigma(\beta t)t^{\gamma p}}{u(t)} \Big)^{\frac{1}{p}} \frac{1}{t} \Big(\int\limits_{\beta t < |x| < 2\beta t} \rho^p(x)dx \Big)^{\frac{1}{p}} \Big(\int\limits_{|x| < t} \rho^{-p'}(x)dx \Big)^{\frac{1}{p'}} \leq$$

$$\leq c_2 t^{\gamma-1} \Big(\int\limits_{\beta t < |x| < 2\beta t} v(x)dx \Big)^{\frac{1}{p}} \Big(\int\limits_{|x| < t} \rho^{-p'}(x)dx \Big)^{\frac{1}{p'}} \leq c_3\overline{B}.$$

$\square$

Lemma 3.4.2. *Let* γ, p, p^* *satisfy the conditions of Lemma 3.4.1. Suppose that* $\rho \in A_{pp^*}(R)$ *and that* σ *and* u *are positive increasing functions on* $(0, \infty)$.

We put $v(x) = \sigma(|x|)\rho^p(x)$ *and* $w(x) = u(|x|)\rho^p(x)$. *If condition* (3.4.1) *is satisfied, then the condition*

$$\overline{B}_1 \equiv \sup_{t>0} \overline{B}_1(t) = \left(\int\limits_{|x|<t} v(x)dx \right)^{\frac{1}{p}} \left(\int\limits_{|x|>t} \frac{w^{1-p'}(x)}{|x|^{(1-\gamma)p'}} dx \right)^{\frac{1}{p'}} < \infty$$

is fulfilled.

Moreover, $\overline{B}_1 \leq b\overline{B}$, *for a positive constant* b.

This Lemma can be proved in the same way as Lemma 3.3.4 .

Theorem 3.4.1. *Let* $0 < \gamma < 1$, $1 < p < \frac{1}{\gamma}$, $p^* = \frac{p}{1-\gamma p}$. *Assume that* $\rho \in A_{pp^*}(R)$. *Suppose that* σ *and* u *are positive increasing functions on* $(0, \infty)$, *and put* $v(x) = \rho^p(x)\sigma(|x|)$, $w(x) = \rho^p(x)u(|x|)$. *The inequality*

$$\|T_\gamma f\|_{L_v^{p\infty}(R)} \leq c\|f\|_{L_w^p(R)}, \quad f \in L_w^p(R), \tag{3.4.2}$$

with a positive constant c *independent of* f, *holds if and only if* (3.4.1) *is satisfied.*

Proof. First prove sufficiency. Let $f \geq 0$. Then we have

$$\|T_\gamma f\|_{L_v^{p\infty}(R)} \leq \left\| \int\limits_{|y|<\frac{|x|}{2}} \frac{f(y)}{|x-y|^{1-\gamma}} dy \right\|_{L_v^{p\infty}(R)} +$$

$$+ \left\| \int\limits_{\frac{|x|}{2}<|y|<2|x|} \frac{f(y)}{|x-y|^{1-\gamma}} dy \right\|_{L_v^{p\infty}(R)} + \left\| \int\limits_{|y|>2|x|} \frac{f(y)}{|x-y|^{1-\gamma}} dy \right\|_{L_v^{p\infty}(R)} =$$

$$= I_1 + I_2 + I_3.$$

If $|y| < \frac{|x|}{2}$, then we have $\frac{|x|}{2} \leq |x-y|$ and consequently using Theorem 1.2.9 we see that

$$I_1 \leq c_1 \left\| |x|^{\gamma-1} \int\limits_{|y|<|x|} f(y)dy \right\|_{L_v^{p\infty}(R)} \leq c_2\overline{B}\|f\|_{L_w^p(R)}.$$

We now estimate I_3. It is easy to verify that if $|y| > 2|x|$, then $\frac{|y|}{2} \leq |x-y|$ and by Corollary 1.1.7 and by Lemma 3.4.2 we have

$$I_3 \leq c_3 \left\| \int\limits_{|y|>|x|} \frac{f(y)}{|y|^{1-\gamma}} dy \right\|_{L_v^{p\infty}(R)} \leq c_4\overline{B}\|f\|_{L_w^p(R)}.$$

It remains to estimate I_2. We have

$$I_2^p = \left\| \int\limits_{\frac{|x|}{2}<|y|<2|x|} \frac{f(y)}{|x-y|^{1-\gamma}} dy \right\|_{L_v^{p\infty}(R)}^p \leq$$

$$\leq \left\| \int\limits_{\frac{|x|}{2}<|y|<2|x|} \frac{f(y)}{|x-y|^{1-\gamma}}dy \right\|_{L_v^p(R)}^p =$$

$$= \int\limits_{-\infty}^{+\infty} v(x) \left(\int\limits_{\frac{|x|}{2}<|y|<2|x|} \frac{f(y)}{|x-y|^{1-\gamma}}dy \right)^p dx.$$

Using Lemma 3.4.1 and Theorem A of Section 3.3 (see also the proof of Theorem 3.3.1) we obtain

$$I_2 \leq c_4 \overline{B} \|f\|_{L_w^p(R)}.$$

Now we show necessity. First we prove that $\int\limits_{|x|<t} w^{1-p'}(x)dx < \infty$ for all $t > 0$. Indeed, let $\int\limits_{|x|<t} w^{1-p'}(x)dx = \infty$. Then $w^{-1/p} \notin L^{p'}(-t,t)$ and consequently there exists a function $g \in L^p(-t,t)$ such that $gw^{-1/p}$ is not integrable on $(-t,t)$. Let $f(x) = |g(x)|w^{-1/p}(x)$ on the interval $(-t,t)$ and $f(x)$ equals to 0 otherwise. If $|x| > t$, then

$$T_\gamma f(x) \geq c_5 \frac{1}{|x|^{1-\gamma}} \int\limits_{|y|<t} g(x)w^{-\frac{1}{p}}(x)dx = \infty.$$

Consequently we have

$$\left(\int\limits_{|x|>t} v(x)dx \right)^{\frac{1}{p}} \leq \left(\int\limits_{\{x:T_\gamma f(x)>\lambda\}} v(x)dx \right)^{\frac{1}{p}} \leq \frac{b_1}{\lambda} \left(\int\limits_{|y|<t} |g(x)|^p dx \right)^{\frac{1}{p}}$$

for all $\lambda > 0$ and therefore

$$\int\limits_{|x|>t} v(x)dx = 0.$$

Now we show that $\overline{B} < \infty$. Let $0 < t < \tau < \infty$ and let

$$f(x) = w^{1-p'}(x)\chi_{(-t,t)}(x).$$

Then if $t < |x| < \tau$, we have

$$T_\gamma f(x) \geq \frac{c_6}{|x|^{1-\gamma}} \int\limits_{|x|<t} w^{1-p'}(x)dx \geq \frac{1}{\tau^{1-\gamma}} \int\limits_{|x|<t} w^{1-p'}(x)dx.$$

Let

$$A_{t\tau} = \frac{1}{\tau^{1-\gamma}} \int\limits_{|x|<t} w^{1-p'}(x)dx.$$

Then

$$\int\limits_{t<|x|<\tau} v(x)dx \le \int\limits_{\{x:T_\gamma f(x)>A_{t\tau}\}} v(x)dx \le \frac{c_7}{A_{t\tau}^p} \int\limits_{|y|<t} w^{1-p'}(y)dy =$$

$$= c_7 \tau^{(1-\gamma)p}\Big(\int\limits_{|y|<t} w^{1-p'}(y)dy \Big)^{1-p}.$$

Therefore we have

$$\frac{1}{\tau^{(1-\gamma)p}}\Big(\int\limits_{t<|x|<\tau} v(x)dx \Big)\Big(\int\limits_{|y|<t} w^{1-p'}(y)dy \Big)^{p-1} \le c_7$$

for all t, τ with $0 < t < \tau < \infty$. Finally we see that $\overline{B} < \infty$. $\square$

We now give examples of appropriate weights.

Example 3.1.1. Let $1 < p < \infty$, $0 < \gamma < \min\{\frac{1}{p}, \frac{1}{p'}\}$ and suppose that

$$v(x) = \begin{cases} |x|^{-\gamma p + p - 1} & \text{if } |x| < e^{-p'}, \\ e^{\gamma pp' - pp' + p' + \beta p'}|x|^\beta & \text{if } |x| \ge e^{-p'}, \end{cases}$$

$$w(x) = \begin{cases} |x|^{p-1}\ln^p\frac{1}{|x|} & \text{if } |x| < e^{-p'}, \\ e^{-p+\alpha p'}(p')^p|x|^\alpha & \text{if } |x| \ge e^{-p'}, \end{cases}$$

where $0 < \beta \le \alpha - \gamma p$ and $\gamma p < \alpha \le \min\{1, p-1\}$. Then for the weight pair (v, w) we have that

$$B = \sup_{t>0}\Big(\int\limits_{|x|>t} \frac{v(x)}{|x|^{(1-\gamma)p}}dx \Big)^{\frac{1}{p}}\Big(\int\limits_{|x|<t/2} w^{1-p'}(x)dx \Big)^{\frac{1}{p'}} < \infty$$

and consequently the operator T_γ is bounded from $L_w^p(R)$ to $L_v^p(R)$.

Example 3.1.2. Let $1 < p < \infty$, $0 < \gamma < \min\{\frac{1}{p}, \frac{1}{p'}\}$. Assume that

$$v(x) = \begin{cases} |x|^{-\gamma p + p - 1}\ln\frac{1}{|x|} & \text{if } |x| < e^\lambda, \\ (-\lambda)e^{-\gamma p\lambda + \lambda p - \lambda - \lambda\beta}|x|^\beta & \text{if } |x| \ge e^\lambda, \end{cases}$$

$$w(x) = \begin{cases} |x|^{p-1}\ln^p\frac{1}{|x|} & \text{if } |x| < e^\lambda, \\ e^{\lambda(p-1)-\alpha\lambda}(-\lambda)^p|x|^\alpha & \text{if } |x| \ge e^\lambda, \end{cases}$$

where $0 < \beta \le \alpha - \gamma p$, $\gamma p < \alpha < \min\{1, p-1\}$, $\lambda = \min\{-p', \frac{1}{\gamma p - p + 1}\}$. Then for v and w condition (3.4.1) is satisfied and thus we have the boundedness of the operator T_γ from $L_w^p(R)$ to $L_v^{p\infty}(R)$; but for these weights $B = \infty$ (see (3.3.2)) and therefore the operator T_γ is not bounded from $L_w^p(R)$ to $L_v^p(R)$.

Corollary 3.4.1. *Let* $1 < p < \infty$, $0 < \gamma < 1$. *Then the classes of weight pairs guaranteeing the weak and strong type two-weight inequalities do not coincide for the operator* T_γ.

3.5. Potentials on bounded intervals

Let

$$T_{\gamma ab} f(x) = \int\limits_a^b \frac{f(y)}{|x - y|^{1-\gamma}} dy,$$

where f is a measurable function on (a, b) and $0 < \gamma < 1$. The following result holds.

Theorem 3.5.1. *Let* $-\infty < a < b < \infty$, $1 < p, q < \infty$ *and* $\frac{1}{p} < \gamma < 1$. *Suppose that* v *is a measurable a.e. positive function on* (a, b). *Then the following statements are equivalent:*
(i) *the operator* $T_{\gamma ab}$ *is bounded from* $L^p(a, b)$ *to* $L^q_v(a, b)$;
(ii) *the operator* $T_{\gamma ab}$ *is compact from* $L_p(a, b)$ *to* $L^q_v(a, b)$;
(iii) v *is integrable on* $[a, b]$, *i.e.*

$$\int\limits_a^b v(x) dx < \infty.$$

Proof. First prove that the condition (iii) guarantees the compactness of the operator $T_{\gamma ab}$.
We have

$$I = \int\limits_a^b v(x) \Big(\frac{1}{|x - y|^{(1-\gamma)p'}} dy \Big)^{\frac{q}{p'}} dx \le$$

$$\le c_1 \int\limits_a^b v(x) \Big(\int\limits_a^x \frac{dy}{(x - y)^{(1-\gamma)p'}} \Big)^{\frac{q}{p'}} dx +$$

$$+ c_1 \int\limits_a^b v(x) \Big(\int\limits_x^b \frac{1}{(y - x)^{(1-\gamma)p'}} dy \Big)^{\frac{q}{p'}} dx = I_1 + I_2.$$

For I_1 we have

$$I_1 = c_2 \int\limits_a^b v(x)(x - a)^{(\gamma-1)q + \frac{q}{p'}} dx \le$$

$$\leq c_2 \Big(\int\limits_a^b v(x)dx \Big)(b-a)^{(\gamma-1)q+\frac{q}{p'}} < \infty.$$

Here we used the fact $\frac{1}{p} < \gamma < 1$.

For I_2 we obtain

$$I_2 = c_3 \int\limits_a^b v(x)(b-x)^{(\gamma-1)q+\frac{q}{p'}}dx \leq$$

$$\leq c_3 \Big(\int\limits_a^b v(x)dx \Big)(b-a)^{(\gamma-1)q+\frac{q}{p'}} < \infty.$$

Thus $I < \infty$ and using Theorem C of Section 2.1 we conclude that the operator $T_{\gamma ab}$ is compact from $L^p(a,b)$ to $L_v^q(a,b)$.

We now show that the boundedness of $T_{\gamma ab}$ implies the validity of condition (iii).

Let $d \in (a,b)$. Then if $f(y) \equiv 1$,

$$\|f\|_{L^p(a,b)} = (b-a)^{\frac{1}{p}}.$$

On the other hand,

$$\|T_{\gamma ab}f\|_{L_v^q(a,b)} \geq \Big(\int\limits_d^b v(x)\Big(\int\limits_a^d \frac{1}{(x-y)^{1-\gamma}}dy \Big)^q dx \Big)^{\frac{1}{q}} \geq$$

$$\geq \Big(\int\limits_d^b v(x)(x-a)^{(\gamma-1)q}dx \Big)^{\frac{1}{q}}(d-a) \geq$$

$$\geq \Big(\int\limits_d^b v(x)dx \Big)^{\frac{1}{q}}(b-a)^{\gamma-1}(d-a).$$

Hence

$$\int\limits_d^b v(x)dx \leq (b-a)^{(1-\gamma)q+1/p}(d-a)^{-q} < \infty.$$

Moreover

$$\|T_{\gamma ab}f\|_{L_v^q(a,b)} \geq \Big(\int\limits_a^d v(x)\Big(\int\limits_d^b (y-x)^{\gamma-1}dy \Big)^q dx \Big)^{\frac{1}{q}} \geq$$

$$\geq \Big(\int\limits_a^d v(x)dx \Big)^{\frac{1}{q}} \Big(\int\limits_d^b (y-a)^{\gamma-1}dy \Big) =$$

$$= c_4 \Big(\int\limits_a^d v(x)dx \Big)^{\frac{1}{q}} [(b-a)^\gamma - (d-a)^\gamma].$$

Thus

$$\int\limits_a^d v(x)dx < \infty.$$

As the compactness of the operator implies boundedness it follows that (iii) $\Rightarrow$ (ii) $\Rightarrow$ (i) $\Rightarrow$ (iii). $\square$

From a duality argument we obtain

Theorem 3.5.2. *Let* $-\infty < a < b < +\infty$, $1 < p, q < \infty$ *and* $\frac{q-1}{q} < \gamma < 1$. *Assume that* w *is measurable a.e. positive function on* (a,b). *Then the following conditions are equivalent:*

(i) *the operator* $T_{\gamma ab}$ *is bounded from* $L_w^p(a,b)$ *to* $L^q(a,b)$;

(ii) *the operator* $T_{\gamma ab}$ *is compact from* $L_w^p(a,b)$ *to* $L^q(a,b)$;

(iii) $w^{1-p'}$ *is integrable on* $[a,b]$ *i.e.*

$$\int\limits_a^b w^{1-p'}(x)dx < \infty.$$

3.6. Two-weight criteria for fractional maximal functions

Now using the results of the preceding sections we state some two-weight criteria for fractional maximal functions defined on the line. We assume that v and w are measurable a.e. positive function on R.

Theorem 3.6.1. *Let* $1/p < \gamma < 1$, $\gamma p - 1 < \beta < p - 1$. *Then for the validity of the inequality*

$$\Big(\int\limits_{-\infty}^{+\infty} (M_\gamma f(x))^p v(x)dx \Big)^{1/p} \leq$$

$$\leq A \Big(\int\limits_{-\infty}^{+\infty} |f(x)|^p |x|^\beta dx \Big)^{1/p}, \quad f \in L_{|x|^\beta}^p(R) \tag{3.6.1}$$

with a constant $A > 0$ which does not depend on f, it is necessary and sufficient that

$$D_1 \equiv \sup_{t>0} \left(\int_{|x|>t} \frac{v(x)}{|x|^{(1-\gamma)p}} dx \right)^{1/p} \cdot t^{1-\frac{1}{p}-\frac{\beta}{p}} < \infty. \tag{3.6.2}$$

In addition, if A is the best constant in (3.6.1) *then we have*

$$A \approx D_1.$$

Proof. Sufficiency follows from Theorem 3.2.1. Let us show that (3.6.2) holds. As it is known (see, e.g., [100]), under the condition (3.6.1) we have

$$D_2 = \sup_{t>0} \left(\int_{-\infty}^{+\infty} v(x)(t+|x|)^{(\gamma-1)p} dx \right)^{1/p} \left(\int_{|x|<t} |x|^{\beta(1-p')} dx \right)^{1/p'} < \infty.$$

For any $k \in Z$ we derive the estimates

$$D_2 \geq \left(\int_{2^k<|x|<2^{k+1}} v(x)(2^k + |x|^{(\gamma-1)p}) dx \right)^{1/p} \times$$

$$\times \left(\int_{|x|<2^k} |x|^{\beta(1-p')} dx \right)^{1/p'} \geq$$

$$\geq c_1 \left(\int_{2^k \leq |x|<2^{k+1}} \frac{v(x)}{|x|^{(1-\gamma)p}} dx \right)^{1/p} 2^{k(-\frac{\beta}{p}+\frac{1}{p'})} \geq$$

$$\geq c_2 \left(\int_{2^k \leq |x|<2^{k+1}} \frac{v(x)}{|x|^{-\gamma p+\beta+1}} dx \right)^{1/p}.$$

Hence if

$$D_3 \equiv \sup_{k \in Z} \left(\int_{2^k \leq |x|<2^{k+1}} \frac{v(x)}{|x|^{-\gamma p+\beta+1}} dx \right)^{1/p},$$

then $D_3 < \infty$.

Consequently according to the results of section 3.2 (see the proof of Theorem 3.2.1) we have $D_1 \leq D_3 < \infty$ and $A \approx D_1$.
$\square$

Analogously, we can prove

Theorem 3.6.2. *Let* $1 < p < \infty$, $\frac{1}{p'} < \gamma < 1$ *and* $-1 < \beta < \gamma p + p - 1$.
Then the inequality

$$\left(\int\limits_{-\infty}^{+\infty} (M_\gamma f(x))^p |x|^\beta dx \right)^{1/p} \le A_1 \left(\int\limits_{-\infty}^{+\infty} |f(x)|^p w(x) dx \right)^{1/p}, \quad (3.6.3)$$

where $f \in L^p_w$, *holds if and only if*

$$\overline{D}_1 \equiv \sup_{t>0} \left(\int\limits_{|x|>t} \frac{w^{1-p'}(x)}{x^{(1-\gamma)p'}} dx \right)^{1/p'} t^{\frac{\beta}{p}+\frac{1}{p}} < \infty.$$

Moreover, for the best constant A_1 *in* (3.6.3) *we have*

$$A_1 \approx \overline{D}_1.$$

Finally we note that the sufficient condition for boundedness of the operator M_γ from $L^p_w(R^n)$ to $L^p_v(R^n)$ were established in [231].

3.7.   Generalized one–sided maximal functions

In this section necessary and sufficient conditions for the boundedness of generalized maximal operators are found. We assume that ν is a positive Borel measure on $(0, \infty)$ and w is a Lebesgue- measurable a.e. positive function on $(0, \infty)$.

Define the following maximal operators:

$$M^+_{\alpha,\beta} f(x) = \sup_{0<x<c} \frac{1}{(c-x)^\beta} \int\limits_x^c \frac{|f(t)|}{(t-x)^{1-\alpha}} dt$$

and

$$M^-_{\alpha,\beta} f(x) = \sup_{0<c<x} \frac{1}{(x-c)^\beta} \int\limits_c^x \frac{|f(t)|}{(x-t)^{1-\alpha}} dt,$$

where $0 < \beta \le \alpha \le 1$.

It is clear, that if $0 < \gamma < 1$, then $M^+_{1,1-\gamma} f(x)$ and $M^-_{1,1-\gamma} f(x)$ are one–sided fractional maximal functions.

The following theorem characterizes the boundedness of the operator $M^-_{\alpha,\beta}$.

Theorem 3.7.1. *Let* $1 < p \le q < \infty$, $0 < \beta < \frac{p-1}{p}$, $\beta + \frac{1}{p} < \alpha \le 1$.
Then the following statements are equivalent:
   (i) $M^-_{\alpha,\beta}$ *is bounded from* $L^p(0, \infty)$ *to* $L^q_\nu(0, \infty)$;

(ii) $M_{\alpha,\beta}^-$ is bounded from $L^p(0,\infty)$ to $L_\nu^{q\infty}(0,\infty)$;
(iii)

$$D = \sup_{t>0} D(t) = \sup_{t>0} \left(\int_t^\infty \frac{1}{x^{(1-\alpha+\beta)q}} d\nu \right)^{1/q} t^{1/p'} < \infty.$$

Moreover, $\|M_{\alpha,\beta}^-\|_{L^p(0,\infty)\to L_\nu^q(0,\infty)} \approx \|M_{\alpha,\beta}^-\|_{L^p(0,\infty)\to L_\nu^{q\infty}(0,\infty)} \approx D$.
Proof. It is easy to verify that

$$M_{\alpha,\beta}^- f(x) \le R_{\alpha-\beta}(|f|)(x)$$

and by Theorem 2.2.1 we conclude that (i) follows from (iii).

Now show that (ii) implies (iii). Let $k \in Z$ and let $f(y) = \chi_{(0,2^{k+1})}(y)$. Then if $x \in [2^k, 2^{k+1})$, we obtain

$$M_{\alpha,\beta}^- f(x) \ge \frac{1}{2^{(k-1)\beta}} \int_{x-2^{k-1}}^x \frac{f(y)}{(x-y)^{1-\alpha}} dy \ge 2^{(k-1)(\alpha-\beta)}.$$

Using the last expression and the boundedness of the operator $M_{\alpha,\beta}^-$ from $L^p(0,\infty)$ to $L_\nu^{q\infty}(0,\infty)$ we have that

$$\left(\int_{[2^k,2^{k+1})} d\nu \right)^{1/q} \le \left(\int_{\{x:\, M_{\alpha,\beta}^- f(x) \ge 2^{(k-1)(\alpha-\beta)}\}} d\nu \right)^{1/q} \le$$

$$\le c_1 2^{(k+1)(\beta-\alpha+1/p)}.$$

From the last inequality we obtain

$$c_1 \ge 2^{(k+1)(\alpha-\beta-1/p)} \left(\int_{[2^k,2^{k+1})} d\nu \right)^{1/q} \ge \left(\int_{[2^k,2^{k+1})} x^{(\alpha-\beta)q-\frac{q}{p}} d\nu \right)^{1/q}.$$

Denote

$$D_1(k) \equiv \left(\int_{[2^k,2^{k+1})} x^{(\alpha-\beta)q-\frac{q}{p}} d\nu \right)^{1/q}.$$

Now we show that

$$D \le c_2 \sup_{k\in Z} D_1(k).$$

Let $t \in (0,\infty)$. Then $t \in [2^m, 2^{m+1})$ for some $m \in Z$. We have

$$D^q(t) \le \left(\int_{[2^m,\infty)} x^{(\alpha-\beta-1)q} d\nu \right) 2^{(m+1)q/p'} =$$

$$= c_4 2^{\frac{mq}{p'}} \sum_{k=m}^{+\infty} \int_{[2^k, 2^{k+1})} x^{(\alpha-\beta-1)q} d\nu \leq$$

$$\leq c_5 D_1^q 2^{\frac{mq}{p'}} \sum_{k=m}^{+\infty} 2^{-\frac{kq}{p'}} \leq c_6 D_1^q$$

and finally we have that $D \leq c_7 D_1$.

As (i) implies (ii) we conclude that (iii) $\Rightarrow$ (i) $\Rightarrow$ (ii)$\Rightarrow$ (iii). $\square$

Using Theorem 2.1.2 we can obtain the following

Theorem 3.7.2. *Let* $1 < p \leq q < \infty$, $0 < \beta < 1/q$, $\beta + 1/q' < \alpha \leq 1$. *Suppose that* w *is a positive Lebesgue-measurable function on* R_+^1. *Then the following statements are equivalent:*

(i) $M_{\alpha,\beta}^+$ *is bounded from* $L_w^p(0, \infty)$ *to* $L^q(0, \infty)$;

(ii) $M_{\alpha,\beta}^+$ *is bounded from* $L_w^p(0, \infty)$ *to* $L^{q\infty}(0, \infty)$;

(iii)

$$\overline{D} = \sup_{t>0} \left(\int_t^\infty \frac{w^{1-p'}(x)}{x^{(1-\alpha+\beta)p'}} dx \right)^{1/p'} t^{1/q} < \infty.$$

Moreover,

$$\|M_{\alpha,\beta}^+\|_{L_w^p(0,\infty)\to L^q(0,\infty)} \approx \|M_{\alpha,\beta}^+\|_{L_w^p(0,\infty)\to L^{q\infty}(0,\infty)} \approx \overline{D}$$

Let $-\infty < a < b \leq \infty$ and $1 < \lambda < \infty$. We say that the measurable function $k : (0, b - a) \to R$ (we assume that $b - a = \infty$ if $b = \infty$) satisfies the condition $V_{1,\lambda}$ ($k \in V_{1,\lambda}$) if there exists a positive constant d such that for all $a < x < b$ the inequality

$$\int_0^{\frac{x-a}{2}} k^{\lambda'}(y) dy \leq d(x-a) k^{\lambda'}\left(\frac{x-a}{2}\right)$$

is fulfilled.

For a locally integrable function f on (a, b) let

$$M_\varphi^- f(x) = \sup_{0 < h < x - a} \frac{1}{\varphi(h)} \int_{x-h}^x |f(t)| dt, \quad x \in (a, b)$$

be its maximal function with respect to a function $\varphi : (0, b - a) \to R$.

Theorem 3.7.3. $-\infty < a < b \leq +\infty$, $1 < p \leq q < \infty$. *Suppose that* φ *is a positive increasing function on* $(0, b - a)$ *and* $\frac{1}{\varphi} \in V_{1,p}$. *Then the following statements are equivalent:*

(i) *the operator M_φ^- is bounded from $L^p(a,b)$ to $L_\nu^q(a,b)$;*
(ii) *the operator M_φ^- is bounded from $L^p(a,b)$ to $L_\nu^{q\infty}(a,b)$;*
(iii)

$$B_\varphi = \sup_{a<t<b} B_\varphi(t) = \sup_{a<t<b} \left(\int_{[t,b)} \varphi^{-q}\left(\frac{x-a}{2}\right) d\nu \right)^{1/q} (t-a)^{1/p'} < \infty.$$

Moreover, there exists positive constants b_1, b_2, b_3 and b_4 depending only on p, q and d such that

$$b_1 B_\varphi \leq \|M_\varphi^-\|_{L^p(a,b)\to L_\nu^q(a,b)} \leq b_2 B_\varphi$$

and

$$b_3 B_\varphi \leq \|M_\varphi^-\|_{L^p(a,b)\to L_\nu^{q\infty}(a,b)} \leq b_4 B_\varphi.$$

Proof. First we show that (i) follows from (iii). Let $x \in (a,b)$ and $0 < h < x - a$. Then we have

$$\int_a^x |f(y)|\varphi^{-1}(x-y)dy \geq \int_{x-h}^x |f(y)|\varphi^{-1}(x-y)dy \geq \frac{1}{\varphi(h)} \int_{x-h}^x |f(y)|dy$$

for all locally integrable f. Therefore using Theorem 2.7.1 we see that (i) follows from (iii).

Now let the operator M_φ^- be bounded from $L^p(a,b)$ to $L_\nu^{q\infty}(a,b)$ and suppose that $-\infty < a < b < \infty$. Assume that $k \in Z$, $k \leq 0$ and $f(y) = \chi_{(a,s_k)}(y)$, where $s_k = a + (b-a)2^k$. Suppose that $x \in [s_{k-1}, s_k)$. Then we have

$$M_\varphi^- f(x) \geq \frac{1}{\varphi((b-a)2^{k-2})} \int_{x-(b-a)2^{k-2}}^x dt = \frac{(b-a)2^{k-2}}{\varphi((b-a)2^{k-2})}.$$

Consequently

$$\left(\int_{[s_{k-1},s_k)} d\nu \right)^{1/q} \leq \left(\int_{\left\{x\in(a,b):\, M_\varphi^- f(x)\geq \frac{(b-a)2^{k-2}}{\varphi((b-a)2^{k-2})}\right\}} d\nu \right)^{1/q} \leq$$

$$\leq c_1 \frac{\varphi((b-a)2^{k-2})}{(b-a)2^{k-2}}((b-a)2^k)^{1/p} = c_2 \varphi((b-a)2^{k-2})((b-a)2^k)^{-1/p'}$$

and therefore

$$c_2 \geq \frac{1}{\varphi((b-a)2^{k-2})}((b-a)2^k)^{1/p'} \left(\int_{[s_{k-1},s_k)} d\nu \right)^{1/q} \geq$$

$$\geq \left(\int\limits_{[s_{k-1},s_k)} \varphi^{-q}\left(\frac{x-a}{2}\right)(x-a)^{q/p'}\,d\nu \right)^{1/q}.$$

Let

$$\overline{B}_\varphi(k) \equiv \left(\int\limits_{[s_{k-1},s_k)} \varphi^{-q}\left(\frac{x-a}{2}\right)(x-a)^{q/p'}\,d\nu \right)^{1/q};$$

then

$$\overline{B}_\varphi \equiv \sup_k \overline{B}_\varphi(k) < \infty.$$

Now we show that $B_\varphi < \overline{B}_\varphi$. Let $k \in (a,b)$; then $t \in [s_{k-1}, s_k)$ for some $k \in Z, k \leq 0$. We have

$$B_\varphi^q(t) \leq \left(\int\limits_{[s_{k-1},b)} \varphi^{-q}\left(\frac{x-a}{2}\right)d\nu \right)[(b-a)2^k]^{q/p'} =$$

$$= [(b-a)2^k]^{q/p'} \sum_{i=k}^{0} \int\limits_{[s_{i-1},s_i)} \varphi^{-q}\left(\frac{x-a}{2}\right)d\nu \leq$$

$$\leq \overline{B}_\varphi^q c_3 [(b-a)2^k]^{q/p'} \sum_{i=k}^{0} [(b-a)2^i]^{-q/p'} \leq c_4 \overline{B}_\varphi^q$$

and thus $B_\varphi \leq c_5 \overline{B}_\varphi$.

Next let $b = \infty$. In this case we take $f(y) = \chi_{(a,a+2^{k+1})}(y)$, where $k \in Z$. If $x \in [a + 2^k, a + 2^{k+1})$, then

$$M_\varphi^- f(x) \geq \frac{1}{\varphi(2^{k-1})} \int\limits_{x-2^{k-1}}^{x} dt = \frac{2^{k-1}}{\varphi(2^{k-1})}.$$

Therefore

$$\left(\int\limits_{[a+2^k,a+2^{k+1})} d\nu \right)^{1/q} \leq c_6\, \varphi(2^{k-1})(2^k)^{-1/p'}$$

and consequently

$$c_7 \geq \left(\int\limits_{[a+2^k,a+2^{k+1})} \varphi^{-q}\left(\frac{x-a}{2}\right)(x-a)^{p'/q}\,d\nu \right)^{1/q}.$$

If we denote

$$\tilde{B}_\varphi(k) \equiv \left(\int\limits_{[a+2^k,a+2^{k+1})} \varphi^{-q}\left(\frac{x-a}{2}\right)(x-a)^{p'/q}\,d\nu \right)^{1/q},$$

then we obtain

$$\widetilde{B}_\varphi = \sup_k \widetilde{B}_\varphi(k) < \infty.$$

The fact that $B_\varphi \leq b_1 \widetilde{B}_\varphi$ with a positive constant b_1 follows analogously. As (ii) follows from (i) we conclude that (iii) $\Rightarrow$ (i) $\Rightarrow$ (ii) $\Rightarrow$ (iii). $\square$

3.8. Potentials with power–logarithmic kernels

Let $-\infty < a < b < +\infty$, $b - a \leq \lambda < \infty$. Suppose that $0 < \gamma \leq 1$ and $\beta \geq 0$. Define the operator $T_{\gamma,\beta}$

$$T_{\gamma,\beta}f(x) = \int\limits_a^b \frac{f(y)}{|x-y|^{1-\gamma}} \ln^\beta \frac{\lambda}{|x-y|} dy$$

where $f : [a,b] \to R$ is a measurable function.

Theorem 3.8.1.   *Let $-\infty < a < b < +\infty$, $b - a \leq \lambda < \infty$ and $1 < p, q < \infty$. Assume that $1/p < \gamma \leq 1$ and $\beta \geq 0$. Suppose that v is a Lebesgue–measurable a.e. positive function on (a,b). Then the following conditions are equivalent:*
(i) *the operator $T_{\gamma,\beta}$ is bounded from $L^p([a,b])$ to $L_v^q([a,b])$;*
(ii) *the operator $T_{\gamma,\beta}$ is compact from $L^p([a,b])$ to $L_v^q([a,b])$;*
(iii) *v is integrable on $[a,b]$ i.e.*

$$\int\limits_a^b v(x)dx < \infty.$$

Proof. First we are going to show that condition (iii) implies (ii). We have

$$I = \int\limits_a^b v(x) \left(\int\limits_a^b \frac{1}{|x-y|^{(1-\gamma)p'}} \ln^{\beta p'} \frac{\lambda}{|x-y|} dy \right)^{\frac{q}{p'}} dx \leq$$

$$\leq c_1 \int\limits_a^b v(x) \left(\int\limits_a^x \frac{1}{(x-y)^{(1-\gamma)p'}} \ln^{\beta p'} \frac{\lambda}{(x-y)} dy \right)^{\frac{q}{p'}} dx +$$

$$+ c_1 \int\limits_a^b v(x) \left(\int\limits_x^b \frac{1}{(y-x)^{(1-\gamma)p'}} \ln^{\beta p'} \frac{\lambda}{(y-x)} dy \right)^{\frac{q}{p'}} dx = I_1 + I_2.$$

For I_1 we have

$$I_1 = c_2 \int\limits_a^b v(x) \left(\int\limits_0^{x-a} \frac{1}{u^{(1-\gamma)p'}} \ln^{\beta p'} \frac{\lambda}{u} du \right)^{\frac{q}{p'}} dx =$$

$$= c_2 \int\limits_a^b v(x) \varphi^{\frac{q}{p'}}(x) dx.$$

Using the fact $\frac{1}{p} < \gamma \le 1$ and $\beta \ge 0$, for arbitrary $x \in (a, b)$ we obtain

$$\varphi(x) = \int\limits_0^{x-a} \frac{1}{u^{(1-\gamma)p'}} \ln^{\beta p'} \frac{\lambda}{u} du =$$

$$= c_3(x - a) \int\limits_0^1 ((x - a)s)^{(\gamma-1)p'} \ln^{\beta p'} \frac{\lambda}{(x - a)s} ds \le$$

$$\le c_3(x - a)^{(\gamma-1)p'+1} \int\limits_0^1 s^{(\gamma-1)p'} \ln^{\beta p'} \frac{\lambda}{x - a} ds +$$

$$+ C_4(x - a)^{(\gamma-1)p'+1} \int\limits_0^1 s^{(\gamma-1)p'} \ln^{\beta p'} \frac{1}{s} ds = \varphi_1(x) + \varphi_2(x).$$

On the other hand,

$$\varphi_1(x) = c_5(x - a)^{(\gamma-1)p'+1} \ln^{\beta p'} \frac{\lambda}{x - a} \le$$

$$\le c_5(x - a)^{(\gamma-1)p'+1} \ln^{\beta p'} \frac{2\lambda}{x - a}$$

and

$$\varphi_2(x) = c_6(x - a)^{(\gamma-1)p'+1} \le c_7(x - a)^{(\gamma-1)p'+1} \ln^{\beta p'} \frac{2\lambda}{x - a}.$$

Consequently

$$\varphi(x) c_7(x - a)^{(\gamma-1)p'+1} \ln^{\beta p'} \frac{2\lambda}{x - a}.$$

Using the last inequality we obtain

$$I_1 \le c_9 \int\limits_a^b v(x)(x - a)^{(\gamma-1)q+q/p'} \ln^{\beta q} \frac{2\lambda}{x - a} dx =$$

$$= c_9 \int\limits_a^b v(x) \psi(x) dx.$$

It is easy to verify that the function ψ increases on the interval (a, λ_1) and decreases on (λ_1, b) where $\lambda_1 = \min\{a + 2\lambda \exp^{\frac{\beta}{1/p-\gamma}}\}$ (if $\lambda_1 = b$, then $(\lambda_1, b) = \emptyset$) and consequently we have

$$\int\limits_a^b v(x)\psi(x)dx \leq \int\limits_a^{\lambda_1} v(x)\psi(x)dx + \int\limits_{\lambda_1}^b v(x)\psi(x)dx \leq$$

$$\leq \psi(\lambda_1) \int\limits_a^{\lambda_1} v(x)dx + \psi(\lambda_1) \int\limits_{\lambda_1}^b v(x)dx = \psi(\lambda_1) \int\limits_a^b v(x)dx < \infty.$$

Now we estimate I_2.

$$I_2 = c_{10} \int\limits_a^b v(x)\Big(\int\limits_0^{b-x} u^{(\gamma-1)p'} \ln^{\beta p'} \frac{\lambda}{u} du\Big)^{\frac{q}{p'}} dx =$$

$$= c_{10} \int\limits_a^b v(x)\varphi_1^{q/p'}(x)dx.$$

It is easy to see that for $x \in (a, b)$ the inequality

$$\varphi_1(x) \leq c_{11}(b - x)^{(\gamma-1)p'+1} \ln^{\beta p'} \frac{2\lambda}{b - x}$$

is fulfilled. Here we used the fact $1/p < \gamma < 1$ and $\beta \geq 0$. Moreover, the function $\varphi_1^{q/p'}$ increases on the interval (a, η) and decreases on (η, b), where $\eta = \max\{a, b - 2\lambda \exp^{\frac{\beta}{1/p-\gamma}}\}$ (if $\eta = a$, then $(a, \eta) = \emptyset$ and in this case $\varphi_1^{q/p'}$ decreases on (a, b)). Consequently we have

$$\int\limits_a^b v(x)\varphi_1^{q/p'}(x)dx \leq \int\limits_a^\eta v(x)\varphi_1^{q/p'}(x)dx + \int\limits_\eta^b v(x)\varphi_1^{q/p'}(x)dx$$

$$\leq \varphi_1^{q/p'}(\eta) \int\limits_a^\eta v(x)dx + \varphi_1^{q/p'}(\eta) \int\limits_\eta^b v(x)d(x) =$$

$$= \varphi_1^{q/p'}(\eta) \int\limits_a^b v(x)dx < \infty.$$

Finally we have that $I < \infty$ and by Theorem C of Section 2.1 we conclude that the operator $T_{\gamma,\beta}$ is compact.

Now we show that (ii) follows from (i). Let $f(x) \equiv 1$. It is clear that

$$\left(\int_a^b v(x) \Big(\int_a^b \frac{f(y) \ln^\beta \frac{\lambda}{|x-y|}}{|x-y|^{1-\gamma}} dy \Big)^q dx \right)^{1/q} \geq$$

$$\geq \left(\int_{\frac{a+b}{2}}^b v(x) \Big(\int_{a+\frac{x-a}{2}}^b \frac{\ln^\beta \frac{\lambda}{x-y}}{(x-y)^{1-\gamma}} dy \Big)^q dx \right)^{1/q} \geq$$

$$\geq c_{11} \left(\int_{\frac{a+b}{2}}^b v(x)(x-a)^{\gamma q} \ln^{\beta q} \frac{2\lambda}{x-a} dx \right)^{1/q} \geq$$

$$\geq c_{12} \left(\int_{\frac{a+b}{2}}^b v(x) dx \right)^{1/q} (b-a)^\gamma \ln^\beta \frac{2\lambda}{b-a}.$$

On the other hand,

$$\|f\|_{L^p[a,b]} = (b-a)^{1/p}$$

and using the boundedness of $T_{\gamma,\beta}$ we obtain that

$$\int_{\frac{a+b}{2}}^b v(x) dx \leq (b-a)^{1/p-\gamma} \ln^{-\beta} \frac{2\lambda}{b-a} < \infty.$$

From the last inequality it follows that

$$\left(\int_a^b v(x) \Big(\int_a^b \frac{f(x)}{|x-y|^{1-\gamma}} \ln^\beta \frac{\lambda}{|x-y|} dy \Big)^q dx \right)^{1/q} \geq$$

$$\geq \left(\int_a^{\frac{b+a}{2}} v(x) \Big(\int_x^b \frac{\ln^\beta \frac{\lambda}{y-x}}{(y-x)^{1-\gamma}} dy \Big)^q dx \right)^{1/q} \geq$$

$$\geq \left(\int_a^{\frac{b+a}{2}} v(x) \right)^{1/q} \Big(\int_{\frac{b+a}{2}}^b \frac{1}{(y-a)^{1-\gamma}} \ln^\beta \frac{\lambda}{y-a} dy \Big).$$

On the other hand,

$$\|f\|_{L^p(a,b)} = (b-a)^{1/p}$$

and by the boundedness of the operator $T_{\gamma,\beta}$ we have

$$\int_a^{\frac{b+a}{2}} v(x) dx < \infty.$$

Thus we have

$$\int_a^b v(x)dx < \infty.$$

As (ii) implies (i) we have that (iii) $\Rightarrow$ (ii) $\Rightarrow$ (i) $\Rightarrow$ (iii). $\square$

From a duality argument we obtain

Theorem 3.8.2. *Let Let* $-\infty < a < b < +\infty$, $b - a \leq \lambda < \infty$. *Suppose that* $1 < p, q < \infty$ *and* $\frac{q-1}{q} < \gamma \leq 1$, $\beta \geq 0$. *We suppose that* w *is a Lebesgue-measurable a.e. positive function on* (a, b). *Then the following statements are equivalent:*

(i) *the operator* $T_{\gamma,\beta}$ *is bounded from* $L_w^p([a, b])$ *to* $L^q([a, b])$;
(ii) *the operator* $T_{\gamma,\beta}$ *is compact from* $L_w^p([a, b])$ *to* $L^q([a, b])$;
(iii) $w^{1-p'}$ *is integrable on* $[a, b]$ *i.e.*

$$\int_a^b w^{1-p'} dx < \infty.$$

Theorem 3.8.3 *Let* $1 < p, q < \infty$, $-\infty < a < b < \infty$. *Assume that* $\frac{1}{p} < \gamma < 1$ *and* v *is a Lebesgue-measurable a.e. positive function on* $(-\infty, +\infty)$. *Then the inequality*

$$\left(\int_{-\infty}^{+\infty} v(x) \left| \int_a^b \frac{f(y)}{|x - y|^{1-\gamma}} dy \right|^q dx \right)^{1/q} \leq$$

$$\leq c \left(\int_a^b |f(x)|^p dy \right)^{1/p}, \quad f \in L^p(a, b), \tag{3.8.1}$$

with a positive constant c *independent of* f *is fulfilled if and only if*

$$B_1 \equiv \left(\int_{\frac{3a-b}{2}}^{\frac{3b-a}{2}} v(x)dx \right)^{1/q} < \infty$$

and

$$B_2 \equiv \left(\int_{|x - \frac{b+a}{2}| > b-a} v(x) |x - (a + b)/2|^{(\gamma-1)q} dx \right)^{1/q} < \infty.$$

Moreover, there exist positive constants $b_i = b_i(a, b, \gamma, p, q)$, $i = 1, 2$, *such that if* c *is the best constant in the inequality* (3.8.1) *then*

$$b_1 B \leq c \leq b_2 B,$$

where $B = \max\{B_1, B_2\}$.

Proof. First we prove sufficiency. Let $f \geq 0$ and

$$I \equiv \left(\int\limits_{-\infty}^{+\infty} v(x) \left(\int\limits_a^b \frac{f(y)}{|x-y|^{1-\gamma}} dy \right)^q dx \right)^{1/q}.$$

Then

$$I \leq \left(\int\limits_{\frac{3a-b}{2}}^{\frac{3b-a}{2}} v(x) \left(\int\limits_a^b \frac{f(y)}{|x-y|^{1-\gamma}} dy \right)^q dx \right)^{1/q} +$$

$$+ \left(\int\limits_{|x-(b+a)/2|>b-a} v(x) \left(\int\limits_a^b \frac{f(y)}{|x-y|^{1-\gamma}} dy \right)^q dx \right)^{1/q} =$$

$$= I_1 + I_2.$$

By Hölder's inequaluity we have

$$I_1 \leq \left(\int\limits_{\frac{3a-b}{2}}^{\frac{3b-a}{2}} v(x) \left(\int\limits_a^b \frac{1}{|x-y|^{(1-\gamma)p'}} dy \right)^{q/p'} dx \right)^{1/q} \|f\|_{L^p([a,b])}.$$

Using the fact that $1/p < \gamma < 1$ we obtain the estimate

$$\eta \equiv \int\limits_a^b \frac{1}{|x-y|^{(1-\gamma)p'}} dy =$$

$$= \int\limits_0^\infty |\{y : |x-y|^{(\gamma-1)p'} > \lambda\} \cap (a,b)| d\lambda \leq$$

$$\int\limits_0^{(b-a)^{(\gamma-1)p'}} (b-a) d\lambda + \int\limits_{(b-a)^{(\gamma-1)p'}}^\infty |\{y : |x-y|^{(\gamma-1)p'} > \lambda\}| d\lambda \equiv$$

$$\equiv \eta_1 + \eta_2.$$

For η_1 we have

$$\eta_1 = (b-a)^{(\gamma-1)p'+1},$$

while

$$\eta_2 = \int\limits_{(b-a)^{(\gamma-1)p'}}^\infty |\{y : |x-y| < \lambda^{\frac{1}{(\gamma-1)p'}}\}| d\lambda =$$

$$\int\limits_{(b-a)^{(\gamma-1)p'}}^{\infty} \lambda^{\frac{1}{(\gamma-1)p'}} d\lambda = c_1(b-a)^{(\gamma-1)p'+1}.$$

Therefore

$$\eta \leq c_2(b-a)^{(\gamma-1)p'+1}.$$

Thus

$$I_1 \leq c_3 B_1 \|f\|_{L^p[a,b]},$$

where c_3 depends on a, b, p, q and γ.

Now we estimate I_2. If $|x - (b+a)/2| > b - a$ and $y \in (a,b)$, then

$$|x - (b+a)/2| \leq |x - y| + |y - (b+a)/2| \leq$$

$$(b-a)/2 \leq \frac{1}{2}|x - (b+a)/2|.$$

Consequently

$$\frac{1}{2}|x - (b+a)/2| \leq |x - y|.$$

Using the last inequality for I_2 we conclude that

$$I_2 \leq \left(\int\limits_{|x-(b+a)/2|>b-a} v(x)|x - (b+a)/2|^{(\gamma-1)q} dx \right)^{1/q} \times$$

$$\times \|f\|_{L^p([a,b])} = c_4 B_2 \|f\|_{L^p([a,b])},$$

where the positive constant c_4 does not depend on f.

Finally we obtain

$$I \leq c_5 B \|f\|_{L^p([a,b])}.$$

Now we prove necessity.

Let $f(y) = 1$ on $[a,b]$. Then

$$I \geq \left(\int\limits_{\frac{a+b}{2}}^{b} v(x)\left(\int\limits_{a}^{x} \frac{1}{(x-y)^{1-\gamma}} dy \right)^q dx \right)^{1/q} =$$

$$= c_6 \left(\int\limits_{\frac{a+b}{2}}^{b} v(x)(x-a)^{\gamma q} dx \right)^{1/q} \geq c_7 \left(\int\limits_{\frac{a+b}{2}}^{b} v(x) dx \right)^{1/q}.$$

Analogously

$$I \geq c_8 \left(\int\limits_{a}^{\frac{a+b}{2}} v(x) dx \right)^{1/q}$$

and

$$I \geq c_9 \left(\int_a^b v(x)dx \right)^{1/q}.$$

Moreover, we have

$$I \geq \left(\int_{\frac{b-a}{2} < |x - \frac{a+b}{2}| < b-a} v(x) \left(\int_a^b \frac{1}{|x - y|^{1-\gamma}} dy \right)^q dx \right)^{1/q} \geq$$

$$\geq c_{10} \left(\int_{\frac{b-a}{2} < |x - \frac{a+b}{2}| < b-a} v(x)|x - (a+b)/2|^{(\gamma-1)q} dx \right)^{1/q} (b - a).$$

In the last inequality we used the fact that if $y \in (a, b)$ and $\frac{b-a}{2} < |x - \frac{b+a}{2}| < b - a$, then $|x - y| \leq |x - \frac{a+b}{2}| + |y - \frac{a+b}{2}| \leq |x - \frac{a+b}{2}| + \frac{a+b}{2} \leq 2|x - \frac{a+b}{2}|$.
Finally we find:

$$I \geq c_{11}B_1.$$

On the other hand, we have

$$I \geq \left(\int_{|x - \frac{a+b}{2}| > b-a} v(x) \left(\int_a^b \frac{1}{|x - y|^{1-\gamma}} dy \right)^q dx \right)^{1/q} \geq c_{12}B_2.$$

So

$$I \geq c_{13}B,$$

where c_{13} depends on a, b, p, q and γ.
Moreover,

$$\|f\|_{L^p([a,b])} = (b - a)^{1/p}$$

and from the inequality (3.8.1) we get

$$B < \infty.$$

□

From a duality argument we derive

Theorem 3.8.4. *Let $1 < p, q < \infty$. Assume that $-\infty < a < b < +\infty$ and $\frac{q-1}{q} < \gamma < 1$. Suppose that w is a Lebesgue-measurable a.e. positive function on $(-\infty, +\infty)$. Then the inequality*

$$\left(\int_a^b \left| \int_{-\infty}^{+\infty} \frac{g(y)}{|x - y|^{1-\gamma}} dy \right|^q dx \right)^{1/q} \leq$$

$$\leq c \left(\int_{-\infty}^{+\infty} |g(y)|^p w(y) dy \right)^{1/p}, \quad g \in L_w^p(R),$$

(3.8.2)

is fulfilled if and only if

$$\widetilde{B}_1 \equiv \left(\int_{\frac{3a-b}{2}}^{\frac{3b-a}{2}} w^{1-p'}(x)dx \right)^{1/p'} < \infty$$

and

$$\widetilde{B}_2 \equiv \left(\int_{|x-(b+a)/2|>b-a} w^{1-p'}(x)\left| x - \frac{a+b}{2} \right|^{(\gamma-1)p'} dx \right)^{1/p'} < \infty.$$

Moreover, there exist positive constants $\widetilde{b}_i = \widetilde{b}_i(a,b,\gamma,p,q)$, $i = 1,2$, such that if c is the best constant in the inequality (3.8.2) then

$$\widetilde{b}_1 \widetilde{B} \le c \le \widetilde{b}_2 \widetilde{B},$$

where $\widetilde{B} = \max\{\widetilde{B}_1, \widetilde{B}_2\}$.

3.9.   Multiple potentials

Let $-\infty < a < b < +\infty$, $-\infty < c < d < +\infty$, $b - a \le \lambda < \infty$ and $d - c \le \eta < \infty$. Suppose that $0 < \gamma_1, \gamma_2 \le 1$ and $\beta_1, \beta_2 \ge 0$ and let

$$If(x,y) = \int_a^b \int_c^d \frac{f(t,\tau) \ln^{\beta_1} \frac{\lambda}{|x-t|} \ln^{\beta_2} \frac{\eta}{|y-\tau|}}{|x-t|^{1-\gamma_1} |y-\tau|^{1-\gamma_2}} dt d\tau,$$

where $f : [a,b] \times [c,d] \to R$ is a measurable function. We shall assume that v and w are Lebesgue-measurable a.e. positive functions on $[a,b] \times [a,b]$, $-\infty \le a < b \le +\infty$, $-\infty \le c < d \le +\infty$.
The following Theorem holds.

Theorem 3.9.1. *Let $-\infty < a < b < +\infty$, $-\infty < c < d < +\infty$, $b - a \le \lambda < \infty$ and $d - c \le \eta < \infty$. Assume that $1 < p,q < \infty$. Let v be Lebesgue-measurable a.e. finite function on $[a,b] \times [c,d]$. Suppose that $1/p < \gamma_1, \gamma_2 \le 1$ and $\beta_1, \beta_2 \ge 0$. Then the operator I is bounded from $L^p([a,b] \times [c,d])$ to $L^q_v([a,b] \times [c,d])$ if and only if*

$$R = \left(\int_a^b \int_c^d v(x,y)dxdy \right)^{1/q} < \infty.$$

Moreover, there exist positive constants $\widetilde{s}_i = \widetilde{s}_i(a,b,c,d,\gamma_1,\gamma_2,\beta_1,\beta_2,p,q)$, $i = 1,2$, such that $\widetilde{s}_2 R \le \|I\| \le \widetilde{s}_1 R$.

Proof. Let $f \geq 0$ and let

$$S \equiv \left(\int\limits_a^b \int\limits_c^d v(x,y)(If(x,y))^q dx dy \right)^{1/q}.$$

Then we have

$$S \leq \left(\int\limits_a^b \int\limits_c^d v(x,y) \left(\int\limits_a^b \int\limits_c^d |x - t|^{(\gamma_1-1)p'} \left(\ln^{\beta_1 p'} \frac{\lambda}{|x-t|} \right) \times \right. \right.$$

$$\left. \left. \times |y - \tau|^{(\gamma_2-1)p'} \ln^{\beta_2 p'} \frac{\eta}{|y-\tau|} dt d\tau \right)^{q/p'} dx dy \right)^{1/q} \|f\|_{L^p([a,b]\times[c,d])} =$$

$$= \tilde{S} \|f\|_{L^p([a,b]\times[c,d])},$$

where

$$\tilde{S} = \left(\int\limits_a^b \int\limits_c^d v(x,y) \left(\int\limits_a^b \int\limits_c^d |x - t|^{(\gamma_1-1)p'} \left(\ln^{\beta_1 p'} \frac{\lambda}{|x-t|} \right) \times \right. \right.$$

$$\left. \left. \times |y - \tau|^{(\gamma_2-1)p'} \ln^{\beta_2 p'} \frac{\eta}{|y-\tau|} dt d\tau \right)^{q/p'} dx dy \right)^{1/q}.$$

Now we estimate $\tilde{S}$.

$$\tilde{S}^q = \int\limits_a^b \int\limits_c^d v(x,y) \left(\int\limits_a^b \varphi_1(|x - t|) dt \right)^{q/p'} \times$$

$$\times \left(\int\limits_c^d \varphi_2(|y - \tau|) d\tau \right)^{q/p'} dx dy,$$

where $\varphi_1(s) = s^{(\gamma_1-1)p'} \ln^{\beta_1 p'} \frac{\lambda}{s}$ and $\varphi_2(s) = s^{(\gamma_2-1)p'} \ln^{\beta_2 p'} \frac{\eta}{s}$.
Moreover we have

$$I = \int\limits_a^b \varphi_1(|x - t|) dt = \int\limits_0^\infty |\{t : \varphi_1(|x - t|) > s\} \cap (a,b)\}| ds =$$

$$= \int\limits_0^{\varphi_1\left(\frac{b-a}{2}\right)} |\{t : \varphi_1(|x - t|) > s\} \cap (a,b)\}| ds +$$

$$+ \int\limits_{\varphi_1\left(\frac{b-a}{2}\right)}^\infty |\{t : \varphi_1(|x - t|) > s\} \cap (a,b)\}| ds = I_1 + I_2.$$

Also

$$I_1 \le \int_0^{\varphi_1\left(\frac{b-a}{2}\right)} (b-a)ds = (b-a)\varphi_1\left(\frac{b-a}{2}\right)$$

while plainly

$$I_2 = \int_{\varphi_1\left(\frac{b-a}{2}\right)}^{\infty} |\{t : |x - t| < \varphi_1^{-1}(s)\}|ds =$$

$$= 2 \int_{\varphi_1\left(\frac{b-a}{2}\right)}^{\infty} \varphi_1^{-1}(s)ds =$$

$$= 2\left(\int_0^{\frac{b-a}{2}} \varphi_1(s)ds - \frac{b-a}{2}\varphi_1((b-a)/2) \right).$$

On the other hand, we have

$$\int_0^{\frac{b-a}{2}} \varphi_1(s)ds = \frac{b-a}{2} \int_0^1 \varphi_1\left(\frac{b-a}{2}u\right)du =$$

$$= \frac{b-a}{2} \int_0^1 \left(\left(\frac{b-a}{2}\right)u\right)^{(\gamma_1-1)p'} \ln^{\beta_1 p'} \frac{2\lambda}{(b-a)u}du \le$$

$$\le b_1(b-a)^{(\gamma_1-1)p'+1} \int_0^1 u^{(\gamma_1-1)p'} \ln^{\beta_1 p'} \frac{2\lambda}{b-a}du +$$

$$+ b_1(b-a)^{(\gamma_1-1)p'+1} \int_0^1 u^{(\gamma_1-1)p'} \ln^{\beta_1 p'} \frac{2\lambda}{u}du \le$$

$$\le b_2(b-a)^{(\gamma_1-1)p'+1} \left(\ln^{\beta p'} \frac{2\lambda}{b-a} + b_3 \right) \le$$

$$\le b_4(b-a)^{(\gamma_1-1)p'+1} \ln^{\beta p'} \frac{2\lambda}{b-a} = b_4\frac{b-a}{2}\varphi\left(\frac{b-a}{2}\right)$$

and consequently

$$I_2 \le b_6\frac{b-a}{2}\varphi_1\left(\frac{b-a}{2}\right).$$

Finally

$$I \le b_7\frac{b-a}{2}\varphi_1\left(\frac{b-a}{2}\right).$$

Analogously it turns out that

$$\int_c^d \varphi_2(|y - \tau|)d\tau \le b_8 \frac{d - c}{2} \varphi_2\left(\frac{d - c}{2}\right)$$

and for $\widetilde{S}$ we have

$$\widetilde{S}^q \le b_9 R^q \left((b - a)(d - c)\right)^{q/p'} \left(\varphi_1\left(\frac{b - a}{2}\right)\varphi_2\left(\frac{d - c}{2}\right)\right)^{q/p'}.$$

Consequently

$$S \le B_{10}R\|f\|_{L^p([a,b]\times[c,d])}.$$

To prove the necessity we take $f(x, y) \equiv 1$. Then due to the boundedness of I we have

$$S = \left(\int_a^b \int_c^d v(x, y)(If(x, y))^q dxdy\right)^{1/q} \ge$$

$$\ge \left(\int_{\frac{a+b}{2}}^b \int_{\frac{c+d}{2}}^d v(x, y) \times\right.$$

$$\times\left(\int_{a+\frac{x-a}{2}}^x \int_{c+\frac{y-c}{2}}^y \frac{\ln^{\beta_1}\frac{\lambda}{x-t}\ln^{\beta_2}\frac{\eta}{y-\tau}dtd\tau}{(x - t)^{1-\gamma_1}(y - \tau)^{1-\gamma_2}}\right)^q \left.dxdy\right)^{1/q} \ge$$

$$\ge c_1 \left(\int_{\frac{a+b}{2}}^b \int_{\frac{c+d}{2}}^d v(x, y)(x - a)^{\gamma_1 q}(\ln^{\beta_1 q}\frac{2\lambda}{x - a}) \times\right.$$

$$\times (y - c)^{\gamma_2 q}\ln^{\beta_2 q}\frac{2\eta}{y - c}dxdy\right)^{1/q} \ge$$

$$\ge c_2 \left(\int_{\frac{a+b}{2}}^b \int_{\frac{c+d}{2}}^d v(x, y)dxdy\right)^{1/q}(b - a)^{\gamma_1} \times$$

$$\times (\ln^{\beta_1}\frac{2\lambda}{b - a})(d - c)^{\gamma_2}\ln^{\beta_2}\frac{2\eta}{d - c}.$$

Moreover

$$S \ge \left(\int_{\frac{a+b}{2}}^b \int_c^{\frac{c+d}{2}} v(x, y) \times\right.$$

$$\times \left(\int\limits_{a+\frac{x-a}{2}}^{x} \int\limits_{y}^{d} \frac{\ln^{\beta_1} \frac{\lambda}{x-t} \ln^{\beta_2} \frac{\eta}{\tau-y} dt d\tau}{(x-t)^{1-\gamma_1}(\tau-y)^{1-\gamma_2}} \right)^{q} dx dy \right)^{1/q} \geq$$

$$\geq c_3 \left(\int\limits_{\frac{a+b}{2}}^{b} \int\limits_{c}^{\frac{c+d}{2}} v(x,y)(x-a)^{\gamma_1 q} \ln^{\beta_1 q} \frac{2\lambda}{x-a} dx dy \right)^{1/q} \times$$

$$\times \left(\int\limits_{\frac{c+d}{2}}^{d} \frac{\ln^{\beta_2} \frac{\eta}{\tau-c} d\tau}{(\tau-c)^{1-\gamma_2}} \right) \geq$$

$$\geq c_4 \left(\int\limits_{\frac{a+b}{2}}^{b} \int\limits_{c}^{\frac{c+d}{2}} v(x,y) dx dy \right)^{1/q} (b-a)^{\gamma_1} \left(\ln^{\beta_1} \frac{2\lambda}{b-a} \right) \times$$

$$\times \left(\int\limits_{\frac{c+d}{2}}^{d} \frac{\ln^{\beta_2} \frac{\eta}{\tau-c} d\tau}{(\tau-c)^{1-\gamma_2}} \right).$$

Analogously we have

$$S \geq c_5 \left(\int\limits_{a}^{\frac{a+b}{2}} \int\limits_{c}^{\frac{c+d}{2}} v(x,y) dx dy \right)^{1/q} \left(\int\limits_{\frac{a+b}{2}}^{b} \frac{\ln^{\beta_1} \frac{\eta}{t-a} dt}{(t-a)^{1-\gamma_1}} \right) \times$$

$$\times \left(\int\limits_{\frac{c+d}{2}}^{d} \frac{\left(\ln^{\beta_2} \frac{\eta}{\tau-c} \right) d\tau}{(\tau-c)^{1-\gamma_2}} \right)$$

and

$$S \geq c_6 \left(\int\limits_{a}^{\frac{a+b}{2}} \int\limits_{\frac{c+d}{2}}^{d} v(x,y) dx dy \right)^{1/q} \left(\left(\ln^{\beta_2} \frac{2\eta}{d-c} \right) (d-c)^{\gamma_2} \right) \times$$

$$\times \left(\int\limits_{\frac{a+b}{2}}^{b} \frac{\ln^{\beta_1} \frac{\eta}{t-a} dt}{(t-a)^{1-\gamma_1}} \right).$$

Thus

$$S \geq c_7 R,$$

where the positive constant c depends on $a, b, c, d, \gamma_1, \gamma_2, p, q, \beta_1, \beta_2$.
On the other hand,

$$S \leq c\|f\|_{L^p(([a,b]\times[c,d])} = ((b-a)(d-c))^{1/p}$$

and so finally we see that

$$R \leq c_8.$$

$\square$

From a duality argument we can obtain

Theorem 3.9.2. *Let* $-\infty < a < b < +\infty$, $-\infty < c < d < +\infty$, $b - a \leq \lambda < \infty$ *and* $d - c \leq \eta < \infty$. *Suppose that* w *is a Lebesgue-measurable a.e. positive function on* $[a, b] \times [c, d]$. *Assume that* $1 < p, q < \infty$, $\frac{q-1}{q} < \gamma_1, \gamma_2 \leq 1$ *and* $\beta_1, \beta_2 \geq 0$. *Then the operator* I *is bounded from* $L_w^p([a, b] \times [c, d])$ *to* $L^q([a, b] \times [c, d])$ *if and only if*

$$\tilde{R} = \left(\int\limits_a^b \int\limits_c^d w^{1-p'} dx dy \right)^{1/p'} < \infty.$$

Moreover, there exist positive constants $\tilde{c}_i = \tilde{c}_i(a, b, c, d, \gamma_1, \gamma_2, \beta_1, \beta_2, p, q)$, $i = 1, 2$, *such that* $\tilde{c}_2 \tilde{R} \leq \|I\| \leq \tilde{c}_1 \tilde{R}$.

Theorem 3.9.3. *let* $a, b, c, d, \gamma_1, \gamma_2, p, q, \beta_1, \beta_2$ *and* v *satisfy the conditions of Theorem 3.9.1. Then the operator* I *is compact from* $L^p([a, b] \times [c, d])$ *to* $L_v^q([a, b] \times [c, d])$ *if and only if* $R < \infty$.

Proof. As the necessity part follows from Theorem 3.9.1 it remains to show sufficiency. As in the proof of Theorem 3.9.1

$$\tilde{S} = \int\limits_a^b \int\limits_c^d v(x,y) \left(\int\limits_a^b \int\limits_c^d |x - t|^{(\gamma_1-1)p'} \left(\ln^{\beta_1 p'} \frac{\lambda}{|x - t|} \right) \times \right.$$

$$\left. \times |y - \tau|^{(\gamma_2-1)p'} \ln^{\beta_2 p'} \frac{\eta}{|y - \tau|} dt d\tau \right)^{q/p'} dx dy \right)^{1/q} < \infty$$

and consequently Theorem C of Section 2.1 shows that the operator I is compact. $\square$

From duality arguments we obtain

Theorem 3.9.4 *Let the conditions of Theorem 3.9.2 be satisfied. Then the operator* I *is compact from* $L_w^p([a, b] \times [c, d])$ *to* $L^q([a, b] \times [c, d])$ *if and only if*

$$\tilde{R} < \infty.$$

Now let $-\infty < a < b < +\infty$, $-\infty < c < d < +\infty$ and $0 < \gamma, \alpha < 1$.
Assume that for measurable $f : [a,b] \times [c,d] \to R$,

$$T f(x,y) = \int\limits_a^b \int\limits_c^d \frac{f(x,\tau)}{|x-t|^{1-\gamma}|y-\tau|^{1-\alpha}} dt d\tau.$$

The following theorem holds.

Theorem 3.9.5. *Let* $1 < p, q < \infty$, $-\infty < a < b < +\infty$ *and* $-\infty < c < d < +\infty$. *Suppose that* $\gamma, \alpha \in (\frac{1}{p}, 1)$. *For the validity of the inequality*

$$\left(\int\limits_{-\infty}^{+\infty} \int\limits_{-\infty}^{+\infty} v(x,y)|Tf(x,y)|^q dxdy \right)^{1/q} \le$$
$$\le c \left(\int\limits_a^b \int\limits_c^d |f(x,y)|^p dxdy \right)^{1/p}, \quad f \in L^p, \tag{3.9.1}$$

where the positive constant c does not depend on f, *it is necessary and sufficient that*

(i)

$$B_1 = \left(\int\limits_{F_1} \int\limits_{F_2} v(x,y) dxdy \right)^{1/q} < \infty,$$

(ii)

$$B_2 = \left(\int\limits_{E_1} \int\limits_{E_2} v(x,y) \left| x - \frac{a+b}{2} \right|^{(\gamma-1)q} \left| y - \frac{c+d}{2} \right|^{(\alpha-1)q} dxdy \right)^{1/q} < \infty,$$

(iii)

$$B_3 = \left(\int\limits_{F_1} \int\limits_{E_2} v(x,y) \left| y - \frac{c+d}{2} \right|^{(\alpha-1)q} dxdy \right)^{1/q} < \infty,$$

(iv)

$$B_2 = \left(\int\limits_{E_1} \int\limits_{F_2} v(x,y) \left| x - \frac{a+b}{2} \right|^{(\gamma-1)q} dxdy \right)^{1/q} < \infty,$$

where, $F_1 = \left\{ x : \left| x - \frac{a+b}{2} \right| < b - a \right\}$, $F_2 = \left\{ y : \left| y - \frac{c+d}{2} \right| < d - c \right\}$,
$E_1 = \left\{ x : \left| x - \frac{a+b}{2} \right| > b - a \right\}$, $E_2 = \left\{ y : \left| y - \frac{c+d}{2} \right| > d - c \right\}$.
Moreover, if c is the best constant in (3.9.1), then there exist positive constants
$b_i = b_i(a,b,c,d,p,q,\gamma,\alpha,)$, $i = 1,2$, *such that*

$$b_2 B \le c \le b_1 B,$$

where $B = \max\{B_1, B_2, B_3, B_4\}$.

Proof. First we prove sufficiency. We may assume that $f \geq 0$. Then

$$I \equiv \left(\int\limits_{-\infty}^{+\infty} \int\limits_{-\infty}^{+\infty} v(x,y)(Tf(x,y))^q dx dy \right)^{1/q} \leq$$

$$\leq \left(\int\limits_{F_1} \int\limits_{F_2} v(x,y)(Tf(x,y))^q dx dy \right)^{1/q} +$$

$$+ \left(\int\limits_{E_1} \int\limits_{E_2} v(x,y)(Tf(x,y))^q dx dy \right)^{1/q} +$$

$$+ \left(\int\limits_{F_1} \int\limits_{E_2} v(x,y)(Tf(x,y))^q dx dy \right)^{1/q} +$$

$$+ \left(\int\limits_{E_1} \int\limits_{F_2} v(x,y)(Tf(x,y))^q dx dy \right)^{1/q} \equiv I_1 + I_2 + I_3 + I_4.$$

For I_1 Hölder's inequality gives

$$I_1 \leq \left(\int\limits_{F_1} \int\limits_{F_2} v(x,y) \times \right.$$

$$\times \left(\int\limits_a^b \int\limits_c^d \frac{1}{|x-t|^{(1-\gamma)p'}|y-\tau|^{(1-\alpha)p'}} dt d\tau \right)^{q/p'} dx dy \right)^{1/q} \times$$

$$\times \|f\|_{L^p([a,b] \times [c,d])} = S \|f\|_{L^p([a,b] \times [c,d])}.$$

Since $\frac{1}{p} < \gamma < 1$ we have

$$\int\limits_a^b \frac{1}{|x-t|^{(1-\gamma)p'}} dt = \int\limits_0^\infty |\{x : |x-t|^{(\gamma-1)p'} > \lambda\} \cap (a,b)| d\lambda =$$

$$= \int\limits_0^{(b-a)^{(\gamma-1)p'}} |\{x : |x-t|^{(\gamma-1)p'} > \lambda\} \cap (a,b)| d\lambda +$$

$$+ \int\limits_{(b-a)^{(\gamma-1)p'}}^\infty |\{x : |x-t|^{(\gamma-1)p'} > \lambda\} \cap (a,b)| d\lambda \leq$$

$$\leq (b-a)^{(\gamma-1)p'+1} + \int\limits_{(b-a)^{(\gamma-1)p'}}^\infty |\{x : |x-t| < \lambda^{\frac{1}{(\gamma-1)p'}}\}| d\lambda =$$

$$= (b-a)^{(\gamma-1)p'+1} + c_1(b-a)^{(\gamma-1)p'+1} = c_2(b-a)^{(\gamma-1)p'+1}.$$

Similarly, as $\frac{1}{p} < \alpha < 1$ we have

$$\int\limits_c^d \frac{1}{|y-\tau|^{(1-\alpha)p'}} d\tau \le c_3(d-c)^{(\alpha-1)p'+1}.$$

Hence

$$\int\limits_a^b \int\limits_c^d \frac{1}{|x-t|^{(1-\gamma)p'}|y-\tau|^{(1-\alpha)p'}} dt d\tau \le c_4(b-a)^{(\gamma-1)p'+1}(d-c)^{(\alpha-1)p'+1},$$

where the positive constant c_4 depends only on γ, α and p. Using the last inequality we obtain

$$S \le c_5 B_1 \|f\|_{L^p[(a,b)\times(c,d)]}.$$

Now we estimate I_3. If $\tau \in (c,d)$ and $y \in E_2$, then we have $|y - \frac{c+d}{2}| \le |y-\tau| + |\tau - \frac{c+d}{2}| \le |y-\tau| + \frac{d-c}{2} \le |y-\tau| + \frac{1}{2}|y - \frac{c+d}{2}|$. Consequently $\frac{1}{2}|y - \frac{c+d}{2}| \le |y-\tau|$.

Using Hölder's inequality, the last inequality and the fact that $\frac{1}{p} < \gamma < 1$, we obtain

$$I_3 \le c_6 \left(\int\limits_{F_1}\int\limits_{E_2} v(x,y)|y - \frac{c+d}{2}|^{(\alpha-1)q} \times \right.$$

$$\times \left(\int\limits_a^b \int\limits_c^d \frac{f(t,\tau)}{|x-t|^{1-\gamma}} dt d\tau \right)^q dx dy \bigg)^{1/q} \le$$

$$\le c_6 \left(\int\limits_{F_1}\int\limits_{E_2} v(x,y) \times \right.$$

$$\times |y - \frac{c+d}{2}|^{(\alpha-1)q} \left(\int\limits_a^b \int\limits_c^d \frac{dt d\tau}{|x-t|^{(1-\gamma)p'}} \right)^{\frac{q}{p'}} dx dy \bigg)^{1/q} \times$$

$$\times \|f\|_{L^p([a,b]\times[c,d])} =$$

$$= c_6(d-c)^{\frac{1}{p'}}(b-a)^{-1+\gamma+\frac{1}{p'}} B_3 \|f\|_{L^p([a,b]\times[c,d])} = \times$$

$$\times c_7 B_3 \|f\|_{L^p([a,b]\times[c,d])}.$$

Analogously, for I_4 we have

$$I_4 \le c_8 B_4 \|f\|_{L^p([a,b]\times[c,d])}.$$

Now, if $\tau \in (c,d)$ and $y \in E_2$, then $\frac{1}{2}\left|y - \frac{c+d}{2}\right| \leq |y - \tau|$. Similarly, if $t \in (a,b)$ and $x \in E_1$, then $\frac{1}{2}\left|x - \frac{a+b}{2}\right| \leq |x - t|$. Using this for I_2 we obtain

$$I_2 \leq c_9 \left(\int\limits_{E_1} \int\limits_{E_2} v(x,y) \left|x - \frac{a+b}{2}\right|^{(\gamma-1)q} \left|y - \frac{c+d}{2}\right|^{(\alpha-1)q} dxdy \right) \times$$

$$\times \left(\int\limits_a^b \int\limits_c^d f(t,\tau)dtd\tau \right) \leq c_{10} B_2 \|f\|_{L^p([a,b]\times[c,d])}.$$

Finally we have that

$$I \leq c_{11} B \|f\|_{L^p([a,b]\times[c,d])},$$

where c_{11} does not depend on f.

Now we prove necessity . Let the inequality (3.9.1) be fulfilled and let $f(x.y) = 1$ on $[a,b] \times [c,d]$. Then we have

$$I = \left(\int\limits_{-\infty}^{+\infty} \int\limits_{-\infty}^{+\infty} v(x,y) \times \right.$$

$$\times \left(\int\limits_a^b \int\limits_c^d \frac{1}{|x-t|^{1-\gamma}|y-\tau|^{1-\alpha}} dtd\tau \right)^q dxdy \right)^{1/q} \geq$$

$$\geq \left(\int\limits_{\frac{a+b}{2}} \int\limits_{\frac{c+d}{2}} v(x,y) \times \right.$$

$$\times \left(\int\limits_a^x \int\limits_c^y \frac{1}{(x-t)^{1-\gamma}(y-\tau)^{1-\alpha}} dtd\tau \right)^q dxdy \right)^{1/q} =$$

$$= c_{12} \left(\int\limits_{\frac{a+b}{2}} \int\limits_{\frac{c+d}{2}} v(x,y)(x-a)^{\gamma q}(y-c)^{\alpha q} dxdy \right)^{1/q} \geq$$

$$\geq c_{13} \left(\int\limits_{\frac{a+b}{2}} \int\limits_{\frac{c+d}{2}} v(x,y)dxdy \right)^{1/q},$$

$$I \geq \left(\int\limits_{\frac{a+b}{2}}^b \int\limits_c^{\frac{c+d}{2}} v(x,y) \times \right.$$

$$\times \left(\int\limits_a^x \int\limits_y^d \frac{1}{(x-t)^{1-\gamma}(\tau-y)^{1-\alpha}} dt d\tau \right)^q dxdy \right)^{1/q} =$$

$$= c_{14} \left(\int\limits_{\frac{a+b}{2}}^b \int\limits_c^{\frac{c+d}{2}} v(x,y)(x-a)^{\gamma q}(d-y)^{\alpha q} dxdy \right)^{1/q} \geq$$

$$\geq c_{15} \left(\int\limits_{\frac{a+b}{2}}^b \int\limits_c^{\frac{c+d}{2}} v(x,y) dxdy \right)^{1/q}.$$

Analogously it turns out that

$$I \geq c_{16} \left(\int\limits_a^{\frac{a+b}{2}} \int\limits_{\frac{c+d}{2}}^d v(x,y) dxdy \right)^{1/q}$$

and

$$I \geq c_{17} \left(\int\limits_{\frac{a+b}{2}}^b \int\limits_{\frac{c+d}{2}}^d v(x,y) dxdy \right)^{1/q}.$$

Consequently

$$I \geq c_{18} \left(\int\limits_a^b \int\limits_c^d v(x,y) dxdy \right)^{1/q}.$$

Moreover,

$$I \geq \left(\int\limits_{M_1} \int\limits_{M_2} v(x,y) \times \right.$$

$$\times \left(\int\limits_a^b \int\limits_c^d \frac{1}{|x-t|^{1-\gamma}|y-\tau|^{1-\alpha}} dt d\tau \right)^q dxdy \right)^{1/q} \geq$$

$$\geq c_{19} \left(\int\limits_{M_1} \int\limits_{M_2} v(x,y)|x - \frac{a+b}{2}|^{\gamma q}|y - \frac{c+d}{2}|^{\alpha q} dxdy \right)^{1/q} \geq$$

$$\geq c_{20} \left(\int\limits_{M_1} \int\limits_{M_2} v(x,y) dxdy \right)^{1/q},$$

where, $M_1 = (\frac{3a-b}{2}, a) \cup (b, \frac{3b-a}{2})$, $M_2 = (\frac{3c-d}{2}, c) \cup (d, \frac{3d-c}{2})$.

Further

$$I \geq \left(\int\limits_{M_1} \int\limits_{\frac{c+d}{2}}^{d} v(x,y) \times \right.$$

$$\times \left(\int\limits_{a}^{b} \int\limits_{c}^{y} \frac{1}{|x-t|^{1-\gamma}(y-\tau)^{1-\alpha}} dt d\tau \right)^{q} dx dy \right)^{1/q} \geq$$

$$\geq c_{21} \left(\int\limits_{M_1} \int\limits_{\frac{c+d}{2}}^{d} v(x,y)|x-(a+b)/2|^{\gamma q}(y-c)^{\alpha q} dx dy \right)^{1/q} \geq$$

$$\geq c_{22} \left(\int\limits_{M_1} \int\limits_{\frac{c+d}{2}}^{d} v(x,y) dx dy \right)^{1/q}.$$

Similarly we find that

$$I \geq c_{23} \left(\int\limits_{M_1} \int\limits_{c}^{\frac{c+d}{2}} v(x,y) dx dy \right)^{1/q},$$

and consequently

$$I \geq c_{24} \left(\int\limits_{M_1} \int\limits_{a}^{b} v(x,y) dx dy \right)^{1/q}.$$

Arguing as above we can obtain that

$$I \geq c_{25} \left(\int\limits_{a}^{b} \int\limits_{M_2} v(x,y) dx dy \right)^{1/q}.$$

Finally we have

$$I \geq c_{26} B_1.$$

Now we show that $I \geq d_1 B_2$. If $t \in (a,b)$ and $x \in E_1$, then we have $|x-t| \leq |x-\frac{a+b}{2}|+|t-\frac{a+b}{2}| \leq |x-\frac{a+b}{2}|+\frac{b-a}{2} \leq |x-\frac{a+b}{2}|+\frac{1}{2}|x-\frac{a+b}{2}| = \frac{3}{2}|x-\frac{a+b}{2}|$.

Analogously, if $\tau \in (c,d)$ and $y \in E_2$, then $|y-\tau| \leq \frac{3}{2}|y-\frac{c+d}{2}|$. Using this we obtain

$$I \geq c_{27} B_2.$$

From the arguments using above we find that $I \geq C_{28} B_3$ and $I \geq c_{29} B_4$, and so

$$I \geq c_{30} B.$$

On the other hand,

$$\|f\|_{L^p([a,b]\times[c,d])} = [(b-a)(d-c)]^{\frac{1}{p}}.$$

Using inequality (3.9.1) we finally have that

$$B < \infty.$$

$\square$

From a duality argument we can obtain the following

Theorem 3.9.6 Let $1 < p, q < \infty$, $-\infty < a < b < +\infty$ and $-\infty < c < d < +\infty$. Assume that $\frac{q-1}{q} < \gamma, \alpha < 1$. The inequality

$$\left(\int\limits_a^b \int\limits_c^d \left| \int\limits_{-\infty}^{+\infty} \int\limits_{-\infty}^{+\infty} \frac{g(t,\tau)}{|x-t|^{1-\gamma}|y-\tau|^{1-\alpha}} dt d\tau \right|^q dx dy \right)^{1/q} \leq$$

$$\leq c \left(\int\limits_{-\infty}^{+\infty} \int\limits_{-\infty}^{+\infty} |g(x,y)|^p w(x,y) dx dy \right)^{1/q}, \tag{3.9.2}$$

where $g \in L^p_w([a,b] \times [c,d])$ and the positive constant c does not depend on g, is fulfilled if and only if

(i)

$$\overline{B}_1 = \left(\int\limits_{F_1} \int\limits_{F_2} w^{1-p'}(x,y) dx dy \right)^{1/p'} < \infty,$$

(ii)

$$\overline{B}_2 = \left(\int\limits_{E_1} \int\limits_{E_2} w^{1-p'}(x,y) \left| x - \frac{a+b}{2} \right|^{(\gamma-1)p'} \left| y - \frac{c+d}{2} \right|^{(\alpha-1)p'} dx dy \right)^{1/p'} < \infty,$$

(iii)

$$\overline{B}_3 = \left(\int\limits_{F_1} \int\limits_{E_2} v(x,y) \left| y - \frac{c+d}{2} \right|^{(\alpha-1)p'} dx dy \right)^{1/p'} < \infty,$$

(iv)

$$\overline{B}_4 = \left(\int\limits_{E_1} \int\limits_{F_2} v(x,y) \left| x - \frac{a+b}{2} \right|^{(\gamma-1)p'} dx dy \right)^{1/p'} < \infty,$$

where F_1, F_2, E_1 and E_2 are defined in Theorem 3.9.5. Moreover, there exist positive constants $\bar{b}_i = \bar{b}_i(\alpha, \gamma, p, q, a, b, c, d)$, $i = 1, 2$ such that if c is the best constant in (3.9.2), then

$$\bar{b}_2\bar{B} \le c \le \bar{b}_1\bar{B},$$

where $\bar{B} = \max\{\bar{B}_1, \bar{B}_2, \bar{B}_3, \bar{B}_4\}$.

Now we consider particular cases of these theorems

Theorem 3.9.7. Let $1 < p, q < \infty$, $0 < a, b < +\infty$. Assume that $\gamma, \alpha \in (\frac{1}{p}, 1)$. Then the inequality

$$\left(\int\limits_{-\infty}^{+\infty}\int\limits_{-\infty}^{+\infty} v(x,y) \Big| \int\limits_{-a}^{a} \int\limits_{-b}^{b} \frac{f(t,\tau)}{|x-t|^{1-\gamma}|y-\tau|^{1-\alpha}} dt d\tau \Big|^q dx dy \right)^{1/q} \le$$

$$\le c \left(\int\limits_{-a}^{a} \int\limits_{-b}^{b} |f(x,y)|^p dx dy \right)^{1/p}, \tag{3.9.3}$$

where $f \in L^p([-a, a] \times [-b, b])$ and the positive constant c does not depend on f, is fulfilled if and only if

(i)

$$A_1 = \left(\int\limits_{-2a}^{2a} \int\limits_{-2b}^{2b} v(x,y) dx dy \right)^{1/q} < \infty,$$

(ii)

$$A_2 = \left(\int\limits_{|x|>2a} \int\limits_{|y|>2b} v(x,y)|x|^{(\gamma-1)q}|y|^{(\alpha-1)q} dx dy \right)^{1/q} < \infty,$$

(iii)

$$A_3 = \left(\int\limits_{-2a}^{2a} \int\limits_{|y|>2b} v(x,y)|y|^{(\alpha-1)q} dx dy \right)^{1/q} < \infty,$$

(iv)

$$A_4 = \left(\int\limits_{|x|>2a} \int\limits_{-2b}^{2b} v(x,y)|x|^{(\gamma-1)q} dx dy \right)^{1/q} < \infty.$$

Moreover there exist positive constants $c_i = c_i(\alpha, b, \alpha, \gamma, p, q)$, $i = 1, 2$, such that if c is the best constant in (3.9.3) then

$$c_2 A \le c \le c_1 A,$$

where $A = \max\{A_1, A_2, A_3, A_4\}$.

Theorem 3.9.8. Let $1 < p, q < \infty$, $0 < a, b < \infty$. Suppose that $\frac{q-1}{q} < \gamma, \alpha < 1$. Then for the validity of the inequality

$$\left(\int\limits_{-a}^{a} \int\limits_{-b}^{b} \left| \int\limits_{-\infty}^{+\infty} \int\limits_{-\infty}^{+\infty} \frac{g(t,\tau)dtd\tau}{|x-t|^{1-\gamma}|y-\tau|^{1-\alpha}} \right|^{q} dxdy \right)^{1/q} \le$$

$$\le c \left(\int\limits_{-\infty}^{+\infty} \int\limits_{-\infty}^{+\infty} |g(x,y)|^{p} w(x,y)dxdy \right)^{1/p}, \tag{3.9.4}$$

with the positive constant c independent of g, $g \in L_w^p(R)$, it is necessary and sufficient that

(i)

$$\overline{A}_1 = \left(\int\limits_{-2a}^{2a} \int\limits_{-2b}^{2b} w^{1-p'}(x,y)dxdy \right)^{1/q} < \infty;$$

(ii)

$$\overline{A}_2 = \left(\int\limits_{|x|>2a} \int\limits_{|y|>2b} w^{1-p'}(x,y)|x|^{(\gamma-1)p'}|y|^{(\alpha-1)p'}dxdy \right)^{1/q} < \infty;$$

(iii)

$$\overline{A}_3 = \left(\int\limits_{-2a}^{2a} \int\limits_{|y|>2b} w^{1-p'}(x,y)|y|^{(\alpha-1)p'}dxdy \right)^{1/q} < \infty;$$

(iv)

$$\overline{A}_4 = \left(\int\limits_{|x|>2a} \int\limits_{-2b}^{2b} w^{1-p'}|x|^{(\gamma-1)p'}dxdy \right)^{1/q} < \infty.$$

Moreover, there exist positive constants $c_i = c_i(a, b, \alpha, \gamma, p, q)$, $i = 1, 2$, such that if c is the best constant in (3.9.4), then

$$\overline{c}_2 \overline{A} \le c \le \overline{c}_1 \overline{A},$$

where $\overline{A} = \max\{\overline{A}_1, \overline{A}_2, \overline{A}_3, \overline{A}_4\}$.

3.10.   One–sided Hörmander–type maximal functions

In this section we discuss the boundedness of one–sided Hörmander- type maximal functions.

For Lebesgue-measurable $f : R_+ \to R$ let

$$M_\alpha^- f(x,t) = \sup_{t<h<x} \frac{1}{h^{1-\alpha}} \int_{x-h}^{x} |f(y)| dy, \quad t < x, \ 0 < \alpha < 1.$$

It is clear, that $M_\alpha^- f(x,t)$ is defined on the triangle $\triangle = \{(x,t) : 0 < t < x < \infty\}$.

Lemma 3.10.1. *Let* $0 < \alpha < 1$, $f \geq 0$. *Then*

$$M_\alpha^- f(x,t) \leq 2^{1-\alpha} T_\alpha f(x,t), \tag{3.10.1}$$

for all $(x,t) \in \triangle$, *where*

$$T_\alpha f(x,t) = \int_{0}^{x} f(y)(x-y+t)^{\alpha-1} dy.$$

Proof. Let $(x,t) \in \triangle$ and let $t < h < x$. Then $h+t < 2h$ and consequently

$$T_\alpha f(x,t) \geq \int_{x-h}^{x} f(y)(x-y+t)^{\alpha-1} dy \geq$$

$$\geq (h+t)^{\alpha-1} \int_{x-h}^{x} f(y) dy \geq 2^{\alpha-1} h^{\alpha-1} \int_{h-h}^{x} f(y) dy$$

for all h with $t < h < x$. Finally we have (3.10.1). $\square$

Let v be a Lebesgue-measurable a.e. finite function on $\triangle$. We shall use the notation: $[r,\infty) \times [0,x) \equiv v_{r,x}$.

Theorem 3.10.1. *Let* $1 < p \leq q < \infty$ *and let* $\alpha \in (1/p, 1)$. *Then the following statements are equivalent*
(i) M_α^- *is bounded from* $L^p(R_+)$ *into* $L_v^q(\triangle)$;
(ii)

$$\mathcal{B}_1 \equiv \sup_{r<0} \left(\int_{r}^{\infty} \left(\int_{0}^{x} x^{(\alpha-1)q} v(x,t) dt \right) dx \right)^{1/q} r^{1/p'};$$

(iii)

$$\mathcal{B}_2 \equiv \sup_{k \in Z} \left(\int_{2^k}^{2^{k+1}} \left(\int_{0}^{x} x^{(\alpha-1)q} x^{q/p'} v(x,t) dt \right) dx \right)^{1/q}.$$

Moreover, $\|M_\alpha^-\| \approx \mathcal{B}_1 \approx \mathcal{B}_2$.

Proof. First we prove the implication (ii) $\Rightarrow$ (i). Let $\tilde{R}_+^2 \equiv R_+ \times R_+$. Let us define the weight function v on $\tilde{R}_+^2 \backslash \Delta$ such that $v(x,t) = v(t,x)$ when $0 < x < t < \infty$. Assume that $\bar{v}$ is the extended function: $\bar{v}(x,t) = v(x,t)$ if $t < x$ and $\bar{v}(x,t) = v(t,x)$ if $x < t$. Suppose that $\bar{v}(x,t) = 0$ when $x = t$. Using Lemma 3.10.1 and Theorem 2.9.1 we obtain that if

$$B \equiv \sup_{r>0} B(r) \equiv \sup_{r>0} \left(\int\limits_r^\infty \int\limits_0^\infty \bar{v}(x,t)(x+t)^{(\alpha-1)q}dxdt \right)^{1/q} r^{1/p'} < \infty,$$

then for $f \geq 0$ we have

$$\left(\int\limits_\Delta (M_\alpha^- f(x,t))^q v(x,t)dxdt \right)^{1/q} \leq$$

$$\leq c_1 \left(\int\limits_\Delta (T_\alpha f(x,t))^q v(x,t)dxdt \right)^{1/q} =$$

$$= c_1/2^{1/q} \left(\int\limits_0^\infty \left(\int\limits_0^x (T_\alpha f(x,t))^q v(x,t)dt \right) dx +\right.$$

$$+ \int\limits_0^\infty \left(\int\limits_0^t (T_\alpha f(t,x))^q v(t,x)dx \right) dt \bigg)^{1/q} =$$

$$= c_1/2^{1/q} \left(\int\limits_0^\infty \left(\int\limits_0^x (T_\alpha f(x,t))^q \bar{v}(x,t)dt \right) dx +\right.$$

$$+ \int\limits_0^\infty \left(\int\limits_x^\infty (T_\alpha f(x,t))^q \bar{v}(x,t)dt \right) dx \bigg)^{1/q} =$$

$$= c_2 \left(\int\limits_\Delta (T_\alpha f(x,t))^q \bar{v}(x,t)dtdx +\right.$$

$$+ \int\limits_{\tilde{R}_+ \backslash \Delta} (T_\alpha f(x,t))^q \bar{v}(x,t)dtdx \bigg)^{1/q} =$$

$$= c_2 \left(\int\limits_{\tilde{R}_+} (T_\alpha f(x,t))^q \bar{v}(x,t)dtdx \right)^{1/q} \leq c_3 \left(\int\limits_{\tilde{R}_+} (f(x))^p dx \right)^{1/p}.$$

Now we show that $B \leq bB_1$ for some positive constant b. Indeed, let $r > 0$.
Then

$$B^q(r) \equiv r^{q/p'} \int_r^\infty \int_0^\infty \bar{v}(x,t)(x+t)^{(\alpha-1)q} dt dx =$$

$$= r^{q/p'} \int_r^\infty \left(\int_0^x \bar{v}(x,t)(x+t)^{(\alpha-1)q} dt \right) dx +$$

$$+ r^{q/p'} \int_r^\infty \left(\int_x^\infty \bar{v}(x,t)(x+t)^{(\alpha-1)q} dt \right) dx \equiv I_1(r) + I_2(r).$$

For $I_1(r)$ we have

$$I_1(r) = r^{q/p'} \int_r^\infty \left(\int_0^x v(x,t)(x+t)^{(\alpha-1)q} dt \right) dx \leq B_1^q.$$

Now we estimate $I_2(r)$. Let $r > 0$. Then $r \in [2^m, 2^{m+1})$ for some $m \in Z$,
and so

$$I_2(r) \leq b_1 2^{mq/p'} \int_{2^m}^\infty \left(\int_x^\infty \bar{v}(x,t)(x+t)^{(\alpha-1)q} dt \right) dx =$$

$$= b_1 2^{mq/p'} \sum_{k=m}^{+\infty} \int_{2^k}^{2^{k+1}} \left(\int_x^\infty v(x,t)(t+x) dt \right) dx \leq$$

$$\leq b_1 \bar{B}_2^q 2^{mq/p'} \sum_{k=m}^{+\infty} 2^{-kq/p'} = b_2 \bar{B}_2^q,$$

where

$$\bar{B}_2 \equiv \sup_{k \in Z} \left(\int_{2^k}^{2^{k+1}} \left(\int_x^\infty v(t,x) x^{q/p'} (t+x)^{(\alpha-1)q} dt \right) dx \right)^{1/q}.$$

Next let $k \in Z$. Then

$$\int_{2^k}^{2^{k+1}} \left(\int_x^\infty v(t,x) x^{q/p'} (t+x)^{(\alpha-1)q} dt \right) dx \leq$$

$$\leq \left(\int_{2^k}^{2^{k+1}} \left(\int_x^\infty v(t,x)(t+x)^{(\alpha-1)q} dt \right) dx \right) 2^{(k+1)q/p'} \leq$$

$$\leq b_3\bigg(\int\limits_{2^k}^{\infty}\bigg(\int\limits_0^t v(t,x)(t+x)^{(\alpha-1)q}dx\bigg)dt\bigg)2^{kq/p'}\leq b_3 B_1^q.$$

Consequently $\bar{B}_2\leq b_4 B_1$ for some constant b_4. Hence

$$\sup_{r>0}I_2(r)\leq b_5 B_1^q.$$

Finally $B\leq bB_1$ for some positive constant b. Thus (ii) $\Rightarrow$ (i). To prove that (ii) $\Rightarrow$ (iii), we take $f_k(y)=\chi_{(0,2^{k+1})}(y)$ for $k\in Z$. Then $\|f\|_{L^p(0,\infty)}=c_0 2^{k/p}$. On the other hand,

$$\|M_\alpha^- f_k\|_{L_v^q(\triangle)}\geq$$

$$\geq\bigg(\int\limits_{2^k}^{2^{k+1}}\bigg(\int\limits_0^{x/2}(M_\alpha^- f_k(x,t))^q v(x,t)dt\bigg)dx\bigg)^{1/q}\equiv J_1(k).$$

Let $h=x/2$. Then as $t<h<x$, we obtain that

$$J_2(k)\geq c_2\bigg(\int\limits_{2^k}^{2^{k+1}}\bigg(\int\limits_0^{x/2}x^{(\alpha-1)q}\bigg(\int\limits_{x/2}^x f_k(y)dy\bigg)^q v(x,t)dt\bigg)dx\bigg)^{1/q}\geq$$

$$\geq c_3\bigg(\int\limits_{2^k}^{2^{k+1}}x^{\alpha q}\bigg(\int\limits_0^{x/2}v(x,t)dt\bigg)dx\bigg)^{1/q}.$$

From the boundedness of M_α^- we get

$$J_2(k)\equiv\int\limits_{2^k}^{2^{k+1}}x^{(\alpha-1)q+q/p'}\bigg(\int\limits_0^{x/2}v(x,t)dt\bigg)dx\leq c_4,$$

where c_4 does not depend on k. Further,

$$\|M_\alpha^- f_k\|_{L_v^q(\triangle)}^q\geq\int\limits_{2^k}^{2^{k+1}}\bigg(\int\limits_{x/2}^x (M_\alpha^- f_k(x,t))^q v(x,t)dt\bigg)dx\equiv J_3(k).$$

Let $t<h<x$. Then we have

$$J_3(k)\geq\int\limits_{2^k}^{2^{k+1}}\bigg(\int\limits_{x/2}^x\bigg(h^{\alpha-1}\int\limits_{x-h}^x f_k(y)dy\bigg)^q v(x,t)dt\bigg)dx=$$

$$= \int\limits_{2^k}^{2^{k+1}} \left(\int\limits_{x/2}^{x} h^{\alpha q} v(x,t) dt \right) dx =$$

$$= \int\limits_{2^k}^{2^{k+1}} \left(\int\limits_{x/2}^{x} h^{(\alpha-1)q} h^q v(x,t) dt \right) dx \geq$$

$$\geq \int\limits_{2^k}^{2^{k+1}} \left(\int\limits_{x/2}^{x} x^{(\alpha-1)q} x^q v(x,t) dt \right) dx.$$

As $\|f_k\|_{L^p(0,\infty)} = c_5 2^{k/q}$, we have

$$J_4(k) \equiv \int\limits_{2^k}^{2^{k+1}} \left(\int\limits_{x/2}^{x} x^{(\alpha-1)q} x^{q/p'} v(x,t) dt \right) dx \leq c_5.$$

Consequently

$$J_2(k) + J_4(k) \leq c_6$$

for every $k \in Z$. Finally $\mathcal{B}_2 \leq c_6$.

It remains to prove that (iii) $\Rightarrow$ (ii). Let $r \in (0,\infty)$. Then $r \in [2^m, 2^{m+1})$ for some $m, m \in Z$. Therefore we have

$$r^{q/p'} \int\limits_{r}^{\infty} \left(\int\limits_{0}^{x} x^{(\alpha-1)q} v(x,t) dt \right) dx \leq$$

$$m \leq c_7 2^{mq/p'} \int\limits_{2^m}^{\infty} \left(\int\limits_{0}^{x} x^{(\alpha-1)q} v(x,t) dt \right) dx =$$

$$= c_8 2^{mq/p'} \sum_{k=m}^{+\infty} \int\limits_{2^k}^{2^{k+1}} \left(\int\limits_{0}^{x} x^{(\alpha-1)q} v(x,t) dt \right) dx \leq$$

$$\leq c_9 \mathcal{B}_2 2^{mq/p'} \sum_{k=m}^{+\infty} 2^{-kq/p'} \leq c_{10} \mathcal{B}_2^q.$$

Consequently $\mathcal{B}_1 \leq c_{11} \mathcal{B}_2$. Finally we have (i) $\Rightarrow$ (iii) $\Rightarrow$ (ii) $\Rightarrow$ (i). $\square$

3.11. Notes and comments on Chapter 3

The characterization of weights ensuring one–weighted estimates for one–sided maximal functions are given in [191] and [262] respectively. The appropriate A_p^+ class of weights is wider than the well-known A_p class. For example, an arbitrary increasing function belongs to A_p^+. The similar problem for the one–sided Hardy–Littlewood maximal function associated to a Borel measure which is finite on bounded intervals has been solved in [5].

The complete description of the class of weights guaranteeing the boundedness of one–sided fractional maximal function (one–sided potentials) from $L_{w^p}^p$ to $L_{w^q}^q$ under the assumption $1 < p < 1/\alpha$ and $1/q = 1/p - \alpha$ is given in [9]. A "geometric" approach to the above–mentioned problem is considered in [193]. In the same paper the case of different weights when $1 < p \leq q < \infty$ is also investigated. The criterion in terms of Sawyer's condition, involving test functions, was given in [193], [181].

In [80] necessary and sufficient conditions on a measure are obtained for the corresponding maximal operator to be of reverse weak $(1,1)$ type.

Two–weight weak type inequalities for generalized one–sided maximal operators are treated in [193], [100]. For a survey of weighted estimates for one–sided operators together with several properties of one–sided weghts we refer to [189].

Concerning the solution of two-weight problems for fractional maximal functions defined on Euclidean space when $1 < p < q < \infty$ we wish to mention [306]. In more general settings, namely in spaces of homogeneous type this problem was solved in [103] (see also [100]), Chapter 4). Later, a new proof of Sawyer's type of criterion was given in [59].

Part of the results presented in this chapter was announced in [201], [157].

Chapter 4

BALL FRACTIONAL INTEGRALS

A new class of fractional integrals connected with balls in R^n was introduced and investigated by B. Rubin in [246] (see also [247]). The special interest in ball fractional integrals (BFI's) arises from the fact that Riesz potentials $I_\alpha \varphi$ over a ball B may be represented by a composition of such integrals. This enables one to derive necessary and sufficient solvability conditions for the equation $I_\alpha \varphi = f$ in Lebesgue spaces with power weights and to construct the solution in closed form.

Our aim is to establish two-weight boundedness (compactness) criteria for BFI's in Lebesgue (Lorentz) spaces. Two–sided estimates of the measure of non–compactness for ball fractional integral operators are also obtained.

4.1. Boundedness criteria

In the present section we find necessary and sufficient conditions for a non-negative σ-finite Borel measure ν which ensure the boundedness of the operator

$$B_+^\alpha(f)(x) = \int\limits_{|y|<|x|} \frac{(|x|^2 - |y|^2)^\alpha}{|x-y|^n} f(y) dy$$

from $L^p(R^n)$ to $L_\nu^q(R^n)$ $(L_\nu^{q\infty}(R^n))$, where $p \in (1,\infty)$, $q \in (0,\infty)$ and $\alpha > \frac{n}{p}$ (When $0 < q < p < \infty$ and $p > 1$, we assume that $d\nu = v(x,t)dxdt$, where v is a measurable a.e. positive function on R^n). The corresponding problem for the operator

$$B_-^\alpha(g)(y) = \int\limits_{|x|>|y|} \frac{(|x|^2 - |y|^2)^\alpha}{|x-y|^n} g(x) d\nu(x)$$

317

is also solved. These operators are called ball fractional integrals.

Some results concerning the operators B_+^α and B_-^α can be found in [246] (see also [247], Chapter 7 and [255], Section Section 29).

Let a be a fixed positive number.

For a measurable function $g : R^n \to R$ define the Hardy–type transform

$$H_a(f)(x) = \int\limits_{|y|<a|x|} f(y)dy.$$

We need the following Lemmas:

Lemma 4.1.1. *Let* $1 < p \le q < \infty$. *Assume that* σ *is a non-negative Borel measure on* R^n *and* w *is a Lebesgue-measurable a.e. positive function on* R^n. *Then the operator* H_a *is bounded from* $L_w^p(R^n)$ *to* $L_\sigma^q(R^n)$ *if and only if*

$$A_a = \sup_{t>0}(\sigma\{x \in R^n : |x| \ge t\})^{\frac{1}{q}} \left(\int\limits_{\{x\in R^n:\, |x|<at\}} w^{1-p'}(x)dx \right)^{\frac{1}{p'}} < \infty,$$

where $p' = p/(p - 1)$. *Moreover, we have* $\|H_a\| \approx A_a$.

This Lemma is a Corollary of Theorem 1.1.6.

Lemma 4.1.2. *Let* $0 < q < p < \infty$ *and let* $p > 1$. *Assume that* v *and* w *are positive, Lebesgue-measurable functions on* R^n. *Then the operator* H_a *is bounded from* $L^p(R^n)$ *to* $L_v^q(R^n)$ *if and only if*

$$A_{1,a} = \left(\int\limits_{R^n} \left(\int\limits_{\{|y|>|x|/a\}} v(y)dy \right)^{\frac{p}{p-q}} \times \right.$$

$$\left. \times \left(\int\limits_{\{y:|y|<|x|\}} w^{1-p'}(y)dy \right)^{\frac{p(q-1)}{p-q}} w^{1-p'}(x)dx \right)^{\frac{p-q}{pq}} < \infty.$$

Moreover, $\|H_a\| \approx A_{1,a}$.

This lemma can be derived from Theorem B of Section 2.1 using polar coordinates. See also [62].

For $\alpha > 0$, let

$$T_\alpha(f)(x) = \int\limits_{|y|<|x|} \frac{f(y)}{|x - y|^{n-\alpha}}dy.$$

To obtain the main results, we need the following lemma.

Lemma 4.1.3. *Let* $1 < p \le q < \infty$ *and let* $\alpha > n/p$. *Assume that* μ *is a non-negative Borel measure on* R^n. *Then the operator* T_α *is bounded from*

$L^p(R^n)$ to $L^q_\mu(R^n)$ if and only if

$$B_1 = \sup_{t>0} \left(\int_{|x|\geq t} |x|^{(\alpha-n)q} d\mu \right)^{1/q} t^{n/p'} < \infty.$$

Moreover, $\|T_\alpha\| \approx B_1$.

Proof. First we show sufficiency. We assume that $\frac{n}{p} < \alpha < n$. Let $f \geq 0$.
Then

$$\left(\int_{R^n} \left(\int_{|y|<|x|} \frac{f(y)}{|x-y|^{n-\alpha}} dy \right)^q d\mu \right)^{\frac{1}{q}}$$

$$\leq c_1 \left(\int_{R^n} \left(\int_{|y|<|x|/2} \frac{f(y)}{|x-y|^{n-\alpha}} dy \right)^q d\mu \right)^{\frac{1}{q}}$$

$$+ c_1 \left(\int_{R^n} \left(\int_{|x|/2<|y|<|x|} \frac{f(y)}{|x-y|^{n-\alpha}} dy \right)^q d\mu \right)^{\frac{1}{q}} \equiv I_1 + I_2.$$

Using Lemma 4.1.1 with respect to the measure $d\sigma = |x|^{(\alpha-n)q} d\mu$, we
obtain

$$I_1 \leq c_2 \left(\int_{R^n} \left(\int_{|y|<|x|} f(y) dy \right)^q |x|^{(\alpha-n)q} d\mu \right)^{\frac{1}{q}} \leq c_3 B_1 \left(\int_{R^n} (f(y))^p dy \right)^{\frac{1}{p}}.$$

Now we estimate I_2. By Hölder's inequality we obtain

$$I_2^q \leq c_4 \int_{R^n} \left(\int_{|x|/2<|y|<|x|} (f(y))^p dy \right)^{\frac{q}{p}} \left(\int_{|x|/2<|y|<|x|} |x-y|^{(\alpha-n)p'} dy \right)^{\frac{q}{p'}} d\mu.$$

On the other hand,

$$\int_{|x|/2<|y|<|x|} \frac{1}{|x-y|^{(n-\alpha)p'}} dy \leq \int_{|y|<|x|} \frac{1}{|x-y|^{(n-\alpha)p'}} dy$$

$$= \int_0^\infty |\{y : |x-y|^{(\alpha-n)p'} > \lambda\} \cap B(0,|x|)| d\lambda \equiv S(x),$$

where $B(0,|x|) \equiv \{y \in R^n : |y| < |x|\}$. We obtain

$$S(x) = \int_0^{|x|^{(\alpha-n)p'}} |\{y : |x-y|^{(\alpha-n)p'} > \lambda\} \cap B(0,|x|)| d\lambda$$

$$+ \int\limits_{|x|^{(\alpha-n)p'}}^{\infty} |\{y : |x - y|^{(\alpha-n)p'} > \lambda\} \cap B(0, |x|)| d\lambda \equiv S_1(x) + S_2(x).$$

For $S_1(x)$ we have

$$S_1(x) \leq \int\limits_0^{|x|^{(\alpha-n)p'}} |B(0, |x|)| d\lambda = c_5 |x|^n |x|^{(\alpha-n)p'} = c_5 |x|^{(\alpha-n)p'+n}.$$

Since $\alpha > n/p$, we see that

$$S_2(x) \leq \int\limits_{|x|^{(\alpha-n)p'}}^{\infty} |\{y : |x - y|^{(\alpha-n)p'} > \lambda\} d\lambda$$

$$= \int\limits_{|x|^{(\alpha-n)p'}}^{\infty} |\{y : |x - y| < \lambda^{\frac{1}{(\alpha-n)p'}}\}| d\lambda = c_6 |x|^{(\alpha-n)p'+n}.$$

We finally obtain that $S(x) \leq c_7 |x|^{(\alpha-n)p'+n}$. Using Hölder's inequality and the last estimate, for I_2^q we derive the estimates:

$$I_2^q \leq c_8 \int\limits_{R^n} \left(\int\limits_{|x|/2 < |y| < |x|} f(y)^p dy \right)^{\frac{q}{p}} |x|^{(\alpha-n)q + \frac{nq}{p'}} d\mu =$$

$$= c_8 \sum_{k \in \mathbf{Z}} \int\limits_{2^k \leq |x| < 2^{k+1}} |x|^{(\alpha-n)q + nq/p'} \left(\int\limits_{|x|/2 < |y| < |x|} (f(y))^p dy \right)^{\frac{q}{p}} d\mu \leq$$

$$\leq c_8 \sum_{k \in \mathbf{Z}} \left(\int\limits_{2^{k-1} < |y| < 2^{k+1}} (f(y))^p dy \right)^{\frac{q}{p}} \left(\int\limits_{2^k \leq |x| < 2^{k+1}} |x|^{(\alpha-n)q + nq/p'} d\mu \right) \leq$$

$$\leq c_9 \sum_{k \in \mathbf{Z}} \left(\int\limits_{2^{k-1} < |y| < 2^{k+1}} (f(y))^p dy \right)^{\frac{q}{p}} \left(\int\limits_{|x| \geq 2^k} |x|^{(\alpha-n)q} d\mu \right) 2^{\frac{knq}{p'}} \leq$$

$$\leq c_{10} B_1^q \|f\|_{L^p(R^n)}^q.$$

Now we show necessity. Let $t > 0$ and let $f(x) = \chi_{\{y:|y|<t/2\}}(x)$. If $|y| < t/2$ and $|x| \geq t$, then $|x - y|^{\alpha-n} \geq b_1 |x|^{\alpha-n}$. Consequently,

$$\left(\int\limits_{R^n} (T_\alpha f(x))^q d\mu \right)^{1/q} \geq \left(\int\limits_{|x| \geq t} \left(\int\limits_{|y| < t/2} \frac{f(y)}{|x - y|^{n-\alpha}} dy \right)^q d\mu \right)^{1/q} \geq$$

$$\geq c_{11}\Big(\int\limits_{|x|\geq t} |x|^{(\alpha-n)q}d\mu\Big)^{1/q} t^n.$$

On the other hand, $\|f\|_{L^p(R^n)} = c_{12}t^{n/p}$ and we find that $B_1 < \infty$. Moreover, $\|T_\alpha\| \geq c_{13}B_1. \square$

Lemma 4.1.4. *Let $0 < q < p < \infty$, $p > 1$ and let $\alpha > \frac{n}{p}$. Suppose that v is Lebesgue-measurable a.e. positive function on R^n. Then the operator T_α is bounded from $L^p(R^n)$ to $L_v^q(R^n)$ if and only if*

$$B_2 = \Big[\int\limits_{R^n}\Big(\int\limits_{|y|>|x|} |y|^{(\alpha-n)q}v(y)dy\Big)^{\frac{p}{p-q}}|x|^{\frac{np(q-1)}{p-q}}dx\Big]^{\frac{p-q}{pq}} < \infty.$$

Morover, $\|T_\alpha\| \approx B_2$.
Proof. Let $f \geq 0$. We have

$$\Big(\int\limits_{R^n}\Big(\int\limits_{|y|<|x|} \frac{f(y)}{|x-y|^{n-\alpha}}dy\Big)^q v(x)dx\Big)^{\frac{1}{q}} \leq$$

$$\leq c_1\Big(\int\limits_{R^n}\Big(\int\limits_{|y|<|x|/2} \frac{f(y)}{|x-y|^{n-\alpha}}dy\Big)^q v(x)dx\Big)^{\frac{1}{q}} \leq$$

$$\leq c_1\Big(\int\limits_{R^n}\Big(\int\limits_{|x|/2<|y|<|x|} \frac{f(y)}{|x-y|^{n-\alpha}}dy\Big)^q v(x)dx\Big)^{\frac{1}{q}} \equiv I_1 + I_2.$$

Using Lemma 4.1.2 we have that $I_1 \leq b\|f\|_{L^p(R^n)}$.
Now we estimate I_2. Let $\frac{n}{p} < \alpha < n$. By Hölder's inequality we have

$$I_2 \leq c_1\Big(\int\limits_{R^n}\Big(\int\limits_{|x|/2<|y|<|x|} (f(y))^p dy\Big)^{\frac{q}{p}} \times$$

$$\times \Big(\int\limits_{|x|/2<|y|<|x|} |x-y|^{(\alpha-n)p'} dy\Big)^{\frac{q}{p'}} v(x)dx\Big)^{\frac{1}{q}}.$$

Since (see the proof of Lemma 4.1.3.),

$$\int\limits_{|x|/2<|y|<|x|} \frac{1}{|x-y|^{(n-\alpha)p'}}dy \leq b_2|x|^{(\alpha-n)p'+n},$$

with a positive constant b_2 independent of x, using Hölder's inequality we obtain

$$I_2^q \leq c_2 \int_{R^n} v(x) \left(\int_{|x|/2<|y|<|x|} (f(y))^p \right)^{q/p} |x|^{(\alpha-n)q+nq/p'} dx \leq$$

$$\leq c_2 \sum_{k\in Z} \left(\int_{2^{k-1}<|y|<2^{k+1}} (f(y))^p dy \right)^{\frac{q}{p}} \left(\int_{2^k<|x|<2^{k+1}} v(x)|x|^{(\alpha-n)q+nq/p'} dx \right) \leq$$

$$\leq c_2 \left(\sum_{k\in Z} \int_{2^{k-1}<|x|<2^{k+1}} (f(y))^p dy \right)^{\frac{q}{p}} \left(\sum_{k\in Z} \left(\int_{2^k<|x|<2^{k+1}} v(x) \times \right. \right.$$

$$\left. \left. \times |x|^{(\alpha-n)q+nq/p'} dx \right)^{\frac{p}{p-q}} \right)^{\frac{p-q}{p}} \leq c_3 \|f\|_{L^p(R^n)}^q S,$$

where

$$S = \left(\sum_{k\in Z} \left(\int_{2^k<|x|<2^{k+1}} v(x)|x|^{(\alpha-n)q+nq/p'} dx \right)^{\frac{p}{p-q}} \right)^{\frac{p-q}{p}}.$$

Moreover,

$$S^{\frac{p}{p-q}} \leq c_4 \sum_{k\in Z} 2^{\frac{knq(p-1)}{p-q}} \left(\int_{2^k<|x|<2^{k+1}} v(x)|x|^{(\alpha-n)q} dx \right)^{\frac{p}{p-q}} \leq$$

$$\leq c_5 \sum_{k\in Z} \int_{2^{k-1}<|y|<2^k} |y|^{\frac{np(q-1)}{p-q}} \left(\int_{|x|>|y|} v(x)|x|^{(\alpha-n)q} dx \right)^{\frac{p}{p-q}} dy = c_5 B_2^{\frac{pq}{p-q}}.$$

Thus,

$$I_2 \leq c_6 \|f\|_{L^p(R^n)} B_2.$$

Now we prove necessity. Assume that $\frac{n}{p} < \alpha < n$ and let the operator T_α be bounded from $L^p(R^n)$ to $L_v^q(R^n)$. If we repeat the arguments used in the proof of Lemma 4.1.3, we get

$$\int_{|x|>t} v(x)|x|^{(\alpha-n)q} dx < \infty$$

for every $t > 0$. Let $m \in Z$. Suppose that $v_m(y) = v(y)\chi_{\{1/m<|z|<m\}}(y)$ and

$$f_m(x) = \left(\int_{|y|>|x|} v_m(y)|y|^{(\alpha-n)q} dy \right)^{\frac{1}{p-q}} (|x| - 1/m)^{\frac{n(q-1)}{p-q}} \chi_{\{1/m<|y|<m\}}(x).$$

Then

$$\|f_m\|_{L^p(R^n)} = \Bigg(\int\limits_{\frac{1}{m}<|x|<m} \Big(\int\limits_{|y|>|x|} v_m(y)|y|^{(\alpha-n)q}dy \Big)^{\frac{p}{p-q}} \times$$

$$\times (|x| - \frac{1}{m})^{\frac{np(q-1)}{p-q}} dx \Bigg)^{\frac{1}{p}} < \infty.$$

On the other hand,

$$\|T_\alpha f_m(x)\|_{L_v^q(R^n)} \geq c_7 \Bigg(\int\limits_{R^n} v(x)|x|^{(\alpha-n)q} \Big(\int\limits_{|y|<|x|} f_m(y)dy \Big)^q dx \Bigg)^{\frac{1}{q}} \geq$$

$$\geq c_7 \Bigg(\int\limits_{\frac{1}{m}<|x|<m} v(x)|x|^{(\alpha-n)q} \Big(\int\limits_{1/m<|y|<|x|} \Big(\int\limits_{|y|<|z|<m} v(z)|z|^{(\alpha-n)q}dz \Big)^{\frac{1}{p-q}} \times$$

$$\times (|y| - 1/m)^{\frac{n(q-1)}{p-q}} dy \Big)^q dx \Bigg)^{\frac{1}{q}} \geq$$

$$\geq c_7 \Bigg(\int\limits_{\frac{1}{m}<|x|<m} v(x)|x|^{(\alpha-n)q} \Big(\int\limits_{|x|<|z|<m} v(z)|z|^{(\alpha-n)q}dz \Big)^{\frac{q}{p-q}} \times$$

$$\times \Big(\int\limits_{1/m<|y|<|x|} (|y| - 1/m)^{\frac{n(q-1)}{p-q}} dy \Big)^q dx \Bigg)^{\frac{1}{q}}$$

$$\geq c_7 \Bigg(\int\limits_{\frac{1}{m}<|x|<m} v(x)|x|^{(\alpha-n)q} \Big(\int\limits_{|x|<|z|<m} v(z)|z|^{(\alpha-n)q}dz \Big)^{\frac{q}{p-q}} \times$$

$$\times (|x| - 1/m)^{\frac{nq(p-1)}{p-q}} dx \Bigg)^{\frac{1}{q}} =$$

$$= c_8 \Bigg(\int\limits_{\frac{1}{m}<|x|<m} \Big(\int\limits_{|x|<|z|<m} v(z)|z|^{(\alpha-n)q}dz \Big)^{\frac{p}{p-q}} (|x| - 1/m)^{\frac{np(q-1)}{p-q}} dx \Bigg)^{\frac{1}{q}}.$$

Using the boundedness of the operator T_α, we obtain

$$\Bigg(\int\limits_{\frac{1}{m}<|x|<m} \Big(\int\limits_{|x|<|z|<m} v(z)|z|^{(\alpha-n)q}dz \Big)^{\frac{p}{p-q}} (|x| - 1/m)^{\frac{np(q-1)}{p-q}} dx \Bigg)^{\frac{1}{q}} \leq c_9,$$

where c_{10} does not depend on m. By Fatou's lemma we finally have that $B_2 < \infty$. Moreover, $B_2 \leq c_{11}\|T_\alpha\|$.

The case $\alpha \geq n$ can be proved analogously. $\Box$

Now we prove the main results of this section

Theorem 4.1.1. *Let* $1 < p \leq q < \infty$ *and let* $\alpha > n/p$. *Then the following statements are equivalent:*
(i) *the operator* B_+^α *is bounded from* $L^p(R^n)$ *to* $L_\nu^q(R^n)$;
(ii) *the operator* B_+^α *is bounded from* $L^p(R^n)$ *to* $L_\nu^{q\infty}(R^n)$;
(iii)

$$D_1 = \sup_{t>0} \left(\int\limits_{|x| \geq t} |x|^{(2\alpha-n)q} d\nu \right)^{1/q} t^{n/p'} < \infty;$$

(iv)

$$D_2 = \sup_{k \in Z} \left(\int\limits_{2^k \leq |x| < 2^{k+1}} |x|^{(2\alpha-n/p)q} d\nu \right)^{1/q} < \infty.$$

Moreover, $\|B_+^\alpha\|_{L^p(R^n) \to L_\nu^q(R^n)} \approx \|B_+^\alpha\|_{L^p(R^n) \to L_\nu^{q\infty}(R^n)} \approx D_1 \approx D_2$.
Proof. First we show that (iii) implies (i). Let $f \geq 0$. Then

$$\left(\int\limits_{R^n} \left(B_+^\alpha f(x) \right)^q d\nu \right)^{\frac{1}{q}} =$$

$$= \left(\int\limits_{R^n} \left(\int\limits_{|y|<|x|} \frac{(|x|-|y|)^\alpha(|x|+|y|)^\alpha}{|x-y|^n} f(y) dy \right)^q d\nu \right)^{\frac{1}{q}} \leq$$

$$\leq c_1 \left(\int\limits_{R^n} \left(\int\limits_{|y|<|x|} |x-y|^{\alpha-n} f(y) dy \right)^q |x|^{q\alpha} d\nu \right)^{\frac{1}{q}}.$$

If we use Lemma 4.1.3 with respect to the measure $d\mu = |x|^{q\alpha} d\nu$, then we obtain

$$\|B_+^\alpha f\|_{L_\nu^q(R^n)} \leq c_2 D_1 \|f\|_{L^p(R^n)}.$$

Now we prove that (ii) $\Longrightarrow$ (iv). Suppose that the operator B_+^α is bounded from $L^p(R^n)$ to $L_\nu^{q\infty}(R^n)$. Let $2^k \leq |x| < 2^{k+1}$ and let $f_k(y) = \chi_{\{|z|<2^{k-1}\}}(y)$. Then,

$$B_+^\alpha f(x) = \int\limits_{|y|<2^{k-1}} \frac{(|x|^2-|y|^2)^\alpha}{|x-y|^n} dy \geq \int\limits_{|y|<2^{k-1}} \frac{(|x|^2-2^{(k-1)2})^\alpha}{|x-y|^n} dy \geq$$

$$\geq \int_{|y|<2^{k-1}} \frac{(|x|^2 - (|x|/2)^2)^\alpha}{|x-y|^n} dy = c_3 |x|^{2\alpha} \int_{|y|<2^{k-1}} \frac{dy}{|x-y|^n}.$$

Moreover, if $2^k \leq |x| < 2^{k+1}$ and $|y| < 2^{k-1}$, then $|x| \leq |x-y| + |y| \leq |x-y| + |x|/2$. Consequently $|x|/2 \leq |x-y|$ and

$$B_+^\alpha f(x) \geq c_4 |x|^{2\alpha-n} 2^{kn} \geq c_5 2^{2k\alpha}.$$

From the boundedness of B_+^α we see that

$$\left(\int_{2^k \leq |x| < 2^{k+1}} d\nu \right)^{\frac{1}{q}} \leq \left(\int_{\{x \in R^n : B_+^\alpha f(x) \geq c_5 2^{2\alpha k}\}} d\nu \right)^{\frac{1}{q}} \leq$$

$$\leq \frac{c_6}{2^{2\alpha k}} \|f\|_{L^p(R^n)} = c_7 2^{nk/p - 2\alpha k}.$$

Therefore

$$\left(\int_{2^k \leq |x| < 2^{k+1}} |x|^{q(2\alpha-n/p)} d\nu \right)^{\frac{1}{q}} \leq c_8$$

for arbitrary $k \in Z$ and hence $D_2 < \infty$. Moreover,

$$\|B_+^\alpha\|_{L^p(R^n) \to L_\nu^{q\infty}(R^n)} \geq c_9 D_2.$$

Now we show that (iv) $\Longrightarrow$ (iii). Let $t > 0$. Then there exists an integer m such that $t \in [2^m, 2^{m+1})$. Denote

$$D_1(t) \equiv \left(\int_{|x| \geq t} |x|^{(2\alpha-n)q} d\nu \right)^{1/q} t^{n/p'}.$$

Then

$$D_1^q(t) \leq \left(\int_{|x| \geq 2^m} |x|^{(2\alpha-n)q} d\nu \right) 2^{(m+1)\frac{nq}{p'}} =$$

$$= b_1 2^{\frac{mnq}{p'}} \sum_{k=m}^{+\infty} \int_{2^k \leq |x| < 2^{k+1}} |x|^{(2\alpha-n)q} d\nu \leq b_2 D_2^q 2^{\frac{mnq}{p'}} \sum_{k=m}^{+\infty} 2^{-\frac{knq}{p'}} \leq b_3 D_2^q.$$

Thus $D_1 \leq b_4 D_2$. As (ii) follows from (i), we finally see that (iii) $\Longrightarrow$ (i) $\Longrightarrow$ (ii) $\Longrightarrow$ (iv) $\Longrightarrow$ (iii). $\square$

Theorem 4.1.2. *Let* $0 < q < p < \infty$, $p > 1$ *and let* $\alpha > \frac{n}{p}$. *Assume that* v *is a Lebesgue-measurable a.e. positive function on* R^n. *Then the operator* B_+^α *is bounded from* $L^p(R^n)$ *to* $L_v^q(R^n)$ *if and only if*

$$
D_3 = \left[\int_{R^n} \left(\int_{|y|>|x|} |y|^{(2\alpha-n)q} v(y) dy \right)^{\frac{p}{p-q}} |x|^{\frac{np(q-1)}{p-q}} dx \right]^{\frac{p-q}{pq}} < \infty.
$$

Moreover, $\|B_\alpha\| \approx D_3$.

Proof. Sufficiency can be obtained in the same way as for Theorem 4.1.1. We only need to use Lemma 4.1.2.

To prove necessity we take $m \in Z$ and $v_m(y) = v(y)\chi_{\{1/m<|z|<m\}}(y)$.

Suppose that $f_m(x) = \left(\int_{|y|>|x|} v_m(y)|y|^{(2\alpha-n)q} dy \right)^{\frac{1}{p-q}} |x|^{\frac{n(q-1)}{p-q}}$. We have

$$
\|f_m\|_{L^p(R^n)} = \left(\int_{R^n} \left(\int_{|y|>|x|} v_m(y)|y|^{(2\alpha-n)q} dy \right)^{\frac{p}{p-q}} |x|^{\frac{np(q-1)}{p-q}} dx \right)^{1/p},
$$

which is finite. On the other hand, using the polar coordinates and integration by parts, we get

$$
\|B_+^\alpha f_m(x)\|_{L_v^q(R^n)} \geq c_1 \left(\int_{R^n} v(x) \left(\int_{|z|>|x|} v_m(z)|z|^{(2\alpha-n)q} dy \right)^{q/(p-q)} \times \right.
$$

$$
\times \left(\int_{|y|<|x|/2} \frac{(|x|^2 - |y|^2)^\alpha}{|x-y|^n} |y|^{\frac{n(q-1)}{p-q}} dy \right)^q dx \right)^{1/q} \geq
$$

$$
\geq c_2 \left(\int_{R^n} v_m(x) \left(\int_{|z|>|x|} v_m(z)|z|^{(2\alpha-n)q} dz \right)^{q/(p-q)} \times \right.
$$

$$
\times |x|^{(2\alpha-n)q} |x|^{\frac{nq(p-1)}{p-q}} dx \right)^{1/q} =
$$

$$
= c_3 \left(\int_{R^n} \left(\int_{|y|>|x|} v_m(y)|y|^{(2\alpha-n)q} dy \right)^{\frac{p}{p-q}} |x|^{\frac{np(q-1)}{p-q}} dx \right)^{\frac{1}{q}}.
$$

From the boundedness of the operator B_+^α and from Fatou's lemma we see that $D_3 < \infty$. $\square$

Remark 4.1.1. Taking into account the methods used above, necessary and sufficient conditions for the boundedness of B_+^α from $L^p(B(0,a))$ into $L_\nu^q(B(0,a))$ can be derived, where $0 < a < \infty$, ν is a Borel measure on the ball $B(0,a)$,

$0 < q < \infty$ and $1 < p < \infty$ (In the case $q < p$ it is assumed that ν is absolutely continuous).

From duality arguments we get the following theorems:

Theorem 4.1.3. *Let* $1 < p \le q < \infty$ *and let* $\alpha > \frac{n(q-1)}{q}$. *Then the following conditions are equivalent:*
(i) B_-^α *is bounded from* $L_\nu^p(R^n)$ *to* $L^q(R^n)$;
(ii)

$$\tilde{D}_1 = \sup_{t>0} \left(\int\limits_{|x|\ge t} |x|^{(2\alpha-n)p'} d\nu \right)^{1/p'} t^{n/q} < \infty.$$

(iii)

$$\tilde{D}_2 = \sup_{k\in Z} \left(\int\limits_{2^k \le |x| < 2^{k+1}} |x|^{(2\alpha-n/q')p'} d\nu \right)^{\frac{1}{p'}} < \infty.$$

Moreover, $\|B_-^\alpha\| \approx \tilde{D}_1 \approx \tilde{D}_2$.

Theorem 4.1.4. *Let* $1 < q < p < \infty$ *and let* $\alpha > \frac{n(q-1)}{q}$. *Assume that* $d\nu = w(x)dx$, *where* w *is a Lebesgue-measurable, a.e. positive function on* R^n. *Then for the boundedness of the operetor* B_-^α *from* $L_w^p(R^n)$ *to* $L^q(R^n)$ *it is necessary and sufficient that*

$$\tilde{D}_3 = \left(\int\limits_{R^n} \left(\int\limits_{|y|>|x|} |y|^{(2\alpha-n)p'} w(y)dy \right)^{\frac{q(p-1)}{p-q}} |x|^{\frac{nq}{p-q}} dx \right)^{\frac{p-q}{pq}} < \infty.$$

Morover, $\|B_-^\alpha\| \approx \tilde{D}_3$.

Next let us consider the case $\alpha < n/p$. Using Lemma 4.1.1 and Sobolev's theorem the following Lemma can be derived in the same way as Lemma 4.1.3 (see also the proof of Theorem 3.3.1).

Lemma 4.1.5. *Let* $1 < p < \infty$ *and* $0 < \alpha < n/p$. *Suppose that* $\tilde{v}$ *and* $\tilde{w}$ *are positive increasing functions on* $(0,\infty)$. *Then for the boundedness of the operator* T_α *from* $L_{\tilde{w}(\cdot)}^p(R^n)$ *into* $L_{\tilde{v}(\cdot)}^p(R^n)$ *it is necessary and sufficient that*

$$C \equiv \sup_{t>0} \left(\int\limits_t^\infty \tilde{v}(\tau)\tau^{(\alpha-n)p+n-1}d\tau \right)^{\frac{1}{p}} \left(\int\limits_0^{t/2} w^{1-p'}(\tau)\tau^{n-1}d\tau \right)^{\frac{1}{p'}} < \infty.$$

Moreover, $\|T_\alpha\| \approx C$.

Using this Lemma we easily obtain the following theorem in the same way as Theorem 4.1.1 was derived from Lemma 4.1.3.

Theorem 4.1.5. *Let* $1 < p < \infty$ *and* $0 < \alpha < n/p$. *Suppose that* $\tilde{v}$ *and* $\tilde{w}$ *are positive increasing functions on* $(0, \infty)$. *Then for the boundedness of the operator* B_+^α *from* $L_{\tilde{w}(\cdot)}^p(R^n)$ *into* $L_{\tilde{v}(\cdot)}^p(R^n)$ *it is necessary and sufficient that*

$$C_1 \equiv \sup_{t>0} \left(\int\limits_t^\infty \tilde{v}(\tau)\tau^{(2\alpha-n)p+n-1}d\tau \right)^{\frac{1}{p}} \left(\int\limits_0^{t/2} w^{1-p'}(\tau)\tau^{n-1}d\tau \right)^{\frac{1}{p'}} < \infty.$$

Moreover, $\|B_+^\alpha\| \approx C_1$.

The next result follows from duality arguments:

Theorem 4.1.6. *Let* $1 < p < \infty$ *and* $0 < \alpha < n/p'$. *Suppose that* $\tilde{v}$ *and* $\tilde{w}$ *are positive decreasing functions on* $(0, \infty)$. *Then for the boundedness of the operator* B_-^α *from* $L_{\tilde{w}(\cdot)}^p(R^n)$ *into* $L_{\tilde{v}(\cdot)}^p(R^n)$ *it is necessary and sufficient that*

$$C_2 \equiv \sup_{t>0} \left(\int\limits_t^\infty \tilde{w}^{1-p'}(\tau)\tau^{(2\alpha-n)p'+n-1}d\tau \right)^{\frac{1}{p}} \left(\int\limits_0^{t/2} v^{1-p'}(\tau)\tau^{n-1}d\tau \right)^{\frac{1}{p'}} < \infty.$$

Moreover, $\|B_-^\alpha\| \approx C_1$.

4.2. Compactness criteria

In this section we consider the compactness of ball fractional integral operators. For the definition of the operators B_+^α and B_-^α see Section 4.1.
First we prove the following Lemma

Lemma 4.2.1. *Let* $1 < p \leq q < \infty$ *and* $\alpha > n/p$. *If*

$$D_1 \equiv \sup_{t>0} \left(\int\limits_{|x|\geq t} |x|^{(2\alpha-n)q}d\nu \right)^{1/q} t^{n/p'} < \infty$$

and $\lim\limits_{a\to 0} D_{1,a} = \lim\limits_{b\to\infty} D_{1,b} = 0$, *where*

$$D_{1,a} = \sup_{0<t<a} \left(\int\limits_{t\leq|x|<a} |x|^{(2\alpha-n)q}d\nu \right)^{\frac{1}{q}} t^{\frac{n}{p'}}$$

and

$$D_{1,b} = \sup_{t>b} \left(\int\limits_{|x|\geq t} |x|^{(2\alpha-n)q}d\nu \right)^{\frac{1}{q}} (t^n - b^n)^{\frac{1}{p'}},$$

then the operator B_+^α is compact from $L^p(R^n)$ to $L_\nu^q(R^n)$.

Proof. Let $0 < a < b < \infty$. Represent B_+^α as follows:

$$B_+^\alpha = \chi_{\{|x|<a\}}B_+^\alpha(f\chi_{\{|y|<a\}}) + \chi_{\{a\le|x|\le b\}}B_+^\alpha(f\chi_{\{|y|<b\}}) +$$
$$+ \chi_{\{|x|>b\}}B_+^\alpha(f\chi_{\{|y|\le b/2\}}) + \chi_{\{|x|>b\}}B_+^\alpha(f\chi_{\{|y|>b/2\}}) +$$
$$+ \tilde{P}_{1a}f + \tilde{P}_{2a}f + \tilde{P}_{3a}f + \tilde{P}_{4a}f.$$

It is easy to see that

$$\tilde{P}_{2a}f = \int\limits_{R^n} \tilde{k}_1(x,y)f(y)dy,$$

where

$$\tilde{k}_1(x,y) = \chi_{\{a\le|z|\le b\}}(x)\chi_{\{|t|<|x|\}}(y)\frac{(|x|^2 - |y|^2)^\alpha}{|x-y|^n}.$$

We have

$$\||\tilde{k}_1(x,y)\|_{L^{p'}(R^n)}\|_{L_\nu^q(R^n)} =$$

$$= \left(\int\limits_{a\le|x|\le b} \left(\int\limits_{|y|<|x|} \frac{(|x|^2 - |y|^2)^{\alpha p'}}{|x-y|^{np'}}dy \right)^{q/p'} d\nu \right)^{1/q} =$$

$$= \left(\int\limits_{a\le|x|\le b} \left(\int\limits_{|y|<|x|} \frac{(|x| - |y|)^{\alpha p'}(|x| + |y|)^{\alpha p'}}{|x-y|^{np'}}dy \right)^{q/p'} d\nu \right)^{1/q} \le$$

$$\le c_1 \left(\int\limits_{a\le|x|\le b} |x|^{\alpha q} \left(\int\limits_{|y|<|x|} |x-y|^{(\alpha-n)p'} dy \right)^{q/p'} d\nu \right)^{1/q} \le$$

$$\le c_2 \left(\int\limits_{a\le|x|\le b} |x|^{(2\alpha-n)q+nq/p'} d\nu \right)^{1/q} < \infty.$$

By Theorem C of Section 2.1 we conclude that $\tilde{P}_{2\alpha}$ is compact. Analogously it can be shown that the operator $\tilde{P}_{3\alpha}$ is compact.

Moreover, using Theorem 4.1.1 restricted to the sets

$$\{x : |x| < a\}, \quad \{x : |x| > b/2\},$$

we have

$$\|\tilde{P}_{1\alpha}\| \le c_3 D_{1,a} \quad \text{and} \quad \|\tilde{P}_{4\alpha}\| \le c_4 D_{1,b/2}.$$

Consequently,

$$\|B_+^\alpha - \tilde{P}_{2\alpha} - \tilde{P}_{3\alpha}\| \leq c_3\|\tilde{P}_{1\alpha}\| + c_4\|\tilde{P}_{4\alpha}\| \to 0$$

as $a \to 0$ and $b \to \infty$ and we finally obtain that B_+^α is compact as it is a limit of compact operators. $\square$

Theorem 4.2.1. Let $1 < p \leq q < \infty$ and let $\alpha > n/p$. Then the following conditions are equivalent :
(i) B_+^α is compact from $L^p(R^n)$ to $L_\nu^q(R^n)$;
(ii) B_+^α is compact from $L^p(R^n)$ to $L_\nu^{q\infty}(R^n)$;
(iii)

$$D_1 = \sup_{t>0} \left(\int_{|x|\geq t} |x|^{(2\alpha-n)q} d\nu \right)^{1/q} t^{n/p'} < \infty$$

and $\lim_{t\to 0} D_1(t) = \lim_{t\to\infty} D_1(t) = 0$, where

$$D_1(t) = \left(\int_{|x|\geq t} |x|^{(2\alpha-n)q} d\nu \right)^{\frac{1}{q}} t^{n/p'};$$

(iv) $D_1 < \infty$ and $\lim_{a\to 0} D_{1,a} = \lim_{b\to\infty} D_{1,b} = 0$;
(v)

$$D_2 = \sup_{k\in Z} \left(\int_{2^k \leq |x| < 2^{k+1}} |x|^{(2\alpha-n/p)q} d\nu \right)^{1/q} < \infty$$

and $\lim_{k\to-\infty} D_2(k) = \lim_{k\to+\infty} D_2(k) = 0$, where

$$D_2(k) = \left(\int_{2^k \leq |x| < 2^{k+1}} |x|^{(2\alpha-n/p)q} d\nu \right)^{\frac{1}{q}}.$$

Proof. From the inequalities

$$D_{1,a} \leq \sup_{t<a} D_1(t), \quad D_{1,b} \leq \sup_{t>b} D_1(t)$$

and from Lemma 4.2.1 we see that (iii) implies (i) .
We now prove that (iii) follows from (i). Let $t > 0$ and let

$$f_t(x) = \chi_{\{|y|<t/2\}}(x)t^{-\frac{n}{p}}.$$

Then it is easy to verify that f_t is weakly convergent to 0 as t tends to 0. Moreover,

$$\|B_+^\alpha f_t\|_{L_\nu^q(R^n)} \geq \left(\int\limits_{|x|\geq t} \left(\int\limits_{|y|<t/2} \frac{(|x|^2-|y|^2)^\alpha}{|x-y|^n} f_t(y)dy \right)^q d\nu \right)^{1/q} \geq$$

$$\geq c_1 \left(\int\limits_{|x|\geq t} |x|^{(2\alpha-n)q}d\nu \right) t^{\frac{n}{p'}} = c_1 D_1(t)$$

and as $B_+^\alpha f_t$ is strongly convergent to 0, it follows that $D_1(t) \to 0$ as $t \to 0$.
If we take the sequence

$$g_t(x) = \chi_{\{|z|>t\}}(x)|x|^{(2\alpha-n)(q-1)} \left(\int\limits_{|y|>t} |y|^{(2\alpha-n)q}d\nu \right)^{-\frac{1}{q'}}$$

and use the fact that the operator B_-^α is compact from $L_\nu^{q'}(R^n)$ to $L^{p'}(R^n)$, we finally obtain that $\lim\limits_{t\to\infty} D_1(t) = 0$.
Let us prove that (v) follows from (ii). Let $k \in Z$ and let

$$f_k(x) = \chi_{\{2^{k-2}\leq|y|<2^{k-1}\}}(x)2^{-\frac{nk}{p}}.$$

Then f_k is weakly convergent to 0 as k tends to $+\infty$ or k tends to $-\infty$. On the other hand,

$$\|B_+^\alpha f_k\|_{L_\nu^{q\infty}(R^n)} \geq \|\chi_{\{y:2^k\leq|x|<2^{k+1}\}}(\cdot)B_+^\alpha f_k(\cdot)\|_{L_\nu^{q\infty}(R^n)} \geq$$

$$\geq c_2\|\chi_{\{y:2^k\leq|x|<2^{k+1}\}}(\cdot)\cdot|^{2\alpha-n}\|_{L_\nu^{q\infty}(R^n)}2^{\frac{kn}{p'}} \geq$$

$$\geq c_3\|\chi_{\{y:2^k\leq|x|<2^{k+1}\}}(\cdot)\|_{L_\nu^{q\infty}(R^n)}2^{k(2\alpha-\frac{n}{p})} \geq$$

$$\geq c_3(\nu\{x:2^k\leq|x|<2^{k+1}\})^{1/q}2^{k(2\alpha-n/p)} \geq$$

$$\geq c_4 \left(\int\limits_{2^k\leq|x|<2^{k+1}} |x|^{(2\alpha-n/p)q}d\nu \right)^{1/q}.$$

Using the fact that $B_+^\alpha f_k$ is strongly convergent to 0, we find that $D_1(k) \to 0$ as $k \to -\infty$ or $k \to +\infty$.

Now we prove that (v) implies (iv). Let $a > 0$. Then there exists an integer $m \in Z$ depending on a such that $2^{m-1} \leq a < 2^m$. Therefore we have

$$D_{1,a} \leq \sup_{0<t<2^m} \left(\int\limits_{t\leq|x|<2^m} |x|^{(2\alpha-n)q}d\nu \right)^{\frac{1}{q}} t^{\frac{n}{p'}} = \sup_{0<t<2^m} D_{1,2^m}(t) = D_{1,2^m},$$

where

$$D_{1,2^m}(t) \equiv \left(\int\limits_{t\leq|x|<2^m} |x|^{(2\alpha-n)q}d\nu \right)^{\frac{1}{q}} t^{\frac{n}{p'}}.$$

Let now $t \in (0, 2^m)$. Then $t \in [2^{i-1}, 2^i)$ for some integer $i \leq m$. We obtain

$$D_{1,2^m}^q(t) \leq \left(\int_{2^{i-1} \leq |x| < 2^m} |x|^{(2\alpha-n)q} d\nu \right) 2^{\frac{nqi}{p'}} =$$

$$= 2^{\frac{nqi}{p'}} \sum_{k=i}^m \int_{2^{k-1} \leq |x| < 2^k} |x|^{(2\alpha-n)q} d\nu =$$

$$= 2^{\frac{nqi}{p'}} \sum_{k=i}^m 2^{-\frac{knq}{p'}} \int_{2^{k-1} \leq |x| < 2^k} |x|^{(2\alpha-n/p)q} d\nu \leq$$

$$\leq c_5 (\sup_{k \leq m} D_2(k-1))^q.$$

Thus $D_{1,2^m} \leq c_6 D_{2,m}$, where $D_{2,m} \equiv \sup_{k \leq m} D_2(k-1)$. If $a \to 0$, then $m \to -\infty$, and $D_{2,m} \to 0$, because $\lim_{k \to -\infty} D_2(k) = 0$. Consequently $\lim_{a \to 0} D_{1,a} = 0$.

Next $t > 0$. Then $t \in [2^m, 2^{m+1})$ for some $m \in Z$. Thus,

$$D_1(t)^q \leq \left(\int_{|x| \geq 2^m} |x|^{(2\alpha-n)q} d\nu \right) 2^{\frac{(m+1)nq}{p'}} =$$

$$= 2^{\frac{(m+1)nq}{p'}} \sum_{k=m}^{+\infty} \left(\int_{2^k \leq |x| < 2^{k+1}} |x|^{(2\alpha-n)q} d\nu \right) \leq$$

$$\leq c_7 2^{\frac{nmq}{p'}} \sum_{k=m}^{+\infty} 2^{-\frac{nkq}{p'}} D_2^q(k) \leq c_8 (\sup_{k \geq m} D_2(k))^q.$$

· Consequently,

$$\lim_{t \to \infty} D_1(t) \leq c_9 \lim_{m \to +\infty} \sup_{k \geq m} D_2(k) = 0.$$

Hence

$$\lim_{b \to +\infty} D_{1,b} \leq \lim_{b \to +\infty} \sup_{t > b} D_1(t) = 0.$$

From Theorem 4.1.1 we have that if $D_2 < \infty$, then $D_1 < \infty$. Using Lemma 4.2.1 and the fact that (i) implies (ii), we finally obtain (iii) $\Longleftrightarrow$ (i) and (iv) $\Longrightarrow$ (i) $\Longrightarrow$ (ii) $\Longrightarrow$ (v) $\Longrightarrow$ (iv). $\square$

The following theorem can be proved analogously.

Theorem 4.2.2. *Let $1 < q < p < \infty$ and let $\alpha > n/p$. Suppose that v is a Lebesgue-measurable a.e. positive function on R^n. The operator B_+^α is* *compact from $L^p(R^n)$ to $L^q_\nu(R^n)$ if and only if*

$$D_3 = \left[\int\limits_{R^n} \left(\int\limits_{|y|>|x|} |y|^{(2\alpha-n)q} v(y) dy \right)^{\frac{p}{p-q}} |x|^{\frac{np(q-1)}{p-q}} dx \right]^{\frac{p-q}{pq}} < \infty.$$

From duality arguments we can obtain the following theorems:

Remark 4.2.1. Analogous results can be obtained for the operator B^α_+ acting from $L^p(B(0,a))$ to $L^q_\nu(B(0,a))$, where $0 < a < \infty$ (We assume that ν is absolutely continuous when $q < p$).

Theorem 4.2.3. *Let p, q and α satisfy the conditions of Theorem 4.1.3. Then the following conditions are equivalent:*

(i) B^α_- *is compact from $L^p_\nu(R^n)$ to $L^q(R^n)$;*

(ii)

$$\tilde{D}_1 = \sup_{t>0} \left(\int\limits_{|x|\geq t} |x|^{(2\alpha-n)p'} d\nu \right)^{1/p'} t^{n/q} < \infty;$$

and $\lim\limits_{t\to 0} \tilde{D}_1(t) = \lim\limits_{t\to\infty} \tilde{D}_1(t) = 0$, where

$$\tilde{D}_1(t) = \left(\int\limits_{|x|\geq t} |x|^{(2\alpha-n)p'} d\nu \right)^{\frac{1}{p'}} t^{n/q};$$

(iii) $\tilde{D}_1 < \infty$ *and $\lim\limits_{a\to 0} \tilde{D}_{1,a} = \lim\limits_{b\to\infty} \tilde{D}_{1,b} = 0$, where*

$$\tilde{D}_{1,a} = \sup_{0<t<a} \left(\int\limits_{t\leq|x|<a} |x|^{(2\alpha-n)p'} d\nu \right)^{\frac{1}{p'}} t^{\frac{n}{q}}$$

and

$$\tilde{D}_{1,b} = \sup_{t>b} \left(\int\limits_{|x|\geq t} |x|^{(2\alpha-n)p'} d\nu \right)^{\frac{1}{p'}} (t^n - b^n)^{\frac{1}{q}};$$

(iv)

$$\tilde{D}_2 = \sup_{k\in Z} \left(\int\limits_{2^k\leq|x|<2^{k+1}} |x|^{(2\alpha-n/q')p'} d\nu \right)^{\frac{1}{p'}} < \infty.$$

and $\lim\limits_{k\to-\infty} \tilde{D}_2(k) = \lim\limits_{k\to+\infty} \tilde{D}_2(k) = 0$, where

$$\tilde{D}_2(k) = \left(\int\limits_{2^k\leq|x|<2^{k+1}} |x|^{(2\alpha-n/q')p'} d\nu \right)^{\frac{1}{p'}}.$$

Theorem 4.2.4. *Let p, q, α, w and ν satisfy the conditions of Theorem 4.1.4 Then the operator B_{-}^{α} is compact from $L_{w}^{p}(R^{n})$ to $L^{q}(R^{n})$ if and only if*

$$\tilde{D}_3 = \left(\int\limits_{R^n} \left(\int\limits_{|y|>|x|} |y|^{(2\alpha-n)p'} w(y)dy \right)^{\frac{q}{p-q}} |x|^{\frac{nq(p-1)}{p-q}} dx \right)^{\frac{p-q}{pq}} < \infty.$$

Remark 4.2.2. From a duality argument necessary and sufficient conditions can be derived for the weight w to ensure the boundedness and compactness of B_{-}^{α} with $d\nu = dx$ from $L_{w}^{p}(R^{n})$ to $L^{q}(R^{n})$.

4.3.   The measure of non–compactness

In the present section we estimate the distance between the ball fractional integral operator

$$B_{+}^{\alpha}(f)(x) = \int\limits_{|y|<|x|} \frac{(|x|^2 - |y|^2)^{\alpha}}{|x-y|^n} f(y)dy$$

and the space of compact operators.

We shall assume that v is a Lebesgue-measurable a.e. positive function on R^n.

The following Lemma holds (see [66], Corollary V.5.4).

Lemma 4.3.1. *Let $1 \leq q \leq \infty$ and $P \in \mathcal{B}(X,Y)$, where $Y = L^{q}(R^{n})$. Then*

$$dist(P, \mathcal{K}(X,Y)) = dist(P, \mathcal{F}_r(X,Y)).$$

We also need the following lemma:

Lemma 4.3.2. *Let $1 \leq q < \infty$ and let $Y = L^{q}(R^{n})$. Suppose that $P \in \mathcal{F}_r(X,Y)$ and $\epsilon > 0$. Then there exist $T \in \mathcal{F}_r(X,Y)$ and $[\alpha, \beta] \subset (0, \infty)$ such that*

$$\|P - T\| < \epsilon$$

and

$$suppTf \subset \{\alpha \leq |x| \leq \beta\}$$

for every $f \in X$.

Proof. As $P \in \mathcal{F}_r(X,Y)$, there exist linearly independent functions u_i ($i = 1, 2, 3, \cdots N$) such that $u_i \in Y$, $\|u_i\|_Y = 1$ and

$$Pf = \sum_{i=1}^{N} c_i(f)u_i$$

for every $f \in X$. For u_i there exist $\varphi_i \in C_0^\infty(R^n)$ such that $\|u_i - \varphi_i\|_Y < \epsilon$, $i = 1, 2, \cdots N$. Let

$$Rf = \sum c_i(f)\varphi_i.$$

Then $R \in \mathcal{F}_r(X, Y)$ and

$$\|Pf - Rf\|_Y \le \sum_{i=1}^N c_i(f)\|\varphi_i - u_i\|_Y$$

$$\le K\epsilon\|f\|_X.$$

Let $supp\ \varphi_i = \{x : \alpha_i < |x| < \beta_i\}$ and put $\alpha = \min \alpha_i$, $\beta = \max \beta_i$. Then

$$suppRf \subset \{x : \alpha < |x| < \beta\}.$$

We shall need the following notation:

$$\overline{B}_+^\alpha f(x) \equiv v^{1/q}(x)B_+^\alpha f(x),$$

$$\tilde{I} \equiv dist\left(B_+^\alpha, \mathcal{K}(X, L_v^q(R^n))\right),$$

$$\overline{I} \equiv dist\left(\overline{B}_+^\alpha, \mathcal{K}(X, L^q(R^n))\right).$$

□

The next lemma follows immediately:

Lemma 4.3.3. *Let* $1 \le q < \infty$. *Then* $\tilde{I} = \overline{I}$.

Theorem 4.3.1. *Let* $1 < p \le q < \infty$ *and let* $\alpha > n/p$. *Assume that* $X = L^p(R^n)$, $Y = L_v^q(R^n)$ *and*

$$D \equiv \sup_{t>0} \left(\int_{|x|\ge t} |x|^{(2\alpha-n)q}v(x)dx \right)^{1/q} t^{n/p'} < \infty.$$

Then there exist positive constants b_1 *and* b_2 *such that*

$$b_1 J \le \tilde{I} \le b_2 J, \qquad\qquad (4.3.1)$$

where $J = \lim_{a\to 0} J^{(a)} + \lim_{b\to\infty} \overline{J}^{(b)}$,

$$J^{(a)} \equiv \sup_{0<t<a} \left(\int_{t<|x|<a} |x|^{(2\alpha-n)q}v(x) \right)^{1/q} t^{n/p'};$$

$$\overline{J}^{(b)} \equiv \sup_{t>b} \left(\int_{|x|\ge t} |x|^{(2\alpha-n)q}v(x)dx \right)^{1/q} (t^n - b^n)^{1/p'}.$$

Proof. The upper estimate of (4.3.1) follows from the proof of Theorem 4.2.1. Let $\lambda > \overline{I}$. Then by Lemma 4.3.3 we have $\lambda \overline{I}$. Using Lemma 4.3.1 we can find $P \in \mathcal{F}_r(X, Y)$ such that

$$\|\overline{B}_+^\alpha - P\| < \lambda.$$

From Lemma 4.3.2 we see that for $\epsilon = (\lambda - \|\overline{B}_+^\alpha - P\|)/2$ there exist $T \in \mathcal{F}_r(X, Y)$, α and β such that

$$\|P - T\| < \epsilon$$

and

$$\operatorname{supp} Tf \subset \{x : \alpha < |x| < \beta\}$$

for all $f \in X$. Consequently for all $f \in X$ the following inequality is fulfilled:

$$\|\overline{B}_+^{(\alpha)} f - Tf\|_Y \leq \lambda \|f\|_X.$$

Hence

$$\int\limits_{|x| \leq \alpha} |\overline{B}_+^\alpha f(x)|^q dx + \int\limits_{|x| \geq \beta} |\overline{B}_+^\alpha f(x)|^q dx \leq \lambda^q \|f\|_X^q. \tag{4.3.2}$$

Let us choose $n \in Z$ such that $2^n < \alpha$. Suppose that $j \in Z, j \leq n - 1$ and $f_j(y) = \chi_{\{|x| < 2^{j-1}\}}(y)$. Then

$$\int\limits_{2^j < |x| \leq 2^{j+1}} |\overline{B}_+^\alpha f(x)|^q dx \geq$$

$$\geq \int\limits_{2^j < |x| \leq 2^{j+1}} v(x) \left(\int\limits_{|y| < 2^{j-1}} \frac{(|x|^2 - |y|^2)^\alpha}{|x - y|^n} dy \right)^q dx \geq$$

$$\geq c_1 \left(\int\limits_{2^j < |x| \leq 2^{j+1}} v(x) |x|^{(2\alpha-n)q} dx \right) 2^{(j-1)qn} \geq$$

$$\geq c_2 \left(\int\limits_{2^j < |x| \leq 2^{j+1}} v(x) |x|^{(2\alpha-n)q} dx \right) 2^{jqn}.$$

On the other hand,

$$\|f_j\|_X = c_3 2^{jn/p}.$$

From (4.3.2) it follows that

$$D_2(j) \equiv c_4 \left(\int\limits_{2^j \leq |x| < 2^{j+1}} v(x) |x|^{(2\alpha-n/p)q} dx \right)^{1/q} < \lambda$$

for every $j \in Z$, $j \leq n - 1$. Consequently for every n such that $2^n < \alpha$ we have

$$c_4 \sup_{j \leq n} D_2(j) \leq \lambda.$$

Hence

$$\lim_{n \to -\infty} \sup_{j \leq n} D_2(j) \leq c_5 \lambda. \tag{4.3.3}$$

Let $a \in (0, \alpha)$. Then $a \in [2^{m-1}, 2^m)$ for some integer $m = m_a$. We have (see the proof of Theorem 4.2.1)

$$J^{(a)} \leq c_6 \sup_{n \leq m} D_2(n) \equiv c_6 D_{2,m}.$$

From the last inequality and from (4.3.3) we obtain

$$\lim_{a \to 0} J^{(a)} \leq c_6 \lim_{m \to -\infty} D_{2,m} \leq c_7 \lambda,$$

where c_7 depends only on p, q and α. As λ is arbitrarily chosen, we conclude that

$$\lim_{a \to 0} J^{(a)} \leq c_7 \overline{I}.$$

Now let $b > \beta$ and let $t > b$. Assume that

$$f(y) = \chi_{(b/2, t/2)}(y).$$

Then $\|f_t\|_X = c_8 (t^n - b^n)^{1/p}$. On the other hand,

$$\int\limits_{|x|>t} v(x) \left(\int\limits_{|y|<|x|} \frac{(|x|^2 - |y|^2)^\alpha}{|x-y|^n} dy \right)^q dx \geq$$

$$\geq c_9 \left(\int\limits_{|x|>t} v(x) |x|^{(2\alpha-n)q} dx \right) (t^n - b^n).$$

From (4.3.2) we conclude that

$$\left(\int\limits_{|x|>t} v(x) |x|^{(2\alpha-n)q} dx \right)^{1/q} (t^n - b^n)^{1/p'} \leq c_{10} \lambda$$

for every $t > b$. As $b > \beta$,

$$\lim_{b \to \infty} \overline{J}^{(b)} \leq c_{10} \lambda.$$

Finally we have that $\lim_{b \to \infty} \overline{J}^{(b)} \leq c_{10} \overline{I}$ and the inequality (4.3.1) is proved. $\square$

4.4. Mapping properties in Lorentz spaces

Here we discuss boundedness and compactness for weighted ball fractional integrals in Lorentz spaces. Let

$$B^\alpha_{+,u_1} f(x) = u_1(x) \int\limits_{\{y \in R^n : |y| < |x|\}} \frac{(|x|^2 - |y|^2)^\alpha}{|x-y|^n} f(y) dy, \quad \alpha > 0, \ x \in R^n,$$

$$B^\alpha_{-,u_2} g(y) \int\limits_{\{x \in R^n : |x| > |y|\}} \frac{(|x|^2 - |y|^2)^\alpha}{|x-y|^n} u_2(x) g(x) d\nu(x), \quad \alpha > 0, \ y \in R^n,$$

be weighted ball fractional integrals, where ν is a positive σ–finite Borel measure on R^n and u_1 and u_2 are ν–measurable a.e. positive functions on R^n.
The following lemma is valid:

Lemma 4.4.1. *Let $1 < r < \infty$, $1 \le s < \infty$ and let $\alpha > n/p$. Then there exists a positive constant c such that for all $x \in R^n$ the following inequality holds*

$$I(x) \equiv \left\| \chi_{\{|x|/2 < |y| < |x|\}}(\cdot) \frac{1}{|x - \cdot|^{n-\alpha}} \right\|_{L^{r's'}(R^n)} \le c|x|^{\alpha - n/r}.$$

Proof. The case $\alpha \ge n$ is trivial. Let us consider the case $n/p < \alpha < n$. We have

$$I(x) \le \left\| \chi_{\{|y| < |x|\}}(\cdot) |x - \cdot|^{\alpha - n} \right\|_{L^{r's'}(R^n)} =$$

$$= \left(s' \int\limits_0^\infty \lambda^{s'-1} \left(|\{y : |y| < |x|, |x-y|^{\alpha-n} > \lambda\}| \right)^{s'/r'} d\lambda \right)^{1/s'} \le$$

$$\le \left(s' \int\limits_0^{|x|^{\alpha-n}} \lambda^{s'-1} \left(|\{y : |y| < |x|, |x-y|^{\alpha-n} > \lambda\}| \right)^{s'/r'} d\lambda \right)^{1/s'} \le$$

$$\le \left(s' \int\limits_{|x|^{\alpha-n}}^\infty \lambda^{s'-1} \left(|\{y : |y| < |x|, |x-y|^{\alpha-n} > \lambda\}| \right)^{s'/r'} d\lambda \right)^{1/s'} \equiv$$

$$\equiv I_1(x) + I_2(x).$$

It is obvious that

$$I_1(x) \le \left(s' \int\limits_0^{|x|^{\alpha-n}} \lambda^{s'-1} |x|^{ns'/r'} d\lambda \right)^{1/s'} \equiv c_1 |x|^{\alpha - n/r}.$$

Further,

$$I_2(x) \leq c_2 \left(s' \int\limits_{|x|^{\alpha-n}}^{\infty} \lambda^{s'-1} \lambda^{\frac{ns'}{(\alpha-n)r'}} d\lambda \right)^{1/s'} = c_3 |x|^{\alpha-n/r}.$$

From the last estimates we obtain the desired result. $\square$

Let

$$I_{\alpha,u_1} f(x) = u_1(x) \int\limits_{|y|<|x|} |x-y|^{\alpha-n} f(y) dy$$

for measurable $f : R^n \to R$ and for $\alpha > 0$.

The following Lemma is proved in the same way as Lemma 4.1.1 using Lemma 4.4.1 and Theorem 1.1.4. Therefore we omit the proof.

Lemma 4.4.2. *Let $1 < r, p < \infty$, $1 \leq s < \infty$ and let $1 < q \leq \infty$. Assume that $\max\{r, s\} \leq \min\{p, q\}$ and $\alpha > n/r$. Then I_{α,u_1} is bounded from $L^{rs}(R^n)$ into $L_\nu^{pq}(R^n)$ if and only if*

$$A \equiv \sup_{t>0} \| | \cdot |^{\alpha-n} u_1(\cdot) \chi_{\{|y|>t\}}(\cdot) \|_{L_\nu^{pq}(R^n)} t^{n/r'} < \infty.$$

Moreover, $\|I_{\alpha,u_1}\| \approx A$.

The following statement is valid:

Theorem 4.4.1. *Let $1 < r, s < \infty$, $1 \leq s < \infty$ and let $1 < q \leq \infty$. Assume also that $\max\{r, s\} \leq \min\{p, q\}$ and $\alpha > n/r$. Then the following statements are equivalent:*
  (i) B_{+,u_1}^α *is bounded from $L^{rs}(R^n)$ into $L_\nu^{pq}(R^n)$;*
  (ii)

$$D_1 \equiv \sup_{t>0} D_1(t) \equiv \| \chi_{\{|y|>t\}}(\cdot) \cdot | \cdot |^{2\alpha-n} u_1(\cdot) \|_{L_\nu^{pq}(R^n)} t^{n/r'},$$

  (iii)

$$D_2 \equiv \sup_{k \in Z} D_2(k) \equiv \sup_{k \in Z} \| \chi_{\{2^k \leq |y| < 2^{k+1}\}}(\cdot) \cdot | \cdot |^{2\alpha-n/r} u_1(\cdot) \|_{L_\nu^{pq}(R^n)}.$$

Moreover, $\|B_{+,u_1}^\alpha\| \approx D_1 \approx D_2$.
Proof. The implication (ii) $\Rightarrow$ (i) follows from the inequality

$$B_{+,u_1}^\alpha f(x) \leq c_1 |x|^\alpha u_1(x) \int\limits_{|y|<|x|} f(y)|x-y|^{\alpha-n} dy$$

and from Lemma 4.4.2. Further, the implications (i) $\Rightarrow$ (iii) and (iii) $\Rightarrow$ (ii) follow in the same way as the proof of Theorem 2.11.1. $\square$

Using Theorem C of Section 2.1 we easily get the following statement:

Theorem 4.4.2. *Let* $1 < r, p, q < \infty$ *and let* $1 \le s < \infty$. *Then the following conditions are equivalent:*

(i) B^{α}_{+,u_1} *is compact from* $L^{rs}(R^n)$ *into* $L^{pq}_{\nu}(R^n)$;

(ii) $D_1 < \infty$ *and* $\lim_{t \to 0} D_1(t) = \lim_{t \to \infty} D_1(t) = 0$;

(iii) $D_1 < \infty$ *and* $\lim_{b \to 0} D^{(b)}_1 = \lim_{c \to \infty} \overline{D}^{(c)}_1 = 0$, *where*

$$D^{(b)}_1 \equiv \sup_{0 < t < b} \|\chi_{\{t < |y| < b\}}(\cdot) \cdot |^{2\alpha - n} u_1(\cdot)\|_{L^{pq}_{\nu}(R^n)} t^{n/r'},$$

$$\overline{D}^{(c)}_1 \equiv \sup_{t > c} \|\chi_{\{|y| > t\}}(\cdot) \cdot |^{2\alpha - n} u_1(\cdot)\|_{L^{pq}_{\nu}(R^n)} (t^n - c^n)^{1/r'};$$

(iii) $D_2 < \infty$ *and* $\lim_{k \to -\infty} D_2(k) = \lim_{k \to +\infty} D_2(k) = 0$.

This theorem is proved in the same way as Theorem 2.12.3: we omit the proof.

From duality arguments we derive the following results concerning B^{α}_{-,u_2}.

Theorem 4.4.3. *Let* $1 < r, p < \infty$, $1 \le s < \infty$ *and* $1 < q \le \infty$. *Assume that* $\max\{r, s\} \le \min\{p, q\}$ *and* $\alpha > n/p'$. *Then the following conditions are equivalent:*

(i) B^{α}_{-,u_2} *is bounded from* $L^{rs}_{\nu}(R^n)$ *into* $L^{pq}(R^n)$;

(ii)

$$E_1 \equiv \sup_{t > 0} E_1(t) \equiv \sup_{t > 0} \|\chi_{\{|y| > t\}}| \cdot |^{2\alpha - n} u_2(\cdot)\|_{L^{r's'}_{\nu}(R^n)} t^{n/p} < \infty;$$

(iii)

$$E_2 \equiv \sup_{k \in Z} E_2(k) \equiv \sup_{k \in Z} \|\chi_{\{2^k \le |y| < 2^{k+1}\}}(\cdot) \cdot |^{2\alpha - n/p'} u_2(\cdot)\|_{L^{r's'}_{\nu}(R^n)} < \infty.$$

Moreover, $\|B^{\alpha}_{-,u_2}\| \approx E_1 \approx E_2$.

Theorem 4.4.4. *Let* $1 < r, s, p, q < \infty$ *with* $\max\{r, s\} \le \min\{p, q\}$ *and let* $\alpha > n/p'$. *Then the following statements are equivalent:*

(i) B^{α}_{-,u_2} *is compact from* $L^{rs}_{\nu}(R^n)$ *into* $L^{pq}(R^n)$;

(ii) $E_1 < \infty$ *and* $\lim_{b \to 0} E^{(b)}_1 = \lim_{c \to \infty} \overline{E}^{(c)}_1 = 0$, *where*

$$E^{(b)}_1 \equiv \sup_{0 < t < b} \|\chi_{\{t < |y| < b\}}(\cdot) \cdot |^{2\alpha - n} u_2(\cdot)\|_{L^{r's'}_{\nu}(R^n)} t^{n/p},$$

$$\overline{E}^{(c)}_1 \equiv \sup_{t > c} \|\chi_{\{|y| > c\}}(\cdot) \cdot |^{2\alpha - n} u_2(\cdot)\|_{L^{r's'}_{\nu}(R^n)} (t^n - c^n)^{1/p};$$

(iii) $E_1 < \infty$ *and* $\lim_{t \to 0} E_1(t) = \lim_{t \to +\infty} E_1(t) = 0$;

(iv) $E_2 < \infty$ and $\lim\limits_{k\to-\infty} E_2(k) = \lim\limits_{k\to+\infty} E_2(k) = 0$.

Let us consider the case $u_1 \equiv 1$. In this case $B^\alpha_{+,u_1} = B^\alpha_+$. We have the following result.

Theorem 4.4.5. *Let* $1 < r, p < \infty$, $1 \leq s < \infty$ *and* $1 < q \leq \infty$. *Assume also that* $\max\{r, s\} \leq \min\{p, q\}$ *and* $\alpha > n/r$. *Then the following statements are equivalent:*

(i) B^α_+ *is bounded from* $L^r(R^n)$ *into* $L^p_\nu(R^n)$;

(ii) B^α_+ *is bounded from* $L^{rs}(R^n)$ *into* $L^{pq}_\nu(R^n)$;

(iii)

$$F_1 \equiv \sup_{t>0} F_1(t) \equiv \sup_{t>0} \left(\int\limits_{\{|x|>t\}} |x|^{(2\alpha-n)p} d\nu(x) \right)^{1/p} t^{n/r'} < \infty;$$

(iv)

$$F_2 \equiv \sup_{k\in Z} F_2(k) \equiv \sup_{k\in Z} \left(\int\limits_{\{2^k \leq |x| < 2^{k+1}\}} |x|^{(2\alpha-n/r)p} d\nu(x) \right)^{1/p} < \infty.$$

Moreover, $\|B^\alpha_+\|_{L^r \to L^p_\nu} \approx \|B^\alpha_+\|_{L^{rs} \to L^{pq}_\nu} \approx F_1 \approx F_2$.

Theorem 4.4.6. *Let* $1 < r, p < \infty$, $1 \leq s < \infty$ *and* $1 < q \leq \infty$. *We assume also that* $\max\{r, s\} \leq \min\{p, q\}$ *and* $\alpha > n/r$. *Then the following conditions are equivalent:*

(i) B^α_+ *is compact from* $L^r(R^n)$ *into* $L^p_\nu(R^n)$;

(ii) B^α_+ *is compact from* $L^{rs}(R^n)$ *into* $L^{pq}_\nu(R^n)$;

(iii) $F_1 < \infty$ *and* $\lim\limits_{b\to 0} F_1^{(b)} = \lim\limits_{c\to\infty} \overline{F}_1^{(c)} = 0$, *where*

$$F_1^{(b)} \equiv \sup_{0<t<b} \left(\int\limits_{\{t<|x|<b\}} |x|^{(2\alpha-n)p} d\nu(x) \right)^{1/p} t^{n/r'},$$

$$\overline{F}_1^{(c)} \equiv \sup_{t>c} \left(\int\limits_{\{|x|>t\}} |x|^{(2\alpha-n)p} d\nu(x) \right)^{1/p} (t^n - c^n)^{1/r'};$$

(iv) $F_1 < \infty$ *and* $\lim\limits_{t\to 0} F_1(t) = \lim\limits_{t\to+\infty} F_1(t) = 0$;

(v) $F_2 < \infty$ *and* $\lim\limits_{k\to-\infty} F_2(k) = \lim\limits_{k\to+\infty} F_2(k) = 0$.

4.5. Notes and comments on Chapter 4

For ball fractional integrals we refer to [245], [247], [255]. In [245–247] two-weight inequalities for ball fractional integrals in case of pairs of certain power weights are derived.

The material of this chapter is taken from [204], [207] (see also the survey [157]).

Chapter 5

POTENTIALS ON R^N

In this chapter we develop a new approach to truncated potentials. We introduce the extension of truncated potentials and prove necessary and sufficient conditions for boundedness from $L^p(R^n)$ into $L_\nu^q(R^n)$, when $1 < p < \infty$, $0 < q < \infty$ and $\alpha > n/p$. A generalization of Sawyer's result [258] is presented. Then a compactness criterion for this operator is proved, and upper and lower estimates of its distance from the class of compact operators are derived.

It is well–known (see, e.g., [2]) that Riesz potential is not compact from $L^p(\Omega)$ to $L^q(\Omega)$ when q is the Sobolev number even in the case of a bounded domain Ω. We provide a condition on a pair of weights (v, w) which ensures the compactness of the Riesz potential from L_w^p to L_v^q when $1 < p < q < \infty$.

The last section is devoted to the compactness property of certain integral transforms with radial kernel.

5.1. Truncated potentials

In the present section, necessary and sufficient conditions are found which guarantee the boundedness (compactness) of the generalized truncated Riesz potential

$$K_\alpha f(x, t) = \int_{|y| \leq 2|x|} (|x - y| + t)^{\alpha - n} f(y) dy, \quad x \in R^n, \ t \in [0, \infty),$$

from $L^p(R^n)$ into $L_\nu^q(R_+^{n+1})$, where $1 < p < \infty$, $0 < q < \infty$ $\alpha > n/p$ and ν is a positive Borel measure on $R_+^{n+1} \equiv R^n \times [0, \infty)$. For $q < p$ we shall assume that $d\nu(x, t) = v(x, t) dx dt$, where v is a Lebesgue-measurable a.e. positive function on R_+^{n+1}.

343

In this section we also propose, necessary and sufficient conditions for the boundedness (compactness) of the operator

$$\widetilde{K}_\alpha g(y) = \int\limits_{|x| \geq \frac{|y|}{2}} (|x - y| + t)^{\alpha - n} g(x, t) d\nu(x, t), \quad y \in R^n,$$

from $L_\nu^p(R_+^{n+1})$ into $L^q(R^n)$ $(1 < p, q < \infty, \ \alpha > \frac{n(q-1)}{q})$.
We also give upper and lower estimates of the distance of the operator K_α from the space of compact operators.

In the sequel we shall use the following notation:

$$U_r \equiv (R^n \setminus B(0, r)) \times [0, \infty),$$

where $B(0, r) \equiv \{x : |x| < r\}$. It is clear that

$$U_r \setminus U_R = (B(0, R) \setminus B(0, r)) \times [0, \infty)$$

for all $0 < r < R < \infty$.
Let

$$Hf(x) = \int\limits_{\{y \in R^n : |y| < |x|\}} f(y) dy.$$

To prove the main results, we need the following lemma.

Lemma 5.1.1. *Let* $1 < p \leq q < \infty$ *and let* μ *be a positive Borel measure on* R_+^{n+1}. *Then the operator* H *is bounded from* $L^p(R^n)$ *into* $L_\mu^q(R_+^{n+1})$ *if and only if*

$$A \equiv \sup (\mu(U_r))^{1/q} r^{n/p'} < \infty, \quad p' = p/(p - 1).$$

Moreover, $\|H\| \approx A$.

Proof. *Sufficiency.* Let $f \geq 0$ and let $f \in L^p(R^n)$. Let us denote

$$I(t) \equiv \int\limits_{|y| < t} f(y) dy.$$

Let

$$\int\limits_{R^n} f(y) dy \in (2^m, 2^{m+1}]$$

for for some integer m. Then there exists a sequence of positive numbers $\{x_k\}_{k=-\infty}^m$ such that $I(x_k) = 2^k$. Moreover,

$$2^k = \int\limits_{x_k < |y| < x_{k+1}} f(y) dy$$

for $k \leq m - 1$. The sequence $\{x_k\}$ is increasing. It is obvious that

$$R^n = \{x : |x| \leq \alpha\} \cup \left(\bigcup_{k \leq m} \{x_k \leq |x| < x_{k+1}\} \right),$$

where $\alpha = \lim_{k \to -\infty} x_k$ and $x_{m+1} = \infty$. If $\int_{R^n} f = \infty$, then we have

$$R^n = \{|x| \leq \alpha\} \cup \left(\bigcup_{k \in Z} \{x_k \leq |x| < x_{k+1}\} \right)$$

(i.e. in this case $m = \infty$). Assume that $t \in [0, \alpha]$. Then $I(t) = 0$. Consequently $I(|x|) = 0$ for $|x| \leq \alpha$. If $t \in [x_k, x_{k+1})$, then $I(t) \leq 2^{k+1}$. Using Hölder's inequality we have

$$\|Hf\|^p_{L^q_\mu(R^{n+1}_+)} \leq \sum_k \|\chi_{U_{x_k} \setminus U_{x_{k+1}}} Hf\|^p_{L^q_\mu(R^{n+1}_+)} \leq$$

$$\leq \sum_k 2^{(k+1)p} \|\chi_{U_{x_k} \setminus U_{x_{k+1}}}\|^p_{L^q_\mu(R^{n+1}_+)} \leq$$

$$\leq 4^p \sum_k 2^{(k-1)p} (\mu(U_{x_k} \setminus U_{x_{k+1}}))^{p/q} =$$

$$= 4^p \sum_k \left(\int_{x_{k-1} < |y| < x_k} f(y) dy \right)^p (\mu(U_{x_k} \setminus U_{x_{k+1}}))^{p/q} \leq$$

$$\leq 4^p \sum_k \left(\int_{x_{k-1} < |y| < x_k} (f(y))^p dy \right) (x_k^n - x_{k-1}^n)^{p-1} (\mu(U_{x_k} \setminus U_{x_{k+1}}))^{p/q} \leq$$

$$\leq 4^p A^p \|f\|^p_{L^p(R^n)}.$$

Necessity. Let $f_r(x) = \chi_{\{|y| < r\}}(x)$, where $r > 0$. Then $\|f_r\|_{L^p(R^n)} = r^{n/q}$. On the other hand,

$$\|Hf_r\|_{L^q_\mu(R^{n+1}_+)} \geq \|\chi_{U_r} Hf_r\|_{L^q_\mu(R^{n+1}_+)} \geq (\mu(U_r))^{1/q} r^n.$$

Consequently $A < \infty$. $\square$

Lemma 5.1.2. *Let* $0 < q < p < \infty$ *and* $p > 1$. *Suppose that* v *is a Lebesgue-measurable a.e. positive function on* R^{n+1}_+. *The operator* H *is bounded from* $L^p(R^n)$ *into* $L^q_v(R^{n+1}_+)$ *if and only if*

$$A_1 \equiv \left(\int_{R^n} \left(\int_{U_{|x|}} v(y, t) dy dt \right)^{\frac{p}{p-q}} |x|^{\frac{np(q-1)}{p-q}} dx \right)^{\frac{p-q}{pq}} < \infty.$$

Moreover, $\|H\| \approx A_1$.

Proof. Let us denote

$$\tilde{v}(y) \equiv \int\limits_0^\infty v(y,t)dt.$$

From the boundedness criterion for the Hardy operator H from $L^p(0,\infty)$ into $L^q_{\tilde{v}}(0,\infty)$ (see Theorem B of Section 2.1) using spherical coordinates we see that the condition $A_1 < \infty$ is equivalent to the boundedness of H from $L^p(R^n)$ into $L^q_{\tilde{v}}(R^n)$. The latter is equivalent to the boundedness of H from $L^p(R^n)$ into $L^q_v(R^{n+1}_+)$. $\square$

Now we prove the main results concerning the boundedness of the operators K_α and $\widetilde{K}_\alpha$. We shall assume that ν is a positive Borel measure on R^{n+1}_+.

Theorem 5.1.1. *Let* $1 < p \le q < \infty$, $\alpha > n/p$. *Then the following conditions are equivalent:*
(i) K_α *is bounded from* $L^p(R^n)$ *into* $L^q_\nu(R^{n+1}_+)$;
(ii)

$$B \equiv \sup_{r>0} \left(\int\limits_{U_r} (|x|+t)^{(\alpha-n)q} d\nu(x,t) \right)^{1/q} r^{n/p'} < \infty; \qquad (5.1.1)$$

(iii)

$$B_1 \equiv \sup_{k \in Z} \left(\int\limits_{U_{2^k} \setminus U_{2^{k+1}}} (|x|+t)^{(\alpha-n)q} |x|^{nq/p'} d\nu(x,t) \right)^{1/q} < \infty. \qquad (5.1.2)$$

Moreover, $\|K_\alpha\| \approx B \approx B_1$.

Proof. First we prove that (ii) $\Rightarrow$ (i). For this we represent $K_\alpha f$ as follows:

$$K_\alpha f(x,t) = \int\limits_{|y|<|x|/2} (|x-y|+t)^{\alpha-n} f(y)dy +$$

$$+ \int\limits_{|x|/2 \le |y| \le 2|x|} (|x-y|+t)^{\alpha-n} f(y)dy \equiv K^{(1)}_\alpha f(x,t) + K^{(2)}_\alpha f(x,t).$$

Consequently

$$\|K_\alpha f\|_{L^q_\nu(R^{n+1}_+)} \le \|K^{(1)}_\alpha f\|_{L^q_\nu(R^{n+1}_+)} + \|K^{(2)}_\alpha f\|_{L^q_\nu(R^{n+1}_+)} \equiv$$

$$\equiv S_1 + S_2.$$

If $|y| < |x|/2$, then $(|x-y|+t)^{\alpha-n} \le b_1(|x|+t)^{\alpha-n}$. Using Lemma 5.1.1 we obtain

$$S_1 \le c_1 \||\cdot|^{\alpha-n} Hf(\cdot)\|_{L^q_\nu(R^{n+1}_+)} \le c_2 B \|f\|_{L^p(R^n)}.$$

Using Hölder's inequality we find that

$$S_2^q \leq \int\limits_{R_+^{n+1}} \left(\int\limits_{\frac{|x|}{2} \leq |y| \leq 2|x|} (f(y))^p dy \right)^{\frac{q}{p}} (\Psi(x,t))^{\frac{q}{p'}} d\nu(x,t),$$

where

$$\Psi(x,t) = \int\limits_{\frac{|x|}{2} \leq |y| \leq 2|x|} (|x-y| + t)^{(\alpha-n)p'} dy.$$

Now we show that there exists a positive constant b_1 such that for any $x \in R^n$ and $t \in [0, \infty)$ the inequality $\Psi(x,t) \leq b_1(|x| + t)^{(\alpha-n)p'}|x|^n$ is fulfilled. Indeed, if $\alpha \geq n$, then this inequality is obvious. Let $\frac{n}{p} < \alpha < n$ and let $t \leq |x|$. Then we have

$$\Psi(x,t) \leq \int\limits_0^\infty |\{y : (|x-y| + t)^{(\alpha-n)p'} > \lambda\} \cap B(0, 3|x|)| d\lambda \leq$$

$$\leq \int\limits_0^{(|x|+t)^{(\alpha-n)p'}} B(0, 3|x|) d\lambda +$$

$$+ \int\limits_{(|x|+t)^{(\alpha-n)p'}}^\infty |\{y : (|x-y| + t)^{(\alpha-n)p'} > \lambda\}| d\lambda \equiv$$

$$\equiv I_1(x,t) + I_2(x,t).$$

For I_1 we have that

$$I_1(x,t) = c_3(|x| + t)^{(\alpha-n)p'}|x|^n.$$

For I_2 we obtain

$$I_2(x,t) \leq \int\limits_{(|x|+t)^{(\alpha-n)p'}}^\infty \left| \left\{ y : |x-y| + t < \lambda^{\frac{1}{(\alpha-n)p'}} \right\} \right| d\lambda \leq$$

$$\leq \int\limits_{(|x|+t)^{(\alpha-n)p'}}^\infty \left| \left\{ y : |x-y| < \lambda^{\frac{1}{(\alpha-n)p'}} \right\} \right| d\lambda =$$

$$= c_4 \int\limits_{(|x|+t)^{(\alpha-n)p'}}^\infty \lambda^{\frac{n}{(\alpha-n)p'}} d\lambda =$$

$$= c_5(|x| + t)^{(\alpha-n)p'+n} \leq c_6(|x| + t)^{(\alpha-n)p'}|x|^n.$$

If $|x| < t$, then

$$\Psi(x,t) \le c_7 t^{(\alpha-n)p'}|x|^n \le c_8(|x|+t)^{(\alpha-n)p'}|x|^n.$$

Finally $\Psi(x,t) \le b_1(|x|+t)^{(\alpha-n)p'}|x|^n$. Further

$$S_2^q \le c_9 \int\limits_{R_+^{n+1}} \left(\int\limits_{\frac{|x|}{2} \le |y| \le 2|x|} (f(y))^p dy \right)^{\frac{q}{p}} \times$$

$$\times (|x|+t)^{(\alpha-n)q}|x|^{nq/p'} d\nu(x,t) =$$

$$= c_9 \sum_{k \in Z} \int\limits_{U_{2^k} \setminus U_{2^{k+1}}} \left(\int\limits_{\frac{|x|}{2} \le |y| \le 2|x|} (f(y))^p dy \right)^{\frac{q}{p}} \times$$

$$\times (|x|+t)^{(\alpha-n)q}|x|^{nq/p'} d\nu(x,t) \le$$

$$\le c_9 \sum_{k \in Z} \left(\int\limits_{2^{k-1} \le |y| \le 2^{k+2}} (f(y))^p dy \right)^{q/p} \times$$

$$\times \int\limits_{U_{2^k} \setminus U_{2^{k+2}}} (|x|+t)^{(\alpha-n)q}|x|^{nq/p'} d\nu(x,t) \le$$

$$\le c_{10} B^q \|f\|_{L^p(R^n)}^q.$$

Now we show that (i) implies (iii). Let $k \in Z$ and $f_k(x) = \chi_{\{y:|y|<2^{k-1}\}}(x)$. Then we have

$$\|K_\alpha f_k\|_{L_\nu^q(R_+^{n+1})} \ge$$

$$\ge c_{11} \left(\int\limits_{U_{2^k} \setminus U_{2^{k+1}}} \left(\int\limits_{|y|<2^{k-1}} (|x-y|+t)^{\alpha-n} dy \right)^q d\nu(x,t) \right)^{1/q} \ge$$

$$\ge c_{12} \left(\int\limits_{U_{2^k} \setminus U_{2^{k+1}}} (|x|+t)^{(\alpha-n)q}|x|^{nq} d\nu(x,t) \right)^{1/q}.$$

On the other hand, $\|f_k\|_{L^p(R^n)} = c_{13} 2^{nk/p}$. Hence the boundedness of K_α implies the condition $B_1 < \infty$.

Let us show that (ii) follows from (iii). Let $r > 0$. Then $r \in [2^m, 2^{m+1})$ for some integer m. We have

$$\left(\int\limits_{U_r} (|x|+t)^{(\alpha-n)q} d\nu(x,t) \right) r^{nq/p'} \le$$

$$\le \left(\int\limits_{U_{2^m}} (|x|+t)^{(\alpha-n)q} d\nu(x,t) \right) 2^{(m+1)nq/p'} =$$

$$= c_{14} 2^{\frac{mnq}{p'}} \sum_{k=m}^{+\infty} \int_{U_{2k} \setminus U_{2k+1}} (|x| + t)^{(\alpha-n)q} d\nu(x,t) \leq$$

$$\leq c_{15} B_1^q 2^{\frac{mnq}{p'}} \sum_{k=m}^{+\infty} 2^{-kqn/p'} \leq c_{16} B_1^q.$$

Consequently $B \leq c_{17} B_1$. Finally (ii) $\Rightarrow$ (i) $\Rightarrow$ (iii) $\Rightarrow$ (ii). $\square$

Theorem 5.1.2. *Let* $0 < q < p < \infty$, $p > 1$ *and* $\alpha > n/p$. *Suppose that* $d\nu(x,t) = v(x,t)dxdt$, *where* v *is a measurable a.e. positive function on* R_+^{n+1}. *Then the operator* K_α *is bounded from* $L^p(R^n)$ *into* $L_v^q(R_+^{n+1})$ *if and only if*

$$D = \left(\int_{R^n} \left(\int_{U_{|x|}} v(y,t)(|y| + t)^{(\alpha-n)q} dy dt \right)^{\frac{p}{p-q}} \right.$$

$$\left. |x|^{\frac{np(q-1)}{p-q}} dx \right)^{\frac{p-q}{pq}} < \infty. \tag{5.1.3}$$

Moreover, $\|K_\alpha\| \approx D$.

Proof. *Sufficiency.* Let $f \geq 0$. As in the proof of Theorem 5.1.1 represent K_α as follows: $K_\alpha f(x,t) = K_\alpha^{(1)} f(x,t) + K_\alpha^{(2)} f(x,t)$. We have

$$\|K_\alpha f\|_{L_v^q(R_+^{n+1})} \leq c_1 \|K_\alpha^{(1)} f\|_{L_v^q(R_+^{n+1})} +$$

$$+ c_1 \|K_\alpha^{(2)} f\|_{L_v^q(R_+^{n+1})} \equiv J_1 + J_2.$$

By virtue of Lemma 5.1.2 for J_1 we find that $J_1 \leq c_2 D \|f\|_{L^p(R^n)}$. By Hölder's inequality we obtain (see also the proof of Theorem 5.1.1)

$$J_2^q \leq c_3 \int_{R_+^{n+1}} v(x,t) \left(\int_{\frac{|x|}{2} \leq |y| \leq 2|x|} (f(y))^p dy \right)^{q/p} \times$$

$$\times (|x| + t)^{(\alpha-n)q} |x|^{nq/p'} dxdt \leq$$

$$\leq c_4 \sum_{k \in Z} \left(\int_{2^{k-1} \leq |y| \leq 2^{k+2}} (f(y))^p dy \right)^{q/p} \times$$

$$\times \int_{U_{2k} \setminus U_{2k+1}} v(x,t)(|x| + t)^{(\alpha-n)q} |x|^{nq/p'} dxdt.$$

Using Hölder's inequality again we find that

$$J_2^q \leq c_4 \left(\sum_{k \in Z} \int_{2^{k-1} \leq |y| \leq 2^{k+2}} (f(y))^p dy \right)^{q/p} \times$$

$$\times \left(\sum_{k \in Z} \left(\int_{U_{2k} \setminus U_{2k+1}} v(x,t)(|x|+t)^{(\alpha-n)q} |x|^{nq/p'} dx dt \right)^{\frac{p}{p-q}} \right)^{\frac{p-q}{p}} \le$$

$$\le c_5 \|f\|_{L^p(R^n)}^q \left(\sum_{k \in Z} J_{2,k} \right)^{(p-q)/p},$$

where

$$J_{2,k} \equiv \left(\int_{U_{2k} \setminus U_{2k+1}} v(x,t)(|x|+t)^{(\alpha-n)q} |x|^{nq/p'} dx dt \right)^{\frac{p}{p-q}}.$$

For $J_{2,k}$ we have

$$J_{2,k} \le c_6 2^{\frac{kn(p-1)q}{p-q}} \left(\int_{U_{2k} \setminus U_{2k+1}} (|x|+t)^{(\alpha-n)q} v(x,t) dx dt \right)^{\frac{p}{p-q}} \le$$

$$\le c_7 \int_{2^{k-1} < |y| < 2^k} |y|^{\frac{knp(q-1)}{p-q}} \left(\int_{U_{|y|}} v(x,t)(|x|+t)^{(\alpha-n)q} dx dt \right)^{\frac{p}{p-q}} dy.$$

Taking into account the last estimate, for J_2 we obtain

$$J_2^q \le c_8 D^q \|f\|_{L^p(R^n)}^q.$$

Necessity. Suppose that K_α is bounded from $L^p(R^n)$ into $L_v^q(R_+^{n+1})$. Then one can easily show that

$$\int_{U_r} v(y,t)(|y|+t)^{(\alpha-n)q} dy dt < \infty$$

for every $r > 0$. Now let $m \in Z$ and

$$f_m(x) = \left(\int_{|y| > |x|} \bar{v}_m(y) dy \right)^{\frac{1}{p-q}} |x|^{\frac{n(q-1)}{p-q}},$$

where

$$\bar{v}_m(y) = \left(\int_0^\infty v(y,t)(|y|+t)^{(\alpha-n)q} dt \right) \chi_{\{\frac{1}{m} < |z| < m\}}(y).$$

From the boundedness of K_α we have that $f_m(x) < \infty$ for every $x \in R^n$. Using spherical coordinates and integration by parts, we obtain

$$\|f_m\|_{L^p(R^n)} = \left(\int_{R^n} \left(\int_{|y| > |x|} \bar{v}_m(y) dy \right)^{\frac{p}{p-q}} |x|^{\frac{np(q-1)}{p-q}} dx \right)^{1/p} =$$

$$= c_9 \left(\int\limits_{R^n} \left(\int\limits_{|y|>|x|} \bar{v}_m(y) dy \right)^{\frac{q}{p-q}} \bar{v}_m(x) |x|^{\frac{nq(p-1)}{p-q}} dx \right)^{1/p} < \infty.$$

On the other hand, again by integration by parts and passing to spherical coordinates, we also find that

$$\|K_\alpha f_m\|_{L_v^q(R_+^{n+1})} \geq$$

$$\geq c_{10} \left(\int\limits_{R_+^{n+1}} \left(\int\limits_{|y|<|x|/2} f_m(y)(|x-y|+t)^{\alpha-n} dy \right)^q v(x,t) dx dt \right)^{1/q} \geq$$

$$\geq c_{11} \left(\int\limits_{R_+^{n+1}} v(x,t) \left(\int\limits_{|y|>|x|} \bar{v}_m(y) dy \right)^{\frac{q}{p-q}} \times \right.$$

$$\left. \times (|x|+t)^{(\alpha-n)q} |x|^{\frac{nq(p-1)}{p-q}} dx dt \right)^{1/q} =$$

$$= c_{11} \left(\int\limits_{R^n} \left(\int\limits_0^\infty v(x,t)(|x|+t)^{(\alpha-n)q} dt \right) \times \right.$$

$$\left. \times \left(\int\limits_{|y|>|x|} \bar{v}_m(y) dy \right)^{\frac{q}{p-q}} |x|^{\frac{nq(p-1)}{p-q}} dx \right)^{1/q} \geq$$

$$\geq c_{11} \left(\int\limits_{R^n} \bar{v}_m(x) \left(\int\limits_{|y|>|x|} \bar{v}_m(y) dy \right)^{\frac{q}{p-q}} |x|^{\frac{nq(p-1)}{p-q}} dx \right)^{1/q} =$$

$$= c_{12} \left(\int\limits_{R^n} \left(\int\limits_{|y|>|x|} \bar{v}_m(y) dy \right)^{\frac{p}{p-q}} |x|^{\frac{np(q-1)}{p-q}} dx \right)^{1/q}.$$

Consequently

$$\left(\int\limits_{R^n} \left(\int\limits_{|y|>|x|} \bar{v}_m(y) dy \right)^{\frac{p}{p-q}} |x|^{\frac{np(q-1)}{p-q}} \right)^{\frac{p-q}{pq}} \leq c_{12}$$

for all $m \in Z$. Finally using Fatou's Lemma we obtain that $D < \infty$. $\square$

From duality arguments we have the following theorems:

Theorem 5.1.3. *Let* $1 < p \leq q < \infty$, $\alpha > \frac{n(q-1)}{q}$. *Then the following conditions are equivalent:*
(i) $\widetilde{K}_\alpha$ *is bounded from* $L_\nu^p(R_+^{n+1})$ *into* $L^q(R^n)$;

(ii)

$$\widetilde{B} \equiv \sup_{r>0} \left(\int_{U_r} (|x| + t)^{(\alpha-n)p'} d\nu(x,t) \right)^{1/p'} r^{n/q} < \infty; \qquad (5.1.4)$$

(iii)

$$\widetilde{B}_1 \equiv \sup_{k\in Z} \left(\int_{U_{2k}\setminus U_{2k+1}} (|x| + t)^{(\alpha-n)p'} |x|^{np'/q} d\nu(x,t) \right)^{1/p'} < \infty. \qquad (5.1.5)$$

Moreover, $\|\widetilde{K}_\alpha\| \approx \widetilde{B} \approx \widetilde{B}_1$.

Theorem 5.1.4. *Let* $1 < q < p < \infty$ *and let* $\alpha > \frac{n(q-1)}{q}$. *Assume that* $d\nu(x,t) = w(x,t)dxdt$, *where* w *is a Lebesgue-measurable a.e. positive function on* R_+^{n+1}. *Then the operator* $\widetilde{K}_\alpha$ *is bounded from* $L_w^p(R_+^{n+1})$ *into* $L^q(R_+^n)$ *if and only if*

$$\widetilde{D} \equiv \left(\int_{R^n} \left(\int_{U_{|x|}} w(y,t)(|y| + t)^{(\alpha-n)p'} dydt \right)^{\frac{q(p-1)}{p-q}} \times \right.$$

$$\left. \times |x|^{\frac{qn}{p-q}} dx \right)^{\frac{p-q}{pq}} < \infty \qquad (5.1.6).$$

Moreover, $\|\widetilde{K}_\alpha\| \approx \widetilde{D}$.

Now we establish weighted criteria for the compactness of the operators K_α and $\widetilde{K}_\alpha$.

Lemma 5.1.3. *Let* $1 < p \le q < \infty$ *and* $\alpha > \frac{n}{p}$. *If* (5.1.1) *holds and* $\lim_{a\to 0} B^{(a)} = \lim_{b\to\infty} B^{(b)} = 0$, *where*

$$B^{(a)} \equiv \sup_{0<r<a} \left(\int_{U_r\setminus U_a} (|x| + t)^{(\alpha-n)q} d\nu(x,t) \right)^{1/q} r^{n/p'},$$

$$B^{(b)} \equiv \sup_{r\ge b} \left(\int_{U_r} (|x| + t)^{(\alpha-n)q} d\nu(x,t) \right)^{1/q} (r^n - b^n)^{1/p'},$$

then the operator K_α *is compact from* $L^p(R^n)$ *into* $L_\nu^q(R_+^{n+1})$.

Proof. Using the notation:

$$P_1 f(x) \equiv \chi_{V_a} K_\alpha(\chi_{B(0,2a)} f),$$

$$P_2 f(x) \equiv \chi_{V_b\setminus V_a} K_\alpha(\chi_{B(0,4b)} f),$$

$$P_3 f(x) \equiv \chi_{U_b} K_\alpha(\chi_{B(0,\frac{b}{2})} f),$$

$$P_4 f(x) \equiv \chi_{U_b} K_\alpha(\chi_{R^n \setminus B(0,\frac{b}{2})} f),$$

where $V_r \equiv B(0,r) \times [0,\infty)$ ($r > 0$) and $0 < a < b < \infty$, we obtain

$$K_\alpha f(x,t) = P_1 f(x,t) + P_2 f(x,t) + P_3 f(x,t) + P_4 f(x,t).$$

Let

$$\bar{k}(x,t,y) = \chi_{V_b \setminus V_a}(x,t)\chi_{\{|z| \leq 2|x|\}}(y)(|x-y|+t)^{\alpha-n}.$$

Then by the inequality

$$\int_{|y| \leq 2|x|} (|x-y|+t)^{(\alpha-n)p'} dy \leq b(|x|+t)^{(\alpha-n)p'}|x|^n,$$

with a positive constant b independent of $x \in R^n$ and $t \geq 0$, we have

$$\| \|\bar{k}(x,t,y)\|_{L^{p'}(R^n)}\|_{L^q_\nu(R^{n+1}_+)} =$$

$$= \left(\int_{V_b \setminus V_a} \left(\int_{|y| \leq 2|x|} (|x-y|+t)^{(\alpha-n)p'} dy \right)^{\frac{q}{p'}} d\nu(x,t) \right)^{1/q} \leq$$

$$\leq c_1 \left(\int_{V_b \setminus V_a} (|x|+t)^{(\alpha-n)q}|x|^{q/p'} d\nu(x,t) \right)^{1/q} < \infty.$$

Consequently due to Theorem C of Section 2.1 we conclude that P_2 is compact. For P_3 we have

$$P_3 f(x,t) = \int_{R^n} \tilde{k}(x,y,t) f(y) dy,$$

where

$$\tilde{k}(x,t,y) = \chi_{U_b}(x,t)\chi_{B(0,\frac{b}{2})}(y)(|x-y|+t)^{\alpha-n}.$$

Moreover

$$\| \|\tilde{k}(x,t,y)\|_{L^{p'}(R^n)}\|_{L^q_\nu(R^{n+1}_+)} < \infty.$$

Using again Theorem A we have that P_3 is compact.

By Theorem 5.1.1 (restricted to the sets $B(0,a)$, $R^n \setminus B(0,b/2)$) we get the estimates:

$$\|P_1\| \leq b_1 B^{(a)} < \infty, \quad \|P_4\| \leq b_2 B^{(b/2)} < \infty. \tag{5.1.1}$$

Therefore

$$\|K_\alpha - P_2 - P_3\| \leq \|P_1\| + \|P_2\| \leq b_3(B^{(a)} + B^{(b/2)}) \to 0 \tag{5.1.2}$$

as $a \to 0$ and $b \to +\infty$. Hence K_α is compact as it is a limit of compact operators. $\square$

Theorem 5.1.5. *Let* $1 < p \le q < \infty$ *and* $\alpha > n/p$. *Then the following conditions are equivalent:*

(i) K_α *is compact from* $L^p(R^n)$ *into* $L^q_\nu(R^{n+1}_+)$;

(ii) (5.1.5) *holds and* $\lim\limits_{a \to 0} B^{(a)} = \lim\limits_{b \to \infty} B^{(b)} = 0$;

(iii) (5.1.5) *holds and* $\lim\limits_{r \to 0} B(r) = \lim\limits_{r \to \infty} B(r) = 0$, *where*

$$B(r) = \left(\int\limits_{U_r} (|x| + t)^{(\alpha-n)q} d\nu(x,t) \right)^{1/q} r^{n/p'};$$

(iv) (5.1.2) *is satisfied and* $\lim\limits_{k \to -\infty} B_1(k) = \lim\limits_{k \to +\infty} B_1(k) = 0$, *where*

$$B_1(k) = \left(\int\limits_{U_{2^k} \setminus U_{2^{k+1}}} (|x| + t)^{(\alpha-n)q} |x|^{nq/p'} d\nu(x,t) \right)^{1/q}.$$

Proof. The fact that (ii) $\Rightarrow$ (i) follows from Lemma 5.1.3. Using the obvious inequalities:

$$B^{(a)} \le \sup_{0 < r < a} B(r)$$

and

$$B^{(b)} \le \sup_{r \ge b} B(r)$$

we conclude that (iii) $\Rightarrow$ (ii). Now we show that (i) $\Rightarrow$ (iii). Let $r > 0$ and let $f_r(x) = \chi_{B(0,r/2)}(x) r^{-n/p}$. Then it is easy to verify that f_r is weakly convergent to 0 as $r \to 0$. On the other hand,

$$\|K_\alpha f_r\|_{L^q_\nu(R^{n+1}_+)} \ge$$

$$\ge \left(\int\limits_{U_r} \left(\int\limits_{|y| < r/2} (|x - y| + t)^{\alpha-n} f(y) dy \right)^q d\nu(x,t) \right)^{1/q} \ge c_1 B(r) \to 0$$

as $r \to 0$, since $K_\alpha f_r$ is strongly convergent to 0. Now let

$$g_r(x,t) = \chi_{U_r}(x,t)(|x| + t)^{(\alpha-n)(q-1)} \left(\int\limits_{U_r} (|y| + t)^{(\alpha-n)q} d\nu(y,t) \right)^{-1/q'}.$$

The sequence g_r is weakly convergent to 0 as $r \to +\infty$. Using the fact that the compactness of K_α from $L^p(R^n)$ into $L^q_\nu(R^{n+1}_+)$ is equivalent to the compactness of $\widetilde{K}_\alpha$ from $L^q_\nu(R^{n+1}_+)$ into $L^{p'}(R^n)$ and taking into account the inequality

$$\|\widetilde{K}_\alpha g_r\|_{L^{p'}(R^n)} \geq B(r),$$

we obtain that $B(r) \to 0$ as $r \to +\infty$. Finally we have that (i) $\Rightarrow$ (iii). To prove that (iv) implies (ii) we take $a > 0$. Then $a \in [2^m, 2^{m+1})$ for some $m \in Z$. Consequently $B^{(a)} \leq \sup\limits_{0<r<2^m} B_{2^m,r} \equiv B^{(2^m)}$, where

$$B_{2^m,r} \equiv \left(\int\limits_{U_r \backslash U_{2^m}} (|x| + t)^{(\alpha-n)q} d\nu(x,t) \right)^{1/q} r^{n/p'}.$$

For $r \in [0, 2^m)$ there exists $j \in Z, j \leq m$, such that $r \in [2^{j-1}, 2^j)$. Hence

$$B^q_{2^m,r} \leq 2^{jnq/p'} \sum_{k=j}^{m} \int\limits_{U_{2^{k-1}} \backslash U_{2^k}} (|x| + t)^{(\alpha-n)q} d\nu(x,t) \leq$$

$$\leq c_3 \left(\sup_{k \leq m} B_1(k-1) \right)^q.$$

From the last estimate we fined that $B^{(2^m)} \leq c_4 S_m$, where

$$S_m \equiv \sup_{k \leq m} B_1(k-1).$$

If $a \to 0$, then $m \to -\infty$. Consequently $S_m \to 0$ as $\lim\limits_{j \to -\infty} B_1(j) = 0$. Therefore $\lim\limits_{a \to 0} B^{(a)} = 0$.

Now let $s > 0$. Then $s \in [2^m, 2^{m+1})$ for some integer m. We have

$$B^q(s) \leq c_5 B^q(2^m) = c_5 2^{\frac{mq}{p'}} \sum_{k=m}^{+\infty} \int\limits_{U_{2^k} \backslash U_{2^{k+1}}} (|x| + t)^{(\alpha-n)q} d\nu(x,t) \leq$$

$$\leq c_6 \left(\sup_{k \geq m} B_1(k) \right)^q.$$

Hence

$$\lim_{s \to +\infty} B(s) \leq c_7 \lim_{m \to +\infty} \left(\sup_{k \geq m} B_1(k) \right) = 0.$$

This means that $\lim\limits_{b \to +\infty} B^{(b)} = 0$. Finally we have that (iv) $\Rightarrow$ (ii). It we take the sequence $f_k(x) = \chi_{B(0,2^{k-1}) \backslash B(0,2^{k-2})}(x) 2^{-kn/p}$, then it is easy to show that f_k is weakly convergent to 0 as $k \to -\infty$ or $k \to +\infty$. Taking into account the inequality

$$\|K_\alpha f_k\|_{L^q_\nu(R^{n+1}_+)} \geq c_8 B_1(k)$$

and using the compactness of K_α from $L^p(R^n)$ into $L^q_\nu(R^{n+1}_+)$, we see that (iv) is valid. So we have that (i) implies (iv). Finally, (iii) $\Leftrightarrow$ (i) and (i) $\Rightarrow$ (iv) $\Rightarrow$ (ii) $\Rightarrow$ (i). $\square$

For the case $q < p$ we have the following

Theorem 5.1.6. *Let* $0 < q < p < \infty$, $p > 1$ *and* $\alpha > n/p$. *Assume that* $d\nu(x,t) = v(x,t)dxdt$, *where* v *is a Lebesgue measurable a.e. positive function on* R^{n+1}_+. *Then* K_α *is compact from* $L^p(R^n)$ *into* $L^q_v(R^{n+1}_+)$ *if and only if* (5.1.3) *is fulfilled.*

This theorem is proved in the same manner as Theorem 5.1.5. It also can be derived from Theorem D of Section 2.1.

From duality arguments we obtain the following results:

Theorem 5.1.7. *Let* $1 < p \leq q < \infty$ *and let* $\alpha > \frac{n(q-1)}{q}$. *Then the following statements are equivalent:*

(i) $\widetilde{K}_\alpha$ *is compact from* $L^p_\nu(R^{n+1}_+)$ *into* $L^q(R^n)$;

(ii) (5.1.4) *is fulfilled and* $\lim_{a \to 0} \widetilde{B}^{(a)} = \lim_{b \to \infty} \widetilde{B}^{(b)} = 0$, *where*

$$\widetilde{B}^{(a)} \equiv \sup_{0 < r < a} \left(\int_{U_r \setminus U_a} (|x| + t)^{(\alpha-n)p'} d\nu(x,t) \right)^{1/p'} r^{n/q};$$

$$\widetilde{B}^{(b)} \equiv \sup_{r > b} \left(\int_{U_r} (|x| + t)^{(\alpha-n)p'} d\nu(x,t) \right)^{1/p'} (r^n - b^n)^{1/q};$$

(iii) (5.1.4) *holds and* $\lim_{r \to 0} \widetilde{B}(r) = \lim_{r \to +\infty} \widetilde{B}(r) = 0$, *where*

$$\widetilde{B}(r) \equiv \left(\int_{U_r} (|x| + t)^{(\alpha-n)p'} d\nu(x,t) \right)^{1/p'} r^{n/q};$$

(iv) (5.1.5) *holds and* $\lim_{k \to -\infty} \widetilde{B}_1(k) = \lim_{k \to +\infty} \widetilde{B}_1(k) = 0$, *where*

$$\widetilde{B}_1(k) \equiv \left(\int_{U_{2^k} \setminus U_{2^{k+1}}} (|x| + t)^{(\alpha-n)p'} |x|^{p'n/q} d\nu(x,t) \right)^{1/p'}.$$

Theorem 5.1.8. *Let* $1 < q < p < \infty$ *and* $\alpha > n/q'$. *Assume that* $d\nu(x,t) = w(x,t)dxdt$, *where* w *is a Lebesgue-measurable a.e. positive function on* R^{n+1}_+. *Then* $\widetilde{K}_\alpha$ *is compact from* $L^p_w(R^{n+1}_+)$ *into* $L^q(R^n)$ *if and only if* $\widetilde{D} < \infty$.

Now we are going to estimate the distance of K_α from the space of compact operators. We shall assume that v is a Lebesgue-measurable a.e. positive function on R_+^{n+1}.

We need the following lemma which can be found in [66] (Corollary 5.4, Chapter V).

Lemma 5.1.4. *Let* $1 \le q < \infty$ *and let* $P \in \mathcal{B}(X, Y)$, *where* $Y = L^q(R_+^{n+1})$. *Then*

$$\text{dist}\,(P, \mathcal{K}(X, Y)) = \text{dist}\,(P, \mathcal{F}_r(X, Y)).$$

The next lemma can be proved in the same way as Lemma V.5.6. in [66] (see also the proof of Lemma 4.3.2).

Lemma 5.1.5. *Let* $1 \le q < \infty$ *and* $Y = L^q(R_+^{n+1})$. *Suppose that* $P \in \mathcal{F}_r(X, Y)$ *and* $\varepsilon > 0$. *Then there exist* $T \in \mathcal{F}_r(X, Y)$ *and* $[\alpha, \beta] \subset (0, \infty)$ *such that* $\|P - T\| < \varepsilon$ *and* $\text{supp} Tf \subset \{\alpha \le |x| \le \beta\} \times [0, \infty)$ *for any* $f \in X$.

The following lemma is obvious:

Lemma 5.1.6. *Let* $1 \le q < \infty$. *Then* $\tilde{I} = \bar{I}$, *where*

$$\tilde{I} \equiv \text{dist}\,(K_\alpha, \mathcal{K}(X, L_v^q(R_+^{n+1}))),$$
$$\bar{I} \equiv \text{dist}\,(K_\alpha', \mathcal{K}(X, L^q(R_+^{n+1})))$$

and

$$K_\alpha' f(x, t) = v^{1/q}(x, t) K_\alpha f(x, t).$$

Theorem 5.1.9. *Let* $1 < p \le q < \infty$ *and* $\alpha > n/p$. *Assume that* $X = L^p(R^n)$ *and* $Y = L_v^q(R_+^{n+1})$. *Also assume that* $B < \infty$ *for* $d\nu(x, t) = v(x, t)dxdt$. *Then there exist positive constants* b_1 *and* b_2, *depending only on* p, q *and* α, *such that*

$$b_1 J \le \text{dist}\,(K_\alpha, \mathcal{K}(X, Y)) \le b_2 J,$$

where

$$J \equiv \lim_{a \to 0} J^{(a)} + \lim_{b \to \infty} J^{(b)},$$

$$J^{(a)} \equiv \sup_{0 < r < a} \left(\int_{U_r \backslash U_a} v(x, t)(|x| + t)^{(\alpha - n)q} dxdt \right)^{1/q} r^{n/p'},$$

$$J^{(b)} \equiv \sup_{r > b} \left(\int_{U_r} v(x, t)(|x| + t)^{(\alpha - n)q} dxdt \right)^{1/q} (r^n - b^n)^{1/p'}.$$

Proof. From (5.1.1) and (5.1.2) and from the arguments of the proof of Lemma 5.1.3 we find that $\tilde{I} \equiv \text{dist}(K_\alpha, \mathcal{K}(X, Y)) \leq b_2 J$. Now let $\lambda > \tilde{I}$. Using Lemma 5.1.6 we have $\lambda > \tilde{I}$. By Lemma 5.1.4 we can find an operator $P : X \to L^q(R_+^{n+1})$ of finite rank such that $\|K_\alpha' - P\| < \lambda$. From Lemma 5.1.5 we conclude that for $\epsilon = (\lambda - \|K_\alpha' - P\|)/2$ there exist $T \in \mathcal{F}_r(X, L^q(R_+^{n+1}))$ and positive numbers α, β, $0 < \alpha < \beta < \infty$, such that $\|P - T\| < \varepsilon$ and $\text{supp} Tf \subset \{\alpha \leq |x| \leq \beta\} \times [0, \infty)$ for all $f \in X$. Hence for all $f \in X$ we have

$$\|K_\alpha' - Tf\| \leq \lambda \|f\|_X.$$

From the last inequality it follows that

$$\int\limits_{\{|x| \leq \alpha\} \times [0,\infty)} |K_\alpha' f(x,t)|^q dx dt + \int\limits_{U_\beta} |K_\alpha' f(x,t)|^q dx dt \leq$$

$$\leq \lambda^q \|f\|_{L^p(R^n)}^q.$$

If $b > \beta$, $r > b$ and $f_r(y) = \chi_{\{b/2 < |x| < r/2\}}(y)$, then $\|f_r\|_{L^p(R^n)}^q = c_1(r^n - b^n)^{q/p}$. On the other hand,

$$\int\limits_{U_r} |K_\alpha' f_r(x,t)|^q dx dt \geq$$

$$\geq \int\limits_{U_r} \left(\int\limits_{B(0,\frac{r}{2}) \setminus B(0,\frac{b}{2})} \frac{1}{(|x-y|+t)^{n-\alpha}} dy \right)^q v(x,t) dx dt \geq$$

$$\geq c_2 \left(\int\limits_{U_r} v(x,t)(|x|+t)^{(\alpha-n)q} dx dt \right) (r^n - b^n)^q.$$

Consequently

$$c_3 \sup_{r>b} \left(\int\limits_{U_r} v(x,t)(|x|+t)^{(\alpha-n)q} dx dt \right)^{1/q} (r^n - b^n)^{1/p'} \leq \lambda.$$

Therefore $c_3 J^{(b)} \leq \lambda$ for $b > \beta$. Finally we find that $c_3 \lim\limits_{b \to \infty} J^{(b)} \leq \lambda$ and since λ is arbitrarily close to $\tilde{I}$ we conclude that $c_3 \lim\limits_{b \to \infty} J^{(b)} \leq \tilde{I}$. Similarly we obtain $c_3 \lim\limits_{a \to 0} J^{(a)} \leq \tilde{I}$. Finally we have $b_1 J \leq \tilde{I}$. $\square$

Remark 5.1.1. Let p, q and α satisfy the conditions of Theorem 5.1.9. Assume that $X = L^p(R^n)$ and $Y = L_v^q(R_+^{n+1})$. It is also assumed that K_α is bounded from $L^p(R^n)$ into $L_v^q(R_+^{n+1})$. Then there exist positive constants $\tilde{b}_1$ and $\tilde{b}_2$

depending only on p, q and α such that the following inequality holds

$$\tilde{b}_1 I \leq \mathrm{dist}\,(K_\alpha, \mathcal{K}(X, Y)) \leq \tilde{b}_2 I,$$

where

$$I = \varlimsup_{k \to -\infty} I(k) + \varlimsup_{k \to +\infty} I(k)$$

and

$$I(k) \equiv \left(\int\limits_{U_{2k} \setminus U_{2k+1}} (|x| + t)^{(\alpha - n)q} |x|^{nq/p'} v(x, t) dx dt \right)^{1/q}.$$

An analogous results for the generalized Riemann-Liouville operator were obtained in Section 2.9.

5.2. Two-weight compactness conditions for Riesz potentials

In this section we give a characterization of pairs of weights which guarantee the boundedness and compactness of Riesz potentials without using the capacity notion.

Let Ω be a domain $\Omega \subseteq R^n$ and let

$$(T_{\gamma, \Omega} f)(x) = \int\limits_{\Omega} \frac{f(y)}{|x - y|^{n - \gamma}} \, dy, \quad 0 < \gamma < n.$$

First we formulate the main results and then prove some of them. The following two theorems are well-known (see for example [100], Chapter 3):

Theorem 5.2.1. *Let* $1 < p < q < \infty$. *The operator* $T_{\gamma, \Omega}$ *is bounded from* $L^p_w(\Omega)$ *into* $L^{q\infty}_v(\Omega)$ *if and only if*

$$\sup_{x \in \Omega,\, r > 0} A_1(x, r, \Omega) < \infty, \tag{5.2.1}$$

where

$$A_1(x, r, \Omega) = \left(\int\limits_{B(x, 2r) \cap \Omega} v(y) dy \right)^{1/q} \left(\int\limits_{\Omega \setminus B(x, r)} \frac{w^{1 - p'}(y)}{|x - y|^{(n - \gamma)p'}} dy \right)^{1/p'},$$

$B(x, r)$ *is a ball in* R^n *with centre* x *and radius* r.

Theorem 5.2.2. *Let* $1 < p < q < \infty$. *The operator* $T_{\gamma, \Omega}$ *is bounded from* $L^p_w(\Omega)$ *into* $L^q_v(\Omega)$ *if and only if* (5.2.1) *and the condition*

$$\sup_{x \in \Omega,\, r > 0} A_2(x, r, \Omega) < \infty \tag{5.2.2}$$

are fulfilled, where

$$A_2(x, r, \Omega) = \Bigg(\int\limits_{B(x,2r) \cap \Omega} w^{1-p'}(y) dy \Bigg)^{1/p'} \Bigg(\int\limits_{\Omega \backslash B(x,r)} \frac{v(y)}{|x - y|^{(n-\gamma)q}} dy \Bigg)^{1/q},$$

Now we present results concerning the boundedness of the operator $T_{\gamma,\Omega}$.

Theorem 5.2.3. *Let* $1 < p < q < \infty$. *Suppose, that* Ω *is a bounded domain. The operator* $T_{\gamma,\Omega}$ *is compact from* $L_w^p(\Omega)$ *into* $L_v^{q\infty}(\Omega)$ *if the conditions*

$$\sup_{x \in \Omega} \sup_{r>0} A_1(x, r, \Omega) < \infty, \qquad (5.2.3)$$

and

$$\lim_{s \to 0} \sup_{x \in \Omega} \sup_{0<r<s} A_1(x, r, \Omega) = 0 \qquad (5.2.4)$$

are fulfilled.

Theorem 5.2.4. *Let* $1 < p < q < \infty$, *and let* Ω *be a bounded domain. The operator* $T_{\gamma,\Omega}$ *is compact from* $L_w^p(\Omega)$ *into* $L_v^q(\Omega)$ *if the conditions:* (5.2.3), (5.2.4),

$$\sup_{x \in \Omega} \sup_{r>0} A_2(x, r, \Omega) < \infty,$$

and

$$\lim_{s \to 0} \sup_{x \in \Omega} \sup_{0<r<s} A_2(x, r, \Omega) = 0$$

are satisfied.

Theorem 5.2.5. *Let* $1 < p < q < \infty$, $\Omega = R^n$. *The operator* $T_{\gamma,\Omega}$ *is compact from* $L_w^p(R^n)$ *into* $L_v^q(R^n)$ *if it is bounded from* $L_w^p(R^n)$ *into* $L_v^q(R^n)$ *and the following conditions are fulfilled:*

$$\lim_{s \to 0} \sup_{0<r<s} A_i(x, r, R^n) = 0 \quad (i = 1, 2),$$

and

$$\lim_{|x| \to \infty} \sup_{0<r<1} A_i(x, r, R^n) = 0 \quad (i = 1, 2).$$

We prove Theorem 5.2.4. Let Ω be a bounded domain and consider the representation:

$$T_\gamma f(x) = \int\limits_\Omega k^t(|x - y|) f(y) \, dy + \int\limits_\Omega k_t(|x - y|) f(y) \, dy,$$

where

$$k^t(u) = \chi_{v_t}(u) |u|^{\gamma-n}, \quad v_t = \{z \in R^n : |z| < t\}$$

$$k_t(u) = \chi_{R^n \setminus v_t}(u) |u|^{\gamma-n}.$$

Denote the corresponding operators by T^t and T_t. To prove the theorem, it is sufficient to show that

$$\lim_{t\to 0} \left\| T^t \right\|_{L_w^p \to L_v^q} = 0$$

and T_t is compact from $L_w^p(\Omega)$ into $L_v^q(\Omega)$ for any $t > 0$.

For given $\varepsilon > 0$ let us take $\delta > 0$ such that

$$\sup_{x\in\Omega} \sup_{0<r<s} (A_1(x,r,\Omega) + A_2(x,r,\Omega)) < \varepsilon$$

when $0 < 2s < \delta$.

Following [2] (p.196) let us consider a partition of R^n into cubes $\{Q\}$ with disjoint interiors, each of side length t for any t such that $3\sqrt{n}t < \delta$. For $f_Q = f\chi_{Q\cap\Omega}$ it is clear that $supp\,(T^t f_Q) \subset 3Q \cap \Omega$, where $3Q$ denotes the cube concentric to Q with side $3t$. We have

$$\left| T^t f(x) \right|^q = \left| \sum_Q T^t f_Q(x) \right|^q \le 3^{n(q-1)} \sum_Q \left| T^t f_Q(x) \right|^q.$$

Therefore

$$\int_\Omega \left| T^t f(x) \right|^q v(x)\,dx \le 3^{n(q-1)} \sum_Q \int_{3Q\cap\Omega} \left| T^t f_Q(x) \right|^q v(x)\,dx.$$

According to Theorem 1 from [30] and Theorem 5.2.2 we have

$$\int_{3Q\cap\Omega} \left| T^t f_Q(x) \right|^q v(x)\,dx \le c\varepsilon \|f\|_{L_w^p(\Omega)}^q.$$

As $q > 1$ we find that

$$\int_\Omega \left| T^t f(x) \right|^q v(x)\,dx \le 3^{n(q-1)} c\varepsilon \|f\|_{L_w^p(\Omega)}^q.$$

From the last inequality we conclude that $\lim_{t\to 0} \|T^t\| = 0$, where $\|T^t\|$ is the operator norm from $L_w^p(\Omega)$ into $L_v^q(\Omega)$.

The compactness of T_t follows from the condition (see Theorem C, Section 2.1.)

$$\int_\Omega \left(\int_\Omega k_t^{p'}(|x-y|)\, w^{1-p'}(y) \right)^{q/p'} v(x)\,dx < \infty$$

which is satisfied according to (5.2.2).

5.3.   Integral transforms with radial kernels

We consider the integral transform

$$K f(x) = \int\limits_{R^n} k(x - y) f(y) dy$$

when the kernel k is radial i.e. $k(x - y) = k(|x - y|)$ and Ω is a bounded domain in R^n. Let $d = \text{diam } \Omega$.

Theorem 5.3.1. *Let $0 \leq \alpha < n$ and let k be a radial decreasing function on $(0, d)$ such that*

$$\int\limits_0^d k(t) \, t^{n-\alpha-1} dt < \infty. \tag{5.3.1}$$

Then K is compact from $L^p(\Omega)$ to $L^q(\Omega)$ if $1 < p < \frac{n}{\alpha}$, $q = \frac{np}{n-\alpha p}$.

Proof. Let for $x \in \Omega$

$$K f(x) = K_r' f(x) + K_r{}'' f(x), \quad r \in (0, d),$$

where

$$K_r' f(x) = \int\limits_\Omega f(y) k(|x - y|) \chi_r(|x - y|) dy,$$

$$K_r'' f(x) = \int\limits_\Omega f(y) k(|x - y|)(1 - \chi_r(|x - y|)) dy$$

and $\chi_r(t) = 1$, when $|t| < r$, and $\chi_r(t) = 0$, when $|t| > r$.

For a bounded domain Ω and any ball $B(x, r)$ we use the notation $D(x, r) \equiv B(x, r) \cap \Omega$, where $x \in \Omega$.

Let us estimate $K_r' f(x)$ as follows:

$$|K_r' f(x)| \leq \sum_{k=0}^\infty \int\limits_{D(x,2^{-k}r) \setminus D(2^{-k-1}r)} k(|x - y|) |f(y)| dy \leq$$

$$\leq \sum_{k=0}^\infty k(2^{-k-1}r) \int\limits_{D(x,2^{-k}r)} |f(y)| dy =$$

$$= \sum_{k=0}^\infty k(2^{-k-1}r) \frac{(2^{-k}r)^{n-\alpha}}{(2^{-k}r)^{n-\alpha}} \int\limits_{D(x,2^{-k}r)} |f(y)| dy \leq$$

$$\leq \sum_{k=0}^\infty k(2^{-k-1}r)(2^{-k}r)^{n-\alpha} \cdot M_\alpha f(x),$$

where M_α is the fractional maximal function:

$$M_\alpha f(x) = \sup_{x \in \Omega, \, r > 0} \frac{1}{|D(x,r)|} \int_{D(x,r)} |f(y)| dy.$$

On the other hand,

$$\sum_{k=0}^\infty k(2^{-k-1}r)(2^{-k}r)^{n-\alpha} \leq c \sum_{k=0}^\infty \int_{2^{-(k+2)}r} k(t) t^{n-\alpha-1} dt =$$

$$= c \int_0^{\frac{r}{2}} k(t) t^{n-\alpha-1} dt \cdot M_\alpha f(x).$$

Hence for $x \in \Omega$ we have

$$|K'_r f(x)| \leq c M_\alpha f(x) \int_0^{\frac{r}{2}} k(t) t^{n-\alpha-1} dt. \qquad (5.3.2)$$

According to the boundedness of operator M_α from $L^p(\Omega)$ to $L^q(\Omega)$ with $q = \frac{np}{n-\alpha p}$, we conclude that

$$\|K'_r f\|_{L^q} \leq c \|f\|_{L^p} \cdot \int_0^{\frac{r}{2}} k(t) t^{n-\alpha-1} dt. \qquad (5.3.3)$$

Due to condition (5.3.1), from (5.3.3) we derive that

$$\lim_{r \to 0} \|K - K''_r\|_{L_p \to L_q} = 0. \qquad (5.3.4)$$

Together with this, since

$$\int_\Omega \left(\int_{\Omega \setminus D(x,r)} k^{p'}(|x-y|) dy \right)^{\frac{q}{p'}} dx \leq k^q(r) |\Omega|^{1 + \frac{q}{p'}},$$

the operators K''_r are compact from L^p to L^q for any $r \in (0,d)$ (see Theorem C of Section 2.1).

Finally from (5.3.4) we conclude that K acts compactly from $L^p(\Omega)$ into $L^q(\Omega)$. $\square$

Corollary 5.3.1. *The Riesz potential*

$$I_\gamma f(x) = \int_{R^n} \frac{f(y)}{|x-y|^{n-\gamma}}, \quad 0 < \gamma < n$$

acts compactly from $L^p(\Omega)$ into $L^q(\Omega)$ with $q < \dfrac{np}{n - \gamma p}$.

Corollary 5.3.2. *The integral transform*

$$T f(x) = \int_a^b \frac{f(y)dy}{|x - y|^{1-\gamma}\left(\ln \frac{1}{|x-y|}\right)^{1+\varepsilon}}, \quad \varepsilon > 0, \quad b - a < 1$$

is compact from $L^p[a, b]$ to $L^q[a, b]$ with $q = \dfrac{p}{1 - \gamma p}$.

Below we discuss the two-weight sufficient conditions for compactness of K assuming that the condition (5.3.1) is satisfied. The sources of such statements are the inequality (5.3.2) and well-known two-weighted boundedness conditions for fractional maximal functions (see [100], Chapter 4).

In the sequel, as above, Ω denotes an arbitrary bounded domain in R^n. The following statements hold:

Theorem 5.3.2. *Let $1 < p < q < \infty$ and let weights v and w satisfy the condition*

$$\sup_{x \in R^n, \, r > 0} (w^{1-p'} D(x, 2r))^{\frac{1}{p'}} \left(\int_{\Omega \backslash D(x,r)} \frac{v(y)}{|x - y|^{(n-\alpha)q}} dy \right)^{1/q} < \infty.$$

Then K acts compactly from $L_w^p(\Omega)$ into $L_v^q(\Omega)$.

Theorem 5.3.3. *Let $1 < p < q < \infty$ and let the measure $w^{1-p'}$ satisfy a reverse doubling condition: there exist constants $\eta_1 > 1$ and $\eta_2 > 1$ such that*

$$w^{1-p'}(B(x, \eta_1 r)) \geq \eta_2 w^{1-p'}(B(x,r))$$

for all $x \in \Omega$ and $r > 0$.
Assume that

$$\sup_{x \in R, \, r > 0} |D(x,r)|^{\frac{\alpha}{n}-1}(vD(x,r))^{\frac{1}{q}}(w^{1-p'}D(x,r))^{\frac{1}{p'}} < \infty,$$

where α is as in condition (5.3.1). Then K is compact from $L_w^p(\Omega)$ to $L_v^q(\Omega)$.

Theorem 5.3.4. *Let $1 < p < \infty$. If for some $s > 1$ the pair of weights (v, w) satisfies the condition*

$$\sup x \in R^n, \, r > 0 |D(x,r)|^{\frac{\alpha}{n}-\frac{1}{p}-\frac{1}{p's}}(vD(x,r))^{\frac{1}{p}}(w^{(1-p')s}D(x,r))^{\frac{1}{p's}} < \infty,$$

then the operator K acts compactly from $L_w^p(\Omega)$ into $L_v^p(\Omega)$.

Remark 5.3.1. The condition (5.3.1) is optimal. For the kernel $K(t) = t^{-n}$ this condition fails and the corresponding operator is not compact.

5.4. Notes and comments on Chapter 5

There is an extensive literature on the operators investigated in this chapter. We mention in particular [275], [195], [1], [33], [2] et al.

Weighted L^p-boundedness criteria for truncated potentials are due to E.Sawyer [258]. We present here a different proof in a more general case. For the historical point of view we mention the two-weighted estimates for fractional integrals in case of power weights obtained in [279] and the sufficient conditions for polynomial weighted inequalities from [288].

The solution of one–weighted problems for fractional integrals is due to [214].The pioneering result concerning the solution of two-weight weak type problem for Riesz potentials is in [263]. More transparent, easy to verify necessary and sufficient conditions for two-weighted weak $L^p - L^q$ ($1 < p < q < \infty$) inequality were found in [88],and then developed in [91]. The solution of strong- type problems when $1 < p < q < \infty$ is in [265] and [100]. For general kernels and structures we refer to [99], [266] and [100].

Some sufficient conditions for the validity of two-weight inequalities were established in [231], [240].

The paper [53] gives necessary and sufficient conditions for the boundedness of the fractional integral operator acting from one weighted Lorentz space to another, the underlying setting being that of a subset of R^n. The conditions are expressed in terms of the rearrangements of the weight functions.

A complete characterization of norm inequalities between Riesz potential $I_\alpha f$ and f in Orlicz space is establish in [51]. For optimal Orlicz -space version of Hedberg's [122] inequality, which allows one to reduce certain problems for I_α to analogous problems for maximal functions, we refer [53].

The cases $p = q$ and $q < p$ have been investigated in [197], [302],

Two-weight weak type problems for generalized potentials was solved in [90]. For the compactness criteria for potential operators in terms of capacities we refer to [2].

It is well-known that Sobolev's embeddings $W_p^s(G) \subset L_q(G)$, $s \in N$, $1 \le p < q < \infty$, for the domain G with regular boundaries are compact for $s - n/p + n/q > 0$. Necessary and sufficient conditions for compact embeddings when $s = 1$ for "bad" domains were established in [195]. These conditions were formulated in terms of capacity and isoperimetric inequalities. Sufficient conditions of embedding compactness in the case $s \ge 2$ were also obtained. However, in [31] compact embedding conditions of weighted and unweighted Sobolev spaces in the domains with regular and non-regular boundaries are presented in very simple geometric terms. More recently in [32] the Sobolev embeddings and corresponding weak type inequalities were established revealing the influence of geometric properties of the domain on the smoothness and summability parameters. The significant impact on the latter results of two-weighted theory for integral transforms with positive kernels developed in [88], [91] should be noted.

Finally we recall the surveys [145] and [157].

This chapter partly relies on the our papers [159], [161], [72].

Chapter 6

FRACTIONAL INTEGRALS
ON MEASURE SPACES

In this chapter we present results concerning the boundedness and compactness of integral transforms generated by various types of fractional integrals.

We give a complete description of those measure spaces with quasi–metrics on which the defined potential-type transforms are bounded in Lebesgue spaces. Sobolev– and Adams–type theorems are established. Boundedness (compactness) conditions in weighted spaces are derived for truncated potentials defined on spaces of nonhomogeneous type.

6.1. Integral transforms on nonhomogeneous spaces

Let (X, d, μ) be a nonhomogeneous space, i.e. a topological space endowed with a locally finite complete measure μ and quasi–metric $d : X \times X \to R^1$ satisfying only the conditions i) - vi) from Definition 1.1.1. We omit the requirement that the measure μ should satisfy the doubling condition. In addition, we shall assume that $\mu(X) = \infty$. First we consider the integral operator

$$\mathcal{K}_0 f(x) = \int\limits_X (d(x,y))^{\gamma-1} f(y) d\mu, \quad 0 < \gamma < 1.$$

Theorem 6.1.1. *Let* $1 < p < q < \infty$ *and let* $0 < \gamma < 1$. *The operator* $\mathcal{K}_0$ *acts boundedly from* $L^p_\mu(X)$ *into* $L^q_\mu(X)$ *if only if there exists a constant* $c > 0$ *such that*

$$\mu B(x,r) \leq cr^\beta, \quad \beta = \frac{pq(1-\gamma)}{pq+p-q}, \tag{6.1.1}$$

for arbitrary balls $B(x,r)$.

To prove this theorem we need some results about the maximal function

$$\widetilde{M}f(x) = \sup_{r>0} \frac{1}{\mu B(x, N_0 r)} \int\limits_{B(x,r)} |f(y)| d\mu,$$

where $N_0 = a_1(1 + 2a_0)$ and the constants a_0 and a_1 are from the definition of a quasi–metric.

Proposition 6.1.1. $\widetilde{M}$ is bounded in $L^p_\mu(X)$, where $1 < p < \infty$.

This can be proved by a well-known method using the following covering lemma.

Lemma A. *Suppose E is a bounded set (i.e.contained in a ball) in X such that for each $x \in E$ there is given a ball $B(x, r(x))$. Then there is a (finite or infinite) sequence of points $x_j \in E$ such that $\{B(x_j, r(x_j))\}$ is a disjoint family of balls and $\{B(x_j, N_0 r(x_j)\}$ is a covering of E.*

For the proof Lemma A see [100], p. 15 (see also [56], p. 623).

Proof of Proposition 6.1.1. For $\lambda > 0$ we set

$$E_\lambda = \{x \in X : \widetilde{M}f(x) > \lambda\}.$$

Let E be a bounded set. Then for arbitrary $x \in E_\lambda \cap E$ there exists a ball $B(x, r(x))$ such that

$$\frac{1}{\mu B(x, N_0 r(x))} \int\limits_{B(x,r(x))} |f(y)| d\mu > \lambda.$$

By Lemma A, from the family $\{B(x, r(x))\}$ we can choose a finite or infinite sequence of balls such that

$$(E_\lambda \cap E) \subset \bigcup_{j=1}^{\infty} B(x_j, N_0 r(x_j)).$$

Further we obtain the estimates

$$\mu(E_\lambda \cap E) \leq \sum_{j=1}^{\infty} (\mu B(x_j, N_0 r(x_j)) \leq \lambda^{-1} \sum_{j=1}^{\infty} \int\limits_{B(x_j, r(x_j))} |f(y)| d\mu \leq$$

$$\leq \lambda^{-1} \int\limits_X |f(x)| d\mu.$$

Thus the weak type inequality

$$\mu\{x : \widetilde{M}f(x) > \lambda\} \leq \lambda^{-1} \left(\int\limits_X |f(x)| d\mu \right)$$

holds.

In addition, it is obvious that $\widetilde{M}$ has strong (∞, ∞) type. Finally using Marcinkiewicz's interpolation theorem we have the boundedness of $\widetilde{M}$ in $L_\mu^p(X)$ for $1 < p < \infty$.

Proof of Theorem 6.1.1. *Necessity.* Let $\mathcal{K}_0$ be bounded from $L_\mu^p(X)$ to $L_\mu^q(X)$:

$$\left(\int\limits_X |\mathcal{K}_0 f(x)|^q d\mu \right)^{1/q} \le c_3 \left(\int\limits_X |f(x)|^p d\mu \right)^{1/p}.$$

In this inequality set $f = \chi_{B(a,r)}$, where $a \in X$ and $r > 0$. We have

$$\left(\int\limits_{B(a,r)} \left(\int\limits_{B(a,r)} (d(x,y))^{\gamma-1} d\mu \right)^q d\mu \right)^{1/q} \le c_3 (\mu B(x,r))^{1/p}. \qquad (6.1.2)$$

When $x \in B(a,r)$ and $y \in B(a,r)$ it is obvious that $d(x,y) \le c_0 r$. Thus from (6.1.2) it follows that

$$r^{\gamma-1} \mu B(a,r)(\mu B(a,r))^{1/q} \le c_4 (\mu B(a,r))^{1/p}.$$

From the last estimate we conclude that (6.1.1) holds.

Sufficiency. Let us introduce the notation

$$\Omega(x) \equiv \sup_{r>0} \frac{\mu B(x,r)}{r^\beta}.$$

For $x \in X$ and $r > 0$ represent $\mathcal{K}_0 f(x)$ by

$$\mathcal{K}_0 f(x) = \int\limits_{B(x,r)} d(x,y)^{\gamma-1} f(y) d\mu + \int\limits_{X \backslash B(x,r)} d(x,y)^{\gamma-1} f(y) d\mu \equiv$$
$$\equiv I_1(x) + I_2(x).$$

Set $D_k = B(x, 2^{-k}r) \backslash B(x, 2^{-k-1}r)$. Then we have

$$|I_1(x)| \le \sum_{k=0}^{\infty} \int\limits_{D_k} (d(x,y))^{\gamma-1} |f(y)| d\mu \le \sum_{k=0}^{\infty} 2^{-k\gamma} r^\gamma 2^k r^{-1} \times$$

$$\times \frac{\mu B(x, N_0 2^{-k}r)}{\mu B(x, N_0 2^{-k}r)} \int\limits_{B(x, 2^{-k}r)} |f(y)| d\mu \le$$

$$\le c_5 r^{\gamma-1+\beta} \sum_{k=0}^{\infty} 2^{-k(\gamma-1+\beta)} \widetilde{M} f(x) \cdot \Omega(x) \le c_6 r^{\gamma-1+\beta} \widetilde{M} f(x) \cdot \Omega(x).$$

The last estimate holds because of the condition $\gamma - 1 + \beta > 0$. Therefore

$$|I_1(x)| \leq c_6 r^{\gamma-1+\beta} \Omega(x) \widetilde{M} f(x).$$

Now let $H_k = B(x, 2^{k+1}r) \backslash B(x, 2^k r)$. Hölder's inequality yields

$$|I_2(x)| \leq \|f\|_{L^p_\mu(X)} \left(\int\limits_{X \backslash B(x,r)} (d(x,y))^{(\gamma-1)p'} d\mu \right)^{1/p'}.$$

From the last inequality we have

$$|I_2(x)| \leq \|f\|_{L^p_\mu(X)} \left(\sum_{k=0}^{\infty} \int\limits_{H_k} (d(x,y))^{(\gamma-1)p'} d\mu \right)^{1/p'} \leq$$

$$\leq c_7 \|f\|_{L^p_\mu(X)} \left(\sum_{k=0}^{\infty} 2^{k(\gamma-1)p'} r^{(\gamma-1)p'} \mu B(x, 2^{k+1}r) \right)^{1/p'} \leq$$

$$\leq c_8 \|f\|_{L^p_\mu(X)} \left(\sum_{k=0}^{\infty} 2^{k(\gamma-1)p'} r^{(\gamma-1)p'} (2^{k+1}r)^\beta \right)^{1/p'} \cdot (\Omega(x))^{1/p'}.$$

Hence

$$|I_2(x)| \leq c_8 \|f\|_{L^p_\mu(X)} r^{\gamma-1+\frac{\beta}{p'}} \left(\sum_{k=0}^{\infty} 2^{k((\gamma-1)p'+\beta)} \right)^{1/p'} (\Omega(x))^{1/p'} \leq$$

$$\leq c_9 \|f\|_{L^p_\mu(X)} r^{\gamma-1+\frac{\beta}{p'}} (\Omega(x))^{1/p'}.$$

$$(6.1.3)$$

The estimetes for I_1 and I_2 imply the following pointwise inequality:

$$|\mathcal{K}_0 f(x)| \leq c_{10}(r^{\gamma-1+\beta} \Omega(x) \widetilde{M} f(x) +$$

$$+ r^{\gamma-1+\beta/p'} (\Omega(x))^{1/p'} \|f\|_{L^p_\mu(X)}).$$

$$(6.1.4)$$

Taking into account condition (6.1.1) and estimate (6.1.4), we deduce that

$$|\mathcal{K}_0 f(x)| \leq c_{11}(r^{\gamma-1+\beta} \widetilde{M} f(x) + r^{\gamma-1+\beta/p'} \|f\|_{L^p_\mu(X)})$$

for arbitrary $x \in X$ and $r > 0$.
In the last inequality we put

$$r = \|f\|^{\frac{p}{\beta}} (\widetilde{M} f(x))^{-\frac{p}{\beta}}.$$

Thus we obtain the estimate

$$|\mathcal{K}_0 f(x)| \leq c_{12} \|f\|_{L^p_\mu(X)}^{\frac{(\gamma-1+\beta)p}{\beta}} (\widetilde{M} f(x))^{1-\frac{(\gamma-1+\beta)p}{\beta}}.$$

$$(6.1.5)$$

From the definition of β we see that

$$q\left(1 - \frac{(\gamma - 1 + \beta)p}{\beta}\right) = p.$$

Using Proposition 6.1.1, from (6.1.5) we conclude that

$$\|\mathcal{K}_0 f\|_{L^q_\mu(X)} \le c_{13}\|f\|_{L^p_\mu(X)}.$$

The proof is complete.

Here an idea from [122] is used.

As an immediate consequence we have

Corollary 6.1.1. *Let* $1 < p < \frac{1}{\gamma}$ *and* $\frac{1}{q} = \frac{1}{p} - \gamma$. *Then* $\mathcal{K}_0$ *acts boundedly from* $L^p_\mu(X)$ *into* $L^q_\mu(X)$ *if and only if*

$$\mu B(x, r) \le cr.$$

This is a statement of Sobolev type for nonhomogeneous spaces.

Theorem 6.1.1 says that if (6.1.1) fails then $\mathcal{K}_0$ is unbounded from $L^p_\mu(X)$ into $L^q_\mu(X)$ for $1 < p < q < \infty$. Nevertheless there exists a weight v such that $\mathcal{K}_0$ acts boundedly from $L^p_\mu(X)$ into $L^q_v(X)$.

Theorem 6.1.2. *Let* $1 < p \le q < \infty$. *Then there exists a constant* $c > 0$ *such that for arbitrary* $f \in L^p_\mu(X)$ *the estimate*

$$\left(\int\limits_X |\mathcal{K}_0 f(x)|^q (\Omega(x))^{\frac{(\gamma - 1)q}{p}} d\mu\right)^{1/q} \le c\|f\|_{L^p_\mu(X)} \tag{6.1.6}$$

holds.

Proof. Note that the inequality (6.1.4) was obtained independently of the condition (6.1.1). Put in (6.1.4)

$$r = \|f\|^{\frac{p}{\beta}} (\widetilde{M}f(x))^{-\frac{p}{\beta}} (\Omega(x))^{-\frac{1}{\beta}}.$$

We have

$$|\mathcal{K}_0 f(x)| \le c\|f\|^{\frac{(\gamma - 1 + \beta)p}{\beta}} (\widetilde{M}f(x))^{1 - \frac{p(\gamma - 1 + \beta)}{\beta}} (\Omega(x))^{\frac{1 - \gamma}{\beta}}.$$

Using again the Proposition 6.1.1 and taking into account that $q(1 - \frac{p(\gamma - 1 + \beta)}{\beta}) = p$, we obtain (6.1.6). $\square$

Together with the operator $\mathcal{K}_0$ we shall also consider the operator

$$\mathcal{K}f(x) = \int\limits_X k(x, y)f(y)d\mu,$$

where the kernel $k : X \times X \to R$ is a positive, measurable function satisfying the condition

$$k(x,y) \le c(\mu B(x, d(x,y)))^{\gamma - 1}, \quad 0 < \gamma < 1,$$

with a constant c independent of x and y from X.

Now we prove

Theorem 6.1.3. *Let* $1 < p < \gamma^{-1}$ *and let* $1 < p < q < \infty$. *Assume that* $\mu B(x, r)$ *is continuous with respect to* r *and* $\mu\{x\} = 0$ *for arbitrary* $x \in X$. *If the pair of measures* (μ, ν) *satisfies the condition*

$$(\nu B(x, 2N_0 r))^{1/q} \le c(\mu B(x, r))^{1/p - \gamma},$$

then the operator $\mathcal{K}$ *acts boundedly from* $L^p_\mu(X)$ *into* $L^q_\nu(X)$.

To prove this we need some auxiliary propositions. Among them is a weak type two-weight estimate for integral transforms with arbitrary positive kernels obtained in [100], Chapter 3, for nonhomogeneous spaces.

Theorem A. *Let* (X, d, μ) *be a nonhomogeneous space. Let* $1 < p < q < \infty$ *and* ν *and* μ *be arbitrary finite measures on* X *such that* $\mu\{x\} = 0$ *for* $x \in X$. *If*

$$\sup (\nu B(x, 2N_0 r))^{1/q} \left(\int_{X \setminus B(x,r)} k^{p'}(x,y) w^{1-p'}(y) d\mu \right)^{1/p'} < \infty, \quad (6.1.7)$$

where the supremum is taken over all $x \in X$ *and* $r > 0$, *then there exists* $c > 0$ *such that for all* $\lambda > 0$ *and* $f \in L^p_w(X)$,

$$\nu\{x \in X : \mathcal{K}f(x) > \lambda\} \le c\lambda^{-q} \left(\int_X |f(x)|^p w(x) d\mu \right)^{q/p}.$$

Note that in this theorem neither the doubling condition nor the continuity of $\mu B(x, r)$ with respect to r is assumed.

In the sequel we shall use the notation

$$\mu B_{xy} \equiv \mu B(x, d(x,y)).$$

Proposition 6.1.2. *Let* $0 < \gamma < 1$ *and* $1 < p < \frac{1}{\gamma}$. *Suppose that* $\mu\{x\} = 0$ *and* $\mu B(x, r)$ *is continuous with respect to* r. *Then there exists a positive constant* c *depending only on* p *and* γ *such that*

$$\left(\int_{X \setminus B(x,r)} (\mu B_{xy})^{(\gamma - 1)p'} d\mu \right)^{1/p'} \le c(\mu B(x, r))^{\gamma - 1/p}. \quad (6.1.8)$$

Proof. Observe that the function $\mu B(x, r)$ is continuous with respect to r if and only if

$$\mu\{y \in X : d(x, y) = t\} = 0$$

for arbitrary $t > 0$ and $x \in X$.

Indeed, when $r_n \downarrow r$ we have

$$\lim_{n \to \infty} \mu B(x, r_n) = \mu B(x, r) + \mu\{y : d(x, y) = r\}.$$

On the other hand,

$$\lim_{n \to \infty} \mu B(x, r_n) = \mu B(x, r)$$

$r_n \uparrow r$.

Let us denote

$$A(x, r) \equiv \int\limits_{X \backslash B(x, r)} (\mu B_{xy})^{(\gamma - 1)p'} d\mu.$$

We have

$$A(x, r) = \int\limits_0^\infty \mu(X \backslash B(x, r) \cap \{y : (\mu B_{xy})^{(\gamma - 1)p'} > \lambda\}) d\lambda =$$

$$= \int\limits_0^{\mu B(x,r)^{(\gamma-1)p'}} + \int\limits_{\mu B(x,r)^{(\gamma-1)p'}}^\infty \equiv A_1(x, r) + A_2(x, r).$$

Now we shall show that $A_2(x, r) = 0$. For given $x \in X$ and $\lambda > 0$ we denote

$$E_\lambda(x) \equiv \{y : (\mu B_{xy})^{(\gamma - 1)p'} > \lambda\}.$$

For $y \in (X \backslash B(x, r)) \cap E_\lambda$ we have $(\mu B_{xy}) < \lambda^{\frac{1}{(\gamma - 1)p'}}$. On the other hand, if $\lambda > (\mu B(x, r))^{(\gamma - 1)p'}$, then $\mu B_{xy} < \mu B(x, r)$. When $y \in X \backslash B(x, r)$ we have $d(x, y) > r$ and therefore $\mu B_{xy} \geq \mu B(x, r)$. Consequently $A_2(x, r) = 0$.

Now we estimate $A_1(x, r)$. First we prove that

$$\mu E_\lambda \leq \lambda^{\frac{1}{(\gamma - 1)p'}}, \tag{6.1.9}$$

when $\lambda < (\mu B(x, r))^{(\gamma - 1)p'}$.

If $E_\lambda = \emptyset$, then (6.1.9) is obvious. Let $E_\lambda \neq \emptyset$. Suppose that

$$t_0 = \sup\{s : \mu B(x, s) < \lambda^{\frac{1}{(\gamma - 1)p'}}\}.$$

We claim that $0 < t_0 < \infty$. If $t_0 = 0$, then $E_\lambda(x) = \{x\}$, for otherwise there exists $y \in E_\lambda(x)$ such that $d(x, y) > 0$ and

$$\mu B_{xy} < \lambda^{\frac{1}{(\gamma-1)p'}}.$$

Thus $\mu E_\lambda(x) = \mu\{x\} = 0 < \lambda^{\frac{1}{(\gamma-1)p'}}$. Consequently $t_0 > 0$. If $t_0 = \infty$ then there exists a sequence $\{t_n\}$ which converges to $+\infty$ and $\mu B(x, t_n) < \lambda^{\frac{1}{(\gamma-1)p'}}$. Hence $\lim_{n\to\infty} \mu B(x, t_n) = \mu X \leq \lambda^{\frac{1}{(\gamma-1)p'}}$, which contradicts the condition $\mu X = \infty$. Thus $0 < t_0 < \infty$.

Now let $z \in E_\lambda(x)$. Then $\mu B_{xz} < \lambda^{\frac{1}{(\gamma-1)p'}}$ and therefore $d(x, z) \leq t_0$. Consequently $z \in \bar{B}(x, t_0)$, where

$$\bar{B}(x, t_0) \equiv \{y : d(x, y) \leq t_0\}.$$

Since $\mu\{y : d(x, y) = t\} = 0$ we obtain

$$\mu E_\lambda(x) \leq \mu\bar{B}(x, t_0) = \mu B(x, t_0) \leq \lambda^{\frac{1}{(\gamma-1)p'}}$$

as $\mu B(x, t_0) \leq \lambda^{\frac{1}{(\gamma-1)p'}}$. This establishes (6.1.9). Further, by the condition $0 < \gamma < \frac{1}{p}$ we get

$$A_1(x, r) \leq \int_0^{\mu B(x,r)^{(\gamma-1)p'}} \lambda^{\frac{1}{(\gamma-1)p'}} d\lambda = c_4(\mu B(x, r))^{(\gamma-1)p'+1}.$$

The last estimate concludes the validity of 6.1.8. □

Proof of Theorem 6.1.3. We pick a number q_1 such that $1 < q_1 < q$ and

$$1 < \frac{q}{p} - \gamma(q - q_1) < q_1.$$

Now define p_1 by the equality

$$\frac{q_1}{p_1} = \frac{q}{p} - \gamma(q - q_1).$$

Then it is easy to verify that $1 < p_1 < q_1$ and $p_1 < p$ since $1 < p < \gamma^{-1}$. Analogously, we can find numbers p_2 and q_2 such that $p < p_2, q < q_2, p_2 < q_2$ and

$$\Big(\frac{1}{p_2} - \gamma\Big)q_2 = \Big(\frac{1}{p} - \gamma\Big)q.$$

Now using Theorem A and Proposition 6.1.2 we observe that the following weak type inequalities are fulfilled:

$$\nu\{x \in X : \mathcal{K}_0 f(x) > \lambda\} \leq c\lambda^{-q_1}\Big(\int_X |f(x)|^{p_1} d\mu\Big)^{\frac{q_1}{p_1}} \qquad (6.1.10)$$

and

$$\nu\{x \in X : \mathcal{K}_0 f(x) > \lambda\} \le c\lambda^{-q_2} \left(\int_X |f(x)|^{p_2} d\mu \right)^{\frac{q_2}{p_2}}. \qquad (6.1.11)$$

Interpolating these inequalities we obtain that $\mathcal{K}_0$ acts boundedly from $L_\mu^p(X)$ into $L_\nu^q(X)$.

Remark 6.1.1. When $X = R$, the assumption in Theorem 6.1.3 we can replace by the condition

$$\nu B(x, r) \le c(\mu B(x, r))^{q(\frac{1}{p} - \gamma)}.$$

The same is true when $X = R^n$ and $\mu B(x, r)$ is a continuous function of r for arbitrary $x \in X$.

These results follow from Remarks 3.1.1 and 3.1.2 in [100] with respect to Theorem A. The piont is that in these cases the theorem mentioned is valid if in condition (6.1.7) we replace $\nu B(x, 2N_0 r)$ by $\nu B(x, r)$.

Theorem 6.1.4. *Let* $1 < p < q < \infty$. *Suppose that the measures* μ *and* ν *satisfy the conditions:*

(i) $\mu B(x, r) \le cr^\beta$ *for some positive* β;

(ii) $\nu B(x, r) \le cr^{(1 - \gamma - \frac{\beta}{p'})q}$, *where* $p' = \frac{p}{p-1}$, $0 < \gamma < 1 - \frac{\beta}{p'}$ *and* $c > 0$ *does not depend on* $x \in X$ *and* $r > 0$.

Then $\mathcal{K}_0$ *is bounded from* $L_\mu^p(X)$ *into* $L_\nu^q(X)$.

Proof. First we show that condition (ii) implies the inequality

$$\nu\{x \in X : \mathcal{K}_0 f(x) > \lambda\} \le c\lambda^{-q} \|f\|_{L_\mu^p(X)}^q.$$

Indeed, note that

$$I(r) \equiv \int_{X \setminus B(x, r)} (d(x, y))^{(\gamma - 1)p'} d\mu(y) =$$

$$= \sum_{k=0}^{\infty} \int_{B(x, 2^{k+1}r) \setminus B(x, 2^k r)} (d(x, y))^{(\gamma - 1)p'} d\mu(y) \le$$

$$\le c_1 \left(\sum_{k=0}^{\infty} 2^{k(\gamma - 1)p'} \mu(B(x, 2^{k+1}r) \setminus B(x, 2^k r)) \right) r^{(\gamma - 1)p'} \le$$

$$\le c_2 \left(\sum_{k=0}^{+\infty} 2^{k(\gamma - 1)p' + \beta k} \right) r^{(\gamma - 1)p' + \beta} \le c_3 r^{(\gamma -)p' + \beta}.$$

In the last inequality we used the condition $\gamma < 1 - \frac{\beta}{p'}$. Consequently, by condition (ii) we conclude that

$$\nu B(x, 2N_0 r) \cdot (I(r))^{\frac{q}{p'}} \leq \nu B(x, 2N_0 r) \cdot r^{(\gamma-1)q + \frac{\beta q}{p'}} \leq c$$

for all $x \in X$ and $r > 0$. In view of of Theorem A the weak (p, q) type inequality is fulfilled.

Now pick q_1 such that $1 < q_1 < q$. For q_1 we can take p_1 with $p_1 < p$, $1 < p_1 < q_1$ and

$$\left(1 - \gamma - \frac{\beta}{p'_1}\right) q_1 = \left(1 - \gamma - \frac{\beta}{p'}\right) q.$$

There exists such p_1, since q_1 is close to q. Analogously we find p_2 and q_2 such that $1 < p < p_2$, $p_2 < q_2$, $1 < q < q_2$ and

$$\left(1 - \gamma - \frac{\beta}{p'_2}\right) q_2 = \left(1 - \gamma - \frac{\beta}{p'}\right) q.$$

Now using Theorem A we conclude that the weak type inequalities (6.1.10) and (6.1.11) are satisfied. Interpolating these inequalities we see that $\mathcal{K}_0$ acts boundedly from $L^p_\mu(X)$ into $L^q_\nu(X)$. $\square$

From the last theorem we can derive the following

Theorem 6.1.5. *Let* $1 < p < q < \infty$. *Then* $\mathcal{K}_0$ *is bounded from* $L^p_\mu(X)$ *into* $L^q_\mu(X)$, *where* $0 < \gamma < 1 - \frac{\beta}{p'}$, *if and only if*

$$\mu B(x, r) \leq cr^\beta, \quad \beta = \frac{pq(1 - \gamma)}{p - q + pq}.$$

Proof. The sufficiency follows from the previous theorem. To prove the necessity we take $f_{x,r}(y) = \chi_{B(x,r)}(y)$. Then $\|f_{x,r}\|_{L^p_\mu(X)} = (\mu B(x, r))^{\frac{1}{p}}$. On the other hand,

$$\|\mathcal{K}_0 f\|_{L^q_\mu(X)} \geq$$

$$\geq \left(\int_{B(x,r)} \left(\int_{B(x,r)} (d(y, z))^{\gamma-1} d\mu(z) \right)^q d\mu(y) \right)^{\frac{1}{q}} \geq c_1 r^{\gamma-1} (\mu B(x, r))^{1 + \frac{1}{q}}.$$

The boundedness of $\mathcal{K}_0$ completes the proof. $\square$

6.2. Theorems of Adams type

In the present section criteria for two-weight inequalities for the operator $\mathcal{K}$ are presented in the setting of spaces of homogeneous type (for the definition of $\mathcal{K}$ see Section 6.1). At the end of the section we consider fractional integrals defined on nonhomogeneous spaces when the reverse doubling condition is assumed.

Everywhere in this section we shall assume that $\mu X = \infty$ and

$$B(x,R) \setminus B(x,r) \neq \emptyset \qquad (*)$$

for all r, R with $0 < r < R < \infty$ and for arbitrary $x \in X$.

Definition 6.2.1. Let $\mu X = \infty$. A measure μ satisfies the reverse doubling condition ($\mu \in RD$) if there exist constants $\eta_1 > 1$ and $\eta_2 > 1$ such that

$$\mu\left(B\left(x,\eta_1 r\right)\right) \geq \eta_2 \mu\left(B\left(x,r\right)\right)$$

for any $x \in X$ and for all $r > 0$.

The following fact is known (see [286], p.11, Lemma 20).

Proposition 6.2.1. *Let* $\mu X = \infty$ *and let* ($*$) *be satisfied. Then if* μ *satisfies the doubling condition, it satisfies the reverse doubling condition as well.*

Theorem 6.2.1. *Let* $1 < p < \min\{1/\gamma, q\}$ *and* (X, d, μ) *be an SHT. Suppose* $\mu\{x\} = 0$ *for arbitrary* $x \in X$. *Assume that the kernel* k *satisfies the condition*

$$k(x,y) \approx (\mu B(x, d(x,y)))^{\gamma - 1}.$$

Then for the boundedness of the operator $\mathcal{K}$ *from* $L^p_\mu(X)$ *to* $L^q_\nu(X)$ *it is necessary and sufficient that*

$$(\nu B(x,r))^{1/q} \leq c(\mu B(x,r))^{\frac{1}{p} - \gamma}$$

for some positive constant c *independent of* $x \in X$ *and* $r > 0$.

To prove this theorem we need the following two lemmas:

Lemma 6.2.1. *Let* (X, d, μ) *be a space of homogeneous type and let* $\lambda > 1$. *Then there exists a positive constant* c *such that*

$$\mu(B(x, \lambda r) \setminus B(x,r)) \leq c\mu B(x,r) \qquad (6.2.1)$$

for arbitrary $x \in X$ *and* $r > 0$.

Proof. The result follows immediately from the doubling condition:

$$\mu B(x, \lambda r) \leq c\mu B(x,r)$$

for some $c > 1$. Indeed

$$\mu(B(x, \lambda r) \setminus B(x,r)) = \mu B(x, \lambda r) - \mu B(x,r) \leq$$
$$\leq c\mu B(x,r) - \mu B(x,r) = (c-1)\mu B(x,r).$$

□

Lemma 6.2.2. *Let* (X, d, μ) *be an SHT. Then* (6.1.8) *holds when* $1 < p < \min \frac{1}{\gamma}$.

Proof. By the assumption, μ satisfies the doubling condition. Therefore by Proposition 6.2.1 the reverse doubling condition is fulfilled, i.e. there exist two constants $\beta_1 < 1$ and $\beta_2 > 1$ such that

$$\mu B(x, r) \leq \beta_1 \mu B(x, \beta_2 r)$$

for arbitrary $x \in X$ and $r > 0$ (see e.g.[286]).

Let $H_k = B(x, \beta_2^{k+1} r) \backslash B(x, \beta_2^k r)$. Using Lemma 6.1.1 we have

$$\int_{X \backslash B(x,r)} \frac{d\mu}{(\mu B(x, d(x, y)))^{(1-\gamma)p'}} = \sum_{k=0}^{\infty} \int_{H_k} \frac{d\mu}{(\mu B(x, d(x, y)))^{(1-\gamma)p'}} \leq$$

$$\leq \sum_{k=0}^{\infty} (\mu B(x, \beta_2^k r))^{(\gamma-1)p'} \mu H_k \leq$$

$$\leq c \sum_{k=0}^{\infty} (\mu B(x, \beta_2^k r))^{(\gamma-1)p'+1}.$$

$$(6.2.2)$$

Now thanks to the reverse doubling condition we have that

$$\sum_{k=0}^{\infty} (\mu B(x, \beta_2^k r))^{(\gamma-1)p'+1} \leq c(\mu B(x, r))^{(\gamma-1)p'+1}.$$

From this estimate and (6.2.2), (6.1.8) follows. □

Proof of Theorem 6.2.1. *Sufficiency.* By Lemma 6.2.2 and doubling condition we conclude that the assumption of Theorem A (Section 6.1) is satisfied. Further using the arguments from the proofs of Theorems 6.1.3 and 6.1.4 we get the boundedness of the operator $\mathcal{K}$ from $L^p_\mu(X)$ to $L^q_\nu(X)$.

Let us prove the necessity. If in the boundedness inequality stand $f(x) = \chi_{B(a,r)}(x)$, $a \in X$, we obtain the estimate

$$\left(\int_{B(a,r)} \left(\int_{B(a,r)} (\mu B(x, d(x, y)))^{\gamma-1} d\mu(y) \right)^q d\nu(x) \right)^{1/q} \leq$$

$$\leq c(\mu B(a, r))^{1/p}.$$

$$(6.2.3)$$

For arbitrary $x \in B(a, r)$ and $y \in B(a, r)$ we have that

$$d(x, y) \leq a_1(d(x, a) + d(a, y)) \leq a_1(a_0 d(a, x) + d(a, y)) \leq Nr,$$

where $N = a_1(a_0 + 1)$. Thus $B(x, d(x, y)) \subset B(x, Nr)$ and

$$(\mu B(x, d(x, y)))^{\gamma-1} \geq (\mu B(x, Nr))^{\gamma-1}.$$

Further, we find a number $N_1 > 1$ such that

$$B(x, Nr) \subset B(a, N_1 r).$$

Indeed, if $z \in B(x, Nr)$ then

$$d(a, z) \le a_1(d(a, x) + d(x, z)) \le N_1 r,$$

where $N_1 = a_1(N + 1)$. Therefore $(\mu B(x, Nr))^{\gamma - 1} \ge (\mu B(a, N_1 r))^{\gamma - 1}$ for any $x \in X$ and $r > 0$. Consequently,

$$(\mu B(x, d(x, y)))^{\gamma - 1} \ge (\mu B(a, N_1 r))^{\gamma - 1}. \tag{6.2.4}$$

Thus, due to (6.2.3) and (6.2.4) we obtain the estimate

$$(\mu B(a, N_1 r))^{\gamma - 1}(\nu B(a, r))^{1/q} \mu B(a, r) \le c(\mu B(a, r))^{1/p}.$$

By virtue of doubling condition from the last estimate we obtain that

$$(\nu B(a, r))^{1/q} \le c_1(\mu B(a, r))^{1/p - \gamma}.$$

This completes the proof of theorem.

From the last theorem we can deduce a statement of Sobolev type for the operator $\mathcal{K}$.

Theorem 6.2.2. *Let* (X, d, μ) *be an SHT and let* $\mu\{x\} = 0$ *for arbitrary* $x \in X$. *Let* $1 < p, q < \infty$ *and let* $\mathcal{K}$ *be an operator with kernel* k *satisfying the condition*

$$k(x, y) \approx \mu B(x, d(x, y))^{\gamma - 1}.$$

Then for the boundedness of $\mathcal{K}$ *from* $L_\mu^p(X)$ *to* $L_\mu^q(X)$ *it is necessary and sufficient that* $\frac{1}{q} = \frac{1}{p} - \gamma$.

Finally we make some remarks concerning the case when the measure μ satisfies the reverse doubling condition (see Definition 6.2.1), weaker than the doubling condition.

Let us consider the operator

$$\tilde{\mathcal{K}}f(x) = \int\limits_X \frac{f(y)}{(\mu B(x, Nd(x, y)))^{1 - \gamma}} d\mu,$$

where $N \ge \eta_1$ and η_1 is the constant from the definition of the reverse doubling condition (see Definition 6.2.1).

Theorem 6.2.3. *Let* (X, d, μ) *be a nonhomogeneous space. Assume also that* $1 < p < q < \infty$, μ *is a measure satisfying the reverse doubling condition* and $\mu\{x\} = 0$ for arbitrary $x \in X$. Then $\tilde{\mathcal{K}}$ acts boundedly from $L^p_\mu(X)$ to $L^q_\nu(X)$ when $1 < p < \frac{1}{\gamma}$ and

$$(\nu B(x, 2N_0 r))^{1/q} \leq (\mu B(x, Nr))^{1/p - \gamma}$$

with a constant c independent of $x \in X$ and $r > 0$, where $N_0 = a_1(1 + 2a_0)$.

The proof is similar to that of proof of Theorem 6.1.3. Let

$$H_k \equiv B(x, \eta_1^{k+1} r) \setminus B(x, \eta_1^k r).$$

Instead of (6.1.8) from the previous section, in the present case we have

$$\sum_{k=0}^{\infty} \int_{H_k} (\mu B(x, Nd(x,y)))^{(\gamma-1)p'} d\mu \leq \sum_{k=0}^{\infty} (\mu B(x, N\eta_1^k r))^{(\gamma-1)p'} \mu H_k \leq$$

$$\leq c_1 \sum_{k=0}^{\infty} (\mu B(x, N\eta_1^k r))^{(\gamma-1)p'} \mu B(x, \eta_1^{k+1} r) \leq$$

$$\leq c_1 \sum_{k=0}^{\infty} (\mu B(x, N\eta_1^k r))^{(\gamma-1)p'+1} \leq c_2 (\mu B(x, Nr))^{(\gamma-1)p'+1}.$$

Corollary 6.2.1. *Let (X, d, μ) be a nonhomogeneous space. Assume that μ satisfies the reverse doubling condition and $\mu\{x\} = 0$ for arbitrary $x \in X$. If $\eta_1 \leq N \leq 2N_0$, then the operator $\tilde{\mathcal{K}}$ acts boundedly from $L^p_\mu(X)$ into $L^q_\mu(X)$ when $1 < p < \frac{1}{\gamma}$ and $\frac{1}{q} = \frac{1}{p} - \gamma$.*

6.3.   Truncated potentials on SHT

In this section we establish the boundedness and compactness of truncated potentials on nonhomogeneous spaces (i.e. on measure spaces with quasi-metric– (X, d, μ), when the doubling condition for μ may fail). As we shall see, the conditions derived here are simultaneously necessary and sufficient when (X, d, μ) is an SHT.

Suppose there exists a point $x_0 \in X$ such that $\mu\{x_0\} = 0$ and let

$$\mu(B(x_0, R) \setminus B(x_0 r)) > 0$$

for all r and R with the condition $0 < r < R < a$, where $a \equiv \sup\{d(x_0, x) : x \in X\}$. Everywhere in this section we shall assume that x_0 is the point mentioned above.

In the sequel we shall use the following notation: $\bar{B}(x, r) \equiv \{y : d(x,y) \leq r\}$, $B(x, (d(x,y)) \equiv B_{xy}$, $\bar{B}(x, (d(x,y)) \equiv \bar{B}_{xy}$.

We define the truncated potential

$$J_\alpha f(x) = \int\limits_{\bar{B}(x_0, 2d(x_0,x))} f(y)(\mu B_{xy})^{\alpha-1} d\mu(y),$$

where $f : X \to R^1$ is a μ-measurable function. For some results concerning the operator J_α on R^n see Chapter 5.

Lemma 6.3.1. *Let (X, d, μ) be a nonhomogeneous space, $1 < p < \infty$ and let $1/p < \alpha < 1$. Assume that $\mu B(x, r)$ is continuous with respect to r for all $x \in X$. Then there exists a positive constant c such that for all $x \in X$ the inequality*

$$I(x) \equiv \int\limits_{\bar{B}(x_0, 2d(x_0,x))} (\mu B_{xy})^{(\alpha-1)p'} d\mu(y) \leq \tag{6.3.1}$$
$$\leq c(\mu \bar{B}(x_0, 2d(x_0, x)))^{(\alpha-1)p'+1}$$

holds.

Proof. Let us put

$$a_x \equiv \mu \bar{B}(x_0, 2d(x_0, x))^{(\alpha-1)p'}, \quad E_\lambda(x) \equiv \{y : (\mu B_{xy})^{(\alpha-1)p'} > \lambda\}.$$

Then

$$I(x) = \int\limits_0^\infty \mu(\bar{B}(x_0, 2d(x_0, x)) \cap E_\lambda(x)) d\lambda = \int\limits_0^{a_x} + \int\limits_{a_x}^\infty \equiv I_1(x) + I_2(x).$$

For $I_1(x)$ we have $I_1(x) \leq c_1 \mu \bar{B}(x_0, 2d(x_0, x))^{(\alpha-1)p'+1}$.

Using the fact that $\mu B(x, r)$ is continuous with respect to r we can prove the inequality $\mu(E_\lambda(x)) \leq \lambda^{\frac{1}{(\alpha-1)p'}}$ (see the proof of Proposition 6.1.2, and (6.1.9)). Consequently,

$$I_2(x) \leq \int\limits_{a_x}^\infty \lambda^{\frac{1}{(\alpha-1)p'}} d\lambda = c_2 a_x^{\frac{1}{(\alpha-1)p'}+1}.$$

The desired result follows. $\square$

The next lemma is proved in the same way as Theorem 1.1.4, and therefore we omit the proof.

Lemma 6.3.2. *Let (X, d, μ) be a nonhomogeneous space and $1 < p \leq q < \infty$. Assume that $a < \infty$ and $\mu\{x : d(x_0, x) = a\} = 0$. Suppose that ν is another measure on X. Then the Hardy-type operator*

$$Hf(x) = \int\limits_{B_{x_0 x}} f(y) d\mu(y)$$

is bounded from $L^p_\mu(X)$ into $L^q_\nu(X)$ if and only if

$$A \equiv \sup_{0<t<a} (\nu(X\backslash\overline{B}(x_0,t)))^{1/q}(\mu\bar{B}(x_0,t))^{1/p'} < \infty.$$

Moreover, $\|H\| \approx A$.

Theorem 6.3.1. Let (X, d, μ) be a nonhomogeneous spece, $1 < p \le q < \infty$ and let $\alpha \in (\frac{1}{p}; 1)$. Assume also that $\mu B(x, r)$ is continuous with respect to r for all $x \in X$. Then the condition

$$B \equiv \sup_{0<t<a} B(t) \equiv$$

$$\equiv \sup_{0<t<a} \left(\int_{X\backslash B(x_0,t)} \left(\mu B\left(x, \frac{d(x_0,x)}{2a_1a_0}\right)\right)^{(\alpha-1)q} d\nu(x) \right)^{1/q} \times \qquad (6.3.2)$$

$$\times (\mu\bar{B}(x_0, \lambda t))^{1/p'} < \infty,$$

where $\lambda = 2(a_1 + a_1^2(a_0 + 2))$, is sufficient for the boundedness of J_α from $L^p_\mu(X)$ to $L^q_\nu(X)$. Moreover, $\|J_\alpha\| \le cB$.

Proof. Let $f \ge 0$. At first we assume that $a = \infty$. Then

$$\|J_\alpha f\|_{L^q_\nu(X)} \le \|J^{(1)}_\alpha f\|_{L^q_\nu(X)} + \|J^{(2)}_\alpha f\|_{L^q_\nu(X)},$$

where

$$J^{(1)}_\alpha f(x) \equiv \int_{F_1(x)} f(y)(\mu B_{xy})^{\alpha-1} d\mu(y),$$

$$J^{(2)}_\alpha f(x) \equiv \int_{F_2(x)} f(y)(\mu B_{xy})^{\alpha-1} d\mu(y)$$

and $F_1(x) \equiv \bar{B}(x_0, \frac{d(x_0,x)}{2a_1})$, $F_2(x) \equiv \bar{B}(x_0, 2d(x_0,x))\backslash\bar{B}(x_0, \frac{d(x_0,x)}{2a_1})$. It is easy to verify that for $y \in F_1(x)$ we have $d(x,y) \ge \frac{d(x_0,x)}{2a_1a_0}$. Consequently, using Lemma 6.3.2 we find that

$$\|J^{(1)}_\alpha f\|_{L^q_\nu(X)} \le c_1 \left(\int_X (\mu B(x, \frac{d(x_0,x)}{2a_1a_0}))^{(\alpha-1)q}(Hf(x))^q d\nu(x) \right)^{1/q} \le$$

$$\le c_2 B\|f\|_{L^p_\mu(X)}.$$

Hölder's inequality yields

$$\|J^{(2)}_\alpha f\|^q_{L^q_\nu(X)} \le \int_X \left(\int_{F_2(x)} (\mu B_{xy})^{(\alpha-1)p'} d\mu(y) \right)^{q/p'} \times$$

$$\times \left(\int_{F_2(x)} (f(y))^p d\mu(y) \right)^{q/p} d\nu(x).$$

By Lemma 6.3.1 we have

$$\|J_\alpha^{(2)}f\|^q_{L^q_\nu(X)} \le c_3 \int\limits_X (\mu\bar{B}(x_0,2d(x_0,x)))^{(\alpha-1)q+q/p'} \times$$

$$\times \left(\int\limits_{F_2(x)} (f(y))^p d\mu(y) \right)^{q/p} d\nu(x) \equiv S.$$

It is easy to establish the following inclusion

$$\bar{B}(x_0,2d(x_0,x)) \subset \bar{B}(x,a_1(a_0+2)d(x_0,x)).$$

Hence,

$$(\mu\bar{B}(x_0,2d(x_0,x)))^{(\alpha-1)q+\frac{q}{p'}} \le$$

$$\le (\mu\bar{B}(x,a_1(a_0+2)d(x_0,x)))^{(\alpha-1)q+\frac{q}{p'}} \le$$

$$\le \left(\mu B\left(x,\frac{d(x_0,x)}{2a_1a_0}\right)\right)^{(\alpha-1)q}(\mu\bar{B}(x,a_1(a_0+2)d(x_0,x)))^{q/p'} \le$$

$$\le \left(\mu B\left(x,\frac{d(x_0,x)}{2a_1a_0}\right)\right)^{(\alpha-1)q}(\mu\bar{B}(x_0,\frac{\lambda}{2}d(x_0,x)))^{q/p'}.$$

Consequently, using the notation $A_k \equiv B(x_0,2^{k+1})\backslash B(x_0,2^k)$, it follows that

$$S = \sum_{k\in Z} c_4 \int\limits_{A_k} \left(\mu B\left(x,\frac{d(x_0,x)}{2a_1a_0}\right)\right)^{(\alpha-1)q} \times$$

$$\times(\mu\bar{B}(x_0,\frac{\lambda}{2}d(x_0,x)))^{q/p'}\left(\int\limits_{F_2(x)} (f(y))^p d\mu(y) \right)^{q/p} d\nu(x) \le$$

$$\le c_4 B^q \sum_{k\in Z} \left(\int\limits_{\bar{B}_{x_0}(2^{k+1})\backslash \bar{B}_{x_0}(2^{k-1}/a_1)} (f(y))^p d\mu \right)^{q/p} \le$$

$$\le c_5 B^q \|f\|^q_{L^p_\mu(X)}.$$

Now let us assume that $a < \infty$. In this case we represent $\|J_\alpha f\|^q_{L^q_\nu(X)}$ as follows:

$$\|J_\alpha f\|^q_{L^q_\nu(X)} = \int\limits_{B(x_0,a/2)} (J_\alpha f(x))^q d\nu(x) +$$

$$+ \int\limits_{X\backslash B(x_0,a/2)} (J_\alpha f(x))^q d\nu(x) \equiv S_1 + S_2.$$

Applying Lemma 6.3.2 we conclude, that $S_1 \leq c_6 B^q \|f\|_{L_\mu^p(X)}^q$.
As above for S_2 we get

$$S_2 \leq \int\limits_{X \setminus B(x_0, a/2)} \left(\int\limits_{F_2(x)} (f(y))^p d\mu(y) \right)^{q/p} \times$$

$$\times \left(\int\limits_{F_2(x)} (\mu B_{xy})^{(\alpha-1)p'} d\mu(y) \right)^{q/p'} d\nu(x) \leq$$

$$\leq c_7 \|f\|_{L_\mu^p(X)}^q \int\limits_{X \setminus B(x_0, a/2)} \left(\mu B\left(x, \frac{d(x_0, x)}{2a_0 a_1}\right) \right)^{(\alpha-1)q} \times$$

$$\times \mu \bar{B}\left(x_0, \frac{\lambda}{2} d(x_0, x)\right) \leq c_7 B^q(a/2) \|f\|_{L_\mu^p(X)}^q \leq$$

$$\leq c_7 B^q \|f\|_{L_\mu^p(X)}^q.$$

□

Now we consider the compactness of J_α.

Theorem 6.3.2. *Let (X, d, μ) be a nonhomogeneous space, $a = \infty$ and let $1 < p \leq q < \infty$. Assume that $\alpha \in (\frac{1}{p}, 1)$. Suppose $\mu B(x, r)$ is continuous with respect to r for all $x \in X$. If* (i) *(6.3.2) holds;* (ii) $\lim\limits_{t \to 0} B(t) = \lim\limits_{t \to \infty} B(t) = 0$, *then J_α is compact from $L_\mu^p(X)$ into $L_\nu^q(X)$.*

Proof. Due to Theorem 6.3.1 it follows that the condition (i) implies the boundedness of J_α. Let $0 < b < c < \infty$. Represent $J_\alpha f$ as follows:

$$J_\alpha f = \chi_{B(x_0, b)} J_\alpha f + \chi_{\bar{B}(x_0, c) \setminus \bar{B}(x_0, b)} J_\alpha(f \cdot \chi_{(x_0, 4c)}) +$$

$$+ \chi_{X \setminus \bar{B}(x_0, c)} J_\alpha(f \cdot \chi_{B(x_0, c/(2a_1))}) + \chi_{X \setminus \bar{B}(x_0, c)} J_\alpha(f \cdot \chi_{X \setminus B(x_0, c/2a_1)}) \equiv$$

$$\equiv P_1 f + P_2 f + P_3 f + P_4 f.$$

Let us put

$$k(x, y) = \chi_{\bar{B}(x_0, c) \setminus \bar{B}(x_0, b)}(x) \chi_{B(x_0, 2d(x_0, x))}(y)(\mu B_{xy})^{\alpha-1}.$$

Then

$$\| \|k(x, y)\|_{L_\mu^{p'}(X)} \|_{L_\nu^q(X)}^q \leq$$

$$\leq \int\limits_{\bar{B}(x_0, c) \setminus \bar{B}(x_0, b)} \left(\int\limits_{F_1(x)} (\mu B_{xy})^{(\alpha-1)p'} d\mu \right)^{q/p'} d\nu(x) +$$

$$+ \int\limits_{\bar{B}(x_0, c) \setminus \bar{B}(x_0, b)} \left(\int\limits_{F_2(x)} (\mu B_{xy})^{(\alpha-1)p'} d\mu \right)^{q/p'} d\nu(x) \equiv S_1 + S_2,$$

where $F_1(x)$ and $F_2(x)$ are defined in the proof of Theorem 6.3.1. Using Lemma 6.3.1, for S_2 we obtain that

$$S_2 \leq c_1 \int_{\bar{B}(x_0,c)\setminus\bar{B}(x_0,b)} (\mu\bar{B}(x_0, 2d(x_0, x)))^{(\alpha-1)q+q/p'} d\nu(x) \leq$$

$$\leq \left(\int_{\bar{B}(x_0,c)\setminus\bar{B}(x_0,b)} (\mu B(x, \frac{d(x_0,x)}{2a_1 a_0}))^{(\alpha-1)q} d\nu(x) \right) \left(\mu\bar{B}(x_0, 2c) \right)^{q/p'} < \infty.$$

If $d(x_0, y) < \frac{d(x_0,x)}{2a_1}$, then $\frac{d(x_0,x)}{2a_0 a_1} \leq d(x,y)$ and for S_1 we have that $S_1 < \infty$. Using Theorem C of Section 2.1 the compactness of P_2 follows. The compactness of P_3 follows similarly.

In view of Theorem 6.3.1 we have the following estimates:

$$\|P_1\| \leq b_1 S^{(b)} \text{ and } \|P_4\| \leq b_2 I^{(c)},$$

where

$$S^{(b)} \equiv \sup_{0<r<b} \left(\int_{B(x_0,b)\setminus B(x_0,r)} \left(\mu B\left(x, \frac{d(x_0,x)}{2a_1 a_0}\right) \right)^{(\alpha-1)q} d\nu(x) \right)^{1/q} \times$$

$$\times (\mu B(x_0, \lambda r))^{1/p'},$$

$$I^{(c)} \equiv \sup_{r>\frac{c}{2a_1}} \left(\int_{X\setminus B(x_0,r)} \left(\mu B\left(x, \frac{d(x_0,x)}{2a_1 a_0}\right) \right)^{(\alpha-1)q} d\nu(x) \right)^{1/q} \times$$

$$\times (\mu B(x_0, \lambda r))^{1/p'}.$$

Hence, taking into account the condition (ii) we finally conclude that

$$\|P - P_2 - P_3\| \leq b_1 S^{(b)} + b_2 I^{(c)} \to 0,$$

when $b \to 0$ and $c \to \infty$. Hence P is a compact operator. This completes the proof of the theorem. $\square$

The next result follows analogously.

Theorem 6.3.3. *Let (X, d, μ) be a nonhomogeneous space, $a < \infty$, $1 < p \leq q < \infty$ and $\alpha \in (\frac{1}{p}, 1)$. Let $\mu B(x, r)$ is continuous with respect to radius for all $x \in X$. Then the following two conditions:*
(i) (6.3.2) is fulfilled;
(ii) $\lim_{t \to 0} B(t) = 0$,
implies the compactness of J_α from $L_\mu^p(X)$ into $L_\nu^q(X)$.
In the sequel we shall need the following

Lemma 6.3.3. *Let* (X, d, μ) *be an SHT,* $1 < p < \infty$ *and let* $\alpha \in (\frac{1}{p}, 1)$. *Then* (6.3.1) *is fulfilled.*

The proof of this lemma is similar to that of Lemma 6.3.1. In this case we use the inequality $\mu(E_\lambda(x)) \leq b\lambda^{\frac{1}{(\alpha-1)p'}}$, where b is the constant from the doubling condition for μ, which is true without the continuity condition for $\mu(x, r)$ with respect to r.

Remark 6.3.1. If (X, d, μ) is an *SHT*, then the condition (6.3.2) is equivalent to the condition

$$D \equiv \sup_{0<t<a} D(t) \equiv \sup_{0<t<a} \left(\int_{X \backslash B(x_0,t)} (\mu B_{x_0 x})^{(\alpha-1)q} d\nu(x) \right)^{1/q} \times$$
$$\times (\mu(B(x_0,t)))^{1/p'} < \infty. \tag{6.3.3}$$

Remark 6.3.2. Let (X, d, μ) be an *SHT* and $a = \infty$. The condition $\lim_{t\to 0} B(t) = \lim_{t\to\infty} B(t) = 0$ (see (6.3.2)) is equivalent to the condition $\lim_{t\to 0} D(t) = \lim_{t\to\infty} D(t) = 0$. If $a < \infty$, then the condition $\lim_{t\to 0} B(t) = 0$ is equivalent to $\lim_{t\to 0} D(t) = 0$.

These remarks follows from the doubling condition.

Theorem 6.3.4. *Let* (X, d, μ) *be an SHT and let* $1 < p \leq q < \infty$. *Let* $\alpha > 1/p$. *Then* J_α *is bounded from* $L^p_\mu(X)$ *into* $L^q_\nu(X)$ *if and only if* (6.3.3) *is satisfied. Moreover,* $\|J_\alpha\| \approx D$.

Proof. Sufficiency follows in the same way as in Theorem 6.3.1 using Lemma 6.3.3 instead of Lemma 6.3.1. We omit the proof. To prove necessity we take $f_t(x) = \chi_{B(x_0, t/(2a_1))}$, where $t \in (0, a)$. Then $\|f_t\|_{L^p_\mu(X)} \leq c_1 (\mu B(x_0, t))^{1/p}$. On the other hand,

$$\|J_\alpha f_t\|_{L^q_\nu(X)} \geq c_1 \left(\int_{X \backslash B(x_0,t)} (\mu(B_{x_0 x}))^{(\alpha-1)q} d\nu(x) \right)^{1/q} \mu B(x_0, t).$$

Consequently, the boundedness of J_α implies that (6.3.3) holds. $\square$

Theorem 6.3.5. *Let* (X, d, μ) *be an SHT and let* $a = \infty$. *Assume that* $1 < p \leq q < \infty$ *and* $\alpha > 1/p$. *Then* J_α *is compact from* $L^p_\mu(X)$ *into* $L^q_\nu(X)$ *if and only if* (i) (6.3.3) *holds; and* (ii) $\lim_{t\to 0} D(t) = \lim_{t\to\infty} D(t) = 0$.

Proof. Sufficiency follows as for Theorem 6.3.2. We only show necessity. Let us take the sequence $f_r(x) = \chi_{B(x_0, r/2a_1)}(\mu B_{x_0}(r/(2a_1)))^{-1/p}$. Then we have

$$\left| \int_X f_r(x) d\mu(x) \right| \leq \left(\int_{B_{x_0}(r/(2a_1))} |\varphi(x)|^{p'} d\mu(x) \right)^{1/p'} \to 0$$

as $r \to 0$.

Hence f_r converges weakly to 0 as $r \to 0$. Further direct computations give us the estimate

$$\|J_\alpha f_r\|_{L^q_\nu(X)} \geq c_1 D(r)$$

which tends to 0 when $r \to 0$, as a compact operator maps a weakly convergent sequence into a strongly convergent one.

To prove the remaining part of the theorem we take

$$g_r(x,r) \equiv \chi_{X \setminus B(x_0,r)}(x)(\mu B_{x_0 x})^{(\alpha-1)(q-1)} \times$$

$$\times \left(\int\limits_{X \setminus B(x_0,r)} (\mu B_{x_0 x})^{(\alpha-1)q} d\nu(x) \right)^{-1/q'}.$$

It easy to verify that this sequence is weakly convergent to 0 as $r \to \infty$. On the other hand,

$$\|J'_\alpha g_r\|_{L^{p'}_\nu(X)} \geq c_2 D(r),$$

where

$$J'_\alpha g(y) \equiv \int\limits_{X \setminus \overline{B}(x_0, d(x_0,y)/2)} g(x) \left(\mu B_{xy} \right)^{\alpha-1} d\nu(x).$$

Since that J_α is compact from $L^p_\mu(X)$ into $L^q_\nu(X)$ if and only if J'_α is compact from $L^{q'}_\nu(X)$ into $L^{p'}_\mu(X)$, we conclude that $D(t) \to 0$ as $t \to \infty$. $\square$

The following theorem is proved just in the same way.

Theorem 6.3.6. *Let (X,d,μ) be an SHT and let $a < \infty$. Let $1 < p \leq q < \infty$. Assume that $\alpha > 1/p$. Then the following statements are equivalent:*
(i) *J_α is compact from $L^p_\mu(X)$ into $L^q_\nu(X)$;*
(ii) *(6.3.3) holds and $\lim\limits_{t \to 0} D(t) = 0$.*

Now let us assume that the measure space (X,d,μ) with quasi–metric apart from the conditions (i) - (iv) of Definition 1.1.1 satisfies the condition

$$\mu B(x,r) \leq cr^\beta$$

for some positive number β, where the positive constant c does not depend on $x \in X$ and $r > 0$.

Let us define the following potential-type operator

$$\bar{J}_\gamma f(x) = \int\limits_{\bar{B}(x_0, \eta d(x_0,x))} f(y)(d(x,y))^{\gamma-1} d\mu(y), \quad 0 < \gamma < 1,$$

where η, $\eta \geq 1$, is some constant. We need the following Lemma:

Lemma 6.3.4. *Let* $1 < p < \infty$ *and let* $\gamma > 1 - \frac{\beta}{p'}$. *Then for all* x, $x \in X$,

$$S(x) \equiv \int\limits_{\bar{B}(x_0, \eta d(x_0, x))} (d(x, y))^{(\gamma-1)p'} d\mu(y) \leq c(d(x_0, x))^{(\gamma-1)p'+\beta},$$

where $c > 0$ *depends only on* p, η *and* γ.

Proof. Let us assume that $1 - \frac{\beta}{p'} < \gamma < 1$. We have

$$S(x) = \int\limits_{0}^{\infty} \mu(\bar{B}(x_0, \eta d(x_0, x)) \cap \{y : (d(x, y))^{(\gamma-1)p'} > \lambda\}) d\lambda =$$

$$= \int\limits_{0}^{(d(x_0, x))^{(\gamma-1)p'}} + \int\limits_{(d(x_0, x))^{(\gamma-1)p'}}^{\infty} \equiv S_1(x) + S_2(x).$$

For $S_1(x)$ we have

$$S_1(x) \leq \int\limits_{0}^{(d(x_0, x))^{(\gamma-1)p'}} \mu \bar{B}(x_0, \eta d(x_0, x)) d\lambda \leq c_1 (d(x_0, x))^{(\gamma-1)p'+\beta}.$$

while for $S_2(x)$ we obtain

$$S_2(x) \leq \int\limits_{(d(x_0, x))^{(\gamma-1)p'}}^{\infty} \mu B\left(x, \lambda^{\frac{1}{(\gamma-1)p'}}\right) d\lambda \leq$$

$$\leq c_2 \int\limits_{(d(x_0, x))^{(\gamma-1)p'}}^{\infty} \lambda^{\frac{\beta}{(\gamma-1)p'}} d\lambda \leq$$

$$\leq c_3 (d(x_0, x))^{\beta+(\gamma-1)p'}.$$

The case $\gamma \geq 1$ follows from the triangle inequality for the quasi–metric d. The lemma is proved. □

We also have

Theorem 6.3.7. *Let* $1 < p \leq q < \infty$ *and let* $\gamma > 1 - \frac{\beta}{p'}$. *Then* $\bar{J}_\gamma$ *is bounded from* $L^p_\mu(X)$ *to* $L^q_\nu(X)$ *if*

$$A_a \equiv \sup_{0<t<a} A_a(t) \equiv \sup_{0<t<a} \left(\int\limits_{X \setminus B(x_0, t)} (d(x_0, x))^{(\gamma-1)q} d\nu(x) \right)^{\frac{1}{q}} t^{\frac{\beta}{p'}} < \infty$$

Theorem 6.3.8. *Let p, q and γ be as in the previous theorem. If*
(a) $A_\infty < \infty$ and $\lim\limits_{t \to 0} A_\infty(t) = \lim\limits_{t \to \infty} A_\infty(t) = 0$ for $a = \infty$;
(b) $A_a < \infty$ and $\lim\limits_{t \to 0} A_a(t) = 0$ for $a < \infty$,
then $\bar{J}_\gamma$ is compact from $L_\mu^p(X)$ to $L_\nu^q(X)$.

These theorems follow just in the same way as Theorems 6.3.1 and 6.3.2.

6.4. Riesz potentials in the half–space

In this section we establish boundedness criteria for the generalized potential operator on a space of homogeneous type (X, d, μ) :

$$T_\gamma f(x, t) = \int\limits_X f(y)(\mu B(x, d(x, y) + t))^{\gamma - 1} d\mu(y),$$

from $L_\mu^p(X)$ into $L_\beta^q(\widehat{X})$ where $0 < \gamma < 1$, $1 < p < q < \infty$ and β is a non-negative measure on $\widehat{X} = X \times [0, \infty)$.
    Let

$$M_\gamma f(x, t) = \sup (\mu(B))^{\gamma - 1} \int\limits_B |f(y)| \, d\mu(y), \ 0 \le \gamma < 1, \ t \ge 0,$$

for μ-measurable $f : X \to R^1$, where the supremum is taken over all balls $B \subset X$ whose radius is not less than $t/2$. For $\gamma = 0$ we denote $M_0 f(x, t)$ by $M f(x, t)$.
    In the sequel we shall use the following notation: $\widehat{B} \equiv \widehat{B}(x, r) \equiv B(x, r) \times [0, 2r)$, where $B(x, r)$ is the ball with center x and radius r.
    We begin with the following Lemma:

Lemma 6.4.1. *Let $0 < \gamma < 1$. Then for every ε, $0 < \varepsilon < \min\{\gamma, 1 - \gamma\}$, there exists $c_\varepsilon > 0$ such that for all $f \in L_{\text{loc}}(X)$ and all (x, t), $(x, t) \in \widehat{X}$, the following inequality holds* •

$$T_\gamma(|f|)(x, t) \le c_\varepsilon (M_{\gamma - \varepsilon} f(x, t) M_{\gamma + \varepsilon} f(x, t))^{\frac{1}{2}}. \tag{6.4.2}$$

Proof. First we show that there exist positive constants c_1 and c_2 such that for all (x, t) there is a ball $B(x, r_0)$ with $r_0 \ge t$ for which the following two-sided inequality is fulfilled:

$$c_1(\mu B(x, r_0))^\varepsilon \le \left(\frac{M_{\gamma + \varepsilon} f(x, t)}{M_{\gamma - \varepsilon} f(x, t)} \right)^{\frac{1}{2}} \le c_2(\mu B(x, r_0))^\varepsilon.$$

To prove (6.4.2) we fix $(x,t) \in \widehat{X}$. Then we can choose the ball $B(x, r_1)$ such that $r_1 > t$ and

$$M_{\gamma+\varepsilon} f(x,t) \leq 2(\mu B(x, r_1))^{\gamma+\varepsilon-1} \int\limits_{B(x,r_1)} |f(y)| \, dy.$$

Consequently

$$\frac{M_{\gamma+\varepsilon} f(x,t)}{M_{\gamma-\varepsilon} f(x,t)} \leq 2 \frac{(\mu B(x, r_1))^{\gamma+\varepsilon-1} \int\limits_{B(x,r_1)} |f(y)| \, d\mu(y)}{(\mu B(x, r_1))^{\gamma-\varepsilon-1} \int\limits_{B(x,r_1)} f(y) \, d\mu(y)} = 2(\mu B(x, r_1))^{2\varepsilon}.$$

Further, there exists a ball $B(x, r_2)$ such that $r_2 > t$ and

$$M_{\gamma-\varepsilon} f(x,t) \leq 2(\mu B(x, r_2))^{\gamma-\varepsilon-1} \int\limits_{B(x,r_2)} |f(y)| \, d\mu(y).$$

Therefore

$$\frac{M_{\gamma+\varepsilon} f(x,t)}{M_{\gamma-\varepsilon} f(x,t)} \geq \frac{1}{2} \frac{(\mu B(x, r_2))^{\gamma+\varepsilon-1} \int\limits_{B(x,r_2)} |f(y)| \, d\mu(y)}{(\mu B(x, r_2))^{\gamma-\varepsilon-1} \int\limits_{B(x,r_2)} f(y) \, d\mu(y)} = \frac{1}{2}(\mu B(x, r_2))^{2\varepsilon}.$$

Let

$$r_0 = \inf\{r_1 : \frac{M_{\gamma+\varepsilon} f(x,t)}{M_{\gamma-\varepsilon} f(x,t)} \leq 2(\mu B(x, r_1))^{2\varepsilon}\}.$$

If $r_0 = t$, then

$$\frac{1}{\sqrt{2}} (\mu B(x, r_0))^{\varepsilon} \leq \left(\frac{M_{\gamma+\varepsilon} f(x,t)}{M_{\gamma-\varepsilon} f(x,t)}\right)^{\frac{1}{2}} \leq \sqrt{2} (\mu B(x, r_0))^{\varepsilon}.$$

Now let us assume that $r_0 > t$. Then it is easy to verify the following inequalities:

$$\frac{\sqrt{2}}{b^{\varepsilon}} (\mu B(x, r_0))^{\varepsilon} \leq \sqrt{2} (\mu B(x, r_0/2))^{\varepsilon} \leq$$

$$\leq \left(\frac{M_{\gamma+\varepsilon} f(x,t)}{M_{\gamma-\varepsilon} f(x,t)}\right)^{\frac{1}{2}} \leq \sqrt{2} (\mu B(x, r_0))^{\varepsilon},$$

where the constant b is from the doubling condition for μ (see Definition 1.1.1). Inequality (6.4.2) is proved.

Further, represent $T_\gamma(|f|)(x,t)$ as follows:

$$T_\gamma(|f|)(x,t) = \int\limits_{B(x,r_0-t)} |f(y)| \, (\mu B(x,\rho(x,y)+t))^{\gamma-1} \, d\mu(y) +$$

$$+ \int\limits_{X\backslash B(x,r_0-t)} |f(y)| \, (\mu B(x,\rho(x,y)+t))^{\gamma-1} \, d\mu(y).$$

First we estimate the second integral. Let $\mu X = \infty$. Assume that $\{r_k\}_{k>0}$ is a sequence of real numbers such that

$$\mu B(x, r_k/2) < 2^k \mu B(x, r_0) \leq \mu B(x, r_k). \tag{6.4.3}$$

Each r_k is derived from r_{k-1} using the doubling condition. Then as $\mu X = \infty$, $\{r_k\}$ is an increasing and infinite sequence.

Let us consider the balls $B_k = B(x, r_k - t)$, $k \geq 0$.

It is easy to see that

$$\int\limits_{X\backslash B(x,r_0-t)} |f(y)| \, \mu B(x, d(x,y)+t) d\mu(y) =$$

$$= \sum_{k=0}^{+\infty} \int\limits_{B_{k+1}\backslash B_k} |f(y)| \, (\mu B(x, d(x,y)+t))^{\gamma-1} \, d\mu(y) \leq$$

$$\leq \sum_{k=0}^{+\infty} (\mu B(x, r_k))^{\gamma-1} \int\limits_{B(x,r_{k+1}-t)} |f(y)| \, d\mu(y).$$

Using (6.4.3) we obtain the estimates

$$\mu B(x, r_k) \geq 2^k \mu B(x, r_0) = \frac{2^{k+1}}{2} \mu B(x, r_0) \geq$$

$$\geq \frac{1}{2} \mu B(x, \frac{r_{k+1}}{2}) \geq \frac{1}{2b} \mu B(x, r_{k+1}).$$

As $\gamma + \varepsilon - 1 < 0$, we observe that

$$\int\limits_{X\backslash B(x_0,r_0-t)} |f(y)| \, \mu B(x, d(x,y)) + t))^{\gamma-1} d\mu(y) \leq$$

$$\leq c \sum_{k=0}^{+\infty} (\mu B(x, r_k))^{-\varepsilon} (\mu B(x, r_{k+1}))^{\gamma+\varepsilon-1} \int\limits_{B(x,r_{k+1})} |f(y)| \, d\mu(y) \leq$$

$$\leq c \sum_{k=0}^{+\infty} 2^{-k\varepsilon} (\mu B(x, r_0))^{-\varepsilon} M_{\gamma+\varepsilon} f(x,t) =$$

$$= c(\mu B(x, r_0))^{-\varepsilon} M_{\gamma+\varepsilon} f(x,t).$$

Using (6.4.2), we have

$$\int\limits_{X\backslash B(x,r_0-t)} |f(y)|\, \mu B(x,d(x,y)+t)^{\gamma-1}d\mu \le c(M_{\gamma-\varepsilon}f(x,t)M_{\gamma+\varepsilon}f(x,t))^{\frac{1}{2}}.$$

Now let $\mu X < \infty$. We can assume that $I \equiv \int\limits_X |f(y)|\, d\mu(y) < \infty$, as if $I = \infty$, then $M_{\gamma\pm\varepsilon}f(x,t) \ge \frac{1}{(\mu X)^{1-\gamma\pm\varepsilon}}\, I = \infty$.

Let $n = \max\{k : 2^k\mu B(x,r_0) < \mu X\}$. Let us define the numbers r_k by (6.4.3) when $0 < k \le n$ and put $B_k \equiv B(x,r_k - t)$, $B_{n+1} \equiv X$. Then

$$\int\limits_{X\backslash B(x,r_0-t)} |f(y)|\, (\mu B(x,\rho(x,y)+t))^{\gamma-1}\, d\mu =$$

$$= \sum_{k=0}^{n} \int\limits_{B_{k+1}\backslash B_k} |f(y)|\, (\mu B(x,\rho(x,y)+t))^{\gamma-1}\, d\mu \le$$

$$\le \sum_{k=0}^{n} (\mu B(x,r_k))^{\gamma-1} \int\limits_{B_{k+1}\backslash B_k} |f(y)|\, d\mu +$$

$$+ \int\limits_{B_{n+1}\backslash B_n} (\mu B(x,\rho(x,y)+t))^{\gamma-1}\, d\mu \le$$

$$\le c \sum_{k=0}^{n-1} 2^{-k\varepsilon}\, (\mu B(x,r_0))^{-\varepsilon}\, M_{\gamma+\varepsilon}f(x,t) +$$

$$+ (\mu B(x,r_n))^{\gamma-1} \int\limits_X |f(y)|\, d\mu \le$$

$$\le c \sum_{k=0}^{n-1} 2^{-k\varepsilon}\, (\mu B(x,r_0))^{-\varepsilon}\, M_{\gamma+\varepsilon}f(x,t) +$$

$$+ (\mu B(x,r_n))^{-\varepsilon}\, (\mu B(x,r_n))^{-\varepsilon} \times$$

$$\times (\mu B(x,r_n))^{\gamma+\varepsilon-1} \int\limits_X |f(y)|\, d\mu.$$

Taking into account the definition of n ,we obtain

$$\mu B(x,r_n) \ge 2^n\mu B(x,r_0) \ge \frac{1}{2}\mu X.$$

Consequently

$$(\mu B(x,r_n))^{\gamma+\varepsilon-1} \int\limits_X |f(y)|\, d\mu \le$$

$$\leq c\,(\mu X)^{\gamma+\varepsilon-1} \int_X |f(y)|\,d\mu \leq c M_{\gamma+\varepsilon} f(x,t).$$

Hence

$$\int_{X\setminus B(x,r_0-t)} |f(y)|\,(\mu B(x,d(x,y)+t))^{\gamma-1}\,d\mu \leq$$

$$\leq c \sum_{k=0}^{n} 2^{-k\varepsilon}\,(\mu B(x,r_0))^{-\varepsilon}\,M_{\gamma+\varepsilon} f(x,t) \leq$$

$$\leq c\,(\mu B(x,r_0))^{-\varepsilon}\,M_{\gamma+\varepsilon} f(x,t) \leq$$

$$\leq c\,(M_{\gamma-\varepsilon} f(x,t) M_{\gamma+\varepsilon} f(x,t))^{\frac{1}{2}}.$$

Now we estimate the integral

$$I(x,r_0,t) \equiv \int_{X\setminus B(x,r_0-t)} |f(y)|\,(\mu B(x,d(x,y)+t))^{\gamma-1}\,d\mu.$$

Let $m_1 = \sup\{k : 2^{-k}\mu B(x,r_0) > \mu\{x\}\}$. It is clear that if $\mu\{x\} = 0$, then $m_1 = \infty$. Consider such r_k that (6.4.3) holds. Note that the sequence $\{r_k\}$ is decreasing. Let

$$m_1 = \sup\{k : 0 \leq k \leq m,\; r_k > t\}.$$

We shall assume that $B_k = B(x,r_k - t)$ for $0 \leq k \leq m$ and $B_{m+1} = \emptyset$.

Suppose that $m < m_1$. Then r_{m+1} is defined and $r_{m+1} \leq t$. Consequently we have

$$\int_{B(x,r_0-t)} |f(y)|\,(\mu B(x,d(x,y)+t))^{\gamma-1}\,d\mu \leq$$

$$\leq \sum_{k=0}^{m} \int_{B_k\setminus B_{k+1}} |f(y)|\,(\mu B(x,d(x,y)+t))^{\gamma-1}\,d\mu \leq$$

$$\leq \sum_{k=0}^{m-1} (\mu B(x,r_{k+1}))^{\gamma-1} \int_{B(x,r_k)} |f(y)|\,d\mu +$$

$$+ (\mu B(x,t))^{\gamma-1} \int_{B(x,r_m)} |f(y)|\,d\mu =$$

$$= \sum_{k=0}^{m-1} (\mu B(x,r_{k+1}))^{\varepsilon}\,(\mu B(x,r_{k+1}))^{\gamma-\varepsilon-1} \int_{B(x,r_k)} |f(y)|\,d\mu +$$

$$+(\mu B(x, r_{m+1}))^{\gamma-1} \int\limits_{B(x,r_m)} |f(y)|\, d\mu =$$

$$= \sum_{k=0}^{m} (\mu B(x, r_{k+1}))^{\varepsilon}\, (\mu B(x, r_{k+1}))^{\gamma-\varepsilon-1} \int\limits_{B(x,r_k)} |f(y)|\, d\mu.$$

Using (6.4.3), we have

$$\mu B(x, r_{k+1}) \geq 2^{-k-1} \mu B(x, r_0) \geq \frac{2^{-k}}{2} \mu B(x, r_0) \geq \frac{1}{2} \mu B(x, \frac{r_k}{2}) \geq$$

$$\geq \frac{1}{2c} \mu B(x, r_k).$$

Therefore ($\gamma - \varepsilon - 1 < 0$, $r_{k+1} < r_k$ and $r_k > t$, $k = 0, \cdots, m$)

$$I(x, r_0, t) \leq c \sum_{k=0}^{m} (\mu B(x, r_k))^{\varepsilon}\, (\mu B(x, r_k))^{\gamma-\varepsilon-1} \int\limits_{B(x,r_k)} |f(y)|\, d\mu \leq$$

$$\leq c \sum_{k=0}^{m} \left(\mu B(x, \frac{r_k}{2}) \right)^{\varepsilon} M_{\gamma-\varepsilon} f(x, t) \leq$$

$$\leq c\, (\mu B(x, r_0))^{\varepsilon}\, M_{\gamma-\varepsilon} f(x, t).$$

If $m = m_1$, then

$$I(x, r_0, t) \leq \sum_{k=0}^{m-1} (\mu B(x, r_{k+1}))^{\gamma-1} \int\limits_{B(x,r_k)} |f(y)|\, d\mu +$$

$$+ (\mu B(x, t))^{\gamma-1} \int\limits_{B(x,r_m)} |f(y)|\, d\mu \leq$$

$$\leq c \sum_{k=0}^{m-1} 2^{-k\varepsilon}\, (\mu B(x, r_0))^{\varepsilon}\, M_{\gamma-\varepsilon} f(x, t) +$$

$$+ (\mu B(x, t))^{\gamma-1} \int\limits_{B(x,r_m)} |f(y)|\, d\mu.$$

Taking into account (6.4.3) we obtain

$$\mu B(x, t) \geq 2^{-m-1} \mu B(x, r_0) = \frac{2^{-m}}{2} \mu B(x, r_0) =$$

$$= \frac{1}{2} \mu B(x, \frac{r_m}{2}) \geq \frac{1}{2b} \mu B(x, r_m).$$

Consequently

$$I(x,r,t) \le c \sum_{k=0}^{m-1} 2^{-\varepsilon k} (\mu B(x,r_0))^\varepsilon M_{\gamma-\varepsilon}f(x,t) +$$

$$+c (\mu B(x,r_m))^\varepsilon (\mu B(x,r_m))^{\gamma-\varepsilon-1} \int_{B(x,r_m)} |f(y)| \, d\mu \le$$

$$\le c \sum_{k=0}^{m-1} 2^{-\varepsilon k} (\mu B(x,r_0))^\varepsilon M_{\gamma-\varepsilon}f(x,t) = c (\mu B(x,r_0))^\varepsilon M_{\gamma-\varepsilon}f(x,t).$$

Using (6.4.2) we finally see that

$$I(x,r_0,t) \le c(M_{\gamma+\varepsilon}f(x,t)M_{\gamma-\varepsilon}f(x,t))^{\frac{1}{2}}.$$

□

Now let us consider the generalized fractional maximal function

$$\widetilde{M_\gamma}f(x,t) = \sup(\mu B)^{-1} \int_B |f(y)| \, d\mu, \ 0 \le \gamma < 1,$$

where the supremum is taken over all balls $B \subset X$ containing x and having radius greater than $\frac{t}{2}$.

The following theorem is well-known (see [100]- Theorem 4.3.1, [98]).

Theorem A. *Let* $0 < \gamma < 1$, $1 < p < q < \infty$. *Then the following statements are equivalent:*

(i) there exists a positive constant c_1 such that for all f, $f \in L^p_\mu(X)$, the following inequality holds

$$\left(\int_{\widehat{X}} \left(\widetilde{M_\gamma}f(x,t) \right)^q d\beta \right)^{\frac{1}{q}} \le c_1 \left(\int_X |f(x)|^p \, d\mu \right)^{\frac{1}{p}};$$

(ii) there exists a positive constant c_2 such that the following inequality is fulfilled

$$\beta\widehat{B} \le c_2(\mu B)^{q(\frac{1}{p}-\gamma)}$$

for all balls $B \subset X$, where $\widehat{X} \equiv X \times [0,\infty)$, $\widehat{B} \equiv \widehat{B}(x,r) \equiv B(x,r) \times [0,2r)$.

Theorem 6.4.1. *Let* $0 < \gamma < 1$ *and let* $1 < p < q < \infty$. *Then the following statements are eqivalent:*

(i) T_γ is bounded from $L^p_\mu(X)$ into $L^q_\beta(X)$;

(ii) there exists a constant $c_1 > 0$ such that for all $\mu-$ measurable sets $E \subset X$,

$$\left(\int\limits_{\widehat{X}} (T_\gamma \chi_E(x,t))^q d\beta(x,t) \right)^{\frac{1}{q}} \leq c_1 \, (\mu E)^{\frac{1}{p}} \,;$$

(iii) there exists a constant c_2 such that for all balls $B \subset X$,

$$\beta \widehat{B} \leq c_2 (\mu B)^{q(\frac{1}{p} - \gamma)}.$$

Proof. It is clear that (i)$\Longrightarrow$(ii). Further,

$$\widetilde{M_\gamma} f(x,t) \leq c T_\gamma(|f|)(x,t),$$

where c does not depend on $(x,t) \in \widehat{X}$.

Indeed, we have

$$T_\gamma f(x,t) \geq \left(\int\limits_{B(x,r)} |f(y)| \, d\mu \right) (\mu B(x, r+t))^{\gamma-1}$$

for all $x \in X$ and $r, t \in [0, \infty)$. If $r \geq \frac{t}{2}$, then $3r \geq r + t$. Because of the doubling condition for μ, we obtain

$$T_\gamma f(x,t) \geq b_1 (\mu B(x,r))^{\gamma-1} \int\limits_{B(x,r)} |f(y)| \, d\mu$$

for all $x \in X$ and $r \geq \frac{t}{2}$. Consequently

$$T_\gamma f(x,t) \geq b_1 M_\gamma f(x,t),$$

where

$$M_\gamma f(x,t) = \sup_{r \geq t/2} (\mu B(x,r))^{\gamma-1} \int\limits_{B(x,r)} |f(y)| \, d\mu.$$

Finally we have

$$T_\gamma f(x,t) = b_1 M_\gamma f(x,t) \geq b_2 \widetilde{M_\gamma} f(x,t).$$

It follows from (ii) that for all balls $B \subset X$ we have

$$\left(\int\limits_{\widehat{B}} \left(\widetilde{M_\gamma} \chi_B(x,t) \right)^q d\beta(x,t) \right)^{\frac{1}{q}} \leq c_1 \, (\mu B)^{\frac{1}{p}}. \qquad (6.4.4)$$

On the other hand, for $(x,t) \in \widehat{B}$ we have $\widetilde{M}_\gamma \chi_B(x,t) \geq (\mu B)^\gamma$. Hence from (6.4.4) we obtain the condition (iii).

Now let us show that (iii)$\Rightarrow$(i). Put

$$q_1 = q \frac{\frac{1}{p} - \gamma}{\frac{1}{p} - \gamma - \varepsilon}, \quad q_2 = q \frac{\frac{1}{p} - \gamma}{\frac{1}{p} - \gamma + \varepsilon}.$$

It is clear that $q_1(\frac{1}{p} - (\gamma + \varepsilon)) = q\left(\frac{1}{p} - \gamma\right)$ and $q_2(\frac{1}{p} - (\gamma - \varepsilon)) = q\left(\frac{1}{p} - \gamma\right)$. Further, $q_1 > q > p$. Moreover, there exists $\varepsilon > 0$ such that $q_2 \geq p$; we can assume that $\varepsilon < \min\{\gamma, 1 - \gamma\}$.

Let $p_1 = \frac{2q_1}{q}$, $p_2 = \frac{2q_2}{q}$. Then $\frac{1}{p_1} + \frac{1}{p_2} = 1$.

Using Lemma 6.4.1 and Hölder's inequality, we obtain

$$\int_{\widehat{X}} |T_\gamma f(x,t)|^q \, d\beta(x,t) \leq \int_{\widehat{X}} (M_{\gamma - \varepsilon} f(x,t))^{\frac{q}{2}} (M_{\gamma + \varepsilon} f(x,t))^{\frac{q}{2}} \, d\beta(x,t) \leq$$

$$\leq \left(\int_{\widehat{X}} (M_{\gamma + \varepsilon} f(x,t))^{\frac{q}{2} p_1} \, d\beta \right)^{\frac{1}{p_1}} \left(\int_{\widehat{X}} (M_{\gamma - \varepsilon} f(x,t))^{\frac{q}{2} p_2} \, d\beta(x,t) \right)^{\frac{1}{p_2}} =$$

$$= \left(\int_{\widehat{X}} (M_{\gamma + \varepsilon} f(x,t))^{q_1} \, d\beta(x,t) \right)^{\frac{q}{2q_1}} \left(\int_{\widehat{X}} (M_{\gamma - \varepsilon} f(x,t))^{q_2} \, d\beta(x,t) \right)^{\frac{q}{2q_2}}.$$

With the help of Theorem A we finally have

$$\int_{\widehat{X}} |T_\gamma f(x,t)|^q \, d\beta(x,t) \leq \left(\int_X |f(y)|^p \, d\mu \right)^{\frac{q}{2p}} \times$$

$$\times \left(\int_X |f(y)|^p \, d\mu \right)^{\frac{q}{2p}} = \left(\int_X |f(y)|^p \, d\mu \right)^{\frac{q}{p}}.$$

$\square$

6.5. Truncated potentials in the half–space

In this section we derive boundedness and compactness criteria for truncated potentials on SHT.

Let (X, d, μ) be an SHT, $\widehat{X} \equiv X \times [0, \infty)$. Suppose that $\hat{\nu}$ is a positive measure on $\widehat{X}$ such that all sets of type $B \times [0, \infty)$ are $\hat{\nu}$ measurable, where B is a ball in X.

We shall use the following notation:

$$B(x,r) \equiv \{y : d(x_0,y) < r\}, \quad \overline{B}(x,r) = \{y : d(x_0,y) \le R\},$$

$$D_r = B(x_0,r) \times [0,\infty), \quad \bar{D}_r = \bar{B}(x_0,r) \times [0,\infty),$$

where $x \in X$ and $r > 0$.

Assume that $x_0 \in X$ is a point such that $\mu\{x_0\} = 0$ and for all r, R with $0 < r < R < a$, $a = \sup\{d(x_0,x) : x \in X\}$, the following condition is satisfied:

$$\mu(B(x_0,R) \setminus B(x_0,r)) > 0. \tag{$*$}$$

Definition 6.5.1. Let (X,d,μ) be an SHT. Let $\mu X = \infty$ and $x_0 \in X$. We say that μ satisfies the reverse doubling condition at the point x_0, if there exist two constants $\eta_1 > 1$ and $\eta_2 > 1$ such that

$$\mu B(x_0, \eta_1 r) \ge \eta_2 \mu B(x_0, r)$$

for all $r > 0$.

Proposition 6.5.1. *Assume that the measure μ satisfies the doubling condition and $(*)$ holds. Then μ satisfies the reverse doubling condition at x_0.*

The proof of this statement follows in the same way as Lemma 20 from [286].

For $R > 0$ let $D_R \equiv B(x_0,R) \times [0,\infty)$ and $\overline{D}_R \equiv \overline{B}(x_0,R) \times [0,\infty)$. Let H be the Hardy- type operator given by

$$H f(x) = \int_{B(x_0,d(x_0,x))} f \, d\mu$$

for any μ-measurable function $f : X \to R^1$.

Lemma 6.5.1. *$\mu(X) < \infty$ if and only if $a < \infty$.*

Proof. Let $\mu(X) < \infty$. Then (see [100], Proposition 1.1.7) there exists $r_0 > 0$ such that $X = B(x_0,r_0)$. In this case $d(x_0,x) < r_0$ for every $x \in X$ and consequently $a \le r_0 < \infty$. Conversely, if $a < \infty$, then $X \subset B(x_0,2a)$ and we have $\mu(X) \le \mu B(x_0,2a) < \infty$. $\square$

Lemma 6.5.2. *Let $1 < p < \infty$ and $\alpha > 1/p$. Then there exists a positive constant c such that for all $x \in X$ the inequality*

$$J(x) \equiv \int_{\bar{B}(x_0,2d(x_0,x))} (\mu B(x,d(x,y)))^{(\alpha-1)p'} d\mu(y) \le$$
$$\le c(\mu B(x_0,d(x_0,x)))^{(\alpha-1)p'+1} \tag{6.5.1}$$

holds.

Proof. Let $\alpha \geq 1$. Then for $d(x_0, y) \leq 2d(x_0, x)$ we have

$$(B(x, d(x, y)))^{\alpha-1} \leq (\mu B(x, a_1(a_0 + 2)d(x_0, x)))^{\alpha-1} \leq$$
$$\leq c_1(\mu B(x, d(x_0, x)))^{\alpha-1} \leq c_2(\mu B(x_0, d(x_0, x)))^{\alpha-1}.$$

Consequently

$$J(x) \leq c_3(\mu B(x_0, d(x_0, x)))^{(\alpha-1)p'+1}.$$

Now let $\frac{1}{p} < \alpha < 1$. Then we have

$$J(x) = \int_0^\infty \mu(\bar{B}(x_0, 2d(x_0, x)) \cap \{y : \mu B(x, d(x, y))^{(\alpha-1)p'} > \lambda\})d\lambda \leq$$

$$\leq \int_0^{(\mu B(x_0, d(x_0, x)))^{(\alpha-1)p'}} \mu \bar{B}(x_0, 2d(x_0, x))d\lambda +$$

$$+ \int_{(\mu B(x_0, d(x_0, x)))^{(\alpha-1)p'}}^\infty \mu\{y : \mu B(x, y))^{(\alpha-1)p'} > \lambda\}d\lambda \equiv$$

$$\equiv J_1(x) + J_2(x).$$

Using the doubling condition for μ (see Definition 1.1.1) we obtain

$$J_1(x) \leq c_4(\mu B(x_0, d(x_0, x)))^{(\alpha-1)p'+1}.$$

Now we prove the inequality

$$\mu(E_\lambda(x)) \leq b\lambda^{\frac{1}{(\alpha-1)p'}} \qquad (6.5.2)$$

for all λ, $\lambda > \mu B(x_0 d(x_0, x))^{(\alpha-1)p'}$ and $x \in X$, where the constant b is from the definition of the doubling condition for the measure μ and

$$E_\lambda(x) \equiv \left\{y : \mu B(x, d(x, y)) < \lambda^{\frac{1}{(\alpha-1)p'}}\right\}.$$

If $E_\lambda(x) = \emptyset$, then the inequality is obvious. Let $E_\lambda(x) \neq \emptyset$ and suppose that

$$t_0 = \sup\left\{s : \mu B(x, s) < \lambda^{\frac{1}{(\alpha-1)p'}}\right\}.$$

First we show that $t_0 > 0$. Indeed, if $t_0 = 0$, then only x belongs to $E_\lambda(x)$. Consequently $\mu\{x\} < \lambda^{\frac{1}{(\alpha-1)p'}}$ (otherwise $E_\lambda(x) = \emptyset$). From this inequality we have $\mu B(x, s) < \lambda^{\frac{1}{(\alpha-1)p'}}$ for some $s > 0$. Consequently $\{s > 0 : d(x, s) < \lambda^{\frac{1}{(\alpha-1)p'}}\} \neq \emptyset$. Hence $t_0 > 0$.

Now we show that $t_0 < \infty$. From the inequality $\lambda^{\frac{1}{(\alpha-1)p'}} < \mu B(x_0, d(x_0, x))$
we have $s < d(x_0, x) < a$ for all s with the condition $\mu B(x_0, s) < \lambda^{\frac{1}{(\alpha-1)p'}}$.
Consequently $t_0 \leq a$. If $\mu(X) < \infty$ then $t_0 < \infty$.

Let $\mu(X) = \infty$ and $t_0 = \infty$. Then there is a sequence $\{t_n\}$ such that $t_n \to \infty$
and $\mu B(x_0, t_n) < \lambda^{\frac{1}{(\alpha-1)p'}}$. Consequently $\mu(X) = \lim_{n\to\infty} \mu B(x_0, t_n) < \infty$.

Let $z \in E_\lambda(x)$. Then $d(x, z) \leq t_0$, i.e. $z \in \bar{B}(x_0, t_0)$. On the other hand,
$\mu B(x, t_0) \leq \lambda^{\frac{1}{(\alpha-1)p'}}$ and we obtain

$$\mu(E_\lambda(x)) \leq \mu \bar{B}(x, t_0) \leq b\mu B(x, t_0) \leq \lambda^{\frac{1}{(\alpha-1)p'}}.$$

The inequality (6.5.2) is proved.
From the inequality (6.5.2) we have

$$J_2(x) \leq b \int_{\mu B(x_0, d(x_0,x))^{(\alpha-1)p'}}^{\infty} \lambda^{\frac{1}{(\alpha-1)p'}} d\lambda = b_1(\mu B(x_0, d(x_0, x)))^{(\alpha-1)p'+1}.$$

Finally we have (6.5.1). □

Lemma 6.5.3. *Let* $\mu(X) = \infty$, $1 < p \leq q < \infty$. *Then the inequality*

$$\left(\int_{\hat{X}} |Hf(x)|^q d\hat{\nu}(x, t) \right)^{1/q} \leq c \left(\int_X |f(x)|^p d\mu(x) \right)^{1/p},$$

where the constant $c > 0$ *is independent of* f, *holds if and only if*

$$A = \sup_{r>0} (\hat{\nu}(\hat{X}\backslash D_r))^{1/q} (\mu B(x_0, r))^{1/p'}.$$

Moreover, $\|H\| \approx A$.

Proof. Using Lemma 6.5.1 we see that $a = \infty$. By the doubling condition
for μ the condition $A < \infty$ is equivalent to

$$\bar{A} = \sup_{r>0} (\hat{\nu}(\hat{X}\backslash \bar{D}_r))^{1/q} (\mu \bar{B}(x_0, r))^{1/p'} < \infty.$$

First we show sufficiency. Let $f \geq 0$ and let $f \in L^p_\mu(X)$. By Hölder's
inequality, $I(s) < \infty$ for every $s > 0$, where $I(s) = \int_{B(x_0,s)} f d\mu$. As $\mu\{x_0\} = 0$, $\lim_{s\to 0} I(s) = 0$. Moreover it is easy to verify that I is left- continuous on
$(0, \infty)$. Now suppose that $S \equiv \int_X f d\mu \in [2^m, 2^{m+1})$ for some $m \in Z$.
Let $s_j = \sup\{s : I(s) \leq 2^j\}$, $j \leq m + 1$. The sequence $\{s_j\}_{j=-\infty}^{m+1}$ is

nondecreasing and $s_{m+1} = \infty$. It is easy to verify that $I(s_i) \leq 2^j$ and $I(s) > 2^j$ when $s > s_j$. Moreover,

$$\int_{\bar{B}(x_0, s_{j+1}) \setminus B(x_0, s_j)} f d\mu \geq 2^j.$$

Let us put $a_0 \equiv \lim_{j \to -\infty} s_j$ and $I_m \equiv \{j : j \leq m, \ s_j < s_{j+1}\}$. It is clear that $I_m \neq \emptyset$ as $m \in I_m$. Moreover, $[0, \infty) = [0, a_0] \cup \left(\cup_{j \in I_m} (s_j, s_{j+1}] \right)$ and consequently we have that $X = F \cup \left(\bigcup_{j \in I_m} E_j \right)$, where $F = \bar{B}(x_0, a_0)$, $E_j = \bar{B}(x_0, s_{j+1}) \setminus \bar{B}(x_0, s_j)$.

If $S = \infty$, then $X = \left(\bigcup_{j \in J} E_j \right) \cup F$, where $J = \{j \in Z : s_j < s_{j+1}\}$. If $s \in (s_j, s_{j+1}]$, then $I(s) \leq 2^{j+1}$ and if $s \in [0, a_0]$, then $I(s) = 0$. We have

$$\left(\int_{\hat{X}} (Hf(x))^q d\hat{\nu}(x, y) \right)^{p/q} \leq$$

$$\leq \left(\sum_{j \in J} \int_{\bar{D}_{s_{j+1}} \setminus \bar{D}_{s_j}} (Hf(x))^q d\hat{\nu}(x, y) \right)^{p/q} \leq$$

$$\leq \sum_{j \in J} \left(\int_{\bar{D}_{s_{j+1}} \setminus \bar{D}_{s_j}} (Hf(x))^q d\hat{\nu}(x, y) \right)^{p/q} \leq$$

$$\leq \sum_{j \in J} 2^{(j+1)p} \left(\int_{E_j \times [0, \infty)} d\hat{\nu}(x, y) \right)^{p/q} =$$

$$= 4^p \sum_j 2^{(j-1)p} (\hat{\nu}(\bar{D}_{s_{j+1}} \setminus \bar{D}_{s_j}))^{p/q} \leq$$

$$\leq 4^p \sum_j \left(\int_{\bar{B}(x_0, s_j) \setminus B(x_0, s_{j-1})} f(x) d\mu(x) \right)^p \times$$

$$\times \hat{\nu}(E_j \times [0, \infty))^{p/q} \leq 4^p \bar{A}^p \sum_j \int_{\bar{B}(x_0, s_j) \setminus B(x_0, s_{j-1})} (f(x))^p d\mu(x) =$$

$$= 4^p \bar{A}^p \int_X (f(x))^p d\mu(x).$$

From the inequality $\tilde{A} \leq c_1 A$ it follows that $\|H\| \leq b_1 A$.

To prove necessity we take $r > 0$. Let $f_r(x) = \chi_{B(x_0,r)}(x)$. Then $\|f_r\|_{L^p_\mu(X)} = (\mu B(x_0, r))^{1/p}$. Moreover,

$$\|Hf_r(\cdot)\|_{L^q_{\hat\nu}(\hat X)} \geq \left(\int_{\hat X \setminus D_r} \left(\int_{d(x_0,y)<d(x_0,x)} f_r(y)d\mu(y) \right)^q d\nu(x,y) \right)^{1/q} \geq$$

$$\geq \hat\nu(\hat X_r)^{1/q}\left(\hat X \setminus D_r \right).$$

Consequently $A < \infty$ and finally we have that $\|H\| \approx A$. $\square$

Analogously we can get the following

Lemma 6.5.4. *Let* $\mu(X) < \infty$, $\mu\{x : d(x_0, x) = a\} = 0$ *and* $1 < p \leq q < \infty$. *Then the inequality*

$$\left(\int_{\hat X} |Hf(x)|^q d\hat\nu(x,t) \right)^{1/q} \leq c\left(\int_X (f(x))^p d\mu(x) \right)^{1/p} \tag{6.5.3}$$

holds if and only if

$$A_1 = \sup_{0<r<a} (\hat\nu(\hat X \setminus D_r))^{1/q}(\mu B(x_0, r))^{1/p'} < \infty.$$

Moreover, if c *is the best constant in* (6.5.3), *then* $c \approx A_1$.

Now let

$$\bar J_\alpha f(x,t) = \int_{\overline B(x_0, 2d(x_0,x))} \frac{f(y)}{(\mu B(x, d(x,y) + t))^{1-\alpha}} d\mu(y), \quad \alpha > 0$$

for $x \in X$, $t \in [0, \infty)$ and μ-measurable $f : X \to R^1$.

Suppose that ν is a σ-finite positive measure on $\hat X = X \times [0, \infty)$ such that all sets of type $B(x, r) \times [0, \infty)$ are ν-measurable.

The following theorem is valid.

Theorem 6.5.1. *Let* $\mu(X) = \infty$ *and let* $1 < p \leq q < \infty$. *Assume that* $\alpha > 1/p$. *Then the following statements are equivalent:*
(i) $\bar J_\alpha$ *is bounded from* $L^p_\mu(X)$ *to* $L^q_\nu(\hat X)$;
(ii)

$$\bar A = \sup_{r>0} \left(\int_{\hat X \setminus D_r} (\mu B(x_0, d(x_0, x) + t))^{(\alpha-1)q} d\nu(x,t) \right)^{1/q} \times$$

$$\times (\mu B(x_0, r))^{1/p'} < \infty;$$

(iii)

$$\bar{A}_1 = \sup_{k \in Z} \bigg(\int_{D_{\eta_1^{k+1}} \setminus D_{\eta_1^k}} (\mu B(x_0, d(x_0 x) + t))^{(\alpha-1)q} \times$$

$$\times (\mu B(x_0, d(x_0, x)))^{q/p'} d\nu(x, y) \bigg)^{1/p'} < \infty,$$

where η_1 is from Definition 6.5.1. Moreover, $\|\bar{J}_\alpha\| \approx \bar{A} \approx \bar{A}_1$.

Proof. As the case $\alpha \geq 1$ is trivial, we consider the case $1/p < \alpha < 1$. Assume that $f \geq 0$. We have

$$\|\bar{J}_\alpha f\|_{L^q_\nu(\hat{X})} \leq$$

$$\leq \bigg(\int_{\hat{X}} \bigg(\int_{B(x_0, \frac{d(x_0, x)}{2a_1})} f(y)(\mu B(x, d(x, y) + t))^{\alpha-1} d\mu(y) \bigg)^q d\nu(x, t) \bigg) +$$

$$+ \bigg(\int_{\hat{X}} \bigg(\int_{\bar{B}(x_0, 2d(x_0, x)) \setminus B(x_0, d(x_0, x)/(2a_1))} f(y) \times$$

$$\times (\mu B(x, d(x, y) + t))^{\alpha-1} d\mu(y) \bigg)^q d\nu(x, t) \bigg)^{1/q} \equiv$$

$$\equiv S_1 + S_2.$$

If $d(x_0, y) < \frac{d(x_0, x)}{2a_1}$, then $\frac{d(x, x)}{2a_1 a_0} \leq d(x, y)$. Hence using the doubling condition for μ we find that

$$\mu B(x, d(x_0, x) + t) \leq \mu B(x, 2a_1 a_0 d(x, y) + t) \leq$$
$$\leq c_1 \mu B(x, d(x, y) + t).$$

In addition, $\mu B(x_0, d(x_0, x) + t) \leq c_2 \mu B(x, d(x_0, x) + t)$. Consequently

$$\mu B(x_0, d(x_0, x) + t) \leq c_3 \mu B(x, d(x, y) + t)$$

and using Lemma 6.5.3 we obtain

$$S_1 \leq c_4 \bigg(\int_{\hat{X}} (\mu B(x_0, d(x_0, x) + t))^{(\alpha-1)q} \times$$

$$\times \bigg(\int_{B(x_0, d(x_0, x))} f(y) d\mu(y) \bigg)^q d\nu(x, t) \bigg)^{1/q} \leq$$

$$\leq c_5 \bar{A} \bigg(\int_X (f(y))^p d\mu(y) \bigg)^{1/p}.$$

Using Hölder's inequality for S_2 we obtain

$$S_2^q \leq \int\limits_{\hat{X}} \left(\int\limits_{E_x} (f(y))^p d\mu(y) \right)^{q/p} \times$$

$$\times \left(\int\limits_{E_x} (\mu B(x, d(x,y) + t))^{(\alpha-1)p'} d\mu(y) \right)^{q/p'} d\nu(x,t),$$

where $E_x = \bar{B}(x_0, 2d(x_0, x)) \backslash B(x_0, d(x_0, x)/(2a_1))$.

Now we claim that there exists a positive constant c_0 such that for all $(x, t) \in \hat{X}$ the following inequality holds:

$$S(x,t) \equiv \int\limits_{E_x} (\mu B(x, d(x,y) + t))^{(\alpha-1)p'} d\mu(y) \leq \qquad (6.5.4)$$

$$\leq c_6 (\mu B(x_0, d(x,y) + t))^{(\alpha-1)p} \mu B(x_0, d(x_0, x)).$$

Let $x \in X$ and let $t \leq d(x_0, x)$. Then by Lemma 6.5.2 we have

$$S(x,t) \leq \int\limits_{E_x} (\mu B(x, d(x,y)))^{(\alpha-1)p'} d\mu(y) \leq$$

$$\leq c_7 (\mu B(x_0, d(x_0, x)))^{(\alpha-1)p'+1} \leq$$

$$\leq c_7 (\mu B(x_0, d(x_0, x) + t))^{(\alpha-1)p'} \mu B(x_0, d(x_0, x) + t) \leq$$

$$\leq c_8 (\mu B(x_0, d(x_0, x) + t))^{(\alpha-1)p'} \mu B(x_0, d(x_0, x)).$$

Now let $t > d(x_0, x)$. Then

$$S(x,t) \leq c_9 \mu B(x_0, d(x_0, x))(\mu B(x, t))^{(\alpha-1)p'} \leq$$

$$\leq c_{10} \mu B(x_0, d(x_0, x))(\mu B(x_0, d(x_0, x) + t))^{(\alpha-1)p'}.$$

(In the last inequality we used the fact $2t > d(x_0, x) + t$ and the doubling condition for μ).

Finally we have (6.5.4). Further,

$$S_2 \leq c_{11} \int\limits_{\hat{X}} (\mu B(x_0, d(x_0, x) + t))^{(\alpha-1)q} \times$$

$$\times \left(\int\limits_{E_x} (f(y))^p d\mu(y) \right)^{q/p} (\mu B(x_0, d(x_0, x)))^{q/p'} d\nu(x,t) =$$

$$= c_{11} \sum_{k \in Z} \int\limits_{D_{2k+1} \backslash D_{2k}} (\mu B(x_0, d(x_0, x) + t))^{(\alpha-1)q} \times$$

$$\times (\mu B((x_0, d(x_0, x)))^{q/p'} \left(\int\limits_{E_x} (f(y))^p d\mu(y) \right)^{q/p} \le$$

$$\le c_{12} \sum_{k \in Z} \left(\int\limits_{\bar{B}(x_0, 2^{k+2}) \setminus \bar{B}(x_0, 2^{k-1}/a_1)} (f(y))^p d\mu(y) \right)^{q/p} \times$$

$$\times \left(\int\limits_{D_{2^{k+1}} \setminus D_{2^k}} (\mu B(x_0, d(x_0, x) + t))^{(\alpha-1)q} d\nu(x, t) \right) \times$$

$$\times (\mu B(x_0, 2^k))^{q/p'} \le c_{13} \bar{A}^q \|f\|^q_{L^p_\mu(X)}.$$

Thus (ii) implies (i).

Now we show that (iii) follows from (i). Let $k \in Z$ and let $f_k(x) = \chi_{B(x_0, \eta_1^{k-1})}(x)$, where η_1 is from Definition 6.5.1. We have

$$\|\bar{J}_\alpha f\|_{L^q_\nu(\hat{X})} \ge$$

$$\ge \left(\int\limits_{D_{\eta_1^{k+1}} \setminus D_{\eta_1^k}} \left(\int\limits_{B(x_0, \eta_1^{k-1})} (\mu B(x, d(x, y) + t))^{\alpha-1} d\mu(y) \right)^q \times \right.$$

$$\left. \times d\nu(x, t) \right)^{1/q} \ge c_{14} \left(\int\limits_{D_{\eta_1^{k+1}} \setminus D_{\eta_1^k}} (\mu B(x_0, d(x_0, x) + t))^{(\alpha-1)q} \times \right.$$

$$\left. \times (\mu B(x_0, d(x_0, x)))^q d\nu(x, t) \right)^{1/q}.$$

On the other hand,

$$\|f_k\|_{L^p_\mu(X)} = (\mu B(x_0, \eta_1^{k-1}))^{\frac{1}{p}}$$

and consequently $\bar{A}_1 < \infty$. Moreover, $\|\bar{J}_\alpha\| \le b_1 \bar{A}_1$.

We claim that there exists $c_{15} > 0$ such that $\bar{A} \le c_{15} \bar{A}_1$. Let $r > 0$. Then $r \in [\eta_1^m, \eta_1^{m+1})$ for some $m \in Z$. Consequently by the doubling condition for μ we have

$$\bar{A}(r) \le \left(\int\limits_{\hat{X} \setminus D_{\eta_1^m}} (\mu B(x_0, d(x_0, x) + t))^{(\alpha-1)q} d\nu(x, t) \right) \times$$

$$\times (\mu B(x_0, \eta_1^{m+1}))^{q/p'} =$$

$$= \left(\sum_{k=m}^{+\infty} \int\limits_{D_{\eta_1^{k+1}} \setminus D^{\eta_1^k}} (\mu B(x_0, d(x_0, x) + t))^{(\alpha-1)q} d\nu(x, t) \right) \times$$

$$\times(\mu B(x_0,\eta_1^{m+1}))^{q/p'} \leq c_{15}\bar{A}_1^q(\mu B(x_0,\eta_1^{m+1}))^{q/p'} \sum_{k=m}^{+\infty} (\mu B(x_0,\eta_1^k))^{-q/p'}.$$

By Proposition 6.5.1 we find that

$$\sum_{k=m}^{+\infty} (\mu B(x_0,\eta_1^k))^{-q/p'} =$$

$$= b_2 \sum_{k=m}^{+\infty} (\mu B(x_0,\eta_1^k))^{-q/p'-1}\mu(B(x_0,\eta_1^{k+1})\backslash B(x_0,\eta_1^k)) \leq$$

$$\leq b_3 \int_{X\backslash B(x_0,\eta_1^m)} (\mu B(x_0,d(x_0,x)))^{-q/p'-1}d\mu(x) \equiv \bar{S}_m.$$

In addition,

$$\bar{S}_m = \int_0^\infty \mu[(X\backslash B(x_0,\eta_1^m)) \cap \{x : (\mu B(x_0,d(x_0,x)))^{-q/p'-1} > \lambda\}]d\lambda =$$

$$= \int_0^{(\mu B(x_0,\eta_1^m))^{-q/p'-1}} \mu\left\{x : \mu B(x_0,d(x_0,x)) < \lambda^{-\frac{1}{q/p'+1}}\right\}d\lambda \leq$$

$$\leq b_4(\mu B(x_0,\eta_1^m))^{-q/p'}.$$

Consequently $\bar{A} \leq b_5\bar{A}_1$.

Finally we obtain (ii) $\Rightarrow$ (i) $\Rightarrow$ (iii) $\Rightarrow$ (ii). $\square$

Analogously we can derive the following theorem (we only need to use Lemma 6.5.4).

Theorem 6.5.2. *Let* $\mu(X) < \infty$, $1 < p \leq q < \infty$ *and* $\alpha \geq 1/p$. *Then the following conditions are equivalent:*

(i) $\bar{J}_\alpha$ *is bounded from* $L_\mu^p(X)$ *to* $L_\nu^q(\hat{X})$;

(ii)

$$\tilde{A} = \sup_{0<r<a} \left(\int_{\hat{X}\backslash D_r} (\mu B(x_0,d(x_0,x)+t))^{(\alpha-1)q}d\nu(x,t) \right)^{1/q} \times$$

$$\times(\mu B(x_0,r))^{1/p'} < \infty;$$

(iii)

$$\tilde{A}_1 = \sup_{k\leq 0} \left(\int_{D_{\eta_1^{k+1}}\backslash D_{\eta_1^k}} (\mu B(x_0,d(x_0,x)+t))^{(\alpha-1)q} \times$$

$$\times (\mu B(x_0, d(x_0, x)))^{q/p'} d\nu(x,t) \Bigg)^{1/q} < \infty.$$

Moreover, $\|\bar{J}\| \approx \tilde{A} \approx \tilde{A}_1$.

Theorem 6.5.3. *Let $\mu(X) = \infty$, $1 < p \le q < \infty$ and $\alpha > 1/p$. Then the following statements are equivalent:*

(i) $\bar{J}_\alpha$ *is compact from $L^p_\mu(X)$ to $L^q_\nu(\hat{X})$;*

(ii) $\bar{A} < \infty$ *and* $\displaystyle\lim_{r \to 0} \bar{A}(r) = \lim_{r \to \infty} \bar{A}(r) = 0$, *where*

$$\bar{A}(r) \equiv \left(\int_{X \backslash D_r} (\mu B(x_0, d(x_0, x) + t))^{(\alpha-1)q} d\nu(x,t) \right)^{\frac{1}{q}} (\mu B(x_0, r))^{\frac{1}{p'}};$$

(iii) $\bar{A}_1 < \infty$ *and* $\displaystyle\lim_{k \to -\infty} \bar{A}_1(k) = \lim_{k \to +\infty} \bar{A}_1(k) = 0$, *where*

$$\bar{A}_1(k) \equiv \left(\int_{D_{\eta_1^{k+1}} \backslash D_{\eta_1^k}} (\mu B(x_0, d(x_0, x) + t))^{(\alpha-1)q} \times \right.$$

$$\left. \times (\mu B(x_0, d(x_0, x)))^{q/p'} d\nu(x,t) \right)^{1/q}.$$

Proof. First we that (ii) implies (i). Let $0 < b < c < \infty$ and represent $\bar{J}_\alpha$ as follows:

$$\bar{J}_\alpha f = \chi_{D_b} \bar{J}_\alpha f + \chi_{\bar{D}_c \backslash D_b} \bar{J}_\alpha (f \chi_{B(x_0, 4c)}) +$$
$$+ \chi_{\hat{X} \backslash \bar{D}_c} J_\alpha (f \chi_{B(x_0, c/2a_1)}) +$$
$$+ \chi_{\hat{X} \backslash \bar{D}_c} J_\alpha (f \chi_{X \backslash B(x_0, c/(2a_1))}) \equiv$$
$$\equiv P_1 f + P_2 f + P_3 f + P_4 f.$$

For P_2 we have

$$P_2 f(x,t) = \int_X k(x,y,t) f(y) d\mu(y),$$

where

$$k(x,y,t) = \chi_{\bar{D}_c \backslash D_b}(x,t)(\mu B(x, d(x,y) + t))^{\alpha-1}$$

if $d(x_0, y) \le 2d(x_0, x)$ and $k(x, y, t) = 0$ if $d(x_0, y) > 2d(x_0, x)$. Using Lemma 6.5.2 we obtain

$$\| \|k(x,y,t)\|_{L^{p'}_\mu(X)} \|^q_{L^q_\nu(\hat{X})} =$$

$$= \int\limits_{\bar{D}_c \backslash \bar{D}_b} \left(\int\limits_{\bar{B}(x_0, 2d(x_0,x))} (\mu B(x, d(x,y) + t))^{(\alpha-1)p'} d\mu(y) \right)^{q/p'} d\nu(x,t) \le$$

$$\le c_1 \int\limits_{\bar{D}_c \backslash \bar{D}_b} (\mu B(x_0, d(x_0,x) + t))^{(\alpha-1)q} (\mu B(x_0, d(x_0,x)))^{q/p'} d\nu(x,t) \le$$

$$\le c_1 \left(\int\limits_{\bar{D}_c \backslash \bar{D}_b} (\mu B(x_0, d(x_0,x) + t))^{(\alpha-1)q} d\nu(x,t) \right) (\mu B(x_0, c))^{q/p'} < \infty.$$

and by Theorem C of Section 2.1 we see that P_2 is compact. Further, we have the following inequalities:

$$\int\limits_{\hat{X} \backslash \bar{D}_c} \left(\int\limits_{\hat{B}(x_0, d/(2a_1))} (\mu B(x, d(x,y) + t))^{(\alpha-1)p'} d\mu(y) \right)^{q/p'} d\nu(x,t) \le$$

$$\le c_2 \left(\int\limits_{\hat{X} \backslash \bar{D}_c} (\mu B(x_0, d(x_0,y) + t))^{(\alpha-1)q} d\nu(x,t) \right) (\mu B(x_0, c))^{q/p'} < \infty.$$

Consequently P_3 is compact.

Repeating the arguments using in the proof of Theorem 6.5.1 we easily get $\|P_1\| \le b_1 S_b$ and $\|P_4\| \le b_2 \bar{S}_{c/(2a_1)}$, where

$$S_b = \sup_{0 < r < b} \left(\int\limits_{D_b \backslash D_r} \mu B(x_0, d(x_0,x)) + t)^{(\alpha-1)q} d\nu(x,t) \right)^{1/q} (\mu B(x_0, r))^{1/p'}$$

and

$$\bar{S}_{c/(2a_1)} = \sup_{r > c/(2a_1)} \left(\int\limits_{\hat{X} \backslash D_r} (\mu B(x_0, d(x_0,x) + t))^{(\alpha-1)q} d\nu(x,t) \right)^{1/q} \times$$

$$\times (\mu(B(x_0, r) \backslash B(x_0, c/(2a_1))))^{1/p'}.$$

Consequently

$$\|P - P_2 - P_3\| \le b_1 S_b + b_2 \bar{S}_{c/(2a_1)} \to 0$$

as $b \to 0$ and $c \to +\infty$.

Now we prove that (i) implies (ii). Let $r > 0$ and let

$$f_r(x) = \chi_{B(x_0, r/(2a_1))} (\mu B(x_0, r/2a_1))^{-1/p}.$$

It is easy to see that the sequence f_r is weakly convergent to 0. Indeed, let $\varphi \in L^{p'}_\mu(X)$. Then using the fact that $\mu\{x_0\} = 0$ we have

$$\left| \int_X f_r(x)\varphi(x)d\mu(x) \right| \le \left(\int_{B(x_0, r/(2a_1))} |\varphi(x)|^{p'} d\mu(x) \right)^{1/p'} \to 0$$

as $r \to 0$. Moreover,

$$\|\bar{J}_\alpha f\|_{L^q_\nu(\hat{X})} \ge$$

$$\ge c_3 \left(\int_{\hat{X} \setminus D_r} (\mu B(x_0, d(x_0, x)) + t)^{(\alpha-1)q} d\nu(x, t) \right)^{1/q} (\mu B(x_0, r))^{1/p'} \to 0$$

as $r \to 0$. Here we used the fact that the compact operator maps a weakly convergent sequence into a strongly convergent one.

Now let $r > 0$ and let

$$g_r(x, t) = \chi_{\hat{X} \setminus D_r}(x, t)(\mu B(x_0, d(x_0, x)) + t)^{(\alpha-1)(q-1)} \times$$

$$\times \left(\int_{\hat{X} \setminus D_r} (\mu B(x_0, d(x_0, y)) + t)^{(\alpha-1)q} d\nu(y, t) \right)^{-1/q'}.$$

Then g_r is weakly convergent to 0 as $r \to +\infty$. Indeed, let $\psi \in L^q_\nu(\hat{X})$. We have

$$\left| \int_{\hat{X}} \psi(x, t)g_r(x, t)d\nu(x, t) \right| \le b_3 \left(\int_{\hat{X} \setminus D_r} (\psi(x, t))^q d\nu(x, t) \right)^{1/q} \to 0$$

as $r \to +\infty$.

Let

$$\bar{I}_\alpha g(y) = \int_{\hat{X} \setminus D_{d(x_0, y)/2}} \frac{g(x, t)}{(\mu B(x, d(x, y)) + t)^{1-\alpha}} d\nu(x, t).$$

It is easy to verify that $\bar{J}_\alpha$ is compact from $L^p_\mu(X)$ to $L^q_\nu(\hat{X})$ if and only if $\bar{I}_\alpha$ is compact from $L^{q'}_\nu(\hat{X})$ to $L^{p'}_\mu(X)$. We have

$$\left(\int_X (\bar{I}_\alpha g(y))^{p'} d\mu(y) \right)^{1/p'} \ge$$

$$\ge \left(\int_{B(x_0, 2r)} \left(\int_{\hat{X} \setminus D_{d(x_0, y)/2}} \frac{g_r(x, t)d\nu(x, t)}{(\mu B(x, d(x, y)) + t)^{1-\alpha}} \right)^{p'} d\mu(y) \right)^{1/p'} \ge$$

$$\geq c_4(\mu B(x_0, 2r))^{1/p'} \left(\int\limits_{\hat{X} \setminus D_r} (\mu B(x_0, d(x_0, x) + t))^{(\alpha-1)q} d\nu(x, t) \right)^{1/q} \geq$$

$$\geq c_5 \overline{A}(r) \to 0,$$

as $r \to +\infty$. We conclude that (i) $\Rightarrow$ (ii).

Now let us prove that (iii) implies (i). As we have seen the compactness of $\bar{J}_\alpha$ follows from the conditions $\overline{A} < \infty$, $\lim\limits_{b \to 0} S_b = 0$ and $\lim\limits_{d \to +\infty} \bar{S}_d = 0$. Let $\tilde{A}_1 < \infty$. Then by Theorem 6.5.1, $\overline{A} < \infty$. Suppose that $b > 0$. Then $b \in [\eta_1^{m-1}, \eta_1^m)$ for some integer m. Consequently $S_b \leq S_{\eta_1^m}$, where

$$S_{\eta_1^m} \equiv \sup_{0 < r < \eta_1^m} S_{\eta_1^m}(r) \equiv \sup_{0 < r < \eta_1^m} \left(\int\limits_{D_{\eta_1^m} \setminus D_r} (\mu B(x_0, d(x_0, x) + t))^{(\alpha-1)q} \times \right.$$

$$\left. \times d\nu(x, t) \right)^{1/q} (\mu B(x_0, r))^{1/p'}.$$

Let $r \in [0, \eta_1^m)$. Then $r \in [\eta_1^{j-1}, \eta_1^j)$ for some integer $j \leq m$. We obtain

$$S_{\eta_1^m}^q(r) \leq \left(\int\limits_{D_{\eta_1^m} \setminus D_{\eta_1^{j-1}}} (\mu B(x_0, d(x_0, x) + t))^{(\alpha-1)q} d\nu(x, t) \right) \times$$

$$\times (\mu B(x_0, \eta_1^j))^{q/p'} =$$

$$= \sum_{k=i}^{m} \left(\int\limits_{D_{\eta_1^k} \setminus D_{\eta_1^{k-1}}} (\mu B(x_0, d(x_0, x) + t))^{(\alpha-1)q} d\nu(x, t) \right) \times$$

$$\times (\mu B(x_0, \eta_1^j))^{q/p'} \leq c_6 (\sup_{k \leq m} \bar{A}_1(k - 1))^q.$$

(In the last inequality we used the reverse doubling condition for μ). Thus

$$S_{\eta_1^m} \leq c_7 \tilde{A}_{1,m},$$

where

$$\tilde{A}_{1,m} \equiv \sup_{k \leq m} \bar{A}_1(k - 1).$$

If $a \to 0$, then $m \to -\infty$. Consequently $\bar{A}_{1,m} \to 0$ (since $\lim\limits_{k \to -\infty} \bar{A}(k) = 0$) and so $\lim\limits_{b \to 0} S_b = 0$.

Now we show that the condition $\lim\limits_{k\to+\infty} \bar{A}_1(k) = 0$ implies $\lim\limits_{r\to+\infty} \bar{A}(r) = 0$.

Let $r > 0$. Then $r \in [\eta_1^m, \eta_1^{m+1})$ for some integer m. By the reverse doubling condition for μ we have

$$(\bar{A}(r))^q \leq c_8 \left(\int\limits_{\hat{X}\backslash D_{\eta_1^m}} (\mu B(x_0, d(x_0, x) + t))^{(\alpha-1)q} d\nu(x, t) \right) \times$$

$$\times (\mu B(x_0, \eta_1^m))^{q/p'} \leq c_7 (\sup_{k\geq m} \bar{A}_1(k))^q.$$

Hence $\lim\limits_{r\to+\infty} \bar{A}(r) = 0$ if $\lim\limits_{k\to+\infty} \bar{A}_1(k) = 0$.

Finally we find that $\lim\limits_{d\to\infty} \bar{A}_{d/(2a_1)} = 0$, if $\lim\limits_{k\to+\infty} \overline{A}_1(k) = 0$.

Let us show that (iii) follows from (i). Let $k \in Z$ and let

$$f_k(x) = \chi_{B(x_0, \eta_1^{k+1}/(2a_1))\backslash B(x_0, \eta_1^{k-2}\backslash(2a_1))}(x)(\mu B(x_0, \eta_1^k))^{-1/p}.$$

Then for $\varphi \in L_\mu^{p'}(X)$ we obtain that

$$\left| \int\limits_X \varphi(x) f_k(x) d\mu(x) \right| \leq$$

$$\leq c_9 \left(\int\limits_{B(x_0, \eta_1^{k-1}/(2a_1))\backslash B(x_0, \eta_1^{k-2}/(2a_1))} |\varphi(x)|^{p'} d\mu(x) \right)^{1/p'} \to 0$$

as $k \to +\infty$ or $k \to -\infty$ (here we used the condition $\mu\{x_0\} = 0$).

On the other hand,

$$\|\bar{J}_\alpha f\|_{L_\nu^q(\hat{X})} \geq$$

$$\geq \left(\int\limits_{D_{\eta_1^{k+1}}\backslash D_{\eta_1^k}} (\mu B(x_0, d(x_0, x) + t))^{(\alpha-1)q} d\nu(x, t) \right)^{1/q} \times$$

$$\times (\mu B(x_0, \eta_1^k))^{1/p'} \to 0$$

as $k \to -\infty$ or $k \to +\infty$. Finally we have (i) $\Rightarrow$ (ii) and (i) $\Rightarrow$ (iii). $\square$

The following statement can be proved analogously:

Theorem 6.5.4. *Let $\mu(X) < \infty$, $1 < p \leq q < \infty$ and $\alpha > 1/p$. Then the following statements are equivalent:*
(i) *$\bar{J}_\alpha$ is compact from $L_\mu^p(X)$ to $L_\nu^q(\hat{X})$;*
(ii) *$\tilde{A} < \infty$ and $\lim\limits_{r\to 0} \tilde{A}(r) = 0$, where*

$$\tilde{A}(r) =$$

$$= \left(\int\limits_{\hat{X}\backslash D_r} (\mu B(x_0, d(x_0, x) + t))^{(\alpha-1)q} d\nu(x,t) \right)^{1/q} (\mu B(x_0, r))^{1/p'} < \infty;$$

(iii) $\tilde{A}_1 < \infty$ and $\lim\limits_{k \to -\infty} \tilde{A}_1(k) = 0$, where

$$\tilde{A}_1(k) = \left(\int\limits_{D_{\eta_1^{k+1}}\backslash D_{\eta_1^k}} (\mu B(x_0, d(x_0, x) + t)))^{(\alpha-1)q} \times \right.$$

$$\left. \times (\mu B(x_0, d(x_0, x)))^{q/p'} d\nu(x,t) \right)^{1/q}.$$

6.6. Two–weight (p, p) type inequalities

In this section we establish necessary and sufficient conditions on weight functions for the validity of two-weight inequalities of strong (weak) type for fractional integrals defined on SHT.

We shall assume that $\mu X = \infty$ and that there exists a point $x_0 \in X$ such that $\mu\{x_0\} = 0$ and

$$\mu(B(x_0, R) \setminus B(x_0, r)) > 0 \qquad\qquad (*)$$

for all r and R such that $0 < r < R < \infty$.

Let us define potential-type operators on X :

$$K_\alpha f(x) = \int\limits_X \frac{f(y)}{\mu B(x, d(x,y))^{1-\alpha}} d\mu, \ \ 0 < \alpha < 1,$$

$$J_\alpha f(x) = \int\limits_{\overline{B}(x_0, 2d(x_0, x))} \frac{f(y)}{\mu B(x, d(x,y))^{1-\alpha}} d\mu, \ \ \alpha > 0.$$

The following theorem holds (see, e.g., [101], [100], Chapter 3):

Theorem A. *Let $0 < \alpha < 1$, $1 < p < 1/\alpha$, $1/p - 1/p^* = \alpha$. Then the operator K_α is bounded from $L^p_{vp}(X)$ into $L^{p^*}_{vp^*}(X)$ if and only if $v \in A_{pp^*}(X)$, that is*

$$\sup \left(\frac{1}{\mu B} \int\limits_B v^{p^*}(x) d\mu \right)^{1/p^*} \left(\frac{1}{\mu B} \int\limits_B v^{-p'}(x) d\mu \right)^{1/p'} < \infty,$$

where the supremum is taken over all balls $B \subset X$.

Recall that a $\mu-$ measurable, locally integrable function $w : X \to R$ which is positive $\mu-$ a.e. is called a weight.

Let $1 < p < \infty$; then $A_p(X)$ is the set of all weights w such that

$$\sup \left(\frac{1}{\mu(B)} \int_B w(x)d\mu \right) \left(\frac{1}{\mu(B)} \int_B w^{-1/(p-1)}(x)d\mu \right)^{p-1} < \infty,$$

where the supremum is taken over all balls $B \subset X$. The Muckenhoupt class $A_1(X)$ is the set of all weights w such that

$$\sup_{B \subset X} \left(\frac{1}{\mu(B)} \int_B w(x)d\mu \right) \left(\text{ess}_{x \in B} \frac{1}{w(x)} \right) < \infty.$$

Further, the maximal operator M is defined by

$$(Mf)(x) = \sup(\mu(B))^{-1} \int_B |f(y)|d\mu(y), \quad x \in X,$$

where the supremum is taken with respect to all balls $B \subset X$ containing x.

We shall need the following Lemmas:

Lemma 6.6.1. *Let $1 < p < \infty$, let $\rho \in A_p(X)$ and suppose $0 < c_1 \leq c_2 < c_3 < \infty$. Then there exists a positive number c such that for any $t > 0$ we have*

$$\int_{B(x_0,c_3t) \backslash B(x_0,c_2t)} \rho(x)d\mu \leq c \int_{B(x_0,c_1t)} \rho(x)d\mu.$$

Proof. By the definition of the maximal function M and the doubling condition we have

$$M\phi(x) \geq \left(\frac{1}{\mu(B(x_0,c_3t))} \int_{B(x_0,c_3t)} |\phi(y)|d\mu \right) \chi_{B(x_0,c_3t) \backslash B(x_0,c_2t)}(x) \geq$$

$$\geq \left(\frac{b_1}{\mu(B(x_0,c_1t))} \int_{B(x_0,c_1t)} |\phi(y)|d\mu \right) \chi_{B(x_0,c_3t) \backslash B(x_0,c_2t)}(x)$$

$$\tag{6.6.1}$$

for any $\phi \in L^p_\rho(X)$. From (6.6.1), in view of the boundedness of the operator M in $L^p_\rho(X)$ (see [286], [278]) we obtain

$$\int_{B(x_0,c_3t) \backslash B(x_0,c_2t)} \left(\frac{1}{\mu(B(x_0,c_1t))} \int_{B(x_0,c_1t)} |\phi(y)|d\mu \right)^p \rho(x)d\mu \leq$$

$$\leq c \int_X |\phi(y)|^p \rho(y)d\mu.$$

If in the last inequality we take $\phi(y) = \chi_{B(x_0,c_1t)}(y)$, the desired estimate follows. $\square$

Lemma 6.6.2. *Let* $0 < \alpha < 1/p$, $1 < p < \infty$, $1/p - 1/p^* = \alpha$. *Assume that* $0 \le c_1 \le c_2 < c_3 < \infty$ *and let* $v \in A_{pp^*}(X)$. *Then there exists a positive constant* c *such that for all* $t > 0$ *the following inequality holds:*

$$\int_{B(x_0,c_3t)\backslash B(x_0,c_2t)} v(x)^{-p'}d\mu \le c \int_{B(x_0,c_1t)} v^{-p'}(x)d\mu. \qquad (6.6.2)$$

Proof. The condition $v \in A_{pp^*}(X)$ is equivalent to the condition $v^{p^*} \in A_{1+p^*/p'}(X)$. Therefore $(v^{p^*})^{-p'/p^*} = v^{-p'}$ belongs to $A_{1+p'/p^*}(X)$ and using Lemma 6.6.1 we obtain (6.6.2). $\square$

Lemma 6.6.3. *Let* $0 < \alpha < 1$, $1 < p < 1/\alpha$, $1/p - 1/p^* = \alpha$. *If* $\rho \in A_{pp^*}(X)$, *then the following condition is fulfilled:*

$$A \equiv \sup_{t>0} \left(\int_{B(x_0,t)} \rho^{p^*}(x)d\mu \right)^{1/p^*} \times$$

$$\times \left(\int_{X\backslash B(x_0,t)} \frac{\rho^{-p'}(x)}{(\mu(B(x_0,d(x_0,x))))^{(1-\alpha)p'}}d\mu \right)^{1/p'} < \infty.$$

Proof. As $\rho \in A_{pp^*}(X)$, by Theorem A we see that K_α is bounded from $L^p_{\rho^p}(X)$ into $L^{p^*}_{\rho^{p^*}}(X)$. Hence (see [101], [100], Chapter 3) $A < \infty$. $\square$

Lemma 6.6.4. *Let* $\rho \in A_p(X)$, *where* $1 < p < \infty$. *Then there exists a positive constant* $c > 0$ *such that for all* $t > 0$ *the following inequality holds:*

$$\int_{d(x_0,x)<2t} \rho(x)d\mu \le c \int_{d(x_0,x)<t} \rho(x)d\mu.$$

Proof. Using Hölder's inequality and the definition of the $A_p(X)$ class, we have

$$\int_{B(x_0,2t)} \rho(x)d\mu \le B(x_0,2t)^p \left(\int_{B(x_0,2t)} \rho^{1-p'}(x)d\mu \right)^{1-p} \le$$

$$\le c\mu B(x_0,t)^p \left(\int_{B(x_0,t)} \rho^{1-p'}(x)d\mu \right)^{1-p} \le$$

$$\le c \left(\int_{B(x_0,t)} \rho(x)d\mu \right) \left(\int_{B(x_0,t)} \rho^{1-p'}d\mu \right)^{p-1} \left(\int_{B(x_0,t)} \rho^{1-p'}d\mu \right)^{1-p} =$$

$$= c \int\limits_{B(x_0,t)} \rho(x)d\mu.$$

$\square$

Lemma 6.6.5. *Let* $0 < \alpha < 1$, $1 < p < 1/\alpha$, $1/p - 1/p^* = \alpha$, $\rho \in A_{pp^*}(X)$ *and let* $v(x) = \sigma(d(x_0, x))\rho(x)$, $w(x) = u(d(x_0, x))\rho(x)$. *Assume that* $\beta \geq 1$. *If*

$$B \equiv \sup_{t>0} \left(\int\limits_{\{d(x_0,x)>t\}} \frac{v(x)}{(\mu(B(x_0, d(x_0, x))))^{(1-\alpha)p}} d\mu \right)^{1/p'} \times$$
$$\times \left(\int\limits_{d(x_0,x)\leq t} w^{1-p'}(x)d\mu \right)^{p-1} < \infty, \tag{6.6.3}$$

then for all $t > 0$ *we have*

$$\sigma(\beta t)(\mu B(x_0, t))^{\alpha p} \leq cu(t), \tag{6.6.4}$$

and the following condition is satisfied:

$$B' \equiv \sup_{t>0} \left(\int\limits_{d(x_0,x)\leq t} v(x)d\mu \right)^{1/p} \times$$
$$\times \left(\int\limits_{d(x_0,x)>t} \frac{w^{1-p'}(x)}{(\mu(B(x_0, d(x_0, x))))^{(1-\alpha)p'}} d\mu \right)^{1/p'} < \infty. \tag{6.6.5}$$

Proof. Let $\eta \geq \eta_1$, where η_1 is the constant from Definition 6.5.1. As before we put

$$\overline{B}(x_0, r) \equiv \{x \in X : d(x_0, x) \leq r\}.$$

By the condition (*) the reverse doubling condition is satisfied at x_0 (see Proposition 6.5.1). Therefore

$$\mu(B(x_0, \eta\beta t) \setminus B(x_0, \beta t)) = \mu(B(x_0, \eta\beta t)) - \mu(B(x_0, \beta t)) \geq$$

$$\geq \mu(B(x_0, \eta\beta t)) - \frac{1}{\eta_2}\mu(B(x_0, \eta\beta t)) \geq$$

$$\geq (1 - \frac{1}{\eta_2})\mu(B(x_0, \eta\beta t)) \geq c_1\mu(B(x_0, \eta\beta t)).$$

Thus

$$\mu(B(x_0, \eta\beta t) \setminus B(x_0, \beta t)) \geq c_1 \mu(B(x_0, \eta\beta t)).$$

By the monotonicity of σ and u, we have

$$\left(\int_{X \setminus B(x_0, t)} v(x)(\mu(B(x_0, d(x_0, x))))^{(\alpha-1)p} d\mu \right) \geq$$

$$\geq \int_{X \setminus B(x_0, \beta t)} \sigma(d(x_0, x))\rho^p(x)(\mu(B(x_0, d(x_0, x))))^{(\alpha-1)p} d\mu$$

$$\geq \int_{B(x_0, \eta\beta t) \setminus B(x_0, \beta t)} \frac{\sigma(d(x_0, x))\rho^p(x)}{(\mu(B(x_0, d(x_0, x))))^{(1-\alpha)p}} d\mu \geq$$

$$\geq c_2 \sigma(\beta t)(\mu(B(x_0, t)))^{(\alpha-1)p} \int_{B(x_0, \eta\beta t) \setminus B(x_0, \beta t)} \rho^p(x) d\mu$$

and

$$\left(\int_{B(x_0, t)} w^{1-p'}(x) d\mu \right)^{p-1} = \left(\int_{B(x_0, t)} u^{1-p'}(d(x_0, x))\rho^{-p'}(x) d\mu \right)^{p-1}$$

$$\geq \frac{1}{u(t)} \left(\int_{B(x_0, t)} \rho^{-p'}(x) d\mu \right)^{p-1}.$$

Using these estimates, Hölder's inequality and Lemma 6.6.1 it follows that

$$\frac{\sigma(\beta t)\mu B(x_0, t)^{\alpha p}}{u(t)} = \frac{\sigma(\beta t)\mu B(x_0, t)^{\alpha p}}{u(t)} \times$$

$$\times \left(\frac{1}{\mu(B(x_0, \eta\beta t) \setminus B(x_0, \beta t))} \int_{B(x_0, \eta\beta t) \setminus B(x_0, \beta t)} \rho(x)\rho^{-1}(x) d\mu \right)^p \leq$$

$$\leq \frac{\sigma(\beta t)\mu B(x_0, t)^{\alpha p}}{u(t)(\mu(B(x_0, \eta\beta t) \setminus B(x_0, \beta t)))^p} \times$$

$$\times \int_{B(x_0, \eta\beta t) \setminus B(x_0, \beta t)} \rho^p(x) d\mu \left(\int_{B(x_0, \eta\beta t) \setminus B(x_0, \beta t)} \rho^{-p'}(x) d\mu \right)^{p-1} \leq$$

$$\leq c_3 \frac{\sigma(\beta t)\mu B(x_0, t)^{(\alpha-1)p}}{u(t)} \int_{B(x_0, \eta\beta t) \setminus B(x_0, \beta t)} \rho^p(x) d\mu \times$$

$$\times \left(\int\limits_{B(x_0,t)} \rho^{-p'}(x)d\mu \right)^{p-1} \le$$

$$\le c_4 \frac{\sigma(\beta t)}{u(t)} \left(\int\limits_{B(x_0,\eta\beta t)\backslash B(x_0,\beta t)} \rho^p(x)(\mu(B(x_0,\eta\beta t)))^{(\alpha-1)p}d\mu \right) \times$$

$$\times \left(\int\limits_{B(x_0,t)} \rho^{-p'}(x)d\mu \right)^{p-1} \le$$

$$\le c_4 \frac{\sigma(\beta t)}{u(t)} \int\limits_{B(x_0,\eta\beta t)\backslash B(x_0,\beta t)} \rho^p(x)\mu(B(x_0,d(x_0,x)))^{(\alpha-1)p}d\mu \times$$

$$\times \left(\int\limits_{B(x_0,t)} \rho^{-p'}(x)d\mu \right)^{p-1} \le$$

$$\le c_5 \left(\int\limits_{X\backslash B(x_0,t)} \frac{v(x)}{(\mu(B(x_0,d(x_0,x))))^{(1-\alpha)p}}d\mu \right) \times$$

$$\times \left(\int\limits_{B(x_0,t)} w^{1-p'}(x)d\mu \right)^{p-1} \le c.$$

This proves the first part of the lemma is proved. Let us establish condition (6.6.5). Using Lemma 6.6.4 we have

$$\int\limits_{d(x_0,x)<2t} \rho^{p^*}(x)d\mu \le c_6 \int\limits_{d(x_0,x)<t} \rho^{p^*}(x)d\mu,$$

where the constant c_6 does not depend on t. Let $t > 0$. Then by Lemma 6.6.3 and by Hölder's inequality we have

$$\left(\int\limits_{\{d(x_0,x)\le t\}} v(x)d\mu \right)^{1/p} \left(\int\limits_{\{d(x_0,x)>t\}} \frac{w^{1-p'}(x)}{\mu B(x_0,d(x_0,x))^{(1-\alpha)p'}}d\mu \right)^{1/p'} \le$$

$$\le c_7 \left(\frac{\sigma(t)}{u(t)} \right)^{1/p} \left(\int\limits_{\{d(x_0,x)\le t\}} \rho^{p^*}(x)d\mu \right)^{1/p^*} \times$$

$$\times \left(\int\limits_{\{d(x_0,x)>t\}} \frac{\rho^{-p'}(x)}{\mu B(x_0,d(x_0,x))^{(1-\alpha)p'}}d\mu \right)^{1/p'} \mu\{d(x_0,x) \le t\}^\alpha \le$$

$$\le c_8 \left(\frac{\sigma(t)}{u(t)} \right)^{1/p} \mu B(x_0,t)^\alpha \le c_9$$

for all $t > 0$. $\square$

Theorem 6.6.1. *Let* $0 < \alpha < 1$, $1 < p < 1/\alpha$ *and* $1/p - 1/p^* = \alpha$. *Assume that* σ *and* u *are positive increasing functions on* $(0, \infty)$ *and* $\rho \in A_{pp^*}(X)$. *We put* $v(x) = \sigma(d(x_0, x))\rho^p(x)$, $w(x) = u(d(x_0, x))\rho^p(x)$. *Then for the boundedness of the operator* K_α *from* $L_w^p(X)$ *into* $L_v^p(X)$ *it is necessary and sufficient that*

$$B \equiv \sup_{t>0} \left(\int\limits_{d(x_0,x)>t} \frac{v(x)}{(\mu B(x_0, d(x_0, x)))^{(1-\alpha)p}} d\mu \right)^{1/p} \times$$

$$\times \left(\int\limits_{d(x_0,x)\leq t} w^{1-p'}(x) d\mu \right)^{1/p'} < \infty.$$

Proof. Let us represent K_α as follows:

$$K_\alpha f(x) = \int\limits_{d(x_0,y)\leq \frac{d(x_0,x)}{2a_1}} \frac{f(y)}{\mu B(x, d(x, y))^{1-\alpha}} d\mu +$$

$$+ \int\limits_{\frac{d(x_0,x)}{2a_1}<d(x_0,y)\leq 2a_1 d(x_0,x)} \frac{f(y)}{\mu B(x, d(x, y))^{1-\alpha}} d\mu +$$

$$+ \int\limits_{d(x_0,y)>2a_1 d(x_0,x)} \frac{f(y)}{\mu B(x, d(x, y))^{1-\alpha}} d\mu \equiv$$

$$\equiv I_1 f(x) + I_2 f(x) + I_3 f(x),$$

where the constant a_1 is from the triangle inequality for the quasi–metric d (see Definition 1.1.1.) Therefore we have

$$\|K_\alpha f\|^p_{L_v^p(X)} \leq c_1 \|I_1 f\|^p_{L_v^p(X)} + c_1 \|I_2 f\|^p_{L_v^p(X)} + c_1 \|I_3 f\|^p_{L_v^p(X)} \equiv$$

$$\equiv S_1 + S_2 + S_3.$$

First we estimate S_1. Note that if $d(x_0, y) \leq \frac{d(x_0,x)}{2a_1}$, then

$$d(x_0, x) \leq a_1(d(x_0, y) + d(y, x)) \leq a_1(d(x_0, y) + a_0 d(x, y)) \leq$$

$$\leq a_1 \left(\frac{d(x_0, x)}{2a_1} + a_0 d(x, y) \right).$$

Hence $\frac{d(x_0,x)}{2a_1 a_0} \leq d(x, y)$ and consequently

$$\mu B(x_0, d(x_0, x)) \leq b_1 \mu B(x_0, d(x, y)). \tag{6.6.6}$$

Further, there exists a positive constant b_2 such that

$$\mu B(x_0, d(x_0, x)) \le b_2 \mu B(x, d(x, y)). \qquad (6.6.7)$$

Indeed, if $z \in B(x_0, d(x, y))$, then

$$d(x, z) \le a_1(d(x, x_0) + d(x_0, z)) \le$$
$$\le a_1(a_0 d(x_0, x) + d(x_0, z)) \le a_1(2a_0^2 a_1 d(x, y) + d(x, y)) =$$
$$= a_1(2a_0^2 a_1 + 1)d(x, y)$$

and so

$$B(x_0, d(x, y)) \subset B(x, a_1(2a_1^2 a_1 + 1)d(x, y))$$

if $d(x_0, x) \le d(x_0, x)/(2a_1)$. From the doubling condition and (6.6.6) we obtain (6.6.7).

Using Theorem 1.1.4 we conclude that

$$S_1 \le c_2 \int\limits_X \frac{v(x)}{(\mu B(x_0, d(x_0, x)))^{(1-\alpha)p}} \left(\int\limits_{B(x_0, d(x_0, x))} |f(y)| d\mu \right)^p d\mu \le$$

$$\le c_3 B^p \int\limits_X |f(x)|^p w(x) d\mu.$$

Now we estimate S_3. First we note that if $d(x_0, y) \ge 2a_1 d(x_0, x)$, then

$$\mu B(x_0, d(x_0, y)) \le b_3 \mu B(x, d(x, y)). \qquad (6.6.8)$$

Indeed, from the inequality

$$d(x_0, y) \le (d(x_0, x) + d(x, y)) \le a_1 \left(\frac{d(x_0, y)}{2a_1} + d(x, y) \right)$$

we have $\frac{d(x_0, y)}{2a_1} \le d(x, y)$. Thus

$$\mu B(x_0, d(x_0, y)) \le b_4 \mu B(x_0, d(x, y)).$$

In addition, it is easy to check that

$$\mu B(x_0, d(x, y)) \le b_5 \mu B(x, d(x, y)).$$

Thus inequality (6.6.8) is proved.

Now we use Corollary 1.1.8 and find that

$$S_3 \le c_3 \int\limits_X v(x) \left(\int\limits_{d(x_0, y) > d(x_0, x)} \frac{|f(y)|}{\mu B(x_0, d(x_0, y))^{1-\alpha}} d\mu \right)^p d\mu \le$$

$$\leq c_5 B^p \int_X |f(x)|^p w(x) d\mu.$$

By Lemma 6.6.5 and by Theorem A we get

$$S_2 = c_4 \sum_{k \in Z} \int_{2^k < d(x_0,x) \leq 2^{k+1}} v(x) \times$$

$$\times \left(\int_{d(x_0,x)/(2a_1) < d(x_0,y) < 2a_1 d(x_0,x)} \frac{|f(y)|}{\mu B(x, d(x,y))^{1-\alpha}} d\mu \right)^p d\mu \leq$$

$$\leq c_4 \sum_{k \in Z} \int_{2^k < d(x_0,x) \leq 2^{k+1}} \sigma(d(x_0,x)) \rho^p(x) \times$$

$$\times \left(\int_X \frac{|f(y)| \chi_{\{2^{k-1}/a_1 < d(x_0,z) < 2^{k+2}a_1\}}(y)}{\mu B(x, d(x,y))^{1-\alpha}} d\mu \right)^p d\mu \leq$$

$$\leq c_4 \sum_{k \in Z} \left(\int_{2^k < d(x_0,x) < 2^{k+1}} (\sigma(d(x_0,x)))^{1/(\alpha p)} d\mu \right)^{\alpha p} \times$$

$$\times \left(\int_{2^k < d(x_0,x) \leq 2^{k+1}} \rho^{p^*}(x) (K_\alpha(f\chi_{\{\frac{2^{k-1}}{a_1} < d(x_0,y) < 2^{k+2}a_1\}})(x))^{p^*} d\mu \right)^{p/p^*} \leq$$

$$\leq c_5 \sum_{k \in Z} \left(\int_{2^k < d(x_0,x) \leq 2^{k+1}} \sigma(d(x_0,x))^{1/(\alpha p)} d\mu \right)^{\alpha p} \times$$

$$\times \left(\int_{\frac{2^{k-1}}{a_1} < d(x_0,x) < 2^{k+2}a_1} \rho^p(x)|f(x)|^p d\mu \right) \leq$$

$$\leq c_6 \sum_{k \in Z} \sigma(2^{k+1}) \mu B(x_0, 2^{k+1})^{\alpha p} \times$$

$$\times \left(\int_{\frac{2^{k-1}}{a_1} < d(x_0,x) < 2^{k+1}a_1} \rho^p(x)|f(x)|^p d\mu \right) \leq$$

$$\leq c_7 \sum_{k \in Z} u\left(\frac{2^{k-1}}{a_1} \right) \left(\int_{\frac{2^{k-1}}{a_1} < d(x_0,x) < 2^{k+2}a_1} \rho^p(x)|f(x)|^p d\mu \right) \leq$$

$$\leq c_7 \sum_{k \in Z} \int_{\frac{2^{k-1}}{a_1} < d(x_0,x) < 2^{k+2}a_1} w(x)|f(x)|^p d\mu \leq c_8 \int_X w(x)|f(x)|^p d\mu.$$

To prove necessity we first show that

$$I(t) = \int\limits_{B(x_0,t)} w^{1-p'}(x)d\mu < \infty$$

for all $t > 0$.

Let $I(t) = \infty$. Then $\left\| \frac{1}{w} \right\|_{L_w^{p'}(B(x_0,t))} = \infty$ and there exists a function $g \geq 0$ such that

$$\|g\|_{L^1(B(x_0,t))} = \infty,$$

and

$$\|g\|_{L_w^p(B(x_0,t))} < \infty.$$

Now let $f(x) = g(x)\chi_{B(x_0,t)}(x)$. Taking into account the inequality

$$\mu B(x, d(x,y)) \leq b_6 \mu B(x_0, d(x_0,x))$$

for x and y with the condition $d(x_0,x) > t$ and $d(x_0,y) < t$, where the constant b_6 is independent of x, we obtain

$$\|K_\alpha f\|_{L_v^p(X)} \geq$$

$$\geq \left(\int\limits_{d(x_0,x)>t} v(x) \left(\int\limits_{d(x_0,y)<t} \frac{g(y)}{(\mu B(x,d(x,y)))^{1-\alpha}} d\mu \right)^p d\mu \right)^{1/p} \geq$$

$$\geq \left(\int\limits_{d(x_0,x)>t} v(x)(\mu B(x_0,d(x_0,x)))^{(\alpha-1)p} d\mu \right)^{1/p} \times$$

$$\times \left(\int\limits_{d(x_0,x)<t} g(x)d\mu \right) = \infty.$$

On the other hand,

$$\int\limits_X |f(x)|^p w(x)d\mu = \int\limits_{B(x_0,t)} g(x)w(x)d\mu < \infty.$$

We conclude that $I(t) < \infty$ for all $t > 0$.

Now let us assume that

$$f(x) = w^{1-p'}(x)\chi_{\{d(x_0,y)<t\}}(x).$$

Then

$$\|K_\alpha f\|_{L_v^p(X)} \geq c_9 \left(\int\limits_{d(x_0,x)>t} v(x)(\mu B(x_0,d(x_0,x)))^{(\alpha-1)p} d\mu \right)^{1/p} \times$$

$$\times \left(\int_{B(x_0,t)} w^{1-p'}(x)d\mu \right).$$

On the other hand,

$$\|f\|_{L_w^p(X)} = \left(\int_{B(x_0,t)} w^{1-p'}(x)dx \right)^{1/p} < \infty.$$

Due to the boundedness of K_α we finally conclude that $B < \infty$. $\square$

Lemma 6.6.6. *Let* $0 < \alpha < 1$, $1 < p < 1/\alpha$, $1/p - 1/p^* = \alpha$, $\rho \in A_{pp^*}(X)$. *Let* $v(x) = \sigma(d(x_0,x))\rho(x)$ *and* $w(x) = u(d(x_0,x))\rho(x)$. *Assume that* $\beta \geq 1$. *If*

$$B_1 \equiv \sup_{0<t<\tau<\infty} \frac{1}{(\mu B(x_0,\tau))^{1-\alpha}} \left(\int_{\{t<d(x_0,x)<\tau\}} v(x)d\mu \right)^{1/p} \times$$

$$\times \left(\int_{d(x_0,x)\leq t} w^{1-p'}(x)d\mu \right)^{1/p'} < \infty,$$

then (6.6.4) *and* (6.6.5) *are fulfilled.*

Proof. Let $\eta \geq \eta_1$, where η_1 is a constant from Definition 6.5.1. As we know, from condition (*) there follows the reverse doubling condition at x_0 (see Proposition 6.5.1). Therefore there exists a positive constant b_1 such that for all $t > 0$ the next inequality holds

$$\mu(B(x_0,\eta\beta t) \setminus B(x_0,\beta t)) \geq b_1\mu(B(x_0,\eta\beta t)).$$

By the monotonicity of σ and u, we have

$$\int_{X\setminus B(x_0,t)} v(x)(\mu(B(x_0,d(x_0,x))))^{(\alpha-1)p}d\mu \geq$$

$$\geq \int_{X\setminus B(x_0,\beta t)} \sigma(d(x_0,x))\rho^p(x)(\mu(B(x_0,d(x_0,x))))^{(\alpha-1)p}d\mu \geq$$

$$\geq \sigma(\beta t) \int_{B(x_0,\eta\beta t)\setminus B(x_0,\beta t)} \frac{\rho^p(x)}{(\mu(B(x_0,d(x_0,x))))^{(1-\alpha)p}}d\mu$$

and

$$\left(\int_{B(x_0,t)} w^{1-p'}(x)d\mu \right)^{p-1} = \left(\int_{B(x_0,t)} u^{1-p'}(d(x_0,x))\rho^{-p'}(x)d\mu \right)^{p-1} \geq$$

$$\geq \frac{1}{u(t)} \left(\int_{B(x_0,t)} \rho^{-p'}(x)d\mu \right)^{p-1}.$$

Using these estimates, Hölder's inequality and Lemma 6.6.1 we find that

$$\frac{\sigma(\beta t)\mu B(x_0,t)^{\alpha p}}{u(t)} \leq$$

$$\leq c_1 \frac{\sigma(\beta t)(\mu B(x_0,\eta\beta t))^{(\alpha-1)p}}{u(t)} \int_{B(x_0,\eta\beta t)\backslash B(x_0,\beta t)} \rho^p(x)d\mu \times$$

$$\times \left(\int_{B(x_0,t)} \rho^{-p'}(x)d\mu \right)^{p-1} \leq c_2$$

for all $t > 0$. The first part of the lemma is proved. Now we show that condition (6.6.5) is satisfied. Using Lemma 6.6.4 we have

$$\int_{d(x_0,x)<2t} \rho^{p^*}(x)d\mu \leq c \int_{d(x_0,x)<t} \rho^{p^*}(x)d\mu,$$

where the constant $c > 0$ does not depend on t. For $t > 0$ by Lemma 6.6.3 and Hölder's inequality we have

$$\left(\int_{\{d(x_0,x)\leq t\}} v^p(x)d\mu \right)^{1/p} \left(\int_{\{d(x_0,x)>t\}} \frac{w^{1-p'}(x)}{\mu B(x_0,d(x_0,x))^{(1-\alpha)p'}}d\mu \right)^{1/p'} \leq$$

$$\leq c_1.$$

$\square$

Theorem 6.6.2. *Let the conditions of Theorem 6.6.1 hold. Then K_α is bounded from $L^p_w(X)$ into $L^{p\infty}_v(X)$ if and only if*

$$B_1 \equiv \sup_{0<t<\tau<\infty} \frac{1}{(\mu B(x_0,\tau))^{1-\alpha}} \left(\int_{t<d(x_0,x)<\tau} v(x)d\mu \right)^{1/p} \times$$

$$\times \left(\int_{d(x_0,x)\leq t} w^{1-p'}(x)d\mu \right)^{1/p'} < \infty.$$

Proof. The sufficiency of this theorem is proved as in the case of Theorem 6.6.1. Therefore we simply prove necessity.

At first we show that

$$I(t) \equiv \int_{B(x_0,t)} w^{1-p'}(x)dx < \infty$$

for all $t > 0$. Indeed, if $I(t) = \infty$ for some $t > 0$, then there exists $g \in L^p(B(x_0,t))$, $g \geq 0$, such that $gw^{-1/p} \notin L^1(B(x_0,t))$. We put $f(y) = g(y)w^{-1/p}(y)\chi_{B(x_0,t)}$. Then for $x \in X \setminus B(x_0,t)(y)$ we have

$$K_\alpha f(x) \geq c_1(\mu B(x_0,d(x_0,x)))^{\alpha-1} \int_{B(x_0,t)} g(y)w^{-1/p}(y)dy = \infty.$$

Therefore

$$S \equiv \int_{X \setminus B(x_0,t)} v(x)d\mu \leq \int_{\{x:|K_\alpha f(x)|>\lambda\}} v(x)d\mu \leq$$

$$\leq \frac{c_2}{\lambda^p} \int_{B(x_0,t)} g(x)^p d\mu.$$

As the constant c_2 does not depend on λ we see that $S = 0$. Now let us take $f(y) = w^{1-p'}(y)\chi_{B(x_0,t)}(y)$. Then for $x \in B(x_0,\tau) \setminus B(x_0,t)$ we have

$$K_\alpha f(x) \geq c_3 \frac{1}{\mu B(x_0,\tau)^{1-\alpha}} I(t).$$

From the boundedness of K_α we obtain the estimates

$$\int_{B(x_0,\tau) \setminus B(x_0,t)} v(x)d\mu \leq \int_{\{x:K_\alpha(x)>c_3 I(t)/\mu B(x_0,\tau)^{1-\alpha}\}} v(x)d\mu \leq$$

$$\leq \frac{c_3(\mu B(x_0,\tau))^{(1-\alpha)p}}{I(t)^p} I(t).$$

$\square$

Using these theorems we can derive the following results for I_α and J_α.

Theorem 6.6.3. Let $0 < \alpha < 1$, $1 < p < 1/\alpha$. Assume that σ and u are positive increasing functions on $(0,\infty)$ and $\rho \in A_{pp^*}(X)$, where $p^* = \frac{p}{1-\alpha p}$.

We put $v(x) = \sigma(d(x_0, x))\rho^p(x)$, $w(x) = u(d(x_0, x))\rho^p(x)$. *Then for the boundedness of the operator* J_α *from* $L^p_w(X)$ *into* $L^p_v(X)$ *it is necessary and sufficient that*

$$\sup_{t>0} \left(\int\limits_{d(x_0,x)>t} \frac{v(x)}{(\mu B(x_0, d(x_0, x)))^{(1-\alpha)p}} d\mu \right)^{1/p} \times$$

$$\times \left(\int\limits_{d(x_0,x)\leq t} w^{1-p'}(x)d\mu \right)^{1/p'} < \infty.$$

Theorem 6.6.4. *Let* $0 < \alpha < 1$, $1 < p < 1/\alpha$. *Assume that* σ *and* u *are positive increasing functions on* $(0, \infty)$ *and* $\rho \in A_{pp^*}(X)$, *where* $p^* = \frac{p}{1-\alpha p}$. *We put* $v(x) = \sigma(d(x_0, x))\rho^p(x)$, $w(x) = u(d(x_0, x))\rho^p(x)$. *Then for the boundedness of the operator* J_α *from* $L^p_w(X)$ *into* $L^{p\infty}_v(X)$ *it is necessary and sufficient that*

$$\sup_{0<t<\tau<\infty} \frac{1}{(\mu B(x_0, \tau))^{1-\alpha}} \left(\int\limits_{t<d(x_0,x)<\tau} v(x)d\mu \right)^{1/p} \times$$

$$\times \left(\int\limits_{d(x_0,x)\leq t} w^{1-p'}(x)d\mu \right)^{1/p'} < \infty.$$

Let G be a homogeneous group with homogeneous norm $x \to r(x)$ and homogeneous dimension Q (see Section 1.4 for the definition and properties of homogeneous groups). It is easy to verify that G with homogeneous norm and Haar measure dx represents a space of homogeneous type. The potential-type operators on G have a form:

$$T_\gamma f(x) = \int\limits_G \frac{f(y)}{r(xy^{-1})^{Q-\gamma}}dx, \quad 0 < \gamma < Q;$$

$$I_\gamma f(x) = \int\limits_{B(e,2r(x))} \frac{f(y)}{r(xy^{-1})^{Q-\gamma}}dx, \quad \gamma > 0.$$

The next statements follow from the results formulated above. The proofs are omitted.

Theorem 6.6.5. *Let* $0 < \gamma < Q$, $1 < p < Q/\gamma$ *and* $p^* = \frac{pQ}{Q-p\gamma}$. *Assume that* σ *and* u *are positive increasing functions on* $(0, \infty)$ *and* $\rho \in A_{pp^*}(G)$. *We* *put* $v(x) = \sigma(r(x))\rho^p(x)$, $w(x) = u(r(x))\rho^p(x)$. *Then for the boundedness of the operator* T_γ *from* $L^p_w(G)$ *into* $L^p_v(G)$ *it is necessary and sufficient that*

$$B \equiv \sup_{t>0} \left(\int_{r(x)>t} \frac{v(x)}{(r(x))^{(Q-\gamma)p}} dx \right)^{1/p} \times$$

$$\times \left(\int_{r(x)<t} w^{1-p'}(x) dx \right)^{1/p'} < \infty.$$

Theorem 6.6.6. *Let the conditions of Theorem 6.6.5 be satisfied. Then* T_γ *is bounded from* $L^p_w(G)$ *into* $L^{p\infty}_v(G)$ *if and only if*

$$B_1 \equiv \sup_{0<t<\tau<\infty} \frac{1}{\tau^{Q-\gamma}} \left(\int_{t<r(x)<\tau} v(x) dx \right)^{1/p} \times$$

$$\times \left(\int_{r(x)<t} w^{1-p'}(x) dx \right)^{1/p'} < \infty.$$

Theorem 6.6.7. *Let* $0 < \gamma < Q$, $1 < p < Q/\gamma$. *Assume that* σ *and* u *are positive increasing functions on* $(0, \infty)$ *and* $\rho \in A_{pp^*}(G)$, *where* $p^* = \frac{pQ}{Q-p\gamma}$. *We put* $v(x) = \sigma(r(x))\rho^p(x)$, $w(x) = u(r(x))\rho^p(x)$. *Then for the boundedness of the operator* I_γ *from* $L^p_w(G)$ *into* $L^p_v(G)$ *it is necessary and sufficient that*

$$\sup_{t>0} \left(\int_{r(x)>t} \frac{v(x)}{(r(x))^{(Q-\gamma)p}} dx \right)^{1/p} \times$$

$$\times \left(\int_{r(x)<t} w^{1-p'}(x) dx \right)^{1/p'} < \infty.$$

Theorem 6.6.8. *Let* $0 < \gamma < Q$, $1 < p < Q/\gamma$ *and* $p^* = \frac{pQ}{Q-p\gamma}$. *Assume that* σ *and* u *are positive increasing functions on* $(0, \infty)$ *and* $\rho \in A_{pp^*}(G)$. *We put* $v(x) = \sigma(r(x))\rho^p(x)$, $w(x) = u(r(x))\rho^p(x)$. *Then for the boundedness of the operator* I_γ *from* $L^p_w(G)$ *into* $L^{p\infty}_v(G)$ *it is necessary and sufficient that*

$$\sup_{0<t<\tau<\infty} \frac{1}{\tau^{Q-\gamma}} \left(\int_{t<r(x)<\tau} v(x) dx \right)^{1/p} \times$$

$$\times \left(\int_{r(x)<t} w^{1-p'}(x) dx \right)^{1/p'} < \infty.$$

Now let us consider the case $\alpha > 1/p$.

Theorem 6.6.9. *Let* $1 < p < \infty$ *and* $1/p < \alpha < 1$. *Then the inequality*

$$\left(\int_X |K_\alpha f(x)| v(x) d\mu(x) \right)^{1/p} \le c \left(\int_X |f(x)|^p (\mu(x_0, d(x_0, x)))^\beta d\mu \right)^{1/p},$$

with a positive constant c independent of f holds if and only if
(i) $\alpha p - 1 < \beta < p - 1$;
(ii)

$$B \equiv \sup_{t>0} \left(\int_{X \setminus B(x_0, t)} \left(\mu B(x_0, d(x_0, x)) \right)^{(\alpha - 1)p} d\mu(x) \right)^{1/p} \times$$

$$\times \left(\mu B(x_0, t) \right)^{1/p' - \beta/p} < \infty.$$

Moreover, $\|K_\alpha\| \approx B$.

Proof. Let us represent K_α as follows

$$K_\alpha f \equiv K_\alpha^{(1)} f + K_\alpha^{(2)} f,$$

where

$$K_\alpha^{(1)} f(x) \equiv \int_{B(x_0, 2a_1 d(x_0, x))} f(y) (\mu B(x, d(x, y)))^{\alpha - 1} d\mu(y),$$

$$K_\alpha^{(2)} f(x) \equiv \int_{X \setminus B(x_0, 2a_1 d(x_0, x))} f(y) (\mu B(x, d(x, y)))^{\alpha - 1} d\mu(y).$$

Consequently we have

$$\|K_\alpha f\|_{L^p_v(X)} \le \|K_\alpha^{(1)}\|_{L^p_v(X)} +$$
$$+ \|K_\alpha^{(2)} f\|_{L^p_v(X)} \equiv S_1 + S_2.$$

Using Corollary 1.1.2, the conditions $\beta < p - 1$, $\alpha > 1/p$, the monotonicity of $\mu B(x, r)$ with respect to r and the inequality

$$\left(\int_{B(x_0, t)} (\mu B(x_0, d(x_0, x)))^{\beta(1 - p')} d\mu \right)^{1/p} \le c_1 (\mu B(x_0, t))^{-\beta/p + 1/p'}$$

with a positive constant c_1 independent of x, we obtain (see also the proof of Theorem 6.3.1)

$$S_1 \le c_2 B \left(\int\limits_X |f(x)|^p (\mu B(x_0, d(x_0, x)))^\beta d\mu \right)^{1/p}.$$

Using Corollary 1.1.8 we find that

$$S_2 \le c_2 B_1 \left(\int\limits_X |f(x)|^p (\mu B(x_0, d(x_0, x)))^\beta d\mu \right)^{1/p},$$

where

$$B_1 \equiv \sup_{t>0} \left(\int\limits_{B(x_0,t)} v(x) d\mu \right)^{1/p} (\mu B(x_0, t))^{-\beta/p+\alpha-1/p}.$$

Further, using the doubling and reverse doubling conditions for μ we have $B_1 \le c_3 B$. Finally we obtain the sufficiency.

Necessity follows in the same way as in the proof of Theorem 3.2.1.

□

6.7. Theorems of Koosis type

In the present section we consider problems of Koosis type for potentials defined on homogeneous groups and on SHT. Analogous problems for classical integral operators have been studied by [45], [249–250], [308], [257], [89] etc. (see also [96] and [100], Chapter 3).

Let G be a homogeneous group with homogeneous norm $x \to r(x)$ and homogeneous dimension Q (see Section 1.4 for the definition of a homogeneous group).

Next, for a measurable function $g : G \to R$ we define the potential on G by

$$T_\alpha g(x) = \int\limits_G \frac{g(y)}{r(xy^{-1})^{Q-\alpha}} dy, \quad 0 < \alpha < Q.$$

The following theorems are well-known.

Theorem A (see [86], p. 188). *Let* $0 < \alpha < 1$, $\frac{1}{q} = 1 - \frac{\alpha}{Q}$. *Then the following inequality is valid:*

$$|\{x : |T_\alpha f(x)| > \lambda\}| \le c \left(\frac{1}{\lambda} \int\limits_G |f(x)| dx \right)^q, \quad f \in L^1(G),$$

where the constant $c > 0$ does not depend on $f \in L^1(G)$ and $\lambda > 0$.

Denote by (X, m) a space with a σ-finite measure m.

Theorem B (see [96], p. 550). *Let $0 < q < p < \infty$, $\frac{1}{r} = \frac{1}{q} - \frac{1}{p}$, and let $A : B \to L^q(X)$ be a sublinear operator (B is a Banach space). If the inequality*

$$\|Af\|_{L^q(X)} \le c\|f\|_B,$$

is fulfilled with $c > 0$ independent of f, then there exists a positive, locally summable function w, $\|w^{-1}\|_{\frac{r}{p}} \le 1$, such that for every $f \in B$ the inequality

$$\int_X |Af(x)|^p w(x) dm(x) \le c^p \|f\|_B^p$$

holds.

Let us quote and prove several lemmas used in proving the basic theorems.

Lemma 6.7.1. *Let $0 < r < 1 < p$, $\frac{1}{\alpha} = \frac{p}{r} - 1$, $\frac{1}{\beta} = p - 1$, and let $A : L^1(G) \to L^r(G)$ be a sublinear operator. If the inequality*

$$\left(\int_G |Af(x)|^r dx \right)^{\frac{1}{r}} \le c_1 \int_G |f(x)| dx, \tag{6.7.1}$$

is fulfilled, where the constant c_1 does not depend on $f \in L^1(G)$, then for every positive $a \in L^\beta(G)$ there exists a positive function $b \in L^\alpha(G)$ such that the following inequality holds:

$$\int_G |Af(x)|^p (b(x))^{-1} dx \le c_2 \int_G |f(x)|^p (a(x))^{-1} dx, \quad f \in L^p_{a-1}(G)$$

where the constant c_2 does not depend on f and $c_2 = c_1^p$.

Proof. Let $a(x) \ge 0$ and $\|a\|_{L^\beta(G)} = 1$. Hölder's inequality implies

$$\int_G |g(x)| dx \le \left(\int_G |g(x)|^p (a(x))^{-1} dx \right)^{\frac{1}{p}} \left(\int_G \alpha^\beta(x) dx \right)^{\frac{1}{\beta p}},$$

that is, $L^p_{a-1}(G) \subset L^1(G)$ and $\| \cdot \|_{L^1(G)} \le \| \cdot \|_{L^p_{a-1}(G)}$. By inequality (6.7.1) we have

$$\|Af\|_{L^r(G)} \le c_1 \|f\|_{L^p_{a-1}(G)}.$$

Using Theorem B and putting $b = w^{-1}$, we have

$$\int_G |Af(x)|^p (b(x))^{-1} dx \le c_1^p \int_G |f(x)|^p (a(x))^{-1} dx$$

for an arbitrary $f \in L^p_{a-1}(G)$, and $\|b\|_\alpha \leq 1 = \|a\|_\beta$.

If $\|a\|_\beta \neq 1$, then we have to consider the function $a_1 = \frac{a}{\|a\|_\beta}$. $\square$

Lemma 6.7.2. *Let* $0 < r < q < \infty$, $\frac{1}{s} = \frac{1}{r} - \frac{1}{q}$, *and let* f *be a non-negative function on* G. *Then the following inequality is fulfilled:*

$$\sup_E \frac{\|f \cdot \chi_E\|_{L^r(G)}}{\|\chi_E\|_{L^s(G)}} \leq \left(\frac{q}{q-r}\right)^{\frac{1}{r}} \sup_{t>0} t|\{x \in G : |f(x)| > t\}|^{\frac{1}{q}},$$

where the supremum is taken over all measurable sets $E \subset G$ *of positive measure.*

Proof. Assuming that

$$\sup_{t>0} t|\{x \in G : f(x) > t\}|^{\frac{1}{q}} = 1,$$

we have

$$\lambda_f(t) = |\{x : f(x) > t\}| \leq t^{-q}$$

for all $t > 0$. This implies that for every set E, $|E| < \infty$, the inequality

$$(\lambda_{(f \cdot \chi_E)})(t) \leq \inf(|E|, t^{-q})$$

is valid.

Now estimate $\|f \cdot \chi_E\|_{L^r(G)}$ as follows:

$$\|f \cdot \chi_E\|_{L^r(G)} = \left(\int_G |f(x)|^r \chi_E(x) dx\right)^{\frac{1}{r}} =$$

$$= \left(\int_0^\infty rt^{r-1}(\lambda_{(f \cdot \chi_E)})(t) dt\right)^{\frac{1}{r}} \leq$$

$$\leq \left\{r \int_0^h t^{r-1}(\lambda_{(f \cdot \chi_E)})(t) dt + r \int_h^\infty t^{r-1}(\lambda_{(f \cdot \chi_E)})(t) dt\right\}^{\frac{1}{r}} \leq$$

$$\leq \left\{r|E| \int_0^h t^{r-1} dt + r \int_h^\infty t^{r-q-1} dt\right\}^{\frac{1}{r}}.$$

Taking $h = |E|^{-\frac{1}{q}}$, we obtain

$$\|f \cdot \chi_E\|_{L^r(G)} \leq |E|^{\frac{1}{s}} \left(\frac{q}{q-r}\right)^{\frac{1}{r}},$$

that is,

$$\frac{\|f \cdot \chi_E\|_{L^r(G)}}{\|\chi_E\|_{L^s(G)}} \leq \left(\frac{q}{q-r}\right)^{\frac{1}{r}}$$

for every measurable $E \subset G, 0 < |E| < \infty$. $\square$

Lemma 6.7.3. *Let* $0 < r < 1$. *Then the inequality*

$$\left(\int_{S_k} |T_\alpha f(x)|^r dx\right)^{\frac{1}{r}} \leq c|S_k|^{\frac{1}{r}} \int_G |f(x)|(1 + r(x))^{\alpha-Q} dx \qquad (6.7.2)$$

is valid, where the constant $c > 0$ *depends only on* Q, α, r, c_0, *and*

$$S_0 = \left\{x : r(x) < \frac{1}{c_0}\right\}, \quad S_k = \left\{x : \frac{2^{k-1}}{c_0} \leq r(x) < \frac{2^k}{c_0}\right\}.$$

Proof. Represent the function f as follows:

$$f(x) = f_1(x) + f_2(x),$$

where $f_1(x) = f \cdot \chi_{B_k}(x), f_2(x) = f(x) - f_1(x)$, and $B_k = \{x; r(x) < 2^{k+1}\}$. For every $x \in S_k$ we have

$$
\begin{aligned}
|T_\alpha f_2(x)| \quad &\leq \int_{r(y) \geq 2^{k+1} > 2c_0 r(x)} \frac{|f(y)|}{r(xy^{-1})^{Q-\alpha}} dy \leq \\
&\leq c_1 \int_{r(y) \geq 2^{k+1}} |f(y)|(1 + r(y))^{\alpha-Q} dy \leq \\
&\leq c_2 \int_G |f(y)|(1 + r(y))^{\alpha-Q} dy,
\end{aligned}
$$

where the constant c_2 depends only on Q, α and c_0.
We obtain

$$\sup_{x \in S_k} |T_\alpha f(x)| \leq c_2 \int_G |f(y)|(1 + r(y))^{\alpha-Q} dy$$

and

$$\left(\int_{S_k} |T_\alpha f_2(x)|^r dx\right)^{\frac{1}{r}} \leq c_2 |S_k|^{\frac{1}{r}} \int_G |f(x)|(1 + r(y))^{\alpha-Q} dy.$$

Take the number q so that $\frac{1}{q} = 1 - \frac{\alpha}{Q}$. Then, by Lemma 6.7.2 and Theorem A, we arrive at

$$\left(\int_{S_k} |T_\alpha f_1(x)|^r dx\right)^{\frac{1}{r}} \leq c_3 |S_k|^{\frac{1}{r}-\frac{1}{q}} \sup_{t>0} t|\{x \in G : |T_\alpha f_1(x)| > t\}|^{\frac{1}{q}} \leq$$

$$\le c_4 |S_k|^{\frac{1}{r}-\frac{1}{q}} \int\limits_{r(x)<2^{k+1}} |f(x)|dx \le$$

$$\le c_5 |S_k|^{\frac{1}{r}} \cdot 2^{-\frac{kQ}{q}} \int\limits_{G} |f(x)| \frac{(1+2^{k+1})^{Q-\alpha}}{(1+r(x))^{Q-\alpha}}\, dx \le$$

$$\le c_6 |S_k|^{\frac{1}{r}} \cdot 2^{-\frac{kQ}{q}} \cdot 2^{k(Q-\alpha)} \int\limits_{G} |f(x)|(1+r(x))^{\alpha-Q}dx =$$

$$= c_6 |S_k|^{\frac{1}{r}} \int\limits_{G} |f(x)|(1+r(x))^{\alpha-Q}dx$$

(the constant c_6 depends only on Q, α, r and c_0).
Finally we obtain (6.7.2). $\square$

Now we formulate the main results:

Theorem 6.7.1. *Let* $0 < \alpha < Q$ *and* $1 < p < \infty$. *Then for the given weight function* w *there exists a weight* v *such that the two-weight inequality*

$$\int\limits_{G} |T_\alpha f(x)|^p v(x)dx \le c_1 \int\limits_{G} |f(x)|^p w(x)dx, \quad f \in L^p_w(G), \qquad (6.7.3)$$

holds if and only if

$$\int\limits_{G} \frac{w^{1-p'}(x)}{(1+r(x))^{(Q-\alpha)p'}}\, dx < \infty.$$

Proof. We first prove sufficiency. Let $r > 1$, $\frac{1}{\lambda} = \frac{p}{r} - 1$, $\frac{1}{\beta} = p - 1$. The operator

$$(T_\alpha)_k g(x) = T_\alpha(g(y)(1+r(y))^{Q-\alpha})(x)\chi_{S_k}(x)$$

is subadditive for every $k \in Z^+$, where

$$S_0 = \left\{ x \in G : r(x) < \frac{1}{c_0} \right\}, \quad S_k = \left\{ x \in G : \frac{2^{k-1}}{c_0} \le r(x) < \frac{2^k}{c_0} \right\}.$$

By Lemma 6.7.3 we have

$$\left(\int\limits_{G} |(T_\alpha)_k f(x)|^r dx \right)^{\frac{1}{r}} \le c_1 2^{\frac{kQ}{r}} \int\limits_{G} |f(x)|dx,$$

where the constant c_1 depends only on r, Q, α and c_0.

Let us take a positive function $a \in L^\beta(G)$. Then by Lemma 6.7.1, there exists a positive function $b_k \in L^\lambda(G)$ such that

$$\|b_k\|_\lambda \leq \|a\|_\alpha$$

and

$$\int_G |(T_\alpha)_k f(x)|^p (b_k(x))^{-1} dx \leq c_1^p 2^{\frac{kQp}{r}} \int_G |f(x)|^p (a(x))^{-1} dx,$$

that is,

$$\int_G |T_\alpha f(x)|^p (b_k(x))^{-1} dx \leq c_1^p 2^{\frac{kQp}{r}} \int_G |f(x)| (a(x))^{-1} (1 + r(x))^{(\alpha-Q)p} dx.$$

Writing $a(x) = (w(x))^{-1} (1 + r(x))^{(\alpha-Q)p}$, we obtain

$$\int_G w^{1-p'}(x) (1 + r(x))^{(\alpha-Q)p'} dx < \infty.$$

Now take $\varepsilon > 0$. Then the inequality

$$\int_{S_k} |(T_\alpha)_k f(x)|^p (b_k(x))^{-1} \cdot 2^{-\frac{kQp}{r}} \cdot 2^{-k\varepsilon} dx \leq$$

$$\leq c_1^p \cdot 2^{-k\varepsilon} \int_G |f(x)|^p w(x) dx$$

holds. Using Fatou's theorem, we get inequality (6.7.3), where

$$v(x) = \sum_{k=0}^\infty 2^{-\varepsilon k} \cdot 2^{-\frac{kQp}{r}} (b_k(x))^{-1} \chi_{S_k}(x).$$

Let $\lambda < \frac{1}{p-1}$. Then for v we have

$$\int_G (v(x))^{-\lambda} (1 + r(x))^{-Qp'} =$$

$$= \sum_{k=0}^\infty 2^{\lambda \varepsilon k} \cdot 2^{\frac{\lambda k Q p}{r}} \int_{S_k} b_k^\lambda(x) (1 + r(x))^{-Qp'} dx \leq$$

$$\leq c_3 \sum_{k=0}^\infty 2^{-Qkp'} \cdot 2^{\lambda \varepsilon k} \cdot 2^{\frac{\lambda k Q p}{r}} \int_{S_k} b_k^\lambda(x) dx \leq$$

$$\leq c_4 \sum_{k=0}^\infty 2^{-k(Qp' - \varepsilon \lambda - \frac{\lambda Q p}{r})}.$$

Taking ε sufficiently small, we obtain

$$\int_G (v(x))^{-\lambda}(1 + r(x))^{-Qp'}dx < \infty \quad \left(\lambda < \frac{1}{p-1}\right). \qquad (6.7.4)$$

Necessity. Now suppose that for the weight w there exists a weight v such that the two-weight inequality (6.7.3) holds. Then for every $t > 0$ we have

$$I_1(t) \equiv \int_{\{r(x)<t\}} w^{1-p'}(x)dx < \infty;$$

$$I_1(t) \equiv \int_{\{r(x)>t\}} w^{1-p'}(x)(r(x))^{(\alpha-Q)p'}dx < \infty.$$

Indeed, let $I_1(t_0) = \infty$ for some t_0. This means that $\|w^{-1}\|_{L_w^{p'}(G)}(B(e,t_0)) < \infty$, where $B(e,t_0) \equiv \{x : r(x) < t_0\}$. Consequently, there exists $g \in L_w^p(B(e,t_0))$ such that $\int_{B(e,t_0)} g(x) = \infty$. Let $f(x) = g(x)\chi_{B(e,t)}$. Then $f \in L_w^p(G)$. On the other hand,

$$\|T_\alpha f(x)\|_{L_v^p(G)} \geq \int_{G\backslash B(e,t_0)} v(x)\left(\int_{B(e,t_0)} (r(xy^{-1}))^{\alpha-Q}f(y)dy\right)^p dx \geq$$

$$\geq c_1 \left(\int_{G\backslash B(e,t_0)} v(x)(r(x))^{(\alpha-Q)p}dx\right) \int_{B(e,t_0)} g(x)dx = \infty,$$

which is impossible. We conclude that $I_1(t_0) < \infty$ for all $t > 0$. Further,

$$\int_G w^{1-p'}(1 + r(x))^{(\alpha-Q)p'}dx \leq \int_{B(e,1)} w^{1-p'}(x)(1 + r(x))^{(\alpha-Q)p}dx +$$

$$+ \int_{G\backslash B(e,1)} w^{1-p'}(x)(1 + r(x))^{(\alpha-Q)p}dx \leq$$

$$\leq I_1(1) + I_2(1) < \infty.$$

$\square$

From duality arguments we can obtain the following result:

Theorem 6.7.2. *Let $1 < p < \infty$ and $0 < \alpha < Q$. Then for the given weight function v there exists a weight function w such that the two-weight inequality*

(6.7.3) *holds for all* $f \in L^p_w(G)$ *if and only if*

$$\int_G \frac{v(x)}{(1 + r(x))^{(Q-\alpha)p}} dx < \infty.$$

Now let (X, d, μ) be a space of homogeneous type SHT. Let us put $a \equiv \sup\{d(x_0, x) : x \in X\}$. We shall assume that for all r, R, with the condition $0 < r < R < \infty$ the following inequality holds

$$\mu(B(x_0, R) \setminus B(x_0, r)) > 0.$$

The following lemma is proved in the same way as Lemma 6.7.1.

Lemma 6.7.4. *Let* $0 < r < 1 < p$, $\frac{1}{\alpha} = \frac{p}{r} - 1$, $\frac{1}{\beta} = p - 1$. *Assume that* $A : L^1(X) \to L^r(X)$ *is a sublinear operator. If the inequality*

$$\left(\int_X |Af(x)|^r d\mu \right)^{\frac{1}{r}} \leq c_1 \int_X |f(x)| d\mu, \quad f \in L^1(X),$$

is satisfied, where the constant c_1 *does not depend on* f, *then for every positive* $a \in L^\beta(X)$ *there exists a positive function* $b \in L^\alpha(X)$ *such that the following inequality holds:*

$$\int_X |Af(x)|^p (b(x))^{-1} d\mu \leq c_2 \int_X |f(x)|^p (a(x))^{-1} d\mu, \quad f \in L^p_{a^{-1}}(X),$$

where the constant c_2 *does not depend on* f, *and* $c_2 = c_1^p$.

We shall need the next result (see, e.g., [101], [100], Chapter 3):

Theorem C. *Let* $0 < \alpha < 1$, $1/q = 1 - \alpha$. *Then the operator* K_α *is bounded from* $L^1(X)$ *into* $L^{q\infty}(X)$.

Lemma 6.7.5. *Let* $0 < r < q < \infty$, $\frac{1}{s} = \frac{1}{r} - \frac{1}{q}$, *and let* f *be a non-negative function on* X. *Then the following inequality holds:*

$$\sup_E \frac{\|f \cdot \chi_E\|_{L^r(X)}}{\|\chi_E\|_{L^s(X)}} \leq \left(\frac{q}{q-r} \right)^{\frac{1}{r}} \sup_{t>0} t\mu\{x \in X : |f(x)| > t\}^{\frac{1}{q}},$$

where the supremum is taken over all μ- *measurable sets* $E \subset X$ *such that* $0 < \mu E < \infty$.

The proof of this lemma is similar to the proof of Lemma 6.7.2.

Lemma 6.7.6. *Let* $x_0 \in X$ *and* $a = \infty$. *Then there exist positive constants* b_1 *and* b_2 *such that for all* $k \in Z$ *the following inequality holds:*

$$b_1 \mu B(x_0, \eta_1^k) \leq \mu E_k \leq b_2 \mu B(x_0, \eta_1^k),$$

where $E_k = B(x_0, \eta_1^{k+1}) \setminus B(x_0, \eta_1^k)$ *and* η_1 *is from the Definition of the reverse doubling condition (see Definition 6.2.1.)*

This lemma is proved using the doubling and reverse doubling conditions for μ.

Lemma 6.7.7. *Let* $x_0 \in X$, $a = \infty$ *and* $\epsilon > 0$. *Then the inequality*

$$\sum_{k=0}^{+\infty} (\mu E_k)^{-\epsilon} \leq c(\mu B(x_0, 1))^{-\epsilon}$$

holds, where c is a positive constant and E_k *is from Lemma 6.7.6.*

Proof. By the reverse doubling condition we have

$$\mu E_k \geq \mu(B(x_0, \eta_1^{k+1}) \setminus B(x_0, \eta_1^k)) \geq \eta_2 \mu B(x_0, \eta_1^k) - \mu B(x_0, \eta_1^k) =$$
$$(\eta_2 - 1)\mu(x_0, \eta_1^k) \geq (\eta_2 - 1)\eta_2^k \mu B(x_0, 1).$$

Hence

$$\sum_{k=0}^{+\infty} (\mu E_k)^{-\epsilon} \leq \sum_{k=0}^{+\infty} \frac{1}{(\eta_2 - 1)^\epsilon} \eta_2^{-\epsilon k} (\mu B(x_0, 1))^{-\epsilon} \leq$$
$$\leq \frac{1}{(\eta_2 - 1)\epsilon} (\mu B(x_0, 1))^{-\epsilon} \sum_{k=0}^{+\infty} \eta_1^{-\epsilon k} = c(\mu B(x_0, 1))^{-\epsilon}.$$

$\square$

Let

$$K_\alpha f(x) = \int_X \frac{f(y)}{(\mu B(x, d(x, y)))^{1-\alpha}}$$

for μ- measurable $f : X \to R$

The following lemma is proved in the same way as Lemma 6.7.3.

Lemma 6.7.8. *Let* $x_0 \in X$, $a = \infty$, $0 < r < 1$. *Then the inequality*

$$\left(\int_{E_k} |K_\alpha f(x)|^r d\mu \right)^{\frac{1}{r}} \leq c(\mu S_k)^{\frac{1}{r}} \int_X |f(x)|(\mu B(x_0, 1 + d(x_0, x)))^{\alpha-1} d\mu,$$

is valid for all $f \in L_w^1(X)$, where $w(x) = (\mu B(x_0, 1 + d(x_0, x)))^{\alpha-1}$ and the constant $c > 0$ depends only on α, r, a_1 and a_0 (a_0 and a_1 are from the definition of SHT, see Section 1.1)

$$E_0 = B(x_0, 1/a_1), \quad E_k = B(x_0, \eta_1^{k+1}/a_1) \setminus B(x_0, \eta_1^k/a_1).$$

Using Lemmas 6.7.4- 6.7.8, we obtain the following result:

Theorem 6.7.3 *Let $x_0 \in X$, $a = \infty$, $0 < \alpha < 1$ and $1 < p < \infty$. Then for a given weight function w there exists a weight v such that the two-weight inequality*

$$\int_X |K_\alpha f(x)|^p v(x) d\mu \le c_1 \int_X |f(x)|^p w(x) d\mu, \quad f \in L_w^p(X), \quad (6.7.5)$$

holds if and only if

$$\int_X \frac{w^{1-p'}(x)}{(\mu B(x_0, 1 + d(x_0, x)))^{(1-\alpha)p'}} d\mu < \infty.$$

From a duality argument we obtain the next theorem:

Theorem 6.7.4. *Let $x_0 \in X$, $a = \infty$, $1 < p < \infty$ and $0 < \alpha < 1$. Then for a given weight function v there exists a weight function w such that the two-weight inequality (6.7.5) holds for all $f \in L_w^p(X)$ if and only if*

$$\int_X \frac{v(x)}{(\mu B(x_0, 1 + d(x_0, x)))^{(1-\alpha)p}} d\mu < \infty.$$

6.8. Fractional maximal functions on SHT

In this section we derive necessary and sufficient conditions for the validity of two-weight inequalities for the fractional maximal functions:

$$M_\gamma f(x) = \sup \frac{1}{(\mu(B))^{1-\gamma}} \int_B |f(y)| d\mu, \quad 0 < \gamma < 1,$$

where the supremum is taken over all balls B containing x;

$$\overline{M}_\gamma f(x) = \sup_{0 < r < \bar{a}\eta_1 d(x_0, x)} \frac{1}{(\mu(B(x, r)))^{1-\gamma}} \int_{B(x,r)} |f(y)| d\mu, \quad 0 < \gamma < 1.$$

(η_1 is the constant from Definition 6.5.1, $\bar{a} = a_1(a_0 + 1)$ and x_0 is a fixed point with the condition $\mu\{x_0\} = 0$).

We shall assume that $a = \sup\{d(x_0, x) : x \in X\} = \infty$ and

$$\mu(B(x_0, R) \setminus B(x_0, r)) > 0$$

for all r and R provided that $0 < r < R < \infty$.

Let ν be an another measure on X such that all balls are ν measurable. We shall use the following notation: $B(x, d(x, y)) \equiv B_{xy}$.

Theorem 6.8.1. *Let $1 < p < \infty$ and let $1/p < \gamma < 1$. Assume also that $\gamma p - 1 < \beta < p - 1$. Then the inequality*

$$\|M_\gamma f\|_{L_\nu^p(X)} \leq c\left(\int\limits_X |f(x)|^p (\mu B_{x_0 x})^\beta d\mu\right)^{1/p}, \quad f \in L^p_{(\mu B_{x_0 x})^\beta}(X),$$

where the positive constant c does not depend on f, holds if and only if

$$B \equiv \sup_{t>0} B(t) \equiv \sup_{t>0}\left(\int\limits_{X \setminus B(x_0, t)} (\mu B_{x_0 x})^{(\gamma-1)p} d\nu(x)\right)^{1/p} \times$$

$$\times (\mu B(x_0, t))^{1/p' - \beta/p} < \infty.$$

Moreover, $\|M_\gamma\| \approx B$.

Proof. Sufficiency follows from Theorem 6.6.9 and from the following inequalities:

$$M_\gamma f(x) \leq c_1 \widetilde{M}_\gamma f(x) \leq c_1 K_\gamma f(x), \tag{6.8.1}$$

where $f \geq 0$, $c_1 > 0$ does not depend on f and $x \in X$ and

$$\widetilde{M}_\gamma f(x) = \sup_{r>0} \frac{1}{(\mu(B(x, r)))^{1-\gamma}} \int\limits_{B(x,r)} f(y) d\mu(y),$$

$$K_\gamma f(x) = \int\limits_X f(y)(\mu B_{xy})^{\gamma-1} d\mu(y).$$

Inequality (6.8.1) follows from the doubling condition. Indeed, let $x \in B \equiv B(x_0, r)$; then if $z \in B$, we have

$$d(x, z) \leq a_1 d(x, x_0) + a_1 d(x_0, z) \leq a_1 a_0 d(x_0, x) + a_1 d(x_0, x) \leq$$
$$a_1 a_0 r + a_1 r = a_1(a_0 + 1)r.$$

On the other hand, if $z \in B(x, a_1(a_0 + 1)r)$, then

$$d(x_0, z) \leq a_1 d(x_0, x) +$$
$$+a_1 d(x, z) \leq a_1 r + a_1^2(a_0 + 1)r = (a_1 + a_1^2(a_0 + 1))r.$$

Consequently $B(x, a_1(a_0+1)r) \subset B(x_0, a_1+a_1^2(a_0+1)r)$. Hence $M_\gamma f(x) \leq c_1 \widetilde{M}_\gamma f(x)$. The inequality $\widetilde{M}_\gamma f(x) \leq K_\gamma f(x)$ is obvious. Now we prove necessity. First we show that

$$B_1 \equiv \sup_{k \in Z} \left(\int_{B(x_0, 2^{k+1}) \setminus B(x_0, 2^k)} (\mu B_{x_0 x})^{\gamma p - \beta - 1} d\nu(x) \right)^{1/p} < \infty.$$

Let $k \in Z$ and let $f_k(x) = \chi_{B(x_0, \eta_1^{k+1})}(x)(\mu B_{x_0 x})^{\beta(1-p')}$, where η_1 is from Definition 6.5.1. Then using the doubling condition we have

$$\|M_\gamma f\|_{L^p_\nu(X)}^p \geq \int_{B(x_0, \eta_1^{k+1}) \setminus B(x_0, \eta_1^{k+1})} (M_\gamma f(x))^p d\nu(x) \geq$$

$$\geq \int_{B(x_0, \eta_1^{k+1}) \setminus B(x_0, \eta_1^k)} \left(\frac{1}{\mu B(x, a_1(a_0+1)\eta_1^{k+1})^{1-\gamma}} \times \right.$$

$$\left. \times \int_{B(x, a_1(a_0+1)\eta_1^{k+1})} f(y) d\mu(y) \right)^p d\nu(x) \geq$$

$$\geq c_2 \int_{B(x_0, \eta_1^{k+1}) \setminus B(x_0, \eta_1^k)} \left(\frac{1}{\mu B(x, \eta_1^{k+1})^{1-\gamma}} \times \right.$$

$$\left. \times \int_{B(x, \eta_1^{k+1})} f(y) d\mu(y) \right)^p d\nu(x) \geq$$

$$\geq c_3 \int_{B(x_0, \eta_1^{k+1}) \setminus B(x_0, \eta_1^k)} (\mu B(x_0, \eta_1^{k+1}))^{\gamma p - \beta p'} d\nu(x) \geq$$

$$\geq c_3 \int_{B(x_0, \eta_1^{k+1}) \setminus B(x_0, \eta_1^k)} (\mu B_{x_0 x})^{\gamma p - \beta p'} d\nu(x)$$

On the other hand,

$$\int_X (f(y))^p (\mu B_{x_0 x})^\beta d\mu(x) =$$

$$= \int_{B(x_0, \eta_1^{k+1})} (\mu B_{x_0 x})^{\beta(1-p')} \leq c_4 (\mu B(x_0, \eta_1^{k+1}))^{\beta(1-p')+1}$$

and finally we see that $B_1 < \infty$.

The inequality $B \leq c_5 B_1$ follows from the reverse doubling condition for μ.

Now let $\eta > 0$ and put

$$\tilde{J}_{\eta, \gamma} f(x) = \int_{B(x_0, \eta d(x_0, x))} (\mu B_{xy})^{\gamma - 1} f(y) d\mu(y), \quad \gamma > 0.$$

□

The next lemma is proved in the same way as Theorem 6.5.1.

Lemma 6.8.1. *Let* $1 < p \leq q < \infty$ *and let* $\gamma > 1/p$. *Then* $\tilde{J}_{\eta, \gamma}$ *is bounded from* $L^p_\mu(X)$ *into* $L^q_\nu(X)$ *if and only if*

$$A \equiv \sup_{t>0} \left(\int_{X \setminus B(x_0, t)} (\mu B_{x_0 x})^{(\gamma - 1)q} d\nu \right)^{1/q} (\mu B_{x_0}(t))^{1/p'} < \infty. \quad (6.8.2)$$

Moreover, $\|\tilde{J}_{\eta, \gamma}\| \approx A$.

Theorem 6.8.2. *Let* $1 < p \leq q < \infty$ *and let* $1/p < \gamma < 1$. *Then* $\overline{M}$ *is bounded from* $L^p_\mu(X)$ *into* $L^q_\nu(X)$ *if and only if* (6.8.2) *is fulfilled. Moreover,* $\|\overline{M}_\gamma\| \approx A$.

Proof. From the inequality

$$\overline{M}_\gamma f(x) \leq \tilde{J}_{2\bar{a} a_1 \eta_1, \gamma} f(x) \quad (x \in X, f \geq 0)$$

and from Lemma 6.8.1 sufficiency follows, where $\bar{a} \equiv a_1(a_0 + 1)$. To prove necessity we take $k \in Z$ and let $f_k(x) = \chi_{B_{x_0}(\eta_1^{k+1})}(x)$. Then

$$\|\overline{M}_\gamma f\|^q_{L^q_\nu(X)} \geq \int_{B(x_0, \eta_1^{k+1}) \setminus B(x_0, \eta_1^{k+1})} (\overline{M}_\gamma f(x))^q d\nu(x) \geq$$

$$\geq \int_{B(x_0, \eta_1^{k+1}) \setminus B(x_0, \eta_1^k)} \left(\frac{1}{\mu B(x, \bar{a}\eta_1^{k+1})^{1-\gamma}} \times \right.$$

$$\left. \times \int_{B(x, \bar{a}\eta_1^{k+1})} f(y) d\mu(y) \right)^q d\nu(x) \geq$$

$$\geq c_1 \int_{B(x_0,\eta_1^{k+1})\backslash B(x_0,\eta_1^k)} \left(\frac{1}{\mu B(x,\eta_1^{k+1})^{1-\gamma}} \int_{B(x,\eta_1^{k+1})} f(y)d\mu(y) \right)^q d\nu(x) =$$

$$= c_1 \int_{B(x_0,\eta_1^{k+1})\backslash B(x_0,\eta_1^k)} (\mu B(x_0,\eta_1^{k+1}))^{\gamma q} d\nu(x) \geq$$

$$\geq c_1 \nu(B(x_0,\eta_1^{k+1}) \backslash B(x_0,\eta_1^k))(\mu B(x_0,\eta q_1^{k+1}))^{\gamma q}.$$

On the other hand,

$$\|f\|_{L_\mu^p(X)}^p = \mu B(x_0,\eta_1^{k+1}).$$

Consequently

$$A_1 \equiv \sup_{k\in Z} \left(\nu(B(x_0,\eta_1^{k+1}) \backslash B(x_0,\eta_1^k)) \right)^{1/q} (\mu B(x_0,\eta_1^{k+1}))^{\gamma-1/p} < \infty.$$

Moreover, using the reverse doubling condition for μ, we have $A \leq c_2 A_1$. The theorem is proved. $\square$

6.9. Weighted estimates in Lorentz spaces

In this section we investigate the mapping properties for the truncated potential in Lorentz spaces defined on SHT- (X,d,μ).

Let $x_0 \in X$ and let $a \equiv \sup\{d(x_0,x) : x \in X\}$. We shall assume that the following condition is satisfied

$$\mu(B(x_0,R) \backslash B(x_0,r)) > 0 \qquad (*)$$

for all r and R with the condition $0 < r < R < \infty$.

Recall that (see Proposition 6.5.1) the condition (*) ensures the reverse doubling condition for μ at the point x_0: there exist constants $\eta_1 > 1, \eta_2 > 1$ such that the inequality

$$\mu B(x_0,\eta_1 r) \geq \eta_2 \mu B(x_0,r)$$

holds for all $r \in (0,a)$.

As before we shall use the following notation: $B(x,d(x,y)) \equiv B_{xy}$, $\{y \in X : d(x_0,y) \leq r\} \equiv \overline{B}(x_0,r)$.

Let φ be a μ measurable μ- a.e. positive function on X and let

$$J_{\alpha,\varphi}f(x) = \varphi(x) \int_{\overline{B}(x_0,2d(x_0,x))} f(y)(\mu B_{xy})^{\alpha-1}d\mu(y), \quad \alpha > 0.$$

If $\varphi \equiv 1$, then $J_{\alpha,\varphi} \equiv J_\alpha$.

The following Lemma can be derived in the same way as in the case of R^n (see [50], [261]).

Lemma 6.9.1. *Let (X, μ) be a σ-finite measure space and let $\{F_k\}$ be a countable family of μ-measurable sets $F_k \subset X$. We assume that $\sum_k \chi_{F_k}(\cdot) \leq c\chi_{\cup_k F_k}(\cdot)$ for some fixed constant $c > 0$. Then*
(a) there exists a positive constant c_1 independent of f such that

$$\sum_k \|f(\cdot)\chi_{F_k}(\cdot)\|^{\lambda}_{L^{rs}_{\mu}(X)} \leq c_1 \|f(\cdot)\chi_{\cup F_k}(\cdot)\|^{\lambda}_{L^{rs}_{\mu}(X)}$$

whenever $\max\{r, s\} \leq \lambda$;
(b) there exists a positive constant c_2 such that

$$\| \sum_k f(\cdot)\chi_{F_k}(\cdot)\|^{\lambda}_{L^{pq}_{\mu}(X)} \leq c_2 \sum_k \|f(\cdot)\chi_{F_k}(\cdot)\|^{\lambda}_{L^{pq}_{\mu}(X)}$$

whenever $0 < \gamma \leq \min\{p, q\}$ and c_2 is independent of f;

The next lemma will also be useful.

Lemma 6.9.2. *Let (X, ν) be a σ-finite measure space and let $E \subset X$ be a ν-measurable set. Suppose that $1 < p, q < \infty$. Further, let f, f_1 and f_2 be ν-measurable functions on X. Then:*
(i)

$$\|\chi_E(\cdot)\|_{L^{pq}_{\nu}(X)} = \left(\nu E\right)^{1/p};$$

(ii)

$$\|f\|_{L^{pq_1}_{\nu}(X)} \leq \|f\|_{L^{pq_2}_{\nu}(X)}$$

for fixed p and $q_2 \leq q_2$;
(iii)

$$\|f_1 f_2\|_{L^{pq}_{\nu}(X)} \leq c \|f_1\|_{L^{p_1 q_1}_{\nu}(X)} \|f_2\|_{L^{p_2 q_2}_{\nu}(X)},$$

where $\frac{1}{p} = \frac{1}{p_1} + \frac{1}{p_2}$, $\frac{1}{q} = \frac{1}{q_1} + \frac{1}{q_2}$.

The first part of this statement was proved in [280], and for (ii), (iii) see, e.g., [130].

We begin with the boundedness of $J_{\alpha,\varphi}$.

Theorem 6.9.1. *Let $\mu X = \infty$, $1 < r, p < \infty$, $1 \leq s < \infty$, $1 < q \leq \infty$ and let $\alpha > 1/r$. We assume that $\max\{r, s\} \leq \min\{p, q\}$ and ν is a σ-finite positive measure on X. Then the following statements are equivalent:*
(i) $J_{\alpha,\varphi}$ is bounded from $L^{rs}_{\mu}(X)$ into $L^{pq}_{\nu}(X)$;
(ii)

$$B_1 \equiv \sup_{t>0} B_1(t) \equiv \sup_{t>0} \|\varphi(x)(\mu B_{x_0 x})^{\alpha-1}\chi_{X \setminus B(x_0,t)}(x)\|_{L^{pq}_{\nu}(X)} \times$$

$$\times (\mu\overline{B}(x_0,t))^{1/r'} < \infty;$$

(iii)

$$B_2 \equiv \sup_{k \in Z} B_1(k) \equiv$$

$$\equiv \|\varphi(x)(\mu B_{x_0 x})^{\alpha - 1/r} \chi_{B(x_0, \eta_1^{k+1}) \setminus B(x_0, \eta_1^k)}(x)\|_{L_\nu^{pq}(X)} < \infty,$$

where η_1 is the constant from Definition 6.5.1.

Moreover, $\|J_{\alpha,\varphi}\| \approx B_1 \approx B_2$.

Proof. The implication (ii) $\Rightarrow$ (i) follows in the same way as in the case of the Lebesgue space using Theorem 1.1.4 , the inequality

$$\|(\mu B_{x_0} \cdot)^{\alpha - 1} \chi_{\overline{B}(x_0, 2d(x_0, x))}(\cdot)\|_{L_\mu^{r' s'}(X)} \leq c \mu B(x_0, d(x_0, x))^{\alpha - 1/r}$$

with a positive constant c independent of $x \in X$, and Lemma 6.9.1.

If we take a function $f_k(x) = \chi_{B(x_0, \eta_1^k/2a_1)}(x)$, then by Lemma 6.9.2 we have $\|f_k\|_{L_\mu^{rs}(X)} = (\mu B(x_0, \eta_1^k/2a_1))^{1/r}$. On the other hand,

$$\|J_{\alpha,\varphi} f_k(\cdot)\|_{L_\nu^{pq}(X)} \geq \|\chi_{B(x_0, \eta_1^{k+1}) \setminus B(x_0, \eta_1^k)}(\cdot) J_{\alpha,\varphi} f_k(\cdot)\|_{L_\nu^{pq}(X)} \geq$$

$$\geq c_1 \|(\mu B_{x_0 x})^\alpha \varphi(x) \chi_{B(x_0, \eta_1^{k+1}) \setminus B(x_0, \eta_1^k)}(x)\|_{L_\nu^{pq}(X)}.$$

From the boundedness of $J_{\alpha,\varphi}$ we obtain the implication (i) $\Rightarrow$ (iii).

It remains to show that (iii) $\Rightarrow$ (ii). For this we take $t > 0$. Then $t \in [\eta_1^{k+1}, \eta_1^k)$ for some integer k. Further, let $1 \leq \sigma \leq \min\{p, q\}$. Then using Lemma 6.9.1 we obtain

$$B_1^\sigma(t) \leq \|\varphi(x)(\mu B_{x_0 x})^{\alpha - 1} \chi_{X \setminus B(x_0, \eta_1^k)}(x)\|_{L_\nu^{pq}(X)}^\sigma (\mu \overline{B}(x_0, \eta_1^{k+1}))^{\sigma/r'} \leq$$

$$\leq \sum_{j=k}^{+\infty} \left\| \varphi(x)(\mu B_{x_0 x})^{\alpha - 1} \chi_{B(x_0, \eta_1^{j+1}) \setminus B(x_0, \eta_1^j)}(x) \right\|_{L_\nu^{pq}(X)}^\sigma \times$$

$$\times (\mu \overline{B}(x_0, \eta_1^{k+1}))^{\sigma/r'} \leq$$

$$\leq c_2 B_2^\sigma (\mu \overline{B}(x_0, \eta_1^{k+1}))^{\sigma/r'} \left(\sum_{j=k}^{+\infty} (\mu B(x_0, \eta_1^j))^{-\sigma/r' - 1} \mu B(x_0, \eta_1^{j-1}) \right) \leq$$

$$\leq c_3 B_2^\sigma (\mu \overline{B}(x_0, \eta_1^{k+1}))^{\sigma/r'} \times$$

$$\times \left(\sum_{j=k}^{\infty} \int_{B(X_0, \eta_1^{j+1}) \setminus B(x_0, \eta_1^j)} (\mu B(x_0, d(x_0, x)))^{-\sigma/r' - 1} d\mu(x) \right) =$$

$$= c_3 B_2^\sigma (\mu \overline{B}(x_0, \eta_1^{k+1}))^{\sigma/r'} \int_{X \setminus B(x_0, \eta_1^k)} (\mu B(x_0, d(x_0, x)))^{-\sigma/r' - 1} \leq c_4 B_2^\sigma.$$

Finally $B_1 \leq c_5 B_2$. $\square$

The next statement can be established in a similar way:

Theorem 6.9.2. *Let* $\mu X < \infty$, $1 < r, p < \infty$, $1 \leq s < \infty$, $1 < q \leq \infty$ *and let* $\alpha > 1/r$. *We assume that* $\max\{r, s\} \leq \min\{p, q\}$. *Let* ν *be a* σ-*finite positive measure on* X. *Then the following statements are equivalent:*
(i) $J_{\alpha,\varphi}$ *is bounded from* $L_\mu^{rs}(X)$ *into* $L_\nu^{pq}(X)$;
(ii)

$$A_1 \equiv \sup_{0 < t < a} A_1(t) \equiv$$

$$\equiv \sup_{0 < t < a} \|\varphi(x)(\mu B_{x_0 x})^{\alpha-1} \chi_{X \setminus B(x_0, t)}(x)\|_{L_\nu^{pq}(X)} (\mu \overline{B}(x_0, t))^{1/r'} < \infty;$$

(iii)

$$A_2 \equiv \sup_{k \in Z_+} A_2(k) \equiv$$

$$\equiv \sup_{k \in Z_+} \|\varphi(x)(\mu B_{x_0 x})^{\alpha-1/r} \chi_{B(x_0, \eta_1^{-k} a) \setminus B(x_0, \eta_1^{-(k+1)} a)}(x)\|_{L_\nu^{pq}(X)} < \infty;$$

where η_1 is a constant from the definition of the reverse doubling condition. Moreover, $\|J_{\alpha,\varphi}\| \approx A_1 \approx A_2$.

Using Theorem C of Section 2.1 we can derive the following results in the same way as in the case of Lebesgue space:

Theorem 6.9.3. *Let* $\mu X = \infty$, $1 < r, p, q < \infty$, $1 \leq s < \infty$ *with* $\max\{r, s\} \leq \min\{p, q\}$ *and let* $\alpha > 1/r$. *Let* ν *be a positive* σ-*finite measure on* X. *Then the following statements are equivalent:*
(i) $J_{\alpha,\varphi}$ *is compact from* $L_\mu^{rs}(X)$ *into* $L_\nu^{pq}(X)$;
(ii) $B_1 < \infty$ *and* $\lim_{t \to 0} B_1(t) = \lim_{t \to \infty} B_1(t) = 0$;
(iii) $B_2 < \infty$ *and* $\lim_{K \to -\infty} B_2(k) = \lim_{k \to +\infty} B_2(k) = 0$.

Theorem 6.9.4. *Let* $\mu X < \infty$, $1 < r, p, q < \infty$, $1 \leq s < \infty$ *with* $\max\{r, s\} \leq \min\{p, q\}$ *and* $\alpha > 1/r$. *Let* ν *be a positive* σ-*finite measure on* X. *Then the following conditions are equivalent:*
(i) $J_{\alpha,\varphi}$ *is compact from* $L_\mu^{rs}(X)$ *into* $L_\nu^{pq}(X)$;
(ii) $A_1 < \infty$ *and* $\lim_{t \to 0} A_1(t) = 0$;
(iii) $A_2 < \infty$ *and* $\lim_{k \to +\infty} A_2(k) = 0$.

Let us now investigate the case when $\varphi \equiv 1$. In this case $J_{\alpha,\varphi} \equiv J_\alpha$.

Theorem 6.9.5. *Let* $\mu X = \infty$ *and let* $1 < r, p < \infty$, $1 \leq s < \infty$, $1 < q \leq \infty$ *with* $\max\{r, s\} \leq \min\{p, q\}$. *Let* $\alpha > 1/r$. *We assume that* ν *is a* σ-*finite positive measure on* X. *Then the following conditions are equivalent:*
(i) J_α *is bounded from* L_μ^{rs} *into* $L_\nu^{pq}(X)$;
(ii) J_α *is bounded from* L_μ^r *into* $L_\nu^p(X)$;

(iii)

$$D_1 \equiv \sup_{t>0} D_1(t) \equiv$$

$$\equiv \sup_{t>0} \left(\int_{X \backslash B(x_0,t)} (\mu B_{x_0 x})^{(\alpha-1)p} d\nu(x) \right)^{1/p} (\mu \overline{B}(x_0,t))^{1/r'} < \infty;$$

(iv)

$$D_2 \equiv \sup_{k \in Z} D_2(k) \equiv$$

$$\equiv \sup_{k \in Z} \left(\nu(B(x_0, \eta_1^{k+1}) \backslash B(x_0, \eta_1^k)) \right)^{1/p} (\mu \overline{B}(x_0, \eta_1^k))^{\alpha-1/r} < \infty,$$

where η_1 is from the definition of the reverse doubling condition.

Moreover, $\|J_\alpha\|_{L_\mu^{rs}(X) \to L_\nu^{pq}(X)} \approx \|J_\alpha\|_{L_\mu^r(X) \to L_\nu^p(X)} \approx D_1 \approx D_2$.

Proof. Using Lemma 6.9.2 we have

$$\|\chi_E\|_{L_\nu^{pq}(X)} = (\nu E)^{1/p},$$

where E is a ν-measurable set in X, $1 < p < \infty$ and $1 < q \le \infty$. From the last equality we obtain (iii) $\Leftrightarrow$ (iv). On the other hand, taking into account Theorem 6.3.5 we have (iii) $\Leftrightarrow$ (ii). Further, by Theorem 6.9.1 (i) $\Leftrightarrow$ (iv). Finally we deduce that (i) $\Leftrightarrow$ (iv) $\Leftrightarrow$ (iii) $\Leftrightarrow$ (ii). $\square$

The next statement follows in an analogous manner:

Theorem 6.9.6 *Let* $\mu X < \infty$, $1 < r, p < \infty$, $1 \le s < \infty$ *and* $1 \le q \le \infty$. *Further, let* $\max\{r, s\} \le \min\{p, q\}$ *and* $\alpha > 1/r$. *We assume that* ν *is a* σ-*finite positive measure on* X. *Then the following statements are equivalent:*

(i) J_α *is bounded from* $L_\mu^{rs}(X)$ *into* $L_\nu^{pq}(X)$;

(ii) J_α *is bounded from* $L_\mu^r(X)$ *into* $L_\nu^p(X)$;

(iii)

$$\overline{D}_1 \equiv \sup_{0 < t < a} \overline{D}_1(t) \equiv \left(\int_{X \backslash B(x_0,t)} (\mu B_{x_0 x})^{(\alpha-1)p} d\nu(x) \right)^{1/p} \times$$

$$\times (\mu \overline{B}(x_0, t))^{1/r'} < \infty;$$

(iv)

$$\overline{D}_2 \equiv \sup_{k \in Z_+} \overline{D}_2(k) \equiv$$

$$\equiv \sup_{k \in Z_+} \left(\nu(B(x_0, \eta_1^{-k} a) \backslash B(x_0, \eta_1^{-(k+1)} a)) \right)^{1/p} \times$$

$$\times (\mu \overline{B}(x_0, \eta_1^{-(k+1)} a))^{\alpha - 1/r} < \infty$$

with the constant η_1 from the definition of the reverse doubling condition. Moreover, $\|J_\alpha\|_{L_\mu^{rs}(X) \to L_\nu^{pq}(X)} \approx \|J_\alpha\|_{L_\mu^{r}(X) \to L_\nu^{p}(X)} \approx D_1 \approx D_2$.

For the compactness of the operator J_α in Lorentz spaces we have the following statements, which follow using the arguments from the proof of Theorems 6.3.2, 6.3.4 and 6.3.5 (see also the proofs of Theorems 6.5.2 and 6.5.3).

Theorem 6.9.7. *Let* $\mu X = \infty$, $1 < r, p, q < \infty$ *and* $1 \leq s < \infty$. *Let* $\max\{r, s\} \leq \min\{p, q\}$. *We assume that* ν *is a* σ-*finite positive measure on* X. *Then the following statements are equivalent:*

(i) J_α *is compact from* $L_\mu^{rs}(X)$ *into* $L_\nu^{pq}(X)$;
(ii) J_α *is compact from* $L_\mu^{r}(X)$ *into* $L_\nu^{p}(X)$;
(iii) $D_1 < \infty$ *and* $\lim\limits_{t \to 0} D_1(t) = \lim\limits_{t \to \infty} D_1(t) = 0$;
(iv) $D_2 < \infty$ *and* $\lim\limits_{k \to -\infty} D_2(k) = \lim\limits_{k \to +\infty} D_2(k) = 0$.

Theorem 6.9.8. *Let* $\mu X < \infty$, $1 < r, p, q < \infty$ *and* $1 \leq s < \infty$. *Let* $\max\{r, s\} \leq \min\{p, q\}$. *We assume that* ν *is a* σ-*finite positive measure on* X. *Then the following statements are equivalent:*

(i) J_α *is compact from* $L_\mu^{rs}(X)$ *into* $L_\nu^{pq}(X)$;
(ii) J_α *is compact from* $L_\mu^{r}(X)$ *into* $L_\nu^{p}(X)$;
(iii) $\overline{D}_1 < \infty$ *and* $\lim\limits_{t \to 0} \overline{D}_1(t) = 0$;
(iv) $\overline{D}_2 < \infty$ *and* $\lim\limits_{k \to +\infty} \overline{D}_2(k) = 0$.

6.10.   Notes and comments on Chapter 6

The cores of the investigations presented in the first part of this chapter are the well-known results of [275] and [1].

In connection with Sections 6.1 and 6.2 we recall that the analogous problems were considered just in Euclidean space in [141–142]. For SHT we refer to [100], Chapter 3.

Theorems of Adams type for generalized potentials defined on SHT are due to [98]. In this work some approaches of [305] were used.

The references concerning Koosis–type theorems for classical potentials can be found in the text of Section 6.

This chapter is essentially based on our recent papers [162], [98], [199]. The results of Sections 6.5, 6.6, 6.6, 6.8 and 6.9 we publish here for the first time.

Chapter 7

SINGULAR NUMBERS

Necessary and sufficient conditions for weighted Volterra integral operators to belong to Schatten–von Neumann ideals are established. Similar problems are solved for Hardy–type transforms and for some operators of potential type defined on R^n and on spaces of homogeneous type. Moreover, two-sided estimates of Schatten–von Neumann norms for these integral transforms are obtained. In the last section we give the asymptotic behaviour of singular numbers of Erdelyi–Köber and Hadamard operators.

7.1. Volterra integral operators

In this section two-sided estimates of Schatten–von Neumann ideal norms for weighted Volterra integral operators are derived.

Let H be a separable Hilbert space and let $\sigma_\infty(H)$ be the class of all compact operators $T : H \to H$, which forms an ideal in the normed algebra $\mathcal{B}$ of all bounded linear operators in H. To construct a Schatten–von Neumann ideal $\sigma_p(H)$ $(0 < p \le \infty)$ in $\sigma_\infty(H)$, the sequence of singular numbers $s_j(T) \equiv \lambda_j(|T|)$ is used, where the eigenvalues $\lambda_j(|T|)$ ($|T| \equiv (T^*T)^{1/2}$) are non–negative and are repeated according to their multiplicity and arranged in decreasing order. A Schatten–von Neumann quasi-norm (norm if $1 \le p \le \infty$) is defined as follows:

$$\|T\|_{\sigma_p(H)} \equiv \Big(\sum_j s_j^p(T) \Big)^{1/p}, \quad 0 < p < \infty,$$

447

with the usual modification if $p = \infty$. Thus we have $\|T\|_{\sigma_\infty(H)} = \|T\|$ and $\|T\|_{\sigma_2(H)}$ is the Hilbert–Schmidt norm given by the formula

$$\|T\|_{\sigma_2(H)} = \left(\int \int |T_1(x,y)|^2 dx dy \right)^{1/2} \tag{7.1.1}$$

for an integral operator

$$Tf(x) = \int T_1(x,y) f(y) dy.$$

We refer, for example, to [105], [164], [235] for more information concerning Schatten–von Neumann ideals.

The next statement is from [105], Chapter III.

Proposition A. *The following equality is valid:*

$$\inf_{K \in \mathcal{B}_n} \|A - K\| = \left(\sum_{j=n+1}^\infty s_j^p(A) \right)^{1/p}, \quad 1 \le p \le \infty,$$

where $\mathcal{B}_n$ is the set of all finite rank operators with rank equal to $n - 1$ and $A \in \sigma_p(H)$.

First we consider the operator

$$K_v f(x) = v(x) \int_0^x f(y)(x-y)^{\alpha-1} dy, \quad x \in (0,a),$$

where v is a measurable function on $(0,a)$, $0 < a < \infty$.

Let us recall some definitions from Section 2.7.

We say that a kernel $k : \{(x,y) : 0 < y < x < a\} \to R_+$ belongs to V ($k \in V$) if there exists a positive constant d_1 such that for all x, y, z with $0 < y < z < x < a$ the inequality

$$k(x,y) \le d_1 k(x,z)$$

holds. Further, $k \in V_\lambda$ ($1 < \lambda < \infty$) if there exists a positive constant d_2 such that for all x, $x \in (0,a)$, the inequality

$$\int_{x/2}^x k^{\lambda'}(x,y) dy \le d_2 x k^{\lambda'}(x,x/2), \quad \lambda' = \frac{\lambda}{\lambda - 1}.$$

is fulfilled.

For some examples of kernels k satisfying the conditions V and V_λ see Section 2.7.

First we investigete the mapping properties of K_v in Lebesgue spaces.

The first part of the next statement follows immediately from Theorem 2.7.1 and Remark 2.7.1. The second part can be derived in the same way as Theorem 2.7.5.

Theorem 7.1.1. *Let* $1 < p \leq q < \infty$, $a = \infty$ *and let* $k \in V \cap V_p$. *Then*

(a) K_v *is bounded from* $L^p(0, \infty)$ *into* $L^q(0, \infty)$ *if and only if*

$$D \equiv \sup_{j \in Z} D(j) \equiv \sup_{j \in Z} \left(\int_{2^j}^{2^{j+1}} k^q(x, x/2) x^{q/p'} |v(x)|^q dx \right)^{\frac{1}{q}} < \infty.$$

Moreover, $\|K_v\| \approx D$.

(b) K_v acts compactly from $L^p(0, a)$ into $L^q(0, a)$ if and only if $D < \infty$ and $\lim_{j \to +\infty} D(j) = \lim_{j \to -\infty} D(j) = 0$.

Analogously we have the following

Theorem 7.1.2. *Let* $1 < p \leq q < \infty$, $a < \infty$ *and let* $k \in V \cap V_p$. *Then*

(a) K_v *is bounded from* $L^p(0, a)$ *into* $L^q(0, a)$ *if and only if*

$$D_a \equiv \sup_{j \geq 0} D_a(j) \equiv \sup_{j \geq 0} \left(\int_{2^{-(j+1)}a}^{2^{-j}a} |v(x)|^q k^q(x, x/2) x^{q/p'} dx \right)^{\frac{1}{q}} < \infty.$$

Moreover, $\|K_v\| \approx D_a$.

(b) K_v acts compactly from $L^p(0, a)$ into $L^q(0, a)$ if and only if $D_a < \infty$ and $\lim_{j \to +\infty} D_a(j) = 0$;

Let $0 < a \leq \infty$, $k : \{(x, y) : 0 < y < x < a\} \to R_+$ be a kernel and let $\overline{k}(x) \equiv xk^2(x, x/2)$. We denote by $l^p(L^2_{\overline{k}}(0, a))$ the set of all measurable functions $g : (0, a) \to R^1$ for which

$$\|g\|_{l^p(L^2_{\overline{k}}(0, \infty))} = \left(\sum_{n \in Z} \left(\int_{2^n}^{2^{n+1}} |g(x)|^2 \overline{k}(x) dx \right)^{p/2} \right)^{1/p} < \infty$$

if $a = \infty$ and

$$\|g\|_{l^p(L^2_{\overline{k}}(0, a))} = \left(\sum_{n=0}^{+\infty} \left(\int_{2^{-(n+1)}a}^{2^{-n}a} |g(x)|^2 \overline{k}(x) dx \right)^{p/2} \right)^{1/p} < \infty$$

if $a < \infty$, with the usual modification for $p = \infty$.

The next statement follows from general interpolation theorem (see, e.g., [299], p. 147 for the interpolation properties of the Schatten classes, and p.127

for the corresponding properties of the sequence spaces. See also [25]– Theorem 5.1.2).

Proposition 7.1.1. Let $0 < a \leq \infty$, $1 \leq p_0, p_1 \leq \infty$, $0 \leq \theta \leq 1$, $\frac{1}{p} = \frac{1-\theta}{p_0} + \frac{\theta}{p_1}$. If T is a bounded operator from $lp_i(L_{\frac{2}{k}}(0,a))$ into $\sigma_{p_i}(L^2(0,a))$, where $i = 0, 1$, then it is also bounded from $l^p(L_{\frac{2}{k}}(0,a))$ into $\sigma_p(L^2(0,a))$. Moreover,

$$\|T\|_{l^p(L_{\frac{2}{k}}) \to \sigma_p(L^2)} \leq \|T\|^{1-\theta}_{l^{p_0}(L_{\frac{2}{k}}) \to \sigma_{p_0}(L^2)} \|T\|^{\theta}_{l^{p_1}(L_{\frac{2}{k}}) \to \sigma_{p_1}(L^2)}.$$

We shall also need the next result from [105], Chapter 3, Section 7.5. It follows also from Lemma 2.11.12 of [235].

Proposition 7.1.2. Let $1 \leq p \leq \infty$ and let $\{f_k\}$, $\{g_k\}$ be orthonormal systems in a Hilbert space H. If $T \in \sigma_p(H)$, then

$$\|T\|_{\sigma_p(H)} \geq \left(\sum_n |\langle Tf_n, g_n \rangle|^p \right)^{1/p}.$$

Now we prove the main results of this section.
In the sequel we shall assume that $v \in L_{\frac{2}{k}}(2^n, 2^{n+1})$ for all $n \in Z$.

Theorem 7.1.3. Let $a = \infty$, $2 \leq p < \infty$ and let $k \in V \cap V_2$. Then K_v belongs to $\sigma_p(L^2(0,\infty))$ if and only if $v \in l^p(L_{\frac{2}{k}}(0,\infty))$. Moreover, there exist positive constants b_1 and b_2 such that

$$b_1 \|v\|_{l^p(L_{\frac{2}{k}}(0,\infty))} \leq \|K_v\|_{\sigma_p(L^2(0,\infty))} \leq b_2 \|v\|_{l^p((L_{\frac{2}{k}}(0,\infty))}.$$

Proof. *Sufficiency.* Note that since $k \in V \cap V_2$ it follows that

$$I(x) \equiv \int_0^x k^2(x,y)dy \leq c\overline{k}(x) \qquad (7.1.2)$$

for some positive constant c independent of x. Indeed, using the condition $k \in V \cap V_2$ we have

$$I(x) = \int_0^{x/2} k^2(x,y)dy + \int_{x/2}^x k^2(x,y)dy \leq c_1\overline{k}(x) + c_2\overline{k}(x) = c_3\overline{k}(x).$$

Consequently, using (7.1.1) and (7.1.2), we find that

$$\|K_v\|_{\sigma_2(L^2(0,\infty))} = \left(\int_0^\infty \int_0^x k^2(x,y)v^2(x)dxdy \right)^{1/2} =$$

$$= \left(\int\limits_0^\infty v^2(x) \left(\int\limits_0^x k^2(x,y)dy \right) dx \right)^{1/2} \le c_1 \left(\int\limits_0^\infty v^2(x)\overline{k}(x)dx \right)^{1/2} =$$

$$= c_4 \left(\sum_{n\in Z} \int\limits_{2^n}^{2^{n+1}} v^2(x)\overline{k}(x)dx \right)^{1/2} = c_4 \|v\|_{l^2(L_{\frac{2}{k}}^2(0,\infty))}.$$

On the other hand, in view of Theorem 7.1.1 we see that there exist positive constants c_5 and c_6 such that

$$c_5 \|v\|_{l^\infty(L_{\frac{2}{k}}^2(0,\infty))} \le \|K_v\|_{\sigma_\infty(L^2(0,\infty))} \le c_6 \|v\|_{l^\infty(L_{\frac{2}{k}}^2(0,\infty))}.$$

Further, Proposition 7.1.1 yields

$$\|K_v\|_{\sigma_p(L^2(0,\infty))} \le c_7 \|v\|_{l^p((L_{\frac{2}{k}}^2(0,\infty))},$$

where $2 \le p < \infty$.

Necessity. Let $K_v \in \sigma_p(L^2(0,\infty))$ and let

$$f_n(x) = \chi_{[2^n,2^{n+1})}(x)2^{-n/2},$$
$$g_n(x) = v(x)x^{1/2}\chi_{[3\cdot 2^{n-1},2^{n+1})}(x)k(x,x/2)\alpha_n^{-1/2},$$

where

$$\alpha_n = \int\limits_{3\cdot 2^{n-1}}^{2^{n+1}} v^2(y)\overline{k}(y)dy.$$

Then it is easy to verify that $\{f_n\}$ and $\{g_n\}$ are orthonormal systems. Further, by virtue of Proposition 7.1.2 (for $p \ge 1$) we have

$$\infty > \|K_v\|_{\sigma_p(L^2(0,\infty))} \ge \left(\sum_{n\in Z} |\langle K_v f_n, g_n \rangle|^p \right)^{1/p} =$$

$$= \left(\sum_{n\in Z} \left(\int\limits_{3\cdot 2^{n-1}}^{2^{n+1}} \left(\int\limits_{2^n}^x 2^{-n/2}k(x,y)dy \right) \times \right. \right.$$

$$\left. \left. \times v^2(x)x^{1/2}k(x,x/2)\alpha_n^{-1/2}dx \right)^p \right)^{1/p} \ge$$

$$\ge c_8 \left(\sum_{n\in Z} \left(\alpha_n^{-1/2} \int\limits_{3\cdot 2^{n-1}}^{2^{n+1}} 2^{-n/2}k(x,x/2) \times \right. \right.$$

$$\times v^2(x)(x-2^n)x^{1/2}dx\bigg)^p\bigg)^{1/p} \geq$$

$$\geq c_9\bigg(\sum_{n\in Z}\bigg(\alpha_n^{-1/2}\int_{3\cdot 2^{n-1}}^{2^{n+1}}\overline{k}(x)v^2(x)dx\bigg)^p\bigg)^{1/p} =$$

$$= c_9\bigg(\sum_{n\in Z}\alpha_n^{p/2}\bigg)^{1/p}.$$

Now let

$$f_n'(x) = \chi_{[3\cdot 2^{n-2},3\cdot 2^{n-1})}(x)(3\cdot 2^{n-2})^{-1/2}$$

and

$$g_n'(x) = v(x)x^{1/2}\chi_{[2^n,3\cdot 2^{n-1})}(x)k(x,x/2)\beta_n^{-1/2},$$

where

$$\beta_n = \int_{2^n}^{3\cdot 2^{n-1}} v^2(y)\overline{k}(y)dy.$$

Then it is easy to verify that $\{f_m'\}$ and $\{g_m'\}$ are orthonormal systems. Further,

$$\infty > \|K_v\|_{\sigma_p(L^2(0,\infty))} \geq \bigg(\sum_{n\in Z}|\langle K_v f_n', g_n'\rangle|^p\bigg)^{1/p} =$$

$$= \bigg(\sum_{n\in Z}\bigg(\int_{2^n}^{3\cdot 2^{n-1}}\bigg(\int_{3\cdot 2^{n-2}}^{x}(3\cdot 2^{n-2})^{-1/2}k(x,y)dy\bigg)\times$$

$$\times v^2(x)x^{1/2}k(x,x/2)\beta_n^{-1/2}dx\bigg)^p\bigg)^{1/p} \geq$$

$$\geq c_{10}\bigg(\sum_{n\in Z}\bigg(\beta_n^{-1/2}\int_{2^n}^{3\cdot 2^{n-1}}2^{-(n-2)/2}k^2(x,x/2)\times$$

$$\times v^2(x)(x-3\cdot 2^{n-2})x^{1/2}dx\bigg)^p\bigg)^{1/p} \geq$$

$$\geq c_{11}\bigg(\sum_{n\in Z}\bigg(\beta_n^{-1/2}\int_{2^n}^{3\cdot 2^{n-1}}\overline{k}(x)v^2(x)dx\bigg)^p\bigg)^{1/p} =$$

$$= c_{11}\bigg(\sum_{n\in Z}\beta_n^{p/2}\bigg)^{1/p},$$

where $p \geq 1$. Consequently

$$\left(\sum_{n \in Z} \left(\int_{2^n}^{2^{n+1}} v^2(x)\overline{k}(x)dx \right)^{p/2} \right)^{1/p} \leq \left(\sum_{n \in Z} (\beta_n + \alpha_n)^{p/2} \right)^{1/p} \leq$$

$$\leq c_{12}\|K_v\|_{\sigma_p(L^2(0,\infty))} + c_{12}\|K_v\|_{\sigma_p(L^2(0,\infty))} \leq$$

$$\leq c_{13}\|K_v\|_{\sigma_p(L^2(0,\infty))} < \infty.$$

$\square$

A result analogous to Theorem 7.1.3 was obtained in [223] for the weighted Riemann–Liouville operator with order greater than $1/2$ (see [224] for the weighted Hardy operator $H_v f(x) = v(x) \int_0^x f(t)dt$).

Let us now consider the case $a < \infty$.

Theorem 7.1.4. *Let* $0 < a < \infty$, $2 \leq p < \infty$ *and let* $k \in V \cap V_2$. *Then* K_v *belongs to* $\sigma_p(L^2(0,a))$ *if and only if* $v \in l^p(L^2_{\frac{1}{k}}(0,a))$. *Moreover, there exists positive constants* b_1 *and* b_2 *such that*

$$b_1\|v\|_{l^p(L^2_{\frac{1}{k}}(0,a))} \leq \|K_v\|_{\sigma_p(L^2(0,a))} \leq b_2\|v\|_{l^p(L^2_{\frac{1}{k}}(0,a))}.$$

Proof. *Sufficiency.* The Hilbert–Schmidt formula (see (7.1.1)) and the condition $k \in V \cap V_2$ yield

$$\|K_v\|_{\sigma_2(L^2(0,a))} =$$

$$\left(\int_0^a v^2(x)\left(\int_0^x k^2(x,y)dy \right)dx \right)^{1/2} \leq$$

$$\leq c_1\left(\sum_{n=0}^{\infty} \int_{2^{-(n+1)}a}^{2^{-n}a} v^2(x)\overline{k}(x)dx \right)^{1/2} = c_1\|v\|_{l^2(L^2_{\frac{1}{k}}(0,a))}.$$

By Theorem 7.1.1 we arrive at

$$\|K_v\|_{\sigma_\infty(L^2(0,a))} \approx \|v\|_{l^\infty(L^2_{\frac{1}{k}}(0,a))}.$$

Using Proposition 7.1.1 we derive

$$\|K_v\|_{\sigma_p(L^2(0,a))} \leq c_2\|v\|_{l^p(L^2_{\frac{1}{k}}(0,a))}$$

when $p \geq 2$.

To prove necessity we take the orthonormal systems of functions defined on $(0, a)$

$$f_n(x) = \chi_{[2^{-(n+1)}a, 2^{-n}a)}(x)(2^{-(n+1)}a)^{-1/2}$$

and

$$g_n(x) = v(x)x^{1/2}\chi_{[3\cdot 2^{-(n+2)}a, 2^{-n}a)}(x)k(x, x/2)\alpha_n^{-1/2},$$

where

$$\alpha_n = \int\limits_{3\cdot 2^{-(n+2)}a}^{2^{-n}a} v^2(y)\overline{k}(y)dy$$

and $n \in Z_+$. Consequently Proposition 7.1.2 yields

$$\infty > \|K_v\|_{\sigma_p(L^2(0,a))} \geq$$

$$\left(\sum_{n=0}^{+\infty} |\langle K_v f_n, g_n\rangle|^p\right)^{1/p} =$$

$$= \left(\sum_{n=0}^{\infty} \left(\int\limits_{3\cdot 2^{-(n+2)}a}^{2^{-n}a} x^{1/2}v^2(x)k(x, x/2) \times\right.\right.$$

$$\left.\left(\int\limits_{2^{-(n+1)}a}^{x} (2^{-(n+1)}a)^{-1/2}k(x, y)dy\right)\alpha_n^{-1/2}dx\right)^p\right)^{1/p} \geq$$

$$\geq c_3\left(\sum_{n=0}^{\infty} \alpha_n^{p/2}\right)^{1/p}.$$

If we take the orthonormal systems

$$f_n'(x) = \chi_{[3\cdot 2^{-(n+3)}a, 3\cdot 2^{-(n+2)}a)}(x)(3\cdot 2^{-(n+3)}a)^{-1/2},$$

$$g_n'(x) = v(x)x^{1/2}\chi_{[2^{-(n+1)}a, 3\cdot 2^{-(n+2)}a)}(x)k(x, x/2)\beta_n^{-1/2},$$

where

$$\beta_n = \int\limits_{2^{-(n+1)}a}^{3\cdot 2^{-(n+2)}a} v^2(y)\overline{k}(y)dy,$$

then we arrive at the estimate

$$\|K_v\|_{\sigma_p(L^2(0,a))} \geq c_4\left(\sum_{n=0}^{\infty} \beta_n^{p/2}\right)^{1/p}.$$

Finally we have the lower estimate for $\|K_v\|_{\sigma_p(L^2(0,a))}$. $\square$

Remark 7.1.1. It follows from the proof of Theorems 7.1.3 and 7.1.4 that the lower estimate of $\|K_v\|_{\sigma_p(L^2(0,a))}$ holds for $1 \leq p \leq \infty$.

The next result gives a manageable equivalent form of the norm of v which appears in Theorem 7.1.3.

Proposition 7.1.3. *Let* $1 \leq p < \infty$. *Then*

$$\|v\|_{l^p(L_{\frac{2}{k}}(0,\infty))} \approx J(v,p),$$

where

$$J(v,p) = \left(\int\limits_0^\infty \left(\int\limits_{x/2}^{2x} v^2(y)k^2(y,y/2)dy \right)^{p/2} x^{p/2-1}dx \right)^{1/p}.$$

Proof. We have

$$\|v\|_{l^p(L_{\frac{2}{k}}(0,\infty))} = \left(\sum_{n \in Z} \left(\int\limits_{2^n}^{2^{n+1}} v^2(x)\overline{k}(x)dx \right)^{p/2} \right)^{1/p} \leq$$

$$\leq \left(\sum_{n \in Z} \left(\int\limits_{2^n}^{2^{n+1}} v^2(x)k^2(x,x/2)dx \right)^{p/2} 2^{(n+1)p/2} \right)^{1/p} =$$

$$= c_1 \left(\sum_{n \in Z} \left(\int\limits_{2^n}^{2^{n+1}} v^2(x)k^2(x,x/2)dx \right)^{p/2} 2^{np/2} \right)^{1/p} \leq$$

$$\leq c_2 \left(\sum_{n \in Z} \int\limits_{2^n}^{2^{n+1}} y^{p/2-1} \left(\int\limits_{2^n}^{2^{n+1}} v^2(x)k^2(x,x/2)dx \right)^{p/2} dy \right)^{1/p} \leq$$

$$\leq c_2 \left(\sum_{n \in Z} \int\limits_{2^n}^{2^{n+1}} y^{p/2-1} \left(\int\limits_{y/2}^{2y} v^2(x)k^2(x,x/2)dx \right)^{p/2} dy \right)^{1/p} =$$

$$= c_2 J(v,p).$$

To prove the reverse inequality we observe that

$$J(v,p) = \left(\sum_{n \in Z} \int\limits_{2^n}^{2^{n+1}} y^{p/2-1} \left(\int\limits_{y/2}^{2y} v^2(x)k^2(x,x/2)dx \right)^{p/2} dy \right)^{1/p} \leq$$

$$\leq \left(\sum_{n \in Z} \left(\int\limits_{2^n}^{2^{n+1}} y^{p/2-1}dy \right) \left(\int\limits_{2^{n-1}}^{2^{n+2}} v^2(x)k^2(x,x/2)dx \right)^{p/2} \right)^{1/p} \leq$$

$$\leq c_3\bigg(\sum_{n\in Z}2^{np/2}\bigg(\int_{2^{n-1}}^{2^n}v^2(x)k^2(x,x/2)dx\bigg)^{p/2}\bigg)^{1/p}+$$

$$+c_3\bigg(\sum_{n\in Z}2^{np/2}\bigg(\int_{2^n}^{2^{n+1}}v^2(x)k^2(x,x/2)dx\bigg)^{p/2}\bigg)^{1/p}+$$

$$+c_3\bigg(\sum_{n\in Z}2^{np/2}\bigg(\int_{2^{n+1}}^{2^{n+2}}v^2(x)k^2(x,x/2)dx\bigg)^{p/2}\bigg)^{1/p}\leq$$

$$c_4\|v\|_{l^p(L^2_{\frac{k}{k}}(0,\infty))}.$$

$\square$

From Theorem 7.1.3 and Proposition 7.1.3 we obviously have

Theorem 7.1.5. *Let* $2\leq p<\infty$ *and let* $k\in V\cap V_\lambda$. *Then*

$$\|K_v\|_{\sigma_p(L^2(0,\infty))}\approx J(v,p).$$

Now we shall investigate the behavious of Schatten–von Neumann norm for the operator

$$\mathcal{K}_v f(x,t)=v(x,t)\int_0^x \mathbf{K}(x,y,t)f(y)dy,\quad \mathbf{K}\geq 0,\ (x,t)\in R_a,$$

where $\widetilde{R}_a\equiv[0,a)\times[0,\infty),\ 0<a\leq\infty$. We shall assume that $v\in L^2(U_{2^n}\setminus U_{2^{n+1}})$ for all $n\in Z$ if $a=\infty$ and $v\in L^2(U_{a\cdot 2^{-(n+1)}}\setminus U_{a\cdot 2^{-n}})$ for $n\in Z_+$ if $a<\infty$, where $U_b\equiv[b,\infty)\times[0,\infty)$.

We recall that the kernel $\mathbf{K}:\{(x,y):0<y<x<a\}\times[0,\infty)\to R_+$ belongs to $\mathcal{V}$ if there exists a positive constant b_1 such that for all x,y,z with $0<y<z<x<a$ and for all $t>0$ the inequality

$$\mathbf{K}(x,y,t)\leq b_1\mathbf{K}(x,z,t)$$

is satisfied. Further, by definition $\mathbf{K}\in\mathcal{V}_\lambda$ $(1<\lambda<\infty)$ if there exists a positive constant b_2 such that for all $x\in(0,a)$, and for all $t>0$ the inequality

$$\int_{x/2}^x \mathbf{K}^{\lambda'}(x,y,t)dy\leq b_2 x\mathbf{K}^{\lambda'}(x,x/2,t)$$

is fulfilled, where $\lambda'=\lambda/(\lambda-1)$.

Let $\overline{K}(x,t) \equiv x\mathbf{K}(x, x/2, t)$. We denote by $l^p(L_{\overline{K}}^2(\widetilde{R}_a))$ ($0 < p \le \infty$, $0 < a \le \infty$) the set of all measurable functions $g : \widetilde{R}_a \to R^1$ such that

$$\|g\|_{l^p(L_{\overline{K}}^2(\widetilde{R}_\infty))} = \left(\sum_{n \in Z} \left(\int_{U_{2^n} \setminus U_{2^{n+1}}} |g(x,t)|^2 \overline{K}(x,t) dx dt \right)^{p/2} \right)^{1/p} < \infty$$

if $a = \infty$ and

$$\|g\|_{l^p(L_{\overline{K}}^2(\widetilde{R}_a))} =$$

$$= \left(\sum_{n=0}^{+\infty} \left(\int_{U_{2^{-(n+1)}a} \setminus U_{2^{-n}a}} |g(x,t)|^2 \overline{K}(x,t) dx dt \right)^{p/2} \right)^{1/p} < \infty$$

if $a < \infty$, with the usual modification for $p = \infty$.

First we characterize the boundedness and compactness of K_v.

Theorem 7.1.6. *Let* $1 < p \le q < \infty$, $a = \infty$ *and let* $\mathbf{K} \in \mathcal{V} \cap \mathcal{V}_p$. *Then*
(a) $\mathcal{K}_v$ *is bounded from* $L^p(0, \infty)$ *into* $L_v^q(\widetilde{R}_\infty)$ *if and only if*

$$E_\infty \equiv \sup_{n \in Z} E_\infty(n) \equiv$$

$$\equiv \left(\int_{U_{2^n} \setminus U_{2^{n+1}}} \mathbf{K}^q(x, x/2, t) x^{q/p'} |v(x,t)|^q dx dt \right)^{1/q} < \infty.$$

Moreover,

$$\|\mathcal{K}_v\| \approx E_\infty;$$

(b) $\mathcal{K}_v$ *is compact from* $L^p(0, \infty)$ *into* $L_v^q(\widetilde{R}_\infty)$ *if and only if* $E_\infty < \infty$ *and*
$\lim_{n \to +\infty} E_\infty(n) = \lim_{n \to -\infty} E_\infty(n) = 0$.

Proof. The first part of this theorem can be derived from Theorem 2.10.1 and Remark 2.10.1. Part (b) follows in the same way as Theorem 2.10.2. $\square$

Analogously, we have

Theorem 7.1.7. *Let* $1 < p \le q < \infty$, $a < \infty$ *and let* $\mathbf{K} \in \mathcal{V} \cap \mathcal{V}_p$. *Then the following statements hold:*
(a) $\mathcal{K}_v$ *is bounded from* $L^p(0, a)$ *into* $L_v^q(\widetilde{R}_a)$ *if and only if*

$$E_a \equiv \sup_{n \in Z_+} E_a(n) \equiv$$

$$\equiv \left(\int_{U_{a \cdot 2^{-(n+1)}} \setminus U_{a \cdot 2^{-n}}} k^q(x, x/2, t) x^{q/p'} |v(x,t)|^q dx dt \right)^{1/q} < \infty.$$

Moreover,

$$\|\mathcal{K}_v\| \approx E_a;$$

(b) $\mathcal{K}_v$ is compact from $L^p(0, a)$ into $L^q_\nu(\widetilde{R}_a)$ if and only if $E_a < \infty$ and $\lim\limits_{n \to +\infty} E_a(n) = 0$.

The next interpolation result in more general cases can be found, for example, in [299], [25], Theorem 5.1.2.

Proposition 7.1.4. *Let* $0 < a \leq \infty$, $1 \leq p_0, p_1 \leq \infty$, $0 \leq \theta \leq 1$, $\frac{1}{p} = \frac{1-\theta}{p_0} + \frac{\theta}{p_1}$. *If T is a bounded operator from* $l^{p_i}(L^2_{\overline{K}}(\widetilde{R}_a))$ *into* $\sigma_{p_i}(L^2(\widetilde{R}_a))$, *where* $i = 0, 1$, *then it is also bounded from* $l^p(L^2(\widetilde{R}_a))$ *into* $\sigma_p(L^2(\widetilde{R}_a))$. *Moreover,*

$$\|T\|_{l^p(L^2_{\overline{K}}) \to \sigma_p(L^2)} \leq \|T\|^{1-\theta}_{l^{p_0}(L^2_{\overline{K}}) \to \sigma_{p_0}(L^2)} \|T\|^\theta_{l^{p_1}(L^2_{\overline{K}}) \to \sigma_{p_1}(L^2)}.$$

Now we are ready to prove

Theorem 7.1.8. *Let* $a = \infty$, $2 \leq p < \infty$ *and let* $\mathbf{K} \in \mathcal{V} \cap \mathcal{V}_2$. *Then* $\mathcal{K}_v \in \sigma_p(L^2(\widetilde{R}_\infty))$ *if and only if* $v \in l^p(L^2_{\overline{K}}(\widetilde{R}_\infty))$. *Moreover,*

$$\|\mathcal{K}_v\|_{\sigma_p(L^2(\widetilde{R}_\infty))} \approx \|v\|_{l^p(L^2_{\overline{K}}(\widetilde{R}_\infty))}.$$

Proof. To prove sufficiency, first note that the condition $\mathbf{K} \in \mathcal{V} \cap \mathcal{V}_2$ implies the inequality

$$\int\limits_0^x \mathbf{K}^2(x, y, t)dy \leq c\overline{\mathbf{K}}(x, x/2, t),$$

with a positive constant $c > 0$ independent of (x, t), $(x, t) \in \widetilde{R}_\infty$.
Hence, using the Hilbert–Schmidt formula (see (7.1.1)) we have

$$\|\mathcal{K}_v\|_{\sigma_2(L^2(\widetilde{R}_\infty))} = \left(\int\limits_0^\infty \int\limits_0^x \int\limits_0^\infty \mathbf{K}^2(x, y, t)v^2(x, t)dxdydt \right)^{1/2} =$$

$$= \left(\int\limits_0^\infty \int\limits_0^\infty v^2(x, t) \left(\int\limits_0^x \mathbf{K}^2(x, y, t)dy \right) dxdt \right)^{1/2} \leq$$

$$\leq c_1 \left(\int\limits_{\widetilde{R}_\infty} v^2(x, t)\overline{\mathbf{K}}(x, t)dxdt \right)^{1/2} =$$

$$= c_1 \left(\sum_{n \in Z} \int_{U_{2n} \setminus U_{2n+1}} v^2(x) \overline{\mathbf{K}}(x,t) dx dt \right)^{1/2} =$$

$$= c_1 \|v\|_{l^2(L^2_{\frac{2}{\mathbf{K}}}(\tilde{R}_\infty))}.$$

By Theorem 7.1.6 we have

$$\|\mathcal{K}_v\|_{\sigma_\infty(L^2(\tilde{R}_\infty))} \approx \|v\|_{l^\infty(L^2_{\frac{2}{\mathbf{K}}}(\tilde{R}_\infty))}.$$

Due to Proposition 7.1.4 we finally obtain

$$\|\mathcal{K}_v\|_{\sigma_p(L^2(\tilde{R}_\infty))} \leq c_2 \|v\|_{l^p(L^2_{\frac{2}{\mathbf{K}}}(\tilde{R}_\infty))}, \quad 2 \leq p < \infty.$$

To prove necessity, we take the orthonormal systems of functions

$$f_n(x) = \chi_{[2^n, 2^{n+1})}(x) 2^{-n/2},$$

$$g_n(x,t) = v(x,t) x^{1/2} \chi_{\Lambda_n}(x,t) \mathbf{K}(x, x/2, t) \lambda_n^{-1/2},$$

where

$$\lambda_n = \int_{\Lambda_n} v^2(y,t) \overline{\mathbf{K}}(y,t) dy dt,$$

$$\Lambda_n = [3 \cdot 2^{n-1}, 2^{n+1}) \times R_+.$$

Taking into account Proposition 7.1.2, we obtain

$$\infty > \|\mathcal{K}_v\|_{\sigma_p(L^2(\tilde{R}_\infty))} \geq \left(\sum_{n \in Z} |\langle \mathcal{K}_v f_n, g_n \rangle|^p \right)^{1/p} =$$

$$= \left(\sum_{n \in Z} \left(\int_{\Lambda_n} \left(\int_{2^n}^x 2^{-n/2} \mathbf{K}(x,y,t) dy \right) \times \right. \right.$$

$$\left. \left. \times v^2(x,t) x^{1/2} \mathbf{K}(x, x/2, t) \lambda_n^{-1/2} dx dt \right)^p \right)^{1/p} \geq$$

$$\geq c_3 \left(\sum_{n \in Z} \left(\lambda_n^{-1/2} \int_{\Lambda_n} 2^{-n/2} \mathbf{K}^2(x, x/2, t) \times \right. \right.$$

$$\left. \left. v^2(x,t)(x - 2^n) x^{1/2} dx dt \right)^p \right)^{1/p} \geq c_4 \left(\sum_{n \in Z} \lambda_n^{p/2} \right)^{1/p}.$$

Now let

$$f'_n(x) = \chi_{[3 \cdot 2^{n-2}, 3 \cdot 2^{n-1})}(x)(3 \cdot 2^{n-2})^{-1/2},$$

$$g'_n(x) = v(x,t)x^{1/2}\chi_{\overline{\Lambda}_n}(x,t)\mathbf{K}(x,x/2,t)(\overline{\lambda}_n)^{-1/2},$$

where

$$\overline{\lambda}_n = \int\limits_{\overline{\Lambda}_n} v^2(y,t)\overline{\mathbf{K}}(y,t)dydt,$$

$$\overline{\Lambda}_n = [2^n, 3 \cdot 2^{n-1}) \times [0, \infty).$$

Then $\{f'_m\}$ and $\{g'_m\}$ are orthonormal systems and using again Proposition 7.1.2 we analogously derive

$$\infty > \|\mathcal{K}_v\|_{\sigma_p(L^2(\widetilde{R}_\infty))} \geq c_5\left(\sum_{n \in Z}(\overline{\lambda}_n)^{p/2}\right)^{1/p},$$

where $p > 0$. From these estimates we finally conclude

$$\|\mathcal{K}_v\|_{\sigma_p(L^2(\widetilde{R}_\infty))} \geq c_6\|v\|_{l^p(L^2_{\frac{1}{\mathbf{K}}}(\widetilde{R}_\infty))}$$

$\square$

The following theorem may be established in a similar manner:

Theorem 7.1.9. *Let $a < \infty$, $2 \leq p < \infty$ and let $\mathbf{K} \in \mathcal{V} \cap \mathcal{V}_2$. Then $\mathcal{K}_v \in \sigma_p(L^2(\widetilde{R}_a))$ if and only if $v \in l^p(L^2_{\frac{1}{\mathbf{K}}}(\widetilde{R}_a))$. Moreover,*

$$\|\mathcal{K}_v\|_{\sigma_p(L^2(\widetilde{R}_a))} \approx \|v\|_{l^p(L^2_{\frac{1}{\mathbf{K}}}(\widetilde{R}_a))}.$$

As the following statement can be derived in the same way as Proposition 7.1.3, we omit the proof.

Proposition 7.1.5. *Let $1 \leq p < \infty$. Then*

$$\|v\|_{l^p(L^2_{\frac{1}{\mathbf{K}}}(\widetilde{R}_\infty))} \approx \overline{J}(v,p),$$

where

$$\overline{J}(v,p) = \left(\int\limits_0^\infty \left(\int\limits_{x/2}^{2x}\int\limits_0^\infty v^2(y,y/2,t)\mathbf{K}^2(y,y/2,t)dydt\right)^{p/2} x^{p/2-1}dx\right)^{1/p}.$$

From Theorem 7.1.8 and Proposition 7.1.5 we have the following

Theorem 7.1.10. *Let $2 \leq p < \infty$ and let $\mathbf{K} \in \mathcal{V} \cap \mathcal{V}_\lambda$. Then*

$$\|\mathcal{K}_v\|_{\sigma_p(L^2(\widetilde{R}_\infty))} \approx \overline{J}(v,p).$$

Let us consider the weighted generalized Riemann–Liouville operator

$$T_{\alpha,v}f(x,t) = v(x,t) \int\limits_0^x (x - y + t)^{\alpha-1} f(y)dy, \quad x > 0, \ \alpha > 0.$$

Theorem 7.1.11. *Let $a = \infty$, $2 \leq p < \infty$ and let $\alpha > 1/2$. Then*

$$\|T_{\alpha,v}\|_{\sigma_p(L^2(\widetilde{R}_\infty))} \approx \|v\|_{l^p(L^2_w(\widetilde{R}_\infty))},$$

where $w(x,t) = x(x + t)^{2\alpha-2}$.

This statement follows immediately from Theorem 7.1.8 when $1/2 < \alpha < 1$ and in the same way as Theorem 7.1.3 (or Theorem 7.1.8) when $\alpha \geq 1$.

7.2.   Riemann–Liouville–type operators

Here we shall deal with the Riemann–Liouville– type operator

$$R_{\alpha,v,w}f(x) = v(x) \int\limits_0^x f(y)(x - y)^{\alpha-1} w(y)dy, \quad \alpha > 0,$$

where v and w are measurable functions on $(0, \infty)$.

We say that the measurable function u belongs to $U_\lambda (1 < \lambda < \infty)$ if there exists a positive constant c such that for all $x \in (0, \infty)$ the following inequality holds:

$$\int\limits_{\frac{x}{2}}^x (x - y)^{(\alpha-1)\lambda'} |u(y)|^{\lambda'} \, dy \leq cx^{(\alpha-1)\lambda'} \int\limits_0^{\frac{x}{4}} |u(y)|^{\lambda'} \, dy, \qquad (7.2.1)$$

where $\lambda' = \frac{\lambda}{\lambda-1}$.

In particular, if u is a decreasing function on $(0, \infty)$ and $\alpha > \frac{1}{\lambda}$, then (7.2.1) is satisfied. Further, if u satisfies the condition

$$M^- \left(|u|^{\lambda'}\right)(x) \leq \frac{c}{x} \int\limits_0^{\frac{x}{4}} |u(y)|^{\lambda'} \, dy,$$

where c is independent of x and

$$M^- g(x) = \sup_{0 < h < x} \frac{1}{h} \int_{x-h}^{x} |g(y)| \, dy,$$

then (7.2.1) holds for u.

From Theorem 2.1.12 we immediately have

Theorem 7.2.1. *Let* $1 < p \leq q < \infty$, $\alpha > 0$ *and let* $w \in U_p$. *Then* $R_{\alpha,v,w}$ *is bounded from* $L^p(0, \infty)$ *into* $L^q(0, \infty)$ *if and only if*

$$B \equiv \sup_{t>0} B(t) \equiv \sup_{t>0} \left(\int_{2t}^{\infty} |v(y)|^q \, y^{(\alpha-1)q} dy \right)^{\frac{1}{q}} \left(\int_{0}^{t} |w(y)|^{p'} \, dy \right)^{\frac{1}{p'}} < \infty.$$

Moreover, $\|R_{\alpha,v,w}\| \approx B$.

In the same way as Theorem 2.1.12 we obtain

Theorem 7.2.2. *Let* $1 < p \leq q < \infty$, $\alpha > 0$ *and let* $w \in U_p$. *Then* $R_{\alpha,v,w}$ *acts boundedly from* $L^p(0, \infty)$ *into* $L^q(0, \infty)$ *if and only if* $B < \infty$ *and* $\lim_{t \to 0} B(t) = \lim_{t \to \infty} B(t) = 0$.

To estimate the Schatten–von Neumann ideal norms of the operator $R_{\alpha,v,w}$, we first prove the following lemma:

Lemma 7.2.1. *Let* $1 \leq p < \infty$ *and* $\alpha > 0$. *Then there exists a positive constant* b_1 *such that the inequality*

$$\|R_{\alpha,v,w}\|_{\sigma_p(L^2(0,\infty))} \geq$$

$$\geq b_1 \sup_{\{a_m\}} \left[\sum_m \left(\int_{a_{m-1}}^{a_m} w^2(y) dy \right)^{p/2} \left(\int_{2a_m}^{2a_{m+1}} \frac{v^2(x)}{x^{(1-\alpha)2}} dx \right)^{p/2} \right]^{1/p},$$

where the supremum is taken over all increasing sequences $\{a_m\}$ *of positive numbers.*

Proof. Let

$$f_m(y) = \alpha_m^{-1} w(y) \chi_{[a_{m-1}, a_m]}(y),$$

$$g_m(y) = \beta_m^{-1} v(y) \chi_{[2a_m, 2a_{m+1}]}(y) y^{\alpha-1},$$

where $\{a_m\}$ is an increasing sequence of positive numbers, and let

$$\alpha_m = \left(\int_{a_{m-1}}^{a_m} w^2(y) dy \right)^{1/2},$$

$$\beta_m = \left(\int\limits_{2a_m}^{2a_{m+1}} v^2(x) x^{(\alpha-1)2} dx \right)^{1/2}.$$

Due to Proposition 7.1.2 we obtain

$$\|R_{\alpha,v,w}\|_{\sigma_p(L^2(0,\infty))} \geq \left(\sum_m |\langle R_{\alpha,v,w} f_m, g_m \rangle|^p \right)^{1/p} =$$

$$= \left(\sum_m \left(\alpha_m^{-1} \beta_m^{-1} \int\limits_{2a_m}^{2a_{m+1}} v^2(x) x^{\alpha-1} \left(\int\limits_{a_{m-1}}^{a_m} (x-y)^{\alpha-1} \times \right. \right. \right.$$

$$\left. \left. \left. \times w^2(y) dy \right) dx \right)^p \right)^{1/p} \geq$$

$$\geq b_1 \left(\sum_m \left(\alpha_m^{-1} \beta_m^{-1} \int\limits_{2a_m}^{2a_{m+1}} v^2(x) x^{(\alpha-1)2} dx \right)^p \times \right.$$

$$\left. \times \left(\int\limits_{a_{m-1}}^{a_m} w^2(y) dy \right)^p \right)^{1/p} =$$

$$= b_1 \left(\sum_m \left(\int\limits_{2a_m}^{2a_{m+1}} v^2(x) x^{(\alpha-1)2} dx \right)^{p/2} \left(\int\limits_{a_{m-1}}^{a_m} w^2(y) dy \right)^{p/2} \right)^{1/p}.$$

□

Theorem 7.2.3. *Let* $2 \leq p < \infty, \alpha > 0$ *and let* $w \in U_2$. *Then there exist positive constants* c_1 *and* c_2 *such that the following two-sided estimate holds:*

$$c_2 A(p) \leq \|R_{\alpha,v,w}\|_{\sigma_p(L^2(0,\infty))} \leq c_1 A(p), \tag{7.2.2}$$

where

$$A(p) = \left(\int\limits_0^\infty \left(\int\limits_0^x w^2(y) dy \right)^{p/2-1} w^2(x) \left(\int\limits_{2x}^\infty \frac{v^2(y)}{y^{(1-\alpha)2}} dy \right)^{\frac{p}{2}} dx \right)^{\frac{1}{p}}.$$

Proof. From Theorem 7.2.1 we have

$$\|R_{\alpha,v,w}\|_{\sigma_\infty(L^q(0,\infty))} \leq b_1 B.$$

On the other hand, the Hilbert–Schmidt formula (see (7.1.1)) yields

$$\|R_{\alpha,v,w}\|_{\sigma_2(L^2(0,\infty))} =$$

$$+ \left(\int_0^\infty v^2(x) \left(\int_0^x (x-y)^{(\alpha-1)2} w^2(y) dy \right) dx \right)^{\frac{1}{2}}. \qquad (7.2.3)$$

Further, we have

$$I(x) \equiv \int_0^x (x-y)^{(\alpha-1)2} w^2(y) dy =$$

$$= \int_0^{\frac{x}{2}} (x-y)^{(\alpha-1)2} w^2(y) dy + \int_{\frac{x}{2}}^x (x-y)^{(\alpha-1)2} w^2(y) dy \equiv$$

$$\equiv I_1(x) + I_2(x).$$

It is easy to see that if $y < \frac{x}{2}$, then $(x-y)^{(\alpha-1)2} \le b_2 x^{(\alpha-1)2}$, where the positive constant b_2 does not depend on x and y. Hence

$$I_1(x) \le b_3 x^{(\alpha-1)2} \int_0^{\frac{x}{2}} w^2(y) dy.$$

In view of the condition $w \in U_2$ we get

$$I_2(x) \le b_4 x^{(\alpha-1)2} \int_0^{\frac{x}{2}} w^2(y) dy,$$

where the positive constant b_4 does not depend on x. Finally we obtain

$$I(x) \le b_5 x^{(\alpha-1)2} \int_0^{\frac{x}{2}} w^2(y) dy.$$

Taking into account the last inequality, from (7.2.3) we derive

$$\|R_{\alpha,v,w}\|_{\sigma_2(L^2(0,\infty))} \le b_6 \left(\int_0^\infty v^2(x) x^{(\alpha-1)2} \left(\int_0^{x/2} w^2(y) dy \right) dx \right)^{\frac{1}{2}} =$$

$$= b_6 \left(\int_0^\infty w^2(y) \left(\int_{2y}^\infty v^2(x) x^{(\alpha-1)2} dx \right) dy \right)^{\frac{1}{2}} \equiv b_6 J.$$

Now choose positive numbers x_k so that

$$\int_0^{x_k} w^2(y)dy = 2^k.\tag{7.2.4}$$

If $\int_0^\infty w^2(y)dy \in [2^m, 2^{m+1})$, for some integer m, then we assume that $x_{m+1} = \infty$. If $\int_0^\infty w^2(y)dy = \infty$, then (7.2.4) holds for all $k \in Z$. It is obvious that

$$\int_{x_{k-1}}^{x_k} w^2(y)dy = 2^{k-1}.\tag{7.2.5}$$

Using the sequence $\{x_k\}$ we obtain

$$J = \left(\sum_k \int_{x_k}^{x_{k+1}} w^2(x)\left(\int_{2x}^\infty \frac{v^2(y)dy}{y^{(1-\alpha)2}}\right)^{\frac{1}{2}} dx\right) \leq$$

$$\leq \left(\sum_k \left(\int_{x_k}^{x_{k+1}} w^2(y)dy\right)\left(\int_{2x_k}^\infty v^2(x)x^{(\alpha-1)2}dx\right)\right)^{\frac{1}{q}} =$$

$$= \left(\sum_k 2^{k+1}\left(\int_{2x_k}^\infty v^2(x)x^{(\alpha-1)2}dx\right)\right)^{\frac{1}{2}} \leq$$

$$\leq 2\left(\sum_k 2^{k-1}\left(\int_{2x_k}^\infty v^2(x)x^{(\alpha-1)2}dx\right)\right)^{\frac{1}{2}} \equiv 2\left(\sum_k \alpha_k^2\right)^{\frac{1}{2}},$$

where

$$\alpha_k \equiv \left[2^{k-1}\left(\int_{2x_k}^\infty v^2(x)x^{(\alpha-1)2}dx\right)\right]^{\frac{1}{2}}.$$

On other hand, if $t > 0$, then $t \in [x_k, x_{k+1})$ for some k. Consequently

$$B(t) \leq \left[\left(\int_{2x_k}^\infty v^2(x)x^{(\alpha-1)2}dx\right)\left(\int_0^{x_{k+1}} w^2(y)dy\right)\right]^{\frac{1}{2}} =$$

$$= 2\left[2^{k-1}\left(\int\limits_{2x_k}^{\infty} v^2(x)x^{(\alpha-1)2}dx\right)\right]^{\frac{1}{2}},$$

where $B(t)$ is from Theorem 7.2.1. Hence $B \leq 2\sup\limits_{k}\alpha_k$.

Thus

$$\|R_{\alpha,v,w}\|_{\sigma_\infty(L^2(0,\infty))} \leq b_7\,\|v\|_E$$

and

$$\|R_{\alpha,v,w}\|_{\sigma_2(L^2(0,\infty))} \leq b_8\,\|v\|_F,$$

where $\|v\|_E \equiv \|\alpha_k\|_{l_\infty}$, $\|v\|_F = \|\alpha_k\|_{l_2}$.
By the interpolation Theorem, we obtain

$$\|R_{\alpha,v,w}\|_{\sigma_p(L^2(0,\infty))} \leq b_9\,\|v\|_{(E,F)}, \quad p \geq 2,$$

where (E,F) is an interpolation space. As $\|v\|_{(E,F)} = \|\alpha_k\|_{l_p}$ (see, e.g., [299]), we finally derive the estimate

$$\|R_{\alpha,v,w}\|_{\sigma_p(L^2(0,\infty))} \leq b_9\left(\sum_k 2^{(k-1)\frac{p}{2}}\left(\int\limits_{2x_k}^{\infty} v^2(x)x^{(\alpha-1)2}dx\right)^{\frac{p}{2}}\right)^{\frac{1}{2}} =$$

$$= b_{10}\left(\sum_k \int\limits_{x_{k-1}}^{x_k} w^2(y)\left(\int\limits_0^{x_{k-1}} w^2(z)dz\right)^{p/2-1} \times \right.$$

$$\left. \times \left(\int\limits_{2x_k}^{\infty} v^2(x)x^{(\alpha-1)2}dx\right)^{\frac{p}{2}}dy\right)^{1/p} \leq$$

$$\leq b_{10}\left(\sum_k \int\limits_{x_{k-1}}^{x_k} w^2(y)\left(\int\limits_0^{y} w^2(z)dz\right)^{p/2-1} \times \right.$$

$$\left. \times \left(\int\limits_{2y}^{\infty} v^2(x)x^{(\alpha-1)2}dx\right)^{\frac{p}{2}}dy\right)^{1/p} =$$

$$= b_{10}\left(\int\limits_0^{\infty} w^2(x)\left(\int\limits_0^{x} w^2(y)dy\right)^{p/2-1}\left(\int\limits_{2x}^{\infty} v^2(y)y^{(\alpha-1)2}dy\right)^{\frac{p}{2}}dx\right)^{1/2},$$

where $2 \leq p < \infty$.
Now we prove the lower estimate of (7.2.2). First note that

$$A(p) = c_1\left(\int\limits_0^{\infty} \overline{v}(2x)\left(\int\limits_{2x}^{\infty} \overline{v}(y)dy\right)^{p/2-1}\left(\int\limits_0^{x} w^2(y)dy\right)^{\frac{p}{2}}dx\right)^{\frac{1}{p}}, \quad (7.2.6)$$

where $\bar{v}(z) \equiv v^2(z)z^{(\alpha-1)2}$ and the positive constant c_1 depends only on p and α. Let $\{x_k\}$ be the sequence of positive numbers defined by (7.2.4). Then using (7.2.5) we have

$$A(2) = c_1 \int\limits_0^\infty \bar{v}(2x)\left(\int\limits_0^x w^2(y)dy\right)dx =$$

$$= c_1 \sum_k \int\limits_{x_{k-1}}^{x_k} \bar{v}(2x)\left(\int\limits_0^x w^2(y)dy\right)dx \le$$

$$\le c_1 \sum_k \left(\int\limits_{x_{k-1}}^{x_k} \bar{v}(2x)dx\right)\left(\int\limits_0^{x_k} w^2(y)dy\right) =$$

$$= c_2 \sum_k \left(\int\limits_{x_{k-1}}^{x_k} \bar{v}(2x)dx\right)\left(\int\limits_{x_{k-2}}^{x_{k-1}} w^2(y)dy\right) =$$

$$= c_3 \sum_k \left(\int\limits_{2x_{k-1}}^{2x_k} \bar{v}(x)dx\right)\left(\int\limits_{x_{k-2}}^{x_{k-1}} w^2(y)dy\right),$$

where c_3 depends only on α and p.
   By Lemma 7.2.1 we have

$$A(2) \le c_4 \|R_{\alpha,v,w}\|_{\sigma_2(L^2(0,\infty))}.$$

Now let $p > 2$. Then we have

$$A^p(p) = c_1 \sum_k \int\limits_{x_{k-1}}^{x_k} \bar{v}(2x)\left(\int\limits_{2x}^\infty \bar{v}(y)dy\right)^{p/2-1}\left(\int\limits_0^x w^2(y)dy\right)^{\frac{p}{2}}dx \le$$

$$\le \sum_k \left(\int\limits_0^{x_k} w^2(y)dy\right)^{\frac{p}{2}}\left(\int\limits_{x_{k-1}}^{x_k} \bar{v}(2x)\left(\int\limits_{2x}^\infty \bar{v}(y)dy\right)^{p/2-1}dx\right) =$$

$$= c_5 \sum_k 2^{kp/2}\left(\int\limits_{x_{k-1}}^{x_k} d\left[\left(-\int\limits_{2x}^\infty \bar{v}(y)dy\right)^{p/2}\right]\right) =$$

$$= c_5 \sum_k 2^{kp/2}\left[\left(\int\limits_{2x_{k-1}}^\infty \bar{v}(y)dy\right)^{p/2} - \int\limits_{2x_k}^\infty \bar{v}(y)dy\right)^{p/2}\right] \le$$

$$\le c_5 \sum_k 2^{kp/2}\left(\int\limits_{2x_{k-1}}^\infty \bar{v}(y)dy\right)^{p/2}.$$

Further, by Hölder's inequality we find that

$$
\int_{2x_{k-1}}^{\infty} \overline{v}(y)dy = \sum_{m=k-1}^{\infty} \int_{2x_m}^{2x_{m+1}} \overline{v}(y)dy =
$$

$$
= \sum_{m\geq k-1} \left(2^{m/2} \int_{2x_m}^{2x_{m+1}} \overline{v}(y)dy \right) 2^{-m/2} \leq
$$

$$
\leq \left(\sum_{m\geq k-1} 2^{mp/4} \left(\int_{2x_m}^{2x_{m+1}} \overline{v}(y)dy \right)^{p/2} \right)^{2/p} \times
$$

$$
\times \left(\sum_{m\geq k-1} 2^{-mp/(2(p-2))} \right)^{(p-2)/p} =
$$

$$
= c_6 2^{-k/2} \left(\sum_{m\geq k-1} 2^{mp/4} \left(\int_{2x_m}^{2x_{m+1}} \overline{v}(y)dy \right)^{p/2} \right)^{2/p}.
$$

Hence taking into account (7.2.5) we have

$$
A^p(p) \leq c_7 \sum 2^{kp/2} 2^{-kp/4} \sum_{m\geq k-1} 2^{mp/4} \left(\int_{2x_m}^{2x_{m+1}} \overline{v}(y)dy \right)^{p/2} =
$$

$$
= c_7 \sum_k 2^{kp/4} \sum_{m\geq k-1} 2^{mp/4} \left(\int_{2x_m}^{2x_{m+1}} \overline{v}(y)dy \right)^{p/2} =
$$

$$
= c_7 \sum_m 2^{mp/4} \left(\int_{2x_m}^{2x_{m+1}} \overline{v}(y)dy \right)^{p/2} \sum_{k\leq m+1} 2^{kp/4} \leq
$$

$$
\leq c_8 \sum_m 2^{mp/2} \left(\int_{2x_m}^{2x_{m+1}} \overline{v}(y)dy \right)^{p/2} =
$$

$$
= c_9 \sum_m \left(\int_{x_{m-1}}^{x_m} w^2(y)dy \right)^{p/2} \left(\int_{2x_m}^{2x_{m+1}} \overline{v}(y)dy \right)^{p/2}.
$$

Due to Lemma 7.2.1 we obtain

$$
A(p) \leq c_{10} \|R_{\alpha,v,w}\|_{\sigma_p(L^2(0,\infty))},
$$

where c_{10} depends only on p and α.

The theorem is proved. □

An alternative lower estimate is given by

Theorem 7.2.4. *Let* $1 \leq p < 2$ *and let* $\alpha > 0$. *Then there exists a positive number* d_1 *such that*

$$\|R_{\alpha,v,w}\|_{\sigma_p(L^2(0,\infty))} \geq d_1 \left(\int\limits_0^\infty \frac{v^2(x)}{x^{(1-\alpha)2}} \left(\int\limits_0^x w^2(y)dy \right)^{p/2} \times \right.$$

$$\left. \times \left(\int\limits_{2x}^\infty v^2(y)y^{(\alpha-1)2}dy \right)^{p/2} dx \right)^{1/2}.$$

Proof. Let $\{a_m\}$ be a sequence of positive numbers such that

$$\int\limits_0^{a_m} w^2(y)dy = 2^m.$$

Then the sequence $\{a_m\}$ is increasing and

$$\int\limits_{a_{m-1}}^{a_m} w^2(y)dy = 2^{m-1}.$$

If $\int\limits_0^\infty w^2(y)dy \in [2^{m_0}, 2^{m_0+1})$ for some $m_0 \in Z$, then we assume that $a_{m+1} = \infty$. Further, using Lemma 7.2.1 and integration by parts twice, we obtain

$$\|R_{\alpha,v,w}\|_{\sigma_p(L^2(0,\infty))} \geq$$

$$\geq b_1 \left[\sum_m \left(\int\limits_{a_{m-1}}^{a_m} w^2(y)dy \right)^{p/2} \left(\int\limits_{2a_m}^{2a_{m+1}} v^2(x)x^{(\alpha-1)2}dx \right)^{p/2} \right]^{1/p} =$$

$$= b_1 \left[\sum_m \left(\int\limits_{2a_m}^{2a_{m+1}} v^2(x)x^{(\alpha-1)2}dx \right)^{p/2} 2^{(m-1)p/2} \right]^{1/p} =$$

$$= b_1 \left[\sum_m \left(\int\limits_{2a_m}^{2a_{m+1}} v^2(x)x^{(\alpha-1)2} \times \right. \right.$$

$$\left. \left. \times \left(\int\limits_x^{2a_{m+1}} v^2(y)y^{(\alpha-1)2}dy \right)^{p/2-1} dx \right) 2^{(m-1)p/2} \right]^{1/p} =$$

$$= b_2 \Bigg[\sum_m \Bigg(\int\limits_{2a_m}^{2a_{m+1}} v^2(x) x^{(\alpha-1)2} \Bigg(\int\limits_{x}^{2a_{m+1}} v^2(y) y^{(\alpha-1)2} dy \Bigg)^{p/2-1} \times$$

$$\times \Bigg(\int\limits_{0}^{a_{m+1}} w^2(y) dy \Bigg)^{p/2} dx \Bigg)\Bigg]^{1/p} \geq$$

$$\geq b_2 \Bigg(\int\limits_{0}^{\infty} v^2(x) x^{(\alpha-1)2} \Bigg(\int\limits_{x}^{\infty} v^2(y) y^{(\alpha-1)2} dy \Bigg)^{p/2-1} \times$$

$$\times \Bigg(\int\limits_{0}^{x/2} w^2(y) dy \Bigg)^{p/2} dx \Bigg)^{1/p} =$$

$$= b_3 \Bigg(\int\limits_{0}^{\infty} \Bigg(\int\limits_{0}^{x/2} w^2(y) dy \Bigg)^{p/2-1} w(x/2) \Bigg(\int\limits_{x}^{\infty} v^2(y) y^{(\alpha-1)2} dy \Bigg)^{p/2} dx \Bigg)^{1/p} =$$

$$= b_4 \Bigg(\int\limits_{0}^{\infty} \Bigg(\int\limits_{0}^{x} w^2(y) dy \Bigg)^{p/2-1} w(x) \Bigg(\int\limits_{2x}^{\infty} v^2(y) y^{(\alpha-1)2} dy \Bigg)^{p/2} dx \Bigg)^{1/p}.$$

□

Now we consider the following operator

$$I_{\varphi,a,v}^{\alpha} = v(x) \int\limits_{a}^{x} (\varphi(x) - \varphi(y))^{\alpha-1} f(y) \varphi'(y) dy \ , \quad \alpha > 0 \ ,$$

where φ is an increasing function on (a, ∞) such that φ' is continous, v is a measurable function on (a, ∞).

In the sequel we shall use the notation: $\overline{g}(x) = g(\varphi^{-1}(x))$, $\widetilde{g}(x) = g(\varphi(x))$.

Using a change of variable it is easy to show that $J_{\varphi,\alpha,v}$ is bounded (compact) from $L_{\varphi'}^p(a, \infty)$ into $L_{\varphi'}^q(a, \infty)$ if and only if $R_{\alpha,\varphi,\overline{v}}$ is bounded (compact) from $L^p(\varphi(a), \varphi(\infty))$ into $L^q(\varphi(a), \varphi(\infty))$, where

$$R_{\varphi,a,\overline{v}}^{\alpha} f(x) = \overline{v}(x) \int\limits_{\varphi(a)}^{x} (x - y)^{\alpha-1} f(y) dy.$$

It follows that from Theorems 2.8.1 and 2.8.3 we obtain the following results:

Theorem 7.2.5. *Let* $1 < p \leq q < \infty$ *and let* $\alpha > 1/p$; *we assume that* $\varphi(\infty) = \infty$.

(a) The following conditions are equivalent:
(i) $I^\alpha_{\varphi,a,v}$ is bounded from $L^p_{\varphi'}(a,\infty)$ into $L^q_{\varphi'}(a,\infty)$;
(ii)

$$B_{\varphi,a} \equiv \sup_{t>a} B_{\varphi,a}(t) \equiv \sup_{t>a} \left(\int_t^\infty |v(x)|^q (\varphi(x)-\varphi(a))^{(\alpha-1)q} \varphi'(x)dx \right)^{1/q} \times$$

$$\times (\varphi(t)-\varphi(a))^{1/p} < \infty;$$

(iii)

$$B^{(1)}_{\varphi,a} \equiv \sup_{k\in Z} B^{(1)}_{\varphi,a}(k) \equiv$$

$$\equiv \sup_{k\in Z} \left(\int_{\varphi^{-1}(\varphi(a)+2^k)}^{\varphi^{-1}(\varphi(a)+2^{k+1})} |v(x)|^q (\varphi(x)-\varphi(a))^{(\alpha-1/p)q} \varphi'(x)dx \right)^{1/q} < \infty.$$

Moreover, $\|J_{\alpha,\varphi,v}\| \approx B_{\varphi,a} \approx B^{(1)}_{\varphi,a}$.

(b) The following conditions are eqivalent:
(i) $I^\alpha_{\varphi,a,v}$ is compact from $L^p_{\varphi'}(a,\infty)$ into $L^q_{\varphi'}(a,\infty)$;
(ii) $B_{\varphi,a} < \infty$ and $\lim_{t\to a+} B_{\varphi,a}(t) = \lim_{t\to\infty} B_{\varphi,a}(t) = 0$;
(iii) $B^{(1)}_{\varphi,a} < \infty$ and $\lim_{k\to-\infty} B^{(1)}_{\varphi,a}(k) = \lim_{k\to+\infty} B^{(1)}_{\varphi,a}(k) = 0$.

Theorem 7.2.6. *Let* $1 < p \le q < \infty$ *and let* $\alpha > 1/p$. *We assume that* $\varphi(\infty) < \infty$.
(a) The following conditions are equivalent:
(i) $I^\alpha_{\varphi,a,v}$ is bounded from $L^p_{\varphi'}(a,\infty)$ into $L^q_{\varphi'}(a,\infty)$;
(ii) $B_{\varphi,a} < \infty$;
(iii)

$$B^{(2)}_{\varphi,a} \equiv \sup_{k\in Z_+} B^{(2)}_{\varphi,a}(k) \equiv$$

$$\equiv \sup_{k\in Z_+} \left(\int_{\varphi^{-1}(c_\varphi(k+1))}^{\varphi^{-1}(c_\varphi(k))} |v(x)|^q (\varphi(x)-\varphi(a))^{(\alpha-1/p)q} \varphi'(x)dx \right)^{1/q} < \infty,$$

where $c_\varphi(k) = \varphi(a) + \frac{\varphi(\infty)-\varphi(a)}{2^k}$.
Moreover, $\|I_{\varphi,\alpha,v}\| \approx B_{\varphi,a} \approx B^{(2)}_{\varphi,a}$.

(b) The following conditions are equaivalent:
(i) $I^\alpha_{\varphi,a,v}$ is compact from $L^p_{\varphi'}(a,\infty)$ into $L^q_{\varphi'}(a,\infty)$;
(ii) $B_{\varphi,a} < \infty$ and $\lim_{t\to a+} B_{\varphi,a}(t) = 0$;

(iii) $B^{(2)}_{\varphi,a} < \infty$ and $\lim\limits_{k\to+\infty} B^{(2)}_{\varphi,a}(k) = 0$.

Having established boundedness and compactness criteria we now turn to the singular numbers of $I^{\alpha}_{\varphi,a,v}$. First we formulate some well-known results for the Riemann–Liouville operator

$$R_{\alpha,a,v}f(x) = v(x) \int\limits_a^x f(y)(x-y)^{\alpha-1}dy, \quad \alpha > 0,$$

where $x \in (a,b)$, $-\infty < a < b \leq \infty$.

Proposition 7.2.1. *Let* $2 \leq p < \infty$, $\alpha > 1/2$. *Then* $\|R_{\alpha,a,v}\|_{\sigma_p(L^2(a,\infty))} \approx I(v,p,\alpha)$ *if* $b = \infty$ *and* $\|R_{\alpha,a,v}\|_{\sigma_p(L^2(a,b))} \approx \overline{I}(v,p,\alpha)$ *if* $b < \infty$, *where*

$$I(v,p,\alpha) = \left(\sum_{k\in Z} \left(\int\limits_{a+2^k}^{a+2^{k+1}} v^2(x)(x-a)^{2\alpha-1}dx \right)^{p/2} \right)^{1/p}$$

and

$$\overline{I}(v,p,\alpha) = \left(\sum_{k\in Z_+} \left(\int\limits_{a+\frac{b-a}{2^{k+1}}}^{a+\frac{b-a}{2^k}} v^2(x)(x-a)^{2\alpha-1}dx \right)^{p/2} \right)^{1/p}.$$

Proof. This follows in the same way as Theorem 7.1.3 for $b = \infty$ and Theorem 7.1.4 for $b < \infty$. □

Theorem A ([223]). *Let* $\alpha > 1/2$ *and* $\frac{1}{\alpha} < p < \infty$. *Then*

$$\|R_{\alpha,v}\|_{\sigma_p(L^2(0,\infty))} \approx \|v\|_{l^p(L^2_{x^{2\alpha-1}}(0,\infty))}.$$

Theorem 7.2.7. *Let* $2 \leq p < \infty$ *and* $\alpha > 1/2$. *Then* $\|I^{\alpha}_{\varphi,a,v}\|_{\sigma_p(L^2_{\varphi'}(a,\infty))} \approx S(v,\varphi,p,\alpha)$ *if* $\varphi(\infty) = \infty$ *and* $\|I^{\alpha}_{\varphi,a,v}\|_{\sigma_p(L^2_{\varphi'}(a,\infty))} \approx \overline{S}(v,\varphi,p,\alpha)$ *if* $\varphi(\infty) < \infty$, *where*

$$S(v,\varphi,p,\alpha) = \left(\sum_{k\in Z} \left(\int\limits_{\varphi^{-1}(\varphi(a)+2^k)}^{\varphi^{-1}(\varphi(a)+2^{k+1})} v^2(x)(\varphi(x)-\varphi(a))^{2\alpha-1}\varphi'(x)dx \right)^{p/2} \right)^{\frac{1}{p}},$$

$$\overline{S}(v,\varphi,p,\alpha) = \left(\sum_{k\in Z_+} \left(\int\limits_{\varphi^{-1}(c_\varphi(k+1))}^{\varphi^{-1}(c_\varphi(k))} v^2(x)(\varphi(x)-\varphi(a))^{2\alpha-1}\varphi'(x)dx \right)^{p/2} \right)^{\frac{1}{p}}.$$

and $c_\varphi(k) = \varphi(a) + \frac{\varphi(\infty) - \varphi(a)}{2^k}$.

Proof. The theorem will be proved if we show that $s_n(I^\alpha_{\varphi,a,v}) = s_n(R^\alpha_{\varphi,a,\bar{v}})$, where by $s_n(T)$ the n-th number of the operator $T : L^2 \to L^2$. As it is well-known the singular numbers of the operator T coincides with the approximation numbers of T (see, e.g., [79].) i.e.

$$s_n(T) = \inf\{\|T - A\| : rank\, A < n\}.$$

Further, by a change of variable we easily obtain the following equalities:

$$\frac{\|I^\alpha_{\varphi,a,v}f - \sum\limits_{j=1}^{n_0-1} c_j(f)u_j\|_{L^q_{\varphi'}(a,\infty)}}{\|f\|_{L^p_{\varphi'}(a,\infty)}} = \frac{\|R^\alpha_{\varphi,a,\bar{v}}\overline{f} - \sum\limits_{j=1}^{n_0-1} c_j(\overline{f})\overline{u}_j\|_{L^q(\varphi(a),\varphi(\infty))}}{\|\overline{f}\|_{L^p(\varphi(a),\varphi(\infty))}}$$

and

$$\frac{\|R^\alpha_{\varphi,a,\bar{v}}f - \sum\limits_{j=1}^{n_0-1} d_j(f)w_j\|_{L^q(\varphi(a),\varphi(\infty))}}{\|f\|_{L^p(\varphi(a),\varphi(\infty))}} = \frac{\|I^\alpha_{\varphi,a,v}\tilde{f} - \sum\limits_{j=1}^{n_0-1} d_j(\tilde{f})\tilde{w}_j\|_{L^q(a,\infty)}}{\|\tilde{f}\|_{L^p_{\varphi'}(a,\infty)}}$$

where $Af(x) = \sum\limits_{j=1}^{n_0-1} c_j(f)u_j(x)$, $A_1f(x) = \sum\limits_{j=1}^{n_0-1} d_j(f)w_j(x)$ are operators with rank $n_0, n_0 < n$.

Hence, $s_n(I^\alpha_{\varphi,a,v}) = s_n(R^\alpha_{\varphi,a,\bar{v}})$ and consequently

$$\left\|I^\alpha_{\varphi,a,v}\right\|_{\sigma_p(L^2_{\varphi'}(a,\infty))} = \left\|R^\alpha_{\varphi,a,\bar{v}}\right\|_{\sigma_p(L^2(\varphi(a),\varphi(\infty)))}.$$

Taking into account Proposition 7.2.1 the theorem follows. $\square$

Analogously, from Theorem A we have the following

Theorem 7.2.8. *Let* $\alpha > 1/2$, $\frac{1}{\alpha} < p < \infty$, $\varphi(a) = 0$, $\varphi(\infty) = \infty$. *Then*

$$\left\|I_{\varphi,a,v}\right\|_{\sigma_p(L^2_{\varphi'}(a,\infty))} \approx \left\|\overline{v}\right\|_{l^p(L^2_{x^{2\alpha-1}}(0,\infty))}.$$

Now let

$$L_vf(x) = v(x)\int\limits_1^x \ln^{\alpha-1}(x/y)f(y)\frac{dy}{y}, \quad \alpha > 0,$$

be an operator of Hadamard type (see Section 2.8), where v is a measurable function on $(1,\infty)$ such that $\overline{v} \in L(2^k, 2^{k+1})$ for all $k \in Z$. From Theorem 7.2.8 we have the following statement:

Theorem 7.2.9. *Let* $\alpha > 1/2$, $\frac{1}{\alpha} < p < \infty$. *Then* $\|L_v\|_{\sigma_p(L^2_{1/y}(1,\infty))} \approx$ $J(v, \alpha, p)$, *where*

$$J(v, \alpha, p) = \left(\sum_{k \in Z} \left(\int\limits_{2^k}^{2^{k+1}} (\bar{v}(x))^2 x^{2\alpha-1} dx \right)^{p/2} \right)^{1/p}.$$

Now we are going to investigate the singular numbers for the operators:

$$(\mathcal{R}_{\alpha,v,\omega_\lambda} f)(x) = v(x) \int\limits_{-\infty}^{x} (x-y)^{\alpha-1} f(y) \omega_\lambda(y) dy,$$

$$(\mathcal{W}_{\alpha,v,\omega_\lambda} f)(x) = v(x) \int\limits_{x}^{+\infty} (y-x)^{\alpha-1} f(y) \omega_\lambda(y) dy,$$

where $\alpha > 0$, $\omega_\lambda(x) \equiv e^{\lambda x}$ ($x, \lambda \in R$) and v is a measurable function on R. First we give criteria for the boundedness (compactness) of $\mathcal{R}_{\alpha,v,\omega_\lambda}$.

Theorem 7.2.11. *Let* $1 < p \le q < \infty$, $\lambda > 0$ *and let* $\alpha \in (1/p, 1)$. *Then* (i) $\mathcal{R}_{\alpha,v,\omega_\lambda}$ *is bounded from* $L^p_{\omega_\lambda}(R)$ *into* $L^q_{v \cdot \omega_\lambda}(R)$ *if and only if*

$$E_\lambda \equiv \sup_{k \in Z} E_\lambda(k) \equiv \sup_{k \in Z} \left(\int\limits_{\frac{k}{\lambda} \ln 2}^{\frac{(k+1)}{\lambda} \ln 2} |v(x)|^q (\omega_\lambda(x))^{q/p'+1} dx \right)^{1/q} < \infty;$$

(ii) $\mathcal{R}_{\alpha,v,\omega_\lambda}$ *is compact from* $L^p_{\omega_\lambda}(R)$ *into* $L^q_{v \cdot \omega_\lambda}(R)$ *if and only if* $E_\lambda < \infty$ and $\lim\limits_{k \to -\infty} E_\lambda(k) = \lim\limits_{k \to +\infty} E_\lambda(k) = 0$.

Proof. It is easy to show that $\mathcal{R}_{\alpha,v,\omega_\lambda}$ is bounded (compact) from $L^p_{\omega_\lambda}(R)$ into $L^q_{\omega_\lambda}(R)$ if and only if $A_{\alpha,\bar{v}}$ is bounded (compact) from $L^p(R)$ into $L^q(R)$, where

$$A_{\alpha,u} f(x) \equiv u(x) \int\limits_{0}^{x} \left(\ln \frac{x}{y} \right)^{\alpha-1} f(y) dy.$$

But using Theorem 7.1.1 and a change of variable we see that $A_{\alpha,v}$ is bounded (compact) if and only if $E_\lambda < \infty$ ($E_\lambda < \infty$ and $\lim\limits_{k \to -\infty} E_\lambda(k) = \lim\limits_{k \to +\infty} E_\lambda(k) = 0$). $\square$

Analogously, the next result for $\mathcal{W}_{\alpha,v,\omega_\lambda}$ holds:

Theorem 7.2.12. *Let* $1 < p \le q < \infty$, $\alpha \in (1/p, 1)$ *and* $\lambda < 0$. *Then*

(a) $W_{\alpha,v,\omega_\lambda}$ is bounded from $L^p_{\omega_\lambda}(R)$ into $L^q_{v\cdot\omega_\lambda}(R)$ if and only if

$$\overline{E}_\lambda \equiv \sup_{k\in Z} \overline{E}_\lambda(k) \equiv \sup_{k\in Z} \left(\int_{\frac{k+1}{\lambda}\ln 2}^{\frac{k}{\lambda}\ln 2} |v(x)|^q (\omega_\lambda(x))^{q/p'+1} dx \right)^{1/q} < \infty;$$

(b) $W_{\alpha,v,\omega_\lambda}$ is compact from $L^p_{\omega_\lambda}(R)$ into $L^q_{v\cdot\omega_\lambda}(R)$ if and only if $\overline{E}_\lambda < \infty$ and $\lim\limits_{k\to-\infty} \overline{E}_\lambda(k) = \lim\limits_{k\to+\infty} \overline{E}_\lambda(k) = 0$.

It is easy to verify that for the operators

$$\mathcal{R}_{\alpha,v,\omega_\lambda} : L^p_{\omega_\lambda}(R) \to L^q_{v\omega_\lambda}(R)$$

and

$$A_{\alpha,\overline{v}} : L^p(R_+) \to L^q(R_+)$$

we have

$$s_j(\mathcal{R}_{\alpha,v,\omega_\lambda}) = s_j(A_{\alpha,\overline{v}}), \quad \lambda > 0,$$

where $A_{\alpha,v}$ is from the proof of Theorem 2.7.11 and $\overline{v} \equiv v(1/\lambda \ln x)$. Moreover,

$$s_j(W_{\alpha,v,\omega_\lambda}) = s_j(A_{\alpha,\overline{v}}),$$

where $\lambda < 0$ and

$$A_{\alpha,u}f(x) = u(x) \int_0^x \left(\ln \frac{x}{y} \right)^{\alpha-1} f(y) dy.$$

Using these arguments, a change of variable and Theorem 7.1.3 we derive the following statement:

Theorem 7.2.13. *Let* $2 \le p < \infty$ *and let* $\alpha \in (1/2, 1)$. *Then*

$$\|\mathcal{R}_{\alpha,v,\omega_\lambda}\|_{\sigma_p(L^2_{\omega_\lambda}(R))} \approx \|v\|_{E(p,\lambda)}$$

if $\lambda > 0$ *and*

$$\|W_{\alpha,v,\omega_\lambda}\|_{\sigma_p(L^2_{\omega_\lambda}(R))} \approx \|v\|_{E(p,\lambda)}$$

if $\lambda < 0$, *where*

$$\|v\|_{E(p,\lambda)} \equiv \left(\sum_{k\in Z} \left(\int_{\frac{k}{\lambda}\ln 2}^{\frac{k+1}{\lambda}\ln 2} |v(x)|^2 \omega_{2\lambda}(x) dx \right)^{p/2} \right)^{1/p}.$$

7.3.   Potential–type operators

In this section we deal with the singular numbers of ball fractional integrals and truncated potentials defined on SHT.

Let (X, d, μ) be an SHT and let w be a $\mu-$measurable a.e. positive function on X. We denote by $l^p(L_w^2(X))$ the set of all $\mu-$measurable functions $\varphi : X \to R^1$ for which

$$\|\varphi\|_{l^p(L_w^2(X))} = \left(\sum_{k \in Z} \left(\int_{B(x_0, 2^{k+1}) \backslash B(x_0, 2^k)} \varphi^2(x) w(x) d\mu(x) \right)^{p/2} \right)^{1/p} < \infty$$

if $\mu X = \infty$ and

$$\|\varphi\|_{l^p(L_w^2(X))} = \left(\sum_{k \in Z_+} \left(\int_{B(x_0, \frac{a}{2^k}) \backslash B(x_0, \frac{a}{2^{k+1}})} \varphi^2(x) w(x) d\mu(x) \right)^{p/2} \right)^{1/p} < \infty$$

when $\mu X < \infty$, where $x_0 \in X$, $a = \sup\{d(x_0, x) : x \in X\}$.

Proposition 7.3.1 ([299]). *Let* $1 \leq p_0, \ p_1 \leq \infty, \ 0 \leq \theta \leq 1, \ \frac{1}{p} = \frac{1-\theta}{p_0} + \frac{\theta}{p_1}$. *If* T *is a bounded operator from* $l^{p_i}(L_w^2(X))$ *to* $\sigma_{p_i}(L_w^2(X))$, *where* $i = 0, 1$, *then it is also bounded from* $l^p(L_w^2(X))$ *to* $\sigma_p(L^2(X))$.

If $X = R^n$ and $w(x) = |x|^\beta$, $\beta \in R^1$, then we set $l^p(L_{|x|^\beta}^2(R^n)) \equiv l^p(L_\beta^2(R^n))$.

For a measurable function $f : R^n \to R^1$ let

$$B_{+,v}^\alpha f(x) = v(x) \int_{|y|<|x|} \frac{\left(|x|^2 - |y|^2\right)^\alpha}{|x-y|^n} f(y) dy, \quad \alpha > 0,$$

be its ball fractional integral, where v is a Lebesgue-measurable function on R^n with $v \in L^2(\{2^n < |y| < 2^{n+1}\})$ for all $n \in Z$.

The next boundedness criterion follows from Theorem 4.1.1.

Theorem 7.3.1. *Let* $1 < p \leq q < \infty$, $\alpha > \frac{n}{p}$. *Then* $B_{+,v}^\alpha$ *acts boundedly from* $L^p(R^n)$ *into* $L^q(R^n)$ *if and only if*

$$F \equiv \sup_{j \in Z} F(j) \equiv \sup_{j \in Z} \left(\int_{2^j < |x| < 2^{j+1}} |v(x)|^q |x|^{q(2\alpha - n/p)} dx \right)^{1/q} < \infty.$$

Moreover, $\left\| B_{+,v}^\alpha \right\| \approx F$.

Compactness can be handled in the same way as Theorem 4.2.1.

Theorem 7.3.2. *Let* $1 < p \leq q < \infty$ *and let* $\alpha > \frac{n}{p}$. *Then* $B^{\alpha}_{+,v}$ *acts compactly from* $L^p(R^n)$ *into* $L^q(R^n)$ *if and only if* $F < \infty$ *and* $\lim\limits_{j \to -\infty} F(j) = \lim\limits_{j \to +\infty} F(j) = 0$.

For the Schatten norms we have

Theorem 7.3.3. *Let* $2 \leq p < \infty$ *and* $\alpha > n/2$. *Then* $B^{\alpha}_{+,v} \in \sigma_p(L^2(R^n))$ *if and only if* $v \in l^p(L^2_{4\alpha-n}(R^n))$. *Moreover, there exist positive constants* b_1 *and* b_2 *such that*

$$b_1\|v\|_{l^p(L^2_{4\alpha-n}(R^n))} \leq \|B^{\alpha}_{+,v}\|_{\sigma_p(L^2(R^n))} \leq b_2\|v\|_{l^p(L^2_{4\alpha-n}(R^n))}.$$

Proof. For sufficiency we use the Hilbert–Schmidt formula (7.1.1) and the condition $\alpha > \frac{n}{2}$. Thus,

$$\|B^{\alpha}_{+,v}\|_{\sigma_2(L^2(R^n))} = \left(\int_{R^n} v^2(x) \left(\int_{|y|<|x|} \frac{\left(|x|^2 - |y|^2\right)^{2\alpha}}{|x-y|^{2n}} dy \right) dx \right)^{\frac{1}{2}} \leq$$

$$\leq c_1 \left(\int_{R^n} |x|^{2\alpha} v^2(x) \left(\int_{|y|<|x|} |x-y|^{(\alpha-n)2} dy \right) dx \right)^{\frac{1}{2}} \leq$$

$$\leq c_2 \left(\int_{R^n} |x|^{4\alpha-n} v^2(x) dx \right)^{\frac{1}{2}} = c_2 \left(\sum_{k=-\infty}^{+\infty} a_k^2 \right)^{\frac{1}{2}},$$

where

$$a_k = \int_{2^k < |y| < 2^{k+1}} |x|^{4\alpha-n} v^2(x) dx.$$

Moreover, using Theorem 7.3.1 we arrive at the following two-sided inequality:

$$c_3\|v\|_{l^\infty(L^2_{4\alpha-n}(R^n))} \leq \left\|B^{\alpha}_{+,v}\right\|_{\sigma_\infty(L^2(R^n))} \leq c_4\|v\|_{l^\infty(L^2_{4\alpha-n}(R^n))}.$$

By Proposition 7.3.1 we conclude that

$$\left\|B^{\alpha}_{+,v}\right\|_{\sigma_p(L^2(R^n))} \leq c_5\|v\|_{l^p(L^2_{4\alpha-n}(R^n))}, \quad 2 \leq p < \infty.$$

Now we prove necessity. For this we take the orthonormal systems $\{f_k\}$ and $\{g_k\}$, where

$$f_k(x) = \chi_{\left\{2^{k-2}<|y|<2^{k-1}\right\}}(x) 2^{-(k-2)n/2} \cdot \lambda_n^{-\frac{1}{2}},$$

$$g_k(x) = \chi_{\{2^k \le |y| < 2^{k+1}\}}(x) \, |x|^{2\alpha - \frac{n}{2}} \, v(x) \alpha_k^{-\frac{1}{2}},$$

$\lambda_n = (2^n - 1)\pi^{n/2}/\Gamma(n/2 + 1)$ and

$$\alpha_k = \int\limits_{2^k \le |x| < 2^{k+1}} v^2(x) \, |x|^{4\alpha - n} \, dx.$$

Then in view of Proposition 7.1.2 we have

$$\infty > \left\| B_{+,v}^\alpha \right\|_{\sigma_p(L^2(R^n))} \ge$$

$$\ge c_6 \left(\sum_{k \in Z} \left(\int\limits_{2^k < |x| < 2^{k+1}} v^2(x) |x|^{2\alpha - \frac{n}{2}} \times \right. \right.$$

$$\left. \left. \times \left(\int\limits_{2^{k-2} < |y| < 2^{k-1}} \frac{(|x|^2 - |y|^2)^\alpha}{|x - y|^n} 2^{-(k-2)n/2} dy \right) dx \right)^p \right)^{\frac{1}{p}} \ge$$

$$\ge c_7 \left(\sum_{k \in Z} \alpha_k^{p/2} \right)^{1/p} = c_7 \, \|v\|_{l^p(L^2_{4\alpha - n}(R^n))}$$

which completes the proof. $\square$

Remark 7.3.1. Let $1 < p \le \infty$. Then there exists a positive constant c such that

$$\left\| B_{+,v}^\alpha \right\|_{\sigma_p(L^2(R^n))} \ge c \, \|v\|_{l^p(L^2_{4\alpha - n}(R^n))} \cdot$$

This follows from the proof of Theorem 7.3.1.

Theorem 7.3.4. *Let* $2 \le p < \infty$ *and let* $\alpha > n/2$. *Then* $B_{+,v}^\alpha \in \sigma_p(L^2(R^n))$ *if and only if*

$$I(v, p, \alpha) \equiv \left(\int\limits_{R^n} \left(\int\limits_{\frac{|x|}{2} < |y| < 2|x|} v^2(y)|y|^{4\alpha - 2n} dy \right)^{p/2} |x|^{np/2 - n} dx \right)^{\frac{1}{p}} < \infty.$$

Moreover,

$$c_1 I(v, p, \alpha) \le \left\| B_{+,v}^\alpha \right\|_{\sigma_p(L^2(R^n))} \le c_2 I(v, p, \alpha)$$

for some positive constants c_1 *and* c_2.

Proof. By Theorem 7.3.3, the statement will be proved if we show that

$$\|v\|_{l^p(L^2_{4\alpha - n}(R^n))} \approx I(v, p, \alpha).$$

Indeed, we have

$$\|v\|_{l^p(L^2_{4\alpha-n}(R^n))} \leq$$

$$\leq \left(\sum_{k\in Z} \left(\int_{2^k<|x|<2^{k+1}} v^2(x)|x|^{4\alpha-2n}dx \right)^{p/2} 2^{(k+1)np/2} \right)^{\frac{1}{p}} =$$

$$= b_1 \left(\sum_{k\in Z} \int_{2^k<|y|<2^{k+1}} |y|^{np/2-n} \times \right.$$

$$\times \left(\int_{\frac{|y|}{2}<|x|<2|y|} v^2(x)|x|^{4\alpha-2n}dx \right)^{p/2} dy \right)^{1/p} =$$

$$= b_1 I(v,p,\alpha).$$

The reverse inequality follows similarly. $\square$

As above, let (X,d,μ) be an SHT and let

$$J_{\alpha,v}f(x) = v(x) \int_{\overline{B}(x_0,2d(x_0,x))} f(y) \left(\mu B(x_0,d(x_0,x))\right)^{\alpha-1} d\mu, \quad \alpha > 0,$$

be a truncated potential, where v is a measurable function on X, $x_0 \in X$ and $\overline{B}(x_0,r) \equiv \{y : d(x_0,x) \leq r\}$. We shall assume that the following condition is satisfied:

$$\mu(B(x_0,R) \setminus B(x_0,r)) > 0 \tag{$*$}$$

for all r and R with $0 < r < R < a$ ($a = \sup\{d(x_0,x) : x \in X\}$).

First we give criteria for the boundedness and compactness properties of $J_{\alpha,v}$. In particular, we have the next result:

Theorem 7.3.5. *Let* $1 < p \leq q < \infty$ *and let* $\mu X = \infty$. *Then*
(a) $J_{\alpha,v}$ *is bounded from* $L^p_\mu(X)$ *into* $L^q_\mu(X)$ *if and only if*

$$G \equiv \sup_{k\in Z} G(k) \equiv$$

$$\equiv \sup_{k\in Z} \left(\int_{B(x_0,2^{k+1})\setminus B(x_0,2^k)} |v(x)|^q (\mu B_{x_0 x})^{q(\alpha-1/p)} d\mu(x) \right)^{1/q} < \infty,$$

where $B_{x_0 x} \equiv B(x_0,d(x_0,x))$. *Moreover,* $\|J_{\alpha,v}\| \approx G$.
(b) $J_{\alpha,v}$ *is compact from* $L^p_\mu(X)$ *into* $L^q_\mu(X)$ *if and only if* $G < \infty$ *and*
$$\lim_{k\to-\infty} G(k) = \lim_{k\to+\infty} G(k) = 0.$$

The first part of this theorem follows from Theorem 6.3.4 and the second part can be derived in the same way as Theorem 6.3.5 (see also Theorems 6.5.1 and 6.5.3).

Analogously we have

Theorem 7.3.6. *Let* $1 < p \le q < \infty$ *and* $\mu X < \infty$. *Then*
(a) $J_{\alpha,v}$ *is bounded from* $L_{\mu}^{p}(X)$ *into* $L_{\mu}^{q}(X)$ *if and only if*

$$G_1 \equiv \sup_{k \in Z} G_1(k) \equiv$$

$$\equiv \sup_{k \in Z_+} \left(\int_{B(x_0,a/2^k) \backslash B(x_0,a/2^{k+1})} |v(x)|^q (\mu B_{x_0 x})^{q(\alpha - 1/p)} d\mu(x) \right)^{1/q} < \infty.$$

Moreover, $\|J_{\alpha,v}\| \approx G_1$.

(b) $J_{\alpha,v}$ *is compact from* $L_{\mu}^{p}(X)$ *into* $L_{\mu}^{q}(X)$ *if and only if* $G_1 < \infty$ *and* $\lim_{k \to +\infty} G_1(k) = 0$.

To investigate the behaviour of the Schatten–von Neumann norms of J_{α} let $v \in L^2(B(x_0, 2^{n+1}) \backslash B(x_0, 2^n))$ for arbitrary $n \in Z$ if $\mu X = \infty$ and $v \in L^2(B(x_0, a/2^n) \backslash B(x_0, a/2^{n+1}))$ for any $n \in Z_+$ if $\mu X < \infty$. We denote $l^p(L_w^2(X))$ by $l^p(L_\beta^2(X))$ if $w(x) = \mu(B(x_0, d(x_0, x)))^\beta$

We have

Theorem 7.3.7. *Let* $\mu X = \infty$, $2 \le p < \infty$, *and let* $\alpha > 1/2$. *Then* $J_{\alpha,v} \in \sigma_p(L^2(X))$ *if and only if* $v \in l^p(L_{2\alpha-1}^2(X))$. *Moreover, the following estimate holds:*

$$c_1 \|v\|_{l^p(L_{2\alpha-1}^2(X))} \le \|J_{\alpha,v}\|_{\sigma_p(L^2(X))} \le c_2 \|v\|_{l^p(L_{2\alpha-1}^2(X))},$$

with some positive constants c_1 *and* c_2.

Proof. Sufficiency is proved in the same way as the previous theorem using the interpolation result, the Hilbert–Schmidt formula and Theorem 7.3.5. Let us show necessity. As in the previous theorems we use Proposition 7.1.2. First note that the condition (∗) implies the reverse doubling condition for μ at the point x_0 (see Proposition 6.5.1), i.e. there exist constants $\eta_1 > 1$, $\eta_2 > 1$ such that

$$\mu B(x_0, \eta_1 r) \ge \eta_2 \mu B(x_0, r)$$

for all $r \in (0, a)$. From this it follows that for arbitrary $n \in Z$ we have

$$\mu \left(B(x_0, \eta_1^n \lambda) \backslash B(x_0, \eta_1^{n-1} \lambda) \right) > 0,$$

where $\lambda > 0$.

Now we take the following orthonormal systems:

$$f_m(x) = \chi_{E_m}(x) a_m^{-1/2},$$

where $a_m = \mu E_m$, $E_m = B(x_0, \eta_1^{m-1}/a_1) \backslash B(x_0, \eta_1^{m-2}/a_1)$ and

$$g_m(x) = v(x)\chi_{F_m}(x)(\mu B_{x_0 x})^{\alpha-1/2}\alpha_m^{-1/2},$$

where

$$\alpha_m = \int\limits_{F_m} v^2(x)(\mu B_{x_0 x})^{2\alpha-1}\,d\mu(x),$$

$B_{x_0 x} \equiv B(x_0, d(x_0, x))$, $F_m = B(x_0, \eta_1^{m+1}) \backslash B(x_0, \eta_1^m)$ and the positive constant a_1 is from the definition of SHT. Further, using the reverse doubling condition for μ we have

$$\infty > \|J_{\alpha,v}\|_{\sigma_p(L^2(X))} = \left(\sum_{m\in Z} |\langle J_{\alpha,v} f_m, g_m\rangle|^p\right)^{1/p} =$$

$$= \left(\sum_{m\in Z} \alpha_m^{-p/2}\left(\int\limits_{F_m} v^2(x)(\mu B_{x_0 x})^{\alpha-1/2} \times\right.\right.$$

$$\left.\left.\times\left(\int\limits_{E_m} (\mu B_{xy})^{\alpha-1}a_m^{-1/2}d\mu(y)\right)d\mu(x)\right)^p\right)^{1/p} \geq$$

$$\geq b_1\left(\sum_{m\in Z} \alpha_m^{-p/2}\left(\int\limits_{F_m} v^2(x)(\mu B_{x_0 x})^{2\alpha-3/2}a_m^{\frac{1}{2}}d\mu(x)\right)^p\right)^{1/p} \geq$$

$$\geq b_2\left(\sum_{m\in Z} \alpha_m^{p/2}\right)^{1/p} = b_2\|v\|_{l^p(L^2_{2\alpha-1}(X))}\,.$$

$\square$

In a similar way we have

Theorem 7.3.8. *Let $\mu X < \infty$, $2 \leq p < \infty$ and let $\alpha > 1/2$. Then $J_{\alpha,v} \in \sigma_p(L^2(X))$ if and only if $v \in l^p(L^2_{2\alpha-1}(X))$. Moreover,*

$$c_1\|v\|_{l^p(L^2_{2\alpha-1}(X))} \leq \|J_{\alpha,v}\|_{\sigma_p(L^2(X))} \leq c_2\|v\|_{l^p(L^2_{2\alpha-1}(X))}$$

for some positive constants c_1 and c_2.

The next criterion follows analogously:

Theorem 7.3.9.   *Let $\mu X = \infty$, $2 \leq p < \infty$ and let $\alpha > 1/2$. Then $J_{\alpha,v} \in \sigma_p(L^2(X))$ if and only if $S(v, p, \alpha) < \infty$, where*

$$S(v, p, \alpha) \equiv \left(\int\limits_X \left(\int\limits_{F_y} v^2(x)(\mu B_{x_0 x})^{2(\alpha-1)}d\mu(x)\right)^{p/2} \times\right.$$

$$\times (\mu B_{x_0 x})^{p/2-1} d\mu(y) \Big)^{1/p},$$

where $B(x_0, \eta_1 d(x_0, y)) \backslash B(x_0, d(x_0, y)/\eta_1)$ *Moreover, there are positive constants* c_1 *and* c_2 *such that*

$$c_1 S(v, p, \alpha) \leq \|J_{\alpha, v}\|_{\sigma_p(L^2(X))} \leq c_2 S(v, p, \alpha).$$

7.4. Hardy–type operators

Let (X, d, μ) be an SHT with $\mu X = \infty$. We assume that there exists a point $x_0 \in X$ with $\mu\{x_0\} = 0$ such that for all $0 < r < R < \infty$ the condition (*) (see Section 7.3) is satisfied.

We recall that (see Proposition 6.5.1) the condition $(*)$ of Section 7.3 ensures the reverse doubling condition at point x_0, i.e. there exists constants η_1 and η_2 with $\eta_1 > 1$, $\eta_2 > 1$ such that for all $r > 0$ the inequality

$$\mu B(x_0, \eta_1 r) \geq \eta_2 \mu B(x_0, r)$$

holds (see Definition 6.5.1). We shall also assume that the measure of the ball of radius r, $\mu B(x_0, r)$, is continous for all $r \geq 0$. This condition (see the proof of Proposition 6.1.2) is equivalent to the condition

$$\mu\{x : d(x_0, x) = r\} = 0$$

for all $r \geq 0$.

In what follows we shall use the notation:

$$B_{x_0 x} \equiv B(x_0, d(x_0, x)),$$

$$E_k \equiv B(x_0, \eta_1^{k+1}) \backslash B(x_0, \eta_1^k),$$

where η_1 is from the definition of the reverse doubling condition.

Now let v be a measurable function on X and let

$$H_v f(x) = v(x) \int\limits_{B_{x_0 x}} f(y) d\mu(y)$$

be the corresponding Hardy–type transform.

We begin with the boundedness and compactness of H_v.

Theorem 7.4.1. *Let* $1 < p \leq q < \infty$ *and let* η_1 *be a constant from the definition of the reverse doubling condition. Then the following conditions are equivalent:*

(i) H_v is bounded from $L_\mu^p(X)$ into $L_\mu^q(X)$;

(ii) $B \equiv \sup_{t>0} B(t) \equiv \sup_{t>0} \left(\int_{X\backslash B(x_0,t)} |v(x)|^q d\mu(x) \right)^{\frac{1}{q}} (\mu B(x_0,t))^{\frac{1}{p'}} < \infty;$

(iii) $B_1 \equiv \sup_{k\in Z} B(k) \equiv \sup_{k\in Z} \left(\int_{E_k} |v(x)|^q d\mu(x) \right)^{\frac{1}{q}} (\mu B(x_0,\eta_1^k))^{\frac{1}{p'}} < \infty.$

Moreover, $\|H_v\| \approx B \approx B_1$.

Proof. The implication (ii) $\Rightarrow$ (i) follows from Theorem 1.1.4 (see also Corollary 1.1.4). If we take the function $f_k(x) = \chi_{E_{k-1}}(x)$ and use the reverse doubling condition, then we easily obtain (iii) from (ii). Now let $t > 0$. Then $t \in \left[\eta_1^m, \eta_1^{m+1} \right)$ for some $m \in Z$. Consequently using the doubling and reverse doubling conditions for μ we get

$$B^q(t) \leq \left(\int_{X\backslash B(x_0,\eta_1^m)} |v(x)|^q \, d\mu(x) \right) \left(\mu B(x_0,\eta_1^{m+1}) \right)^{q/p'} =$$

$$= c_1 \left(\mu B(x_0,\eta_1^m) \right)^{\frac{q}{p'}} \sum_{k=m}^{+\infty} \int_{E_k} |v(x)|^q \, d\mu(x) \leq$$

$$\leq c_1 B_1^q \left(\mu B(x_0,\eta_1^m) \right)^{\frac{q}{p'}} \sum_{k=m}^{+\infty} \left(\mu B(x_0,\eta_1^k) \right)^{-\frac{q}{p'}} \leq$$

$$\leq c_2 B_1^q \left(\mu B(x_0,\eta_1^m) \right)^{\frac{q}{p'}} \int_{X\backslash B(x_0,\eta_1^m)} (\mu B_{x_0 x})^{-\frac{q}{p'}-1} \, d\mu(x) \leq$$

$$\leq c_3 B_1^q.$$

Hence (iii) $\Rightarrow$ (ii). $\square$

Theorem 7.4.2. *Let* $1 < p \leq q < \infty$. *Then the following conditions are equivalent:*

(i) H_v *is compact from* $L_\mu^p(X)$ *into* $L_\mu^q(X)$;
(ii) $B < \infty$ *and* $\lim_{t\to 0} B(t) = \lim_{t\to\infty} B(t) = 0;$
(iii) $B_1 < \infty$ *and* $\lim_{k\to -\infty} B_1(k) = \lim_{k\to +\infty} B_1(k).$

Proof. That (ii) $\Rightarrow$ (i) follows from Theorem 1.1.9. To prove the implication (i) $\Rightarrow$ (iii) we take the sequence $g_k(x) = \chi_{F_k}(x) (\mu F_k)^{-\frac{1}{p}}$, where

$$F_k = \{y \in X : \eta_1^{k-2}/a_1 < d(x_0,y) < \eta_1^{k-1}/a_1\}.$$

Then g_k converges weakly to 0 as $k \to +\infty$ or $k \to -\infty$. On the other hand, using the doubling and reverse doubling conditions we obtain

$$\|H_v g_k\|_{L_\mu^q(X)} \geq c_1 \left(\int_{E_k} |v(x)|^q \, d\mu(x) \right)^{\frac{1}{q}} \left(\mu B(x_0,\eta_1^k) \right)^{\frac{1}{p'}}$$

which converges to 0 as $k \to +\infty$ or $k \to -\infty$.

The implication (iii) $\Rightarrow$ (ii) follows in the standard way (see also the proofs of Theorems 2.1.5 and 6.5.3 for details). $\square$

Now we investigate the singular numbers of H_v. Let w be a μ- measurable, a.e. positive function on X. We say that the μ- measurable function g belongs to $l^p\left(L^2_w(X)\right)$, $1 \leq p \leq \infty$, if

$$\|g\|_{l^p(L^2_w(X))} = \left(\sum_k \left(\int_{E_k} g^2(x)w(x)d\mu(x) \right)^{p/2} \right)^{\frac{1}{p}} < \infty$$

for $p < \infty$ and

$$\|g\|_{l^\infty(L^2_w(X))} = \sup_{k \in Z} \left(\int_{E_k} g^2(x)w(x)d\mu(x) \right)^{1/2} < \infty$$

for $p = \infty$.

The next interpolation theorem holds (see i.e. [299], pp. 127, 147).

Propositon 7.4.1. *Let* $1 \leq p_0$, $p_1 \leq \infty$, $0 \leq \theta \leq 1$, $\frac{1}{p} = \frac{1-\theta}{p_0} + \frac{\theta}{p_1}$. *If the operator* T *is bounded from* $l^{p_i}\left(L^2_w(X)\right)$ *into* $\sigma_{p_i}\left(L^2_\mu(X)\right)$, $i = 0, 1$, *then it is also bounded from* $l^p\left(L^2_w(X)\right)$ *into* $\sigma_p\left(L^2_\mu(X)\right)$. *Moreover,*

$$\|T\|_{l^p(L^2_w(X)) \to \sigma_p(L^2_\mu(X))} \leq \|T\|^{1-\theta}_{l^{p_0}(L^2_w(X)) \to \sigma_{p_0}(L^2_\mu(X))} \|T\|^{\theta}_{l^{p_1}(L^2_w) \to \sigma_{p_1}(L^2_\mu(X))} \cdot$$

For convenience we shall use the notationn

$$l^p(L^2_w(X)) \equiv l^p(L^2_{\mu B_{x_0 x}}(X)),$$

for $w(x) \equiv \mu B_{x_0 x}$.

Theorem 7.4.3. *Let* $2 \leq p < \infty$. *Then* $\sigma_p\left(L^2(X)\right) \approx l^p(L^2_{\mu B_{x_0 x}}(X))$.

Proof. To prove this theorem we again use the arguments from the proof of Theorems of Sections 7.1 and 7.3. By Theorem 7.4.1 we have

$$\|H_v\|_{\sigma_\infty\left(L^2_\mu(X)\right)} \approx \|v\|_{l^\infty(L^2_{\mu B_{x_0 x}}(X))} \cdot$$

On the other hand, due to the Hilbert–Schmidt formula (see (7.1.1)) for H_v we obtain

$$\|H_v\|_{\sigma_2(L^2_j \mu(X))} = \left(\int_X v^2(x)\mu(B_{x_0 x})d\mu(x) \right)^{1/2} =$$

$$= \left(\sum_k \left(\int_{E_k} v^2(x) \mu(B_{x_0 x}) d\mu(x) \right)^{2/2} \right)^{1/2} = \|v\|_{l^2(L^2_{\mu B_{x_0 x}}(X))} \cdot$$

Taking into account Proposition 7.4.1 we see that

$$\|H_v\|_{\sigma_p(L^2_\mu(X))} \leq c_1 \|v\|_{l^p(L^2_{\mu B_{x_0 x}}(X))}, \quad 2 \leq p < \infty,$$

with a positive constant c_1 depending only on p.

To prove the reverse inequality we take the orthonormal systems $\{f_m\}$ and $\{g_m\}$, where

$$f_m(x) = (\mu E_{m-1})^{-\frac{1}{2}} \chi_{E_{m-1}}(x);$$

$$g_m(x) = v(x) \chi_{E_m}(x) \left(\int_{E_m} v^2(y) \mu_{B_{x_0 y}} d\mu(y) \right)^{-\frac{1}{2}} (\mu B_{x_0 x})^{\frac{1}{2}}.$$

In view of Proposition 7.1.2 we get

$$\left(\sum_k |\langle H_v f_k, g_k \rangle|^p \right)^{1/p} \geq c_2 \left(\sum_k \left(\int_{E_k} v^2(y) \mu(B_{x_0 x}) d\mu(x) \right)^{p/2} \right)^{1/p} =$$

$$= \|v\|_{l^p(L^2_{\mu B_{x_0 x}}(X))},$$

where $1 \leq p < \infty$. $\square$

Let (X, μ) be a measure space and let $x_0 \in X$. We assume that φ and ψ are real non-negative functions on X such that

$$\{x \in X : \varphi(x) \in (\alpha, \beta)\} \neq \emptyset,$$

$$\{x \in X : \psi(x) \in (\alpha, \beta)\} \neq \emptyset$$

for all α and β with $0 < \alpha < \beta < \infty$. Let us assume also that $\mu\{x_0\} = 0$ and $\mu\{x : \varphi(x) = R\} = \{x : \psi(x) = R\} = 0$ for all $R > 0$.

Let

$$H_{v,w} f(x) = v(x) \int_{\{y: \varphi(y) < \psi(x)\}} f(y) w(y) d\mu(y)$$

for μ - measurable function $f \to R$, where v and w are μ- measurable functions on X.

In the sequel we shall use the notation:

$$E_\varphi(a_k) \equiv \{x : a_k < \varphi(x) < a_{k+1}\}, \quad E_\psi(a_k) \equiv \{x : a_k < \psi(x) < a_{k+1}\},$$

where $\{a_k\}$ is an increasing sequence of positive numbers.

We are going to estimate the Schatten–von Neumann norms of the operator $H_{v,w}$.

The following statement holds:

Theorem 7.4.4. *Let* $1 \le p < \infty$. *Then*

$$\|H_{v,w}\|_{\sigma_p(L^2_\mu(X))} \approx J_{\varphi,\psi},$$

where

$$J_{\varphi,\psi} \equiv \sup_{\{a_m\}} \left\{ \sum_m \left(\int_{E_\varphi(a_{k-1})} w^2(y)d\mu(y) \right)^{p/2} \left(\int_{E_\psi(a_k)} v^2(x)d\mu(x) \right)^{p/2} \right\}^{\frac{1}{p}},$$

where the supremum is taken over all sequences $\{a_m\}$ *of positive numbers.*

Proof. First we prove necessity. Let us take the sequence of positive numbers $\{a_m\}$ and the orthonormal systems:

$$f_m(y) = \alpha_m^{-1} w(y)\chi_{E_\varphi(a_{m-1})}(y),$$

$$g_m(x) = \beta_m^{-1} v(x)\chi_{E_\psi(a_m)}(x),$$

where

$$\alpha_m = \left(\int_{E_\varphi(a_{m-1})} w^2(y)dy \right)^{\frac{1}{2}},$$

$$\beta_m = \left(\int_{E_\psi(a_m)} v^2(y)dy \right)^{\frac{1}{2}}.$$

Using Proposition 7.1.2 we derive

$$\|H_{v,w}\|_{\sigma_p(L^2(X))} \ge \left(\sum_m |\langle H_{v,w}f_m, g_m\rangle|^p \right)^{1/p} =$$

$$= \left(\sum_m (\beta_m\alpha_m)^{-p} \left(\int_{E_\psi(a_m)} v^2(x)d\mu(x) \right)^p \left(\int_{E_\varphi(a_{m-1})} w^2(y)d\mu(y) \right)^p \right)^{1/p} =$$

$$= \left(\sum_m \alpha_m^p \beta_m^p \right)^{1/p}.$$

To prove sufficiency, we use the arguments from the proof of Theorem 4 in [78]. Assume that $H_{v,w}$ is compact. Then by Theorem 1.1.5 we have that

$$\|P_c H_{v,w} P_c\| \to 0 \quad as \quad c \to 0,$$

$$\|Q_b H_{v,w} Q_b\| \to 0 \ as \ b \to \infty,$$

where

$$(P_c H_{v,w} P_c)(f) = \chi_{\{x:\ \psi(x)<c\}} H_{v,w} \left(f \cdot \chi_{\{y:\ \varphi(y)<c\}} \right),$$

$$(Q_b H_{v,w} Q_b)(f) = \chi_{\{x:\ \psi(x)>b\}} H_{v,w} \left(f \cdot \chi_{\{y:\ \psi(y)>b\}} \right).$$

Let $\varepsilon < \|H_{v,w}\|$. Then we choose disjoint intervals $I_k = [a_k, a_{k+1})$ such that $0 = a_1 < a_2 < ... < a_N < a_{N+1} = \infty$ and

$$\|P_{I_k} H_{v,w} P_{I_k}\| = \varepsilon, \quad k = 1, 2, \cdots, N-2, N, \tag{7.4.1}$$

$$\|P_{N-1} H_{v,w} P_{N-1}\| \le \varepsilon. \tag{7.4.2}$$

where

$$(P_{I_k} H_{v,w} P_{I_k})(f) = \chi_{\{x:\ \psi(x)\in I_k\}} H_{v,w} \left(f \cdot \chi_{\{y:\ \varphi(y)\in I_k\}} \right).$$

Let P_N be an operator defined as follows

$$P_N f(x) = \sum_{k=1}^{N} \chi_{I_k}(x) v(x) \int_{\{y:\ \varphi(y)<a_k\}} f(y) w(y) d\mu(y).$$

Then it is clear that $\operatorname{rank} P_N = N$. Further, taking into account a restricted version of Theorem 1.1.4 (or Corollary 1.1.4), (7.4.1) and (7.4.2) we obtain

$$\|(H_{v,w} - P_N)f\|^2_{L^2_\mu(X)} = \left\| \sum_{k=1}^{N} (P_{I_k} H_{v,w} P_{I_k})f \right\|^2_{L^2_\mu(X)} =$$

$$= \sum_{k=1}^{N} \|(P_{I_k} H_{v,w} P_{I_k})f\|^2_{L^2_\mu(X)} \le \varepsilon^2 \sum_{k=1}^{N} \|f \cdot \chi_{I_k}\|^2_{L^2_\mu(X)} = \varepsilon^2 \|f\|^2_{L^2_\mu(X)}.$$

Consequently

$$\|H_{v,w} - P_N\| \le \varepsilon.$$

Since

$$s_j(T) = \inf \{\|T - P\| \ : \ \operatorname{rank} P < j\},$$

we get

$$s_1(H_{v,w}) \le ... \le s_N(H_{v,w}) \le \varepsilon.$$

Hence by Theorem 1.1.4 (or Corollary 1.1.4) restricted to an interval and Proposition A from Section 7.1 we derive an upper bound for $\|H_{v,w}\|_{\sigma_p(L^2_\mu(X))}$ by applying (7.4.1). $\square$

7.5.   Asymptotic behaviour of singular and entropy numbers

In this section we investigate the asymptotic behaviour of singular and entropy numbers of the following integral operators:

$$I_{\alpha,\sigma}f(x) = \frac{1}{\Gamma(\alpha)} \int\limits_{0}^{x} (x^{\sigma} - y^{\sigma})^{\alpha-1} f(y) dy, \quad x > 0, \; \alpha > 0, \; \sigma > 0,$$

and

$$H_{\alpha}f(x) = \frac{1}{\Gamma(\alpha)} \int\limits_{1}^{x} \left(\ln \frac{x}{y} \right)^{\alpha-1} f(y) dy, \quad x > 1, \; \alpha > 0.$$

We get singular–value decompositions of these operators in some weighted L^2 spaces.

The statements which we prove in this section mainly rely on results obtained in [84], [61], [108-110], where the asymptotic behaviour of the singular numbers of fractional integrals was investigated (see also [223] for Riemann–Liouville operators $R_{\alpha,v}f(x) = v(x)R_{\alpha}f(x)$ of order $\alpha > 1/2$).

Analogous problems for the Hardy operator were studied in [16], [67] (see also [68], [70] for weighted Hardy operators).

In [34–35] some results were obtained concerning asymptotics of singular numbers of certain pseudo–differential operators of general types.

Let A and B be infinite- dimensional Hilbert spaces. It is known that if $K : A \to B$ is an injective compact linear operator, then there exists

(a) an orthonormal basis $\{u_j\}_{Z_+}$ in A;

(b) an orthonormal basis $\{v_j\}_{Z_+}$ in B;

(c) a nonincreasing sequence $\{s_j(K)\}_{Z_+}$ of positive numbers with limit 0 as $j \to +\infty$ such that

$$Ku_j = s_j(K)v_j, \quad j \in Z_+.$$

The numbers $s_j(K)$ are known as singular numbers or $s-$ numbers of the operator K; the system $\{s_j(K), u_j, v_j\}_{j \in Z_+}$ is called a singular system of K. If $A = B$, then $s_j(K) = \lambda_j(|K|)$, where $|K| = (K^*K)^{1/2}$ and $\lambda_j(|K|)$ is an eigenvalue of $|K|$ (see Section 7.1). For the operator K the singular value decomposition

$$Kf = \sum_{j=0}^{\infty} s_j(K)(f, u_j)_A v_j, \quad f \in A$$

is valid.

Let us first consider the Riemann–Liouville operator

$$R_\alpha f(x) = \frac{1}{\Gamma(\alpha)} \int_0^x (x-y)^{\alpha-1} f(y) dy, \quad x > 0, \ \alpha > 0.$$

The following result is well-known (see [Go-V1]):

Theorem A. *Let $\alpha > 0$, $\beta > -1$, $\varphi(t) = t^{-\beta} e^{-t}$, $\psi(t) = t^{-(\alpha+\beta)} e^{-t}$. Then the singular system $\{s_j(R_\alpha), u_j, v_j\}_{j\in Z_+}$ of the operator $R_\alpha : L^2_\varphi(R_+) \to L^2_\psi(R_+)$ is given by*

$$s_n(R_\alpha) = \left(\frac{\Gamma(n+\beta+1)}{\Gamma(n+\alpha+\beta+1)} \right)^{1/2}, \qquad (7.5.1)$$

$$u_n(t) = \left(\frac{n!}{\Gamma(n+\beta+1)} \right)^{1/2} t^\beta L_n^{(\beta)}(t),$$

$$v_n(t) = \left(\frac{n!}{\Gamma(n+\alpha+\beta+1)} \right)^{1/2} t^{\alpha+\beta} L_n^{(\alpha+\beta)}(t);$$

and $s_n(R_\alpha)/n^{-\alpha/2} \to 1$ as $n \to \infty$, where $L_n^{(\gamma)}$ is the Laguerre polynomial

$$L_n^{(\gamma)}(x) = \sum_{k=0}^n (-1)^k \binom{n+\gamma}{n-k} \frac{x^k}{k!}, \quad \gamma > -1, \ n \in Z_+.$$

We also have

Theorem B ([108]). *Let $\alpha > 0$, $\lambda > \alpha - 1/2$, $\lambda \neq 0$. Then the operator $R_\alpha : L^2_\varphi(R_+) \to L^2_\psi(R_+)$, where $\varphi(x) = x^{1/2-\lambda}(1+x)^{2\alpha}$, $\psi(x) = x^{1/2-\lambda-\alpha}$, has the following singular system:*

$$s_n(R_\alpha) = \left(\frac{\Gamma(n+\lambda-\alpha+1/2)}{\Gamma(n+\lambda+\alpha+1/2)} \right)^{1/2}, \qquad (7.5.2)$$

$$u_n(t) = 2^\lambda a_n t^{\lambda-1/2}(1+t)^{-\lambda-\alpha-1/2} C_n^\lambda\left(\frac{1-t}{1+t} \right),$$

$$v_n(t) = 2^\lambda b_n t^{\lambda+\alpha-1/2}(1+t)^{-\lambda-\alpha-3/2} P_n^{(\lambda-\alpha-1/2,\lambda+\alpha-1/2)}\left(\frac{1-t}{1+t} \right),$$

where

$$a_n = \left(\frac{2^{2\lambda-1}(n+\lambda)n!}{\pi\Gamma(n+2\lambda)} \right)^{1/2} \Gamma(\lambda),$$

$$b_n = \left(\frac{2^{1-2\lambda}(n+\lambda)n!\Gamma(n+2\lambda)}{\Gamma(n+\lambda-\alpha+1/2)\Gamma(n+\lambda+\alpha+1/2)} \right)^{1/2},$$

$C_n^\lambda(t)$ is the Gegenbauer polynomial

$$C_n^\lambda(t) = \frac{1}{\Gamma(\alpha)} \sum_{j=0}^{[n/2]} (-1)^j \frac{\Gamma(\alpha + n - j)}{j!(n - 2j)!} (2t)^{n-2j},$$

and $P_m^{(\alpha,\beta)}$ is the Jacobi polynomial

$$P_n^{(\alpha,\beta)}(t) = 2^{-n} \sum_{m=0}^{n} \binom{n + \alpha}{m} \binom{n + \beta}{n - m} (t - 1)^{n-m}(t + 1)^m, \quad n \in Z_+.$$

Moreover, $\lim_{n\to\infty} s_n(R_\alpha)/n^{-\alpha} = 1$.

Yet another known result is

Theorem C ([110]). *The singular numbers of the operator $R_\alpha : L^2(0, 1) \to L_{x-\gamma}^2(0, 1)$ have the following asymptotics:*

$$s_n(R_\alpha) \approx n^{-\alpha}, \quad 0 \le \gamma < \alpha.$$

When $\gamma = 0$, the upper estimate in the previous statement was derived in [84], while the lower estimate was given in [61].

The next lemma is immediate.

Lemma 7.5.1. *Let φ, ψ, v and w be measurable a.e. positive functions on a measurable set $\Omega \subseteq R_+$. Then an operator A is compact from $L_\varphi^2(\Omega)$ to $L_\psi^2(\Omega)$ if and only if the operator $A_1 f(x) = v^{1/2}(x)A(fw^{-1/2})(x)$ is compact from $L_{\varphi w^{-1}}^2(\Omega)$ to $L_{\psi v^{-1}}^2(\Omega)$.*

Taking into account the definition of the singular system of the operator, we easily derive the next statement.

Lemma 7.5.2. *Let v and w be a.e. positive measurable functions on a measurable set $\Omega \subseteq R_+$. The system $\{s_j(A), u_j, v_j\}_{j \in Z_+}$ is a singular system for the operator $A : L_\varphi^2(\Omega) \to L_\psi^2(\Omega)$ if and only if the operator $A_1 : L_{\varphi w^{-1}}^2(\Omega) \to L_{\psi v^{-1}}^2(\Omega)$ has the singular system $\{s_j(A_1), w^{1/2}u_j, v^{1/2}v_j\}_{j \in Z_+}$, where*

$$A_1 f(x) = v^{1/2}(x)A(fw^{-1/2})(x) \quad and \quad s_j(A_1) = s_j(A).$$

Let

$$\mathcal{I}_{\alpha,\sigma} f(x) = \frac{1}{\Gamma(\alpha)} \int_0^x (x^\sigma - y^\sigma)^{\alpha-1} y^{\sigma-1} f(y) dy, \quad \alpha > 0, \ \sigma > 0, \ x > 0.$$

From the definition of compactness we easily deduce

Lemma 7.5.3. *Let* $\alpha > 0$, $\sigma > 0$ *and let* $\Omega = (0,1)$ *or* $\Omega = (0,\infty)$. *Assume that* v *and* w *are measurable a.e. positive functions on* Ω. *Then the operator* $\mathcal{I}_{\alpha,\sigma}$ *is compact from* $L^2_w(\Omega)$ *to* $L^2_v(\Omega)$ *if and only if* R_α *is compact from* $L^2_W(\Omega)$ *to* $L^2_V(\Omega)$, *where* $W(x) = w(x^{1/\sigma})x^{1/\sigma-1}$, $V(x) = v(x^{1/\sigma})x^{1/\sigma-1}$.

We shall also need the next statement:

Lemma 7.5.4. *Let* $\alpha > 0$, $\sigma > 0$ *and let* v *and* w *be measurable a.e. positive functions on* Ω, *where* $\Omega = (0,\infty)$ *or* $\Omega = (0,1)$. *Then for the singular system* $\{s_j(\mathcal{I}_{\alpha,\sigma}\}, \overline{u}_j, \overline{v}_j\}_{j\in Z_+}$ *of the operator* $\mathcal{I}_{\alpha,\sigma} : L^2_w(\Omega) \to L^2_v(\Omega)$ *we have* $s_j(\mathcal{I}_{\alpha,\sigma}) = \sigma^{-1}s_j(R_\alpha)$, $\overline{u}_j(x) = \sigma^{1/2}u_j(x^\sigma)$, $\overline{v}_j(x) = \sigma^{1/2}v_j(x^\sigma)$, *where* $\{s_j(R_\alpha), u_j, v_j\}_{j\in Z_+}$ *is a singular system for the operator* $R_\alpha : L^2_W(0,\infty) \to L^2_V(0,\infty)$, *with* $W(x) = w(x^{1/\sigma})x^{1/\sigma-1}$ *and* $V(x) = v(x^{1/\sigma})x^{1/\sigma-1}$.

Proof. Let $\Omega = (0,\infty)$. Using the change of variable $y = t^{1/\sigma}$, we have

$$(\mathcal{I}_{\alpha,\sigma}\overline{u}_j)(x) = \frac{1}{\Gamma(\alpha)} \int_0^x (x^\sigma - y^\sigma)^{\alpha-1}y^{\sigma-1}\overline{u}_j(y)dy =$$

$$= \frac{\sigma^{1/2}}{\Gamma(\alpha)} \int_0^x (x^\sigma - y^\sigma)^{\alpha-1}u_j(y^\sigma)y^{\sigma-1}dy =$$

$$= \frac{\sigma^{-1/2}}{\Gamma(\alpha)} \int_0^{x^\sigma} (x^\sigma - t)^{\alpha-1}u_j(t)dt =$$

$$= \sigma^{-1/2}(R_\alpha u_j)(x^\sigma) = s_j(R_\alpha)\sigma^{-1/2}v_j(x^\sigma) = \sigma^{-1}s_j(R_\alpha)\overline{v}_j(x).$$

A change of variable also yields

$$\int_0^\infty \overline{v}_j(x)\overline{v}_i(x)v(x)dx = \sigma \int_0^\infty v_j(x^\sigma)v_i(x^\sigma)V(x^\sigma)x^{\sigma-1}dx =$$

$$= \int_0^\infty v_j(x)v_i(x)V(x)dx = \delta_{ij},$$

where δ_{ij} denotes Kronecker's symbol.

Analogously, we have

$$\int_0^\infty \overline{u}_j(x)\overline{u}_i(x)w(x)dx = \int_0^\infty u_j(x)u_i(x)W(x)dx = \delta_{ij},$$

Hence $\{\overline{v}_j\}$ and $\{\overline{u}_j\}$ are orthonormal systems in $L_v^2(R_+)$ and $L_w^2(R_+)$ respectively.

The case $\Omega = (0, 1)$ follows in a similar way. $\square$

Theorem 7.5.1. *Let* $\alpha > 0$, $\sigma > 0$ *and* $0 \le \gamma < \alpha$. *Then there exist positive constants* c_1 *and* c_2 *depending on* α, σ *and* γ *such that for the singular numbers of the operator* $I_{\alpha,\sigma} : L_{x^{1-\sigma}}^2(0,1) \to L_{x^{\sigma-1-\gamma\sigma}}^2(0,1)$ *we have* $s_n(I_{\alpha,\sigma}) \approx n^{-\alpha}$.
Proof. By Lemma 7.5.2 we have that $s_j(I_{\alpha,\sigma}) = s_j(\mathcal{I}_{\alpha,\sigma})$, where $\mathcal{I}_{\alpha,\sigma}$ acts from $L_{x^{\sigma-1}}^2(0,1)$ to $L_{x^{\sigma-1-\gamma\sigma}}^2(0,1)$, while Lemma 7.5.4 yields $s_j(\mathcal{I}_{\alpha,\sigma}) = 1/\sigma s_j(R_\alpha)$, where R_α is a Riemann–Liouville operator acting from $L^2(0,1)$ to $L_{x^{-\gamma}}^2(0,1)$. Theorem C completes the proof. $\square$

Theorem 7.5.2. *Let* $\alpha > 0$, $\sigma > 0$, $\lambda > \alpha - 1/2$ *and* $\lambda \ne 0$. *Assume that* $w(x) = x^{1-\sigma/2-\sigma\lambda}(1 + x^\sigma)^{2\alpha}$, $v(x) = x^{3\sigma/2-\sigma\lambda-\sigma\alpha-1}$. *Then the operator* $I_{\alpha,\sigma} : L_w^2(0,\infty) \to L_v^2(0,\infty)$ *has a singular system* $\{s_n(I_{\alpha,\sigma}), \overline{u}_n, \overline{v}_n\}_{n\in Z_+}$, *where*

$$s_n(I_{\alpha,\sigma}) = 1/\sigma \left(\frac{\Gamma(n + \lambda - \alpha + 1/2)}{\Gamma(n + \lambda + \alpha + 1/2)} \right)^{1/2},$$

$$\overline{u}_n(x) = \sigma^{1/2} 2^\lambda x^{\sigma-1} a_n x^{\sigma(\lambda-1/2)} (1 + x^\sigma)^{-\lambda-\alpha-1/2} C_n^\lambda \left(\frac{1 - x^\sigma}{1 + x^\sigma} \right),$$

$$\overline{v}_n(x) = \sigma^{1/2} 2^\lambda b_n x^{\sigma(\lambda+\alpha-1/2)} (1+x^\sigma)^{-\lambda-\alpha-3/2} P_n^{(\lambda-\alpha-1/2,\lambda+\alpha-1/2)} \left(\frac{1 - x^\sigma}{1 + x^\sigma} \right)$$

$C_n^\lambda(x)$ *and* $P_n^{(\alpha,\beta)}$ *are Gegenbauer and Jacobi polynomials, respectively (see Theorem B) and* a_n, b_n *are constants defined in Theorem B. Moreover,*

$$\lim_{n\to\infty} s_n(I_{\alpha,\sigma})/n^{-\alpha} = 1/\sigma.$$

Proof. Lemma 7.5.2 implies that the singular system

$$\{s_m(I_{\alpha,\sigma}), \overline{u}_m, \overline{v}_m\}_{m\in Z_+}$$

of the map $I_{\alpha,\sigma} : L_w^2(0,\infty) \to L_v^2(0,\infty)$ coincides with the singular system

$$\{s_m(\mathcal{I}_{\alpha,\sigma}), \widetilde{u}_m, \widetilde{v}_m\}_{m\in Z_+}$$

of the map $\mathcal{I}_{\alpha,\sigma} : L_W^2(0,\infty) \to L_V^2(0,\infty)$, where $W(x) = w(x)x^{2(\sigma-1)}$, $V(x) = v(x)$, $\widetilde{u}_m(x) = x^{1-\sigma}u_m(x)$, $\widetilde{v}_m(x) = \overline{v}_m(x)$. Further, by Lemma 7.5.4 we have that the operator $R_\alpha : L_\varphi^2(0,\infty) \to L_\psi^2(0,\infty)$ ($\varphi(x) = x^{1/2-\lambda}(1 + x)^{2\alpha}$, $\psi(x) = x^{1/2-\lambda-\alpha}$) has a singular system

$$\{s_m(R_\alpha), u_m, v_m\}_{m\in Z_+},$$

where

$$s_m(R_\alpha) = \sigma s_m(\mathcal{I}_{\alpha\sigma}) \approx m^{-\alpha}, \quad \bar{u}_m(x) = \sigma^{1/2} x^{\sigma-1} u_m(x^\sigma),$$

$$\bar{v}_m(x) = \sigma^{1/2} v_m(x^\sigma).$$

$\square$

Analogously, we have the following

Theorem 7.5.3. *Let* $\alpha > 0$, $\sigma > 0$, $\beta > -1$, $w(y) = y^{-\sigma\beta-\sigma+1}e^{-y^\sigma}$ *and* $v(y) = y^{-\sigma(\alpha+\beta)+\sigma-1}e^{-y^\sigma}$. *Then the operator* $I_{\alpha,\sigma} : L_w^2(0,\infty) \to L_v^2(0,\infty)$ *has a singular system*

$$\{s_m(I_{\alpha,\sigma}), \bar{u}_m, \bar{v}_m\}_{m \in Z_+}$$

defined by

$$s_n(I_{\alpha,\sigma}) = 1/\sigma \left(\frac{\Gamma(n+\beta+1)}{\Gamma(n+\alpha+\beta+1)} \right)^{1/2},$$

$$\bar{u}_n(x) = \sigma^{1/2} x^{\sigma-1} \left(\frac{n!}{\Gamma(n+\beta+1)} \right)^{1/2} x^{\sigma\beta} L_n^{(\beta)}(x^\sigma),$$

$$\bar{v}_n(x) = \sigma^{1/2} \left(\frac{n!}{\Gamma(n+\alpha+\beta+1)} \right)^{1/2} x^{\sigma(\alpha+\beta)} L_n^{(\alpha+\beta)}(x^\sigma),$$

where $L_n^{(\gamma)}(x)$ *is a Laguerre polynomial (see Theorem A). Moreover,*

$$\lim_{n\to\infty} s_n(I_{\alpha,\sigma})/n^{-\alpha/2} = 1/\sigma.$$

Now we consider the operator of Hadamard's type H_α.
We first need the next

Lemma 7.5.5. *Let* $\alpha > 0$ *and* (v,w) *be a pair of weights defined on* $(1,\infty)$. *Then* $\{s_m(L_\alpha), \bar{u}_m, \bar{v}_m\}_{m \in Z_+}$ *is a singular system for the operator* $L_\alpha : L_w^2(1,\infty) \to L_v^2(1,\infty)$, *where*

$$L_\alpha f(x) = \frac{1}{\Gamma(\alpha)} \int_1^x \left(\ln\frac{x}{y} \right)^{\alpha-1} f(y) \frac{dy}{y},$$

if and only if the Riemann–Liouville operator $R_\alpha : L_W^2(0,\infty) \to L_V^2(0,\infty)$ *has a singular system* $\{s_m(R_\alpha), \tilde{u}_m, \tilde{v}_m\}_{m \in Z_+}$, *where* $W(x) = w(e^x)e^x$, $V(x) = v(e^x)e^x$, $s_m(R_\alpha) = s_m(L_\alpha)$, $\tilde{u}_m(x) = \bar{u}_m(e^x)$, $\tilde{v}_m(x) = \bar{v}_m(e^x)$.
Proof. Using the change of variable $y = e^z$ we have

$$(L_\alpha \overline{u}_m)(x) = \frac{1}{\Gamma(\alpha)} \int_1^x \left(\ln \frac{x}{y} \right)^{\alpha-1} \overline{u}_m(y) \frac{dy}{y} =$$

$$= \frac{1}{\Gamma(\alpha)} \int_0^{\ln x} (\ln x - z)^{\alpha-1} \widetilde{u}_m(z) dz = (R_\alpha \widetilde{u}_m)(\ln x) = \widetilde{v}(\ln x) s_j(R_\alpha).$$

On the other hand,

$$\int_0^\infty \widetilde{u}_i(x) \widetilde{u}_j(x) W(x) dx = \int_0^\infty \overline{u}_i(e^x) \overline{u}_j(e^x) w(e^x) e^x dx =$$

$$= \int_1^\infty \overline{u}_i(y) \overline{u}_j(y) w(y) dy = \delta_{ij},$$

where δ_{ij} is Kronecker's symbol.
Similarly, we have

$$\int_0^\infty \widetilde{v}_i(x) \widetilde{v}_j(x) V(x) dx = \int_1^\infty \overline{v}_i(y) \overline{v}_j(y) v(y) dy = \delta_{ij}.$$

Finally we conclude that $\{s_m(L_\alpha), \overline{u}_m, \overline{v}_m\}_{m \in Z_+}$ is a singular system for the operator R_α. $\square$

Lemmas 7.5.2 and 7.5.5 yield

Theorem 7.5.4. *Let* $\alpha > 0$, $\beta > -1$, $w(x) = \ln^{-\beta} x$, $v(x) = x^{-2} \ln^{-(\alpha+\beta)} x$. *Then the operator* $H_\alpha : L_w^2(1, \infty) \to L_v^2(1, \infty)$ *has a singular system*

$$\{s_n(H_\alpha), \widetilde{u}_n, \widetilde{v}_n\}_{n \in Z_+},$$

where

$$s_n(H_\alpha) = s_n(R_\alpha)$$

($s_m(R_\alpha)$ *is defined by* (7.5.1)),

$$\widetilde{u}_n(x) = x^{-1} \left(\frac{n!}{\Gamma(n+\beta+1)} \right)^{1/2} L_n^{(\beta)}(\ln x) \ln^\beta x,$$

$$\widetilde{v}_n(x) = \left(\frac{n!}{\Gamma(n+\alpha+\beta+1)} \right)^{1/2} L_n^{(\alpha+\beta)}(\ln x) \ln^{\alpha+\beta} x,$$

and $L_n^{(\gamma)}$ *is a Laguerre polynomial. Moreover,* $\lim_{n \to \infty} s_n(H_{/\alpha l})/n^{-\alpha/2} = 1$.

Theorem 7.5.5. *Let* $\lambda > \alpha - 1/2$, $\lambda \neq 0$. *Then the operator* H_α : $L_w^2(1, \infty) \to L_v^2(1, \infty)$ *has a singular system* $\{s_n(H_\alpha), \tilde{u}_m, \tilde{v}_n\}_{m \in Z_+}$, *where* $v(x) = x^{-1} \ln^{1/2-\lambda-\alpha} x$, $w(x) = (1 + \ln x)^{2\alpha} x \ln^{1/2-\lambda} x$,

$$s_n(H_\alpha) = s_n(R_\alpha)$$

$(s_n(R_\alpha)$ *is defined by* (7.5.2)*)*,

$$\tilde{u}_n(x) = 2^\lambda a_n (1 + \ln x)^{-\lambda-\alpha-1/2} C_n^\lambda \left(\frac{1 - \ln x}{1 + \ln x} \right) x^{-1} \ln^{\lambda-1/2} x,$$

$$v_n(x) = 2^\lambda b_n (1 + \ln x)^{-\lambda-\alpha-3/2} P_n^{(\lambda-\alpha-1/2, \lambda+\alpha-1/2)} \times$$
$$\times \left(\frac{1 - \ln x}{1 + \ln x} \right) \ln^{\lambda+\alpha-1/2} x,$$

and C_n^λ *and* $P_n^{(\alpha,\beta)}$ *are Gegenbauer and Jacobi polynomials respectively. Moreover,* $\lim\limits_{n \to \infty} s_n(H_\alpha)/n^{-\alpha} = 1$.

Definition 7.5.1. Let X and Y be Banach spaces and let T be a bounded linear map of X to Y. Then for all $k \in N$, the kth entropy number $e_k(T)$ of T is defined by

$$e_k(T) = \inf\{\varepsilon > 0 : T(U_X) \subset \cup_{j=1}^{2^{k-1}} (b_i + \varepsilon U_Y)$$

$$for \ some \ b_1, \cdots, b_{2^{k-1}} \in Y\},$$

where U_X and U_Y are the closed unit balls in X and Y respectively.

It is easy to verify that

$$\|T\| = e_1(T) \geq e_2(T) \geq \cdots \geq 0.$$

For other properties of the entropy numbers see, e.g., [79].

It is known (see, e.g., [44]), that if T is a compact linear map of a Hilbert space X into a Hilbert space Y, then $s_n(T) \approx n^{-\lambda}$ if and only if $e_n(T) \approx n^{-\lambda}$. Hence we can get asymptotics of the entropy numbers for the operators $I_{\alpha,\sigma}$ and H_α. In particular, Theorems 7.5.1, 7.5.2 and 7.5.3 yield

Proposition 7.5.1. *Let* $\alpha > 0$ *and* $\sigma > 0$. *Then the following statements are valid:*
  (a) *If* $0 \leq \gamma < \alpha$, *then the asymptotic formula*

$$e_n(I_{\alpha,\sigma}) \approx n^{-\alpha} \qquad (7.5.3)$$

holds for the operator $I_{\alpha,\sigma} : L_{x^{1-\sigma}}^2(0,1) \to L_{x^{\sigma-1-\gamma\sigma}}^2(0,1)$.

(b) Assume that $\lambda > \alpha - 1/2$ and $\lambda \neq 0$. Then the asymptotic formula (7.5.3) is valid for the map $I_{\alpha,\sigma} : L_w^2(0,\infty) \to L_v^2(0,\infty)$, where $w(x) = x^{-\sigma/2-\sigma\lambda+1}(1+x^\sigma)^{2\alpha}$ and $v(x) = x^{3\sigma/2-\sigma\lambda-\sigma\alpha-1}$.

(c) For the entropy numbers $e_n(I_{\alpha,\sigma})$ of the operator $I_{\alpha,\sigma} : L_w^2(0,\infty) \to L_v^2(0,\infty)$ $(w(y) = y^{-\sigma\beta-\sigma+1}e^{-y^\sigma}, v(y) = y^{-\sigma(\alpha+\beta)+\sigma-1}e^{-y^\sigma}, \beta > -1)$ we have

$$e_n(I_{\alpha,\sigma}) \approx n^{-\alpha/2}.$$

Let $T : L_w^2 \to L_v^2$ be a compact linear operator. We shall denote by $n(t,T)$ the distribution function of singular numbers for the operator T, i.e.

$$n(t,T) \equiv \sharp\{k : s_k(T) > t\}.$$

Theorem 7.5.6. *Let $\alpha > 1/2$ and $\sigma > 0$. Assume that v is a measurable a.e. positive function of $(0,\infty)$ satisfying the condition*

$$\sum_{k\in Z} \left(\int_{2^{k/\sigma}}^{2^{(k+1)/\sigma}} v(y)y^{(2\alpha-1)\sigma}dy \right)^{1/(2\alpha)} < \infty. \qquad (7.5.4)$$

Then for the operator $I_{\alpha,\sigma} : L_w^2(R_+) \to L_v^2(R_+)$, where $w(x) = x^{1-\sigma}$, the asymptotic formula

$$\lim_{t\to 0} t^{1/\alpha}n(t,I_{\alpha,\sigma}) = \frac{\sigma^{-1/\alpha+1}}{\pi} \int_0^\infty v^{1/(2\alpha)}(y)y^{(1-\sigma)(1/(2\alpha)-1)}dy$$

holds.

Proof. Condition (7.5.4) implies that

$$\sum_{k\in Z} \left(\int_{2^k}^{2^{k+1}} \bar{v}^2(y)y^{2\alpha-1}dy \right)^{1/(2\alpha)} < \infty,$$

where $\bar{v}(x) \equiv [v(x^{1/\sigma})x^{1/\sigma-1}]^{1/2}$. By virtue of Theorem 1 from [223] we have that for the operator $R_{\alpha,\bar{v}} : L^2(R_+) \to L^2(R_+)$, where $R_{\alpha,\bar{v}}f(x) \equiv \bar{v}(x)R_\alpha f(x)$, the asymptotic formula

$$\lim_{t\to 0} t^{1/\alpha}n(t,R_{\alpha,\bar{v}}) = \pi^{-1} \int_{R_+} \bar{v}^{1/\alpha}(x)dx$$

holds. Further, using Lemmas 7.5.1, 7.5.2 and 7.5.3 we find that $s_k(R_{\alpha,\bar{v}}) = \sigma \cdot s_k(I_{\alpha,\sigma})$. Consequently

$$\lim_{t\to 0} t^{1/\alpha}n(t,I_{\alpha,\sigma}) = \sigma^{-1/\alpha}\lim_{t\to 0} t^{1/\alpha}n(t,R_{\alpha,\bar{v}}) =$$

$$= \sigma^{-1/\alpha}\frac{1}{\pi}\int\limits_0^\infty (\bar{v}(x))^{1/\alpha}dx = \frac{\sigma^{-1/\alpha+1}}{\pi}\int\limits_0^\infty (v(y))^{1/(2\alpha)}y^{(1-\sigma)(1/(2\alpha)-1)}\,dy.$$

$\square$

Now we prove

Theorem 7.5.7. *Let* $\alpha > 1/2$ *and* $\sigma > 0$. *Suppose that* v *is a measurable a.e. positive function on* $(0,1)$ *satisfying the condition*

$$\sum_{k\in Z}\left(\int\limits_{a_k}^{a_{k+1}} v(x)x^{-\sigma+2\alpha\sigma}(1-x^\sigma)^{-1}dx\right)^{1/(2\alpha)} < \infty, \qquad (7.5.5)$$

where $a_k = (2^k/(2^k+1))^{1/\sigma}$. *Then for the operator* $I_{\alpha,\sigma}$ *acting from* $L_w^2(0,1)$ *into* $L_v^2(0,1)$, *where* $w(x) = (1-x^\sigma)^{2\alpha}x^{1-\sigma}$, *we have*

$$\lim_{t\to 0} t^{1/\alpha}n(t,I_{\alpha,\sigma}) = \frac{\sigma^{-1/\alpha+1}}{\pi}\int\limits_0^1 v^{1/(2\alpha)}(x)x^{(1-\sigma)(1/(2\alpha)-1)}(1-x^\sigma)^{-1}dx.$$

Proof. Using Lemmas 7.5.1-7.5.4 we have that $s_n(I_{\alpha,\sigma}) = 1/\sigma s_n(R_\alpha)$, where R_α is the Riemann–Liouville operator acting from $L_{w_1}^2(0,1)$ into $L_{v_1}^2(0,1)$, with $w_1(x) = w(x^{1/\sigma})x^{1-1/\sigma}, v_1(x) = v(x^{1/\sigma})x^{1/\sigma-1}$. Further, by the change of variable $x = y/(1-y)$ we find that the operator $\bar{R}_\alpha : L_{w_2}^2(R_+) \to L_{v_2}^2(R_+)$ has singular numbers $s_n(\bar{R}_\alpha) = \sigma s_n(I_{\alpha,\sigma})$, where $w_2(x) = w_1(x/(x+1))(x+1)^{-2}$, $v_2(x) = v_1(x/(x+1))(x+1)^{-2}$ and $\bar{R}_\alpha f(x) = \psi(x)R_\alpha(f\varphi)(x)$ with $\psi(x) = (x+1)^{-\alpha+1}$, $\varphi(x) = (x+1)^{-1-\alpha}$. Hence the singular numbers of the Riemann–Liouville operator $R_\alpha : L_{w_3}^2(R_+) \to L_{v_3}^2(R_+)$ satisfy $s_n(R_\alpha) = \sigma s_n(I_{\alpha,\sigma})$, where $w_3(x) = w_2(x)(x+1)^{2\alpha+2} = 1$ and $v_3(x) = v_2(x)(x+1)^{2-2\alpha}$. Further, condition (7.5.5) implies

$$\sum_{k\in Z}\left(\int\limits_{2^k}^{2^{k+1}} v_3(y)y^{2\alpha-1}dy\right)^{1/(2\alpha)} < \infty.$$

Thus taking into account Theorem 1 from [223] we arrive at

$$\lim_{t\to 0} t^{1/\alpha}n(t,I_{\alpha,\sigma}) = \sigma^{-1/\alpha}\lim_{t\to 0} t^{1/\alpha}n(t,R_\alpha) =$$

$$= \sigma^{-1/\alpha}\frac{1}{\pi}\int\limits_0^\infty v_3^{1/(2\alpha)}(x)dx =$$

$$= \frac{\sigma^{-1/\alpha+1}}{\pi}\int\limits_0^1 (v(y))^{1/(2\alpha)}y^{(1-\sigma)(1/(2\alpha)-1)}(1-y^\sigma)^{-1}dy.$$

In the last equality we used a change of variable twice. $\square$

Finally we have

Theorem 7.5.8. *Let $\alpha > 1/2$ and let v be a measurable a.e. positive function on $(1, \infty)$ satisfying the condition*

$$\sum_{k \in Z} \left(\int_{a_k}^{a_{k+1}} v(x) \ln^{2\alpha-1} x dx \right)^{1/(2\alpha)} < \infty, \quad a_k = e^{2^k}. \qquad (7.5.6)$$

Then for the operator $H_\alpha : L_w^2(1, \infty)$ into $L_v^2(1, \infty)$, where $w(x) = e^x$, we have the asymptotic formula

$$\lim_{t \to 0} t^{1/\alpha} n(t, H_{\alpha,\sigma}) = \frac{1}{\pi} \int_1^\infty v^{1/(2\alpha)}(x) x^{1/(2\alpha)-1} dy. \qquad (7.5.7)$$

Proof. Taking into account Lemmas 7.5.2 and 7.5.5 we see that $s_n(R_\alpha) = s_n(H_\alpha)$, where R_α is the Riemann–Liouville operator acting from $L^2(R_+)$ into $L_{v_1}^2(R_+)$, $v_1(x) = v(e^x)e^x$. By condition (7.5.6), Theorem 1 from [N-S] and the change of variable $x = e^y$ we conclude that (7.5.7) holds. $\square$

7.6.   Notes and comments on Chapter 7

Some estimates from above of eigen– and singular values of integral operators in terms of properties of their kernels may be found in, e.g., [34], [164] (see also monograph [66]).

Let T be the unit circle and φ be a bounded function on **T**. The Hankel operator $H_\varphi : H^2 \to H_-^2$ $(H_-^2 = L^2 \ominus H^2)$ is defined as follows

$$H_\varphi f = P_- \varphi f, \quad f \in H^2,$$

where H^2 is the Hardy class and P_- is an orthogonal projection from L^2 to H_-^2. The function φ is called a symbol of the Hankel operator H_φ.

Sharp two–sided sharp estimates of singular numbers of the Hankel operators H_φ were established in [229], [230], [243]. Analogous problems for the weighted Hardy operator were studied in [67–68]. For integral operators with Oinarov [225] kernels see [78].

Necessary and sufficient conditions for the weighted Hardy operator to belong to Schatten–von Neumann ideals were obtained in [224]. Closely related results had been derived in [36].

Two -sided estimates of Schatten–von Neumann norms for the Riemann–Liouville operators were established in [223], while analogous problems in the case of two weights, when the order of integration is greater than one, were solved in [78].

The cut off result at $p = 1$ which shows that the weighted Hardy operator can not be of trace class was obtained in [224], [244]. For similar problems in the case of commutators of singular integral operators see [132] and for the Hankel operator see, e.g., [11].

Chapter 8

SINGULAR INTEGRALS

The question of the boundedness of integral transforms defined on spaces of homogeneous type (SHT) arises naturally when studying boundary–value problems for partial differential equations with variable coefficients. For example, when the underlying domain is strongly pseudo–convex, one is led to use the concept of the Heisenberg group (and more general structures) as a model for the boundary of the domain in the theory of functions of several complex variables. Such problems indicate a strong need for structures more general than spaces of functions on Euclidean space. The space domain might, for instance, be most conveniently endowed with a quasi–metric induced by a differential operator or tailored to suit the kernel of a given integral operator (see [278], Chapters I, XII and XIII).

On the other hand, it is well–known that the solubility of boundary–value problems for elliptic partial differential equations in domains with non-smooth boundaries depends crucially on the geometry of the boundary. In [137], Chapter IV it is shown that the presence of angular points (involving cusps) can result in non-existence or non-uniqueness of solutions of Dirichlet and Neumann problems for harmonic functions from Smirnov classes and boundary functions in appropriate Lebesgue spaces. In this connection, two-weight inequalities for singular integrals with pairs of weights like those considered in the sequel enable one to identify, for the boundary functions, the weighted Lebesgue spaces for which the problem becomes soluble.

Two-weight inequalities of strong type with monotonic weights for Hilbert transforms have been established in [215]. Analogous problems for singular integrals in Euclidean spaces were considered in [117] and were generalised in [113] for singular integrals on Heisenberg groups. For Calderón–Zygmund singular integrals, conditions for a pair of radial weights ensuring the validity of two–weight inequalities of strong type have been obtained by the first two authors [71] (see also [239]) and generalised for homogeneous groups and on spaces of homogeneous type with some additional assumptions by the last two authors [150–151], [155–156]. Moreover, these last papers contain weak type inequalities as well.

In [52] necessary and sufficient conditions governing two-weight inequalities for certain singular integrals are established in terms of the rearrangements of the weights.

In this chapter we investigate two-weight problems for singular integrals. For a certain class of pairs of weights criteria for weak and strong type inequalities for classical singular integrals are established. Then these results for singular integrals defined on spaces of nonhomogeneous type are generalized. The consideration of spaces of homogeneous type enable us to extend the class of admissible pairs of weights in appropriate inequalities and to derive, as corollaries, two-weight estimates for Cauchy–Szegö projections. Finally we present Koosis–type theorems for singular integrals defined on homogeneous groups.

8.1. Two–weight strong–type estimates

In this section we derive various two-weight inequalities of strong type. The sufficient conditions given are optimal in the sense that for Hilbert transforms these are also necessary, see for example [215] and [71]. Some very special cases of the problems discussed in this section were studied in [100], Chapter IX.

Let (X, d, μ) be a space of homogeneous type (see Section 1.1 for the definition). There are numerous interesting examples of SHT, such as Euclidean space with an anisotropic distance and Lebesgue measure, any compact C^∞ Riemannian manifold with the Riemannian metric and volume, and the boundary of any bounded Lipschitz domain in R^n with the induced Euclidean metric and Lebesgue measure (see [100], Chapter 1 for some other examples of SHT).

We shall assume that there exists a point x_0 such that for every r and R with $0 < r < R < a$ the following condition is satisfied:

$$\mu(B(x_0, R) \setminus B(x_0, r)) > 0, \qquad (*)$$

where $a = \sup\{d(x_0, x) : x \in X\}$.

For the definition of the $A_p(X)$ classes see Chapter 6.

Theorem A ([128]). *Let $1 < p < \infty$. If $w \in A_p(X)$ then the operator K is bounded in $L_w^p(X)$.*

Now we pass to the definition of singular integrals on SHT (see for example [56], [278]). Let $k : (X \times X) \setminus \{(x, x); x \in X\} \to R$ be a measurable function satisfying the conditions:

$$|k(x,y)| \leq \frac{c}{\mu(B(x,d(x,y)))},$$

for all $x, y \in X$, $x \neq y$, and

$$|k(x_1, y) - k(x_2, y)| + |k(y, x_1) - k(y, x_2)| \leq$$

$$\leq c\omega \left(\frac{d(x_2, x_1)}{d(x_2, y)} \right) \frac{1}{\mu(B(x_2, d(x_2, y)))},$$

for every $x_1, x_2, y \in X$ such that $d(x_2, y) > bd(x_1, x_2)$. Here ω is a positive, non-decreasing function on $(0, \infty)$, satisfying the well-known Δ_2-condition (that is, $\omega(2t) \leq c\omega(t)$ for all $t > 0$ and some $c > 0$ independent of t) and the Dini condition

$$\int_0^1 \frac{\omega(t)}{t} dt < \infty.$$

We assume as well that for some p_0, $1 < p_0 < \infty$, and all $f \in L_\mu^{p_0}(X)$ the limit

$$Kf(x) = \lim_{\varepsilon \to 0+} \int_{X \backslash B(x,\varepsilon)} k(x,y)f(y)d\mu$$

exists a.e. and that the operator K is bounded in $L_\mu^{p_0}(X)$. For the definition of singular integrals and other remarks see [278], Chapter I, pp.29-36 and also [100], p.295. The boundedness of the operator K under the above-mentioned conditions for some $p_0 \in (1, \infty)$ guarantees the existence of the principal value Kf a.e. for $f \in L_\mu^p(X)$ and its boundedness in $L_\mu^p(X)$ for all $p \in (1, \infty)$ (see, for example, [278], Chapter V, 6.17, p.223).

Theorem A ([128]). *Let* $1 < p < \infty$. *If* $w \in A_p(X)$ *then the operator* K *is bounded in* $L_w^p(X)$.

Lemma 8.1.1. *Let* $1 < p < \infty$, *suppose that* $\mu\{x_0\} = 0$, *let* w *be a weight function on* X, *let* $\rho \in A_p(X)$ *and suppose that the following conditions are satisfied:*

(1) there exists an increasing function σ on $(0, 4a_1a)$, such that for some positive constant c_1,

$$\rho(x)\sigma(2a_1d(x_0, x)) \leq c_1w(x) \quad a.e.;$$

(2) for arbitrary t, $0 < t < a$,

$$\int_{B(x_0,t)} w^{1-p'}(x)d\mu < \infty.$$

Then $K\phi(x)$ exists $\mu-$ a.e. for any $\varphi \in L_w^p(X)$.

Proof. Let $0 < \alpha < \frac{a}{a_1}$ and put

$$S_\alpha = \left\{ x \in X : d(x_0, x) \geq \frac{\alpha}{2} \right\}.$$

Let us suppose $\phi \in L_w^p(X)$. Then

$$\phi(x) = \phi_1(x) + \phi_2(x), \tag{8.1.1}$$

where $\phi_1 = \phi\chi_{S_\alpha}$ and $\phi_2 = \phi - \phi_1$. Using condition (8.1.1) it is easy to see that

$$\int\limits_X |\phi_1(x)|^p \rho(x)d\mu = \frac{\sigma\left(\frac{\alpha}{2}\right)}{\sigma\left(\frac{\alpha}{2}\right)} \int\limits_{S_\alpha} |\phi(x)|^p \rho(x)d\mu$$

$$\leq \frac{1}{\sigma\left(\frac{\alpha}{2}\right)} \int\limits_{S_\alpha} |\phi(x)|^p \rho(x)\sigma\left(2a_1 d(x_0, x)\right) d\mu$$

$$\leq \frac{c_1}{\sigma\left(\frac{\alpha}{2}\right)} \int\limits_{S_\alpha} |\phi(x)|^p w(x)d\mu < \infty$$

for arbitrary α, $0 < \alpha < \frac{a}{a_1}$. Consequently for such α, $K\phi_1 \in L_\rho^p(X)$ (see Theorem A) and $K\phi_1(x)$ exists $\mu-$a.e. on X.

Now let x be such that $d(x_0, x) > \alpha a_1$ (the constant a_1 appears in Definition 1.1.1). If $y \in X$ and $d(x_0, y) < \frac{\alpha}{2}$ then

$$d(x_0, x) \leq a_1\left(d(x_0, y) + d(y, x)\right) \leq a_1\left(d(x_0, y) + a_0 d(x, y)\right).$$

Hence

$$d(x, y) \geq \frac{1}{a_0 a_1}d(x_0, x) - \frac{1}{a_0}d(x_0, y) \geq \frac{\alpha}{a_0} - \frac{\alpha}{2a_0} = \frac{\alpha}{2a_0}.$$

In addition

$$\mu\left(B\left(x_0, d(x, y)\right)\right) \leq c\mu\left(B(x, d(x, y))\right).$$

In fact for $z \in B\left(x_0, d(x, y)\right)$ we have

$$d(x, z) \leq a_1\left(d(x, x_0) + d(x_0, z)\right) \leq a_1\left(d(x, x_0) + d(x, y)\right).$$

On the other hand,

$$d(x, x_0) \leq a_1\left(d(x, y) + d(y, x_0)\right) \leq a_1\left(d(x, y) + a_0 d(x_0, y)\right)$$

$$\leq a_1\left(d(x, y) + \frac{a_0\alpha}{2}\right) \leq a_1\left(d(x, y) + a_0^2 d(x, y)\right) =$$

$$= a_1\left(1 + a_0^2\right) d(x, y)$$

and

$$d(x,z) \leq a_1 \left(1 + a_1 \left(1 + a_0^2\right)\right) d(x,y).$$

Hence

$$B\left(x_0, d(x,y)\right) \subset B\left(x, a_1 \left(1 + a_1 \left(1 + a_0^2\right)\right) d(x,y)\right).$$

By the doubling condition (vii) above we conclude that

$$\mu\left(B\left(x_0, d(x,y)\right)\right) \leq c_1 \mu\left(B\left(x, d(x,y)\right)\right). \tag{8.1.2}$$

For $K\phi_2$ by (8.1.2) and the Hölder inequality we have

$$|K\phi_2(x)| = \left| \int_X \phi_2(y) k(x,y) d\mu \right| \leq$$

$$\leq c_2 \int_{B\left(x_0, \frac{\alpha}{2}\right)} \frac{|\phi(y)|}{\mu\left(B\left(x, d(x,y)\right)\right)} d\mu$$

$$\leq c_3 \int_{B\left(x_0, \frac{\alpha}{2}\right)} \frac{|\phi(y)|}{\mu\left(B\left(x_0, d(x,y)\right)\right)} d\mu \leq$$

$$\leq \frac{c_4}{\mu\left(B\left(x_0, \frac{\alpha}{2a_0}\right)\right)} \int_{B\left(x_0, \frac{\alpha}{2}\right)} |\phi(y)| \, d\mu$$

$$\leq \frac{c_5}{\mu B\left(x_0, \frac{\alpha}{2a_0}\right)} \left(\int_{B\left(x_0, \frac{\alpha}{2}\right)} |\phi(y)|^p \, w(y) d\mu \right)^{1/p} \times$$

$$\times \left(\int_{B\left(x_0, \frac{\alpha}{2}\right)} w^{1-p'}(y) d\mu \right)^{1/p'} < \infty.$$

Thus $K\phi(x)$ is absolutely convergent for arbitrary x such that $d(x_0, x) > \alpha a_1$. We can take α arbitrarily small and as $\mu\{x_0\} = 0$ we conclude that $K\phi_2(x)$ converges absolutely μ–a.e. on X. By (8.1.1), $K\phi(x)$ exists a.e. on X. $\square$

Theorem 8.1.1. *Let* $1 < p < \infty$, *suppose that* $\mu\{x_0\} = 0$, *let* σ *be a positive continuous increasing function on* $(0, 4a_1 a)$, *let* $\rho \in A_p(X)$ *and suppose that* w *is a weight function on* X. *Let* $v(x) = \sigma\left(d(x_0, x)\right) \rho(x)$ *and suppose that the following conditions are fulfilled:*

(i) there exists $c > 0$ such that

$$\sigma\left(2a_1 d(x_0, x)\right)\rho(x) \le cw(x) \quad \mu - a.e.; \tag{8.1.3}$$

(ii)

$$\sup_{0<t<a}\left(\int\limits_{X\backslash B(x_0,t)} \frac{v(x)}{\left(\mu\left(B\left(x_0, d(x_0, x)\right)\right)\right)^p}\right) \times$$

$$\times\left(\int\limits_{B(x_0,t)} w^{1-p'}(x)d\mu\right)^{p-1} < \infty. \tag{8.1.4}$$

Then there is a constant $c > 0$ such that for any $f \in L_w^p(X)$ we have

$$\int\limits_X |Kf(x)|^p v(x)d\mu \le c\int\limits_X |f(x)|^p w(x)d\mu. \tag{8.1.5}$$

Proof. Without loss of generality we can suppose that σ may be represented by

$$\sigma(t) = \sigma(0+) + \int\limits_0^t \phi(\tau)d\tau, \quad \phi \ge 0.$$

In fact there exists a sequence of absolutely continuous functions σ_n such that $\sigma_n(t) \le \sigma(t)$ and $\lim_{n\to\infty} \sigma_n(t) = \sigma(t)$ for any $t \in (0, 4a_1 a)$. For such functions we may take

$$\sigma_n(t) = \sigma(0+) + n\int\limits_0^t \left[\sigma(\tau) - \sigma\left(\tau - \frac{1}{n}\right)\right]d\tau.$$

We have

$$\int\limits_X |Kf(x)|^p v(x)d\mu = \sigma(0+)\int\limits_X |Kf(x)|^p \rho(x)d\mu +$$

$$+ \int\limits_X |Kf(x)|^p \rho(x)\left(\int\limits_0^{d(x_0,x)} \phi(t)dt\right)d\mu = I_1 + I_2.$$

If $\sigma(0+) = 0$ then $I_1 = 0$. If $\sigma(0+) \ne 0$, by Theorem A and (8.1.3) we find that

$$I_1 \le c\sigma(0+)\int\limits_X |f(x)|^p \rho(x)d\mu \le$$

$$\le c_1 \int\limits_X |f(x)|^p \rho(x)\sigma\left(2a_1 d\left(x_0, x\right)\right)d\mu \le c\int\limits_X |f(x)|^p w(x)dx. \tag{8.1.6}$$

After changing the order of integration in I_2 we have

$$I_2 = \int_0^a \phi(t) \left(\int_{\{x:d(x_0,x)>t\}} |Kf(x)|^p \rho(x)d\mu \right) dt \le$$

$$\le c_2 \int_0^a \phi(t) \left(\int_{\{x:d(x_0,x)>t\}} \rho(x) \left| \int_{\{y:d(x_0,y)>\frac{t}{2a_1}\}} f(y)k(x,y)d\mu \right|^p d\mu \right) dt +$$

$$+ c_2 \int_0^a \phi(t) \left(\int_{\{x:d(x_0,x)>t\}} \rho(x) \left| \int_{\{y:d(x_0,y)\le\frac{t}{2a_1}\}} f(y)k(x,y)d\mu \right|^p d\mu \right) dt =$$

$$= I_{21} + I_{22}.$$

Using the boundedness of K in $L_\rho^p(X)$ (see Theorem A) we obtain

$$I_{21} \le c_3 \int_0^a \phi(t) \left(\int_{\{y:d(x_0,y)>\frac{t}{2a_1}\}} |f(y)|^p \rho(y)d\mu \right) dt \le$$

$$\le c_3 \int_X |f(y)|^p \rho(y) \left(\int_0^{2a_1 d(x_0,y)} \phi(t)dt \right) d\mu \le \qquad (8.1.7)$$

$$\le c_3 \int_X |f(y)|^p \rho(y)\sigma\left(2a_1 d(x_0,y)\right) d\mu \le$$

$$\le c_4 \int_X |f(y)|^p w(y)d\mu.$$

Now we estimate I_{22}. When $d(x_0,x) > t$ and $d(x_0,y) \le \frac{t}{2a_1}$ we have

$$d(x_0,x) \le a_1 \left(d(x_0,y) + d(y,x) \right) \le a_1 \left(d(x_0,y) + a_0 d(x,y) \right) \le$$

$$\le a_1 \left(\frac{t}{2a_1} + a_0 d(x,y) \right) \le a_1 \left(\frac{d(x_0,x)}{2a_1} + a_0 d(x,y) \right).$$

Hence

$$\frac{d(x_0,x)}{2a_1 a_0} \le d(x,y)$$

and

$$\mu\left(B\left(x, d(x_0,x)\right)\right) \le b_1 \mu \left(B \left(x, \frac{d(x_0,x)}{2a_1 a_0} \right) \right) \le \qquad (8.1.8)$$

$$\le b_2 \mu\left(B\left(x, d(x,y)\right)\right).$$

As in the preceding Lemma 8.1.1 we conclude that

$$\mu\left(B\left(x_0, d\left(x_0, x\right)\right)\right) \le b_3 \mu\left(B\left(x, d\left(x_0, x\right)\right)\right)$$

and therefore from (8.1.8) we have

$$\mu\left(B\left(x_0, d\left(x_0, x\right)\right)\right) \le b_4 \mu\left(B\left(x, d\left(x, y\right)\right)\right). \tag{8.1.9}$$

Using (8.1.9) we derive the inequalities

$$I_{22} \le c_2 \int_0^a \phi(t) \left(\int_{\{x:d(x_0,y)>t\}} \rho(x) \times \right.$$

$$\times \left(\int_{\{y:d(x_0,y)\le\frac{t}{2a_1}\}} \frac{|f(y)|}{\mu(B(x,d(x,y)))} \right)^p d\mu \right) dt \le$$

$$\le c_5 \int_0^a \phi(t) \left(\int_{\{x:d(x_0,x)>t\}} \frac{\rho(x)}{(\mu(B(x_0,d(x_0,x))))^p} \right) \times$$

$$\times \left(\int_{B(x_0,t)} |f(y)|d\mu \right)^p dt.$$

It is easy to see that for any s, $0 < s < a$, we have

$$\int_s^a \phi(t) \left(\int_{\{x:d(x_0,x)>t\}} \frac{\rho(x)}{(\mu\left(B\left(x_0, d\left(x_0, x\right)\right)\right))^p} d\mu \right) dt \le$$

$$\le \int_{\{x:d(x_0,x)\ge s\}} \frac{\rho(x)}{(\mu\left(B\left(x_0, d\left(x_0, x\right)\right)\right))^p} \left(\int_s^{d(x_0,x)} \phi(t)dt \right) d\mu \le$$

$$\le \left(\int_{\{x:d(x_0,x)\ge s\}} \frac{\rho(x)\sigma\left(d\left(x_0, x\right)\right)}{(\mu\left(B\left(x_0, d\left(x_0, x\right)\right)\right))^p} \right) d\mu.$$

Now applying Theorem 1.2.1 we conclude that from (8.1.4) there follows the inequality

$$I_{22} \le c_6 \int_X |f(x)|^p w(x)d\mu.$$

Finally from the last estimate, (8.1.6) and (8.1.7) we obtain (8.1.5). $\square$

Corollary 8.1.1. Let $1 < p < \infty$, suppose that $\mu\{x_0\} = 0$, σ and u be positive, increasing functions on $(0, 4a_1a)$, let $\rho \in A_p(X)$ and put $v(x) =$

$\sigma\left(d\left(x_0, x\right)\right)\rho(x)$, $w(x) = u\left(d\left(x_0, x\right)\right)\rho(x)$. *Then the inequality* (8.1.5) *holds under the following two conditions:*

(i) there exists a positive number b such that

$$\sigma(2a_1 t) \leq bu(t) \tag{8.1.10}$$

(ii) for any $t \in (0, 2a)$;

$$\sup_{0 < t < a} \left(\int_{X \setminus B(x_0, t)} \frac{v(x)}{\left(\mu(B(x_0, d(x_0, x)))\right)^p} d\mu \right) \times$$
$$\times \left(\int_{B(x_0, t)} w^{1-p'}(x) d\mu \right)^{p-1} < \infty. \tag{8.1.11}$$

Now we are going to consider the case when the weight on the left side is decreasing. We shall assume that throughout the rest of this section

$$a = \sup\left\{d\left(x_0, x\right) : x \in X\right\} = \infty.$$

According to Lemma 6.5.1 this condition is equivalent to the condition $\mu X = \infty$.

Lemma 8.1.2. *Let* $1 < p < \infty$, *let* σ *be a positive decreasing function on* $(0, \infty)$, *let* $\rho \in A_p(X)$ *and let* w *be a weight function. Suppose the following conditions are fulfilled:*

(i) there exists a positive constant b such that

$$\sigma\left(\frac{d\left(x_0, x\right)}{2a_1}\right)\rho(x) \leq bw(x) \quad a.e. \tag{8.1.12}$$

(ii)

$$\int_{X \setminus B(x_0, t)} w^{1-p'}(x)\left(\mu\left(B\left(x_0, d\left(x_0, x\right)\right)\right)\right)^{-p'} d\mu < \infty \tag{8.1.13}$$

for any $t > 0$.

Then $K\phi(x)$ exists μ−a.e. for arbitrary $\phi \in L_w^p(X)$.

Proof. Fix arbitrarily $\alpha > 0$ and let

$$S_\alpha = \left\{x : d\left(x_0, x\right) \geq \alpha\right\}.$$

Write

$$\phi(x) = \phi_1(x) + \phi_2(x),$$

where $\phi_1(x) = \phi(x)\chi_{S_\alpha}(x)$ and $\phi_2(x) = \phi(x) - \phi_1(x)$.

For ϕ_2 we have

$$
\int\limits_X |\phi_2(x)|^p \, \rho(x)d\mu = \frac{\sigma(\alpha)}{\sigma(\alpha)} \int\limits_{B(x_0,\alpha)} |\phi(x)|^p \, \rho(x)d\mu \le
$$

$$
\le \frac{1}{\sigma(\alpha)} \int\limits_{B(x_0,\alpha)} |\phi(x)|^p \, \rho(x)\sigma\left(d\left(x_0,x\right)\right)d\mu \le
$$

$$
\le \frac{c_1}{\sigma(\alpha)} \int\limits_{B(x,\alpha)} |\phi(x)|^p \, w(x)d\mu < \infty.
$$

Hence $\phi_2 \in L^p_\rho(X)$ and so $K\phi_2 \in L^p_\rho(X)$. From this we see that $K\phi_2(x)$ exists almost everywhere.

Now let $x \in X$ be such that $d\left(x_0,x\right) < \frac{\alpha}{2a_1}$. If $d\left(x_0,y\right) \ge \alpha$, then

$$
d\left(x_0,y\right) \le a_1\left(d\left(x_0,x\right) + d\left(x,y\right)\right)
$$

and so

$$
d(x,y) \ge \frac{1}{a_1}d\left(x_0,y\right) - d\left(x_0,x\right) \ge \frac{\alpha}{a_1} - \frac{\alpha}{2a_1} = \frac{\alpha}{2a_1}.
$$

Moreover, it is easy to prove that

$$
\mu\left(B\left(x_0,d\left(x_0,y\right)\right)\right) \le b_1\mu\left(B\left(x,d\left(x,y\right)\right)\right).
$$

From these inequalities we obtain estimates for ϕ_1 :

$$
|K\phi_1(x)| \le c_1 \int\limits_{S_\alpha} \frac{|\phi(y)|}{\mu\left(B\left(x,d\left(x,y\right)\right)\right)}d\mu \le
$$

$$
\le c_2 \int\limits_{S_\alpha} \frac{|\phi(y)|}{\mu\left(B\left(x_0,d\left(x_0,y\right)\right)\right)}d\mu \le
$$

$$
\le c_2 \left(\int\limits_{S_\alpha} |\phi(y)|^p \, w(y)d\mu \right)^{1/p} \times
$$

$$
\times \left(\int\limits_{S_\alpha} w^{1-p'} \left(\mu\left(B\left(x_0,d\left(x_0,y\right)\right)\right)\right)^{-p'} d\mu \right)^{1/p'} < \infty.
$$

As we may take α arbitrarily large we conclude that $T\phi_1(x)$ and consequently $T\phi(x)$, exists a.e. $\square$

Theorem 8.1.2. *Let* $1 < p < \infty$, *suppose that* $\mu\{x_0\} = 0$, *let* σ *be a positive continuous decreasing function on* $(0,\infty)$, *let* $\rho \in A_p(X)$, $v(x) =$

$\sigma\left(d\left(x_0, x\right)\right)\rho(x)$ *and suppose that w is a weight function. Assume that the following two conditions are fulfilled:*

(i) there exists a positive c such that

$$\sigma\left(\frac{d\left(x_0, x\right)}{2a_1}\right)\rho(x) \le cw(x); \tag{8.1.14}$$

(ii)

$$\sup_{t>0}\left(\int_{B(x_0,t)} v(x)d\mu\right) \times$$
$$\times\left(\int_{X\backslash B(x_0,t)} w^{1-p'}(x)(\mu(B(x_0, d(x_0, x))))^{-p'}d\mu\right)^{p-1} < \infty. \tag{8.1.15}$$

Then the inequality (8.1.5) holds.

Proof. Without loss of generality we suppose that σ is representable as

$$\sigma(t) = \sigma(+\infty) + \int_t^\infty \phi(\tau)d\tau.$$

In fact there exists a sequence of decreasing absolutely continuous functions such that $\sigma_n(t) \le \sigma(t)$ and $\lim_{n\to\infty} \sigma_n(t) = \sigma(t)$ for any t. For example, we may take

$$\sigma_n(t) = \sigma(+\infty) + n\int_t^\infty \left[\sigma(\tau) - \sigma\left(\tau + \frac{1}{n}\right)\right]d\tau.$$

It is easy to see that

$$\sigma_n(t) = n\int_t^{t+\frac{1}{n}} \sigma(\tau)d\tau.$$

Moreover, $\lim_{n\to\infty} \sigma_n(t) = \sigma(t)$ for any t by virtue of the continuity of σ. On the other hand $\sigma_n(t) \le \sigma(t)$ for any $t > 0$. Hence

$$\int_X |Kf(x)|^p v(x)dx = \sigma(+\infty)\int_X |Kf(x)|^p \rho(x)d\mu +$$
$$+ \int_X |Kf(x)|^p \left(\int_{d(x_0,x)}^\infty \phi(t)dt\right)d\mu = I_1 + I_2.$$

If $\sigma(+\infty) = 0$ then $I_1 = 0$. But if $\sigma(+\infty) \neq 0$, by virtue of the boundedness of K in $L_\rho^p(X)$ we have

$$I_1 \leq c_1 \sigma(+\infty) \int\limits_X |f(x)|^p \rho(x) d\mu \leq$$

$$\leq c_1 \int\limits_X |f(x)|^p \rho(x) \sigma\left(\frac{d(x_0, x)}{2a_1}\right) d\mu \leq c_2 \int\limits_X |f(x)|^p w(x) d\mu. \qquad (8.1.16)$$

Now we pass to I_2 :

$$I_2 \leq \int\limits_0^\infty \phi(t) \left(\int\limits_{B(x_0,t)} |Kf(x)|^p \rho(x) d\mu \right) dt \leq$$

$$\leq c_3 \int\limits_0^\infty \phi(t) \left(\int\limits_{B(x_0,t)} \rho(x) \left| \int\limits_{B(x_0,2a_1 t)} f(y) k(x,y) d\mu \right|^p d\mu \right) +$$

$$+ c_3 \int\limits_0^\infty \phi(t) \left(\int\limits_{B(x_0,t)} \rho(x) \left| \int\limits_{X \backslash B(x_0,2a_1 t)} f(y) k(x,y) d\mu \right|^p d\mu \right) dt =$$

$$= I_{21} + I_{22}.$$

Again since K is bounded in $L_\rho^p(X)$ we obtain:

$$I_{21} \leq c_4 \int\limits_0^\infty \phi(t) \left(\int\limits_{B(x_0,2a_1 t)} \left| f(y) \right|^p \rho(y) d\mu \right) dt =$$

$$= c_4 \int\limits_X \left| f(y) \right|^p \rho(y) \left(\int\limits_{\frac{d(x_0,x)}{2}}^\infty \phi(t) dt \right) d\mu \leq \qquad (8.1.17)$$

$$\leq c_5 \int\limits_X \left| f(y) \right|^p w(y) d\mu.$$

It remains to estimate I_{22}. When $x \in B(x_0, t)$ and $y \in X \backslash B(x_0, 2a_1 t)$ we have

$$\mu(B(x_0, d(x_0, y))) \leq b\mu(B(x, d(x, y))).$$

In fact,

$$d(x_0, y) \leq a_1 (d(x_0, x) + d(x, y)) \leq$$

$$\leq a_1(t + d(x,y)) \leq a_1 \left(\frac{d(x_0, y)}{2a_1} + d(x, y) \right).$$

Hence

$$\frac{d(x_0, y)}{2a_1} \leq d(x, y)$$

and

$$\mu\left(B\left(x, d\left(x_0, y\right)\right)\right) \le b\mu\left(B\left(x, d\left(x, y\right)\right)\right). \tag{8.1.18}$$

In addition

$$\mu\left(B\left(x_0, d\left(x_0, y\right)\right)\right) \le b\mu\left(B\left(x, d\left(x_0, y\right)\right)\right). \tag{8.1.19}$$

If $z \in B\left(x_0, d\left(x_0, y\right)\right)$, then

$$d\left(x, z\right) \le a_1\left(d\left(x, x_0\right) + d\left(x_0, z\right)\right) \le a_1\left(a_0 d\left(x_0, x\right) + d\left(x_0, y\right)\right) \le$$
$$\le a_1\left(a_0 t + d\left(x_0, y\right)\right) \le a_1\left(a_0 \frac{d\left(x_0, y\right)}{2a_1} + d\left(x_0, y\right)\right) =$$
$$= a_1\left(\frac{a_0}{2a_1} + 1\right) d\left(x_0, y\right).$$

From this using the doubling condition we obtain (8.1.18). Using inequalities (8.1.18) and (8.1.19) we derive the estimates:

$$I_{22} \le c_5 \int_0^\infty \phi(t)\left(\int_{B(x_0,t)} \rho(x)d\mu\right) \times$$

$$\times \left(\int_{X\backslash B(x_0,2a_1t)} |f(y)|\mu(B(x_0, d(x_0, y)))d\mu\right)^p dt \le$$

$$\le c_5 \int_0^\infty \phi(t)\left(\int_{B(x_0,t)} \rho(x)d\mu\right) \times$$

$$\times \left(\int_{\{y:d(x_0,y)>t\}} |f(y)|\mu(B(x_0, d(x_0, y)))d\mu\right)^p dt.$$

In addition

$$\int_0^s \phi(t)\left(\int_{B(x_0,t)} \rho(x)d\mu\right) dt = \int_{B(x_0,s)} \rho(x)\left(\int_{d(x_0,x)}^s \phi(t)dt\right) d\mu.$$

Now application of Theorem 8.1.2 and inequalities (8.1.16), (8.1.17) give the desired inequality (8.1.5). $\square$

Corollary 8.1.2. *Let* $1 < p < \infty$, $\mu\{x_0\} = 0$, *let* σ *and* u *be positive decreasing functions on* $(0, \infty)$ *with* σ *continuous, let* $\rho \in A_p(X)$, *put* $v(x) = \sigma\left(d\left(x_0, x\right)\right)\rho(x)$, $w(x) = u\left(d\left(x_0, x\right)\right)\rho(x)$ *and suppose that the following two conditions are fulfilled:*

(i) there exists a positive number b_1 such that

$$\sigma\left(\frac{t}{2a_1}\right) \leq b_1 u(t) \qquad (8.1.20)$$

for any $t > 0$;

(ii)

$$\sup_{t>0}\left(\int_{B(x_0,t)} v(x)d\mu\right) \times$$

$$\times\left(\int_{X\backslash B(x_0,t)} w^{1-p'}(x)(\mu(B(x_0,d(x_0,x))))^{-p'}d\mu\right)^{p-1} < \infty. \qquad (8.1.21)$$

Then (8.1.5) holds.

In the sequel we shall investigate the cases when the condition (8.1.11) ((8.1.21)) implies (8.1.10) ((8.1.20)).

We now recall the definition of the reverse doubling condition: Let $\mu X = \infty$. A measure μ satisfies the reverse doubling condition ((RD) condition) at the point x_0 if there exist constants $\eta_1 > 1$ and $\eta_2 > 1$ such that

$$\mu\left(B\left(x_0,\eta_1 r\right)\right) \geq \eta_2 \mu\left(B\left(x_0,r\right)\right)$$

for all $r > 0$ (see Definition (6.5.1)).

As a measure with the doubling condition by the condition (*) satisfies the reverse doubling condition at x_0 as well (see Proposition 6.5.1), we are able to show that from (8.1.11) ((8.1.21)) follows (8.1.10) ((8.1.20)).

Theorem 8.1.3 *Let $1 < p < \infty$, suppose that $\mu\{x_0\} = 0$, let σ and u be positive increasing functions on $(0,\infty)$ with σ continuous, let $\rho \in A_p(X)$, put $v(x) = \sigma\left(d\left(x_0,x\right)\right)\rho(x)$, $w(x) = u\left(d\left(x_0,x\right)\right)\rho(x)$ and suppose that*

$$\sup_{t>0}\left(\int_{X\backslash B(x_0,t)} \frac{v(x)}{(\mu(B(x_0,d(x_0,x))))^p}d\mu\right) \times$$

$$\times\left(\int_{B(x_0,t)} w^{1-p'}(x)d\mu\right)^{p-1} < \infty. \qquad (8.1.22)$$

Then (8.1.5) holds.

Proof. By Corollary 8.1.1 it is sufficient to prove that (8.1.22) implies that given $\beta > 1$, there is a positive constant b such that

$$\sigma\left(\beta t\right) \leq b_1 u(t) \qquad (8.1.23)$$

for all $t > 0$.

Let $\eta \geq \eta_1 > 0$, where η_1 is as in the definition of the (RD) condition. In view of (RD) we have

$$\mu(B(x_0, \eta\beta t)) \setminus (B(x_0, \beta t)) \geq$$

$$\geq \mu(B(x_0, \eta\beta t)) - \frac{1}{\eta_2}\mu(B(x_0, \eta\beta t)) \geq (1 - \frac{1}{\eta_2})\mu(B(x_0, \eta\beta t)).$$

Hence

$$\mu\left(B\left(x_0, \eta\beta t\right)\right) \setminus \left(B\left(x_0, \beta t\right)\right) \geq b_2\mu\left(B\left(x_0, \eta\beta t\right)\right). \tag{8.1.24}$$

Since σ and u are monotone, we have

$$\left(\int_{X \setminus B(x_0, t)} v(x)\left(\mu\left(B\left(x_0, d\left(x_0, x\right)\right)\right)\right)^{-p} d\mu\right) \geq$$

$$\geq \int_{X \setminus B(x_0, \beta t)} \sigma\left(d\left(x_0, x\right)\right) \rho(x)\left(\mu\left(B\left(x_0, d\left(x_0, x\right)\right)\right)\right)^{-p} d\mu \geq$$

$$\geq \int_{B(x_0, \eta\beta t) \setminus B(x_0, \beta t)} \frac{\sigma\left(d\left(x_0, x\right)\right) \rho(x)}{\left(\mu\left(B\left(x_0, d\left(x_0, x\right)\right)\right)\right)^p} d\mu \geq \tag{8.1.25}$$

$$\geq \sigma(\beta t) \int_{B(x_0, \eta\beta t) \setminus B(x_0, \beta t)} \frac{\rho(x)}{\left(\mu\left(B\left(x_0, d\left(x_0, x\right)\right)\right)\right)^p} d\mu$$

and

$$\left(\int_{B(x_0, t)} w^{1-p'}(x) d\mu\right)^{p-1} \geq$$

$$\geq \frac{1}{u(t)}\left(\int_{B(x_0, t)} \rho^{1-p'}(x) d\mu\right)^{p-1}. \tag{8.1.26}$$

Using Hölder's inequality, Lemma 6.6.1, (8.1.24), (8.1.25) and (8.1.26) it follows that

$$\frac{\sigma(\beta t)}{u(t)} = \frac{\sigma(\beta t)}{u(t)}\left(\frac{1}{\mu(B(x_0, \eta\beta t) \setminus B(x_0, \beta t))} \times\right.$$

$$\times \int_{B(x_0, \eta\beta t) \setminus B(x_0, \beta t)} \rho^{1/p}(x) \rho^{-1/p}(x) d\mu\bigg)^p \leq$$

$$\leq \frac{\sigma(\beta t)}{u(t)(\mu(B(x_0, \eta\beta t) \setminus B(x_0, \beta t)))^p} \int\limits_{B(x_0,\eta\beta t)\setminus B(x_0,\beta t)} \rho(x)d\mu \times$$

$$\times \left(\int\limits_{B(x_0,\eta\beta t)\setminus B(x_0,\beta t)} \rho^{1-p'}(x)d\mu \right)^{p-1} \leq$$

$$\leq b_1 \frac{\sigma(\beta t)}{u(t)(\mu(B(x_0, \eta\beta t)))^p} \int\limits_{B(x_0,\eta\beta t)\setminus B(x_0,\beta t)} \rho(x)d\mu \times$$

$$\times \left(\int\limits_{B(x_0,t)} \rho^{1-p'}(x)d\mu \right)^{p-1} \leq$$

$$\leq b_1 \frac{\sigma(\beta t)}{u(t)} \left(\int\limits_{B(x_0,\eta\beta t)\setminus B(x_0,\beta t)} \rho(x)(\mu(B(x_0, \eta\beta t)))^{-p}d\mu \right) \times$$

$$\times \left(\int\limits_{B(x_0,t)} \rho^{1-p'}(x)d\mu \right)^{p-1} \leq$$

$$\leq b_2 \frac{\sigma(\beta t)}{u(t)} \left(\int\limits_{B(x_0,\eta\beta t)\setminus B(x_0,\beta t)} \rho(x)(\mu(B(x_0, d(x_0, x))))^{-p}d\mu \right) \times$$

$$\times \left(\int\limits_{B(x_0,t)} \rho^{1-p'}(x)d\mu \right)^{p-1} \leq b_2 \left(\int\limits_{X\setminus B(x_0,t)} \frac{v(x)}{(\mu(B(x_0, d(x_0, x))))^p}d\mu \right) \times$$

$$\times \left(\int\limits_{B(x_0,t)} w^{1-p'}(x)d\mu \right)^{p-1} \leq c.$$

Finally by Corollary 8.1.1 we obtain (8.1.5). □

Analogously, we can prove

Theorem 8.1.4. *Let* $1 < p < \infty$, *suppose* $\mu\{x_0\} = 0$, *let* σ *and* u *be positive decreasing functions,* σ *be continuous,* $\rho \in A_p(X)$ *and put* $v(x) = \sigma(d(x_0, x))\rho(x)$, $w(x) = u(d(x_0, x))\rho(x)$. *Suppose that*

$$\sup_{t>0} \left(\int\limits_{B(x_0,t)} v(x)d\mu \right) \times$$

$$\times \left(\int\limits_{X\setminus B(x_0,t)} \frac{w^{1-p'}(x)}{(\mu(B(x_0, d(x_0, x))))^{p'}}d\mu \right)^{p-1} < \infty. \qquad (8.1.27)$$

Then (8.1.5) *holds.*

We shall now discuss the following question: if a pair (σ, u) of positive increasing (decreasing) functions satisfies the condition (8.1.22) ((8.1.27)) with $\rho \equiv 1$, then for which functions $\rho \in A_p(X)$ does (8.1.5) remain valid? It is evident that not all ρ in $A_p(X)$ have this property. Nevertheless we have

Theorem 8.1.5 *Let* $1 < p < \infty$*, let* $\mu\{x_0\} = 0$*. Let* σ *and* u *be positive increasing functions on* $(0, \infty)$*, with* σ *continuous. If*

$$
\sup_{t>0} \left(\int_{X \backslash B(x_0, t)} \frac{\sigma(d(x_0, x))}{(\mu(B(x_0, d(x_0, x))))^p} d\mu \right) \times
$$
$$
\times \left(\int_{B(x_0, t)} u^{1-p'}(d(x_0, x)) \, d\mu \right)^{p-1} < \infty
\tag{8.1.28}
$$

and $\rho \in A_1(X)$*, then we have*

$$
\int_X |Kf(x)|^p \sigma(d(x_0, x))\rho(x)d\mu \le
$$
$$
\le c \int_X |f(x)|^p u(d(x_0, x))\rho(x)d\mu.
\tag{8.1.29}
$$

Proof. Let η_1 and η_2 be as in the definition of condition (RD). Then we have

$$
\mu(B(x_0, \eta_1^{k+1}t) \backslash B(x_0, \eta_1^k t)) \ge (\eta_2 - 1) \mu\left(B\left(x_0, \eta_1^k t\right)\right)
$$

for any non-negative integer k.

From this and the doubling condition for μ we derive

$$
\int_{B(x_0, \eta_1^{k+1}t) \backslash B(x_0, \eta_1^k t)} \rho(x) \left(\mu\left(B\left(x_0, d\left(x_0, x\right)\right)\right)\right)^{-p} d\mu \le
$$
$$
\le \frac{b_1}{\left(\mu\left(B\left(x_0, \eta_1^k t\right)\right)\right)^{p-1}} \frac{1}{\mu\left(B\left(x_0, \eta_1^{k+1}t\right)\right)} \int_{B(x_0, \eta_1^{k+1}t)} \rho(x)d\mu.
\tag{8.1.30}
$$

Now using (8.1.30) and the $A_1(X)$ condition for ρ we obtain the following estimates, in which

$$
A_k = \operatorname*{ess\,sup}_{x \in B(x_0, \eta_1^{k+1}t)} \frac{1}{\rho(x)} \left(\int_{B(x_0, t)} u^{1-p'}(d(x_0, x)) \, d\mu \right)^{p-1} :
$$

$$\int\limits_{X\setminus B(x_0,t)} \frac{\sigma(d(x_0,x))\rho(x)}{(\mu(B(x_0,d(x_0,x))))^p}d\mu \times$$

$$\times \left(\int\limits_{B(x_0,t)} u^{1-p'}(d(x_0,x))\rho^{1-p'}(x)d\mu \right)^{p-1} \le$$

$$\le A_k \sum_{k=0}^{\infty} \int\limits_{B(x_0,\eta_1^{k+1}t)\setminus B(x_0,\eta_1^k t)} \frac{\sigma(d(x_0,x))\rho(x)}{(\mu(B(x_0,d(x_0,x))))^p}d\mu \le$$

$$\le A_k b_2 \sum_{k=0}^{\infty} \frac{\sigma(\eta_1^{k+1}t)}{(\mu(B(x_0,\eta_1^k t)))^{p-1}} \frac{1}{(\mu(B(x_0,\eta_1^{k+1}t)))} \int\limits_{B(x_0,\eta_1^{k+1}t)} \rho(x)d\mu \le$$

$$\le b_3 \sum_{k=0}^{\infty} \int\limits_{B(x_0,\eta_1^{k+2}t)\setminus B(x_0,\eta_1^{k+1}t)} \frac{\sigma(d(x_0,x))}{(\mu(B(x_0,d(x_0,x))))^p}d\mu \times$$

$$\times \left(\int\limits_{B(x_0,t)} u^{1-p'}(d(x_0,x))d\mu \right)^{p-1} \le b_4.$$

Finally with the help of Theorem 8.1.3 we obtain (8.1.29). $\square$

Theorem 8.1.6. *Let* $1 < p < \infty$, $\mu\{x_0\} = 0$, *let* σ *and* u *be decreasing functions on* $(0,\infty)$ *with* σ *continuous, and suppose that*

$$\sup_{t>0} \left(\int\limits_{B(x_0,t)} \sigma(d(x_0,x))d\mu \right) \times$$

$$\times \left(\int\limits_{X\setminus B(x_0,t)} \frac{u^{1-p'}(d(x_0,x))}{(\mu(B(x_0,d(x_0,x))))^{p'}}d\mu \right)^{p-1} < \infty. \qquad (8.1.31)$$

Then if $\rho \in A_1(X)$ *we have the inequality*

$$\int\limits_X |Kf(x)|^p \sigma(d(x_0,x))\rho^{1-p}(x)d\mu$$

$$\le c \int\limits_X |f(x)|^p u(d(x_0,x))\rho^{1-p}(x)d\mu \qquad (8.1.32)$$

with a constant c *independent of* f.

Proof. Again let η_1 and η_2 be as in the definition of the reverse doubling condition,. By the A_1 condition for ρ, the doubling and the (RD) conditions and (8.1.31) we obtain the following chain of inequalities:

$$\left(\int\limits_{B(x_0,t)} \sigma\left(d\left(x_0,x\right)\right)\rho^{1-p}(x)d\mu \right)^{p'-1} \times$$

$$\times \left(\int\limits_{X\setminus B(x_0,t)} \frac{u^{1-p'}(d(x_0,x))\rho^{(1-p)(1-p')}(x)}{(\mu(B(x_0,d(x_0,x))))^{p'}} d\mu \right) =$$

$$= \left(\int\limits_{B(x_0,t)} \sigma(d(x_0,x))\rho^{1-p}(x)d\mu \right)^{p'-1} \times$$

$$\times \sum_{k=0}^{\infty} \int\limits_{B(x_0,\eta_1^{k+1}t)\setminus B(x_0,\eta_1^k t)} \frac{u^{1-p'}(d(x_0,x))\rho(x)}{(\mu(B(x_0,d(x_0,x))))^{p'}} d\mu \le$$

$$\le b_1 \left(\int\limits_{B(x_0,t)} \sigma\left(d\left(x_0,x\right)\right)d\mu \right)^{p'-1} \sum_{k=0}^{\infty} \frac{u^{1-p'}\left(\eta_1^{k+1}t\right)}{(\mu\left(B\left(x_0,\eta_1^k t\right)\right))^{p'-1}} \le$$

$$\le b_2 \left(\int\limits_{B(x_0,t)} \sigma\left(d\left(x_0,x\right)\right)d\mu \right)^{p'-1} \sum_{k=0}^{\infty} u^{1-p'}\left(\eta_1^{k+1}t\right) \times$$

$$\times \int\limits_{B(x_0,\eta_1^{k+2}t)\setminus B(x_0,\eta_1^{k+1}t)} \frac{d\mu}{(\mu\left(B\left(x_0,d\left(x_0,x\right)\right)\right))^{p'}} \le$$

$$\le b_2 \left(\int\limits_{B(x_0,\eta_1 t)} \sigma\left(d\left(x_0,x\right)\right)d\mu \right)^{p'-1} \times$$

$$\times \int\limits_{X\setminus B(x_0,\eta_1 t)} \frac{u^{1-p'}(d(x_0,x))}{(\mu(B(x_0,d(x_0,x))))^{p'}} d\mu \le b_3.$$

As $\rho \in A_1(X)$ it follows that $\rho^{1-p} \in A_p(X)$, and by Theorem 8.1.4 the last estimation leads to the desired result. $\square$

Now we give some examples of weights.

Let $\Gamma \subset \mathbf{C}$ be a connected rectifiable curve and let ν be arc-length measure on Γ. By definition, Γ is regular if

$$\nu\left(\Gamma \cap B(z,r)\right) \le cr$$

for every $z \in \mathbf{C}$ and all $r > 0$.

For r smaller than half the diameter of Γ, the reverse inequality

$$\nu\left(\Gamma \cap B(z,r)\right) \ge r$$

holds for all $z \in \Gamma$. Equipped with ν and the Euclidean metric, the regular curve becomes an SHT.

The associated kernel in which we are interested is

$$k(z, w) = \frac{1}{z - w}.$$

The Cauchy integral

$$S_\Gamma f(t) = \int\limits_\Gamma \frac{f(\tau)}{t - \tau} d\nu(\tau)$$

is the corresponding singular operator.

The above-mentioned kernel in the case of regular curves is a Calderón–Zygmund kernel. As was proved by David [60], a necessary and sufficient condition for continuity of the operator S_Γ in $L_p(\Gamma)$ ($1 < p < \infty$) is that Γ is regular.

From the results obtained in the preceding section we can derive several two-weight estimates for S_Γ.

Definition 8.1.1. A measurable, almost everywhere positive function w on Γ is said to be in the class $A_p(\Gamma)$ if

$$\sup_{z \in \Gamma, r > 0} \frac{1}{\nu(B(z, r) \cap \Gamma)} \int\limits_{B(z,r) \cap \Gamma} w(t) d\nu \times$$

$$\times \left(\frac{1}{\nu(B(z, r) \cap \Gamma)} \int\limits_{B(z,r) \cap \Gamma} w^{1-p'}(t) d\nu \right)^{p-1} < \infty. \qquad (8.1.33)$$

It is known (see for example [60]) that for the continuity of S_Γ in $L_w^p(\Gamma)$, $1 < p < \infty$, when Γ is regular, it is necessary and sufficient that $w \in A_p(\Gamma)$.

Since for regular curves the measure ν satisfies the reverse doubling condition as well, we derive from Theorem 8.1.3

Proposition 8.1.1. *Let* $1 < p < \infty$, *let* Γ *be an unbounded regular curve, and let* $t_0 \in \Gamma$. *Let* σ *and* u *be positive increasing functions on* $(0, \infty)$ *with* σ *continuous, let* $\rho \in A_p(\Gamma)$ *and put* $v(t) = \sigma(|t - t_0|) \rho(t)$, $w(t) = u(|t - t_0|) \rho(t)$. *If*

$$\sup_{r > 0} \int\limits_{\Gamma \backslash B(t_0, r)} \frac{v(t)}{(\nu(B(t_0, |t - t_0|) \cap \Gamma))^p} d\nu \left(\int\limits_{B(t_0, r) \cap \Gamma} w^{1-p'}(t) d\nu \right)^{p-1} < \infty,$$

then the inequality

$$\int\limits_\Gamma |S_\Gamma f(t)|^p v(t) d\nu \leq c \int\limits_\Gamma |f(t)|^p w(t) d\nu$$

holds with a constant c independent of $f \in L^p_w(\Gamma)$. *A corresponding version of Theorem 8.1.4 holds for* S_Γ.

From the results of the last section we can obtain two-weight inequalities in more general situations than the case just considered.

Let Γ be a subset of R^n which is an s-set $(0 \leq s \leq n)$ in the sense that there is a Borel measure μ on R^n such that (i) $\mathrm{supp}\mu = \Gamma$; (ii) there are positive constants c_1 and c_2 such that for all $x \in \Gamma$ and all $r \in (0, 1)$,

$$c_1 r^s \leq \mu(B(x, r) \cap \Gamma) \leq c_2 r^s.$$

It is known (see [301], Theorem 3.4) that μ is equivalent to the restriction of Hausdorff s-measure $\mathcal{H}_s$ to Γ; we shall thus identify μ with $\mathcal{H}_s | \Gamma$.

Given $x \in \Gamma$, put $\Gamma(x, r) = B(x, r) \cap \Gamma$. By definition, if $1 < p < \infty$, $\rho \in A_p(\Gamma)$ if

$$\sup_{x \in \Gamma, r > 0} \frac{1}{\mathcal{H}_s(\Gamma(x, r))} \int_{\Gamma(x,r)} \rho(y) d\mathcal{H}_s(y) \times$$

$$\times \left(\frac{1}{\mathcal{H}_s(\Gamma(x, r))} \int_{\Gamma(x,r)} \rho^{1-p'}(z) d\mathcal{H}_s(z) \right)^{p-1} < \infty.$$

Let K_Γ be a Calderón–Zygmund singular integral defined on an s-set Γ. Since $\mathcal{H}_s|_\Gamma$ satisfies condition (RD) we have, for example, the following

Proposition 8.1.2. *Let* $1 < p < \infty$ *and* $x_0 \in \Gamma$. *Let* σ *and* u *be positive increasing functions on* $(0, \infty)$ *with* σ *continuous. Let* $\rho \in A_p(\Gamma)$ *and put* $v(x) = \sigma(|x - x_0|) \rho(x)$, $w(x) = u(|x - x_0|) \rho(x)$. *Suppose that*

$$\sup_{r > 0} \left(\int_{R^n \setminus \Gamma(x_0, r)} \frac{v(x)}{|x - x_0|^{sp}} d\mathcal{H}_s(x) \right) \left(\int_{\Gamma(x_0, r)} w^{1-p'}(y) d\mathcal{H}_s(y) \right)^{p-1} < \infty.$$

Then there is a constant c such that for all $f \in L^p_w(\Gamma)$,

$$\int_\Gamma |K_\Gamma f(x)|^p v(x) d\mathcal{H}_s(x) \leq c \int_\Gamma |f(x)|^p w(x) d\mathcal{H}_s(x).$$

It is clear that other direct consequences of the results of previous sections may be formulated in the setting of s-sets. Note that (see [301], 4.9) since the Cantor set in R^n is an s-set, where

$$s = \frac{\log(3^n - 1)}{\log 3},$$

we can obtain two-weight estimates for singular integrals on a Cantor set in R^n.

Now we provide several examples in which the conditions guaranteeing two-weight estimates for singular integrals defined on SHT are satisfied.

Let $x_0 \in X$ be such that $\mu \{x_0\} = 0$. Then the function

$$w(x) = (\mu (B(x_0, d(x_0, x))))^{\alpha}$$

belongs to $A_p(X)$ if, and only if, $-1 < \alpha < p - 1$ (see [100]). For this weight we have the one-weight inequality

$$\int_X |Kf(x)|^p \, w(x) d\mu \le c \int_X |f(x)|^p \, w(x) d\mu.$$

For simplicity let us consider SHT for which $\mu(B(x, r)) \sim r$. From the results of previous sections we deduce

Proposition 8.1.3. *Let $1 < p < \infty$. Suppose also that $a < \infty$. Then there exists a positive constant $c > 0$ such that the inequalities*

$$\int_X |Kf(x)|^p \, (d(x_0, x))^{p-1} \, d\mu \le c \int_X |f(x)|^p \, (d(x_0, x))^{p-1} \log^p \frac{a}{d(x_0, x)} d\mu$$

and

$$\int_X |Kf(x)|^p \frac{d\mu}{d(x_0, x) \log^{p-1} \frac{a}{d(x_0, x)}} \le c \int_X |f(x)|^p \frac{d\mu}{d(x_0, x)}$$

hold.

Remerk 8.1.1. All results of this section remain valid if we omit the continuity of σ, but require that $\mu B(x_0, r)$ is continuous with respect to r, where x_0 is the same fixed point as above. This follows from the fact that in this case the sequence $\sigma_n(d(x_0, x))$, where σ_n are absolutely continuous functions (see the proof of Theorem 8.1.1.), converges to $\sigma(d(x_0, x))$ a.e. on X. Recall that the continuity of $\mu B(x_0, r)$ with respect to r is equivalent to the condition: $\mu\{x : d(x_0, x) = r\} = 0$ for all $r > 0$.

8.2.   Weak–type estimates

In the present section we establish two-weight weak-type inequalities for singular integrals defined on an SHT- (X, d, μ).

We shall assume that there exists a point $x_0 \in X$ such that

$$a \equiv \sup\{d(x_0, x) : x \in X\} = \infty$$

and

$$\mu(B(x_0, R) \setminus B(x_0, r)) > 0$$

for all r and R with the condition $0 < r < R < \infty$.

Recall that by Lemma 6.5.1 we have that $\mu(X) = \infty$ is equivalent to $a = \infty$.

Assume that K is a Calderón–Zygmund operator (see Section 8.1 for the definition).

We need the following Lemmas.

Lemma 8.2.1. *Let* $\mu\{x_0\} = 0$, *let* w *be a weight function on* X *and let* $\rho \in A_1(X)$. *Suppose that the following conditions are fulfilled:*
(i) there exists a positive increasing function σ on $(0, \infty)$ such that

$$\sigma(2a_1 d(x_0, x)) \rho(x) \le b_1 w(x) \quad a.e.,$$

for some positive constant b_1, and with the constant a_1 from the definition of SHT (see Section 1.1).
(ii)

$$\operatorname*{ess\,sup}_{x \in B(x_0, t)} \frac{1}{w(x)} < \infty$$

for any $t > 0$.

Then $Kf(x)$ exists a.e. on X for any $\phi \in L^1_w(X)$.

Proof. Fix $\alpha > 0$. Let

$$S_\alpha = X \setminus B\left(x_0, \frac{\alpha}{2}\right)$$

and given $\phi \in L^1_w(X)$, put

$$\phi(x) = \phi_1(x) + \phi_2(x)$$

where $\phi_1(x) = \phi(x)\chi_{S_\alpha}(x)$, $\phi_2(x) = \phi(x) - \phi_1(x)$. For ϕ_1 we have

$$\int_X |\phi_1(x)| \rho(x) d\mu =$$

$$= \frac{\sigma(\frac{\alpha}{2})}{\sigma(\frac{\alpha}{2})} \int_{S_\alpha} |\phi(x)| \rho(x) d\mu \le \frac{1}{\sigma(\frac{\alpha}{2})} \int_{S_\alpha} |\phi(x)| \rho(x) \sigma(d(x_0, x)) d\mu \le$$

$$\le \frac{c_1}{\sigma(\alpha/2)} \int_{S_\alpha} |\phi(x)| w(x) d\mu < \infty.$$

Consequently $\phi_1 \in L^1_\rho(X)$ and so due to the weak one-weight inequality $K\phi_1$ belongs to weak $L^1_\rho(X)$ (see [128], [100]). Hence $K\phi_1$ exists a.e..

Now we shall show that $K\phi_2(x)$ converges absolutely on the set $\{x : d(x_0, x) > \alpha a_1\}$. For $d(x_0, y) < \frac{\alpha}{2}$ and $d(x_0, x) > \alpha$ we have $d(x, y) \geq \frac{\alpha}{2a_0}$ and

$$\mu(B(x_0, d(x, y))) \leq c_2 \mu(B(x, d(x, y))).$$

(See the proof of Lemma 8.1.1.) Then

$$|K\phi_2(x)| \leq c_2 \int\limits_{B(x_0, \alpha/2)} \frac{|\phi(y)|}{\mu(B(x, d(x, y)))} d\mu \leq$$

$$\leq c_3 \frac{1}{\mu(B(x, \alpha))} \int\limits_{B(x_0, \alpha/2)} |\phi(y)| d\mu \leq$$

$$\leq \frac{c_3}{\mu(B(x_0, \alpha))} \int\limits_{B(x_0, \alpha)} |\phi(y)| w(y) d\mu \left(\operatorname*{ess\,sup}_{x \in B(x_0, \alpha)} \frac{1}{w(x)} \right) < \infty.$$

In view of the arbitrariness of α we conclude that $K\phi_2(x)$ is convergent and $K\phi(x)$ exists a.e. $\square$

Analogously we can prove

Lemma 8.2.2. *Let w be a weight function on X and let $\rho \in A_1(X)$. Suppose that*

(i) *there exists a positive decreasing function σ on $(0, \infty)$ such that*

$$\sigma\left(\frac{d(x_0, x)}{2a_1}\right) \rho(x) \leq c w(x) \quad a.e.;$$

(ii)

$$\operatorname*{ess\,sup}_{x \in X \setminus B(x_0, t)} \frac{1}{w(x) \mu(B(x_0, d(x_0, x)))} < \infty$$

for any $t > 0$.

Then $K\phi(x)$ exists a.e. for arbitrary $\phi \in L_w^1(X)$.

We shall need the following proposition which follows from Fubini's theorem

Proposition 8.2.1. *Let v and w be non-negative measurable functions respectively on $(0, \infty)$ and X. If*

$$\sup_{0 < t < a} \left(\int\limits_0^t v(\tau) d\tau \right) \operatorname*{ess\,sup}_{\{x : d(x_0, x) > t/2\}} \frac{1}{w(x)} < \infty$$

then we have, for all $f \in L_w^1(X)$,

$$\int\limits_0^a v(t) \left| \int\limits_{\{x : d(x_0, x) > t\}} f(x) d\mu \right| dt \leq c \int\limits_X |f(x)| w(x) d\mu$$

with a constant c independent of f.

Theorem 8.2.1. *Let $\mu\{x_0\} = 0$, σ be a positive continuous increasing function on $(0, \infty)$, let $\rho \in A_1(X)$, w be a weight function on X and put $v(x) = \sigma(d(x_0, x))\rho(x)$. Suppose the following two conditions are satisfied: there exists a positive constant b_1 such that for $\mu-$ almost all $x \in X$,*

$$\rho(x)\sigma(2a_1 d(x_0, x)) \le b_1 w(x); \tag{8.2.1}$$

and

$$\sup_{\tau > t}\left(\frac{1}{\mu(B(x_0, \tau))}\int\limits_{\{x \in X : t < d(x_0, x) < \tau\}} v(x)d\mu\right) \times$$

$$\times \operatorname*{ess\,sup}_{d(x_0, x) \le t}\frac{1}{w(x)} < \infty. \tag{8.2.2}$$

Then there exists $c > 0$ such that for any $\lambda > 0$ and $f \in L_w^1(X)$ we have

$$\int\limits_{\{x \in X : |Kf(x)| > \lambda\}} v(x)d\mu \le \frac{c}{\lambda}\int\limits_X |f(x)|w(x)d\mu. \tag{8.2.3}$$

Proof. Put $\{x \in X : |Kf(x)| > \lambda\} = H_\lambda$. We assume that $\sigma(t) = \sigma(0+) + \int_0^t \psi(\tau)d\tau$, $\psi \ge 0$. Then

$$\int\limits_{H_\lambda} v(x)d\mu = \int\limits_{H_\lambda} \sigma(0+)\rho(x)d\mu + \int\limits_{H_\lambda} \rho(x)\left(\int\limits_0^{d(x_0, x)} \psi(t)dt\right)d\mu = I_1 + I_2.$$

Using a weak-type one-weight inequality for K (see [128], [100] Section 7.3) and condition (8.2.1) we derive

$$I_1 \le \frac{c_1\sigma(0+)}{\lambda}\int\limits_X |f(x)|\rho(x)d\mu \le$$

$$\le \frac{c_1}{\lambda}\int\limits_X |f(x)|\rho(x)\sigma(2a_1 d(x_0, x))d\mu \le \frac{c_2}{\lambda}\int\limits_X |f(x)|w(x)d\mu. \tag{8.2.4}$$

Now we estimate I_2. Let

$$S \equiv \left\{x \in X : \left|\int\limits_{X \setminus B(x_0, t/2a_1)} k(x, y)f(y)d\mu\right| > \frac{\lambda}{2}\right\},$$

$$S_1 \equiv \left\{x \in X : \left|\int\limits_{B(x_0, t/2a_1)} k(x, y)f(y)d\mu\right| > \frac{\lambda}{2}\right\}.$$

Then

$$I_2 = \int_0^\infty \psi(t) \left(\int_{\{x:d(x_0,x)>t\}} \rho(x)\chi_{H_\lambda} d\mu \right) dt \le$$

$$\le \int_0^\infty \psi(t) \left(\int_{\{x:d(x_0,x)>t\}} \rho(x)\chi_S d\mu \right) dt +$$

$$+ \int_0^\infty \psi(t) \left(\int_{\{x:d(x_0,x)>t\}} \rho(x)\chi_{S_1} d\mu \right) dt \equiv$$

$$\equiv I_{21} + I_{22}.$$

Since $\rho \in A_1(X)$ we have

$$I_{21} \le \int_0^\infty \psi(t) \left(\int_S \rho(x) d\mu \right) dt \le$$

$$\le \frac{c_3}{\lambda} \int_0^\infty \psi(t) \left(\int_{X/B(x_0,t/2a_1)} |f(x)|\, \rho(x) d\mu \right) dt = \tag{8.2.5}$$

$$= \frac{c_3}{\lambda} \int_X \rho(x)\, |f(x)| \left(\int_0^{2a_1 d(x_0,x)} \psi(t) dt \right) d\mu \le$$

$$\le \frac{c_4}{\lambda} \int_X |f(x)|\, w(x) d\mu.$$

Further we note that for $d(x_0, x) > t$ and $d(x_0, x) \le \frac{t}{2a_1}$ the inequality

$$\mu(B(x_0, d(x_0, x))) \le c_5 \mu(B(x, d(x, y)))$$

holds (see the proof of Theorem 8.1.1). By virtue of the last inequality we obtain the estimates

$$I_{22} \le \int_0^\infty \psi(t) \left(\int_{\{d(x_0,x)>t\}} \rho(x)\chi_{\{x \in X : c_5 Pf(x) > \lambda\}} d\mu \right) dt =$$

$$= \int_{\{x \in X : c_5 Pf(x) > \lambda\}} \rho(x) \left(\int_0^{d(x_0,x)} \psi(t) dt \right) d\mu \le$$

$$\le \int_{\{x \in X : c_5 Pf(x) > \lambda\}} v(x) d\mu,$$

where

$$Pf(x) = \frac{1}{\mu(B(x_0, d(x_0, x)))} \int_{B(x_0, d(x_0, x))} f(y) dy.$$

By Theorem 1.2.9 and condition (8.2.2) we obtain

$$I_{22} \leq \frac{c_6}{\lambda} \int_X |f(x)| \, w(x) d\mu. \tag{8.2.6}$$

Finally (8.2.4), (8.2.5) and (8.2.6) lead to (8.2.3). $\square$

Theorem 8.2.2. *Let σ be a positive continuous decreasing function on $(0, \infty)$, let $\rho \in A_1(X)$ and let w be a weight function on X. Suppose the following two conditions hold:*
(i) *there exists a positive constant b such that*

$$\rho(x) \sigma \left(\frac{d(x_0, x)}{2a_1} \right) \leq b w(x)$$

a.e. on X, (ii)

$$\sup_{t>0} \left(\int_{B(x_0, t)} v(x) d\mu \right) \sup_{x \in X \backslash B(x_0, t/2)} \frac{1}{w(x) \mu(B(x_0, d(x_0, x)))} < \infty.$$

Then the inequality (8.2.3) is true.

The proof of this theorem is based on Proposition 8.2.1 and some aspects of the proof of Theorem 8.1.2.

Corollary 8.2.1. *Let $\mu\{x_0\} = 0$. Let σ and u be positive increasing functions on $(0, \infty)$ with σ continuous, let $\rho \in A_1(X)$ and put $v(x) = \sigma(d(x_0, x)) \rho(x)$, $w(x) = u(d(x_0, x)) \rho(x)$. Suppose the following conditions are satisfied:*
(i) *there exists a positive constant b_1 such that*

$$\sigma(2a_1 t) \leq b_1 u(t) \tag{8.2.7}$$

for all $t > 0$;
(ii)

$$\sup_{\tau > t} \left(\frac{1}{\mu(B(x_0, \tau))} \int_{\{x \in X : t < d(x_0, x) < \tau\}} v(x) d\mu \right) \times$$
$$\times \operatorname*{ess\,sup}_{\{x : d(x_0, x) \leq t\}} \frac{1}{w(x)} < \infty. \tag{8.2.8}$$

Then the inequality (8.2.3) holds.

Corollary 8.2.2.. *Let* μ, σ, u, ρ, v *and* w *be as in Corollary 8.2.1 except that* σ *and* u *are decreasing rather than increasing. Suppose also that*
(i) *there exists a positive constant* b *such that*

$$\sigma\left(\frac{t}{2a_1}\right) \le bu(t)$$

for all $t > 0$;
(ii)

$$\sup_{t>0}\left(\int_{B(x_0,t)} v(x)d\mu\right) \operatorname*{ess\,sup}_{x\in X\setminus B(x_0,t/2)} \frac{1}{w(x)\mu\left(B\left(x_0,d\left(x_0,x\right)\right)\right)} < \infty.$$

Then (8.2.3) *holds.*

Theorem 8.2.3 *Let* $\mu\{x_0\} = 0$, *let* σ *and* u *be positive increasing functions on* $(0,\infty)$ *with* σ *continuous and suppose that* $\rho \in A_1(X)$. *If further* $v(x) = \sigma\left(d\left(x_0,x\right)\right)\rho(x)$, $w(x) = u\left(d\left(x_0,x\right)\right)\rho(x)$ *and* (8.2.8) *is satisfied, then* (8.2.3) *holds.*

Proof. By virtue of Corollary 8.2.1 it is sufficient to prove the implication (8.2.8) $\implies$ (8.2.7).

Let $\beta \ge 1$, $\eta \ge \eta_1 > 1$, where η_1 is the constant from the reverse doubling condition which is satisfied at point x_0 (see Proposition 6.5.1). As $\rho \in A_1(X)$ we know that $\rho \in A_p(X)$ for any $p > 1$. Now using Hölder's inequality, Lemma 6.6.1 and the argument of the proof of Theorem 8.1.3 we derive the chain of inequalities

$$\frac{\sigma(\beta t)}{u(t)} \le c_1 \frac{\sigma(\beta t)}{u(t)\left(\mu\left(B\left(x_0,\eta\beta t\right)\right)\right)^p} \int_{B(x_0,\eta\beta t)\setminus B(x_0,\beta t)} \rho(x)d\mu \times$$

$$\times\left(\int_{B(x_0,\eta\beta t)\setminus B(x_0,\beta t)} \rho(x)^{1-p'}d\mu\right)^{p-1} \le$$

$$\le c_2 \frac{\sigma(\beta t)}{u(t)(\mu(B(x_0,\eta\beta t)))^p} \times$$

$$\times\int_{B(x_0,\eta\beta t)\setminus B(x_0,\beta t)} \rho(x)d\mu\left(\int_{B(x_0,\frac{t}{2})} \rho(x)^{1-p'}d\mu\right)^{p-1} \le$$

$$\le c_2 \frac{\sigma(\beta t)}{\left(\mu\left(B\left(x_0,\eta\beta t\right)\right)\right)^p} \int_{B(x_0,\eta\beta t)\setminus B(x_0,\beta t)} \rho(x)d\mu \times$$

$$times \left(\int\limits_{B(x_0,\frac{t}{2})} w(x)^{1-p'} d\mu \right)^{p-1} \leq$$

$$\leq c_2 \frac{\sigma(\beta t)}{(\mu\,(B\,(x_0,\eta\beta t)))^p} \left(\int\limits_{B(x_0,\eta\beta t)\backslash B(x_0,\beta t)} \rho(x)d\mu \right) \times$$

$$\times \operatorname*{ess\,sup}_{x\in B(x_0,\frac{t}{2})} \frac{1}{w(x)} (\mu\,(B\,(x_0,\eta\beta t)))^{p-1} \leq$$

$$\leq c_3 \frac{1}{\mu\,(B\,(x_0,\eta\beta t))} \left(\int\limits_{\left\{x:\frac{t}{2}<d(x_0,x)<\eta\beta t\right\}} v(x)d\mu \right) \operatorname*{ess\,sup}_{x\in B(x_0,\frac{t}{2})} \frac{1}{w(x)} \leq c_4.$$

The theorem is proved. □

Theorem 8.2.4. *Let σ and u be positive decreasing functions on $(0,\infty)$ with σ continuous, let $\rho \in A_1(X)$ and put $v(x) = \sigma\,(d\,(x_0,x))\,\rho(x)$ and put $w(x) = u\,(d\,(x_0,x))\,\rho(x)$. Suppose that*

$$\sup_{t>0} \left(\int\limits_{B(x_0,t)} v(x)d\mu \right) \operatorname*{ess\,sup}_{x\in X\backslash B(x_0,t/2)} \frac{1}{w(x)\mu\,(B\,(x_0,d\,(x_0,x)))} < \infty.$$

Then (8.2.3) holds.

Now let us consider the following operator

$$Lf(x) \equiv (p.v) \int\limits_{B(x_0,2a_1d(x_0,x))\backslash B(x_0,d(x_0,x)/2a_1)} k(x,y)f(y)d\mu(y),$$

where k ia a Calderón–Zygmund kernel, x_0 is a fixed point of X and a_1 is from the Definition of SHT (see Section 1.1). In addition, we shall assume that

$$L_\alpha f(x) = a(x,\alpha)Lf(x),$$

where $a(x,\alpha) \equiv \mu B(x_0,d(x_0,x))^\alpha$.

We have the following theorem (for the Hilbert transform see [8])

Theorem 8.2.5. *Let $\mu\{x_0\} = 0$, $w \in A_1(X)$ and let $\alpha \in R$. Then there exists a constant $c_\alpha > 0$ such that for all $\lambda > 0$ and for all $f \in L_w^1(X)$ the following inequality holds*

$$\int\limits_{S_\lambda} w(x)a(x,\alpha)^{-1}d\mu \leq \frac{c}{\lambda}\|f\|_{L_w^1(X)},$$

where $S_\lambda \equiv \{x \in X : |L_\alpha f(x)| > \lambda\}$.

Proof. Let $2^n < d(x_0, x) \leq 2^{n+1}$, where $n \in Z$. Then

$$
Lf(x) = \int\limits_{B(x_0, 2^{n+2}a_1) \setminus B(x_0, 2^{n-1}/a_1)} f(y)k(x, y)d\mu -
$$

$$
- \int\limits_{B(x_0, d(x_0,x)/2a_1) \setminus B(x_0, 2^{n-1}/a_1)} f(y)k(x, y)d\mu -
$$

$$
- \int\limits_{B(x_0, 2^{n+2}a_1) \setminus B(x_0, 2a_1 d(x_0,x))} f(y)k(x, y)d\mu \equiv I_{1,n} - I_{2,n} - I_{3,n}.
$$

We have

$$
\int\limits_{S_\lambda} w(x)a(x, \alpha)^{-1}d\mu =
$$

$$
= \sum_{n \in Z} \int\limits_{S_\lambda} \chi_{B(x_0, 2^{n+1}) \setminus B(x_0, 2^n)}(x)w(x)(a(x, \alpha))^{-1}d\mu \equiv \sum_n I_n.
$$

Further

$$
I_n \leq \int\limits_{\{x:a(x,\alpha)|I_{1,n}(x)|>\lambda/3\}} w(x)\chi_{B(x_0, 2^{n+1}) \setminus B(x_0, 2^n)}(x)a(x, \alpha)^{-1}d\mu +
$$

$$
+ \int\limits_{\{x:a(x,\alpha)|I_{2,n}(x)|>\lambda/3\}} w(x)\chi_{B(x_0, 2^{n+1}) \setminus B(x_0, 2^n)}(x)a(x, \alpha)^{-1}d\mu +
$$

$$
+ \int\limits_{\{x:a(x,\alpha)|I_{3,n}(x)|>\lambda/3\}} w(x)\chi_{B(x_0, 2^{n+1}) \setminus B(x_0, 2^n)}(x)a(x, \alpha)^{-1}d\mu \equiv
$$

$$
\equiv J_{1,n} + J_{2,n} + J_{3,n}.
$$

Now let $al \geq 0$. Then using the one-weight weak type inequality for K we obtain

$$
J_{1,n} \leq \mu B(x_0, 2^n)^{-al} \times
$$

$$
\times \int\limits_{\{x:\mu(x_0,2^{n+1})^\alpha |I_{1,n}(x)|>\lambda/3\}} w(x)\chi_{B(x_0, 2^{n+1}) \setminus B(x_0, 2^n)}(x)a(x, \alpha)^{-1}d\mu \leq
$$

$$
\leq \frac{c_1}{\lambda} \int\limits_{B(x_0, 2^{n+2}a_1) \setminus B(x_0, 2^{n-1}/a_1)} |f(x)|w(x)d\mu.
$$

The case $\alpha < 0$ is proved analogously.

If $2^n \leq d(x_0,x) < 2^{n+1}$ and $d(x_0,y) \leq \frac{d(x_0,x)}{2a_1}$, then $\frac{d(x_0,x)}{2a_0 a_1} \leq d(x,y)$. Moreover, $\mu B(x_0, d(x_0,x)) \leq b_1 \mu B(x, d(x_0,x))$ and for $J_{2,n}$ we have

$$J_{2,n} \leq \frac{c_3}{\lambda} \int\limits_{B(x_0,2^{n+1})\setminus B(x_0,2^n)} w(x) \times$$

$$\times \Big| \int\limits_{B\left(x_0, d(x_0,x)/(2a_1)\right)\setminus B(x_0, 2^{n-1}/a_1)} f(y)k(x,y)d\mu \Big| d\mu \leq$$

$$\leq \frac{c_4}{\lambda} \int\limits_{B(x_0,2^{n+1})\setminus B(x_0,2^n)} \frac{w(x)}{\mu B(x_0, d(x_0,x))} \times$$

$$\times \Big(\int\limits_{B\left(x_0, d(x_0,x)/(2a_1)\right)\setminus B(x_0, 2^{n-1}/a_1)} |f(y)|d\mu(y) \Big) d\mu(x) \leq$$

$$\leq \frac{c_4}{\lambda} \int\limits_{B(x_0,2^{n+1})\setminus B(x_0,2^n)} \frac{w(x)}{\mu B(x_0, d(x_0,x))} d\mu \times$$

$$\times \int\limits_{B(x_0,2^n/a_1)\setminus B(x_0,2^{n-1}/a_1)} |f(y)|d\mu \leq$$

$$\leq \frac{c_5}{\lambda\mu(B(x_0,2^{n+1})} \Big(\int\limits_{B(x_0,2^{n+1})} w(x)d\mu(x) \Big) \Big(\operatorname*{ess\,sup}_{x\in B(x_0,2^{n+2})} \frac{1}{w(x)} \Big) \times$$

$$\times \int\limits_{B(x_0,2^n/a_1)\setminus B(x_0,2^{n-1}/a_1)} |f(y)|w(y)d\mu(y) \leq$$

$$\leq \frac{c_6}{\lambda} \int\limits_{B(x_0,2^n/a_1)\setminus B(x_0,2^{n-1}/a_1)} |f(y)|w(y)d\mu.$$

Now we are going to estimate $J_{3,n}$. For x and y with the conditions $x \in B(x_0, 2^{n+1}) \setminus B(x_0, 2^n)$, $d(x_0,y) > 2a_1 d(x_0,x)$ we have

$$\mu B(x_0, d(x_0,y)) \leq b_2 \mu B(x, d(x,y))$$

and consequently

$$J_{3,n} \leq \frac{c_7}{\lambda} \int\limits_{B(x_0,2^{n+1})\setminus B(x_0,2^n)} w(x)d\mu(x) \times$$

$$\times \Big(\int\limits_{B(x_0,2^{n+2}a_1)\setminus B(x_0,2^{n-1}a_1)} |f(y)|(\mu B(x_0, d(x_0,y)))^{-1}d\mu(y) \Big) \leq$$

$$\leq \frac{c_8}{\lambda}\mu B(x_0, 2^{n+3}a_1)^{-1}\left(\int_{B(x_0, 2^{n+3}a_1)} w(x)d\mu\right) \operatorname*{ess\,sup}_{x \in B(x_0, 2^{n+3}a_1)} \frac{1}{w(x)} \times$$

$$\times \left(\int_{B(x_0, 2^{n+2}a_1)\backslash B(x_0, 2^{n-1}a_1)} |f(y)|w(y)d\mu(y)\right) \leq$$

$$\leq \frac{c_9}{\lambda} \int_{B(x_0, 2^{n+2}a_1)\backslash B(x_0, 2^{n-1}a_1)} |f(y)|w(y)d\mu(y).$$

$\square$

Theorem 8.2.6. *Let* $\mu\{x_0\} = 0$, $\rho \in A_1(X)$ *and* $\alpha \leq 0$. *Then there exists a constant* $c_d > 0$ *such that for all* $\lambda > 0$ *and for all* $f \in L^1_w(X)$ *the following inequality holds*

$$\int_{E_\lambda} w(x)a(x,\alpha)^{-1}d\mu \leq \frac{c}{\lambda}\|f\|_{L^1_w(X)},$$

where $E_\lambda \equiv \{x \in X : a(x,\alpha)|Kf(x)| > \lambda\}$.

Proof. It is obvious that

$$|Kf(x)| \leq |Lf(x)| + \int_{B(x_0, d(x_0,x)/(2a_1))} \frac{|f(y)|}{\mu B(x, d(x,y))}d\mu +$$

$$+ \int_{X\backslash B(x_0, 2a_1 d(x_0,x))} \frac{|f(y)|}{\mu B(x, d(x,y))}d\mu =$$

$$= |Lf(x)| + I_1f(x) + I_2f(x).$$

First we estimate $I_1f(x)$. When $d(x_0,y) \leq \frac{d(x_0,x)}{2a_1}$ then

$$\mu B(x_0, d(x_0,x)) \leq b_1\mu B(x, d(x,y)).$$

Consequently

$$I_1f(x) \leq \frac{c_1}{\mu B(x_0, d(x_0,x))} \int_{B(x_0, d(x_0,x))} |f(x)|d\mu.$$

Now we estimate $I_2f(x)$. Note that for y with the condition $d(x_0,y) \geq 2a_1 d(x_0,x)$ we have

$$d(x_0,x) \leq a_1 d(x_0,x) + a_1 d(x,y) \leq$$

$$\leq a_1\frac{d(x_0,y)}{2a_1} + a_1 d(x,y) = d(x_0,y)/2 + a_1 d(x,y).$$

Hence $\frac{d(x_0,y)}{2a_1} \le d(x,y)$. Therefore

$$\mu B(x_0, d(x_0, y)) \le b_1 \mu B(x, d(x, y)).$$

So, we obtain

$$I_2 f(x) \le c_2 \int_{B(x_0, d(x_0, x))} |f(y)| (\mu B(x_0, d(x_0, y)))^{-1} d\mu(y).$$

Using the fact $w \in A_1(X)$ and Theorem 8.2.5, we have

$$\int_{S_\lambda} a(x, d)^{-1} w(x) d\mu \le \frac{c_3}{\lambda} \int_X |f(y)| w(x) d\mu.$$

Now let us establish the estimate

$$\int_{\{x : a(x,\alpha)\mu B(x_0, d(x_0,x))^{-1} \int_{B(x_0, d(x_0,x))} |f(y)| d\mu > \lambda\}} a(x, \alpha)^{-1} w(x) d\mu \le$$

$$\le \frac{c_4}{\lambda} \int_X |f(y)| w(y) d\mu.$$

This inequality follows from the obvious estimate ($\alpha \le 0$)

$$\frac{1}{\mu B(x_0, b)^{1-\alpha}} \left(\int_{a < d(x_0,x) < b} a(x, d)^{-1} d\mu(x) \right) \le$$

$$\le \frac{1}{\mu B(x_0, b)} \left(\int_{x \in B(x_0, b)} w(y) d\mu \right) \operatorname*{ess\,sup}_{B(x_0, b)} \frac{1}{w(x)} < \infty$$

and from Theorem 1.2.9.
In addition

$$\int_{\{x : a(x,\alpha) \int_{X \setminus B(x_0, d(x_0,x))} \frac{|f(y)|}{\mu B(x_0, d(x_0,y))} d\mu(y) > \lambda\}} a(x, \alpha)^{-1} w(x) d\mu \le$$

$$\le \frac{c_5}{\lambda} \int_X w(x) \left(\int_{X \setminus B(x_0, d(x_0,x))} \frac{|f(y)|}{\mu B(x_0, d(x_0,y))} d\mu(x) \right) d\mu(y) \le$$

$$\le \frac{c_5}{\lambda} \int_X |f(y)| (\mu B(x_0, d(x_0, y)))^{-1} \left(\int_{B(x_0, d(x_0,y))} w(x) d\mu(x) \right) d\mu(y) \le$$

$$\leq \frac{c_6}{\lambda} \int\limits_X |f(y)|(\mu B(x_0, 2(x_0, y)))^{-1} \bigg(\int\limits_{B(x_0, 2d(x_0,y))} w(x)d\mu(x) \bigg) d\mu(y) \leq$$

$$\leq \frac{c_7}{\lambda} \int\limits_X |f(y)|(\operatorname*{ess\,inf}_{B(x_0, 2d(x_0,y))} w(x))d\mu \leq$$

$$\leq \frac{c_7}{\lambda} \int\limits_X |f(y)|w(y)d\mu.$$

The theorem is proved. $\square$

Theorem 8.2.7. *Let* $1 \leq p < \infty$, $d \leq 0$ *and let* σ *be an increasing positive continuous fuction on* $(0, \infty)$. *Assume that* w *is a weight on* X *and* $\rho \in A_p(X)$. *We put* $v(x) = \sigma(d(x_0, x))\rho(x)$. *If the following two conditions are satisfied:*
1) there exist a positive constant b such that

$$\sigma(2a_1 d(x_0, x))\rho(x) \leq cw(x);$$

2)

$$\sup_{\tau > t} \frac{1}{\mu B(x_0, \tau)} \bigg(\int\limits_{B(x_0,\tau)\backslash B(x_0,t)} v(x)d\mu(x) \bigg)^{1/p} \times$$

$$\times \bigg(\int\limits_{d(x_0,x)\leq t} w(x)d\mu(x) \bigg)^{1/p'} < \infty$$

for $p > 1$ and

$$\sup \tau > t \frac{1}{\mu B(x_0, \tau)} \bigg(\int\limits_{B(x_0,\tau)\backslash B(x_0,t)} v(x)d\mu(x) \bigg) \operatorname*{ess\,sup}_{d(x_0,x)\leq t} \frac{1}{w(x)} < \infty$$

for $p = 1$,
then there exists a positive constant c_α such that for all $\lambda > 0$ and $f \in L_w^p(X)$,

$$\int\limits_{\{x:\mu B(x_0,d(x_0,x))^\alpha |Kf(x)|>\lambda\}} \mu B(x_0, d(x_0, x))^{-\alpha p} v(x)d\mu \leq$$

$$\leq \frac{c}{\lambda^p} \int\limits_X |f(x)|^p w(x)d\mu.$$

Proof. Without loss of generality we can assume that

$$\sigma(t) = \sigma(0+) + \int\limits_0^t \phi(\tau)d\tau, \quad \phi \geq 0.$$

Let us put

$$H_\lambda \equiv \{x : \mu B(x_0, d(x_0, x))^\alpha |Kf(x)| > \lambda\},$$

where $a(x, \alpha) \equiv \mu B(x_0, d(x_0, x))^\alpha$.
Then we have (recall that $a(x, \alpha) \equiv \mu B(x_0, d(x_0, x))^\alpha$).

$$\int_{H_\lambda} a(x, \alpha)^{-p} v(x) d\mu = \sigma(0+) \int_{H_\lambda} a(x, \alpha)^{-p} \rho(x) d\mu +$$

$$+ \int_{H_\lambda} \Big(\int_0^{d(x_0,x)} \phi(t) dt \Big) a(x, \alpha)^{-p} \rho(x) d\mu \equiv I_1 + I_2.$$

If $\sigma(0+) = 0$, then $I_1 = 0$ otherwise using a one-weight strong (or weak) (p, p) type inequality for K if $p > 1$ and Theorem 8.2.6 if $p = 1$, we obtain that

$$I_1 \leq \frac{c_1}{\lambda^p} \sigma(0+) \|f\|^p_{L^p_\rho(X)} \leq \frac{c_2}{\lambda^p} \|f\|^p_{L^p_w(X)}.$$

Now we estimate I_2.

$$I_2 = \int_0^\infty \phi(t) \Big(\int_{\{x:d(x_0,x) > t, a(x,\alpha)|Kf(x)| > \lambda\}} a(x, \alpha)^{-p} \rho(x) d\mu(x) \Big) dt \leq$$

$$\leq \int_0^\infty \phi(t) \Big(\int_{\{x:d(x_0,x) > t, a(x,\alpha)|Kf_{1,t}(x)| > \lambda\}} a(x, \alpha)^{-p} \rho(x) d\mu \Big) dt +$$

$$+ \int_0^\infty \phi(t) \Big(\int_{\{x:d(x_0,x) > t, a(x,\alpha)|Kf_{2,t}(x)| > \lambda\}} a(x, \alpha)^{-p} \rho(x) d\mu \Big) dt \equiv$$

$$\equiv I_{21} + I_{22},$$

where $f_{1,t}(x) = f(x) \cdot \chi_{d(x_0,y) > t/(2a_1)}(x)$ and $f_{2,t} = f(x) - f_{1,t}(x)$.
Using again a strong (or weak) (p, p) type inequality if $p > 1$ and Theorem 8.2.6 for $p = 1$, we find that

$$I_{21} \leq \frac{c_3}{\lambda^p} \int_0^\infty \phi(t) \Big(\int_{\{x:d(x_0,x) > t/(2a_1)\}} |f(x)|^p \rho(x) d\mu \Big) dt =$$

$$= \frac{c_3}{\lambda^p} \int_X |f(x)|^p \rho(x) \Big(\int_0^{2a_1 d(x_0,x)} \phi(t) dt \Big) d\mu \leq$$

$$\leq \frac{c_4}{\lambda^p} \int_X |f(x)|^p w(x) d\mu.$$

Let $d(x_0, x) > t$ and $y \in d(x_0, t/(2a_1))$. Then

$$\mu B(x_0, d(x_0, x)) \leq b_1 \mu B(x, d(x, y)).$$

Therefore

$$I_{22} \leq \int\limits_0^\infty \phi(t) \times$$

$$\times \left(\int\limits_{\substack{\{x: d(x_0,x)>t, \mu B(x_0,d(x_0,x))^{\alpha-1} \\ \int\limits_{B(x_0,d(x_0,x))} f(y)d\mu > \frac{\lambda}{c_5}\}}} \frac{\rho(x)}{a(x,\alpha)^p} d\mu \right) dt =$$

$$= \int\limits_{\substack{\{x: \mu B(x_0,d(x_0,x))^{\alpha-1} \\ \int\limits_{B(x_0,d(x_0,x))} f(y)d\mu > \frac{\lambda}{c_5}\}}} \frac{v(x)}{\mu(B(x_0, d(x_0, x))^{\alpha p}} d\mu.$$

As $d \leq 0$, we have the estimate

$$\mu B(x_0, b)^{(\alpha-1)p} \int\limits_{B(x_0,b)\backslash B(x_0,a)} v(x)a(x,\alpha)^{-p} d\mu \leq$$

$$\leq \mu B(x_0, b)^{-p} \int\limits_{B(x_0,b)\backslash B(x_0,a)} v(x)d\mu.$$

Hence using Theorem 1.2.9 we finally obtain

$$I_{22} \leq \frac{c_6}{\lambda^p} \int\limits_X |f(x)|^p w(x)d\mu.$$

$\square$

Theorem 8.2.8. *Let* $1 \leq p < \infty$, $\alpha \leq 0$ *and let* σ *and* u *be positive increasing fuctions on* $(0, \infty)$. *Suppose that* σ *is continuous and that* $\rho \in A_p(X)$. *We put* $v(x) = \sigma(d(x_0, x))\rho(x)$, $w(x) = u(d(x_0, x))\rho(x)$. *Assume that*

$$\sup_{\tau>t} \frac{1}{\mu B(x_0, \tau)} \left(\int\limits_{B(x_0,\tau)\backslash B(x_0,t)} v(x)d\mu(x) \right)^{1/p} \left(\int\limits_{d(x_0,x)\leq t} w(x)d\mu(x) \right)^{1/p'} < \infty$$

for $p > 1$ *and*

$$\sup_{\tau>t} \frac{1}{\mu B(x_0, \tau)} \left(\int\limits_{B(x_0,\tau)\backslash B(x_0,t)} v(x))d\mu(x) \right) \operatorname*{ess\,sup}_{d(x_0,x)\leq t} \frac{1}{w(x)} < \infty$$

for $p = 1$.

Then there exists a positive constant c_α such that for all $\lambda > 0$ and $f \in L^p_w(X)$ the following inequality holds

$$\int_{\{x:\mu B(x_0,d(x_0,x))^\alpha |Kf(x)|>\lambda\}} \mu B(x_0, d(x_0, x))^{-\alpha p} v(x) d\mu \le$$

$$\le \frac{c}{\lambda^p} \int_X |f(x)|^p w(x) d\mu.$$

This theorem is a consequence of Theorem 8.2.7. We only need to use the fact that the reverse doubling condition for μ at point x_0 and Lemma 6.6.1 implies the inequality

$$\sigma(2a_1 t) \le bu(t).$$

see also the proof of Theorem 8.1.3.

As the following theorem can be derived in the same way as Theorem 8.2.7, we omit the proof.

Theorem 8.2.9. *Let $\alpha \le 0$ and let σ and u be positive decreasing fuctions on $(0, \infty)$, with the condition that σ is continuous. Assume that $\rho \in A_1(X)$. We put $v(x) = \sigma(d(x_0, x))\rho(x)$, $w(x) = u(d(x_0, x))\rho(x)$. If*

$$\sup_{t>0} \left(\int_{B(x_0,2t)} v(x) d\mu(x) \right) \operatorname*{ess\,sup}_{d(x_0,x) \ge t} \frac{1}{w(x)\mu B(x_0, d(x_0, x))} < \infty,$$

then there exists a positive constant c_α such that for all $\lambda > 0$ and $f \in L^1_w(X)$,

$$\int_{\{x:a(x,\alpha)|Kf(x)|>\lambda\}} a(x, \alpha)^{-1} v(x) d\mu \le \frac{c}{\lambda} \int_X |f(x)| w(x) d\mu$$

$(a(x, \alpha) \equiv \mu B(x_0, d(x_0, x))^\alpha)$.

For the rest of this section we shall assume that (X, d, μ) satisfies the condition

$$\mu B(x_0, r) \approx r^\delta$$

for some positive δ.

Theorem 8.2.10. *Let $\rho \in A_1(X)$, $\alpha \ge 1$. Then there exists a constant $c_\alpha > 0$ such that for all $\lambda > 0$ and for all $f \in L^1_w(X)$ the following inequality holds:*

$$\int_{E_\lambda} w(x) d(x_0, x)^{-\alpha\delta} d\mu \le \frac{c}{\lambda} \|f\|_{L^1_w(X)},$$

where $E_\lambda \equiv \{x \in X : \mu d(x_0, x)^{d\delta}|Kf(x)| > \lambda\}$.

Proof. As in the proof of Theorem 8.2.6 we have

$$|Kf(x)| \le |Lf(x)| + I_1 f(x) + I_2 f(x),$$

where $I_1 f(x)$ and $I_2 f(x)$ are the same expressions as in that proof. By theorem 8.2.5 we have

$$\int_{\{x:d(x_0,x)^{\delta\alpha}|Lf(x)|>\lambda\}} d(x_0, x)^{-\delta\alpha} w(x) d\mu \le \frac{c_1}{\lambda} \int_X |f(x)| w(x) d\mu.$$

We claim that

$$\int_{\{x:d(x_0,x)^{\delta(d-1)} \int_{B(x_0,d(x_0,x))} |f(y)|d\mu > \lambda\}} d(x_0, x)^{-\delta\alpha} w(x) d\mu \le$$

$$\le \frac{c_2}{\lambda} \int_X |f(x)| w(x) d\mu.$$

As $\delta(\alpha - 1) > 0$, we have

$$\frac{1}{t^{\delta(1-\alpha)}} \left(\int_{d(x_0,x)>t} v(x) d(x_0, x)^{-\delta\alpha} d\mu \right) =$$

$$= \frac{c_3}{t^{\delta(\alpha-1)}} \left(\int_{d(x_0,x)>t} v(x) \left(\int_{d(x_0,x)}^{\infty} \tau^{-\alpha\delta-1} d\tau \right) d\mu \right) =$$

$$= \frac{c_3}{t^{\delta(1-\alpha)}} \left(\int_t^{\infty} \tau^{-\alpha\delta-1} \left(\int_{t<d(x_0,x)<\tau} v(x) d\mu \right) d\tau \right) \le$$

$$\le \frac{c_3}{t^{\delta(1-\alpha)}} \left(\int_t^{\infty} \tau^{-\alpha\delta-1+\alpha} \left(\frac{1}{\tau^\delta} \int_{B(x_0,\tau)} v(x) d\mu \right) d\tau \right) \le$$

$$\le \frac{c_4}{t^{\delta(1-d)}} \left(\int_t^{\infty} \tau^{-\alpha\delta-1+\delta} \left(\operatorname*{ess\,inf}_{B(x_0,\tau)} w(x) \right) d\tau \right) \le$$

$$\le \frac{c_6}{t^{\delta(1-\alpha)}} \left(\operatorname*{ess\,inf}_{x \in B(x_0,t)} w(x) \right) \left(\int_t^{\infty} \tau^{-\alpha\delta-1+\delta} d\tau \right) =$$

$$= c_7 \operatorname*{ess\,inf}_{x \in B(x_0,t)} w(x).$$

for all $t > 0$. Therefore by Theorem 1.2.14 we obtain the desired inequality.

Further

$$\int\limits_{\substack{\{x:c_8 d(x_0,x)^{-\delta\alpha} \int\limits_{d(x_0,y)>d(x_0,x)} \frac{|f(y)|}{d(x_0,y)^\delta}d\mu>\lambda\}}} w(x)d(x_0,x)^{-\delta d}d\mu \le$$

$$\le \frac{c_5}{\lambda}\int\limits_X w(x)\bigg(\int\limits_{d(x_0,y)>d(x_0,x)} \frac{|f(y)|}{d(x_0,y)^\delta}d\mu(y)\bigg)d\mu(x) =$$

$$= \frac{c_5}{\lambda}\int\limits_X \frac{|f(y)|}{d(x_0,x)^\delta}\bigg(\int\limits_{d(x_0,x)<d(x_0,y)} w(x)d\mu(x)\bigg)d\mu(y) =$$

$$= \frac{c_6}{\lambda}\int\limits_X \frac{|f(y)|}{2d(x_0,x)^\delta}\bigg(\int\limits_{d(x_0,x)<2d(x_0,y)} w(x)d\mu(x)\bigg)d\mu(y) \le$$

$$\le \frac{c_7}{\lambda}\int\limits_X |f(y)|\bigg(\operatorname*{ess\,inf}_{x\in B(x_0,2d(x_0,y))} w(x)\bigg)d\mu(y) \le$$

$$\le \frac{c_7}{\lambda}\int\limits_X |f(y)|w(y)d\mu.$$

$\square$

Theorem 8.2.11. *Let* $1 \le p < \infty$, $\alpha \ge 1$ *and let* σ *and* u *be increasing positive functions on* $(0,\infty)$. *Suppose that* σ *is continuous and* $\rho \in A_p(X)$. *We put* $v(x) = \sigma(d(x_0,x))\rho(x)$, $w(x) = u(d_(x_0,x))\rho(x)$. *If*

$$\sup_{\tau>t} \frac{1}{\tau^\delta}\bigg(\int\limits_{B(x_0,\tau)\setminus B(x_0,t)} v(x)d\mu(x)\bigg)^{1/p}\bigg(\int\limits_{d(x_0,x)\le t} w(x)d\mu(x)\bigg)^{1/p'} < \infty$$

for $p > 1$ *and*

$$\sup_{\tau>t} \frac{1}{\tau^\delta}\bigg(\int\limits_{B(x_0,\tau)\setminus B(x_0,t)} v(x)d\mu(x)\bigg) \operatorname*{ess\,sup}_{d(x_0,x)\le t} \frac{1}{w(x)} < \infty$$

for $p = 1$, *then there exists a positive constant* c_α *such that for all* $\lambda > 0$ *and* $f \in L_w^p(X)$,

$$\int\limits_{\{x:\mu d(x_0,x)^{\delta\alpha}|Kf(x)|>\lambda\}} d(x_0,x)^{-\delta\alpha p}v(x)d\mu \le \frac{c}{\lambda^p}\int\limits_X |f(x)|^p w(x)d\mu.$$

Proof. This theorem is proved in the same way as Theorem 8.2.8. According to Theorem 1.2.14 we only need to verify the following condition

$$\sup_{t>0} \frac{1}{t^{\delta(1-\alpha)}}\bigg(\int\limits_{\{d(x_0,x)>t\}} \frac{v(x)}{(d(x_0,x))^{-\delta p\alpha}}d\mu\bigg) \times$$

$$\times \left(\int\limits_{\{x:d(x_0,x)\leq t\}} w^{1-p'}(x)d\mu \right)^{p-1} < \infty.$$

Indeed

$$\frac{1}{t^{\delta(1-\alpha)p}} \int\limits_{\{x:d(x_0,x)>t\}} v(x)d(x_0,x)^{-\delta\alpha p}d\mu \times$$

$$\times \left(\int\limits_{d(x_0,x)\leq t} w^{1-p'}(x)d\mu(x) \right)^{p-1} =$$

$$= \frac{c_1}{t^{\delta(1-\alpha)p}} \int\limits_{\{x:d(x_0,x)>t\}} v(x) \left(\int\limits_{d(x_0,x)}^{\infty} \tau^{-\alpha p\delta-1}d\tau \right)d\mu \times$$

$$\times \left(\int\limits_{\{x:d(x_0,x)\leq t\}} w^{1-p'}(x)d\mu(x) \right)^{p-1} =$$

$$= \frac{c_1}{t^{\delta(1-\alpha)p}} \int\limits_t^{\infty} \tau^{-\alpha p\delta-1} \left(\int\limits_{t<d(x_0,x)<\tau} v(x)d\mu \right)d\tau \times$$

$$\times \left(\int\limits_{\{x:d(x_0,x)\leq t\}} w^{1-p'}(x)d\mu(x) \right)^{p-1} \leq$$

$$\leq \frac{c_1}{t^{\delta(1-\alpha d)p}} \int\limits_t^{\infty} \tau^{-\alpha p\delta-1+p\delta} \left(\tau^{-\delta p} \int\limits_{t<d(x_0,x)<\tau} v(x)d\mu \right)d\tau \times$$

$$\times \left(\int\limits_{\{x:d(x_0,x)\leq t\}} w^{1-p'}(x)d\mu(x) \right)^{p-1} \leq$$

$$\leq c_2 \frac{1}{t^{p\delta-\delta\alpha p}} t^{-\alpha p\delta+p\delta} = c_2.$$

Analogously we can handle the case $p = 1$. $\square$

The following theorem is proved in the same manner.

Theorem 8.2.12. *Let $\alpha \geq 1$ and let σ and u be a decreasing continuous function on $(0,\infty)$. Assume that $\rho \in A_1(X)$. We put $v(x) = \sigma(d(x_0,x))\rho(x)$, $w(x) = u(d(x_0,x))\rho(x)$. If*

$$\sup_{t>0} \left(\int\limits_{B(x_0,2t)} v(x)d\mu(x) \right) \operatorname{ess\,sup}_{d(x_0,x)\geq t} \frac{1}{w(x)d(x_0,x)^\delta} < \infty$$

then there exists a positive constant c_α such that for all $\lambda > 0$ and $f \in L^1_w(X)$,

$$\int\limits_{\{x:d(x_0,x)^{\delta\alpha}|Kf(x)|>\lambda\}} d(x_0,x)^{-\delta\alpha} v(x) d\mu \leq \frac{c}{\lambda} \int\limits_X |f(x)| w(x) d\mu.$$

8.3. Singular integrals on nonhomogeneous spaces

In the present section we present weighted inequalities for Calderón–Zygmund singular integrals defined on nonhomogeneous spaces.

Let X be a metric space with metric d. Suppose that μ is a measure on X satisfing the condition

$$\mu\overline{B}(x,r) \leq cr^\alpha, \quad x \in X, \quad r > 0,$$

for some $\alpha > 0$, where $\overline{B}(x,r) \equiv \{y : d(x,y) \leq r\}$ (the doubling condition for μ need not be valid and may fail).

Let the kernel k satisfy the following conditions:
1)
$$|k(x,y)| \leq c_1 d(x,y)^{-\alpha},$$

for all $x, y \in X, x \neq y$;
2) there exist $c_2 > 0$ and $\varepsilon \in (0,1]$ such that

$$\max\{|k(x,y) - k(x_1,y)|, |k(y,x) - k(y,x_1)|\} \leq c_2 \frac{d(x,x_1)^\varepsilon}{d(x,y)^{\alpha+\varepsilon}}$$

whenever $d(x,x_1) \leq 2^{-1} d(x,y)$, $x \neq y$. Assume also that the integral operator

$$Tf(x) = \lim_{\varepsilon \to 0} \int\limits_{X \setminus \overline{B}(x_0,\epsilon)} f(y) k(x,y) d\mu$$

acts boundedly in $L^2_\mu(X)$.

We have the following theorem.

Theorem A([221]). *The operator T is bounded in $L^p_\mu(X)$ for $1 < p < \infty$ and is of weak type $(1,1)$.*

As before we shall assume that $a \equiv \sup\{d(x_0,x) : x \in X\}$, where x_0 is a fixed point of X and

$$\mu(B(x_0,R) \setminus B(x_0,r)) > 0$$

whenever $0 < r < R < a$. Further, suppose that $\mu\{x : d(x_0,x) = a\} = 0$.

Lemma 8.3.1. *Let* $\mu\{x_0\} = 0$, $1 < p < \infty$ *and let* w *be a weight function on* X. *Assume the following two conditions are satisfied:*

(i) *there exists a positive increasing function* v *on* $(0, 4a)$ *such that for almost all* $x \in X$ *the inequality*

$$v(2d(x_0, x)) \le b_1 w(x)$$

holds, where the positive constant b_1 does not depend on x;

(ii)

$$I(t) \equiv \int\limits_{B(x_0, t)} w^{1-p'}(x) d\mu < \infty$$

for all $t > 0$.

Then $T\varphi(x)$ exists μ- a.e. for any $\varphi \in L^p_w(X)$.

Proof. Let $0 < \alpha < a$ and put

$$S_\beta \equiv \{x \in X : d(x_0, x) \ge \beta/2\}.$$

Suppose $\varphi \in L^p_w(X)$ and represent φ as follows:

$$\varphi(x) = \varphi_1(x) + \varphi_2(x),$$

where $\varphi_1 = \varphi \chi_{S_\beta}$ and $\varphi_2 = \varphi - \varphi_1$. By condition (i) we see that

$$\int\limits_X |\varphi_1(x)|^p \, d\mu = \frac{v\left(\frac{\beta}{2}\right)}{v\left(\frac{\beta}{2}\right)} \int\limits_{S_\beta} |\varphi(x)|^p \, d\mu$$

$$\le \frac{1}{v\left(\frac{\beta}{2}\right)} \int\limits_{S_\beta} |\varphi(x)|^p \, v\left(2d\left(x_0, x\right)\right) d\mu \le$$

$$\le \frac{c_1}{v\left(\frac{\beta}{2}\right)} \int\limits_{S_\beta} |\varphi(x)|^p \, w(x) d\mu < \infty$$

for arbitrary β, $0 < \beta < a$. Consequently $T\varphi_1 \in L^p_\mu(X)$ and according to Theorem A, $T\varphi_1(x)$ exists μ−a.e. on X.

Now let x be such that $d(x_0, x) > \beta$. If $y \in X$ and $d(x_0, y) < \frac{\beta}{2}$, then

$$d(x_0, x) \le d(x_0, y) + d(x, y) \le \frac{d(x_0, x)}{2} + d(x, y).$$

Hence $\beta < \frac{d(x_0, x)}{2} \le d(x, y)$ and we obtain

$$|K\varphi_2(x)| = \left| \int\limits_X \varphi_2(y)k(x,y)d\mu \right| \leq$$

$$\leq c_2 \int\limits_{B(x_0,\frac{\beta}{2})} \frac{|\varphi(y)|}{d(x,y)^\alpha}d\mu \leq$$

$$\leq c_3 \int\limits_{B(x_0,\frac{\beta}{2})} \frac{|\varphi(y)|}{d(x,y)^\alpha}d\mu \leq$$

$$\leq c_4\beta^{-\alpha} \int\limits_{B(x_0,\frac{\beta}{2})} |\varphi(y)|d\mu$$

$$\leq c_4\beta^{-\alpha} \left(\int\limits_{B(x_0,\frac{\beta}{2})} |\varphi(y)|^p w(y)d\mu \right)^{1/p} \times$$

$$\times \left(\int\limits_{B(x_0,\frac{\beta}{2})} w^{1-p'}(y)d\mu \right)^{1/p'} < \infty.$$

Thus $T\varphi(x)$ is absolutely convergent for arbitrary x such that $d(x_0,x) > \beta$. As we can take β arbitrarily small and $\mu\{x_0\} = 0$, we conclude that $T\varphi(x)$ exists $\mu-$ a.e. on X. $\square$

Theorem 8.3.1. *Let* $1 < p < \infty$, *and let* $\mu\{x_0\} = 0$. *Assume that* v *is a positive increasing continuous function on* $(0, 4a)$. *Suppose that* w *is a weight on* X. *and that following two conditions hold:*

(i) *there exists a constant* $b_1 > 0$ *such that the inequality*

$$v(2d(x_0,x)) \leq b_1 w(x)$$

is fulfilled for μ-*a.a.* $x \in X$

(ii)

$$\sup_{0<t<a} \left(\int\limits_{X\backslash B(x_0,t)} \frac{v(d(x_0,x))}{d(x_0,x)^{\alpha p}}d\mu \right)^{1/p} \left(\int\limits_{B(x_0,t)} w^{1-p'}(x)d\mu \right)^{1/p'} < \infty.$$

Then T *is bounded from* $L^p_w(X)$ *to* $L^p_{v(d(x_0,\cdot))}(X)$.

Proof. Without loss of generality we can suppose that v has the form

$$v(t) = v(0+) + \int\limits_0^t \phi(\tau)d\tau, \quad \phi \geq 0.$$

We have

$$\int\limits_X |Tf(x)|^p v(d(x_0, x)) d\mu = v(0+) \int\limits_X |Tf(x)|^p d\mu +$$

$$+ \int\limits_X |Tf(x)|^p \left(\int\limits_0^{d(x_0,x)} \phi(t)dt \right) d\mu \equiv I_1 + I_2.$$

If $v(0+) = 0$, then $I_1 = 0$. If $v(0+) \neq 0$, by Theorem A we obtain

$$I_1 \leq c_1 v(0+) \int\limits_X |f(x)|^p \, d\mu \leq$$

$$\leq c_1 \int\limits_X |f(x)|^p v\, (d\,(x_0, x))\, d\mu \leq c_2 \int\limits_X |f(x)|^p \, w(x)dx.$$

Using change of the order of integration in I_2, we have

$$I_2 = \int\limits_0^a \phi(t) \left(\int\limits_{\{x:d(x_0,x)>t\}} |Tf(x)|^p \, d\mu \right) dt \leq$$

$$\leq c_3 \int\limits_0^a \phi(t) \left(\int\limits_{\{x:d(x_0,x)>t\}} \left| \int\limits_{\{y:d(x_0,y)>\frac{t}{2}\}} f(y)k(x,y)d\mu \right|^p d\mu \right) dt +$$

$$+ c_3 \int\limits_0^a \phi(t) \left(\int\limits_{\{x:d(x_0,x)>t\}} \left| \int\limits_{\{y:d(x_0,y)\leq\frac{t}{2}\}} f(y)k(x,y)d\mu \right|^p d\mu \right) dt$$

$$\equiv I_{21} + I_{22}.$$

Using again the boundedness of T in $L_\mu^p(X)$ we obtain

$$I_{21} \leq c_4 \int\limits_0^a \phi(t) \left(\int\limits_{\{y:d(x_0,y)>\frac{t}{2}\}} |f(y)|^p \, d\mu \right) dt =$$

$$= c_4 \int\limits_X |f(y)|^p \left(\int\limits_0^{2d(x_0,y)} \phi(t)dt \right) d\mu \leq$$

$$\leq c_4 \int\limits_X |f(y)|^p v\, (2a_1 d\,(x_0, y))\, d\mu \leq$$

$$\leq c_5 \int\limits_X |f(y)|^p \, w(y) d\mu.$$

Now let us estimate I_{22}. When $d(x_0, x) > t$ and $d(x_0, y) \leq \frac{t}{2}$ we have

$$d(x_0, x) \leq d(x_0, y) + d(y, x) = d(x_0, y) + d(x, y) \leq$$

$$\leq \frac{t}{2} + d(x, y) \leq \frac{d(x_0, x)}{2} + d(x, y).$$

Hence

$$\frac{d(x_0, x)}{2} \leq d(x, y).$$

Consequently

$$I_{22} \leq c_6 \int\limits_0^a \phi(t) \left(\int\limits_{\{x: d(x_0, y) > t\}} \left(\int\limits_{\{y: d(x_0, y) \leq \frac{t}{2}\}} \frac{|f(y)|}{d(x, y)^\alpha} d\mu \right)^p d\mu \right) dt \leq$$

$$\leq c_7 \int\limits_0^a \phi(t) \left(\int\limits_{\{x: d(x_0, x) > t\}} \frac{1}{d(x_0, x)^{\alpha p}} d\mu \right) \left(\int\limits_{B(x_0, t)} |f(y)| d\mu \right)^p dt.$$

It is easy to see that for any s, $0 < s < a$, we have

$$\int\limits_s^a \phi(t) \left(\int\limits_{\{x: d(x_0, x) > t\}} \frac{1}{d(x_0, x)^{\alpha p}} d\mu \right) dt \leq$$

$$\leq \int\limits_{\{x: d(x_0, x) \geq s\}} \frac{1}{d(x_0, x)^{\alpha p}} \left(\int\limits_s^{d(x_0, x)} \phi(t) dt \right) d\mu$$

$$\leq \int\limits_{\{x: d(x_0, x) \geq s\}} \frac{v(d(x_0, x))}{d(x_0, x)^{\alpha p}} d\mu.$$

Finally, using Corollary 1.2.1, we obtain

$$I_{22} \leq c_8 \int\limits_X |f(x)|^p \, w(x) d\mu.$$

□

Now we consider the case of decreasing weights. Here we shall assume that $a = \infty$.

Lemma 8.3.2. *Let* $1 < p < \infty$, *let* w *be a weight function on* X. *Suppose the following conditions are fulfilled:*
(i) *there exists a positive decreasing function* v *on* $(0, \infty)$ *such that*

$$v\left(\frac{d(x_0, x)}{2}\right) \le cw(x) \quad a.e.;$$

(ii)

$$\int\limits_{X \setminus B(x_0, t)} w^{1-p'}(x)(d(x_0, x))^{-\alpha p'} d\mu < \infty$$

for all $t > 0$. *Then* $T\varphi(x)$ *exists* $\mu-$*a.e. for arbitrary* $\varphi \in L_w^p(X)$.

Proof. Fix arbitrarily $\beta > 0$ and let

$$S_\beta = \{x : d(x_0, x) \ge \beta\}.$$

Represent φ as follows:

$$\varphi(x) = \varphi_1(x) + \varphi_2(x),$$

where $\varphi_1(x) = \varphi(x)\chi_{S_\beta}(x)$ and $\varphi_2(x) = \varphi(x) - \varphi_1(x)$.
For φ_2 we have

$$\int\limits_X |\varphi_2(x)|^p d\mu = \frac{v(\beta)}{v(\beta)} \int\limits_{B(x_0, \beta)} |\varphi(x)|^p d\mu \le$$

$$\le \frac{1}{v(\beta)} \int\limits_{B(x_0, \beta)} |\varphi(x)|^p v(d(x_0, x)) d\mu \le$$

$$\le \frac{c_1}{v(\beta)} \int\limits_{B(x, \beta)} |\varphi(x)|^p w(x) d\mu < \infty.$$

Consequently $\varphi_2 \in L_\mu^p(X)$ and by Theorem A we have $T\varphi_2 \in L_\mu^p(X)$. Hence $T\varphi_2(x)$ exists a. e. on X.
Now let $x \in X$ and let $d(x_0, x) < \beta/2$. If $d(x_0, y) \ge \beta$, then

$$\frac{d(x_0, y)}{2} \ge d(x, y).$$

Using this inequality we obtain

$$|T\varphi_1(x)| \le c_2 \int\limits_{S_\beta} \frac{|\varphi(y)|}{d(x, y)^\alpha} d\mu \le$$

$$\leq c_3 \int\limits_{S_\beta} \frac{|\varphi(y)|}{d(x_0, y)^\alpha} d\mu \leq$$

$$\leq c_3 \left(\int\limits_{S_\beta} |\varphi(y)|^p w(y) d\mu \right)^{1/p} \times$$

$$\times \left(\int\limits_{S_\beta} w^{1-p'}(y) d(x_0, y)^{-\alpha p'} d\mu \right)^{1/p'} < \infty.$$

As we may take β arbitrarily large, we conclude that $T\varphi(x)$ exists a.e. $\square$

Theorem 8.3.2. *Let* $1 < p < \infty$ *and let* v *be a positive continuous decreasing function on* $(0, \infty)$. *Suppose that* w *is a weight function on* X *and the following conditions are satisfied:*

(i)

$$v(d(x_0, x)/2) \leq cw(x)$$

for almost all x;

(ii)

$$\sup_{t>0} \left(\int\limits_{B(x_0, t)} v(d(x_0, x)) d\mu \right)^{1/p} \times$$

$$\times \left(\int\limits_{X \backslash B(x_0, t)} w^{1-p'}(x) d(x_0, x)^{-\alpha p'} d\mu \right)^{1/p'} < \infty.$$

Then the operator T is bounded from $L^p_w(X)$ to $L^p_{v(d(x_0, \cdot))}(X)$.

Proof. Without loss of generality we can represent v as

$$v(t) = v(+\infty) + \int\limits_t^\infty \phi(\tau) d\tau, \quad \phi \geq 0.$$

Further

$$\int\limits_X |Tf(x)|^p v(d(x_0, x)) dx = v(+\infty) \int\limits_X |Tf(x)|^p d\mu +$$

$$+ \int\limits_X |Tf(x)|^p \left(\int\limits_{d(x_0, x)}^\infty \phi(t) dt \right) d\mu \equiv I_1 + I_2.$$

If $v(+\infty) = 0$, then $I_1 = 0$. But if $v(+\infty) \neq 0$, then by virtue of the boundedness of T in $L^p_\mu(X)$ we have

$$I_1 \leq c_1 v(+\infty) \int_X |f(x)|^p d\mu \leq$$

$$\leq c_1 \int_X |f(x)|^p v(d(x_0, x)) d\mu \leq c_2 \int_X |f(x)|^p w(x) d\mu.$$

Now we pass to I_2 :

$$I_2 = \int_0^\infty \phi(t) \left(\int_{B(x_0,t)} |Tf(x)|^p d\mu \right) dt \leq$$

$$\leq c_3 \int_0^\infty \phi(t) \left(\int_{B(x_0,t)} |Tf_t^{(1)}(x)|^p d\mu \right) dt +$$

$$+ c_3 \int_0^\infty \phi(t) \left(\int_{B(x_0,t)} |Tf_t^{(2)}(x)|^p d\mu \right) dt = I_{21} + I_{22},$$

where $f_t^{(1)} = f \cdot \chi_{B(x_0,2t)}$ and $f_t^{(2)} = f - f_t^{(1)}$. Using Theorem A again we have

$$I_{21} \leq c_4 \int_0^\infty \phi(t) \left(\int_{B(x_0,2t)} |f(y)|^p d\mu \right) dt =$$

$$= c_4 \int_X |f(y)|^p \left(\int_{\frac{d(x_0,x)}{2}}^\infty \phi(t) dt \right) d\mu \leq c_5 \int_X |f(y)|^p w(y) d\mu.$$

It remains to estimate I_{22}. If $x \in B(x_0, t)$ and $y \in X \setminus B(x_0, 2t)$, then

$$\frac{d(x_0, y)}{2} \leq d(x, y).$$

Consequently

$$I_{22} \leq c_5 \int_0^\infty \phi(t) \left(\int_{B(x_0,t)} d\mu \right) \left(\int_{X \setminus B(x_0, 2a_1 t)} |f(y)| d(x_0, y)^{-\alpha} d\mu \right)^p dt.$$

Moreover

$$\int_0^s \phi(t) \left(\int_{B(x_0,t)} d\mu \right) dt = \int_{B(x_0,s)} \left(\int_{d(x_0,x)}^s \phi(t) dt \right) d\mu \leq$$

$$\leq \int\limits_{B(x_0,s)} v(d(x_0,x))d\mu$$

and by Corollary 1.2.2 we finally conclude that T is bounded. $\square$

Now we shall deal with weak type inequalities. Using Theorem A, the following Lemma can be proved in the same way as Lemma 8.3.1.

Lemma 8.3.3. *Let $\mu\{x_0\} = 0$ and let w be a weight function on X. Suppose that the following conditions are fulfilled:*
(i) there exists a positive increasing function v on $(0, \infty)$ such that for some positive b_1 the inequality

$$v(2d(x_0,x)) \leq b_1 w(x)$$

holds for almost all $x \in X$; (ii)

$$\operatorname*{ess\,sup}_{x \in B(x_0,t)} \frac{1}{w(x)} < \infty$$

for any $t > 0$.
Then $T\varphi(x)$ exists for $\varphi \in L_w^1(X)$ almost everywhere on X.

Lemma 8.3.4. *Let w be a weight on X. Suppose that*
(i) there exists a positive decreasing function v on $(0, \infty)$ such that

$$v\left(\frac{d(x_0,x)}{2}\right) \leq b_1 w(x),$$

for μ- a.a $x \in X$;
(ii) for any $t > 0$

$$\operatorname*{ess\,sup}_{x \in X \setminus B(x_0,t)} \frac{1}{w(x)d(x_0,x)^\alpha} < \infty.$$

Then if $\varphi \in L_w^1(X)$, $T\varphi(x)$ exists almost everywhere on X.

Using Fubini's theorem we easily derive a following proposition:

Proposition 8.3.1. *Let v and w be a weight functions respectively on $(0, \infty)$ and X. Then the condition*

$$\sup_{t>0} \left(\int_0^t v(\tau)d\tau \right) \operatorname*{ess\,sup}_{d(x_0,x)>t/2} \frac{1}{w(x)} < \infty$$

implies the boundedness of the Hardy–type operator

$$Hf(t) = \int\limits_{d(x_0,x)>t} f(y)d\mu \quad \text{from} \quad L_w^1(X) \quad \text{to} \quad L_v^1(R_+).$$

Theorem 8.3.3. *Let $\mu\{x_0\} = 0$, $1 < p < \infty$ and let v be a positive increasing continuous function on $(0, \infty)$. Suppose that w is a weight function on X. Let the following two conditions be satisfied:*
(i) *there exists a positive constant c such that for almost all $x \in X$*

$$v(2d(x_0, x)) \leq cw(x);$$

(ii)

$$\sup_{\tau > t} \tau^{-\alpha} \left(\int_{B(x_0,\tau)\backslash B(x_0,t)} v(d(x_0, x))d\mu(x) \right)^{1/p} \times$$

$$\times \left(\int_{d(x_0,x)\leq t} w(x)d\mu(x) \right)^{1/p'} < \infty.$$

Then T is bounded from $L^p_w(X)$ to $L^{p\infty}_{v(d(x_0,\cdot))}(X)$.

Proof. Let us denote $H_\lambda \equiv \{x \in X : |Tf(x)| > \lambda\}$. If we represent v as $v(t) = v(0+) + \int\limits_0^t \psi(\tau)d\tau$, $\psi \geq 0$, then

$$\int_{H_\lambda} v(d(x_0, x))d\mu = \int_{H_\lambda} v(0+)d\mu +$$

$$+ \int_{H_\lambda} \left(\int_0^{d(x_0,x)} \psi(t)dt \right) d\mu = I_1 + I_2.$$

Using Theorem A we derive

$$I_1 \leq \frac{c_1 v(0+)}{\lambda^p} \int_X |f(x)|^p d\mu \leq \frac{c_1}{\lambda^p} \int_X |f(x)|^p v(2a_1 d(x_0, x))d\mu \leq$$

$$\leq \frac{c_2}{\lambda^p} \int_X |f(x)|^p w(x)d\mu.$$

Now we estimate I_2. Let

$$S = \left\{ x \in X : \left| \int_{X\backslash B(x_0,t/2a_1)} k(x,y)f(y)d\mu \right| > \frac{\lambda}{2} \right\},$$

$$S_1 = \left\{ x \in X : \left| \int_{B(x_0,t/2a_1)} k(x,y)f(y)d\mu \right| > \frac{\lambda}{2} \right\}.$$

Then we obtain

$$I_2 = \int_0^\infty \psi(t)\left(\int_{\{x:d(x_0,x)>t\}} \chi_{H_\lambda}d\mu \right)dt \le$$

$$\le \int_0^\infty \psi(t)\left(\int_{\{x:d(x_0,x)>t\}} \chi_S d\mu \right)dt \le$$

$$+ \int_0^\infty \psi(t)\left(\int_{\{x:d(x_0,x)>t\}} \chi_{S_1} d\mu \right)dt \equiv$$

$$\equiv I_{21} + I_{22}.$$

Using Theorem A again we find that

$$I_{21} \le \int_0^\infty \psi(t)\left(\int_S d\mu \right)dt \le$$

$$\le \frac{c_3}{\lambda^p} \int_0^\infty \psi(t)\left(\int_{X\setminus B(x_0,t/2)} |f(x)|^p d\mu \right)dt =$$

$$= \frac{c_3}{\lambda^p} \int_X |f(x)|^p \left(\int_0^{2d(x_0,x)} \psi(t)dt \right)d\mu \le$$

$$\le \frac{c_4}{\lambda^p}\|f\|^p_{L^p_w(X)}.$$

Further, if $d(x_0,x) > t$ and $d(x_0,x) \le \frac{t}{2}$, then $\frac{d(x_0,x)}{2} \le d(x,y)$ and by Theorem 1.2.9 we obtain

$$I_{22} \le$$

$$\le \int_0^\infty \psi(t)\left(\int_{\{d(x_0,x)>t\}} \chi_{\{y\in X:c_5 d(x_0,x)^{-\alpha} \int_{B(x_0,d(x_0,x))} f(y)d\mu(y)>\lambda\}}(x)d\mu \right)dt$$

$$= \int_{\left\{x\in X:c_5 d(x_0,x)^{-\alpha} \int_{B(x_0,d(x_0,x))} f(y)d\mu(y)>\lambda\right\}} \left(\int_0^{d(x_0,x)} \psi(t)dt \right)d\mu \le$$

$$\le \frac{c_6}{\lambda^p}\|f\|^p_{L^p_w(X)}.$$

□

The following theorems are proved analogously.

Theorem 8.3.4. *Let* $\mu\{x_0\} = 0$, $1 < p < \infty$ *and let* v *be a positive increasing continuous function on* $(0, \infty)$. *Let* w *be a weight function on* X *and let the following conditions be fulfilled:*
(i) *there exists a positive constant* c *such that for almost all* $x \in X$,

$$v(2d(x_0, x)) \leq cw(x);$$

(ii)

$$\sup_{\tau > t} \tau^{-\alpha} \left(\int_{B(x_0, \tau) \backslash B(x_0, t)} v(d(x_0, x)) d\mu(x) \right) \operatorname*{ess\,sup}_{d(x_0, x) \leq t} \frac{1}{w(x)} < \infty.$$

Then T is bounded from $L^1_w(X)$ to $L^{1\infty}_{v(d(x_0, x))}(X)$.

Theorem 8.3.5. *Let* $1 < p < \infty$ *and let* v *be a positive decreasing continuous function on* $(0, \infty)$. *Assume that* w *is a weight function on* X *and the following conditions are fulfilled:*
(i)

$$v(\frac{d(x_0, x)}{2}) \leq cw(x)$$

for almost all $x \in X$;
(ii)

$$\sup_{t > 0} \left(\int_{B(x_0, t)} v(d(x_0, x)) d\mu(x) \right) \operatorname*{ess\,sup}_{d(x_0, x) \geq t/2} \frac{1}{w(x)d(x_0, x)^\alpha} < \infty.$$

Then T is bounded from $L^1_w(X)$ to $L^{1\infty}_{v(d(x_0, x))}(X)$.

Remark 8.3.1. The results of this section are also valid if we do not require the continuity of v, but assume that $\mu B(x_0, r)$ is continuous with respect to r, where x_0 is the same fixed point as above.

8.4.   Two–weight inequalities for the Hilbert transform

In the present section two-weight criteria are established for the Hilbert transform

$$Hf(x) = p.v. \int_{-\infty}^{+\infty} f(t)(x - t)^{-1} dt.$$

First we consider the strong- type inequalities.

Theorem 8.4.1. *Let* $1 < p < \infty$ *and let* σ *and* u *be positive increasing functions on* $(0, \infty)$. *Suppose that* $\rho \in A_p(R)$. *We put* $v(x) = \sigma(|x|)\rho(x)$, $w(x) = u(|x|)\rho(x)$. *The inequality*

$$\|Hf\|_{L^p_v(R)} \leq c\|f\|_{L^p_w(R)}, \tag{8.4.1}$$

with a constant c *independent of* f, $f \in L^p_w(R)$, *holds if and only if*

$$\sup_{t>0} \left(\int_{|x|>t} \frac{v(x)}{|x|^p} dx \right) \left(\int_{|x|<t} w^{1-p'}(x) dx \right)^{p-1} < \infty. \tag{8.4.2}$$

Proof. The sufficiency follows in the same way as in the case of Theorem 8.1.2 taking into account Remark 8.1.1 (see also [100], Chapter 9). In this case we do not require the continuity of σ.

Now let us prove necessity. First we show that

$$I \equiv \int_{|x|<t} w^{1-p'}(x) dx < \infty$$

for all $t > 0$. Indeed, suppose that $I(t) = \infty$ for some $t > 0$. Then there exists $g \in L^p(-t, t)$ such that $\int_{\{|\cdot|<t\}} gw^{-1/p} = \infty$. Let us assume that $f_t(y) = g(y)w^{-1/p}(y)\chi_{\{|\cdot|<t\}}(y)$. Then we have

$$\|Hf_t\|_{L^p_v(R)} \geq \|\chi_{\{|\cdot|>t\}}Hf_t\|_{L^p_v(R)} \geq$$

$$\geq c_1 \left(\int_{|x|>t} v(x)|x|^{-p} dx \right)^{1/p} \int_{-t}^{t} g(y)w^{-1/p}(y) dy = \infty.$$

On the other hand,

$$\|f_t\|_{L^p_w(R)} = \int_{-t}^{t} g(x) dx = \infty.$$

Thus from the inequality 8.4.1 we conclude that $I(t) < \infty$ for all $t > 0$.

To prove that condition 8.4.2 holds we put $f_t(y) = w^{1-p'}(y)\chi_{(-t,t)}(y)$ in 8.4.1. Then it is obvious that

$$\|Hf_t\|_{L^p_v(R)} \geq \|\chi_{|\cdot|>t}Hf_t\|_{L^p_v(R)} \geq$$

$$\geq c_1 \left(\int_{|x|>t} v(x)|x|^{-p} dx \right)^{1/p} \int_{|x|<t} w^{1-p'}(x) dx.$$

In addition,

$$\|f_t\|_{L^p_w(R)} = \int\limits_{|x|<t} w^{1-p'} dx < \infty$$

and consequently from the boundedness of H we obtain condition (8.4.2). $\square$

The following theorem is proved analogously (it can be obtained also from duality arguments).

Theorem 8.4.2. *Let* $1 < p < \infty$ *and let* σ *and* u *be positive decreasing functions on* $(0,\infty)$. *Suppose that* $\rho \in A_p(R)$. *We put* $v(x) = \sigma(|x|)\rho(x)$, $w(x) = u(|x|)\rho(x)$. *Then inequality 8.4.1 holds if and only if*

$$\sup_{t>0} \left(\int\limits_{|x|<t} v(x)dx \right) \left(\int\limits_{|x|>t} w^{1-p'}|x|^{-p'}(x)dx \right)^{p-1} < \infty$$

For $n = 1$ in the case when the function ρ is even, Theorems 8.4.1 and 8.4.2 were proved in [71] (see also [100], Chapter 9).

The following result is a Corollary of Theorem 8.4.1 and is proved in the same way as Theorem 8.1.5.

Theorem 8.4.3. *Let* $1 < p < \infty$ *and let* v *and* w *be positive increasing functions on* $(0,\infty)$. *Suppose that* $\rho \in A_1(R)$. *If for all* $f \in L^p_w(R)$

$$\int\limits_R |Hf(x)|^p v(|x|)dx \le c \int\limits_R |f(x)|^p w(|x|)dx, \qquad (8.4.3)$$

with a constant c independent of f, then there exists a positive constant c_1, *such that the inequality*

$$\int\limits_R |Hf(x)|^p v(|x|)\rho(x)dx \le c_1 \int\limits_R |f(x)|^p w(|x|)\rho(x)dx$$

holds for all $f \in L^p_w(R)$.

Theorem 8.4.4. *Let* $1 < p < \infty$. *Suppose that* v *and* w *are positive decreasing functions on* $(0,\infty)$. *Suppose also that* $\rho \in A_1(R)$. *If the inequality* (8.4.3) *holds, then there exists a positive constant* c_1 *such that the inequality*

$$\int\limits_R |Hf(x)|^p v(|x|)\rho^{1-p}(x)dx \le$$

$$\le c_1 \int\limits_R |f(x)|^p w(|x|)\rho^{1-p}(x)dx, \quad f \in L^p_{w|\cdot|\rho^{1-p}(\cdot)}(R),$$

holds .

The following example is taken from [71]

Example 8.4.1. Let $1 < p < \infty$, $w(t) = t^{(p-1)} \ln^p \frac{1}{t}$, $v(t) = t^{p-1}$ for $0 < t \leq b$. Let $w(t) = (b^{p-1-\beta} \ln^p \frac{1}{b})t^{\beta}$, $v(t) = b^{p-1-\alpha}t^{\alpha}$ when $t > b$, where $0 < \alpha \leq \beta < p - 1$, $b = e^{-p'}$, $p' = \frac{p}{p-1}$. Then for the pair $(v(| \cdot |), w(| \cdot |))$ the inequality (8.4.3) holds.

In [100] an appropriate example for the decreasing weights is presented as well.

Now we consider weak- type inequalities for H. We have the following Theorem:

Theorem 8.4.5. *Let* $1 < p < \infty$, *let* σ *and* u *be positive increasing functions on* $(0, \infty)$ *and suppose that* $\rho \in A_p(R)$. *We put* $v(x) = \sigma(|x|)\rho(x)$, $w(x) = u(|x|)\rho(x)$. *The inequality*

$$\int\limits_{\{x \in R: |Hf(x)| > \lambda\}} v(x)dx \leq \frac{c}{\lambda^p}\|f\|^p_{L^p_w(R)}, \qquad (8.4.4)$$

holds with a constant c independent of $f \in L^p_w(R)$ *and* $\lambda > 0$, *if and only if*

$$\sup_{\tau > t} \frac{1}{\tau^p} \int\limits_{t < |x| < \tau} v(x)dx \Big(\int\limits_{|x| < t} w^{1-p'}(x)dx \Big)^{p-1} < \infty.$$

Proof. Sufficiency follows as in the case of Theorem 8.2.8. For necessity we first prove that

$$I(t) \equiv \int\limits_{|x| < t} w^{1-p'}(x)dx < \infty$$

for arbitrary $t > 0$. Indeed if $I(t) = \infty$ for some $t > 0$ then there exists $g \in L^p(-t, t)$ such that $\int\limits_{|\cdot| < t} gw^{-1/p} = \infty$. Put $f_t(y) = g(y)w^{-1/p}(y)$ when $|y| < t$ and, $f(y) = 0$ for $|y| \geq t$. Then for $|x| > t$ we have

$$|Hf_t(x)| = \int\limits_{-t}^{t} \frac{g(y)w^{-1/p}(y)}{|x - y|}dy \geq (2|x|)^{-1} \int\limits_{-t}^{t} g(y)w^{-1/p}(y)dy = \infty.$$

Therefore from the two-weight inequality we derive the estimate

$$I_1(t) \equiv \int\limits_{|x|>t} v(x)dx \leq \int\limits_{\{x:\,|Hf(x)|>\lambda\}} v(x)dx \leq$$

$$\leq \frac{c_1}{\lambda^p} \int\limits_{-t}^{t} |g(x)|^p dx$$

and as c_1 does not depend on λ, we have that $I_1(t) = 0$.
Now let

$$f(y) = w^{1-p'}(y)\chi_{(-t,t)}(y).$$

Then for $t < |x| < s$ we have

$$|Hf(x)| = \int\limits_{-t}^{t} \frac{w^{1-p'}(y)}{|x-y|}dy \geq \frac{1}{2s}\int\limits_{-t}^{t} w^{1-p'}(y)dy = \frac{1}{2s}I(t).$$

Hence by (8.4.4) we obtain the estimate

$$\int\limits_{t<|x|<s} v(x)dx \leq \int\limits_{\{x:\,|Hf(x)|\geq \frac{I(t)}{2s}\}} v(x)dx \leq \frac{c_2 s^p}{I^p(t)} \cdot I(t).$$

This establishes necessity. □

The next two theorems follow analogously and so the proofs are omitted.

Theorem 8.4.6. *Let σ and u be positive increasing functions on $(0,\infty)$ and let $\rho \in A_1(R)$. We put $v(x) = \sigma(|x|)\rho(x)$, $w(x) = u(|x|)\rho(x)$. For the validity of the inequality*

$$\int\limits_{\{x\in R:\,|Hf(x)|>\lambda\}} v(x)dx \leq \frac{c}{\lambda}\|f\|_{L^1_w(R)}, \qquad (8.4.5)$$

with a positive constant c independent of $f \in L^1_w(R)$ and $\lambda > 0$, it is necessary and sufficient that

$$\sup_{\tau>t}\left(\frac{1}{\tau^n}\int\limits_{t<|x|<\tau} v(x)dx\right) \operatorname{ess\,sup}_{|x|<t} \frac{1}{w(x)} < \infty.$$

Theorem 8.4.7. *Let σ and u be positive decreasing functions on $(0,\infty)$ and let $\rho \in A_1(R)$. We put $v(x) = \sigma(|x|)\rho(x)$, $w(x) = u(|x|)\rho(x)$. For the validity of the inequality* (8.4.5) *it is necessary and sufficient that*

$$\sup_{t>0}\left(\int\limits_{|x|<t} v(x)dx\right) \operatorname{ess\,sup}_{|x|<t} \frac{1}{w(x)|x|} < \infty.$$

Example 8.4.2. Let $1 < p < \infty$ and suppose that

$$w(t) = \left\{ \begin{array}{l} t^{p-1} \ln^p \frac{1}{t}, \quad \text{for} \quad 0 < t \le b (b^{p-1-\beta} \ln^p \frac{1}{b}) t^\beta, \quad \text{for} \quad t > b \end{array} \right.$$

$$v(t) = \left\{ \begin{array}{ll} t^{p-1} \ln \frac{1}{t}, & \text{for} \quad 0 < t \le b \\ (b^{p-1-\alpha} \ln \frac{1}{b}) t^\alpha, & \text{for} \quad t > b, \end{array} \right.$$

where $0 < \alpha \le \beta < p-1, b = e^{-p'}, p' = \frac{p}{p-1}$. Then for the pair $(v(|\cdot|), w(|\cdot|))$ the two-weight weak type inequality

$$\int\limits_{\{x \in R : |Hf(x)| > \lambda\}} v(|x|) dx \le \frac{c}{\lambda^p} \|f\|^p_{L^p_{w(|\cdot|)}(R)}$$

holds but H is not bounded from $L^p_{w(|\cdot|)}(R)$ into $L^p_{v(|\cdot|)}(R)$.

In [152] (see also [100], Chapter 9) the two-weight (p, q) type inequalities for the Hilbert transform in the case $q < p$ are established.

8.5. Estimates in Lorentz spaces

In the present section optimal sufficient conditions on weight functions ensuring the validity of weight inequalities for singular integrals in Lorenz spaces defined on spaces of homogeneous and nonhomogeneous type are obtained.

Analogous problems in Lebesgue spaces were studied in Sections 8.1- 8.4.

Let (X, d, μ) be an SHT (see Definition 1.1.1). As in the previous sections we suppose that there exists a point $x_0 \in X$ such that if $a = \sup\{d(x_0, x) : x \in X\}$, then

$$\mu\{x_0\} = \mu\{x \in X : d(x_0, x) = a\} = 0;$$

and for arbitrary r and R with condition $0 < r < R < a$ the folowing inequality is fulfilled:

$$\mu(B(x_0, R) \setminus B(x_0, r))) > 0. \tag{$*$}$$

Note, that the condition $a = \infty$ is equivalent to the condition $\mu(X) = \infty$(see Lemma 6.5.1) and if $a = \infty$, then automatically we have

$$\mu\{x \in X : d(x_0, x) = a\} = 0.$$

Recall that for the space of nonhomogeneous type we shall assume the measure space with quasi–metric (X, d, μ) satisfies the conditions (i)-(v) of Definition 1.1.1 , i.e. the doubling condition is not assumed and may not hold.

Let

$$H_{\varphi\psi} f(x) = \varphi(x) \int\limits_{\{y : d(x_0, y) < d(x_0, x)\}} f(y) \psi(y) w(y) d\mu(y)$$

and

$$H'_{\varphi\psi} f(x) = \varphi(x) \int\limits_{\{y:d(x_0,\,y)>d(x_0,x)\}} f(y)\psi(y)w(y)d\mu(y).$$

The following proposition is valid (see Corollary 1.1.2 for $a = \infty$. The case $a < \infty$ can be obtained similarly):

Proposition A. *Let* $r = s = 1$ *or* $r \in (0, \infty]$ *and* $s \in [1, \infty]$, $p = q = 1$ *or* $p \in (1, \infty)$ *and* $q \in [1, \infty]$; *also*, $\max\{r, s\} \leq \min\{p, q\}$. *The operator* $H_{\varphi\psi}$ *is bounded from* $L_w^{rs}(X)$ *to* $L_v^{pq}(X)$ *if and only if*

$$A = \sup_{0<t<a} \|\varphi(\cdot)\chi_{\{d(x_0,\,y)\,>\,t\}}(\cdot)\|_{L_v^{pq}(X)} \times$$
$$\times \|\psi(\cdot)\chi_{\{d(x_0,\,y)\leq\,t\}}(\cdot)\|_{L_w^{r's'}(X)} < \infty.$$

Moreover, $\|H_{\varphi\psi}\| \approx A$.

From duality arguments we have the following result:

Proposition B. *Let* $r = s = 1$ *or* $r \in (0, \infty]$ *and* $s \in [1, \infty]$, $p = q = 1$ *or* $p \in (1, \infty)$ *and* $q \in [1, \infty]$; *also*, $\max\{r, s\} \leq \min\{p, q\}$. *The operator* $H'_{\varphi\psi}$ *is bounded from* $L_w^{rs}(X)$ *to* $L_v^{pq}(X)$ *if and only if*

$$A_1 = \sup_{0<t<a} \|\varphi(\cdot)\chi_{\{d(x_0,\,y)\leq t\}}(\cdot)\|_{L_v^{pq}(X)} \times$$
$$\times \|\psi(\cdot)\chi_{\{d(x_0,y)>t\}}(\cdot)\|_{L_w^{r's'}(X)} < \infty.$$

Moreover, $\|H'_{\varphi\psi}\| \approx A$.

First we establish weighted inequalities for singular integrals . Let us recall the definition of a singular integral on an SHT.

Let $k : X \times X \backslash \{(x, x) : x \in X\} \to R$ be a measurable function satisfying the condition

$$|k(x, y)| \leq \frac{c}{\mu B(x, d(x, y))},$$

for all $x, y \in X$, $x \neq y$, and

$$|k(x_1, y) - k(x_2, y)| + |k(y, x_1) - k(y, x_2)| \leq$$
$$\leq c\omega\left(\frac{d(x_2,\, x_1)}{d(x_2,\, y)}\right)\frac{1}{\mu(B(x_2, d(x_2,\, y)))},$$

for all $x_1, x_2, y \in X$ provided that $d(x_2, y) > bd(x_1, x_2)$, where ω is a positive non-decreasing function on $(0, \infty)$ satisfying the Δ_2- condition ($\omega(2t) \leq c\omega(t)$) and the Dini condition

$$\int\limits_0^1 \frac{\omega(t)}{t}dt < \infty.$$

We shall also suppose that there exists p_0, $1 < p_0 < \infty$, such that the limit

$$Kf(x) = \lim_{\varepsilon \to 0+} \int_{X \setminus B(x,\varepsilon)} k(x,y)f(y)d\mu$$

exists a.e. on X for all $f \in L^{p_0}(X)$, and that the operator K is bounded in $L_\mu^{p_0}(X)$.

Theorem A ([147], p.207). *Let $1 < p, q < \infty$. If $w \in A_p(X)$, then the operator K is bounded in $L_w^{pq}(X)$. If the Hilbert transform*

$$Hf(x) = p.v. \int_{-\infty}^{+\infty} f(t)(x-t)^{-1}dt$$

is bounded in $L_w^{pq}(R)$, then $w \in A_p(R)$.

We have the following Lemma.

Lemma 8.5.1. *Let $1 < s \le p < \infty$, $\rho \in A_p(X)$. Suppose that for weight functions w and w_1 the following conditions are satisfied:*

(1) *there exist positive increasing function σ defined on $(0, 4a_1 a)$ and positive constant b such that for every $x \in X$ the following inequality holds:*

$$\sigma(d(x_0, x))\rho(x) \le bw(x)w_1(x);$$

(2) *for all $t, 0 < t < a$, we have*

$$\left\| \frac{1}{w(\cdot)\, w_1(\cdot)} \chi_{\{d(x_0,y) \le t\}}(\cdot) \right\|_{L_w^{p's'}(X)} < \infty.$$

Then $Kg(x)$ exists a.e. on X for any g satisfying the condition

$$\|g(\cdot)w_1(\cdot)\|_{L_w^{ps}(X)} < \infty.$$

Proof. Let us fix some α satisfying $0 < \alpha < \frac{a}{a_1}$ and suppose

$$S_\alpha = \{x \in X : d(x_0, x) \ge \frac{\alpha}{2}\}.$$

Assume that $\|g(\cdot)w_1(\cdot)\|_{L_w^{ps}(X)} < \infty$. Introduce g in the following way :

$$g(x) = g_1(x) + g_2(x),$$

where $g_1(x) = g(x) \cdot \chi_{S_a}(x)$, $g_2(x) = g(x) - g_1(x)$.
For g_1 we have.

$$\int\limits_X |g_1(x)|^p \rho(x)d\mu = \frac{\sigma(\alpha/2)}{\sigma(\alpha/2)} \int\limits_{S_a} |g(x)|^p \rho(x)d\mu \le$$

$$\le \frac{1}{\sigma(\alpha/2)} \int\limits_{S_a} |g(x)|^p \rho(x)\sigma(2a_1 d(x_0,x))d\mu \le$$

$$\le \frac{b}{\sigma(\alpha/2)} \int\limits_{S_a} |g(x)|^p w_1^p(X)w(x)d\mu =$$

$$= \frac{b_1}{\sigma(\alpha/2)} \|g(\cdot)w_1(\cdot)\|_{L_w^p(X)}^p \le \frac{b}{\sigma(\alpha/2)} \|g(\cdot)w_1(\cdot)\|_{L_w^{ps}(X)}^p.$$

(In the latter inequality we have used Lemma 6.9.1 part (ii)). Hence $g_1 \in L_\rho^p(X)$ and, according to Theorem A of Section 8.1, $Kg \in L_\rho^p(X)$ and consequently $Kg(x)$ exists a.e. on X.

Now, let $d(x_0,x) > \alpha a_1$ and let $d(x_0,y) < \alpha/2$. We have

$$d(x_0,x) \le a_1(d(x_0,y) + d(y,x)) \le a_1(d(x_0,y) + a_0 d(x,y)).$$

Hence

$$d(x,y) \ge \frac{d(x_0,x)}{a_1 a_0} - \frac{1}{a_0}d(x_0,y) \ge \frac{\alpha}{a_0} - \frac{\alpha}{2a_0} = \frac{\alpha}{2a_0}.$$

Moreover

$$B(x_0, d(x,y)) \subset B(x, a_1(1 + a_1(1 + a_0^2))d(x,y))$$

and consequently we obtain the inequality

$$\mu B(x_0, \alpha/2) \le c_1 \mu B(x, d(x,y)),$$

where c_1 is independent of x. Hence for Kg_2, using the latter estimate and the Hölder's inequality, we have

$$|Kg_2(x)| = \left| \int\limits_X g(y)k(x,y)\,d\mu \right| \le c_2 \int\limits_{B(x_0,\alpha/2)} \frac{|g(y)|}{\mu B(x,d(x,y))}\,d\mu \le$$

$$\le \frac{c_3}{\mu B(x_0, \frac{\alpha}{2c_0})} \int\limits_{B(x_0,\alpha/2)} |g(y)|\,d\mu \le$$

$$\le \frac{c_3}{\mu B(x_0, \frac{\alpha}{2c_0})} \int\limits_{B(x_0,\alpha/2)} |g(y)| \frac{1}{w(y)w_1(y)}w_1(y)w(y)d\mu \le$$

$$\le \frac{c_3}{\mu B(x_0, \frac{\alpha}{2c_0})} \left\| \chi_{B(x_0,\frac{\alpha}{2})}(\cdot)\frac{1}{w(\cdot)w_1(\cdot)} \right\|_{L_w^{p's'}(X)} \times$$

$$\times \|g(\cdot)w_1(\cdot)\|_{L_w^{ps}(X)} < \infty.$$

Thus $Kg_2(x)$ converges absolutely for all x with condition $d(x_0, x) > \alpha a_1$. Since we can choose α arbitrarily small and $\mu\{x_0\} = 0$, we conclude that $Kg(x)$ exists μ- a.e. on X. $\square$

From Lemma 8.5.1 it is easy to derive

Lemma 8.5.2. *Let* $1 < s \le p < \infty$. *If u and u_1 are positive increasing functions on $(0, 4a_1 a)$ and*

$$\left\| \frac{1}{u(d(x_0, \cdot))u_1(d(x_0, \cdot))} \chi_{\{d(x_0, y) \le t\}}(\cdot) \right\|_{L^{p's'}_{u(d(x_0, \cdot))}(X)} < \infty$$

for all t with $0 < t < a$, then for arbitrary g satisfying the condition

$$\|\varphi(\cdot)u_1(d(x_0, \cdot))\|_{L^{ps}_{u(d(x_0, \cdot))}(X)} < \infty,$$

$K\varphi(x)$ *exists a.e. on* X.

The following lemma is proved in the same way as Lemma 8.5.1.

Lemma 8.5.3. *Let* $a = \infty$, $1 < s \le p < \infty$, $\rho \in A_p(X)$. *Suppose the weight functions w and w_1 satisfy the conditions:*

(1) there exists a decreasing positive function σ on $(0, \infty)$ such that for almost all $x \in X$ we have

$$\sigma(d(x_0, \cdot))\rho(x) \le bw(x)w_1^p(x),$$

where the positive constant b does not depend on $x \in X$;
(2) if for every $t > 0$

$$\left\| \frac{\mu B(d(x_0, \cdot))^{-1}}{w(\cdot)w_1(\cdot)} \chi_{\{d(x_0, y) > t\}}(\cdot) \right\|_{L^{p's'}_w(X)} < \infty.$$

Then $Kg(x)$ exists a.e. on X for arbitrary g with $\|g(\cdot)w_1(\cdot)\|_{L^{ps}_w(X)} < \infty$.

From Lemma 8.5.3 we have the following lemma:

Lemma 8.5.4. *Let* $a = \infty$, $1 < s \le p < \infty$. *Assume that u and u_1 are positive decreasing functions on $(0, \infty)$ and the condition*

$$\left\| \frac{\mu B(d(x_0, \cdot))^{-1}}{u(d(x_0, \cdot))u_1(d(x_0, \cdot))} \chi_{\{d(x_0, y) > t\}}(\cdot) \right\|_{L^{p's'}_{u(d(x_0, \cdot))}(X)} < \infty,$$

holds for any $t > 0$. Then $Kg(x)$ exists a.e. on X for arbitrary g with $g(\cdot)u_1(d(x_0, \cdot)) \in L^{ps}_{u(d(x_0, \cdot))}(X)$.

Next we prove weighted inequalities for the operator K.

Theorem 8.5.1. *Let $1 < s \leq p \leq q < \infty$, let w be a weight function on X, suppose that σ is a positive increasing continuous function on $(0, 4a_1a)$, $\rho \in A_p(X)$ and $v(x) = \sigma(d(x_0, x))\rho(x)$. Suppose the following conditions are satisfied:* a) *there exists a positive constant b, such that the inequality*

$$\sigma(2a_1 d(x_0, x))\rho(x) \leq bw(x)$$

holds for almost every $x \in X$;
  b)

$$B \equiv \sup_{0 < t < a} \|\mu B(x_0, d(x_0, \cdot))^{-1} \chi_{\{d(x_0, y) > t\}}\|_{L_v^{pq}(X)} \times$$

$$\times \|\frac{1}{w(\cdot)} \chi_{\{d(x_0, y) \leq t\}}(\cdot)\|_{L_w^{p's'}(X)} < \infty.$$

Then there exists a positive constant c such that

$$\|Kf(\cdot)\|_{L_v^{pq}(X)} \leq c\|f(\cdot)\|_{L_w^{ps}(X)} \tag{8.5.1}$$

for all $f \in L_w^{ps}(X)$.

Proof. First, let us assume that σ is of the kind

$$\sigma(t) = \sigma(0+) + \int_0^t \varphi(\tau)d\tau, \quad \varphi \geq 0;$$

then we have

$$\|Kf(\cdot)\|_{L_v^{pq}(X)} \leq$$

$$\leq c_1 \left(q \int_0^\infty \lambda^{q-1} \left(\int_{\{x \in X : |Kf(x)| > \lambda\}} \rho(x)\sigma(0+)d\mu \right)^{\frac{q}{p}} d\lambda \right)^{\frac{1}{q}} +$$

$$+c_1 \left(q \int_0^\infty \lambda^{q-1} \left(\int_{\{x \in X : |Kf(x)| > \lambda\}} \rho(x) \left(\int_0^{d(x_0, x)} \varphi(t)dt \right) d\mu \right)^{\frac{q}{p}} d\lambda \right)^{\frac{1}{q}} \equiv$$

$$\equiv I_1 + I_2.$$

In the case, where $\sigma(0+) = 0$, we have $I_1 = 0$; in the other case, by Theorem A and Lemma 6.9.2 (part(ii)), we have

$$I_1 = c_1 \sigma(0+)^{\frac{1}{p}} \|Kf(\cdot)\|_{L_p^{pq}(X)} \leq c_2 \sigma^{\frac{1}{p}}(0+) \|f(\cdot)\|_{L_p^{pq}(X)} \leq$$

$$\leq c_2 \sigma^{\frac{1}{p}}(0+) \|f(\cdot)\|_{L_p^{ps}(X)} \leq c_2 \|f(\cdot)\|_{L_w^{ps}(X)}.$$

We etimate I_2. Set

$$f_{1t}(x) = f(x)\chi_{\{d(x_0,x) > \frac{t}{2a_1}\}}, \quad f_{2t}(x) = f(x) - f_{1t}(x).$$

We have

$$I_2 = c_1 \bigg(q \int_0^\infty \lambda^{q-1} \bigg(\int_0^a \varphi(t) \times$$

$$\bigg(\int_{\{x:d(x_0,x)>t,|Kf(x)|>\lambda\}} \rho(x)d\mu \bigg) dt \bigg)^{\frac{q}{p}} d\lambda \bigg)^{\frac{1}{q}} \leq$$

$$\leq c_3 \bigg(q \int_0^\infty \lambda^{q-1} \bigg(\int_0^a \varphi(t) \times$$

$$\times \bigg(\int_{\{x:d(x_0,x)>t\}} \chi_{\{x:|Kf_{1t}(x)|>\frac{\lambda}{2}\}} \rho(x)d\mu \bigg) dt \bigg)^{\frac{q}{p}} d\lambda \bigg)^{\frac{1}{q}} +$$

$$+ c_3 \bigg(q \int_0^\infty \lambda^{q-1} \bigg(\int_0^a \varphi(t) \times$$

$$\times \bigg(\int_{\{x:d(x_0,x)>t\}} \chi_{\{x:|Kf_{2t}(x)|>\frac{\lambda}{2}\}} \rho(x)d\mu \bigg) dt \bigg)^{\frac{q}{p}} d\lambda \bigg)^{\frac{1}{q}} \equiv$$

$$\equiv I_{21} + I_{22}.$$

Applying the Minkowki's inequality twice ($\frac{q}{p} \geq 1, \frac{p}{s} \geq 1$) and using Theorem A and Lemma 6.9.2 (part(ii)), we obtain

$$I_{21} \leq c_4 \bigg(\int_0^a \varphi(t) \bigg(\int_0^\infty \lambda^{q-1} \bigg(\int_{\{x \in X:|Kf_{1t}(x)|>\frac{\lambda}{2}\}} \rho(x)d\mu \bigg)^{\frac{p}{q}} d\lambda \bigg)^{\frac{q}{p}} dt \bigg)^{\frac{1}{p}} \leq$$

$$\leq c_5 \bigg(\int_0^a \varphi(t) \|f_{1t}(\cdot)\|_{L_\rho^{pq}(X)}^p \, dt \bigg)^{\frac{1}{p}} \leq c_5 \bigg(\int_0^a \varphi(t) \|f_{1t}(\cdot)\|_{L_\rho^{ps}(X)}^p \, dt \bigg)^{\frac{1}{p}} \leq$$

$$\leq c_5 bigg(\int_0^\infty \lambda^{s-1} \bigg(\int_{\{x \in X:|f(x)|>\lambda\}} \rho(x) \bigg(\int_0^{2a_1 d(x_0,x)} \varphi(t)dt \bigg) d\mu \bigg)^{\frac{s}{p}} d\lambda \bigg)^{\frac{1}{s}} \leq$$

$$\leq c_5 \|f(\cdot)\|_{L_w^{ps}(X)} \leq c_6 \|f(\cdot)\|_{L_w^{ps}(X)}.$$

Next we shall estimate I_{22}. Note, that if $d(x_0, x) > t$ and $d(x_0, y) < \frac{t}{2a_1}$, then

$$d(x_0, x) \leq a_1 \left(d(x_0, y) + d(y, x) \right) \leq$$

$$\leq a_1 \left(d(x_0, y) + a_0 d(x, y) \right) \leq a_1 \left(\frac{t}{2a_1} + a_0 d(x, y) \right) \leq$$

$$\leq a_1 \left(\frac{d(x_0, x)}{2a_1} + a_0 d(x, y) \right).$$

Hence

$$\frac{d(x_0, x)}{2a_1 a_0} \leq d(x, y),$$

and we also have

$$\mu \left(B(x, d(x_0, x)) \right) \leq b\mu \left(B\left(x, \frac{d(x_0, x)}{2a_1 a_0} \right) \right) \leq b\mu \left(B(x, d(x, y)) \right).$$

Moreover, it is easy to show that

$$\mu B \left(x_0, d(x_0, x) \right) \leq b_1 \mu B \left(x, d(x_0, x) \right)$$

We conclude that

$$\mu B((x_0, d(x_0, x))) \leq b_2 \mu B(x, d(x, y)).$$

Taking into account this inequality and using Proposition A we have

$$I_{22} \leq c_7 \left\| \frac{1}{\mu B(x_0, d(x_0, \cdot))} \int_{\{d(x_0, y) < d(x_0, \cdot)\}} |f(y)| dy \right\|_{L_v^{pq}(X)} \leq$$

$$\leq c_9 \left\| f(\cdot) \right\|_{L_w^{ps}(X)}.$$

Now, let σ be a positive continuous, but not absolutely continuous, increasing function on $(0, 4a_1 a)$; then there exists a sequence of absolutely continuous functions σ_n such that $\sigma_n(t) \leq \sigma(t)$ and $\lim_{n \to \infty} \sigma_n(t) = \sigma(t)$ for arbitrary $t \in (0, 4a_1 a)$. For these functions, we can take $\sigma_n(t) = \sigma(0+) + n \int_0^t \left[\sigma(\tau) - \sigma\left(\tau - \frac{1}{n} \right) \right] d\tau$.

Put $v_n(x) = \rho(x) \sigma_n(d(x_0, x))$; then $B_n < B$, where

$$B_n \equiv \sup_{t>0} \left\| \chi_{\{d(x_0, y) > t\}}(\cdot) \mu \left(B\left(x_0, d(x_0, \cdot) \right) \right)^{-1} \right\|_{L_{v_n}^{pq}(X)} \times$$

$$\times \left\| \frac{1}{w(\cdot)} \chi_{\{d(x_0, y) \leq t\}}(\cdot) \right\|_{L_w^{p's'}(X)}.$$

By virtue of what was been proved, if $B < \infty$, then the following inequality holds:

$$\|Kf(\cdot)\|_{L^{pq}_{v_n}(X)} \leq c \|f(\cdot)\|_{L^{ps}_w(X)},$$

where the constant $c > 0$ does not depend on n.

By passing to the limit as $n \to \infty$, we obtain inequality (8.5.1). $\square$

Using the representation $\sigma(t) = \sigma(+\infty) + \int_t^\infty \psi(\tau)d\tau$, where $\sigma(+\infty) = \lim_{t \to \infty} \sigma(t)$ and $\psi \geq 0$ on $(0, \infty)$ and Proposition B, we obtain the following result.

Theorem 8.5.2. *Let $a = \infty$, $1 < s \leq p \leq q < \infty$, suppose that σ is a positive decreasing continuous function on $(0, \infty)$. Assume that $\rho \in A_p(X)$ and $v(X) = \sigma(d(x_0, x))\rho(x)$. If the conditions:*
a) *there exists a positive constant b such that the inequality*

$$\rho(x)\sigma\left(\frac{d(x_0, x)}{2a_1}\right) \leq bw(x)$$

is true for almost every $x \in X$;
b)

$$B' = \sup_{t>0} \left\|\chi_{\{d(x_0,y) \leq t\}}(\cdot)\right\|_{L^{pq}_v(X)} \times$$

$$\times \left\|\frac{(\mu B(x_0, d(x_0, x)))^{-1}}{w(\cdot)}\chi_{\{d(x_0,y)>t\}}(\cdot)\right\|_{L^{p's'}_w(X)} < \infty$$

are fulfilled. Then inequality (8.5.1) holds.

Now let us consider particular cases of Theorems 8.5.1 and 8.5.2.

Theorem 8.5.3. *Let $a = \infty$, $1 < p \leq q < \infty$, suppose σ_1 and σ_2 are positive, increasing functions on $(0, \infty)$, let σ_1 be a continuous function and suppose that $\rho \in A_p(X)$. We put $v(x) = \sigma_2(d(x_0, x))\rho(x)$, $w(x) = \sigma_1(d(x_0, x))\rho(x)$. If*

$$\sup_{t>0} \|(\mu B(x_0, d(x_0, \cdot)))^{-1}\chi_{\{d(x_0,y)>t\}}(\cdot)\|_{L^{pq}_v(X)} \times$$

$$\times \left\|\frac{1}{w(\cdot)}\chi_{\{d(x_0,y) \leq t\}}(\cdot)\right\|_{L^{p'}_w(X)} < \infty,$$

then there exists a positive constant c such that

$$\|Kf(\cdot)\|_{L^{pq}_v(X)} \leq c \|f(\cdot)\|_{L^p_w(X)}$$

holds for all $f \in L_w^p(X)$.

Proof. By Theorem 8.5.1 it is sufficient to show that there exists a positive constant b such that $\sigma_2(2a_1t) \leq b\sigma_1(t)$ for all $t \in (0, \infty)$.

From the doubling condition (see (vii) in Definition 1.1.1) and the condition (*) it follows that the measure μ satisfies the reverse doubling condition at the point x_0 (see Proposition 6.5.1). In other words there exist constants $\eta_1 > 1$ and $\eta_2 > 1$ such that

$$\mu(B(x_0, \eta_1 r)) \geq \eta_2 \mu(B(x_0, r)),$$

for all $r > 0$.

Applying Hölder's inequality and using Lemma 6.6.1, the reverse doubling condition and the fact that $\rho^{1-p'} \in A_{p'}(X)$, we obtain

$$\frac{\sigma_2(2a_1t)}{\sigma_1(t)} \leq (\mu(B(x_0, 2a_1\eta_1 t) \setminus B(x_0, 2a_1 t)))^{-p} \times$$

$$\times \left(\int_{B(x_0, 2a_1\eta_1 t) \setminus B(x_0, 2a_1 t)} \rho(x)d\mu \right) \times$$

$$\times \left(\int_{B(x_0, 2a_1\eta_1 t) \setminus B(x_0, 2a_1 t)} \rho^{1-p'}(x)d\mu \right)^{p-1} \frac{\sigma_2(2a_1t)}{\sigma_1(t)} \leq$$

$$\leq \left(\left(1 - \frac{1}{\eta_2}\right) \mu B(x_0, 2a_1\eta t) \right)^{-p} \left(\int_{B(x_0, 2a_1\eta_1 t) \setminus B(x_0, 2a_1 t)} \rho(x)d\mu \right) \times$$

$$\times \left(\int_{B(x_0, t)} \rho^{1-p'}(x)d\mu \right)^{p-1} \frac{\sigma_2(2a_1t)}{\sigma_1(t)} \leq$$

$$\leq c_1 (\mu B(x_0, 2a_1\eta t))^{-p} \left\| \chi_{(B(x_0, 2a_1\eta_1 t) \setminus B(x_0, 2a_1 t))}(\cdot) \right\|_{L_v^{pq}(X)}^p \times$$

$$\times \left\| \frac{1}{w(\cdot)} \chi_{B(x_0, t)}(\cdot) \right\|_{L_w^{p'}(X)}^p \leq$$

$$\leq c_2 \left\| (\mu B(x_0, d(x_0, \cdot)))^{-1} \chi_{\{d(x_0, y) > t\}}(\cdot) \right\|_{L_v^{pq}(X)}^p \times$$

$$\times \left\| \frac{1}{w(\cdot)} \chi_{B(x_0, t)}(\cdot) \right\|_{L_w^{p'}(X)}^p \leq c.$$

$\square$

Analogous theorems for singular integrals on homogeneous groups proved in [156] and for singular integrals on SHT in Lebesgue space, see Section 8.1.

Theorem 8.5.4. *Let* $a = \infty$, $1 < s \le p \le q < \infty$; *suppose* σ_1, σ_2, u_1 *and* u_2 *are weight functions defined on* X. *Let* $\rho \in A_p(X)$ *and suppose* $v = \sigma_2 \rho$, $w = \sigma_1 \rho$. *Assume that the following conditions are fulfilled:*

1) there exists positive constant b such that for all $t > 0$

$$\sup_{F_t} \sigma_2^{\frac{1}{p}}(x) \sup_{F_t} u_2(x) \le b \inf_{F_t} \sigma_1^{\frac{1}{p}}(x) \inf_{F_t} u_1(x);$$

holds, where $F_t = \{x \in X : \frac{t}{a_1} \le d(x_0, x) < 8a_1 t\}$;

2)

$$\sup_{t>0} \left\| u_2(\cdot) \left(\mu \left(B \left(x_0, d\left(x_0, \cdot \right) \right) \right) \right)^{-1} \chi_{\{d(x_0,t)>t\}}(\cdot) \right\|_{L_v^{pq}(X)} \times$$

$$\times \left\| \frac{1}{u_1(\cdot) w(\cdot)} \chi_{\{d(x_0,y)\le t\}}(\cdot) \right\|_{L_w^{p's'}(X)} < \infty;$$

3)

$$\sup_{t>0} \left\| u_2(\cdot) \chi_{\{d(x_0,t)\le t\}}(\cdot) \right\|_{L_v^{pq}(X)} \times$$

$$\times \left\| \frac{(\mu B \left(x_0, d \left(x_0, \cdot \right) \right))^{-1}}{u_1(\cdot) w(\cdot)} \chi_{\{d(x_0,y)>t\}}(\cdot) \right\|_{L_w^{p's'}(X)} < \infty.$$

Then

$$\| u_2(\cdot) K f(\cdot) \|_{L_v^{pq}(X)} \le c \| u_1(\cdot) f(\cdot) \|_{L_w^{ps}(X)}, \quad u_1 f \in L_w^{pq}(X), \quad (8.5.2)$$

where the positive constant c does not depend on f.

Proof. Let

$$E_k \equiv B\left(x_0, 2^{k+1}\right) \setminus B\left(x_0, 2^k\right), \quad G_{k,1} \equiv B\left(x_0, 2^{k-1}/a_1\right),$$

$$G_{k,2} \equiv B\left(x_0, a_1 2^{k+2}\right) \setminus B\left(x_0, 2^{k-1}/a_1\right), \quad G_{k,3} \equiv X \setminus B\left(x_0, a_1 2^{k+2}\right).$$

We obtain

$$\| u_2(\cdot) K f(\cdot) \|_{L_v^{pq}(X)}^p \le c_1 \left\| \sum_{k \in Z} u_2(\cdot) K(f\chi_{G_{k,1}})(\cdot) \chi_{E_k}(\cdot) \right\|_{L_v^{pq}(X)}^p +$$

$$+ c_1 \left\| \sum_{k \in Z} u_2(\cdot) K(f\chi_{G_{k,2}})(\cdot) \chi_{E_k}(\cdot) \right\|_{L_v^{pq}(X)}^p +$$

$$+ c_1 \left\| \sum_{k \in Z} u_2(\cdot) K(f\chi_{G_{k,3}})(\cdot) \chi_{E_k}(\cdot) \right\|_{L_v^{pq}(X)}^p \equiv$$

$$\equiv c_1 \left(S_1^p + S_2^p + S_3^p \right).$$

Now we estimate S_1. Note that

$$d(x_0, y) < \frac{2^{k-1}}{a_1} \leq \frac{d(x_0, x)}{2a_1}$$

when $x \in E_k$, and $y \in G_{k,1}$. From the latter inequality we have

$$\mu B(x, d(x_0, x)) \leq b_1 \mu B(x, d(x, y)).$$

Indeed,

$$d(x_0, x) \leq a_1 (d(x_0, y) + d(y, x)) \leq a_1 \left(\frac{d(x_0, x)}{2a_1} + a_0 d(x, y) \right).$$

Hence

$$\frac{1}{2a_1 a_0} d(x_0, x) \leq d(x, y).$$

Correspondingly,

$$\mu B(x, d(x_0, x)) \leq b_2 \mu B\left(x, \frac{d(x_0, x)}{2a_1 a_0}\right) \leq b_2 \mu B(x, d(x, y)).$$

It is easy to see that

$$\mu B(x_0, d(x_0, x)) \leq b_3 \mu B(x, d(x_0, x))$$

and, finally, we obtain

$$\mu B(x_0, d(x_0, x)) \leq b_4 \mu B(x, d(x, y)).$$

By considering this last inequality we have

$$|K(f\chi_{G_{k1}})(x)| \leq b_5 \int_X \frac{|f(y)| \chi_{G_{k1}}(y)}{\mu B(x, d(x, y))} d\mu \leq$$

$$\leq \frac{b_6}{\mu B(x_0, d(x_0, x))} \int_{B(x_0, d(x_0, x))} |f(y)| d\mu,$$

when $x \in E_k$ and by Proposition A we obtain

$$S_1^p \leq c_2 \left\| u_2(\cdot) \left(\mu B(x_0, d(x_0, \cdot))\right)^{-1} \int_{B(x_0, d(x_0, x))} |f(y)| \, d\mu \right\|_{L_v^{pq}(X)}^p \leq$$

$$\leq c_3 \|u_1(\cdot) f(\cdot)\|_{L_w^{ps}(X)}^p.$$

Now, we estimate S_3^p. It is easy to check that if $x \in E_k$ and $y \in G_{k,3}$, then $d(x_0, y) \le d(x, y)$ and

$$\mu B(x_0, d(x_0, y)) \le b_7 \mu B(x, d(x, y)).$$

By virtue of Proposition B we obtain

$$S_3^p \le c_4 \left\| u_2(\cdot) \int_{\{d(x_0, y) > d(x_0, x)\}} \frac{|f(y)|}{\mu B(x_0, d(x_0, y))} d\mu \right\|_{L_v^{pq}(X)}^p \le$$

$$\le c_5 \left\| u_1(\cdot) f(\cdot) \right\|_{L_w^{pq}(X)}^p.$$

We estimate S_2^p. By Lemma 6.9.1 (part(b)) we have

$$S_2^p \le \sum_{k \in u} \left\| u_2(\cdot) K(f \chi_{G_{k,2}})(\cdot) \chi_{E_k}(\cdot) \right\|_{L_v^{pq}(X)}^p \equiv \sum_{k \in u} S_{k,2}^p.$$

We shall use the following notation:

$$u_{2,k} \equiv \sup_{x \in E_k} u_2(x), \quad \sigma_{2,k} \equiv \sup_{x \in E_k} \sigma_2(x), \quad u_{1,k} \equiv \inf_{x \in G_{k,2}} u_1(x),$$

$$\sigma_{1,k} \equiv \inf_{x \in G_{k,2}} \sigma_1(x).$$

By Theorem A and Lemma 6.9.2 we have

$$S_{k,2} \le u_{2,k} \sigma_{2,k}^{\frac{1}{p}} \| K(f \chi_{G_{k,2}})(\cdot) \|_{L_\rho^{pq}(X)} \le$$

$$\le c_6 u_{2,k} \sigma_{2,k}^{\frac{1}{p}} \| f(\cdot) \chi_{G_{k,2}}(\cdot) \|_{L_\rho^{pq}(X)} \le$$

$$\le c_6 u_{2,k} \sigma_{2,k}^{\frac{1}{p}} \left\| f(\cdot) \chi_{G_{k,2}}(\cdot) \right\|_{L_\rho^{ps}(X)} \le$$

$$\le c_7 u_{1,k} \sigma_{1,k}^{\frac{1}{p}} \left\| f(\cdot) \chi_{G_{k,2}}(\cdot) \right\|_{L_\rho^{ps}(X)}^p \le$$

$$\le c_8 \| u_1(\cdot) f(\cdot) \chi_{G_{k,2}}(\cdot) \|_{L_w^{ps}(X)}.$$

Using the Lemma 6.9.1 (part(i)) we finally obtain:

$$S_2^p \le c_9 \| u_1(\cdot) f(\cdot) \|_{L_w^{ps}(X)}.$$

$\square$

Remark 8.5.1. It is easy to verify, that Theorem 8.5.4 is still valid if we replace 1) by the condition
1')

$$\sup_{E_x} \sigma_2(y) \sup_{E_x} u_2^p(y) \le \bar{b} \sigma_1(x) u_1^p(x),$$

where $\bar{b} > 0$ does not depend on $x \in X$ and where

$$E_x = \left\{ y : \frac{d(x_0, x)}{4a_1} \le d(x_0, y) < 4a_1 d(x_0, x) \right\}.$$

Indeed, we have

$$S_{k,2} \le \sigma_{2,k}^{\frac{1}{p}} u_{2,k} \| T(f \chi_{G_{k,2}})(\cdot) \|_{L_\rho^{pq}(X)} \le$$

$$\le \sigma_{2,k}^{\frac{1}{p}} u_{2,k} \| T(f \chi_{G_{k,2}})(\cdot) \|_{L_\rho^p(X)} \le$$

$$\le b_1 \sigma_{2,k}^{\frac{1}{p}} u_{2,k} \| f(\cdot) \chi_{G_{k,2}}(\cdot) \|_{L_\rho^p(X)} =$$

$$= b_1 \int\limits_{G_{k,2}} \left(\sup_{2^k \le d(x_0, y) < 2^{k+1}} \sigma_2(y) \right) \times$$

$$\times \left(\sup_{2^k \le d(x_0, y) < 2^{k+1}} u_2^p(y) \right) |f(x)|^p \rho(x) d\mu \right)^{\frac{1}{p}} \le$$

$$\le b_1 \left(\int\limits_{G_{k,2}} \left(\sup_{E_x} \sigma_2(y) \right) \left(\sup_{E_x} u_2^p(y) \right) |f(x)|^p \rho(x) d\mu \right)^{\frac{1}{p}} \le$$

$$\le b_2 \left(\int\limits_{G_{k,2}} \sigma_1(x) u_1^p(x) \rho(x) |f(x)|^p d\mu \right)^{\frac{1}{p}} =$$

$$= b_2 \| f(\cdot) u_1(\cdot) \chi_{G_{k,2}}(\cdot) \|_{L_w^p(X)} \le$$

$$\le b_2 \| f(\cdot) u_1(\cdot) \chi_{G_{k,2}}(\cdot) \|_{L_w^{sp}(X)}.$$

In [239] it is proved that conditions 2) and 3) of Theorem 8.5.4 are also necessary for equality (8.5.2) to be fulfilled when K is a Hilbert transform.

Theorem 8.5.5. *Let $\mu(X) = \infty$, $1 < s \le p \le q$, and suppose φ_1, φ_2, and v are positive increasing functions on $(0, \infty)$. If the condition*

$$\sup_{t>0} B(t) \equiv \sup_{t>0} \| \varphi_2(d(x_0, \cdot))(\mu B(x_0, d(x_0, x)))^{-1} \times$$

$$\chi_{\{d(x_0,y)>t\}}(\cdot) \times \| L_{v(d(x_0,\cdot))}^{pq}(X) \right\| \frac{1}{\varphi_1(d(x_0, \cdot))} \chi_{\{d(x_0,y) \le t\}}(\cdot) \right\|_{L^{p's'}(X)} < \infty,$$

is fulfilled, then the following weighted inequality holds:

$$\| Kf(\cdot) \varphi_2(d(x_0, \cdot)) \|_{L_{v(d(x_0,\cdot))}^{pq}(X)} \le c \| f(\cdot) \varphi_1(d(x_0, \cdot)) \|_{L^{ps}(X)}. \quad (8.5.3)$$

Proof. First, let us prove the inequality

$$\varphi_2(8a_1 t) v^{\frac{1}{p}}(8a_1 t) \le b_1 \varphi_1 \left(\frac{t}{a_1} \right),$$

where the positive constant b_1 does not depend on $t > 0$. Indeed, by Lemma 6.9.2 (part (i)) and the reverse doubling condition for μ we have

$$c \geq B(t) \geq$$

$$\geq \left\| \varphi_2(d(x_0,\cdot))(\mu B(x_0, d(x_0,\cdot)))^{-1}\chi_{\{t<d(x_0,y)<\eta_1 t\}}(\cdot) \right\|_{L^{ps}_{v(d(x_0,\cdot))}(X)} \times$$

$$\times \left\| \frac{1}{\varphi_1(d(x_0,\cdot))}\chi_{\{d(x_0,y)\leq\frac{t}{8a_1^2}\}}(\cdot) \right\|_{L^{p's'}(X)} \geq$$

$$\geq c_1(\mu B(x_0, \eta_1 t))^{-1}\varphi_2(t)\left(\int\limits_{\{t<d(x_0,y)<\eta_1 t\}} v(d(x_0,y))d\mu \right)^{\frac{1}{p}} \times$$

$$\times \varphi_1^{-1}(\frac{t}{8a_1^2})(\mu B(x_0, t/(8a_1^2)))^{1/p} \geq$$

$$\geq c_2\varphi_2(t)v^{1/p}(t)\mu B(x_0,t)^{-1}\mu B(x_0,t)^{1/p'}\varphi_1^{-1}(\frac{t}{8a_1^2})\mu B(x_0,t)^{1/p'} =$$

$$= c_2\frac{\varphi_2(t)v^{1/p}(t)}{\varphi_1(\frac{t}{8a_1^2})}.$$

Now we show that the following condition is fulfilled:

$$\sup_{t>0} B_1(t) \equiv \sup_{t>0} \left\| \varphi_2(d(x_0,\cdot))\chi_{\{d(x_0,y)\leq t\}}(\cdot) \right\|_{L^{pq}_v(X)} \times$$

$$\times \left\| \frac{\mu B(x_0, d(x_0,\cdot))^{-1}}{\varphi_1(d(x_0,\cdot))}\chi_{\{d(x_0,y)>t\}}(\cdot) \right\|_{L^{p's'}(X)} < \infty.$$

Indeed, by the monotonic property of the functions φ_1, φ_2, and v and by Lemma 6.9.2 (part(i)), we obtain:

$$B_1(t) \leq c_3\varphi_2(t)(\mu B(x_0,t))^{1/p}v^{1/p}(t)\varphi_1^{-1}(\frac{t}{8a_1^2}) \times$$

$$\times \left\| (\mu B(x_0, d(x_0,\cdot))^{-1}\chi_{\{d(x_0,y)>t\}}(\cdot) \right\|_{L^{p's'}(X)}.$$

On the other hand, we have

$$\left\| (\mu B(x_0, d(x_0,\cdot))^{-1}\chi_{\{d(x_0,y)>t\}}(\cdot) \right\|_{L^{p's'}(X)} =$$

$$= \left(s' \int\limits_0^\infty \lambda^{s'-1}(\mu\{x : (\mu B(x_0, d(x_0,x)))^{-1} > \lambda\} \cap \right.$$

$$\left. \cap \{d(x_0,x) \geq t\}))^{\frac{s'}{p'}}d\lambda \right)^{\frac{1}{s'}} \leq$$

$$= \left(s' \int_0^{(\mu B(x_0,t))^{-1}} \lambda^{s'-1} (\mu\{x : \mu B(x_0, d(x_0,x)) < \lambda^{-1}\})^{\frac{s'}{p'}} d\lambda \right)^{\frac{1}{s'}} \le$$

$$\le c_4 \left(\int_0^{(\mu B(x_0,t))^{-1}} \lambda^{s'-1} \lambda^{-\frac{s'}{p'}} d\lambda \right)^{\frac{1}{s'}} =$$

$$= c_5 (\mu B(x_0, t))^{-1/p}.$$

Here we have used the inequality (see the proof of Proposition 6.1.2)

$$\mu\{x : \mu B(x_0, d(x_0,x)) < \lambda^{-1}\} \le b\lambda^{-1},$$

where the positive constant b is from the doubling condition for μ. Thus, we obtain

$$B_1(t) \le c_6 \varphi_2(t) v^{1/p}(t) \varphi_1^{-1}(t/8a_1^2) \le c_7$$

for arbitrary $t > 0$.

By Theorem 8.5.2 we conclude that inequality (8.5.3) holds. □

From now in this chapter we shall consider singular integrals on spaces of nonhomogeneous type.

Let (X, d, μ) be a measure and a metric space with a measure μ and a metric d (i.e. $a_0 = 1$, $a_1 = 1$ and (vii) fail in Definition 1.1.1) and satisfying the condition

$$\mu\overline{B}(x,r) \le r^\alpha, \quad x \in X, \; r > 0.$$

for some $\alpha > 0$, where $\overline{B}(x,r) \equiv \{y : d(x,y) \le r\}$.

Recall that $a \equiv \sup\{d(x_0, x) : x \in X\}$, where x_0 is a fixed point of X and $B(x_0, R) \setminus B(x_0, r) \ne \emptyset$ for all $0 < r < R < a$.

Let us mention that the condition $\mu\{x\} = 0$ for all $x \in X$ is automatically satisfied.

Assume that T is a singular integral operator on (X, d, μ) (see Section 8.3.).

From Theorem A of Section 8.3 and from an interpolation theorem (see, e.g., [280], Theorem 3.15) we have the following theorem

Theorem C. *Let* $1 < p, q < \infty$. *Then* T *is bounded in* $L^{pq}(X)$.

The next Lemmas are obtained in the same way as in the homogeneous case.

Lemma 8.5.5. *Let* $1 < s \le p < \infty$. *Suppose the weight functions* w *and* w_1 *satisfy the conditions*

(1) *there exists an increasing function* v *on* $(0, 4a)$ *such that the inequality*

$$v(d(x_0,x)) \le bw(x)w_1(x)$$

holds for almost all $x \in X$;

(2) for every t, $0 < t < a$, the norm

$$\left\| \frac{1}{w(\cdot)\, w_1(\cdot)} \chi_{\{d(x_0,y) \leq t\}}(\cdot) \right\|_{L_w^{p'\, s'}(X)}$$

is finite.

Then $Tg(x)$ exists a.e. on X for any g satisfying the condition

$$\|g(\cdot)w_1(\cdot)\|_{L_w^{ps}(X)} < \infty.$$

Lemma 8.5.6. *Let* $1 < s \leq p < \infty$. *Suppose also that* u *and* u_1 *are positive increasing functions on* $(0, 4a_1 a)$ *and*

$$\left\| \frac{1}{u(d(x_0,\cdot))u_1(d(x_0,\cdot))} \chi_{\overline{B}(x_0,t)}(\cdot) \right\|_{L_{u(d(x_0,\cdot))}^{p'\, s'}(X)} < \infty$$

for all t *satisfying the condition* $0 < t < a$. *Then for arbitrary* φ *with*

$$\|\varphi(\cdot)u_1(d(x_0,\cdot))\|_{L_{u(d(x_0,\cdot))}^{ps}(X)} < \infty,$$

$T\varphi(x)$ *exists a.e. on* X.

The next lemmas also hold:

Lemma 8.5.7. *Let* $a = \infty$ *and* $1 < s \leq p < \infty$. *Suppose that for weights* w *and* w_1 *the following conditions are satisfied:*

(1) *there exists a decreasing positive function* v *on* $(0, \infty)$ *such that*

$$v(d(x_0,\cdot)) \leq bw(x)w_1^p(x)$$

for almost all $x \in X$;

(2) *for every* $t > 0$,

$$\left\| \frac{d(x_0,\cdot)^{-\alpha}}{w(\cdot)w_1(\cdot)} \chi_{X \setminus \overline{B}(x_0,t)}(\cdot) \right\|_{L_w^{p'\, s'}(X)} < \infty.$$

Then $Tg(x)$ exists a.e. on X for arbitrary g satisfying $\|g(\cdot)w_1(\cdot)\|_{L_w^{ps}(X)} < \infty$,

From the previous lemmas we can easily obtain

Lemma 8.5.8. *Let* $a = \infty$, $1 < s \leq p < \infty$. *Suppose also that for the decreasing functions* u *and* u_1 *on* $(0, \infty)$ *the following condition is satisfied:*

$$\left\| \frac{d(x_0,\cdot)^{-\alpha}}{u(d(x_0,\cdot))u_1(d(x_0,\cdot))} \chi_{X \setminus \overline{B}(x_0,t)}(\cdot) \right\|_{L_{u(d(x_0,\cdot))}^{p'\, s'}(X)} < \infty,$$

for all $t > 0$. *Then* $Tg(x)$ *exists a.e. on* X *for* g *satisfying the condition*

$$g(\cdot)u_1(d(x_0, \cdot)) \in L^{ps}_{u(d(x_0,\cdot))}(X).$$

Using Proposition A and Theorem C we obtain the following results in the same way as Theorem 8.5.1.

Theorem 8.5.6. *Let* $1 < s \le p \le q < \infty$ *and let* w *be a weight function on* X. *Assume that* v *is a positive increasing continuous function on* $(0, 4a)$. *Suppose also that the following two conditions are satisfied:*
1) there exists a positive constant c such that the inequality

$$v(2a_1 d(x_0, x)) \le cw(x)$$

holds for almost every $x \in X$;
2)

$$\sup_{0<t<a} \|(d(x_0, \cdot))^{-\alpha}\chi_{X\setminus\overline{B}(x_0,t)}\|_{L^{pq}_v(X)} \times$$

$$\times \left|\frac{1}{w(\cdot)}\chi_{\overline{B}(x_0,t)}(\cdot)\right|_{L^{p's'}_w(X)} < \infty.$$

Then the operator T is bounded from $L^{ps}_w(X)$ to $L^{pq}_{v(d(x_0,\cdot))}(X)$.

Theorem 8.5.7. *Let* $a = \infty$, *and let* $1 < s \le p \le q < \infty$; *suppose that* v *is a positive decreasing continuous function on* $(0, \infty)$. *Assume also that:*
(i) there exists a positive constant b such that for almost every $x \in X$ the following inequality is true:

$$v\left(\frac{d(x_0, x)}{2a_1}\right) \le bw(x);$$

(ii)

$$\sup_{t>0} \left\|\chi_{\overline{B}(x_0,t)}(\cdot)\right\|_{L^{pq}_v(X)} \times$$

$$\times \left\|\frac{(d(x_0, \cdot))^{-\alpha}}{w(\cdot)}\chi_{X\setminus\overline{B}(x_0,t)}(\cdot)\right\|_{L^{p's'}_w(X)} < \infty.$$

Then T acts boundedly from $L^{ps}_w(X)$ to $L^{pq}_{v(d(x_0,\cdot))}(X)$.

Finally we formulate

Theorem 8.5.8. *Let* $a = \infty$, $1 < s \le p \le q < \infty$. *Suppose that* v, w, u_1 *and* u_2 *are weights on* X. *Assume that the following conditions are fulfilled:*

1) there exists a positive constant b such that

$$\sup_{F_t} v^{\frac{1}{p}}(x) \sup_{F_t} u_2(x) \leq b \inf_{F_t} w^{\frac{1}{p}}(x) \inf_{F_t} u_1(x)$$

holds for all $t > 0$, where $F_t = \{x \in X : \frac{t}{a_1} \leq d(x_0, x) < 8a_1 t\}$;

2)

$$\sup_{t>0} \left\| u_2(\cdot)(d(x_0, \cdot))^{-\alpha} \chi_{X \setminus \overline{B}(x_0, t)}(\cdot) \right\|_{L_v^{pq}(X)} \times$$

$$\times \left\| \frac{1}{u_1(\cdot)w(\cdot)} \chi_{\overline{B}(x_0, t)}(\cdot) \right\|_{L_w^{p's'}(X)} < \infty;$$

3)

$$\sup_{t>0} \left\| u_2(\cdot) \chi_{\overline{B}(x_0, t)}(\cdot) \right\|_{L_v^{pq}(X)} \times$$

$$\times \left\| (d(x_0, \cdot))^{-\alpha} (u_1(\cdot)w(\cdot))^{-1} \chi_{X \setminus \overline{B}(x_0, t)}(\cdot) \right\|_{L_w^{p's'}(X)} < \infty.$$

Then

$$\| u_2(\cdot)Tf(\cdot) \|_{L_v^{pq}(X)} \leq c \| u_1(\cdot)f(\cdot) \|_{L_w^{ps}(X)},$$

where the positive constant c does not depend on f.

Now we give some applications of the results derived above.

Let $\Gamma \subset C$ be a connected rectifiable curve and let ν be arc-length measure on Γ.

As we know (see Section 8.1.), $\Gamma \subset C$ is called regular if

$$\nu(\Gamma \cap B(z, r)) \leq c_0 r$$

for every $z \in \Gamma$ and all $r > 0$. For r smaller than half the diameter of Γ, the reverse inequality: $\nu(\Gamma \cap B(z, r)) \geq r$ holds for all $z \in C$. Equipped with ν and the Euclidean metric, the regular curve becomes an SHT.

The Cauchy integral

$$S_\Gamma f(t) = \int_\Gamma \frac{f(\tau)}{t - \tau} d\nu(\tau)$$

is the corresponding singular operator.

Let $\overline{\Gamma}(z, r) = \{t \in C : |t - z| \leq r\} \cap \Gamma$. From Theorem 8.5.3 there follows

Proposition 8.5.1. *Let $1 < p \leq q < \infty$, let $\nu(\Gamma) = \infty$ and $t_0 \in \Gamma$. Let σ_1 and σ_2 be a positive, increasing functions on $(0, \infty)$, with σ_1 continuous, let $\rho \in A_p(\Gamma)$ and $\nu(t) = \sigma_2(|t - t_0|)\rho(t)$, $w(t) = \sigma_1(|t - t_0|)\rho(t)$. If*

$$\sup_{r>0} \left\| |\cdot - t_0|^{-1} \chi_{\Gamma \setminus \overline{\Gamma}(t_0, r)}(\cdot) \right\|_{L_v^{pq}(\Gamma)} \left\| \frac{1}{w(\cdot)} \chi_{\overline{\Gamma}(t_0, r)}(\cdot) \right\|_{L_w^p(\Gamma)} < \infty,$$

then the inequality:

$$\|S_\Gamma f(\cdot)\|_{L_v^{pq}(\Gamma)} \le c \|f(\cdot)\|_{L_w^p(\Gamma)}$$

holds, with a constant c independent of $f \in L_w^p(\Gamma)$.
From Theorem 8.5.5 we derive

Proposition 8.5.2. *Let $\nu(\Gamma) = \infty$ and $1 < s \le p \le q < \infty$, let φ_1, φ_2 and v be positive, increasing functions on $(0, \infty)$ and let $t_0 \in \Gamma$. If*

$$\sup_{r>0} \left\| \varphi_2 \left(|\cdot - t_0| \right) |\cdot - t_0|^{-1} \chi_{\Gamma \setminus \overline{\Gamma}(t_0,r)}(\cdot) \right\|_{L_{v(|\cdot - t_0|)}^{pq}(\Gamma)} \times$$

$$\times \left\| \frac{1}{\varphi_1 \left(|\cdot - t_0| \right)} \chi_{\overline{\Gamma}(t_0,r)}(\cdot) \right\|_{L^{p's'}(\Gamma)} < \infty,$$

then there exists a positive constant c such that

$$\|S_\Gamma f(\cdot)\varphi_2 \left(|\cdot - t_0| \right)\|_{L_{v(|\cdot - t_0|)}^{pq}(\Gamma)} \le c \|f(\cdot)\varphi_1 \left(|\cdot - t_0| \right)\|_{L^{ps}(\Gamma)} \quad (8.5.4)$$

holds for all f with $\|f(\cdot)\varphi_1 \left(|\cdot - t_0| \right)\|_{L^{ps}(\Gamma)} < \infty$.

Now let $\Gamma \subset R^n$ be an $s-$ set $(0 \le s \le n)$ in R^n with the conditions:
(i) supp $\mu = \Gamma$;
(ii) there exist positive constants c_1 and c_2 such that

$$c_1 r^s \le \mu \left(B \left(x, r \right) \cap \Gamma \right) \le c_2 r^s \quad (8.5.5)$$

for arbitrary $x \in \Gamma$ and $r \in (0, 1)$.

In thit case the measure μ is referred as s-measure.
As we mentioned in 8.1 μ is equivalent to the restriction of Hausdorff s-measure H_s to Γ. We shall thus identify μ with $H_s|_\Gamma$. For $x \in \Gamma$ and $r > 0$ let $B(x,r) \cap \Gamma = \Gamma(x,r)$. From (8.5.5) follows, that the doubling condition

$$H_s \left(\Gamma \left(x, 2r \right) \right) \le b H_s \Gamma \left(x, r \right)$$

holds with a constant b independent of $x \in \Gamma$ and of $r, 0 < r \le 1$.
Let K_Γ be a singular integral operator. Assume $a \equiv \sup \{|x - x_0| : x \in \Gamma\}$ where $x_0 \in \Gamma$. We have, for example, the following

Proposition 8.5.3. *Let $1 < \lambda \le p \le q < \infty$, and let σ be a positive increasing continuous function on $(0, 4a_1 a)$. Let $\rho \in A_p(X)$ and let w be a weight function on Γ, put $v(x) = \sigma \left(2 |x - x_0| \right) \rho(x)$ Suppose:*
a) There exists positive constant b such that

$$\sigma \left(2 |x - x_0| \right) \rho(x) \le bw(x)$$

for a.e. $x \in \Gamma$;

b)

$$\sup_{0<t<a} \left\| |x - x_0|^{-s} \chi_{\{x \in \Gamma : |x-x_0|>t\}}(\cdot) \right\|_{L_v^{pq}(\Gamma)} \times$$

$$\times \left\| \frac{1}{w(\cdot)} \chi_{\{x \in \Gamma : |x-x_0| \leq t\}}(\cdot) \right\|_{L_w^{p'\lambda'}(\Gamma)} < \infty.$$

Then

$$\|K_\Gamma f(\cdot)\|_{L_v^{pq}(\Gamma)} \leq c \|f(\cdot)\|_{L_w^{p\lambda}(\Gamma)}.$$

Proposition 8.5.4. *Let* $1 < p \leq q < \infty$, $H_s(\Gamma) = \infty$, Γ *be a global s-set, i.e.* $H_s(B(x,r)) \approx r^s$ *for all* $r > 0$. *Let* σ_1 *and* σ_2 *be positive, increasing functions on* $(0, \infty)$, *with* σ_1 *continuous. Let* $\rho \in A_p(\Gamma)$, $v(x) = \sigma_2(|x - x_0|)\rho(x)$, $w(x) = \sigma_1(|x - x_0|)\rho(x)$. *For* v *and* w *suppose that*

$$\sup_{t>0} \left\| |\cdot - t_0|^{-s} \chi_{\{x \in \Gamma : |x-x_0|>t\}}(\cdot) \right\|_{L_v^{pq}(\Gamma)} \times$$

$$\times \left\| \frac{1}{w(\cdot)} \chi_{\{x \in \Gamma | |x-x_0| \leq t\}}(\cdot) \right\|_{L_w^{p'}(\Gamma)} < \infty.$$

Then

$$\|K_\Gamma f(\cdot)\|_{L_v^{pq}(\Gamma)} \leq c \|f(\cdot)\|_{L_w^p(\Gamma)}$$

holds for arbitrary $f \in L_w^p(\Gamma)$.

Let $(X, d\mu)$ be an SHT such that the condition $\mu(B(x,r)) \approx r$ holds (if $a < \infty$, then we assume that this condition is fulfilled for $0 < r \leq 1$). It is known (see Section 8.1), that if $\mu(X) < \infty$, then

$$\int_X |Kf(x)|^p (d(x_0, x))^{p-1} d\mu \leq c \int_X |f(x)|^p (d(x_0, x))^{p-1} \log^p \frac{b}{d(x_0, x)} d\mu$$

where $x_0 \in X$, $1 < p < \infty$, and $b = 8a_1 a e^{p'}$ and the positive constant c does not depend on f.

Example 8.5.1. Let $1 < p < q < \infty$, and let $a < \infty$; put $v(t) = t^{p-1}$, $w(t) = t^{p-1} \ln^\gamma \frac{b}{t}$, then $t \in (0, 4a_1 a)$, where $b = 8a_1 a e^{\frac{\gamma}{p-1}}$, $\gamma = \frac{p}{q} + p - 1$, and a_1 is the positive constant from Definition 1.1. Then from Theorem 8.5.1 it follows that

$$\|Kf(\cdot)\|_{L_{v(d(x_0,\cdot))}^{pq}(X)} \leq c \|f(\cdot)\|_{L_{w(d(x_0,\cdot))}^p(X)}, \tag{8.5.6}$$

where the positive constant c does not depend on f.

Example 8.5.2. Let $1 < p < q < \infty$, $a = \infty$, $v(t) = t^{p-1}$ when $0 < t \le 1$ and $v(t) = t^{\alpha}$ when $t > 1$, and let $w(t) = t^{p-1} \ln^{\gamma} \frac{2e^{\frac{\gamma}{p-1}}}{t}$ when $t \le 1$, and $w(t) = t^{\beta} \ln^{\beta} \left(2e^{\frac{\gamma}{p-1}}\right)$ when $t > 1$, where $\gamma = \frac{p}{q} + p - 1$, $0 < \alpha \le \beta < p - 1$. Then inequality (8.5.6) holds.

An appropriate example for the conjugate function

$$\tilde{f}(x) = \frac{1}{\pi} \int\limits_{-\pi}^{\pi} \frac{f(x+t)}{2tg\frac{t}{2}} dt$$

is presented in [146].

Remark 8.5.2. The main results of this section are also valid if we do not require the continuity of σ in the homogeneous case and the continuity of v if (X, d, μ) is a space of nonhomogeneous type, but suppose that $\mu B(x_0, r)$ is continuous with respect to r.

8.6. Cauchy–Szegö projection

In the complex space $\mathbf{C}^{n+1}$ let us consider the upper half–space

$$U^n = \left\{ z \in \mathbf{C}^{n+1} : Im \ z_{n+1} > \sum_{j=1}^{n} |z_j|^2 \right\}.$$

The boundary of U^n is

$$bU^n = \left\{ z \in \mathbf{C}^{n+1} : Im \ z_{n+1} = \sum_{j=1}^{n} |z_j|^2 \right\}.$$

We come now to the Heisenberg group, which gives the translation of the domain U^n. Abstractly this group consists of the set

$$\mathbf{C}^n \times R = \{[\zeta, t] : \zeta \in \mathbf{C}^n, \ t \in R\}$$

with the multiplication law

$$[\zeta, t] \odot [\eta, s] = [\zeta + \eta, t + s + 2 \ Im \ (\zeta, \bar{\eta})] \qquad (8.6.1)$$

It is easy to see that the law (8.6.1) makes $\mathbf{C}^n \times R$ into a group whose identity is the origin $[0, 0]$ and where the inverse is given by $[\zeta, t]^{-1} = [-\zeta, -t]$. The couple $(\mathbf{C}^n \times R, \odot)$ is the Heisenberg group and will be denoted by H^n (see [278]).

We can identify the Heisenberg group with bU^n via its action on the origin

$$H^n \ni [\zeta, t] \to \left(\zeta, t + i |\zeta|^2\right) \in bU^n \qquad (8.6.2)$$

Since the Heisenberg group H^n preserves the boundary bH^n, and since we have identified bH^n with H^n, we get as a result an action of H^n on itself. Because the general element of bU^n is of the form $h(0)$ for some $h \in H^n$, the action of another $h_1 \in H^n$ maps it to $(h_1 h)(0)$, and so the action of H^n on itself is simply by left translation:

$$h_1 : h \to h_1 h.$$

The Cauchy–Szegö integral may be viewed as the orthogonal projection of $L^2(H^n)$ onto its subspace of boundary values of holomorphic functions.

The identification of bU^n with H^n by (8.6.2) allows us to transport the Haar measure dh on H^n to a measure $d\beta$ on bU^n; that is, we have the integration formula

$$\int_{bU^n} F(z) d\beta(z) = \int_{C^n \times R^1} F(z', t + i |z'|^2) dz' dt$$

for (say) continuous F of compact support. With these measures we can define the space $L^2(H^n) = L^2(bU^n)$.

For function F on U^n, we write F_ε for its "vertical translate":

$$F_\varepsilon(z) = F(z + \varepsilon i), \quad \text{where } i = (0, 0, ..., 0, i).$$

If $\varepsilon > 0$, then F_ε is defined on bU^n.

The space $\mathcal{H}^2(U^n)$ consists of all functions F holomorphic on U^n, for which

$$\|F\|_{\mathcal{H}^2} \equiv \left(\sup_{\varepsilon > 0} \int_{bU^n} |F_\varepsilon(z)|^2 \, d\beta(z) \right)^{\frac{1}{2}} < \infty. \qquad (8.6.3)$$

The space $\mathcal{H}^2(U^n)$ is a Hilbert space and its element can be identified with their boundary values,

Theorem A ([278]). *Let $F \in \mathcal{H}^2(U^n)$. Then*

(i) there exists an $F^b \in L^2(bU^n)$ so that $F(z + \varepsilon i)\big|_{bU^n} \to F^b$ in the $L^2(bU^n)$ norm, as $\varepsilon \to 0$;

(ii) the space of F^b so obtained is closed subspace of $L^2(bU^n)$;

(iii) $\left\| F^b \right\|_{L^2(bU^n)} = \|F\|_{\mathcal{H}^2}$.

We shall now define the Cauchy–Szegö kernel $s(z, w)$ for the domain U^n. This is the function defined on $U^n \times U^n$ with the following properties:

(a) For each $w \in U^n$, the function $z \to s(z, w)$ is holomorphic for $z \in U^n$ and belongs to $\mathcal{H}^2(U^n)$. This allows us to define, for each $w \in U^n$, the boundary function $s^b(z, w)$ (which is defined for almost all $z \in bU^n$).

(b) The kernel s is symmetric, in the sense that

$$s(t, w) = \overline{s(w, t)}$$

for each $w \in U^n$.

(c) The kernel s satisfies the reproducing property

$$F(z) = \int_{bU^n} s(z, w) F^b(w) d\beta(w), \quad z \in U^n,$$

whenever $F \in \mathcal{H}^2(U^n)$.

Let T denotes the Cauchy–Szegö projection operator: it is the orthogonal projection from $L^2(U^n)$ to the subspace of functions $\left\{ F^b \right\}$ that are boundary values of functions $F \in \mathcal{H}^2(U^n)$. In other words, for each $f \in L^2(U^n)$, we have that $Tf = F^b$ for some $F \in \mathcal{H}^2(U^n)$. Moreover, $Tf = F^b$ and T is self adjoint, i.e. $T^* = T$.

For fixed $z \in U^n$ we have that $TF(z) = F(z)$ is well defined and the kernel $s(z, w)$ will be determined by the representation

$$F(z) = \int_{bU^n} s(z, w) f(w) d\beta(w). \tag{8.6.4}$$

The mapping $f \to Tf$ assigns to each element $f \in L^2(bU^n)$ another element of the form $Tf = F^b$ for some $F \in \mathcal{H}^2(U^n)$.

As a consequence of (8.6.4) and Theorem A we have that

$$Tf(z) = \lim_{\varepsilon \to 0} \int_{bU^n} s(z + \varepsilon i, w) f(w) d\beta(w), \quad z \in bU^n, \tag{8.6.5}$$

where the limit exists in the $L^2(bU^n)$ norm.

We now use the identification of bU^n with H^n. Then

$$Tf(x) = \lim_{\varepsilon \to 0} \int_{H^n} K_\varepsilon(y^{-1} \odot x) f(y) dy, \quad f \in L^2(H^n)$$

where the limit is taken in $L^2(H^n)$ and $K_\varepsilon(x) = c[t + i\,|\zeta|^2 + i\varepsilon]^{-n-1}, x = [\zeta, t], \quad c = (2i)^{n+1}\, c_n = 2^{n-1} i^{n+1} n! / \pi^{n+1}$. Since $K_\varepsilon(x) = -\frac{\partial}{\partial t}(\frac{c}{n}[t + i\,|\zeta|^2 + i\varepsilon]^{-n})$, and the function $[t + i\,|\zeta|^2]^{-n}$ is locally integrable on H^n, we see that the distribution K is given by

$$K = -\frac{\partial}{\partial t}(\frac{c}{n}[t + i\,|\zeta|^2]^{-n}),$$

and that this distribution equals the function

$$c(t + i|\zeta|^2)^{-n-1}$$

away from the origin. Another immediate consequence of (8.6.5) is that we can write

$$Tf(x) = \int_{H^n} K(x,y)f(y)dy, \qquad (8.6.6)$$

where $K(x,y) = K(y^{-1}x)$ for $x \neq y$ (i.e. $y^{-1}x \neq 0$). In particular, (8.6.6) holds for $f \in L^2(H^n)$ with support in a compact set, for every x outside the support of f.

We next consider the quasi-norm ρ given by

$$\rho(x) = \rho([\zeta, t]) = \max(|\zeta|, |t|^{\frac{1}{2}}).$$

Note that $\rho(x^{-1}) = \rho(-x) = \rho(x)$ and $\rho(\delta \circ x) = \delta\rho(x)$, where

$$\delta \circ x = [\delta\zeta, \delta^2 t].$$

In addition, the function ρ satisfies a quasi–triangle inequality

$$\rho(x \odot y) \leq c\left(\rho(x) + \rho(y)\right).$$

We define the quasi–distance $\rho(x,y) = \rho(y^{-1} \odot x)$ on H^n. Further ρ is clearly symmetric. It is clear that

$$\rho(x,y) \leq c\left(\rho(x,z) + \rho(y,z)\right)$$

for every $x, y, z \in H^n$.

Let $v(x,y) = |B(y,\delta)| = c'\delta^{2n+2} = c'\rho(x,y)^{2n+2}$.

Observe also that the kernel

$$K(x) = c(i|\zeta|^2 + t)^{-n-1}$$

of the Cauchy–Szegö projection satisfies the condition $|K(x)| \approx \rho(x)^{-2n-2}$. If we write $K(x,y) = K(y^{-1} \odot x)$, then we have that $|K(x,y)| \approx \rho(x,y)^{-2n-2}$; in particular

$$|K(x,y)| \leq \frac{A}{v(x,y)}.$$

Also, we have

$$|K(x,y) - K(x,y_0)| \leq A\frac{\rho(y,y_0)}{\rho(x,y_0)}v(x,y_0)^{-1}$$

wherever $\rho(x,y_0) \geq \bar{c}\rho(y,y_0)$, for some appropriately large constant $\bar{c}$. Note that K is a formally self-adjoint; i.e. $K(x,y) = \overline{K}(y,x)$.

Theorem B ([278]). *Let* $1 < p < \infty$. *Then the Cauchy–Szegö projection has an extension to a bounded operator on* $L^p(H^n)$.

Theorem C ([86], [278]). *There exists a positive constant c such that if F is a nonnegative function on* $(0, \infty)$, *then*

$$\int\limits_{H^n} F(\rho(x))dx = c \int\limits_0^\infty F(r)r^{2n+1}dr,$$

where dx is a Haar measure on H^n.

This theorem is a particular case of Theorem A from Section 1.4.

Definition 8.6.1. *Let* $1 < p < \infty$. *We say that* $w \in A_p(H^n)$ *if*

$$\sup \left(\frac{1}{|B|} \int\limits_B w(x)dx \right) \left(\frac{1}{|B|} \int\limits_B w^{1-p'}(x)dx \right)^{p-1},$$

where the supremum is taken over all balls $B \subset H^n$ *and* $|B|$ *denotes the Haar measure of B. Further,* $w \in A_1(H^n)$ *if there exists a positive constant c such that for all balls* $B \subset H^n$ *the following inequality holds*

$$\frac{1}{|B|} \int\limits_B w(x)dx \leq c \operatorname*{ess\,sup}_{x \in B} w(x).$$

As H^n with a Haar measure and quasi–distance represents an SHT, we have the following

Theorem D. *Let* $1 < p < \infty$. *Then the operator* T *is bounded on* $L^p_w(H^n)$ *if* $w \in A_p(H^n)$.

As a consequence of results obtained in Section 8.1, we derive the following result:

Theorem 8.6.1. *Let* $1 < p < \infty$. *Suppose that* σ *and* u *are positive increasing functions on* $(0, \infty)$ *and* $\eta \in A_p(H^n)$. *We put* $v(x) = \eta(x)\sigma(\rho(x))$, $w(x) = \eta(x)u(\rho(x))$. *If*

$$\sup_{t>0} \left(\int\limits_{\rho(x)>t} v(x) \, (\rho(x))^{-(2n+2)p} \, dx \right)^{\frac{1}{p}} \left(\int\limits_{\rho(x)<t} w^{1-p'}(x)dx \right)^{\frac{1}{p'}} < \infty,$$

then T *is bounded from* $L^p_w(H^n)$ *into* $L^p_v(H^n)$.

Theorem 8.6.2. *Let* $1 < p < \infty$. *Assume that* σ *and* u *are positive decreasing functions on* $(0, \infty)$. *Suppose that* $\eta \in A_p(H^n)$. *We put* $v(x) =$

$\eta(x)\sigma(\rho(x))$ and $w(x) = \eta(x)u(\rho(x))$. *If*

$$\sup_{t>0}\left(\int_{\rho(x)<t} v(x)dx\right)^{\frac{1}{p}}\left(\int_{\rho(x)>t} w^{1-p'}(x)\rho(x)^{-(2n+2)p'}dx\right)^{\frac{1}{p'}} < \infty,$$

then T is bounded from $L_w^p(H^n)$ *into* $L_v^p(H^n)$.

Theorem 8.6.3. *Let* $1 < p < \infty$. *Let* σ *and* u *be positive increasing functions on* $(0,\infty)$. *Suppose that* $\eta \in A_1(H)$. *If*

$$\sup_{t>0}\left(\int_{\rho(x)>t} \frac{\sigma(\rho(x))}{\rho(x)^{(2n-2)p}}dx\right)^{1/p} \times$$

$$\times\left(\int_{\rho(x)<t} u^{1-p'}(\rho(x))dx\right)^{1/p'} < \infty,$$

then there exists a positive constant c such that for all f, $f \in L_{u(\rho(\cdot))\eta(\cdot)}(H^n)$, *the inequality*

$$\int_{H^n} |Tf(x)|^p \sigma(\rho(x))\eta(x)dx \le c\int_{H^n} |f(x)|^p u(\rho(x))\eta(x)dx$$

is fulfilled.

Theorem 8.6.4. *Let* $1 < p < \infty$. *Let* σ *and* u *be positive decreasing functions on* $(0,\infty)$. *Assume that*

$$\sup_{t>0}\left(\int_{\rho(x)<t} \sigma(\rho(x))dx\right)^{\frac{1}{p}}\left(\int_{\rho(x)>t} \frac{u^{1-p'}(\rho(x))}{\rho(x)^{(2n-2)p'}}dx\right)^{\frac{1}{p'}} < \infty$$

and $\eta \in A_1(H^n)$. *Then we have the inequality*

$$\int_{H^n} |Tf(x)|^p \sigma(\rho(x))\eta^{1-p}(\rho(x))\,dx \le c\int_{H^n} |f(x)|^p u(\rho(x))\eta^{1-p}(x)dx$$

with a constant c independent of f, $f \in L_{u(\rho(\cdot))\eta^{1-p}(\cdot)}^p(H^n)$.

Next we consider the case $q < p$. In this situation we have the following results, which can be obtained from the two-weight inequalities for singular integrals defined on homogeneous groups (see, e.g., [151], [100], Chapter 9).

Theorem 8.6.5. *Let* $1 < q < p < \infty$. *Suppose that* v *and* w *are a positive increasing functions on* $(0,\infty)$. *If*

$$\int_0^\infty\left[\left(\int_t^\infty v(\tau)\tau^{2n+1-(2n+2)q}d\tau\right)\left(\int_0^{\frac{t}{2}} w^{1-p'}(\tau)\tau^{2n+1}d\tau\right)^{q-1}\right]^{\frac{p}{p-q}} w^{1-p'} \times$$

$$\times (t/2)t^{2n+1}dt < \infty,$$

then the operator T is bounded from $L^p_{w(\rho(\cdot))}(H^n)$ into $L^q_{v(\rho(\cdot))}(H^n)$.

Theorem 8.6.6. *Let $1 < q < p < \infty$ and let v and w be positive decreasing functions on $(0, \infty)$. If*

$$\int\limits_0^\infty \left[\left(\int\limits_0^{\frac{t}{2}} v(\tau)\tau^{2n-1}d\tau \right) \times \right.$$

$$\left. \times \left(\int\limits_t^\infty w^{1-p'}(\tau)\tau^{-1-\frac{2n+2}{p-1}}d\tau \right)^{q-1} \right]^{\frac{p}{p-q}} w^{1-p'}(t)t^{-1-\frac{2n+2}{p-1}}dt < \infty,$$

then T is bounded from $L^p_{w(\rho(\cdot))}(H^n)$ into $L^q_{v(\rho(\cdot))}(H^n)$.

8.7.   Singular integrals via Clifford analysis

In this section we establish two- weight estimates for higher–dimensional singular integrals in the framework of Clifford analysis.

The study of the L^2- boundedness for singular integrals on Lipschitz surfaces was introduced by Murray in [216] for small Lipschitz constant. Analogous problem for all Lipschitz constants was considered by McIntosh in [198] (see also [172- 173], [211], [252]).

The L^2- boundedness problem for singular double- layer potential operator was solved by Calderón [4] for small Lipschitz constants and by Coifman, McIntosh and Meyer [55] for arbitrary Lipschitz constants.

Let $\mathcal{A}_m$ denote real or complex Clifford algebra, i.e. $\mathcal{A}_m$ is a 2^m- dimensional real or complex algebra (with identity e_0) generated by the basis vectors $e_0, e_1, \cdots, e_m$ satisfying the conditions:

$$e_j^2 = -e_0,$$

$$e_j e_k = -e_k e_j$$

when $1 \le j \le m, 1 \le k \le m$ and $j \ne k$.

It is assumed that $R^{m+1} = R \oplus R^m$ is embedded in $\mathcal{A}_m$. We denote by $|x|$ an Euclidean norm of $x = x_0 e_0 + X \in R^{m+1}$, where $X \in R^m$. The conjugate $\overline{x}$ of $x = x_0 e_0 + X \in R^{m+1}$ is $\overline{x} = x_0 e_0 - X \in R^{m+1}$.

In the sequel we shall denote by Σ the Lipschitz surface consisting of all points

$$x = g(X)e_0 + X \in R^{m+1},$$

where $X \in R^m$ and g is a real-valued Lipschitz function which satisfies the condition

$$\|Dg\|_\infty \equiv \operatorname*{ess\,sup}_{X \in R^m} \left\{ \sum_{j=1}^{m} \left| \frac{\partial g}{\partial x_j}(X) \right|^2 \right\}^{1/2} \leq \tan \omega < \infty$$

where $0 \leq \omega < \pi/2$.

We denote by C_Σ the singular Cauchy integral operator, defined for almost all $x \in \Sigma$ by

$$\left(C_\Sigma f\right)(x) \equiv \lim_{\epsilon \to 0} \frac{2}{\sigma_m} \int\limits_{\{y \in \Sigma : |x-y| > \epsilon\}} \frac{\overline{x-y}}{|x-y|^{m+1}} n(y) f(y) dS_y,$$

where dS_y is the surface measure $(dS_y = \left(1+|Dg(Y)|^2\right)^{1/2} dY, y = g(Y)e_0 + Y, Y \in R^m)$, $n(y)$ is the unit normal:

$$n(y) = \frac{e_0 - Dg(Y)}{(1+|Dg(Y)|^2)^{1/2}}$$

and $f : \Sigma \to \mathcal{A}_m$ is an $\mathcal{A}_m$ algebra–valued function (we assume that $\|\cdot\|$ is a norm in $\mathcal{A}_m$).

A measurable, locally integrable, a.e. positive function $w : \Sigma \to R$ is called a weight.

We shall need the Muckenhoupt class $A_p(\Sigma)$ $(1 < p < \infty)$ which is the set of all weights w such that

$$\sup_B \left(\frac{1}{|B|} \int\limits_B \overline{w}(X) dX \right) \left(\frac{1}{|B|} \int\limits_B \overline{w}^{1-p'}(X) dX \right)^{p-1} < \infty \quad \left(p' = \frac{p}{p-1} \right),$$

where the supremum is taken over all balls B in R^m and $\overline{w}(x) = w(g(X), X)$, $X \in R^m$. The Muckenhoupt class $A_1(\Sigma)$ is the set of all weights w such that

$$\sup_B \left(\frac{1}{|B|} \int\limits_B \overline{w}(X) dX \right) \operatorname*{ess\,sup}_{X \in B} \frac{1}{\overline{w}(X)} < \infty,$$

where again the supremum is taken over all balls $B \subset R^m$.

For weight w we denote by $L_w^p(\Sigma)$, $1 \leq p < \infty$, the space of all $\mathcal{A}_m$ algebra–valued functions f on Σ with finite norm

$$\|f\|_{L_w^p(\Sigma)} = \left(\int\limits_\Sigma \|f(x)\|^p w(x) dS_x \right)^{1/p}.$$

The following statements hold (see, e.g., [211], [128], [100], Chapter 5):

Theorem A. *Let* $1 < p < \infty$. *Then the integral operator* C_Σ *is bounded in* $L^p_w(\Sigma)$ *for* $w \in A_p(\Sigma)$.

Theorem B. *Let* $w \in A_1(\Sigma)$. *Then there exists a positive constant c such that for all* $\lambda > 0$ *and* $f \in L^1_w(\Sigma)$,

$$\int\limits_{\{x\in\Sigma:\|C_\Sigma f(x)\|>\lambda\}} w(x)dS_x \le \frac{c}{\lambda}\|f\|_{L^1_w(\Sigma)}.$$

Now we pass to two-weight inequalities for the operator C_Σ. We shall denote by $\nu(E)$ a surface measure of the measurable set $E \subset \Sigma$.

Theorem 8.7.1. *Let* $1 < p < \infty$, *suppose that* $\eta_0 \in \Sigma$ *and that* $\nu(\Sigma) = \infty$. *Let* σ *and* u *be positive increasing functions on* $(0, \infty)$ *and let* $\rho \in A_p(\Sigma)$. *We put* $v(x) = \sigma(|x - \eta_0|)\rho(x)$, $w(x) = u(|x - \eta_0|)\rho(x)$. *If*

$$\sup_{t>0} \left(\int\limits_{\Sigma\setminus D(\eta_0,t)} v(x)|x|^{-mp}dS_x \right) \left(\int\limits_{D(\eta_0,t)} w^{1-p'}(x)dS_x \right)^{p-1} < \infty, \quad (8.7.1)$$

where $D(\eta_0, t) \equiv \Sigma \cap B(\eta_0, t)$, $B(\eta_0, t) \equiv \{x \in \Sigma : |x - \eta_0| < t\}$, *then the operator* C_Σ *is bounded from* $L^p_w(\Sigma)$ *to* $L^p_v(\Sigma)$.

Theorem 8.7.2. *Let* $1 < p < \infty$ *and let* $\eta_0 \in \Sigma$. *Suppose that* $\nu(\Sigma) = \infty$. *Let* σ *and* u *be positive decreasing functions on* $(0, \infty)$ *and let* $\rho \in A_p(\Sigma)$. *We put* $v(x) = \sigma(|x - \eta_0|)\rho(x)$, $w(x) = u(|x - \eta_0|)\rho(x)$. *If*

$$\sup_{t>0} \left(\int\limits_{D(\eta_0,t)} v(x)dS_x \right) \left(\int\limits_{\Sigma\setminus D(\eta_0,t)} w^{1-p'}(x)|x|^{-p'm}dS_x \right)^{p-1} < \infty, \quad (8.7.2)$$

then the operator C_Σ *is bounded from* $L^p_w(\Sigma)$ *to* $L^p_v(\Sigma)$.

From the results of Section 8.5 it follows that conditions (8.7.1) and (8.7.2) are necessary in the case of the Hilbert transform H.

Similar results for singular integrals defined on SHT were presented in Sections 8.1 and 8.2.

Now we provide several examples of weights guaranteeing the boundedness of the operator C_Σ in weighted spaces (For the Hilbert transform H see Section 8.4)

Example 8.7.1. Let $1 < p < \infty$, $\eta_0 \in \Sigma$ and let $b \equiv diam\,\Sigma < \infty$. Let $v(x) = |x - \eta_0|^{(p-1)m}$, $w(x) = |x - \eta_0|^{(p-1)m} \ln^p \frac{4b}{|x-\eta_0|}$. Then the operator C_Σ is bounded from $L^p_w(\Sigma)$ to $L^p_v(\Sigma)$.

Example 8.7.2. Let $1 < p < \infty$ and $\eta_0 \in \Sigma$. Suppose that $b < \infty$. Let $v(x) = |x - \eta_0|^{-m} \ln^{-p} \frac{4b}{|x - \eta_0|}$, $w(x) = |x - \eta_0|^{-m}$. Then the operator C_Σ is bounded from $L_w^p(\Sigma)$ to $L_v^p(\Sigma)$.

Analogous results for the singular integral operator on Σ

$$Tf(x) = p.v. \int_\Sigma f(y)k(x,y)dS_y, \quad x \in \Sigma,$$

hold, where the kernel k satisfies the following conditions:

(i) there exists a positive constant c such that for all x y with $y - x \notin \Gamma_a$ (Γ_a is the upright circular cone in the upper–half space R_+^{m+1} having aperture a, $0 < a < \pi/2 - arctg(\|\nabla g\|_\infty)$, and whose vertex is at the origin) the inequality

$$\|k(x,y)\| \leq c|x - y|^{-m},$$

holds;

(ii) for any element $\xi \in \mathcal{A}_m$ there exists $\epsilon = \epsilon(\xi) > 0$ such that for any $x \in R^{m+1}$

$$D < k(\cdot, x), \xi >= 0 \quad \text{on} \quad R^{m+1} \setminus (-\Gamma_a - \epsilon + x),$$

$$< k(\cdot, x), \xi > D = 0 \quad \text{on} \quad R^{m+1} \setminus (\Gamma_a + \epsilon + x).$$

We note that results formulated above remain valid if we replace C_Σ by its scalar part (the singular double- layer potential operator)

$$C_{\Sigma 0}f(x) = \frac{2}{\sigma_m} \lim_{\epsilon \to 0} \int_{\{y \in \Sigma : |x-y| > \epsilon\}} \frac{< x - y, n(y) >}{|x - y|^{m+1}} f(y)dS_y.$$

For real- valued functions all these statements follow from the results obtained in Sections 8.1 and 8.2. In the case of algebra–valued functions these results can be obtained just in the same way as in the previous sections.

8.8. Theorems of Koosis type

In the present section we consider problems of Koosis type for singular integrals defined on homogeneous groups and spaces of homogeneous type (SHT).

Analogous problems for Calderón–Zygmund singular integrals have been studied in [45], [249- 250], [308] (see also [96]).

The definition of a homogeneous group G is presented in Section 1.4. We shall assume that $r(x)$ is a homogeneous norm of $x \in G$ and Q is a homogeneous dimension of G.

For measurable $f : G \to R$ let

$$Tf(x) \equiv \lim_{\varepsilon \to 0} \int_{G \setminus B(x,\varepsilon)} k(xy^{-1}) f(y) dy,$$

be a singular integral, where the kernel $k : G \to R$ satisfies the following conditions:

(1)
$|k(x)| \leq \frac{c}{r(x)^Q}$ for arbitrary $x \in G$;

(2) k is homogeneous of order $-Q$, i.e.,

$$K(\delta_t x) = t^{-Q} K(x),$$

for arbitrary $x \in G$ and $t > 0$;

(3)
$$\int_{S_G} K(x) d\sigma(x) = 0, \quad \text{where } S_G = \{x \in G : r(x) = 1\};$$

(4)
modulus of continuity

$$\omega(k, \delta) = \sup \left\{ |k(x) - k(y)| : x, y \in S_G, r(xy^{-1}) \leq \delta \right\}$$

of the function k satisfies the condition

$$\int_0^1 \frac{\omega(k, t)}{t} dt < \infty.$$

As it is known (see, e.g., [278], [128], [100]) the operator T is bounded from $L^1(G)$ into $L^{1\infty}(G)$.

We begin with the following Lemma:

Lemma 8.8.1. *Let $0 < r < 1$ and let T be a singular integral operator. Then for $f \in L^1_{(1+r(\cdot))^{-Q}}(G)$ the following inequality holds:*

$$\left(\int_{S_k} |Tf(x)|^r dx \right)^{\frac{1}{r}} \leq c|S_k|^{\frac{1}{r}} \int_G |f(x)|(1 + r(x))^{-Q} dx, \qquad (8.8.1)$$

where

$$S_0 = \left\{ x \in G : r(x) < \frac{1}{c_0} \right\},$$

$$S_k = \left\{ x \in G : \frac{2^{k-1}}{c_0} \leq r(x) < \frac{2^k}{c_0} \right\}, \quad k = 1, 2, \ldots,$$

and the constant $c > 0$ does not depend on f and k (c_0 is a constant from the triangle inequality for the homogeneous norm r).

Proof. Represent f as follows:

$$f(x) = f_1(x) + f_2(x),$$

where $f_1(x) = f \cdot \chi_{B_k}(x)$, $f_2(x) = f(x) - f_1(x)$, and $B_k = \{x \in G : r(x) < 2^{k+1}\}$.

If $r(y) > 2c_0 r(x)$, then $r(xy^{-1}) \geq \frac{1}{2c_0} r(y)$, and for every $x \in S_k$ we have

$$|Tf_2(x)| = \left| \int_{r(y) \geq 2^{k+1} > 2c_0 r(x)} k(xy^{-1}) f(y) \right| \leq$$

$$\leq c_1 \int_{r(y) \geq 2^{k+1} > 2c_0 r(x)} \frac{|f(y)|}{r(xy^{-1})^Q} dy \leq$$

$$\leq c_1 (2c_0)^Q \int_{r(y) \geq 2^{k+1}} \frac{|f(y)|}{r(y)^Q} dy \leq$$

$$\leq c_2 \int_{r(y) \geq 2^{k+1}} |f(y)| (1 + r(y))^{-Q} dy,$$

where the constant c_2 depends only on Q and c_0.

Thus

$$\sup_{x \in S_k} |Tf_2(x)| \leq c_2 \int_G |f(y)| (1 + r(y))^{-Q} dy$$

and

$$\left(\int_{S_k} |Tf_2(x)|^r dx \right)^{\frac{1}{r}} \leq c_2 |S_k|^{\frac{1}{r}} \int_G |f(y)| (1 + r(y))^{-Q} dx.$$

By virtue of the weak-type inequality for the operator T and by Lemma 6.7.2 we have

$$\left(\int_{S_k} |Tf_1(x)|^r dx \right)^{\frac{1}{r}} \leq c_3 |S_k|^{\frac{1}{r}-1} \sup_{t>0} t |\{x \in G : |Tf_1(x)| > t\}| \leq$$

$$\leq c_4 |S_k|^{\frac{1}{r}-1} \int_{r(x) < 2^{k+1}} |f(x)| dx \leq$$

$$\leq c_5 |S_k|^{\frac{1}{r}} \int_G |f(x)| (1 + r(y))^{-Q} dx.$$

Finally we obtain (8.8.1). $\square$

Theorem 8.8.1. *Let* $1 < p < \infty$ *and let* T *be a singular integral operator. Let*

$$\int_G \frac{w^{1-p'}(x)}{(1 + r(x))^{Qp'}}\, dx < \infty \quad \left(p' = \frac{p}{p-1}\right).$$

Then there exists a weight function v *such that for all* $f \in L_w^p(G)$,

$$\int_G |Tf(x)|^p v(x)dx \le c \int_G |f(x)|^p w(x)dx,$$

where the constant $c > 0$ *is independent of* f *and the function* v *satisfies the condition*

$$\int_G (v(x))^{-\lambda}(1 + r(x))^{-Qp'} dx < \infty \quad \left(\lambda < \frac{1}{p-1}\right). \tag{8.8.2}$$

Proof. Let $r < 1$, $\frac{1}{\lambda} = \frac{p}{r} - 1$ and $\frac{1}{\beta} = p - 1$. For every $k \in Z^+$ the operator

$$T_k g(x) = T(g(y)(1 + r(y))^Q)(x)\chi_{S_k}(x),$$

where

$$S_k = \left\{x \in G : \frac{2^{k-1}}{c_0} \le r(x) < \frac{2^k}{c_0}\right\}, \quad S_0 = \left\{x \in G : r(x) < \frac{1}{c_0}\right\},$$

is sublinear, and by Lemma 8.8.1 we have

$$\left(\int_G |T_k f(x)|^r dx\right)^{\frac{1}{r}} \le c_1 2^{\frac{kQ}{r}} \int_G |f(x)| dx,$$

where the constant c_1 depends only on r, c_0 and Q.

Let us take a positive function $a \in L^\beta(G)$. Then by Lemma 6.7.1 there exists a positive function $b_k \in L^\lambda(G)$ such that

$$\|b_k\|_\lambda \le \|a\|_\beta$$

and

$$\int_G |T_k f(x)|^p (b_k(x))^{-1} dx \le c_1^p 2^{\frac{kQp}{r}} \int_G |f(x)|^p (a(x))^{-1} dx.$$

Thus

$$\int_{S_k} |Tf(x)|^p (b_k(x))^{-1} dx \le c_1^p 2^{\frac{kQp}{r}} \int_G |f(x)|^p (a(x))^{-1}(1 + r(x))^{-Qp} dx.$$

Denote $a(x) = (w(x))^{-1}(1 + r(x))^{-Qp}$. Then we have

$$\int_G w^{1-p'}(x)(1 + r(x))^{-Qp'} dx < \infty.$$

Let $\varepsilon > 0$. Then the inequality

$$\int_{S_k} |Tf(x)|^p (b_k(x))^{-1} \cdot 2^{-\frac{kQp}{r}} \cdot 2^{-k\varepsilon} dx \le c_1^p \cdot 2^{-k\varepsilon} \int_G |f(x)|^p w(x) dx$$

is valid. Applying Fatou's theorem, we obtain

$$\int_G |Tf(x)|^p v(x) dx \le c_2 \int_G |f(x)|^p w(x) dx,$$

where

$$v(x) = \sum_{k=0}^{\infty} 2^{-\varepsilon k} \cdot 2^{-\frac{kQp}{r}} (b_k(x))^{-1} \chi_{S_k}(x).$$

For v we have

$$\int_G (v(x))^{-\lambda}(1 + r(x))^{-Qp'} =$$

$$= \sum_{k=0}^{\infty} 2^{\lambda \varepsilon k} \cdot 2^{\frac{\lambda k Qp}{r}} \int_{S_k} b_k^{\lambda}(x)(1 + r(x))^{-Qp'} dx \le$$

$$\le c_3 \sum_{k=0}^{\infty} 2^{-Qkp'} \cdot 2^{\lambda \varepsilon k} \cdot 2^{\frac{\lambda k Qp}{r}} \int_{S_k} b_k^{\lambda}(x) dx \le$$

$$\le c_4 \sum_{k=0}^{\infty} 2^{-k(Qp' - \varepsilon\lambda - \frac{\lambda Qp}{r})}.$$

Taking ε sufficiently small, we obtain condition (8.2.2). $\square$

Now we formulate an analogous theorem for a singular integral operator K defined on SHT (for the definition of K see Section 8.1) which can be derived in the same way as Theorem 8.8.1. In this case we use the inequality of weak $(1,1)$ type for the operator K.

Theorem 8.8.2. *Let $x_0 \in X$ and $\mu X = \infty$. Suppose that $1 < p < \infty$. If*

$$\int_X \frac{w^{1-p'}(x)}{(\mu B(x_0, 1 + d(x_0, x)))^{p'}} d\mu < \infty,$$

then for a given weight function w there exists a weight v such that the two-weight inequality

$$\int_X |Kf(x)|^p v(x)d\mu \le c_1 \int_X |f(x)|^p w(x)d\mu, \quad f \in L^p_w(X),$$

holds.

8.9. Notes and comments on Chapter 8

For the weight theory of singular integral operators we refer to [96], [278], [147] and [100], where the reader can find many references.

The approach to one-weight inequalities for the Hilbert transforms with general weights was developed in the pioneering work [131].

For two-weight estimates in the case of discrete Hilbert transforms we refer to [94], [95], [92], [93].

More recently, in [219–221] and [291–295] a Calderón–Zygmund theory was developed in a separable metric spaces X endowed with a nonnegative "n-dimensional" Borel measure μ, i.e., a measure satisfying

$$\mu(B(x,r)) \le r^n$$

for all $x \in X$ and $r > 0$. This enables one to obtained in greater generality such basic results as the weak L^1-boundedness, the L^2-boundedness, Cotlar's inequality, $T(1)$ and $T(2)$ theorems. We refer to [294].

For a more modern approach to two-weighted problems for singular integrals we refer to [57–58], [297].

This chapter gives some of the results obtained in [71], [74], [153], [156–157], [199]. Some results of Sections 8.2, 8.8 are published here for the first time.

Chapter 9

MULTIPLIERS OF FOURIER TRANSFORMS

In this chapter weighted Triebel–Lizorkin spaces are defined in a general setting. The two–weighted criteria for fractional and singular integrals derived in the previous chapters enable us to develop a new approach to the theory of multipliers of Fourier transforms. For (L^p, L^q) multipliers we establish two–weight estimates involving weight functions which do not necessarily belong to the class A_p. It should be noted that the derived conditions are not only sufficient but also necessary for a whole class of multipliers under consideration.

A spacial section contains examples of pairs of weight functions ensuring two–weight inequalities for multipliers.

9.1. Weighted Triebel–Lizorkin spaces

Let $S(R^n)$ be the Schwartz space of rapidly decreasing functions (see [270]). For $\varphi \in S(R^n)$ the Fourier transform $\hat{\varphi}$ is defined by

$$\hat{\varphi}(\lambda) = (2\pi)^{-\frac{n}{2}} \int\limits_{R^n} \varphi(x) e^{-i\lambda x} dx;$$

let $\check{\varphi}$ denote the inverse Fourier transform of φ. For the Fourier transform and its inverse the notation $F(\varphi)$ and $F^{-1}(\varphi)$ respectively will be also used.

The Fourier transform determines a topological isomorphism of the space S into itself.

Let S' be the space of tempered distributions, i.e. a space of linear bounded functionals over $S(R^n)$. In the sequel the Fourier transforms in the framework of the theory of S'-distributions will be considered.

Now we give a definition of a weighted Triebel–Lizorkin space in a general setting.

593

Let $\{m_j\}_{j=-\infty}^{\infty}$ be a two-sided increasing sequence of positive numbers such that $\lim_{j\to-\infty} m_j = 0$ and $\lim_{j\to+\infty} m_j = +\infty$. Let $\mathcal{I}$ be the collection of all intervals $(m_j, m_{j+1}]$ and $[-m_{j+1}, -m_j)$, $j \in Z$. Any interval of this type we shall denote by I. It is clear that $\bigcup_I I = R^1\backslash\{0\}$. Now considering n similar decompositions of $R^1\backslash\{0\}$ by the sets

$$\Gamma_j^i = [m_{j,i}, m_{j+1,i}) \cup (-m_{j+1,i}, -m_{j,i}), \quad j \in Z, \quad i = 1, \cdots, n,$$

we denote by J the collection of all intervals of the form

$$J = I_1^{(1)} \times \ldots \times I_n^{(n)},$$

where $I_j^{(i)}$ is an arbitrary one-dimensional interval of the above-mentioned type. This gives a decomposition of $R^1\backslash\{0\} \times \ldots \times R^1\backslash\{0\}$.

Let $\{\beta_{j,i}\}_{j=-\infty}^{\infty}$, $i = 1, \cdots, n$, be sequences of positive numbers which for arbitrary i, $i = 1, \cdots, n$, satisfy the following conditions:

(i)

$$\sum_{j=-\infty}^{0} \beta_{j,i}(m_{j+1,i} - m_{j,i}) < \infty; \tag{9.1.1}$$

(ii) there exists some ε, $0 < \varepsilon < 1$, such that

$$\sum_{j=-\infty}^{0} \beta_{j,i} m_{j,i}^{\varepsilon} < \infty; \tag{9.1.2}$$

(iii) there exists some natural k such that

$$\sum_{j=1}^{\infty} \frac{\beta_{j,i}}{m_{j,i}^k} < \infty. \tag{9.1.3}$$

Now let

$$A = \Gamma_{j_1}^{i_1} \times \ldots \times \Gamma_{j_m}^{i_m}, \quad j_m \in Z, \quad i_l \in \{1, \ldots, n\}.$$

Put

$$\beta_A = \prod_{l=1}^{n} \beta_{j_l, i_l}.$$

For $\varphi \in S(R^n)$ let

$$\varphi_A = F^{-1}(\chi_J \hat{\varphi}).$$

Suppose now that $1 < p$, $\theta < \infty$. If for some locally finite regular measure ν and for any $\varphi \in S(R^n)$ the quantity

$$|\varphi, F_{\beta,\nu}^{p,\theta}| = \left(\int_{R^n} \left(\sum_A \beta_A^\theta |\varphi_A(x)|^\theta \right)^{p/\theta} d\nu \right)^{1/p} \tag{9.1.4}$$

is finite, then the completion of $S(R^n)$ with respect to the norm will be called a weighted Triebel–Lizorkin space and denoted by $F_{\beta,\nu}^{p,\theta}(R^n)$. For unweighted Triebel–Lizorkin spaces we refer to [299], [179].

Proposition 9.1.1. *Let* $1 < p, \theta < \infty$ *and suppose that*

$$(1 + |x|)^{-p(1-\varepsilon)} \in L_\nu^1(R^n) \qquad (9.1.5)$$

for the constant ε from (2.2). Then (9.1.4) is finite for arbitrary $\varphi \in S(R^n)$.

Proof. To avoid awkward computations we give a proof for $n = 1$; the case $n > 1$ may be handled in a similar manner. Without loss of generality we can consider a function $\varphi \in S(R^1)$ whose Fourier transform $\hat{\varphi}$ vanishes for $\lambda < 0$. Let $I_j = [m_j, m_{j+1}]$, $j \in Z$, and let $\varphi_j = \varphi_{I_j}$. We have

$$|\varphi, F_{\beta,\nu}^{p,\theta}| = \left(\int_{R^1} \left(\sum_{j=-\infty}^{\infty} |\varphi_j(x)|^\theta \beta_j^\theta \right)^{p/\theta} d\nu \right)^{1/p} \leq$$

$$\leq \left(\int_{R^1} \left(\sum_{j=-\infty}^{\infty} |\varphi_j(x)| \beta_j \right)^p d\nu \right)^{1/p} \leq$$

$$\leq \left(\int_{|x|<1} \left(\sum_{j=-\infty}^{0} \beta_j |\varphi_j(x)| \right)^p d\nu \right)^{1/p} +$$

$$+ \left(\int_{|x|>1} \left(\sum_{j=-\infty}^{0} \beta_j |\varphi_j(x)| \right)^p d\nu \right)^{1/p} +$$

$$+ \left(\int_{|x|<1} \left(\sum_{j=1}^{\infty} \beta_j |\varphi_j(x)| \right)^p d\nu \right)^{1/p} +$$

$$+ \left(\int_{|x|>1} \left(\sum_{j=1}^{\infty} \beta_j |\varphi_j(x)| \right)^p d\nu \right)^{1/p} =$$

$$= I_1 + I_2 + I_3 + I_4.$$

It is obvious that

$$I_1 \leq c \sum_{j=-\infty}^{0} \beta_j (m_{j+1} - m_j) \left(\int_{-1}^{1} d\nu \right)^{1/p} < \infty$$

by condition (9.1.1).

Integration by parts leads to the estimate

$$\sqrt{2\pi} |\varphi_j(x)| \leq |\hat{\varphi}(m_{j+1}) - \hat{\varphi}(m_j)| \frac{1}{|x|} +$$

$$+\hat\varphi(m_j)\frac{|e^{im_j x}-1|}{|x|}+\frac{1}{|x|}\int\limits_{m_j}^{m_{j+1}}|\hat\varphi(\lambda)|d\lambda.$$

Hence

$$I_2\le c\Bigg(\bigg(\int\limits_{|x|>1}\Big(\sum_{j=-\infty}^{0}\beta_j(m_{j+1}-m_j)\Big)^p\frac{d\nu}{|x|^p}\bigg)^{1/p}+$$

$$+\bigg(\int\limits_{|x|>1}\Big(\sum_{j=-\infty}^{0}\beta_j|e^{im_j x}-1|\Big)^p\frac{d\nu}{|x|^p}\bigg)^{1/p}+$$

$$+\bigg(\int\limits_{|x|>1}\Big(\sum_{j=-\infty}^{0}\beta_j\int\limits_{m_j}^{m_{j+1}}|\hat\varphi(\lambda)|d\lambda\Big)^p\frac{d\nu}{|x|^p}\bigg)^{1/p}\Bigg)=I_2^{(1)}+I_2^{(2)}+I_2^{(3)}.$$

Since $(1+|x|)^{-p}\in L_\nu^1(R^1)$ and the condition (9.1.1) is satisfied, we conclude that $I_2^{(1)}$ is finite. For $I_2^{(2)}$ we derive

$$I_2^{(2)}\le c\sum_{j=-\infty}^{0}\beta_j\bigg(\int\limits_{|x|>1}\frac{|e^{im_j x}-1|}{|x|^p}d\nu\bigg)^{1/p}\le$$

$$\le c\sum_{j=-\infty}^{\infty}\beta_j\bigg(\int\limits_{|x|>1}\frac{|\sin m_j\frac{x}{2}|^\varepsilon}{|x|^p}d\nu\bigg)^{1/p}\le$$

$$\le c\sum_{j=-\infty}^{0}\beta_j m_j^{p\varepsilon}\bigg(\int\limits_{|x|>1}\frac{d\nu}{|x|^{-p(1-\varepsilon)}}\bigg)^{1/p}<\infty.$$

The boundedness of $\hat\varphi$ and the condition (9.1.2) imply that $I_2^{(3)}$ is finite. Further since $\hat\varphi\in S(R^1)$ and $|\hat\varphi(\lambda)|\le c\lambda^{-(k+1)}$ (where k is as in (9.1.3)) we have by (9.1.3),

$$I_3\le c\sum_{j=1}^{\infty}\frac{\beta_j}{m_j^k}\bigg(\int\limits_{-1}^{1}d\nu\bigg)^{1/p}<\infty.$$

Integrating by parts and using the estimates $|\hat\varphi(\lambda)|\le c|\lambda|^{-k}$ and $|\hat\varphi'(\lambda)|\le c\lambda^{-(k+1)}$, we get that

$$I_4\le c\bigg(\int\limits_{|x|>1}\Big(\sum_{j=1}^{\infty}\beta_j|\hat\varphi(m_{j+1})|+|\hat\varphi(m_j)|\Big)^p\frac{d\nu}{|x|^p}\bigg)^{1/p}+$$

$$+c\bigg(\int\limits_{|x|>1}\Big(\sum_{j=1}^{\infty}\beta_j\int\limits_{m_j}^{m_{j+1}}|\hat\varphi'(\lambda)d\lambda\Big)^p\frac{d\nu}{|x|^p}\bigg)^{1/p}\le$$

$$\leq c \sum_{j=1}^{\infty} \frac{\beta_j}{m_j k} \left(\int\limits_{|x|>1} \frac{d\nu}{|x|^p} \right)^{1/p} < \infty,$$

thanks to (9.1.1), (9.1.3) and (9.1.5).

Summarizing all these estimates we conclude that (9.1.4) is finite for arbitrary $\varphi \in S(R^1)$. $\square$

For an absolutely continuous measure $d\nu = w(x)dx$, where w is a locally integrable a.e. positive function, we write $F_{\beta,w}^{p,\theta}$ instead of $F_{\beta,\nu}^{p,\theta}$. The function w, as usual, will be called a weight function. If $w \equiv 1$, then we use the notation $L_{\beta,w}^{p,\theta} \equiv L_{\beta}^{p,\theta}$.

It is easy to see that the space $F_{\beta,\nu}^{p,\theta_1}$ is continuously imbedded into $F_{\beta,\nu}^{p,\theta_2}$ when $\theta_1 \leq \theta_2$ because of the inequality

$$\left(\sum_{j=1}^{\infty} a_j^{\theta_2} \right)^{1/\theta_2} \leq \left(\sum_{j=1}^{\infty} a_j^{\theta_1} \right)^{1/\theta_1}.$$

The spaces $F_{\beta,\nu}^{p,\theta}$ ($1 < p, \theta < \infty$) are Banach spaces and each $f \in F_{\beta,\nu}^{p,\theta}$ can be regarded as an element of S'.

Remark. When $\theta = 2$, $\beta_A = 1$, the decomposition is lacunary and ν is absolutely continuous with weight function w, the norm (9.1.4) is equivalent to the $L_w^p(R^n)$ norm, thanks to the weighted version of the Littlewood-Paley theorem (see [169]) when the weight function w satisfies the condition A_p:

$$\sup \frac{1}{|J|} \int_J w(x)dx \left(\frac{1}{|J|} \int_J w^{1-p'}(x)dx \right)^{p'-1} < \infty,$$

where the supremum is taken over all $n-$ dimensional boxes J with sides parallel to the coordinate axes.

Proposition 9.1.2. *Let ν be an arbitrary locally finite, regular measure. Then $S(R^n)$ is dense in $L_\nu^p(R^n)$.*

Proof. It is sufficient to prove that $C_0^\infty(R^n)$ is dense in $L_w^p(R^n)$. Given $\varphi \in L_\nu^p(R^n)$ and $\varepsilon > 0$, choose a continuous function g with compact support such that

$$\|f - g\|_{L_\nu^p(R^n)} < \frac{\varepsilon}{2}$$

(see [64] Lemma IV.8.19).

Let ψ be a non-negative, infinitely differentiable function supported in the unit ball of R^n with total integral equal to 1.

Define

$$\psi_t(x) = t^{-n}\psi(\frac{x}{t}), \quad t > 0.$$

It is easy to see that $\psi_t * g \in C_0^\infty(R^n)$ for all $t > 0$ and $\psi_t * g \to g$ as $t \to 0$ uniformly on compact subset of R^n. If B is a large ball containing the support of g in its interior, choose t so small that

$$\|g - \psi_t * g\|_\infty < \frac{\epsilon}{2} \left(\int_B d\nu \right)^{-1/p}.$$

Then

$$\|f - \psi_t * g\|_{L_\nu^p} \le \|f - g\|_{L_\nu^p} + \|g - \psi_t * g\|_{L_\nu^p} < \epsilon.$$

Hence $C_0^\infty(R^n)$ and therefore $S(R^n)$ is dense in $L_\nu^p(R^n)$. $\square$

9.2. Two–weight multipliers in Triebel–Lizorkin spaces

Let X and Y be two function space on R^n with norms $\| \cdot \|_X$ and $\| \cdot \|_Y$ respectively. Assume that $S(R^n)$ is dense in both X and Y.

Definition 9.2.1. A distribution $m \in S'$ is called an (X, Y) multiplier if for the operator K defined by the Fourier transform equation

$$\hat{\mathcal{K}}f = m\hat{f}, \quad f \in S(R^n), \tag{9.2.1}$$

there exists a constant c such that

$$\|\mathcal{K}f\|_Y \le c\|f\|_X$$

for all $f \in S(R^n)$.

In this case we write $m \in \mathcal{M}(X, Y)$. The number $\sup\limits_{\|\varphi\|_X=1} \|F^{-1}(m\hat{\varphi})\|_Y$ is the norm of the (X, Y)-multiplier m.

In the sequel we shall need the following definitions of weight classes, the weights being defined on R.

Definition 9.2.2. Let $\alpha \in (0, 1)$. We say that the weight pair (v, w) belongs to the class $U_\alpha^{q,p}$ if $1 < p < q < \infty$ and for v and w conditions (2.2.2) and (2.2.3) from Section 2.2 are satisfied. Further, the weight pair (v, w) belongs to $\widetilde{U}_\alpha^{q,p}$ if $1 < p < q < \infty$ and conditions (2.2.4) and (2.2.5) from Section 2.2 hold.

Definition 9.2.3. Let $0 < \alpha < 1$.
(i) The pair of weight functions (v, w) belongs to the class $W_\alpha^{q,p}$ if $1 < p < q < \infty$, $n = 1$, $\Omega = R$ and (v, w) satisfies conditions (5.2.1) and (5.2.2) from Section 5.2;

(ii) Let $1 < p < \infty$, $n = 1$ and let I_α be the Riesz potential on R. We say that the weight function v belongs to the class V_α^p if $I_\alpha v \in L_{loc}^{p'}(R^n)$ and the condition

$$I_\alpha(I_\alpha v)^{p'}(x) \le c(I_\alpha v)(x) < \infty, \quad p' = p/(p-1),$$

is satisfied for almost all $x \in R$;

(iii) The weight $v \in \Gamma_\alpha^{q,p}$ if $n = 1$ and the condition

$$\int\limits_{-\infty}^{+\infty} \left(\int\limits_0^\infty \left[\frac{1}{r^{1-\alpha p}} \int\limits_{I(x,r)} v(y)dy \right]^{p'-1} \frac{dr}{r} \right)^{\frac{q(p-1)}{p-q}} v(x)dx < \infty$$

holds, where $I(x,r)$ is the interval $(x - r, x + r)$.

The results of Chapters 2 and 5 show that the equality

$$W_\alpha^{q,p} = U_\alpha^{q,p} \cap \tilde{U}_\alpha^{q,p}$$

holds.

On the other hand, for $w \equiv 1$ we have (see Section 2.3)

$$W_\alpha^{q,p} = U_\alpha^{q,p} = \tilde{U}_\alpha^{q,p}.$$

Definition 9.2.4. Let $0 < \alpha < 1$. The weight function v on R is said to be of the class B_α^p (resp. $\tilde{B}_\alpha^p$) if $W_\alpha(v) \in L_{loc}^{p'}(R)$ and

$$W_\alpha(W_\alpha v^{p'})(x) \le c(W_\alpha v)(x) < \infty$$

a.e. on R (resp. $R_\alpha(v) \in L_{loc}^{p'}(R)$ and

$$(R_\alpha(R_\alpha v^{p'})(x) \le c(R_\alpha v)(x) < \infty \text{ a.e. on } R),$$

where by W_α and R_α, the Weyl and Riemann–Liouville operators defined on R are denoted respectively.

Now as in the previous section let $\mathcal{I}$ be a decomposition of R with corresponding numbers β_j satisfying conditions (9.1.1), (9.1.2) and (9.1.3).

We have the following statements:

Theorem 9.2.1. *Let $1 < p < q < \infty$, $1 < \theta < \infty$ and $(v, w) \in W_\alpha^{q,p}$. Let $\mathcal{K}$ be defined by (3.1), where the function m is represented in an arbitrary interval $I \in \mathcal{I}$ as*

$$m(\lambda) = \int\limits_{-\infty}^{\lambda} \frac{d\mu_I}{(\lambda - t)^\alpha}, \quad 0 < \alpha < 1, \quad (9.2.2)$$

and the μ_I are finite measures for which

$$\sup_{I \in \mathcal{I}} var\mu_I = M < \infty. \qquad (9.2.3)$$

Then $m \in \mathcal{M}(F_{\beta,w}^{q,\theta}, F_{\beta,v}^{q,\theta})$ and, moreover,

$$|\mathcal{K}f, F_{\beta,v}^{q,\theta}| \leq cM|f, F_{\beta,w}^{p,\theta}|, \qquad (9.2.4)$$

where c does not depend on f and m.

In the sequel by $B(x, r)$ we understand the interval $[x - r, x + r]$.

Theorem 9.2.1'. *Let $1 < p < q < \infty$, $1 < \theta < \infty$. Assume that the measure ν on R satisfies the condition*

$$\nu B(x, r) \leq cr^{q(\frac{1}{p} - \alpha)}, \quad 0 < \alpha < 1,$$

with the constant c independent of x and r. Then for any measurable function m satisfying the conditions of the previous theorem we have that $m \in \mathcal{M}(F_\beta^{q,\theta}, F_{\beta,\nu}^{p,\theta})$.

In the sequel in this section we shall a priori assume that

$$(1 + |x|)^{-q+\varepsilon} \in L_v^1(R)$$

and

$$(1 + |x|)^{-p+\varepsilon} \in L_w^1(R)$$

for some $\varepsilon > 0$.

Note that if $w \in A_p(R)$, then the last condition is satisfied.

The following statements also hold:

Theorem 9.2.2. *Let $1 < p < q < \infty$, $1 < \theta < \infty$ and $(v, w) \in \cup_\alpha^{q,p}$. Suppose that a function m in each I, $I \in \mathcal{I}$, is defined by the formula*

$$m(\lambda) = \int_{-\infty}^{\lambda} \frac{d\mu_I(t)}{(\lambda - t)^\alpha} + e^{i\alpha\pi} \int_{\lambda}^{\infty} \frac{d\mu_I(t)}{(\lambda - t)^\alpha}, \quad 0 < \alpha < 1, \qquad (9.2.5)$$

where finite measures μ_I satisfy (9.2.3). Then $m \in \mathcal{M}(F_{\beta,w}^{p,\theta}, F_{\beta,v}^{q,\theta})$ and (9.2.4) holds.

Theorem 9.2.3. *Let $1 < p < q < \infty$, $1 < \theta < \infty$ and let $(v, w) \in \tilde{U}_\alpha^{q,p}$. Let a measurable function m be represented in each $I \in \mathcal{I}$ by*

$$m(\lambda) = \int_{-\infty}^{\lambda} \frac{d\mu_I(t)}{(\lambda - t)^\alpha} + e^{-i\alpha\pi} \int_{\lambda}^{\infty} \frac{d\mu_I(t)}{(t - \lambda)^\alpha}, \quad 0 < \alpha < 1, \qquad (9.2.6)$$

where μ_I satisfies (9.2.3). Then $m \in \mathcal{M}(F_{\beta,w}^{p,\theta}, F_{\beta,v}^{q,\theta})$ and (9.2.4) holds.

Theorem 9.2.4. *Let $1 < p, \theta < \infty$ and $\alpha < 1/p$. Assume that $v \in B_\alpha^p$ (resp. $v \in \tilde{B}_\alpha^p$). Then for the function m represented in each $I \in \mathcal{I}$ by (9.2.5) (resp. 9.2.6) we have $m \in \mathcal{M}(F_\beta^{p,\theta}, F_{\beta,v}^{p,\theta})$.*

Theorem 9.2.5. *Let $1 \le q < p < \infty$, $1 < \theta < \infty$ and $v \in \Gamma_\alpha^{q,p}$, $0 < \alpha < 1$. Then for the functions m defined by (9.2.2) and satisfying the condition (9.2.3) we have $m \in \mathcal{M}(F_\beta^{p,\theta}, F_{\beta,w}^{q,\theta})$.*

Theorem 9.2.6. *Let $1 < p, \theta < \infty$ and let $v \in V_\alpha^p$. Then the function m from Theorem 9.2.1 is a $(F_\beta^{p,\theta}, F_{\beta,v}^{p,\theta})$ multiplier.*

Remark 9.2.1. Let $\theta = 2$ and suppose that in addition to the above-mentioned conditions for a pair of weights (v, w), we have $v \in A_q(R)$ and $w \in A_p(R)$. Then the foregoing theorems give multiplier statements for (L_w^p, L_v^q).

Proposition 9.2.1. *If for some pair of weights (v, w) and all functions m of type (9.2.2) with condition (3.3) belong to $\mathcal{M}(L_w^p, L_v^q)$ then $(v, w) \in W_\alpha^{q,p}$.*

The same is true for other multipliers and appropriate classes of pairs (v, w).

The proofs of all the theorems formulated above are carried out essentially by the same method, that is, by the representation of the operator under consideration as a composition of certain elementary transformations.

Let

$$x_+^\alpha = \begin{cases} x^\alpha, & x > 0 \\ 0, & x < 0 \end{cases}$$

and

$$x_-^\alpha = \begin{cases} 0, & x > 0 \\ |x|^\alpha, & x < 0 \end{cases}$$

We consider the following distributions:

$$l(\lambda) = \lambda_+^{-\alpha},$$
$$h(\lambda) = \lambda_+^{-\alpha} + e^{i\alpha\pi}\lambda_-^{-\alpha}$$

and

$$\gamma(\lambda) = \lambda_+^{-\alpha} + e^{-i\alpha\pi}\lambda_-^{-\alpha},$$

where $0 < \alpha < 1$.

It is known that their Fourier preimages are given by

$$\check{l}(x) = A(\alpha)x_+^{\alpha-1} + B(\alpha)x_-^{\alpha-1},$$

$$\check{h} = C(\alpha)x_+^{\alpha-1}$$

and

$$\check{\gamma}(x) = D(\alpha)x_-^{\alpha-1},$$

where

$$A(\alpha) = \frac{1}{\sqrt{2\pi}}e^{-i\alpha\frac{\pi}{2}}\Gamma(1-\alpha), \quad B(\alpha) = \frac{1}{\sqrt{2\pi}}e^{i\alpha\frac{\pi}{2}}\Gamma(1-\alpha),$$

$$C(\alpha) = \frac{\sqrt{2\pi}}{\Gamma(\alpha)}e^{i\alpha\frac{\pi}{2}}, \quad D(\alpha) = \frac{\sqrt{2\pi}}{\Gamma(\alpha)}e^{-i\alpha\frac{\pi}{2}}.$$

(see [97] p.172).

Lemma 9.2.1. *After completion with respect to the norm of $L_w^p(R)$, the mapping $\varphi \to \psi$ defined on $S(R)$ by the Fourier transform equation*

$$\hat{\psi}(\lambda) = h(\lambda)\hat{\varphi}(\lambda)$$

generates a bounded operator from $L_w^p(R)$ to $L_v^q(R)$ $(1 < p < q < \infty)$ if $(v, w) \in U_\alpha^{q,p}$.

Proof. The convolution of φ with the preimage of h, i.e. with $C(\alpha)x_+^{\alpha-1}$, gives the Riemann–Liouville operator $\mathcal{R}^\alpha$ on R (see [255], Theorem 7.1). By the assumptions $(v, w) \in U_\alpha^{q,p}$ it acts boundedly from L_w^p into L_v^q (see Theorem 2.2.1).

Similar propositions hold for the other "elementary multipliers" l and γ and for appropriate classes of pairs of weights. Henceforth the proofs will be given only for h.

Let us consider the family of operators $\mathcal{R}_\alpha^t$ defined by the Fourier transform equation

$$\hat{\psi}(\lambda) = h(\lambda - t)\hat{\varphi}(\lambda), \quad \varphi \in S(R), \quad t \in R.$$

Since the shift of t in the Fourier image corresponds to the multiplication by e^{itx} of the Fourier image, the norms of $\mathcal{R}_\alpha^t$ coincide with that of $\mathcal{R}_\alpha$. $\square$

Theorem 9.2.7. *Let $1 < p < q < \infty$ and let $(v, w) \in U_\alpha^{q,p}$. Suppose that a function m is defined by the formula*

$$m(\lambda) = \int\limits_{-\infty}^{\lambda} \frac{d\mu(t)}{(\lambda - t)^\alpha} + e^{i\alpha\pi} \int\limits_{\lambda}^{\infty} \frac{d\mu(t)}{(t - \lambda)^\alpha}, \quad 0 < \alpha < 1, \qquad (9.2.7)$$

where μ is a finite measure on R. Then the operator $\mathcal{K}$ acts boundedly from $L_w^p(R)$ into $L_v^q(R)$.

Proof. It is easy to see that m is a regular tempered distribution. Indeed, since the images of $\varphi \in S$ by the Riemann–Liouville and Weyl operators are bounded functions we have

$$\left| \int_R m(\lambda)\varphi(\lambda)d\lambda \right| = \left| \int_R \left(\int_t^\infty \frac{\varphi(\lambda)d\lambda}{(\lambda-t)^\alpha} + e^{i\alpha\pi} \int_{-\infty}^t \frac{\varphi(\lambda)d\lambda}{(t-\lambda)^\alpha} \right) d\mu(t) \right| \leq$$

$$\leq c \text{ var } \mu < \infty.$$

Therefore $m\widehat{\varphi} \in L^1(R)$. By definition of the operator $\mathcal{K}$ and the function m we have

$$\mathcal{K}\varphi(x) = \frac{1}{\sqrt{2\pi}} \int_R m(\lambda)\widehat{\varphi}(\lambda)e^{i\lambda x}d\lambda =$$

$$= \frac{1}{\sqrt{2\pi}} \left(\int_R \left(\int_R h(\lambda-t)d\mu(t) \right) \widehat{\varphi}(\lambda)e^{i\lambda x}d\lambda \right).$$

Changing the order of integration we get

$$\mathcal{K}\varphi(x) = \frac{1}{\sqrt{2\pi}} \int_R \left(\int_R h(\lambda-t)\widehat{\varphi}(\lambda)e^{i\lambda x}d\lambda \right) d\mu(t) = \frac{1}{\sqrt{2\pi}} \int_R \mathcal{R}_\alpha^t d\mu(t).$$

From Lemma 9.2.1 and the previous remark with respect to $\mathcal{R}_\alpha^t$ it follows that

$$\|\mathcal{K}f\|_{L_v^q} \leq \frac{1}{\sqrt{2\pi}} \int_R \|\mathcal{R}_\alpha^t \varphi\|_{L_v^q} d\mu(t) \leq cM\|\varphi\|_{L_w^p},$$

where M is the total variation of μ. $\square$

The next two statements follow in the same manner as the previous theorem. Therefore proofs are omitted.

Theorem 9.2.8. *Let $1 < p < \infty$ and let $v \in B_\alpha^p$. Then the operator $\mathcal{K}$ defined by (9.2.1), where m is defined by (9.2.7), is bounded from L^p to L_v^p.*

Theorem 9.2.9. *Let $1 < p < \infty$ and let $v \in \tilde{B}_\alpha^p$. Then the operator $\mathcal{K}$ with*

$$m(\lambda) = \int_{-\infty}^\lambda \frac{d\mu(t)}{(\lambda-t)^\alpha} + e^{-i\alpha\pi} \int_\lambda^\infty \frac{d\mu(t)}{(t-\lambda)^\alpha}, \quad 0 < \alpha < 1,$$

acts boundedly from L^p into L_v^p.

Now we shall deal with Fourier multipliers in weighted Lebesgue spaces of vector-valued functions with values in $l^\theta (1 < \theta < \infty)$.

Definition 9.2.5. Let $1 < p < \infty$, $1 \leq \theta < \infty$. By $L_v^p(l^\theta)$ is denoted the set of vector-valued functions $f(x) = \{f_j(x)\}_{j=1}^\infty$, $x \in R$, with measurable components and with finite norm

$$|f, L^p_\nu(l^\theta)| = \left(\int\limits_R \left(\sum_{j=1}^\infty |f_j(x)|^\theta \right)^{p/\theta} d\nu(x) \right)^{1/p}.$$

It is well-known that $L^p_\nu(l^\theta)$ is a Banach space (see [64], p.162). Further, it is evident that if $f \in L^p_\nu(l^\theta)$ then $f_j \in L^p_\nu$ for all $j \in N$.

Let $\vec{S}$ be the set of all vector-valued functions $\varphi = (\varphi_1, \varphi_2, \cdots)$ where $\varphi_j \in S$, $j \in N$.

Note that the set $\vec{S} \cap L^p_\nu(l^\theta)$ is dense in $L^p_\nu(l^\theta)$ for $1 < p$, $\theta < \infty$ and an arbitrary measure ν.

Indeed, let $f \in L^p_\nu(l^\theta)$ and a positive ε be given. By Proposition 9.1.2 for any j we can choose φ_j such that

$$\|f_j - \varphi_j\|_{L^p_\nu} < \frac{\varepsilon}{2^{j/p}} \quad \text{when } 1 < p \le \theta$$

and

$$\|f_j - \varphi_j\|_{L^p_\nu} < \frac{\varepsilon}{2^{j/\theta}} \quad \text{when } p > \theta.$$

When $1 < p \le \theta$ for $\varphi = (\varphi_1, \varphi_2, \cdots)$ we have

$$|f - \varphi, L^p_\nu(l^\theta)| = \left(\int\limits_R \left(\sum_{j=1}^\infty |f_j - \varphi_j|^\theta \right)^{p/\theta} d\nu \right)^{1/p} \le$$

$$\le \left(\int\limits_R \sum_{j=1}^\infty |f_j - \varphi_j|^p d\nu \right)^{1/p} = \left(\sum_{k=1}^\infty \|f_j - \varphi_j\|_{L^p_\nu} \right)^{1/p} < \varepsilon.$$

If $p > \theta$ then it is sufficient to use Jessen's inequality ([119], p.182):

$$\|f - \varphi\|_{L^p_\nu(l^\theta)} = \left(\int\limits_R \left(\sum_{j=1}^\infty |f_j - \varphi_j|^\theta \right)^{p/\theta} d\nu \right)^{1/p} \le$$

$$\le \left(\sum_{j=1}^\infty \left(\int\limits_R |f_j - \varphi_j|^p d\nu \right)^{\frac{\theta}{p}} \right)^{1/p} = \left(\sum_{j=1}^\infty \|f_j - \varphi_j\|_{L^p_\nu}^\theta \right)^{1/p} < \varepsilon.$$

Thus we see that $\varphi \in \vec{S} \cap L^p_\nu(l^\theta)$ and $\|f - \varphi\|_{L^p_\nu(l^\theta)} < \varepsilon$. The Fourier transform of the vector-valued function $f \in L^p_w(l^\theta)$ is defined by $\widehat{f} = Ff = \left\{ \widehat{f}_j \right\}_{j=1}^\infty$.

Recall that Ff_j is defined by means of the Fourier transform of distributions.

The convolution of the vector-valued function $f \in L^p_v(l^\theta)$ with a tempered distribution $h \in S'$ is considered coordinate-wise:

$$h * f = \{h * f_j\}_{j=1}^\infty.$$

The following equality for the Fourier transform of a convolution holds:

$$F(h * f) = F(h) \cdot F(f).$$

Lemma 9.2.2. *The transform $\varphi \mapsto \psi$ defined by the equality*

$$\psi(\lambda) = h(\lambda)\widehat{\varphi}(\lambda), \quad \varphi \in \overrightarrow{S},$$

generates a bounded operator from $L^p_w(l^\theta)$ to $L^q_v(l^\theta)$ when $1 < p < q < \infty$, $1 < \theta < \infty$ and the pair of weights (v, w) belongs to $\cup^{q,p}_\alpha$.

The proof is analogous to that of Lemma 9.2.1. It should only be noted that

$$\left\| \int_{-\infty}^{x} \frac{f_j(y)}{(x-y)^{1-\alpha}} dy \right\|_{l^\theta} \leq \int_{-\infty}^{x} \frac{\|f_j(y)\|_{l^\theta}}{(x-y)^{1-\alpha}} dy.$$

Theorem 9.2.10. *Let $1 < p < q < \infty$, $1 < \theta < \infty$ and let the pair of weights (v, w) belongs to $\cup^{q,p}_\alpha$. Then for the transform $\mathcal{K}$ defined by*

$$\widehat{\mathcal{K}f}(\lambda) = m(\lambda)\widehat{\varphi}(\lambda), \quad \varphi \in \overrightarrow{S},$$

where $m(\lambda)$ is represented by (9.2.7), we have

$$|\mathcal{K}\varphi, L^q_w(l^\theta)| \leq cM|\varphi, L^p_w(l^\theta)|,$$

with a constant c independent of f and m. The operator $\mathcal{K}$ is extendable to a bounded operator from $L^p_w(l^\theta)$ to $L^q_v(l^\theta)$.

Proof. If we consider the superposition in $L^p_w(l^\theta)$ defined by

$$\widehat{\psi}(\lambda) = h(\lambda - t)\widehat{\varphi}(\lambda), \quad \varphi \in \overrightarrow{S}, \ t \in R,$$

the norm of this operator coincides with its value when $t = 0$. The rest of the proof is the same as in Theorem 9.2.7. $\square$

Theorem 9.2.11. *Let $1 < p < q < \infty$, $1 < \theta < \infty$ and let $(v, w) \in U^{q,p}_\alpha$. Assume that the measurable functions m_j are defined by*

$$m_j(\lambda) = \int_{-\infty}^{\lambda} \frac{d\mu_j(t)}{(\lambda - t)^\alpha} + e^{i\alpha\pi} \int_{\lambda}^{\infty} \frac{d\mu_j(t)}{(t - \lambda)^\alpha} \qquad (9.2.8)$$

where μ_j are finite measures for which

$$\sup_j \ var \ \mu_j < \infty. \qquad (9.2.9)$$

Then the operator $\mathcal{K}$ defined on $\vec{S}$ by the Fourier-transform equation

$$\widehat{\mathcal{K}\varphi} = \{\mu_j \varphi_j\}_{j=1}^{\infty}, \quad \varphi \in \vec{S}, \tag{9.2.10}$$

is extendable to a bounded operator from $L_w^p(l^\theta)$ to $L_v^q(l^\theta)$.

The proof of Theorem 9.2.11 will be divided into several steps.

Let us consider the transform $T(\lambda)$, $\lambda = \{\lambda_1, \lambda_2, ..., \lambda_n, ...\}$, $\lambda_j \in R$, defined on $\vec{S}$:

$$T(\lambda) : f_j \longrightarrow g_j, \ \widehat{g}_j(\lambda) = h(\lambda - \lambda_j)\widehat{f}_j(\lambda), \ j = 1, 2, ..., \ \lambda \in R,$$

where as above

$$h(\lambda) = \lambda_+^{-\alpha} + e^{i\alpha\pi}\lambda_-^{\alpha\pi} \text{ and } \breve{h}(x) = C(\alpha)x_+^{\alpha-1}.$$

Lemma 9.2.3. *The transform $T(\lambda)$ is extendable to a bounded operator from $L_w^p(l^\theta)$ to $L_v^q(l^\theta)$ under the conditions that $1 < p < q < \infty$, $1 < \theta < \infty$ and with $(v, w) \in \cup_\alpha^{q,p}$.*
Proof. The operator $T(\lambda)$ can be represented as

$$T(\lambda) = L(\lambda)T(0)L(-\lambda),$$

where the operator L is defined by

$$\{f_j(x)\}_{j=1}^{\infty} \to \left\{e^{i\lambda x}f_j(x)\right\}_{j=1}^{\infty}.$$

It is clear that $T(0)$ is the Riemann–Liouville operator and since the operator L is an isometry of $L_w^p(l^\theta)$ the desired result follows from Theorem 9.2.10. Further we truncate the operator T and for any given n consider the operator

$$T(t_1, \cdots, t_n) : f \to g, \ \widehat{g}_j(\lambda) = h(\lambda - t_j)\widehat{f}_j(\lambda),$$

when $j \le n$ and $\widehat{g}_j(\lambda) = \widehat{f}_j(\lambda)$ when $j > n$. $\square$

Lemma 9.2.4. *Let $1 < p < q < \infty$, $1 < \theta < \infty$ and let $(v, w) \in \cup_\alpha^{q,p}$. Let m_j be defined by (9.2.8). Assume that the transform $\mathcal{K}_n$ is determined in $L_w^p(l^\theta)$ by*

$$\mathcal{K}_n f = G = \{G_j\},$$

$$\widehat{G}_j = \begin{cases} m\widehat{f}_j, & j \le n, \\ \widehat{f}_j, & j > n. \end{cases}$$

Then there exists a positive constant c independent of f and n such that

$$\left|\mathcal{K}_n f, L_v^q(l^\theta)\right| \le cM \left|f, L_w^p(l^\theta)\right|.$$

Proof. Without loss of generality we assume that the measures μ_j are positive and normalized to 1.

For $j \leq n$ we have

$$G_j(x) = \frac{1}{\sqrt{2\pi}} \int\limits_R m(\lambda) \widehat{f}_j(\lambda) e^{i\lambda x} d\lambda =$$

$$= \frac{1}{\sqrt{2\pi}} \int\limits_R \left(\int\limits_R h(\lambda - t) d\mu_j \right) \widehat{f}_j(\lambda) e^{i\lambda x} d\lambda =$$

$$= \frac{1}{\sqrt{2\pi}} \int\limits_R \left(\int\limits_R h(\lambda - t) \widehat{f}_j(\lambda) e^{i\lambda x} d\lambda \right) d\mu_j(t_j) =$$

$$= \frac{1}{\sqrt{2\pi}} \int\limits_R \cdots \int\limits_R \left(\int\limits_R h(\lambda - t_j) \widehat{f}_j(\lambda) e^{i\lambda x} d\lambda \right) d\mu_1(t_1) \cdots d\mu_n(t_n).$$

Hence $\mathcal{K}_\backslash$ can be represented in the form

$$\mathcal{K}_n f(x) = \frac{1}{\sqrt{2\pi}} \int\limits_R \cdots \int\limits_R T(t_1, \cdots, t_n) d\mu_1(t_1) \cdots d\mu_n(t_n).$$

Therefore

$$|\mathcal{K}_n f, L_v^q(l^\theta)| \leq c|T(t_1, \cdots, t_n) f, L_v^q(l^\theta)|.$$

Applying Lemma 9.2.3 we can see that $\mathcal{K}_n$ is bounded from $L_w^p(l^\theta)$ to $L_v^q(l^\theta)$ with an upper estimate of the norm independent of n. $\square$

Proof of Theorem 9.2.9. First of all we show that $\lim\limits_{n \to \infty} \mathcal{K}_n f = \mathcal{K} f$ exists in the sense of convergence in the norm of $L_v^q(l^\theta)$ for arbitrary $f \in \vec{S} \cap L_w^p(l^\theta)$.

Let

$$\mathcal{K}_{n_1} f - \mathcal{K}_n f = \{\psi_j\}_{j=1}^\infty, \quad n > n_1.$$

By definition of $\mathcal{K}_n$ and Lemma 9.2.4:

$$|\mathcal{K}_{n_1} f - \mathcal{K}_{n_2} f, L_v^q(l^\theta)| \leq cM \left\| \left(\sum_{j=n_1}^n |f_j(x)|^\theta \right)^{1/\theta} \right\|_{L_w^p}.$$

The right-hand side tends to zero. The proof of the theorem now follows from the uniform boundedness principle (see [64], p.73).

The following statements can be proved analogously:

Theorem 9.2.12. *Let* $1 < p < q < \infty$, $1 < \theta < \infty$ *and let* $(v, w) \in \tilde{U}_\alpha^{q,p}$. *Let the functions* m_j *be defined by*

$$m_j(\lambda) = \int\limits_{-\infty}^{\lambda} \frac{d\mu_j(t)}{(\lambda - t)^\alpha} + e^{-i\alpha\pi} \int\limits_{\lambda}^{\infty} \frac{d\mu_j(t)}{(t - \lambda)^\alpha}, \qquad (9.2.11)$$

where the m_j *are finite measures satisfying condition* (9.2.9). *Then the operator* $\mathcal{K}$ *defined by* (9.2.10) *is extendable to a bounded operator from* $L_w^p(l^\theta)$ *to* $L_v^q(l^\theta)$ *and*

$$\|\mathcal{K}\|_{L_w^p(l^\theta) \to L_v^q(l^\theta)} \leq cM.$$

Theorem 9.2.13. *Suppose that the functions* m_j *are representable in the form*

$$m_j(\lambda) = \int\limits_{-\infty}^{\lambda} \frac{d\mu_j(t)}{(\lambda - t)^\alpha}$$

with the condition (9.2.9). *Then:*

i) *If* $1 < p < q < \infty$, $1 < \theta < \infty$ *and the pair* (v, w) *belongs to the class* $W_\alpha^{q,p}$ ($v \in V_\alpha^p$), *then the operator* $\mathcal{K}$ *is bounded from* $L_w^p(l^\theta)$ *to* $L_v^q(l^\theta)$ (*acts boundedly from* $L^p(l^\theta)$ *to* $L_v^p(l^\theta)$);

ii) *If* $1 \leq q < p < \infty$, $1 < \theta < \infty$ *and* $v \in \Gamma_\alpha^{q,p}$ *then the operator* $\mathcal{K}$ *acts boundedly from* $L^p(l^\theta)$ *into* $L_v^q(l^\theta)$.

Below the intervals of decomposition forming $\mathcal{I}$ we regard as enumerated by $\{I_j\}_{j=1}^\infty$.

Proposition 9.2.1. *Let* m_j *be the functions for which the operator*

$$\widehat{Tf_j} = m_j \widehat{f_j}$$

is bounded from $L_w^p(l^\theta)$ *to* $L_v^q(l^\theta)$ *for the weight pair* $(v, w) \in \cup_\alpha^{q,p}$ *and for* p *and* q *with* $1 < p < q < \infty$. *Let the function* m *be defined by* $m(\lambda) = m_j(\lambda)$, $\lambda \in I_j$. *Then the operator* $\mathcal{K}$ *defined by* (9.2.1) *is bounded from* $F_w^{p,\theta}$ *to* $F_v^{q,\theta}$.

The proof is evident in view of the equality

$$|\mathcal{K}f, F_{\beta,v}^{q,\theta}| = |Tf_\beta, L_v^q(l^\theta)|,$$

where $f_\beta = \{\beta_j f_j\}$.

Now Theorem 9.2.2 immediately follows from Proposition 9.2.1 and Theorem 9.2.11.

9.3. $\left(F_{\beta,w}^{q,\theta}, F_{\beta,v}^{p,\theta}\right)$ multipliers. The case $1 < q \le p < \infty$

Let us consider the function

$$\theta(\lambda) = \begin{cases} 1, & \text{when } \lambda > 0 \\ 0, & \text{when } \lambda < 0. \end{cases}$$

It is known (see [97]) that the tempered distribution $\check{\theta}$ is given by the equality

$$\check{\theta}(x) = \sqrt{2\pi}\left(\delta(x)/2 + (2\pi i)^{-1}1/x\right), \qquad (9.3.1)$$

where δ is the Dirac function.

Let the transform $\varphi \mapsto \psi$ be defined on $S(R^1)$ by the Fourier equation

$$\hat{\psi} = \theta\hat{\varphi}.$$

This equality corresponds to the convolution

$$\psi = \check{\theta} * \varphi.$$

according to (9.3.1) the latter leads to the Hilbert transform

$$Hf(x) = \frac{1}{\pi}\int_{-\infty}^{\infty}\frac{f(y)}{x-y}dy.$$

The considerations of previous sections together with two-weighted estimates for singular integrals proved in [71] and [152] (see also Section 8.4) enable us to prove assertions about $(F_{\beta,w}^{p,\theta}, F_{\beta,v}^{p,\theta})$ $(1 < p < \infty)$ and $(F_{\beta,w}^{p}, F_{\beta,v}^{q})$ $(1 < q < p < \infty)$ multipliers.

Definition 9.3.1. A pair of weights (v, w) belongs to a_p $(1 < p < \infty)$, if $v(x) = \sigma(|x|)\rho(x)$, $w(x) = u(|x|)\rho(x)$, $\rho \in A_p(R^1)$, σ and u are increasing functions on $(0, \infty)$ and

$$\sup_{t>0}\int_{|x|>t}\frac{v(x)}{|x|^p}dx\left(\int_{|x|<t}w^{1-p'}(x)dx\right)^{p-1} < \infty.$$

Definition 9.3.2. A pair (v, w) is said to be of class b_p $(1 < p < \infty)$, if σ and u decrease on $(0, \infty)$, $\rho \in A_p(R^1)$ and

$$\sup_{t>0}\int_{|x|<t}v(x)dx\left(\int_{|x|>t}\frac{w^{1-p'}(x)}{|x|^{p'}}dx\right)^{p-1} < \infty.$$

In [71] it is proved that if $(v, w) \in a_p \cup b_p$, $1 < p < \infty$, then H is bounded from L_w^p into L_v^q.

Theorem 9.3.1. $1 < p$, $\theta < \infty$ *and let* $(v, w) \in a_p \cup b_p$. *Suppose that the function m is expressed in any interval I of the decomposition $\mathbf{I}$ by the following form*

$$m(\lambda) = \int\limits_{-\infty}^{\lambda} d\mu \quad (\lambda \in I),$$

where the positive measures μ_I satisfy the condition

$$\sup_{I} var \ \mu_I < \infty.$$

Then $m \in \mathcal{M}(F_{\beta,w}^{p,\theta}, F_{\beta,v}^{p,\theta})$.

Theorem 9.3.2. *Let* $1 < q < p < \infty$, $1 < \theta < \infty$. *If the pair of even, increasing on $(0, \infty)$ weight functions satisfies the condition*

$$\int\limits_{0}^{\infty} \left[\left(\int\limits_{t}^{\infty} v(x) x^{-nq} dx \right) \left(\int\limits_{0}^{t/2} w^{1-p'}(x) dx \right)^{q-1} \right]^{\frac{p}{p-q}} \cdot w^{1-p'}(\frac{t}{2}) dt < \infty,$$

then the function m from the previous theorem is an $\left(F_{\beta,w}^{p,\theta}, F_{\beta,v}^{q,\theta} \right)$ *multiplier.*

These statements follow in the same manner as the above-formulated corresponding theorems.

9.4. Multipliers in weighted spaces with mixed norms

Let $\bar{p} = (p_1, \cdots, p_n)$, $1 < p_i < \infty$, $i = 1, \cdots, n$. Put $\bar{w} \equiv (w_1, \cdots, w_n)$, where $w_i = w_i(x_i)$ $(i = 1, \cdots, n)$ are weight functions defined on R^1.

By definition $L_{\bar{w}}^{\bar{p}}(R^n)$ is the space of functions $f : R^n \to R^1$ with the condition

$$\|f\|_{L_{\bar{w}}^{\bar{p}}} =$$

$$= \left(\int\limits_{-\infty}^{\infty} d\nu_1 \left(\int\limits_{-\infty}^{\infty} d\nu_2 ... \left(\int\limits_{-\infty}^{\infty} |f(x)|^{p_n} d\nu_n \right)^{\frac{1}{p_n}} \right)^{\frac{1}{p_2} \cdot p_1} \right)^{\frac{1}{p_1}} < \infty \quad (9.4.1)$$

The definition of $F_{\beta,\bar{w}}^{\bar{p},\theta}$ is similar to that given in Section 9.2. We have to use the norm (9.4.1).

Let $\mathcal{J}$ be a decomposition of $R^n \backslash \{0\}$ of the type defined in Section 9.1 and let $J \in \mathcal{J}$.

Theorem 9.4.1. *Let* $1 < p_i < q_i < \infty$, $1 < \theta < \infty$ *and let* $(v_i, w_i) \in W_{\alpha_i}^{q_i, p_i}(R^1)$, $i = 1, \cdots, n$. *Suppose that a function* m *is represented in the form*

$$m(\lambda) = \int_{-\infty}^{\lambda_1} \cdots \int_{-\infty}^{\lambda_n} \prod_{j=1}^{n} (\lambda_j - t_j)^{-\alpha_j} \, d\mu_J, \quad \lambda \in J, \tag{9.4.2}$$

where μ_J *are finite measures and*

$$\sup_{J \in J} \text{var } \mu_J \leq M. \tag{9.4.3}$$

Then $m \in \mathcal{M}\left(F_{\beta,\overline{w}}^{\overline{p},\theta}, F_{\beta,\overline{v}}^{\overline{q},\theta}\right)$.

For the rest of this section we shall assume that

$$(1 + |x|)^{-q_i + \epsilon} \in L_{v_i}^1(R^1)$$

and

$$(1 + |x|)^{-p_i + \epsilon} \in L_{w_i}^1(R^1)$$

for some $\epsilon > 0$.

Theorem 9.4.2. *Let* $1 < p_i < q_i < \infty$, $1 < \theta < \infty$ *and let* $(v_i, w_i) \in U_{\alpha_i}^{q_i, p_i}$, *where* $0 < \alpha_i < 1$ $(i = 1, \cdots, n)$. *Assume that the function* m *in each* J *is represented by*

$$m(\lambda) = \int_{-\infty}^{\infty} \cdots \int_{-\infty}^{\infty} \prod_{i=1}^{n} \left(\frac{1}{(\lambda_i - t_i)_+^{\alpha_i}} + e^{i\alpha_i \pi} \frac{1}{(t_i - \lambda_i)^{\alpha_i}} \right) d\mu_J(t), \tag{9.4.4}$$

where μ_J, $J \in \mathcal{J}$ *satisfy the condition* (9.4.3). *Then* $m \in \mathcal{M}\left(F_{\beta,\overline{w}}^{\overline{p},\theta}, F_{\beta,\overline{v}}^{\overline{q},\theta}\right)$.

Theorem 9.4.3. *Let* $1 < p_i < q_i < \infty$, $1 < \theta < \infty$ *and* $(v_i, w_i) \in \widetilde{U}_{\alpha_i}^{q_i, p_i}$, $(i = 1, \cdots, n)$. *Let the function* m *be represented as*

$$m(\lambda) = \int_{-\infty}^{\infty} \cdots \int_{-\infty}^{\infty} \prod_{i=1}^{n} \left(\frac{1}{(\lambda_i - t_i)_+^{\alpha_i}} + e^{-i\alpha_i \pi} \frac{1}{(t_i - \lambda_i)_-^{\alpha_i}} \right) d\mu_J(t)$$

under the condition (9.4.3). *Then* $m \in \mathcal{M}(F_{\overline{w}}^{\overline{p},\theta}, F_{\overline{v}}^{\overline{q},\theta})$.

The idea of the proof is similar to the one-dimensional case; however, we should make some remarks.

Let

$$l(\lambda) = \prod_{i=1}^{n} (\lambda_i)_+^{-\alpha},$$

$$h(\lambda) = \prod_{i=1}^{n} \left((\lambda_i)_+^{-\alpha_i} + e^{i\alpha_i \pi} (\lambda_i)_-^{-\alpha_i} \right)$$

and

$$\gamma(\lambda) = \prod_{i=1}^{n} \left((\lambda_i)_+^{-\alpha_i} + e^{-i\alpha_i \pi} (\lambda_i)_-^{-\alpha_i} \right).$$

It is well-known (see [271]) that the Fourier transform of a direct product is given by direct product of Fourier images of factors.

The Fourier preimage of l is a linear combination of the products

$$(x_1)_+^{\alpha_1 - 1} \times \cdots \times (x_i)_+^{\alpha_i - 1} \times (x_{i+1})_-^{\alpha_{i+1} - 1} \times \cdots \times (x_n)_-^{\alpha_n - 1}.$$

Analogously,

$$\breve{h}(x) = \prod_{j=1}^{n} C(\alpha_j)(x_j)_+^{\alpha_j - 1}$$

and

$$\breve{\gamma}(x) = \prod_{j=1}^{n} D(\alpha_j)(x_j)_-^{\alpha_j - 1}.$$

These lead to the integral operators:

$$I_{\overline{\alpha}} f(x) = \int_{R^n} \prod_{j=1}^{n} |x_j - y_j|^{\alpha_j - 1} f(y) dy,$$

$$\mathcal{R}_{\overline{\alpha}} f(x) = \int_{-\infty}^{x_1} \cdots \int_{-\infty}^{x_n} \prod_{j=1}^{n} (x_j - y_j)^{\alpha_j - 1} f(y) dy$$

and

$$\mathcal{W}_{\overline{\alpha}} f(x) = \int_{x_1}^{+\infty} \cdots \int_{x_n}^{+\infty} \prod_{j=1}^{n} (y_j - x_j)^{\alpha_j - 1} f(y) dy,$$

where $0 < \alpha_j < 1$ $(i = 1, \cdots, n)$, $\overline{\alpha} = (\alpha_1, \cdots, \alpha_n)$.

Now making $n-$ fold applications of appropriate one–dimensional two– weighted inequalities and using Minkowski's inequality we derive

Proposition 9.4.1. *Let* $1 < p_i < q_i < \infty$, $(i = 1, \cdots, n)$. *The following statements hold:*

i) *If* $(v_i, w_i) \in W_{\alpha_i}^{q_i, p_i}$ $(i = 1, \cdots, n)$, *then* $I_{\overline{\alpha}}$ *is bounded from* $L_{\overline{w}}^{\overline{p}}(R^n)$ *to* $L_{\overline{v}}^{\overline{q}}(R^n)$.

ii) *If* $(v_i, w_i) \in U_{\alpha_i}^{q_i, p_i}$ $(i = 1, \cdots, n)$, *then* $\mathcal{R}_{\overline{\alpha}}$ *is bounded from* $L_{\overline{w}}^{\overline{p}}(R^n)$ *to* $L_{\overline{v}}^{\overline{q}}(R^n)$.

iii) When $(v_i, w_i) \in \tilde{U}_{\alpha_i}^{q_i, p_i}$ $(i = 1, \cdots, n)$, then $\mathcal{W}_{\overline{\alpha}}$ acts boundedly from $L_{\overline{w}}^{\overline{p}}(R^n)$ to $L_{\overline{v}}^{\overline{q}}(R^n)$.

From Theorem 9.4.1 it follows the two-weighted version of Mikhlin–Lizorkin type multipliers.

Theorem 9.4.4. *Let m be continous outside the coordinate planes and have there continous derivatives*

$$\frac{\partial^k m}{\partial \lambda_1^{k_1} \cdots \partial \lambda_n^{k_n}}, \quad 0 \le k_1 + k_2 + \cdots + k_n = k \le n, \quad k_j = 0, 1.$$

Moreover, assume that

$$\left| \lambda_1^{k_1 + \alpha_1} \cdots \lambda_n^{k_n + \alpha_n} \frac{\partial^k m}{\partial \lambda_1^{k_1} \cdots \partial \lambda_n^{k_n}} \right| \le M.$$

Then the following statements hold:

i) When $1 < p_i < q_i < \infty$, $0 < \alpha_i < 1$, $i = 1, \cdots, n$, $1 < \theta < \infty$ and $(v_i, w_i) \in W_{\alpha_i}^{q_i, p_i}$, then $m \in \mathcal{M}(F_{\beta, \overline{w}}^{\overline{q}, \theta}, F_{\beta, \overline{v}}^{\overline{p}, \theta})$.

ii) If $p_i = q_i$, $\alpha_i = 0$, $j = 1, \cdots, n$, $1 < \theta < \infty$ and (v_i, w_i) satisfy the condition of Theorem 9.4.1, then $m \in \mathcal{M}(F_{\beta, \overline{w}}^{\overline{p}, \theta}, F_{\beta, \overline{v}}^{\overline{p}, \theta})$.

Finally, in addition, if $w_i \in A_{p_i}(R^1)$ and $v_i \in A_{q_i}(R^1)$ $(i = 1, \cdots, n)$, then we obtain $\left(L_{\overline{w}}^{\overline{p}}, L_{\overline{v}}^{\overline{q}} \right)$ multiplier statements. The n−dimensional weighted version of the Littlewood-Paley theorem must be applied. The proof of the last result (see [169]) works in weighted Lebesgue spaces with mixed norms as well.

9.5. Examples

Here on the basis of the previous sections' results we derive various examples of pairs of weights ensuring validity of two-weight estimates for appropriate multipliers.

Proposition 9.5.1. *Let $1 < p$, $\theta < \infty$ and m be a function of the form*

$$m(\lambda) = \int\limits_{-\infty}^{\lambda} d\mu_I \quad \lambda \in I,$$

where I are intervals of decomposition of R^1 and the finite measures μ_I are such that

$$\sup var \, \mu_I = M < \infty.$$

Assume that

$$w(x) = \begin{cases} |x|^{p-1} \ln^p \frac{1}{|x|} & when \quad |x| \le e^{-p'} \\ (p')^p e^{-p+\beta p} |x|^\beta l; & when \quad |x| > e^{-p'}, \end{cases}$$

$$v(x) = \begin{cases} |x|^{p-1} & when \quad |x| \le e^{-p'} e^{-\gamma p' - p} \\ |x|^\gamma & when \quad |x| > e^{-p'}, \end{cases}$$

where $0 < \gamma \le \alpha < p - 1$. *Then* $m \in \mathcal{M}(F_{\beta,w}^{p,\theta}, F_{\beta,v}^{p,\theta})$.

It is easy to show that the pair (v, w) satisfies the condition a_p from Definition 9.3.1 and thus from Theorem 9.3.1 we obtain Proposition 9.5.1.

Note that these weights do not belong to A_p class. On the other hand, the conditions

$$\int_{R^1} w(x) (1 + |x|)^{-p+\varepsilon} dx < \infty \tag{9.5.1}$$

and

$$\int_{R^1} v(x) (1 + |x|)^{-p+\varepsilon} dx < \infty \tag{9.5.2}$$

are satisfied, so that $S \subset F_{\beta,w}^{p,\theta} \cap F_{\beta,v}^{p,\theta}$.

Proposition 9.5.2. *Let* $1 < p < \frac{1}{\alpha}$, $\frac{1}{q} = \frac{1}{p} - \alpha$. *Assume that a function* m *is given by the formula*

$$m(\lambda) = \int_{-\infty}^{\lambda} \frac{d\mu_I}{(\lambda - t)^\alpha},$$

where μ_I *satisfy the condition indicated in previous Proposition. If the function* w *satisfies the* A_{pq} *condition, i.e.*

$$\sup \left(\frac{1}{|I|} \int_I w^q(x) dx \right)^{1/q} \left(\frac{1}{|I|} \int_I w^{-p'}(x) dx \right)^{1/p'} < \infty, \tag{9.5.3}$$

where supremum is taken over all one-dimensional intervals I, *then*

$$m \in \mathcal{M}(F_{\beta,w^p}^{p,\theta}, F_{\beta,w^q}^{q,\theta}).$$

In particular, if $0 < \beta < p - 1$, $w(x) = |x|^\beta$ *and* $v(x) = |x|^{\frac{\beta q}{p}}$ *then the condition (9.5.3) is satisfied and, consequently,* $(v, w) \in W_\alpha^{q,p}$.

Example 9.5.1. Suppose that $1 < p < q < \infty$, $1/p < \alpha < \frac{1}{q'}$.

Let

$$w(x) = \begin{cases} |x|^{p-1} \ln^p \frac{1}{|x|} & \text{when } |x| \le e^{-p'} \\ e^{p'\lambda} (p')^p |x|^\lambda & \text{when } |x| > e^{-p'} \end{cases}$$

$$v(x) = \begin{cases} |x|^\gamma & \text{when } |x| \le e^{-p'} \\ |x|^\beta e^{p'(\beta-\gamma)} & \text{when } |x| > e^{-p'}, \end{cases}$$

where $0 < \gamma = q - q\alpha - 1$, $0 < \lambda < p - 1$, $0 < \beta < (1-\alpha)q + \frac{\lambda q}{p} - \frac{q}{p'} - 1$.
Then the function m from the previous theorem belongs to $\mathcal{M}(F_{\beta,w}^{p,\theta}, F_{\beta,v}^{q,\theta})$.

The above-defined pair (v, w) satisfies the condition (5.2.1) and (5.2.2) (for $n = 1$ and $\Omega = R^1$) of Section 5.2.

Note also that weight functions v and w satisfy the conditions (9.5.1) and (9.5.2) for p and q respectively.

Example 9.5.2. Let $\alpha \in (0,1)$, $1 - p < \mu_0 < p/q - \alpha p$, $\alpha p - 1 \le \mu_1 < p - 1$, $q/p - 1 < \varepsilon < q/p - 1 + q$, $\gamma = -\alpha q + q/p - \varepsilon - \mu_0 q/p$, $w(x) = (1 + |x|)^{-\mu_0 - \mu_1} |x|^{\mu_1}$, $v(x) = (1 + |x|)^\gamma |x|^{-q/p+\varepsilon}$. Then $\mathcal{R}_\alpha$ is bounded from $L_w^p(R)$ to $L_v^q(R)$ (see [255], p. 93) and $w \in A_p(R), v \in A_q(R)$.

9.6. Notes and comments on Chapter 9

The well–known Marcinkiewicz, Mikhlin and Hörmander multiplier theorems have been extended by several authors (see, e.g., [175], [228], [176–177], [179], [26–27], etc.).

General theorems on (L^p, L^q) multipliers of Fourier integrals and their applications to embedding theory for spaces of differentiable functions of several variables are presented in [178], [180]. For further generalizations we refer to [149]. An improvment of Hörmander's multiplier theorem in terms of spaces of fractional smoothness is obtained in [26]. On the basis of integral representations of functions certain spaces of differentiable functions are studied in [27] in terms of which sufficient conditions are established for Fourier integral multipliers in $L^p(R^n)$ when $|1/2 - 1/p| < 1/q$ for some $q > 2$.

Finally we recall monograph [196], where the Fourier multipliers in various spaces of differentiable functions are studied and applications to the theory of differential and integral operators are presented.

Chapter 10

PROBLEMS

In this section we list some problems which seem to have resisted solution up to now.

1) To give a complete description of weights w for which the inclusion

$$S(R^n) \subset L_w^p(R^n) \tag{10.1}$$

holds.

It is obvious that the condition

$$\int_{R^n} \frac{w(x)}{(1+|x|)^{pk}} dx < \infty$$

for some k, is sufficient for the above–mentioned inclusion.

2) Let Φ be the Lizorkin space, i.e. the set of all functions $\varphi \in S$ for which all moments vanish. The question is to give a full characterization of weights w for which Φ is dense in L_w^p. For $w \in A_p$ this density holds (see, e.g., [254]).

3) Let $1 < p < \infty$ and let $w^{1-p'} \notin L(0, a)$ for some $a > 0$. Find necessary and sufficient conditions for the functions $f \in L_w^p \cap L_{loc}$ to have their Hardy transforms in $L_w^p(R)$ (see Notes and comments to Chapter 1).

4) Let k be a positive kernel defined on R^n. The problem is to give a complete description of all k and all those pairs of weights (v, w) for which the operator

$$\mathcal{K}f(x) = \int_{R^n} k(x, y)f(y)dy$$

is bounded from L_w^p to L_v^q, $1 < p \leq q < \infty$. For some classes of kernels k and $1 < p < q < \infty$ see [266] and [100].

617

In particular, to solve two–weight problems for

$$I_\alpha^{(n)} f(x) = \int_{R^n} \prod_{k=1}^{n} |x_k - y_k|^{\alpha_k - 1} f(y) dy, \quad 0 < \alpha_k < 1 \ (k = 1, 2, \cdots, n).$$

5) To find a complete description of those pairs of Young functions (Ψ, Φ) for which $R_\alpha(W_\alpha)$ acts boundedly from L_Φ into L_Ψ, where L_Φ and corresponding L_Ψ denote Orlicz spaces (For Riesz potentials see [51], [53]).

6) To give a characterization of all pairs of weights (v, w) for which

$$\left\| \int_{|y| < 2|x|} f(y)|x - y|^{\alpha - 1} dy \right\|_{L_v^q(R)} \leq c \|f\|_{L_w^p(R)},$$

where $1 < p < \infty$ and $\alpha > 0$.

Let $0 \leq 1/p - 1/q \leq \alpha$. When $\alpha > 1/p$ the solution of this problem is presented in Section 5.1 (see also [258]). For the case $w \in A_p \cap RH_p$, $\beta = \alpha - (1/p - 1/q)$ and $v(x) = |x|^{-\beta q} w^{q/p}$ we refer to [288]. Here RH_p denotes the reverse Hölder condition of order $r > 1$.

7) To find the best constant c in the inequality

$$\|R_\alpha f(x)\|_{L_v^q} \leq c \|f\|_{L_w^p}, \quad 1 < p \leq q < \infty, \ 0 < \alpha < 1.$$

For $\alpha \geq 1$ see [188], [269].

8) For the Volterra integral operator considered in Section 2.7 to establish two–sided estimates for the Schatten–von Neumann ideal norm $\| \cdot \|_{\sigma_p}$ when $1 < p < 2$ (For $p \geq 2$ see Chapter 7).

9) To derive two–sided estimates of the measure of noncompactness for the Riemann–Liouville operator R_α acting from $L^{rs}(R_+)$ into $L_v^{pq}(R_+)$ where $1 < r, s, p < \infty, 1 < q \leq \infty, \max\{r, s\} \leq \min\{p, q\}$ and $\alpha > 1/p$ (For the case of Lebesgue space see Section 2.1).

10) Find a complete description for weights ensuring one–weighted estimates for singular integrals on metric measure space when the condition

$$\mu B(x, r) \leq c r^\alpha, \quad \alpha > 0,$$

is satisfied.

11) To derive boundedness (compactness) criteria from $L^p(R^n)$ to $L_v^q(R^n)$ for ball fractional integrals and truncated potentials when $0 < \alpha < n/p$ (For $\alpha > n/p$ see Chapters 4 and 5).

12) Assume that

$$\int_{R^n} \frac{w^{1-p'}(x)}{(1 + |x|)^{p'(n-\alpha)}} dx = +\infty.$$

The problem is to find necessary and sufficient conditions for the function $f \in L_w^p(R^n) \cap L^1(R^n)$ to have its Riesz potential $I_\alpha f$ in $L_w^p(R^n)$ when $1 < p < \infty$.

13) Let L^* be the set of functions f for which $(1 + |x|)^{-n} f(x) \in L^1(R^n)$. Assume that

$$\int_{R^n} \frac{w^{1-p'}(x)}{(1 + |x|)^{p'n}} dx = +\infty.$$

The question is: what are necessary and sufficient conditions for the function $f \in L_w^p \cap L^*$ to have Riesz transforms R_j $(j = 1, \cdots, n)$ in $L_w^p(R^n)$ when $1 < p < \infty$. We note some results (see [115]) concerning this problem for the Hilbert transform

$$Hf(x) = \int_{-\infty}^{+\infty} \frac{f(y)}{x - y} dy.$$

Denote by $Z_p, p > 0$, the family of functions f such that

$$Z_p(f) = \left(\int_0^\infty \frac{dy}{y} \left| \int_R {}^y f(x) dx \right|^p \right)^{1/p} < \infty,$$

where ${}^y f$ denotes the upper cut of f. The following statements hold:

Theorem A. *Let $f \in L_{|x|^\alpha}^p(R) \cap L^*$, where $p > 1$ and $np < \alpha < np - 1 + p$, $n \in N$. Then $Hf \in L_{|x|^\alpha}^p(R)$ if and only if*

$$\int_R t^k f(t) dt = 0, \quad k = 0, \cdots, n - 1.$$

Moreover, if the last condition holds, then

$$\|Hf\|_{L_{|x|^\alpha}^p(R)} \le c \|f\|_{L_{|x|^\alpha}^p(R)}.$$

Theorem B. *Let $f \in L_{|x|^\alpha}^p(R) \cap L^*$, where $p > 1$ and $\alpha = p - 1$. Then $Hf \in L_{|x|^{p-1}}^p(R)$ if and only if $f \in Z_p$ and*

$$\|Hf\|_{L_{|x|^{p-1}}^p(R)} \le c_p \left[\|f\|_{L_{|x|^{p-1}}^p(R)} + Z_p(f) \right],$$

$$Z_p(f) \le c_p \left[\|f\|_{L_{|x|^{p-1}}^p(R)} + \|Hf\|_{L_{|x|^{p-1}}^p(R)} \right].$$

14) For a given $s \in (0, \alpha]$, does there exist a pair of weights v and w such that singular numbers of the Riemann–Liouville operator $R_\alpha : L_w^2(\Omega) \to L_v^2(\Omega)$ where $\Omega = [0, 1]$ or $\Omega = R_+$ have the asymptotics $a_n(R_\alpha) \approx n^{-s}$? (see [108]).

15) For a given $m \in S'$ to find necessary and sufficient conditions on a weight w (v) for the existence of an almost everywhere positive function v (w) such that

$$\|F^{-1}(m\hat{\varphi})\|_{L_v^p} \leq c\|\varphi\|_{L_w^p}, \quad \varphi \in S,$$

holds, with a constant c independent of φ.

16) Let the operator

$$H_{v,w}f(x) = v(x) \int_0^x f(y)w(y)dy \qquad (10.2)$$

be compact from $L^p(R_+)$ to $L^q(R_+)$, when $1 < p, q < \infty$.
i) To study the existence of

$$\lim_{n \to \infty} na_n(H_{v,w})$$

(For the cases of $p = q = 2$ see [68] and for $1 < p = q < \infty$ we refer to [82].
ii) If $p = 2$, then it is known that

$$\lim_{n \to \infty} na_n(H_{v,w}) = \frac{1}{\pi} \int_0^\infty w(x)v(x)dx$$

(see [68]).
Later in [82] it was proved that

$$\lim_{n \to \infty} na_n(H_{v,w}) = \alpha_p \int_0^\infty w(x)v(x)dx,$$

where a positive constant α_p depends on p (For ralated topics when $p, q \in (1, \infty)$ see [174]).

The corresponding problem for fractional integrals is open.

Another interesting problem is to determine in more detail the asymptotic behaviour of the approximation numbers of $H_{v,w}$. Some progress in this derection, for the case $p = q = 2$, may be found in [70].

17) To establish the boundedness (compactness) criteria for the generalized Riemann–Liouville operator

$$\overline{R}_\alpha f(x,t) = \int_{-\infty}^x (x - y + t)^{\alpha-1} f(y)dy, \quad (x,t) \in R \times R_+, \ \alpha \in (0,1),$$

from $L_w^p(R)$ to $L_v^q(R \times R_+)$ (for the Riemann-Liouville operator R_α see Section 2.9).

18) Let $\Gamma \subset C$ be a connected rectifiable curve and let ν be arc–length measure on Γ. Consider the Cauchy singular integral on Γ:

$$S_\Gamma f(t) = \int_\Gamma \frac{f(\tau)}{t - \tau} d\nu(\tau).$$

In [60] it has been proved that S_Γ is bounded in $L^p(\Gamma)$ $(1 < p < \infty)$ if and only if Γ is regular i.e.,

$$\nu(\Gamma \cap B(z,r)) \leq cr, \quad r > 0,$$

where $B(z,r)$ is a ball with center $z \in \Gamma$ and radius r and the constant c does not depend on z and r.

The problem is to find necessary and sufficient condition for Γ to be S_γ bounded from $L^p(\Gamma)$ to $L^q(\Gamma)$, where $q < p$ (for some sufficient conditions see [138]).

19) Let the operator $H_{v,w}$ (see (2)) be compact from $L^p(R_+)$ to $L^q(R_+)$.

i) The problem is to derive necessary and sufficient conditions for which

$$I(s) \equiv \left(\sum_{n=0}^{\infty} a_n^s(H_{v,w}) \right)^{1/s} < \infty, \quad 0 < s < \infty,$$

where $a_n(H_{v,w})$ is the n–th approximation number of $H_{v,w}$, and to establish two–sided estimates of $I(s)$ in terms of v, w, p, q and s. If $p = q = 2$ and $w \equiv 1$ this problem was solved in [224] and for the case of two weights we refer to [78]. For $1 < p = q < \infty$ and $s \in (1, \infty)$ see [185].

ii) To solve the analogous problem for fractional integrals (For the Riemann–Liouville operator R_α when $p = q = 2$, $\alpha > 1/s$, $w \equiv 1$ we refer to [223] and for the Volterra–type integral operators see Chapter 7).

20) To derive necessary and sufficient conditions for weight functions v and w defined on SHT– (X, d, μ) which guarantee the boundedness of the Hardy-type operator

$$H_\mu f(x) = \int_{B(x_0, d(x_0, x))} f(y) d\mu(y) \qquad (10.3)$$

from $L_w^p(X)$ to $L_v^q(X)$, where $1 < q < p < \infty$.

21) To establish a weighted criterion for compactness and to estimate the measure of noncompactness of the operator H_μ (see (3)) acting from $L_w^p(X)$ into $L_v^q(X)$, where $1 \leq p \leq q < \infty$ without the requirement

$$\mu\{x \in X : d(x_0, x) = t\} = 0, \quad t > 0.$$

22) Let φ and ψ be positive functions on SHT– (X, d, μ) such that $\varphi(x) < \psi(x)$ μ–a.e. on X. Let us consider the operator

$$T_{\varphi,\psi} f(x) = \int_{B(x_0, \psi(x)) \backslash B(x_0, \varphi(x))} f(y) d\mu(y), \quad x_0 \in X.$$

The problem is to derive boundedness (compactness) criteria for the operator $T_{\varphi,\psi}$ from $L_w^p(X)$ into $L_v^q(X)$, where $1 \leq p, q < \infty$.

References

[1] D. R. Adams, *A trace inequality for generalized potentials*, Studia Math. **48**(1973), 99–105.

[2] D.R. Adams and L.I. Hedberg, *Function spaces and potential theory, Springer–Verlag, Berlin,* 1996.

[3] D.R. Adams and M. Pierre, *Capacitary strong type estimates in semilinear problemns.* Ann. Inst. Fourier (Grenoble) **41**(1991), 117–135.

[4] K.F.Andersen, Weighted inequalities for maximal functions in spaces of homogeneous type with applications to nonisotropic fractional integrals. *In: General Inequalities 5,* (5–*th Int. Conf. Oberwohlfach 1986) Internat. Schriffreihe Num. Mat.* **80**, *Birkhäuser, Basel–Boston,* 1987, 117–129.

[5] K.F. Andersen, Weighted inequalities for maximal functions associated with general measures, *Trans. Amer. Math. Soc.***236**(1991), No.2, 907–920.

[6] K. Andersen, Weighted generalized Hardy inequalities for nonincreasing functions, *Canad. J. Math.***43**(1991), 1121–1135.

[7] K. F. Andersen and H. P. Heinig, Weighted norm inequalities for certain integral operators. *SIAM J. Math. Anal.* **14**(1983), No. 4, 834–844.

[8] K.F.Andersen and B. Muckenhoupt, Weighted weak type Hardy inequalities with applications to Hilbert transforms and maximal functions, *Studia Math.***72**(1982), 9–26.

[9] K.F. Andersen and E. T. Sawyer, Weighted norm inequalities for the Riemann–Liouville and Weyl fractional operators. *Trans. Amer. Math. Soc.***308**(1988), No.2, 547–558.

[10] T. Ando, On the compactness of integral operators, *Indag. Math.* (N.S.) **24**(1962), 235–239.

[11] J. Arazy, S. Fisher and J. Peetre, Hankel operators on weighted Bergmann spaces, *Amer. J. Math.***110**(1988), 989–1055.

[12] M. Arino and B. Muckenhoupt, Maximal functions on classical Lorentz spaces and Hardy's inequality with weights for nonincreasing functions. *Trans. Amer. Math. Soc.***320**(1990), 727–735.

[13] P. Baras and M. Pierre, Critére d'existence de solutions positives pour des équations semi-linéaires non monotones. *Ann. Inst. H. Poincaré Anal. Non Linéaire* **2**(1985), 185–212.

[14] D. I. Bashaleishvili, Mathematical software of authomated systems of design and control. **IV** (Russian), *Tbilisi State University Publishing House, Tbilisi,* 1993.

[15] E. N. Batuev and V. D. Stepanov, Weighted inequalities of Hardy type. *Siberian Math. J.* **30**(1989), 8–15.

[16] J. Baumeister, Stable solution of inverse problems, *Vieneg and Sohn, Brunschweig/ Weisbaden,* 1987.

[17] G. Bennett, Some elementary inequalities, *Quart. J. Math. Oxford* (2),**38** (1987), 401–425.

[18] G. Bennett, Some elementary inequalities, II, *Quart. J. Math. Oxford* (2),**39** (1988), 385–400.

[19] G. Bennett, Some elementary inequalities, III, *Quart. J. Math. Oxford* (2),**42** (1991), 149–174.

[20] G. Bennett and R. Sharply, Interpolation of operators, *Pure and Appl. Math.* **129**, *Academic press*, 1988.

[21] E.I. Berezhnoi, Exact constants of operators on cones in the ideal spaces. (Russian) *Tr. Mat. Inst. Steklova* **204** (1993), 173–186.

[22] E. I. Berezhnoi, Two–weighted estimates for the Hardy–Littlewood maximal function in ideal Banach spaces. *Proc. Amer. Math. Soc.* **127**(1999), No. 1, 79–87.

[23] J. Bergh, Hardy's inequality–A complement. *Math. Z.* **202**(1989), 147–149.

[24] J. Bergh, V. I. Burenkov and L. E. Persson, Best constants in reversed Hardy's inequalities for quasi–monotonous functions. *Acta Sci. Math.* (*Szeged*) **59**(1993), 221–239.

[25] J. Bergh and J. Löfström, Interpolation spaces: An introduction, *Grundlehren Math. Wiss.*, Vol. **223**, *Springer–Verlag, Berlin*, 1976.

[26] O. V. Besov, On Hörmander's theorem on Fourier multipliers (Russian) *Tr. Mat. Inst. Steklova* **173**(1966). *English Transl. Proc. Steklov Inst. Mat.*, 1987, Issue 4, 1–12.

[27] O. V. Besov, Application of integral representations of functions to interpolation of spaces of differentiable functions and Fourier multipliers. (Russian) *Tr. Math. Inst. Steklova* **187**(1989), *English Transl. Proc. Steklov Inst. Math.* 1990, Issue 3, 1–12

[28] O. V. Besov, Weighted spaces of differentiable functions with variable smoothness. (Russian) *Tr. Mat. Inst. Steklova.* **210**(1995),31–40.

[29] O. V. Besov. Embeddings of function spaces with variable smoothnesss.(Russian) *Dokl. Acad. Nauk* **347**(1996), No. 1, 7–10.

[30] O. V. Besov. Embeddings of differential function spaces with variable smoothness. (Russian) *Tr. Mat. Inst. Steklova* **214**(1997), 25–58.

[31] O. V. Besov, On compact embedding of weighted Sobolev spaces on the domain with irregular boundaries. (Russian) *Tr. Mat. Inst. Steklova* **231**, 2001, 72–93.

[32] O. V. Besov, Sobolev imbedding theorem for the domains with irregular boundaries. (Russian) *Mat. Sb.***1992**(2001), No. 3, 1–26.

[33] O. V. Besov, V. P. Ilin and S. M. Nikolski, Integral representation of functions and embedding theorems, I, II, *Willey, New-York*, 1979.

[34] M. Sh. Birman and M. Solomyak, Estimates for the singular numbers of integral operators (Russian). *Uspekhi Mat. Nauk.* **32** (1977), No.1, 17–82. English transl. *Russian Math. Surveys* **32**(1977).

[35] M. Sh. Birman and M. Solomyak, Asymptotic behaviour of the spectrum differential operators with anisotropically homogeneous symbols. *Vestnik Leningrad. Univ.* (Russian) **13**(1977), 13–21, English Transl. *Vestnik Leningrad Univ. Math.* **10**(1982).

[36] M. Sh. Birman and M. Solomyak, Estimates for the number of negative eigenvalues of the Schrödinger operator and its generalizations, *Adv. Soviet Math.* **7**(1991), 1–55.

[37] S. Bloom and R. Kerman, Weighted norm inequalities for operators of Hardy type, *Proc. Amer. Math. Soc.* **113**(1991), 135–141.

[38] J. S. Bradley, Hardy inequality with mixed norms. *Canad. Math. Bull.* **21**(1978), 405–408.

[39] M. Sh. Braverman, On a class of operators, *J. London Math. Soc.* **(2)47**(1993), 119–128.

[40] V. I. Burenkov, On the best constants in Hardy's inequality for $0 < p < 1$ for monotone functions. (Russian) *Tr. Mat. Inst. Steklova* **194** (1992),58–62.

[41] V. I. Burenkov and M. L. Gol'dman, Calculation of the norm of a positive operator on a cone of monotone functions. (Russian) *Tr. Mat. Inst. Steklova* **210**(1995), 65–89.

[42] V. I. Burenkov and M.S.Tujakbaev, On Fourier integral multipliers in weighted L^p spaces with exponential weights. (Russian) *Trudy Math. Inst. Steklov.* **204**(1993), 81–112.

[43] A. P. Caldéron, Cauchy integrals on Lipschitz curves and related operators, *Proc. Natl. Acad. Sci. USA* **74**(1977), 1324–1327.

[44] B. Carl, Inequalities of Bernstein–Jackson type and the degree of compactness of operators in Banach spaces. *Ann. Inst. Fourier* **35**(1985), 79–118.

[45] L. Carleson and P. Jones, Weighted norm inequalities and a theorem of Koosis. *Mittag–Leffler Inst., Rep.* **12**(1981), No. 2,

[46] A. Carpinteri and F. Mainardy (eds), Fractal calculus in continuum mechanics, *Spinger, Verlag, Wien,* 1997.

[47] M. J. Carro and J. Soria, Boundedness of some integral operators, *Canad. J. Math.* **45**(1993), 1155–1166

[48] R.S.Chisholm and W.N. Everitt, On bounded integral operators in the space of integrable-square functions, *Proc. Roy. Soc. Edinburgh* **A69** (1971), 199–204.

[49] M. Christ, A $T(b)$ theorem with remarks on analytic capacity and Cauchy integral. *Colloq. Math.* **61**(1990), 601–628.

[50] H. -M. Chung, R. A. Hunt and D. S. Kurtz, The Hardy–Littlewood maximal function on $L(p,q)$ spaces with weights, *Indiana Univ. Math. J.* **31**(1982), 109–120;

[51] A. Cianchi, Strong and Weak type inequalities for some classical operators in Orlicz spaces. *J. London Math. Soc.* **(2) 60**(1999) 187–202.

[52] A. Cianchi and D.E.Edmunds, On fractional integration in weighted Lorentz spaces. *Quarterly J. Math. Oxford Ser.* **(2)48**(1997), 439–451.

[53] A. Cianchi and B. Stroffolini, An extension of Hedberg's convolution inequality. *J. Math. Anal. Appl.* **227**(1998), 168–186.

[54] F. Cobos, Interpolation theory and measures related to operator ideals. In: *Nonlinear Analysis, Function Spaces and Applications* 6, 93–118, *Proceedings, Mathematical Instituete Czech Academy of Sciences, Olimpa Press*, 1998.

[55] R.R. Coifman, A. McIntosh and Y. Meyer, L'intégral de Cauchy définit un opérateur borné sur L^2 pour les courbes Lipschitziennes, *Ann. Math.* **116**(1982), 361–387.

[56] R. R. Coifman, G. Weiss, Extensions of Hardy spaces and their use in analysis. *Bull. Amer. Math. Soc.* **83**(1977), 569–645.

[57] M. Cotlar and C. Sadosky, On the Helson–Szegö theorem and a related class of modified Toeplitz kernels. *In Harmonic Analysis in Euclidean Spaces, ed. by G. Weiss and S. Wainger, Proc. Symp. Pure Math.* 35, *Amer. Math. Soc., Providence, R.I.*, 1979, 383–407.

[58] M. Cotlar and C. Sadosky, On some L^p version of the Helson–Szegö theorem, *Conference on Harmonic Analysis in honor of Antony Zygmund (Chicago, 1981)*, Vol.1, *ed. by Beckner et.al., Wadsworth Math. Ser., Wadsworth, Belmond, CA*, 1983, 306–317.

[59] D. Cruz–Uribe SFO, New proof of two–weight norm inequalities for the maximal operator. *Georgian Math. J.* 7(2000), No.1, 33–42.

[60] G. David, Opérateurs intégraux singuliers sur certaines courbes du plan complexe. *Ann. Sci. École. Norm. Sup.* (4)17(1984), 157–189.

[61] M. R. Dostenić, Asymptotic behaviour of the singular value of fractional integral operators. *J. Math. Anal. Appl.* **175**(1993), 380–391.

[62] P. Drabek, H. Heinig and A. Kufner, Higher dimentional Hardy inequality, 123, 3–16, *General Inequalities VII, International Series of Numerical Mathematics, Birkäuser Verlag, Basel* 1997.

[63] P. Drabek, A. Kufner and F. Nikolosi, Nonlinear elliptic equations singular and degenerate case, *University of West Bohemia in Plzen*, 1966.

[64] N. Dunford and J.T. Schwartz, Linear operators. Part I: General theory. *Interscience Publ. New York- London*, 1958.

[65] D. E. Edmunds, Recent developments concerning entropy and approximation numbers. *In: Nonlinear Analysis, Function spaces and Applications*, 5, 33–76, *Olimpia Press, Prague*, 1994.

[66] D. E. Edmunds and W.D. Evans, Spectral theory and differential operators. *Oxford Univ. Press, Oxford*, 1987.

[67] D. E. Edmunds, W. D. Evans and D. J. Harris, Approximation numbers of certain Volterra integral operators. *J. London Math. Soc.* 37(1988), No. 2, 471–489.

[68] D. E. Edmunds, W. D. Evans and D. J. Harris, Two-sided estimates of the approximation numbers of certain Volterra integral operators. *Studia Math.* **124**(1997), No. 1, 59–80.

[69] D. E. Edmunds, P. Gurka and L. Pick. Compactness of Hardy–type integral operators in weighted Banach function spaces. *Studia Math.* **109**(1994), No. 1, 73–90.

[70] D. E. Edmunds, R. Kerman and J. Lang, Remainder estimates for the approximation numbers of weighted Hardy operators acting on L^2. *J. d' Analyse Math.* (to appear).

[71] D. E. Edmunds and V. Kokilashvili, Two-weight inequalities for singular integrals. *Canadian Math. Bull.* **38**(1995), 119–125.

[72] D.E. Edmunds and V. Kokilashvili, Two–weight compactness criteria for potential type operators, *Proc. A. Razmadze Math. Inst.* **117**(1998), 123–125.

[73] D. E. Edmunds, V. Kokilashvili and A. Meskhi, Boundedness and compactness of Hardy-type operators in Banach function spaces. *Proc. A. Razmadze Math. Inst.* **117** (1998), 7–30.

[74] D. E. Edmunds, V. Kokilashvili and A. Meskhi, Two–weight estimates for singular integrals defined on spaces of homogeneous type. *Canad. J. Math.***52**(2000), No. 3, 468–502.

[75] D. E. Edmunds, V. Kokilashvili and A. Meskhi, Weight inequalities for singular integrals defined on homogeneous and nonhomogeneous type spaces. *Georgian Math. J* **8**(2001), No. 1. 33–59.

[76] D. E. Edmunds, V. Kokilashvili and A. Meskhi, On Fourier multipliers in weighted Triebel–Lizorkin spaces, *J. Ineq. Appl.* (*to appear.*)

[77] D. E. Edmunds and V. D. Stepanov, The measure of noncompactness and approximation numbers of certain Volterra integral operators, *Math. Ann.* **298**(1994), 41–66.

[78] D. E. Edmunds and V. D. Stepanov, On the singular numbers of certain Volterra integral operators. *J. Funct. Anal.* **134**(1995), No. 1, 222–246.

[79] D. E. Edmunds and H. Triebel, Function spaces, entropy numbers, differential operators.*Cambridge University Press, Cambridge*, 1996.

[80] L. Ephremidze, On reverse weak (1,1) type inequalities for the maximal operators with respect to arbitrary measures. *Real Anal. Exchange.* **24**(2)(1998/9), 761–764.

[81] W. D. Evans, D. J. Harris and J. Lang, On the approximation numbers of Hardy-type operators in L^∞ and L^1, *Studia Math.***130**(1998), No. 2, 171–192.

[82] W. D. Evans, D. J. Harris and J. Lang, The approximation numbers of Hardy–type operators on trees. *Preprint,* 2000.

[83] W. D. Evans, D. J. Harris and L. Pick, Weighted Hardy and Poincare inequalities on trees.*Proc. London Math. Soc.* **52**(1995), 121–136.

[84] V. Faber and G. M. Wing, Singular values of fractional integral operators: a unification of theorems of Hille, Tamarkin, and Chang, *J. Math. Anal. Appl.* **120**(1986), 745–760.

[85] T. M. Flett, A note on some inequalities. *Proc. Glasgow Math. Assoc.* **4**(1958), No.1, 7–15.

[86] G. B. Folland and E. M. Stein, Hardy spaces on homogeneous groups. *Princeton University Press and University of Tokyo Press, Princeton, New Jersey, 1982.*

[87] M. Frazier, B. Jawerth and G. Weiss, Littlewood-Paley theory and the study of function spaces. *Regional Conference Series in Mathematics,* Vol. **79**, *Amer. Math. Soc.* Providence, RI, 1991.

[88] M. Gabidzashvili, Weighted inequalities for anisotropic potentials. (Russian) *Trudy Tbilissk. Mat. Inst. Razmadze Akad. Gruz. SSR* **82**(1986), 25-36.

[89] M. Gabidzashvili, Two-weight norm inequalities for fractional maximal functions. In *Integral Operators and Boundary Properties of Functions, Fourier Series. Nova Science Publishers, Inc., New York,* 1992, 1–17.

[90] M. Gabidzashvili, I. Genebashvili and V. Kokilashvili, Two-weight inequalities for generalized potentials. (Russian) *Trudy Mat. Inst. Steklov* **194**(1992), 89–96. English transl. *Proc. Steklov Inst. Math.* **94**(1993), Issue 4, 91–99.

[91] M. Gabidzashvili and V. Kokilashvili, Two weight weak type inequalities for fractional type integrals, *Preprint,* No. **45**, *Mathematical Institute Czech Acad. Sci., Prague,* 1989.

[92] I. Gabisonija, Two–weighted inequalities for a discrete Hilbert transforms (for decreasing sequences). *Bull. Georgian Acad. Sci.* **159**(1999), No 1, 9–10.

[93] I. Gabisonija, Two–weighted inequalities for a discrete operators. *Proc. A. Razmadze Math. Inst.* **117**(1998), 144–146.

[94] I. Gabisonija and A. Meskhi, Two–weight inequalities for a descrete Hardy operators and Hilbert transform. *Bull. Acad. Georgia.* **158**(1998), No. 2.

[95] I. Gabisonija and A. Meskhi, Two–weighted inequalities for a discrete Hilbert transform. *Proc. A.Razmadze Math. Inst.* **116**(1998), 107–122.

[96] J. Garsia–Cuerva and J. L. Rubio de Francia. Weighted norm inequalities and related topics. *North–Holland–Amsterdam, New-York, Oxford* 1985.

[97] I. M. Gelfand and G. E. Shilov, *Verallgemeneirte Functionen (Distribution) I, Veb Deutsche Verlag der Wissenschaften, Berlin,* 1960.

[98] I. Genebashvili, Carleson measures and potentials defined on the spaces of homogeneous type. *Bull. Georgian Acad. Sci.* **135**(1989), No. 3, 505–508.

[99] I. Genebashvili, A. Gogatishvili and V. Kokilashvili, Solution of two-weight problems for integral transforms with positive kernels, *Georgian Math. J.* **3**(1996), No. 1, 319–342.

[100] I. Genebashvili, A. Gogatishvili, V. Kokilashvili and M. Krbec, Weight theory for integral transforms on spaces of homogeneous type, *Pitman Monographs and Surveys in Pure and Applied Mathematics,* **92**, *Longman, Harlow,* 1998.

[101] I. Genebashvili and V. Kokilashvili, Weighted norm inequalities for fractional maximal functions and integrals defined on the spaces of homogeneous type. *Bull. Georgian Acad. Sci.* **143**(1991), 13–15.

[102] A. Gogatishvili and V. Kokilashvili, Criteria of weighted inequalities for integral transforms defined on homogeneous type spaces. In: *Topological Vector Spaces, Algebras and Related Areas.* Longman, Pitman Research Notes in Mathematics, **316**(1994), 251–262.

[103] A. Gogatishvili and V. Kokilashvili, Criteria of two–weight strong type inequalities for fractional maximal functions. *Georgian Math. J.* 3(1996), No.5, 423–446.

[104] A. Gogatishvili and J. Lang, The generalized Hardy operator with kernel and variable integral limits in Banach function spaces. *J. Inequal. Appl.*4(1999), No.1, 1–16.

[105] I. C. Gohberg and M. G. Krein, Introduction to the theory of linear non-selfadjoint operators. *Amer. Math. Soc.* Transl. **128**, *Amer. Math. Society, Providence, RI,* 1969.

[106] M. L. Gol'dman, On integral inequalities on the set of functions with some properties of monotonicity.*Teubner Texte zur Math.* **133**, 1993, 274–279.

[107] R. Gorenflo and F. Mainardi, Essentials of fractional calculus, *Preprint submitted to MaPhySto Center, Preliminary version, 25 January, 2000.*

[108] R. Gorenflo and S. Samko, On the dependence of asymptotics of $s-$ numbers of fractional integration operators on weight functions. *Preprint No. A − 3/96, Freie Universität, Berlin, Fachbereich Mathematik und Informatik, Serie A, Mathematik.*

[109] R. Gorenflo and Vu Kim Tuan, Asymptotics of singular values of fractional integration operators. *Preprint No. A- 3/94, Freie Universität, Berlin Fachbereich Mathematik und Informatik, Serie A, Mathematik.*

[110] R. Gorenflo and Vu Kim Tuan, Singular value decomposition of fractional integration operators in L_2- spaces with weights, *J. Inv. Ill– Posed Problems* 3(1995), No.1, 1–9.

[111] K.-G. Grosse-Erdmann, The blocking technique, weighted mean operators and Hardy's inequality, *Lecture Notes in Math.* **1679**, *Springer, Berlin,* 1998.

[112] M. Guer, Problem III, an illustrative problem on Abel's integral equation. In : *C.T. Herman, et. al. (eds.) Basic Methods of Tomography and Inverse Problems,* p. 643–667. *Adam Hilger, Bristol and Philadelphia,* 1987.

[113] V. Guliev, Two–weight L_p inequality for singular integral operator on Heisenberg groups. *Georgian Math. J.* 1(1994), No. 4, 367–376.

[114] P. Gurka, Generalized Hardy's inequality, *Cas. Pestovani Mat.* **109**(1984), 194–203.

[115] R. I. Gurielashvili, On the Hilbert transform. *Anal. Math.* 13(1987), 121–137.

[116] R. I. Gurielashvili, On the Fourier coefficients of functions of the class $L^p(|x|^{p-1}dx)$, $p > 1$. (Russian) *Trudy Tbiliss. Mat. Inst A. Razmadze* **98**(1991), 19–51.

[117] E. G. Gusseinov, Singular integrals in the space of function summable with monotone weight (Russian). *Mat. Sb.* **132(174)**(1977), No.1, 28–44.

[118] G. H. Hardy and J. E. Littlewood, Some properties of fractional integrals, I, *Math. Z.* 27(1928), No.4, 565–606.

[119] G. H. Hardy, J. E. Littlewood and G. Polya, Inequalities.*Cambridge University Press,* 1934.

[120] K. Hansson, Imbedding theorems of Sobolev type in potential theory. *Math. Scand.* **45**(1979), 77–102.

[121] D. J. Harris, On the approximation numbers of Hardy–type operators. *In Function Spaces, Differential Operators and Nonlinear Analysis. M. Krbec et. al. (eds.), Prometheus Publishing House, Prague,* 1996, 216–220.

[122] L. Hedberg, On certain convolution inequalities. *Proc. Amer. Math. Soc.***36**(1972), 505–510.

[123] H. P. Heinig, Weighted inequalities in Fourier analysis. *Nonlinear Analysis, Function Spaces and Appl.* 4, *Proc. Spring School,* 1990, 42–85, *Teubner-Verlag, Leipzig,* 1990.

[124] H.P. Heinig and A. Kufner, Hardy operators of monotone functions and sequences in Orlicz spaces. *J. London Math. Soc.***53**(1996), 256–270.

[125] H. Heinig and G. Sinnamon, Mapping proprties of integral averaging operator. *Studia Math.***129**(1998), No.2, 157–177.

[126] R. Hilfer (Editor), Applications of fractional calculus in physics. *World Scientific, Singapore,* 2000.

[127] E. Hille and R.S. Phillips, Functional analysis and semi–groups. *AMS Colloq. Publ. Providence,* 1957.

[128] S. Hoffman, Weighted norm inequalities and vector–valued inequalities for certain rough operators, *Indiana Univ. Math. J.* **42**(1993), 1–14.

[129] L. Hörmander, Estimates for translation invariant operators in L^p–spaces. *Acta Math.***104**(1960), 93–140.

[130] R. A. Hunt, on $L(p, q)$ spaces. *L'enseign. Math.* **12**(1966), 249–276;

[131] R. A. Hunt, B. Muckenhoupt and R. Wheeden, Weighted norm inequalities for the conjugate function and Hilbert transform. *Trans. Amer. Math. Soc.* **176**(1973), 227–251.

[132] S. Janson and T. Wolff, Schatten classes and commutators of singular integral operators, *Ark. Mat.***20**(1982), 301–310.

[133] R. K. Juberg, The measure of noncompactness in L^p for a class of integral operators, Indiana Univ. Math. J.**23**(1974), 925–936.

[134] R. K. Juberg, Measure of noncompactness and interpolation of compactness for a class of integral transformations, *Duke Math. J.***41**(1974), 511–525.

[135] L. P. Kantorovich and G. P. Akilov, Functional Analysis, *Pergamon, Oxford,* 1982.

[136] N. K. Karapetiantz and B. S. Rubin, Local properties of fractional integrals and BMO spaces on the real line. (Russian), *Deposit in VINITI, Rostov,*1985, p. 43.

[137] G. Khuskivadze, V. Kokilashvili and V. Paatashvili, Boundary value problems for analytic and harmonic functions in domains with nonsmooth boundaries. Applications to conformal mappings. *Mem. Differential Equations Math. Phys., Tbilisi,* 1998, 195p.

[138] G. Khuskivadze and V. Paatashvili, On the boundedness of the Cauchy singular operator S_Γ in the case, where Γ is a countable family of concentric circumferences.*Proc. A. Razmadze Math. Inst.* **124**(2000), 107–114.

[139] V. M. Kokilashvili, On Hardy's inequalities in weighted spaces. (Russian) *Soobsch. Akad. Nauk Gruz. SSR* **96**(1979), 37–40.

[140] V. M. Kokilashvili, On weighted Lizorkin–Triebel spaces. Singular integrals, multipliers, imbedding theorems. (Russian)*Tr. Mat. Inst. Steklova* **161**(1983), Engllish Transl. in *Proc. Steklov Inst. Math.* **3**(1984), 135–162.

[141] V. Kokilashvili, Fractional integrals on curves. *Trudy Tbiliss. Mat. Inst. Razmadze* **95**(1989), 56–70.

[142] V. Kokilashvili, Weighted estimates for classical integral operators, *Nonlinear Analysis, Function spaces and Appl. IV, Teubner–Verlag, Leipzig,* 1990, 86–103.

[143] V. Kokilashvili, Potentials on thin sets. (Russian) *In: Function Spaces and Appl. Diff. Equat. Izd. Univ. Druzbi Narodov, Moskva,* 1992, 25–47

[144] V. Kokilashvili, Riesz potentials in weighted Lorentz spaces. In: *Continuum Mechanics and Related Problems in Analysis,* 383–389, *Proc. Int. Symp., Metsniereba, Tbilisi,* 1993.

[145] V. Kokilashvili, New aspects in weight theory and applications. *Function Spaces, Differential Operators and Nonlinear Analysis, M. Krbec et.al.(eds), Paseky nad Jizerou, September* 3-9 , 1995, *Prometheus Publishing House, Prague,* 1996, 51–70.

[146] V. Kokilashvili, On two–weight inequalities for conjugate functions. *Proc. A. Razmadze Math. Inst.* **112**(1997), 141–142.

[147] V. Kokilashvili and M. Krbec, Weighted inequalities in Lorentz and Orlicz spaces. *World Scientific, Singapore, New Jersey, London , Hong Kong,* 1991.

[148] V. Kokilashvili and A. Kufner, Fractional integrals on spaces of homogeneous type. *Comment. Math. Univ. Carolinae* **30**(1989), No. 3, 511–523.

[149] V. Kokilashvili and P. Lizorkin, Two–weight estimates for multipliers, and embedding theorems. (Russian) *Dokl. Akad. Nauk* **336**(1994), No.4, English Transl. *Russian Acad. Sci. Dokl. Math.* **49**(1994), No.3, 515–519.

[150] V. Kokilashvili and A. Meskhi, Two–weight estimates for singular integrals defined on homogeneous groups. (Russian) *Doklady Acad. Nauk* **354**(1997), No 3. 301–303. English transl. in *Doklady Mathematics,* **55**(1997), No 3, 362–364.

[151] V. Kokilashvili and A. Meskhi, Two weight inequalities for singular integrals defined on homogeneous groups. *Proc. A.Razmadze Math. Inst.* **112**(1997), 57–90.

[152] V. Kokilashvili and A. Meskhi, Weighted inequalities for the Hilbert transform and multipliers of Fourier transforms. *J. Ineq. Appl.* **1**(1997), No 3, 239–252 .

[153] V. Kokilashvili and A. Meskhi, Two–weight inequalities for singular integrals on the Lorentz spaces defined on homogeneous groups. *Proc. A. Razmadze Math. Inst.* **112**(1997), 138–140 .

[154] V. Kokilashvili and A. Meskhi, Two-weight inequalities for the Hardy operator on the Lorentz spaces defined on homogeneous groups. *Proc. A. Razmadze Math. Inst.* **112** (1997), 138–140.

[155] V. Kokilashvili and A. Meskhi, Two-weight inequalities for Hardy type transforms and singular integrals defined on homogeneous type spaces. *Proc. A. Razmadze Math. Inst.* **114**(1997), 119–123.

[156] V. Kokilashvili and A. Meskhi, Two-weight inequalities for integral operators in Lorentz spaces defined on homogeneous groups. *Georgian Math. J.* **6**(1999), No 1, 65–82.

[157] V. Kokilashvili and A. Meskhi, Boundedness and compactness criteria for some classical integral operators. *In: Lecture Notes in Pure and Applied Mathematics*, **213**, " *Function Spaces V"*, *Proceedings of the Conference, Poznań, Poland* (*Ed. H. Hudzik and L. Skrzypczak*), 279–296, New York, Bazel, Marcel Dekker, 2000.

[158] V. Kokilashvili and A. Meskhi, Norms of positive operators on some cones of functions defined on measure spaces, *Proc. A. Razmadze Math. Inst.* **122** (2000), 59–78.

[159] V. Kokilashvili, A. Meskhi, On the boundedness and compactness of generalized truncated potentials. *Bull. Tbilisi Int. Centre Math. Inf.* **4**(2000), 28–31.

[160] V. Kokilashvili and A. Meskhi, Criteria for the boundedness and compactness for integral transforms with power–logarithmic kernels, *Anal. Math.* **27**(2001), No.3, 173–185.

[161] V. Kokilashvili and A. Meskhi, Boundedness and compactness criteria for the generalized truncated potentials (Russian), *Tr. Mat. Inst. Steklova* **232**(2001), 164–178. Engl. Transl. *In Proc. Steklov Inst. Math.* **232**(2001), 157–171.

[162] V. Kokilashvili and A. Meskhi, Fractional integrals on measure spaces, *Fract. Calc. Appl. Anal.* **4**(2001), No. 1, 1–24.

[163] V. Kokilashvili and A. Meskhi, On a trace inequality for one–sided potentials and applications to the solvability of nonlinear integral equations. *Georgian Math. J.* **8** (2001), No.3, 521-536.

[164] H. König, Eigenvalue distribution of compact operators. *Birkhäuser, Boston*, 1986.

[165] M. A. Kransnoselskii, P.P. Zabreiko, E.I. Pustilnik and P.E. Sobolevskii, Integral operators in spaces of summable functions. (Russian) *Nauka, Moscow*, 1966, English transl. *Noordhoft International Publishing, Leiden*, 1976.

[166] M. Krbec, B. Opic, L. Pick, and J. Rakosnik, Some recent results on Hardy type operators in weighted fucntion spaces and related topics. In: *Proceedings of the International Spring School. Function Spaces, Differential Operators and Nonlinear Analysis. Teubner Texte zur Mathematik*, Band **133**, Stuttgart, Leipzig, 158–182.

[167] A. Kufner, O. John and S. Fucik, Function spaces. *Noordoff Int. Publ., Leyden* 1977.

[168] A. Kufner and L. E. Persson, Integral equations with weights. *Preprint*, 2000.

[169] D. S. Kurtz, Littlewood–Paley and multiplier theorems in weighted L^p spaces, *Trans. Amer. Math. Soc.* **259**(1980), 235–263.

[170] D. S. Kurtz and R. L. Wheeden, Results on weighted norm inequalities for multipliers. *Trans. Amer. Math. Soc.* **255**(1979), 343–362.

[171] Sh. Lai, Weighted norm inequalities for general operators on monotone functions. *Trans. Amer. Math. Soc.***340**(1993), 811–836.

[172] C. Li, A. McIntosh and T. Qian, Clifford algebras, Fourier transforms and singular convolution operators on Lipschitz surfaces, *Rev. Mat. Iberoamericana.* **10**(1994), 665–721.

[173] C. Li, A. McIntosh and S. Semmes, Convolution singular integrals on Lipschitz surfaces, *J. Amer. Math. Soc.* **5**(1992), No. 3, 455–481.

[174] M. A. Lifshits and W. Linde, Approximation and entropy numbers of Volterra operators with applications to Brownian motion.*Preprint Math/ Inf /*99 / 27, *Universität Jena, Germany,* 1999.

[175] W. Littman, Multipliers in L^p and interpolation. *Bull. Amer. Math. Soc.***71**(1966), 765–766.

[176] P.I. Lizorkin, (L_p, L_q)–multipliers of Fourier integrals. (Russian) *Dokl. Akad. Nauk SSSR* **152**(1963), 808–811, Engl. transl. in *Soviet Math. Dokl.* **4** (1963), 1420–1424.

[177] P. Lizorkin, On multipliers of Fourier integrals in the spaces $L_{p,\theta}$. (Russian) *Tr. Mat. Inst. Steklova* **89**(1967), 231–248. English Transl. in *Proc. Steklov Inst. Math.* **89**(1967), 269–190.

[178] P. Lizorkin, Generalized Liouville differentiation and multiplier's method in embedding theory of differentiable functions. (Russian) *Tr. Mat. Inst. Steklova* **105**(1969), 89–167.

[179] P. Lizorkin, Properties of functions from the space $\Lambda^r_{p,\theta}$ (Russian), *Tr. Mat. Inst. Steklova* **131** (1974), 158–181.

[180] P. Lizorkin, On the theory of multipliers. (Russian), *Tr. Mat. Inst. Steklova* **173**(1986), 149–163.

[181] M. Lorente, A characterization of two–weight norm inequalities for one–sided operators of fractional type. *Canad. J. Math.* **49**(1997), No. 5, 1010–1033.

[182] M. Lorente and A. de la Torre, Weighted inequalities for some one–sided operators. *Proc. Amer. Math. Soc.* **124**(1996), 839–848.

[183] E. Lomakina and V. Stepanov, On the compactness and approximation numbers of Hardy–type integral operators in Lorentz spaces. *J. London Math. Soc.*(2)**53**(1996), No. 170, 369–382.

[184] E. Lomakina and V. Stepanov, On the Hardy–type integral operators in Banach function spaces. *Preprint,* 1997.

[185] E. Lomakina and V. Stepanov, On asymptotic behaviour of the approximation numbers and estimates of Schatten–von Neumann norms of the Hardy–type integral operators. *Preprint,* 2000.

[186] W. A. J. Luxemburg, Banach function spaces, *Thesis, Delft,* 1955.

[187] W. A. J. Luxemburg and A. C. Zaanen, Compactness of integral operators in Banach function spaces, *Math. Ann.* **149**(1963), 150–180.

[188] V. M. Manakov, On the best constant in weighted inequalities for Riemann–Liouville integrals. *Bull. London Math. Soc.* **24** (1992), No.5, 442–448.

[189] F. J. Martin–Reyes, Weights, one–sided operators, singular integrals, and ergodic theorems. In: *Nonlinear Analysis, Function Spaces and Applications*, Vol **5**, *Proceedings*, M. Krbec et. al. (eds), *Prometheus Publishing House, Prague*, 1994, 103–138.

[190] F. J. Martin–Reyes and P. Ortega, On weighted weak type inequalities for modified Hardy operators, *Proc. Amer. Math. Soc.* **126**(1998), 1739–1746.

[191] F. J. Martin–Reyes, P. Ortega Salvador and A. de la Torre, Weighted inequalities for one–sided maximal functions, *Trans. Amer. Math. Soc.* **319** (1990), 514–534.

[192] F. J. Martin–Reyes and E. Sawyer, Weighted inequalities for Riemann–Liouville fractional integrals of order one and greater, *Proc. Amer. Math. Soc.* **106**(1989), 727–733.

[193] F. J. Martin–Reyes and A. de la Torre, Two–weight norm inequalities for fractional one–sided maximal operators. *Proc. Amer. Math. Soc.* **117**(1993), 483–489.

[194] P. Mattila, Rectifiability, analytic capacity, and singular integrals. *Doc. Math. J. DMV*, 657–664.

[195] V. G. Maz'ya, Sobolev spaces, *Springer, Berlin*, 1985.

[196] V. G. Maz'ya and T.O. Shaposhnikova, Multipliers in the spaces of differentiable functions. (Russian) *Izdatelstvo Leningradskogo Universiteta, Leningrad*, 1986.

[197] V. G. Maz'ya, I. E. Verbitsky, Capacitary inequalities for fractional integrals, with applications to partial differential equations and Sobolev multipliers. *Ark. Mat.* **33**(1995), 81–115.

[198] A. McIntosh, Clifford algebras and the higher dimentional Cauchy integral, approximation and function spaces, *Banach Center Publ., Warsaw*, vol. **22**(1989), 253–267.

[199] A. Meskhi, Koosis type theorems for singular integrals and potentials defined on homogeneous groups. *Proc. A. Razmadze Math. Inst.* **114**(1997), 59–70.

[200] A. Meskhi, Solution of some weight problems for the Riemann–Liouville and Weyl operators, *Georgian Math. J.* **5**(1998), No. 6, 565–574.

[201] A. Meskhi, Boundedness and compactness weighted criteria for Riemann–Liouville and one–sided maximal operators, *Proc. A.Razmadze Math. Inst.* **117** (1998), 126–128.

[202] A. Meskhi, Weighted inequalities for Riemann–Liouville transform. *Proc. A. Razmadze Math. Inst.* **117** (1998), 151–153.

[203] A. Meskhi, Boundedness and compactness criteria for the generalized Riemann–Liouville operator. *Proc. A. Razmadze Math. Inst.* **121**(1999). 161–162.

[204] A. Meskhi, On the boundedness and compactness of ball fractional integral operators. *Fract. Calc. Appl. Anal.* **3**(2000), No.1, 13–30.

[205] A. Meskhi, Criteria for the boundedness and compactness of integral transforms with positive kernels, *Proc. Edinburgh Math. Soc.* **44**(2001), 267–284.

[206] A. Meskhi, Criteria of the boundedness and compactness for generalized Riemann–Liouville operator. *Real Anal. Exchange,* **26** (2000/2001), No.1, 217–236.

[207] A. Meskhi, On the measure of non–compactness and singular numbers for the Volterra integral operators. *Proc. A. Razmadze Math. Inst.* **123**(2000), 162–165.

[208] A. Meskhi, On the singular numbers for some integral operators. *Revista Mat. Comp.* **14**(2001), No.2, 379–393.

[209] A. Meskhi, Asymptotic behaviour of singular and entropy numbers for some Riemann–Liouville type operators. *Georgian Math. J.* **8**(2001), No.2, 157–159.

[210] A. Meskhi, Criteria of the boundedness and compactness for the Riemann–Liouville type discrete operators. *Proc. A. Razmadze Math. Inst.* **126**(2001), 119–121.

[211] M. Mitrea, Clifford Wavelets, singular integrals and Hardy spaces, *Lecture Notes in Math.*, **1575**, *Springer, Berlin,* 1994.

[212] B. Muckenhoupt, Hardy's inequality with weights. *Studia Math.* **44**(1972), 31–38.

[213] B. Muckenhoupt, Weighted norm inequalities for the Hardy maximal function.*Trans. Amer. Math. Soc.***165**(1972), 207–226.

[214] B. Muckenhoupt and R. Wheeden, Weighted norm inequalities for fractional integrals. *Trans. Amer. Math. Soc.***192**(1974), 261–276.

[215] B. Muckenhout and R. L. Wheeden, Two–weight function norm inequalities for the Hardy–Littlewood maximal function and the Hilbert transform. *Studia Math.* **55**(1976), No.3, 279–294.

[216] M. Murrey, The Cauchy integral, Calderon commutators and conjugations of singular integrals in R^n. *Trans. Amer. Math. Soc.* **289**(1985), 497–518.

[217] E. A. Myasnikov, L. E. Persson and V. D. Stepanov, On the best constants in certain integral inequalities for monotone functions. *Dept. Math. McMaster Univ. Preprint,* 1993.

[218] M. A. Najmark, Lineare differential operatoren. *Akademie Verlag, Berlin,* 1960.

[219] F. Nazarov, S. Treil and A. Volberg, Cauchy integral and Calderon–Zygmund operators on nonhomogeneous spaces, *Internat. Math. Res. Notices* **15**(1997), 703–726.

[220] F. Nazarov, S. Treil and A. Volberg, Perfect hair, *Preprint.* 1998.

[221] F. Nazarov, S. Treil and A. Volberg, Weak type estimates and Cotlar inequalities for Calderon–Zygmund operators on nonhomogeneous spaces. *Internat. Math. Res. Notices* **9**(1998), 463–487.

[222] C. J. Neugebauer, Weighted norm inequalities for operators of monotone functions. *Publ. Math.* **35**(1991), 429–447.

[223] J. Newman and M. Solomyak, Two–sided estimates on singular values for a class of integral operators on the semi–axis. *Integral Equations Operator Theory* **20**(1994), 335–349.

[224] K. Nowak, Schatten ideal behavior of a generalized Hardy operator. *Proc. Amer. Math. Soc.* **118**(1993), 479–483.

[225] R. Oinarov, Two–sided estimates of certain classes of integral operators, (Russian) *Tr. Mat. Inst. Steklova* **204**(1993), 240–250.

[226] B. Opic, On the distance of the Riemann–Liouville operators from compact operators. *Proc. Amer. Math. Soc.* **122**(1994), No. 2, 495–501.

[227] B. Opic and A. Kufner, Hardy-Type inequalities. *Pitman Research Notes in Math. Series* **219**, *Longman Sci. and Tech., Harlow,* 1990.

[228] J. Peetre, Applicatioons de la théorie des espaces dinterpolation class lánalyse harmonique. *Riserche Mat.* **15**(1966), 3–36.

[229] V. V. Peller, Hankel operators of class S_p and their applications (rational approximation, Gaussian processes, the problem of majorizing operators). (Russian) *Mat. Sb.* **41**(1980), 538–581. English transl. *Math. USSR Sbornik* **41**(1982), 443–472.

[230] V. V. Peller, A descreption of Hankel opertors of class for $p > 0$, an investigation of the rate of rational approximation, and other applications. (Russian)*Mat. Sb.* **122** (1983), 481–510, English transl. *Math. USSR Sbornik* **50**(1985), 465–494.

[231] C. Perez, Two–weighted norm inequalities for potential and fractional type maximal operators. *Indiana Univ. Math. J.* **43**(1994), 31–45.

[232] C. Perez Weighted norm inequalities for singular integral operators, *J. London Math. Soc.* **2**(49)(1994), 296–308.

[233] C. Perez, Banach function spaces and the two–weight problem for maximal functions. *In: Function Spaces, Differential Operators and Nonlinear Analysis, M. Krbec et. al. (eds), Paseky nad Jizerou,* 1995, 1–16, *Prometheus Publishing House, Prague,* 1996, 141–158.

[234] L. E. Persson, Generalizations of some classical inequalities and their applications. *Teubner Texte zur Math.* **119**, 1990, 127–148.

[235] A. Pietsch, Operator ideals. *North–hulland, Amsterdam,* 1980.

[236] A. Pietsch, Eigenvalues and s–numbers. *Cambridge Univ. Press, Cambridge,* 1987.

[237] D. V. Prokhorov, On the boundedness of a class of integral operators. *J. London Math. Soc.* **61**(2000), No.2, 617–628.

[238] Y. Rakotondratsimba, Weighted norm inequalities for Riemann–Liouville fractional integrals of order less than one.*Z. Anal. Anwend***16** (1997), No.4, 801–824.

[239] Y. Rakotondratsimba, On boundedness of classical operators on weighted Lorentz spaces. *Georgian Math. J.* **5**(1998), No. 2, 177–200.

[240]   Y. Rakotondratsimba, Two weight norm inequality for the fractional maximal operator and the fractional integral operator. *Publ. Math.*, **24**(1998), 81–101.

[241]   Y. Rakotondratsimba, Two–dimentional discrete Hardy operators in weighted Lebesgue spaces with mixed norms, *Proc. A. Razmadze Math. Inst.* **123**(2000), 117–134.

[242]   S. D. Riemenschneider, Compactness of a class of Volterra operators. *Tôhoku Math. J.* (2) **26**(1974), 285–387.

[243]   R. Rochberg, Toeplitz and Hankel operators, wavelets, NWO sequences, and almost diagonalization of operators (*W.B. Arveson and R.D. Douglas, eds.*), *Proc. Sympos. Pure Math.* Vol. **51**, *Amer. Math. Soc., Providence, RI*, 1990, 425–444.

[244]   R. Rochberg, Eigenvalue estimates for Calderon–Toeplitz operators, *Preprint*, 1990.

[245]   B. Rubin, Fractional integrals and Riesz potentials with radial kernels in spaces with power weights (Russian), *Izv. AN Arm. SSR, Math.* **21**(1986), No.5, p. 488.

[246]   B. S. Rubin, Fractional integrals and weakly singular integral equations of the first kind in the n–dimensional ball. *J. Anal. Math.* **63**(1994), 55–102.

[247]   B. S. Rubin, Fractional integrals and potentials. *Addison Wesley Longman, Essex , U.K.*, 1996.

[248]   B. S. Rubin, Inversion of exponential $k-$ plane transforms, *Preprint.*

[249]   J. L. Rubio de Francia, Boundedness of maximal functions and singular integrals in weighted L^p spaces. *Proc. Amer. Math. Soc.* **83**(1981), 673–679.

[250]   J. L. Rubio de Francia, Weighted norm inequalities and vector valued inequalities. *Lecture Notes in Math.* **908**(1982), 86–101.

[251]   W. Rudin, Functional Analysis, *McGraw–Hill Book Company, New York,* 1973.

[252]   J. Ryan (Editor), Clifford algebras in analysis and related topics, *CRC Press, Boca Raton, FL,* 1996.

[253]   A. Saginashvili, Integral equations arising in the control theory of the complex systems. *Georgian Math. J.* **8**(2001), No. 3.

[254]   S. G. Samko, Hypersingular integrals and their applications. *Gordon & Breach Sci. Publ., Series "Analytic Methods and Special Functions",* **5**, 2000.

[255]   S. G. Samko, A.A. Kilbas and O.I. Marichev, Fractional integrals and derivatives. Theory and applications. *Gordon and Breach Science Publishers, London- New York,* 1993.

[256]   E. Sawyer, A characterization of a two–weight norm inequality for maximal operators. *Studia Math.* **75**(1982), 1–11.

[257]   E. T. Sawyer, Two–weight norm inequalities for certain maximal and integral operators. *Lecture Notes in Math.* **908**(1982), 102–127.

[258]   E. Sawyer, Multipliers of Besov and power–weighted L^2 spaces. *Indiana Univ. Math. J.* **33**(1984), No. 3. 353–366.

[259] E. T. Sawyer, A two–weight weak type inequality for fractional integrals, *Trans. Amer. Math. Soc.* **281**(1984), 339–345.

[260] Sawyer, Weighted inequalities for the two–dimentional Hardy operator. *Studia Math.* **82**(1985), 1–11.

[261] E.T. Sawyer, Weighted Lebesgue and Lorentz norm inequalities for the Hardy operator, *Trans. Amer. Math. Soc.* **281**(1984), 329–337.

[262] E. Sawyer, Weighted inequalities for the one sided Hardy–Littlewood maximal functions. *Trans. Amer. Math. Soc.* **297**(1986), 53–61.

[263] E. Sawyer, A characterization of two weight norm inequalities for fractional and Poisson integrals, *Trans. Amer. Math. Soc.* **308**(1988), 533–545.

[264] E. Sawyer, Boundedness of classical operators on classical Lorentz spaces. *Studia Math.* **96**(1990), 429–447.

[265] E.T. Sawyer and R.L.Wheeden, Weighted inequalities for fractional integrals on Euclidean and homogeneous spaces. *Amer. J. Math.* **114**(1992), 813–874.

[266] E. T. Sawyer, R. L. Wheeden and S. Zhao, Weighted norm inequalities for operators of potential type and fractional maximal functions, *Potential Anal.*, **5**(1996), 523–580.

[267] H. J. Schmeisser and H. Triebel, Topics in Fourier analysis and function spaces. *Wiley, Chichester,* 1987.

[268] T. Schott, Pseudodifferential operators in function spaces with exponential weights. *Math. Nachr.* **2000**(1999), 119–149.

[269] T. Schott, Ungleichungen von Hardy type. *Dissertation zur Erlangung des akademischen Grades doctor rerum naturalium,* 1991, 93 S. *Erfurt, Mülhausen, Päd. Hochsch. FB Math.*

[270] J. Schwartz, A remark on inequalities of Calderon–Zygmund type for vector–valued functions. *Comm. Pure Appl. Math.* **14**(1960), No. 7, 785–799.

[271] L. Schwartz, Theory of distributions. **I-II.** *Paris,* 1980–1981.

[272] G. Sinnamon Weighted Hardy and Opial–type inequalities. *J. Math. Anal. Appl.* **160**(1991), 434–445.

[273] G. Sinnamon and V. D. Stepanov, The weighted Hardy inequality: new proofs and the case $p = 1$, *J. London Math. Soc.* **54**(1996), 89–101.

[274] S. L. Sobolev, On a theorem of functional analysis, (Russian) *Mat. Sb.* **4(46)**(1938), 471–497. English Transl. in *Amer. Math. Soc. Transl.*(2)**34**(1963), 39–68.

[275] S. L. Sobolev, Some applications of functional analysis in mathematical physics. (Russian), *Nauka, Moscow,* 1988.

[276] V. A. Sollonnikov, Simple proof of the Hardy–Littlewood inequalities for fractional integrals. *Vestnik Leningrad. Univ., Ser. Mat. Mech. Astr.* **13**(1962), 150–153.

[277] M. Solomyak, Estimates for the approximation numbers of the weighted Riemann–Liouville operator in the space L_p. *Oper. Theory, Adv. Appl.* **113**(2000), 371–383.

[278] E. M. Stein, Harmonic analysis: real variable methods, orthogonality and oscillatory integrals, *Princeton University Press, Princeton, New Jersey,* 1993.

[279] E. M. Stein and G. Weiss, Fractional integrals on n–dimensional Euclidean spaces. *J. Math. Mech.* **7**(1958), No.4, 503–514.

[280] I. M. Stein and G. Weiss, Introduction to Fourier analysis on Euclidean spaces, *Princeton, New Jersey, Princeton University Press,* 1971.

[281] V. D. Stepanov, Weigted inequalities of Hardy type for higher derivatives, and their applications. (Russian) *Soviet Math. Dokl.,* **38**(1989), 389–393.

[282] V. D. Stepanov, Two–weight estimates for the Riemann–Liouville operators, (Russian), *Izv. Akad. Nauk SSSR,* **54** (1990), No. 3, 645–656.

[283] V. D. Stepanov, The weighted Hardy's inequality inequality for nonincreasing functions. *Trans. Amer. Math. Soc.* **338**(1993), 173–186.

[284] V. D. Stepanov, Weighted norm inequalities for integral operators and related topics. *In: Nonlinear Analysis, Function spaces and Applications,* **5**, 139–176, *Olimpia Press, Prague,* 1994.

[285] V. D. Stepanov, On the lower bounds for Schatten–von Neumann norms of certain Volterra integral operators. *J. London Math. Soc.* **61**(2000), No. 2, 905–922.

[286] J. O. Strömberg and A. Torchinsky, Weighted Hardy spaces, *Lecture Notes in Math.* **1381**, Springer Verlag, Berlin, 1989.

[287] J.O. Strömberg and R. L. Wheeden, Fractional integrals on weighted H^p and L^p spaces. *Trans. Amer. Math. Soc.* **287**(1985), No.1, 293–321.

[288] J.O. Strömberg and R. L. Wheeden, Kernel estimates for fractional integrals with polynomial weights. *Studia Math.* **84**, 133–157.

[289] C. A. Stuart, The measure of non–compactness of some linear integral operators. *Proc. R. Soc. Edinburgh* Sect. **A71**(1973), 167–179.

[290] G. Talenti, Osservazioni sopra una classe di disuguaglianze. *Rend. Sem. Mat. Fiz. Milano* **39**(1969), 171–185.

[291] X. Tolsa, Cotlar's inequality and existence of principal values for the Cauchy integral without the doubling condition, *J. Riene Angrew. Math.* **502**(1998), 199–235.

[292] X. Tolsa, Curvature of measures, Cauchy singular integral and analytic capacity. *Ph. D. Thesis, Universitat Autonoma de Barselona,* 1998.

[293] X. Tolsa, L^2–boundedness of the Cauchy integral operator for continuous measures, *Duke Math. J.* **98**(1999), No.2, 269–304.

[294] X. Tolsa, Littlewood–Paley theory and the $T(1)$ theorem with non–doubling measures. *Preprint, Department of Mathematics, Chalmers University of Technology, Göteborg University, Göteborg, Sweden,* 2000.

[295] X. Tolsa, Weighted norm inequalities for Calderón–Zygmund operators without doubling conditions. *Preprint*, 2001.

[296] G. Tomaselli, A class of inequalities. *Boll. Un. Mat. Ital.* **21**(1969), 622–631.

[297] S. Treil, A. Volberg and D. Zheng, Hilbert transform, Toeplitz operators and Hankel opertors, and invariant A_∞ weights, *Rev. Mat. Iberoamericana* **13**(1997), No.2, 319–360.

[298] H. Triebel, Fourier analysis and function spaces, *Teubner Verlag, Leipzig,* 1977.

[299] H.Triebel, Interpolation theory, function spaces, differential operators. *North–Holland, Amsterdam,* 1978 (Second ed. *Barth, Heidelberg,* 1995).

[300] H. Triebel, Theory of function spaces. *Birkhäuser, Basel,* 1983.

[301] H. Triebel, Fractals and spectra related to Fourier analysis and function spaces. *Birkhäuser, Basel,* 1997.

[302] I. E. Verbitsky, Superlinear equalities, potential theory and weighted norm inequalities, In: *Nonlinear Analysis, Function Spaces and Applications* 6, *Proceedings, Mathematical Instituete Czech Academy of Sciences, Olimpia Press,* 1998.

[303] I. E. Verbitsky and R.L. Wheeden, Weighted estimates for fractional integrals and applications to semilinear equatuions. *J. Funct. Anal.* **129**(1995), 221–241.

[304] I.E. Verbitsky and R.L. Wheeden, Weighted norm inequalities for integral operators. *Trans. Amer. Math. Soc.***350**(1998), 3371–3391.

[305] G. Welland, Weighted norm inequalities for fractional integrals. *Proc. Amer. Math. Soc.***51**(1975), 143–148.

[306] R. L. Wheeden, A characterization of some weighted norm inequalities for the fractional maximal functions. *Studia Math.***107**(1993), 251–272.

[307] R. L. Wheeden and J. M. Wilson, Weighted norm estimates for gradients of half–space extensions, *Indiana Univ. Math. J.* **44**(1995), No. 3, 917–969.

[308] W. S. Young, Weighted norm inequalities for the Hardy–Littlewood maximal functions. *Proc. Amer. Math. Soc.* **85**(1982), No. 1, 24–26.

[309] X. P. Zheng, J. Y. Mo and P. X. Cai, Simultaneous application of spline wavelet and Riemann–Liouville transform filtration in electroanalytical chemistry. *Anal. Commun.* **35**(1998), 57–59

[310] A. Zygmund, Trigonometric series, Vol. I, II, *Cambridge, Cambridge University press,* 1959.

Index

641